Industrial Research and Innovation in Business

The International Library of Critical Writings in Business History

Series Editor: Geoffrey Jones
Professor of Business History,
University of Reading

1. The Growth of Multinationals
Mira Wilkins
2. Government and Business
Steven W. Tolliday
3. Mergers and Acquisitions
Gregory P. Marchildon
4. Antitrust and Regulation
Giles H. Burgess Jr
5. The Rise of Big Business
Barry E. Supple
6. Marketing (Volumes I and II)
Stanley C. Hollander and Kathleen M. Rassuli
7. Coalitions and Collaboration in International Business
Geoffrey Jones
8. Business Elites
Youssef Cassis
9. Technology Transfer and Business Enterprise
David J. Jeremy
10. Big Business in Mining and Petroleum
Christopher Schmitz
11. Organizational Capability and Competitive Advantage
William Lazonick and William Mass
12. Entrepreneurship and the Growth of Firms (Volumes I and II)
Harold C. Livesay
13. Family Business
Mary B. Rose
14. Industrial Research and Innovation in Business
David E.H. Edgerton

Future titles will include:

Human Resources and the Firm in International Perspective
Robert Fitzgerald and Christopher Rowley

Women in Business
Mary A. Yeager

The Rise and Fall of Mass Production
Steven W. Tolliday

Industrial Research and Innovation in Business

Edited by

David E.H. Edgerton

Reader in the History of Technology,
Head of the Centre for the History of Science, Technology and Medicine
Imperial College, London, UK

THE INTERNATIONAL LIBRARY OF CRITICAL WRITINGS IN BUSINESS HISTORY

An Elgar Reference Collection
Cheltenham, UK • Brookfield, US

Published by
Edward Elgar Publishing Limited
8 Lansdown Place
Cheltenham
Glos GL50 2HU
UK

Edward Elgar Publishing Company
Old Post Road
Brookfield
Vermont 05036
US

A catalogue record for this book
is available from the British Library

Library of Congress Cataloging in Publication Data
Industrial research and innovation in business : edited by D. E. H.
Edgerton
(International library of critical writings in
business history : 14)
Includes bibliographical references and index.
1. Technological innovations—Economic aspects—History—20th
century. 2. Research. Industrial—History—20th century.
3. Schumpeter. Joseph Alois, 1883–1950. I. Edgerton, D. E. H. ,
1959– . II. Series.
HC79.T415412 1996
338 .064—dc20 96–25420
CIP

ISBN 1 85898 258 8

Printed in Great Britain by Galliard (Printers) Ltd, Great Yarmouth

Contents

Acknowledgements

The editor and publishers wish to thank the authors and the following publishers who have kindly given permission for the use of copyright material.

Academic Press Limited for article: John Cantwell (1995), 'The Globalisation of Technology: What Remains of the Product Cycle Model', *Cambridge Journal of Economics*, **19** (1), February, 155–74.

Blackwell Publishers Ltd for article: D.E.H. Edgerton and S.M. Horrocks (1994), 'British Industrial Research and Development before 1945', *Economic History Review*, **XLVII** (2), 213–38.

Frank Cass & Co. Ltd for article: Jonathan Liebenau (1984), 'Industrial R&D in Pharmaceutical Firms in the Early Twentieth Century', *Business History*, **XXVI** (3), November, 329–46.

Council of the British Society for the History of Science for article: Ernst Homburg (1992), 'The Emergence of Research Laboratories in the Dyestuffs Industry, 1870–1900', *British Journal for the History of Science*, **25**, March, 91–111.

Elsevier Science B.V. for article: Yukiko Fukasaku (1992), 'Origins of Japanese Industrial Research: Prewar Government Policy and In-house Reserch at Mitsubishi Nagasaki Shipyard', *Research Policy*, **21**, 197–213.

Harvard Business School and the President and Fellows of Harvard College for articles: Leonard S. Reich (1977), 'Research, Patents, and the Struggle to Control Radio: A Study of Big Business and the Uses of Industrial Research', *Business History Review*, **LI** (2), Summer, 208–35; David C. Mowery (1984), 'Firm Structure, Government Policy, and the Organization of Industrial Research: Great Britain and the United States, 1900–1950', *Business History Review*, **58** (4), Winter, 504–31.

Harvard University Press and the President and Fellows of Harvard College for excerpt: Jacob Schmookler (1972), 'The Size of Firm and the Growth of Knowledge', in Zvi Griliches and Leonid Hurwicz (eds), *Patents, Invention, and Economic Change: Data and Selected Essays*, 36–46.

Harwood Academic Publishers GmbH for article: Ulrich Marsch (1994), 'Strategies for Success: Research Organization in German Chemical Companies and IG Farben until 1936', *History and Technology*, **12**, 23–77.

National Institute of Economic and Social Research for article: C. Freeman (1963), 'The Plastics Industry: A Comparative Study of Research and Innovation', *National Institute Economic Review*, **26**, November, 22–62.

Oxford University Press for excerpts: Richard Nelson (1992), 'The Roles of Firms in Technical Advance: A Perspective from Evolutionary Theory', in Giovanni Dosi, Renato Giannetti and Pier Angelo Toninelli (eds), *Technology and Enterprise in a Historical Perspective*, Chapter 5, 164–84; Keith Pavitt (1992), 'Some Foundations for a Theory of the Large Innovating Firm', in Giovanni Dosi, Renato Giannetti and Pier Angelo Toninelli (eds), *Technology and Enterprise in a Historical Perspective*, Chapter 7, 212–28; David A. Hounshell (1992), 'Continuity and Change in the Management of Industrial Research: The Du Pont Company, 1902–1980', in Giovanni Dosi, Renato Giannetti and Pier Angelo Toninelli (eds), *Technology and Enterprise in a Historical Perspective*, Chapter 8, 231–60.

Oxford University Press, Inc. for excerpt: Hiroyuki Odagiri and Akira Goto (1993), 'The Japanese System of Innovation: Past, Present, and Future', in Richard R. Nelson (ed.), *National Innovation Systems: A Comparative Analysis*, 76–114.

Sage Publications, Inc. for article: David F. Noble (1978), 'Social Choice in Machine Design: The Case of Automatically Controlled Machine Tools, and a Challenge for Labor', *Politics and Society*, **8** (3–4), 313–47.

Sage Publications Ltd for articles: Michael Sanderson (1972), 'Research and the Firm in British Industry, 1919–39', *Science Studies*, **2** (2), April, 107–51; Michael Aaron Dennis (1987), 'Accounting for Research: New Histories of Corporate Laboratories and the Social History of American Science', *Social Studies of Science*, **17**, 479–518.

University of Chicago Press for articles: John J. Beer (1958), 'Coal Tar Dye Manufacture and the Origins of the Modern Industrial Research Laboratory', *Isis*, **49**, Part 2 (156), June, 123–31; Georg Meyer-Thurow (1982), 'The Industrialization of Invention: A Case Study from the German Chemical Industry', *Isis*, **73** (268), September, 363–81; Walter G. Vincenti (1984), 'Technological Knowledge without Science: The Innovation of Flush Riveting in American Airplanes, ca. 1930–ca. 1950', *Technology and Culture*, **25** (3), July, 540–76; Mark Clark (1993), 'Suppressing Innovation: Bell Laboratories and Magnetic Recording', *Technology and Culture*, **34** (3), July, 516–38; Eric Schatzberg (1994), 'Ideology and Technical Choice: The Decline of the Wooden Airplane in the United States, 1920–1945', *Technology and Culture*, **35** (1), January, 34–69.

Steven W. Usselman for his own article: (1993), 'IBM and its Imitators: Organizational Capabilities and the Emergence of the International Computer Industry', *Business and Economic History*, **22** (2), Winter, 1–35.

Every effort has been made to trace all the copyright holders but if any have been inadvertently overlooked the publishers will be pleased to make the necessary arrangement at the first opportunity.

In addition the publishers wish to thank the Library of the London School of Economics and Political Science and the Marshall Library of Economics, Cambridge University for their assistance in obtaining these articles.

Introduction

David E.H. Edgerton

It is commonplace to acknowledge the role of science and technology, and research and innovation specifically, in transforming the twentieth century world. Most of such accounts, however, focus on scientific and engineering research in general, or technological systems and paradigms, or on government-funded initiatives. Much less attention has been given to business, despite its obvious contribution to innovation, and its even more obvious role in producing and using new technologies. We have, for example, no general account of the role of industrial research and innovation in the twentieth century. But such accounts as we have tend to stress industrial *research*, rather than industrial *innovation*. It is quite wrong to equate the two things; indeed an industrial research focused account of innovation has given us a misleading account of industrial innovation in general since much industrial innovation has not come from research at all. It is also important not to identify the history of industrial science and technology with the history of industrial innovation, since only a fraction of the scientists and engineers employed in industry, even in large firms, have been concerned with innovation.

In a few paragraphs of *Capitalism, Socialism and Democracy*, published in 1942, Joseph Schumpeter placed the bureaucratization and routinization of research and innovation by the giant corporation at the centre of his historical sociology of modern capitalism. Schumpeter wrote about industrial research in the wake of arguments that the stream of new technical ideas had been washed up, and that technological development was being retarded by industrial monopolies which restrained research and bought up patents which threatened their interests. Against this picture Schumpeter argued that large firms did more research than an equivalent number of smaller firms (the 'Schumpeter hypothesis'), and much more importantly, that innovation was changing qualitatively.

> The first thing a modern concern does as soon as it feels it can afford to is to establish a research department every member of which knows that his bread and butter depends on his success in devising improvements. This practice does not obviously suggest aversion to technological progress. Nor can we in reply be referred to the cases in which patents acquired by business concerns have not been used promptly or not been used at all. For there may be perfectly good reasons for this; for example, the patented process may turn out to be no good or at least not to be in shape to warrant application on a commercial basis. Neither the inventors themselves nor the investigating economists or government officials are unbiased judges of this, and from their remonstrances or reports we may easily get a very distorted picture. (pp. 96–7)

But the twist in this tale was that capitalism would be laid to rest by its success in industrial research and not – as its critics generally held – by its failure. The key point in Schumpeter's argument was the routinization, mechanization, automatization of progress, represented by the research department. For Schumpeter the management of a 'stationary economy', and therefore socialism of a sort, was quite possible; the difficulty was to combine management with innovation. But Schumpeter argued that:

> Progress itself may be mechanised as well as the management of a stationary economy, and this mechanisation of progress may affect entrepreneurship and capitalist society nearly as much as the cessation of economic progress would. (p. 131) ... [entrepreneurship] is already losing importance and is bound to lose it at an accelerating rate in the future even if the economic process itself of which entrepreneurship was the prime mover went on unabated. For, on the one hand, it is much easier now than it has been in the past to do things that lie outside familiar routine – innovation itself is being reduced to routine. Technological progress is increasingly becoming the business of teams of trained specialists who turn out what is required and make it work in predictable ways. (p. 132)

For Schumpeter, what could once only be 'visualised in a flash of genius' could now be 'strictly calculated' (p. 132). It is out of these considerations that Schumpeter made his famous comment that:

> if capitalist evolution – 'progress' – ceases or becomes completely automatic, the economic basis of the industrial bourgeoisie will be reduced eventually to wages such as are paid for current administrative work ... Since capitalist enterprise, by its very achievements, tends to automatise progress, we conclude that it tends to make itself superfluous – to break to pieces under the pressure of its own success. The perfectly bureaucratised giant industrial unit not only ousts the small or medium-sized firm and 'expropriates' its owners, but in the end it also ousts the entrepreneur and expropriates the bourgeoisie as a class which in the process stands to lose not only its income but also what is infinitely more important, its function. The true pacemakers of socialism were not the intellectuals or agitators who preached it but the Vanderbilts, Carnegies and Rockefellers. This result may not in every respect be to the taste of Marxian socialists of a more popular (Marx would have said, vulgar) description. But so far as prognosis goes, it does not differ from theirs. (p. 134)

Today we are a long way from such thoughts, but Schumpeter's iconoclastic vision of the role of industrial research in undermining traditional capitalism, and making socialism possible, suggests at the very least a series of interesting questions about the history of industrial research. The questions that Schumpeter asked have hardly found much echo in the journal literature, but the trend he and so many others noticed has been the subject of enquiry.

However, a 'Schumpeterian' perspective has been very influential in recent studies of industrial innovation. A particularly important feature has been the linking of accounts of technological innovation to that of firms and their behaviour. Such a focus is very clear in the methodological and theoretical articles by Richard Nelson (Chapter 1) and Keith Pavitt (Chapter 2) which open this volume. It is important to bear in mind that the Schumpeterian emphasis on the efficacy of large firms in promoting innovation has been challenged, although this is sometimes forgotten. Chapter 3 by Jacob Schmookler, which draws on detailed empirical work on the USA, raises some very pertinent issues, which the history of industrial research needs to address more clearly.

More generally, readers will note in the following a perhaps surprising lack of cross reference between different literatures, and a lack of debate about the big questions between different theoretical traditions. The engagement between empirical and theoretical perspectives is also sometimes surprisingly weak. It is also important to note how historical accounts of industrial innovation and research have changed over time, and that they have changed in relation to differing contemporary accounts of the sources of innovation. It is for these reasons that I have included a number of papers which go over the same empirical ground, especially in the cases of the origins of industrial research, and the history of British industrial research before 1945.

From Coal Tar Dyes to IG Farbenindustrie

The study of industrial research is the study of the creation of novelty. We should not be too tempted by the thesis that the creation of all novelty became routinized through the research laboratory as Schumpeter ingeniously suggested. Firstly, the great bulk of industrial invention and innovation took place outside the research laboratory. Secondly, we should not assume that the research laboratory did indeed routinize innovation. As Hounshell's study of Du Pont (Chapter 10) shows, it was very difficult to get innovation to order.

The origins of the novel form of novelty creation, the industrial research laboratory, are usually held to be in the German dyestuffs industry in the late nineteenth century. In Chapter 4 J.J. Beer makes the point that in the dyestuffs industry there was a built in demand for novelty: fashions in colours changed rapidly. Beer's short study (which was expanded into a book) was the pioneering historical study of R&D. More recently, Meyer-Thurow (Chapter 5) looked in detail at the case of Bayer. The focus of his paper is the 'industrialization of invention'. He notes that it was only in 1891 that Bayer opened a laboratory dedicated to research, and that the development of research was an evolutionary process arising out of the extensive use of chemists by Bayer in non-research functions. Only some 20 per cent of Bayer chemists were in research by the early part of the twentieth century, and in 1909 research expenditures accounted for well under 1 per cent of turnover (as they were in the case of Du Pont). Homburg (Chapter 6) looks at the origins of research in a much greater sample of dye firms. He notes important examples of British and French industrial research in the 1860s, but argues that the idea of research was formally institutionalized in the German dye firms of the early 1880s. Furthermore he sees the establishment of research as being an act of management, which often brought in academic chemists to the new research laboratories. Ulrich Marsch (Chapter 7), in by far the most comprehensive study of the research efforts of IG Farben, which succeeded the great German dye firms, notes that too much attention has been given to the main laboratories of IG. The ten or so such laboratories had around 300 chemists but approximately 1000 chemists were engaged in research out of a total of around 2000. Marsch insists that as well as looking at the main laboratories it is necessary to understand the role of the development and testing laboratories. It is worth noting that by comparison with the pre-war Bayer, IG had a far higher proportion of its chemists in research, and a much higher research intensity. It was almost certainly the largest research spending firm in the world.

Meyer-Thurow makes the point that when they set up research laboratories the German dye firms were already large and wealthy. He goes on to note that after the mid-1880s, when research was established in the main firms, there were no new significant entrants into the German dye industry. This argument bears some expansion. It appears to be true not only of Germany, but of the world as a whole (excluding Switzerland). This is not surprising in that the Germans dominated *world* and not just German markets in synthetic dyestuffs. This helps explain why no research intensive dye firms were established elsewhere before 1914, and puts the complaints of the lack of such development in Britain in a somewhat different light. It is also worth pointing out that there were no such firms in the USA either, though there appear to be no complaints about this. The establishment of fine organic chemicals research in Britain and the USA took place during World War I in quite different economic circumstances.

The Research Laboratory and the US Corporate Economy

The innovation of the corporate research laboratory is almost universally credited to the Germans. We should note however, Homburg's caveat in Chapter 6, and the case of Thomas Edison's Menlo Park laboratory (1876) and his West Orange laboratory (1887), though these are sometimes dismissed as invention factories, even though the latter was to serve as a kind of research laboratory to General Electric before 1900. However, there is little doubt in the literature that the essentially German innovation was developed and most effectively diffused by US businesses.

The historiography of US industrial research usually places its origins as late as 1900, with the establishment of the General Electric Laboratory. Other significant laboratories were established by the explosives firm Du Pont (1902 and 1903) the telephone company AT&T (around 1911) and the photographic giant Eastman Kodak (around 1912). All these firms were very large, and with very strong market positions; each were already very innovative in 'science-based' technologies; two of them were multinationals already operating in Europe. We should stress that they did *not* need *research* laboratories to achieve this dominance. Indeed it appears that one of the main factors leading to the establishment of research in these firms was potential threats from European innovations which *did not* themselves come out of industrial research laboratories: a colour photographic system in the case of Eastman Kodak, the Nernst lamp (the rights to which had been acquired by the German firm AEG) in the case of GE, and the radio in the case of AT&T. It was a potential threat to existing market control which prompted research rather than a desire to increase the rate of innovation. In the case of Du Pont, research started to improve high explosives manufacture, and then, in a separate laboratory, was intended to be at the core of Du Pont's diversification strategy, pursued under pressure from anti-trust legislation. But until the 1930s Du Pont relied on external sources of technology.

Leonard Reich's study of AT&T (Chapter 8) stresses that research was not just devoted to innovation, but to controlling the innovation of others by patents. He makes the forceful point, barely acknowledged elsewhere in the literature, that research spending is not an indicator of innovation in such circumstances. His work also suggests very strongly that Schumpeter was far too sanguine about corporate control of research. Such a conclusion is strongly supported in Chapter 9 by Mark Clark's study of AT&T's refusal to market its impressive magnetic recorders.

David Hounshell's study (Chapter 10), summarizing his book, written with John K. Smith (1988) on Du Pont's research over nearly a century, brings out very well the changing relations between research and corporate policy, in a very particular legal environment. It is worth noting that Du Pont's R&D expenditure increased from less than 1 per cent before 1914, to around 3 per cent in the inter-war years, and around 7 per cent after World War II. Du Pont's spending was well above the average for manufacturing but mirrors the general trend. One particular point of interest is the changing place of 'fundamental research' within Du Pont. Such research was started in polymer chemistry in 1927 and led to neoprene and nylon. Thereafter, 'new nylons' proved difficult to find.

A great deal of attention has been given to the 'fundamental' or 'basic' aspect of the research effort of firms. In his review of the literature Michael Dennis (Chapter 11) takes issue with the image of industrial research either as representing a 'university in exile' or 'prostituted science'. His essential point is that the late nineteenth and early twentieth centuries saw the development of *research* for the first time in both universities and firms. In one case research was linked to teaching, in the other to the needs of the firm. But in both cases research was highly organized and focused. The ideal of 'pure science' was not a reality with which industrial research should be compared.

So far we have been considering research and researchers in chemistry and in physics. Indeed these two scientific disciplines dominate consideration of industrial research. Engineers barely figure in the articles discussed above. That in itself is an issue worth thinking about. It bears on the point that industrial research and innovation are not the same thing. The lack of literature on this topic makes Vincenti's pioneering analysis (Chapter 12) especially important. Vincenti takes an important innovation which owed nothing to science, nor to engineering science, nor even to a conceptual breakthrough; it was what Vincenti calls a production-centred innovation. Eric Schatzberg (Chapter 13), also focusing on aviation, argues a thesis which is also of general applicability. He shows that there was no technical reason why wood could not have remained the main material for airframe production. The deeply ingrained preference for metal he attributes to the idea that metal was the material of the future. The general point that arises is that we should not retrospectively argue that superseded technologies were massively inferior, or even inferior at all. This is important because we too often forget the existence of real historical alternatives when discussing technology. The pioneering study of the role of ideological, or if one likes, non-technical factors, in technical development in the twentieth century is David Noble's study (Chapter 14) of numerically-controlled and record-playback controlled machine tools. Noble opens with a critique of the idea of technological determinism which he holds dominates thinking about the relations of technology and society. Contrary to one of the premises of technological determinism he holds that technology is socially shaped.

International and Comparative Studies

After World War II there were few doubts about the technical progressiveness of capitalism at least when aided by the state. Christopher Freeman's pioneering study of research and development, invention and innovation in the plastics industry (Chapter 15) remains a vital source for information on technical change in that important industry from the inter-war years into the early 1960s. The great power of US corporations is very evident from the study. One important issue thrown up by comparative studies is the question whether differences are to be explained in terms of national characteristics, or rather in terms of the characteristics of particular firms in each economy. Some studies insist that it is differences in firms which are most significant. Thus in Chapter 16 Liebenau argues that the medical and scientific resources available in the USA, Britain and Germany around the turn of the century were similar, and yet research in pharmaceutical industries developed in very different ways. He attributes this to the differing character of firms operating in this sector. More formally, in Chapter 17 Usselman shows that resources of particular firms fundamentally shaped their ability to succeed in the computer industry: IBM started from a very powerful position in data-processing which it was able to build on; its competitors generally did not. The general argument is that US corporations got it right, and the British stand as an example of how not to do it. John Cantwell's empirical study of the patenting of the world's largest innovators (Chapter 18) shows very clearly that their innovative activity was much more internationalized than such a nationalistic emphasis implies. Again, the role of firms, rather than nations, is highlighted.

The Not-So-Peculiar British Case

Other studies focus on the comparative performance of nations, while drawing on the argument that the structure of individual firms is important. Most such comparative studies are concerned to explain allegedly poor British economic performance. Mowery's influential comparative account of US and British industrial research (Chapter 19) seeks to explain what he sees as British backwardness in R&D (which he sees as a powerful cause of the British decline) in terms of British corporate structure, British government policy and British education. What he finds is that in the USA in-house industrial R&D dominated, while in Britain the cooperative industrial research association had an important role. His picture of British industry was different from that given in a much earlier paper by Sanderson (Chapter 20). It is also worth comparing their estimates (using different measures) of the relative R&D efforts of British and US business. The paper by Edgerton and Horrocks (Chapter 21) challenges the Mowery interpretation of British industrial R&D and makes the point that no evidence has been adduced for the common view that British industrial R&D was behind German in the inter-war years. The British case is a very important one in the twentieth century. Unfortunately we do not have, at least in the journal literature, a detailed study of the research effort of a particular British firm, certainly by comparison with the US case. This does not mean that British firms should be ignored, especially not the largest British R&D spender – Imperial Chemical Industries.

Industrial Innovation and the Rise of Japan After 1945

Although by the 1920s the USA was by far and away the largest spender on industrial R&D, Britain was probably second, with Germany not far behind. France, Japan and Italy were not in the contest. In Chapter 22 Fukasaku shows, however, that Japanese firms were doing a surprisingly large amount of R&D in the inter-war years. After World War II, Japanese firms increased their spending on innovation very dramatically. In the 1960s and 1970s, Japanese firms overtook the absolute spending on R&D of first Britain and then Germany, although it is important to remember that their spending is still below that of US firms. Japan is an essentially interesting case because it highlights the fact that high rates of growth do not correlate with high levels of R&D spending. But it also suggests very strongly that in order to stay rich, if not to become so, firms do invest hugely in R&D.

References

Hounshell, D.A. and Smith, J.K. Jr (1988), *Science and Corporate Strategy: Du Pont R&D, 1902–80*, NY: Cambridge University Press.
Schumpeter, J. (1942), *Capitalism, Socialism and Democracy*, London: Allen & Unwin, 1976 edition.

Part I
Theoretical Perspectives

Excerpt from *Technology and Enterprise in a Historical Perspective*, Giovanni Dosi, Renato Giannetti and Pier Angelo Toninelli (eds.)

[1]

5

The Roles of Firms in Technical Advance: A Perspective from Evolutionary Theory

Richard Nelson

1. FIRMS AS PLAYERS IN AN EVOLUTIONARY GAME

The invitation to prepare a paper on 'the roles of firms in technical advance' encouraged me to return to unfinished business. In our *An Evolutionary Theory of Economic Change* (1982), Sidney Winter and I laid out a theoretical framework designed to illuminate the processes of technical advance in modern capitalist economies, and the connections between the workings of those processes and such variables as productivity growth at an industry or aggregative level, the distribution of firm sizes, the shapes of diffusion curves, etc. That book was written at two quite different levels of abstraction, and we were well aware that we had developed the analysis unevenly.

One portion of the book, and this is the part that has been most referred to by economists, consisted of a family of formal evolutionary models, each concerned with illuminating or explaining a particular set of issues, all united by a common broad theoretical point of view. Within these models firms were treated in a highly stylized manner, and industry details were similarly stripped down, with just enough detail to enable the model to run.

In other chapters of the book, we sketched but did not develop particularly far two matters: a theory of the behaviour and capabilities of firms that is much finer in texture than the one contained in our formal models, and a characterization of the institutional richness and cross-industry variety that marks modern capitalism. It is interesting that these parts of our book have been the ones most referred to by sociologists, and by scholars of technical advance who come to the topic not tightly bound by any disciplinary perspective.

We were and are quite self-conscious about the differences in the levels of abstraction involved. Indeed we argued that a solid theoretical understanding of technical advance needed to be developed at several levels. For some purposes, like exploring whether the observed time series of macroeconomic output and inputs is compatible with the notion that technical advance is an evolutionary process, the high level of abstraction of our formal models is appropriate, and more detail mainly adds clutter and detracts from the clarity (transparency) of the analysis. For other purposes, like analysis of technical advance in the computer industry, or comprehending the differences between the computer industry and the pharmaceutical industry, a finer-grained treatment is clearly needed. We concluded our book by promising to do further work on the finer-grained theoretical issues.

When I received the invitation to write this chapter I had already done considerable work fleshing out the analysis of institutional complexity and cross-industry differences (see Nelson, 1987). I am now ready to get back to work on a more detailed theory of the firm appropriate to an evolutionary theory of technical advance.

1.1. Theories of the Firm: Those Where the Focus is on the Game, and Those Where the Focus is on the Players

On reflecting on the kind of theory I wanted to develop, and on the collections of quite disparate writings that purport to be theories about the firm, it became clear to me that a critical distinction was whether the theoretical focus was on phenomena that are the result of the behaviour of firms in a certain context, or on firms *per se*; put another way, on a game in which firms are critical actors, or on characteristics of the actors. What I am trying to do here is to contribute to the development of theory that will enable better understanding of what firms are doing, but my basic interest is in the 'game' of technical advance.

Thus I would like to distinguish the theory I am trying to develop from those in various research traditions where the focus is on particular activities that go on in firms, or characteristics of firms that lead to their good or poor performance. The internal organization of firms and the processes of decision-making have received attention as topics of interest on their own account, and as variables that influence firm performance. So too have the sources of conflict and mechanisms of conflict resolution. Scholars have studied how to design a firm so that it will be flexible in an unpredictable environment, and have examined what it takes to profit from R. & D. While the writings in question are extremely diverse, I lump them together here because they have as a common denominator lack of interest in variables and mechanisms that are defined at a higher level than individual firms, except in so far as these influence the well-being of firms, what goes on in firms, or what the firm should do. If it is recognized that the firm is involved in a game with other actors, the focus of attention is not on the general character of the game, or on game outcomes broadly described, but on the particular player.[1] While I intend my analysis to be conformable with these writings, and consider conformability an important check on the plausibility of my theorizing, my objective is a different one.

These research traditions have little intellectual contact with other collections of theories of the firm where the central interest is one the behaviour of a system containing many firms, and where individual firms are of concern mainly as players in a game whose outcome determines macro-behaviour and performance. Firms are not of interest *per se*, and are basically treated as anonymous. They are modelled in a stark stylized way with the attention limited to dimensions strictly relevant to macro-outcomes. The theory of the firm contained in neo-classical economic theory fits this mould. The name 'firm' is given to the individual profit-maximizing entities presumed to be engaged in a certain kind of economic game, whose outcomes can be characterized in terms of some kind of equilibrium. The interest is in certain aggregate characteristics of the equilibrium, like price, or industry output, or employment.

The formal evolutionary models developed by Sidney Winter and myself are of course concerned with a different kind of game from most neo-classical models, and this game—technical advance—is the one I am concerned with here. However, the treatment of firms in our models and in neo-classical ones is similar in many respects. Our interest is in what happens to variables defined at the level of an industry or economy. The only attributes of firms' behaviour that are modelled are those that bear on these matters. And in our formal models, as in neo-classical ones, firms are anonymous.

However, there is one important difference between evolutionary models as we have developed them and neo-classical ones. It is that in the logic of evolutionary modelling what individual firms do often matters significantly. Since we treat technical advance as involving a significant stochastic element, and since individual firms make different 'draws', best-practice technology at any time can be identified as having been found (or created) by particular firms in our model. And, given the stochastic nature of things, it is not inevitable that had one firm not found a technology, another would have. Thus while the firms in our model are formally treated as anonymous, a recount of a 'history' generated by our models generally requires a description of what particular entities did.

The theory of the firm contained in our evolutionary models represents a start on a theory of the firm that enables one to understand the roles of firms in technical advance. The models are of the right 'game', and have some of the right variables in them, innovative and imitative R. & D. for example, and suggest some interesting distinctions, for example between industries where technical advance is 'cumulative' in the sense that one technical advance sets the stage for the next one, and those where new science, arrived at independently of concerns about new technology, provides the seeds for technical advance. It also highlights the fact that such a theory must recognize differences between firms and not cover them up. However, Winter and I not only acknowledged but emphasized that if one wants a theory of the firm that is useful in illuminating the role of firms in particular technological episodes or histories, the formalism contained in our models is too stark. In other parts of our book we laid out starting-places for a more detailed analysis.

I return here to try to develop that analysis further. In doing so now I have the advantage of many interesting theoretical and empirical articles and books that were not available when I was last at that task.[2] However, my interest continues to be in the 'game' of technical advance, and the more macroeconomic measures of the outcome of that game. My theorizing about firms reflects my belief that one cannot understand technical advance in a particular field unless one has a good theory of the firm so that one can understand what they do and don't do. But, on the other hand, one cannot characterize what one wants to understand about the players until one has characterized the game in some detail.

1.2. Key Aspects of Technical Advance in Capitalist Economies

There are three aspects that I would like to highlight here.[3] They will come as no news to historians of technical advance, but they define the game within which the firms are players.

In the first place, in capitalist economies technical advance is set up as an evolutionary process. A central purpose of Nelson and Winter (1982) was to argue that empirical studies of technical advance, historical and other, demonstrated this overwhelmingly, and to draw from this some implications for modelling. Of course we are not alone in calling attention to the evolutionary nature of technical advance. The parlance is common among historians, and Joseph Schumpeter observed that 'in dealing with capitalism one is dealing with an evolutionary process'.

Schumpeter's model is in fact a good starting-place for the general discussion. In his *Capitalism, Socialism, and Democracy* (1950), the key actors are for-profit business firms, equipped with industrial research laboratories, in rivalrous competition with each other. The arena of competition is set, on the one side, by public scientific and technological knowledge, and on the other, by various mechanisms that enable firms to keep proprietary the new technologies they create, at least to some degree, and for a while. The former lends power to industrial R. & D. The latter makes it potentially profitable. Indeed in this context if a firm's competitors engage in R. & D., it may have no choice other than to do so itself. The result is a flow of new products and processes into actual use in the market-place, with *expost* selection on these and on firms themselves. This is clearly a description of an evolutionary process and, as noted, Schumpeter called it such.

Evolutionary processes in general, and technical change in particular, are inherently wasteful, at least with the vision of hindsight. Compared with any kind of an ideal, R. & D. allocation is inefficient. There is bound to be duplication or near-duplication of effort. Economies of scale and scope that might be achieved through R. & D. co-ordination are often missed. Certain kinds of

R. & D. that would have high social value simply may not be done because they would not yield proprietary advantage, or because no one is minding the overall portfolio. Also, because technology is to a considerable extent proprietary one can see many enterprises operating inefficiently, even failing, sometimes at considerable social cost, for want of access to the best technology.

It is something of a puzzle, therefore, as to why capitalist innovation systems have performed so well compared with systems in the socialist countries that have relied to a much greater extent on central planning and co-ordination mechanisms. I have argued at length elsewhere that the basic reason undoubtedly resides in the uncertainty that surrounds technical advance, and differences of intuition and insight among experts (see Nelson, 1987). In such a context, the attempt to plan and co-ordinate will almost certainly be ineffectual, and probably counter-productive. What the capitalist system does is to permit and encourage a variety of approaches, with *expost* selection. This is certainly wasteful, but has proved wonderfully powerful.

Also, the second aspect of the capitalist innovation system I want to highlight serves to cut down somewhat on the waste. In capitalist economies technology, or aspects of it, is partially a public good. Publically financed science, which is an important part of the system, surely is. But so too are aspects of industrial technology created through privately funded R. & D.

Schumpeter's analysis clearly recognizes that, while proprietary gains and the hope for them provide the key incentive, there are limits on the ability of firms to block access to the fruits of their R. & D. Sooner or later competitors will be able to imitate, or invent around, the initial innovation. The fact that technology ultimately becomes public has several benefits. First, it ensures that restrictions on access are limited and that a healthy share of the benefits go to users. Second, knowledge of the new innovation provides a base and a spur for further innovation by others. Third, by facilitating subsequent competition, the dangers that a company can build a wide and durable industry monopoly out of a particular innovation are kept under control. However, all of these benefits will come to naught if fear of rapid imitation damps incentives to innovate in the first place.

Some analysts have seen the tension here in terms of a trade-off.

Access to an innovation by competitors should not be so rapid or complete as to dull incentives to innovate. On the other hand, the stronger the restrictions on access and the longer they last, the higher the social costs in terms of less-than-optimal use.

However, while the trade-off notion is illuminating, it takes attention away from the third feature of the technical-change game that I want to highlight. In capitalist economies firms are only one of the institutional actors. There are universities, professional societies, and government agencies in the system. There are public moneys as well as private. Many things that firms cannot be expected to do on their own because the returns are not readily appropriable are picked up by other institutions or are publically financed or both. Also, there is a variety of modes whereby firms can and do work co-operatively, rather than in rivalry.

The picture I have sketched is quite complex, both in terms of the processes involved, and in terms of the institutional actors. But that is because the technical-change game in capitalist economies is a complex one, with many different types of moves, and many different types of players. In order to develop a theory that comes to grips with the roles of firms one must have a comprehension of the game within which firms play.

Within this context, there is a large variety of questions a theory of the firm might address. Here I begin to explore two topics: first, understanding the central role played by R. & D. done by firms that basically make their profit by making and selling products; second, understanding what is private and what is public about technology in modern capitalism.

2. THE CENTRAL ROLE OF INDUSTRIAL R. & D.

During the first Industrial Revolution, invention was largely the province of men working alone, or occasionally in pairs or small groups. As a rule the inventive efforts of these men were taken on their own initiative, with no particular obligation to a company where they might otherwise be employed. In some cases the inventive work seems to have been associated with advancing the capabilities of the company employing the men, or paying them explicitly for their inventive work, but until the last quarter of the nineteenth century one cannot recognize anything that resembles modern industrial research and development laboratories.

Towards the end of the nineteenth century, special R. & D. laboratories tied to particular business firms and dedicated to advancing their product and process technologies began to grow up, first in industries where chemical processes or products were important, and later in those producing products based on electrical technology. Since that time industrial research and development has increasingly been the principal source of the design, problem-solving, and development work leading to new industrial products and processes, across a broad front of industries.[4]

Free-lance individual inventors have played a diminishing role. While in the early decades of the twentieth century, free-standing research and development laboratories, which contracted to do work for industrial firms, played a significant role in some industries, these kinds of institutions have also become less important. Though, as we shall see shortly, university-based research plays an important role in technical advance in modern capitalist economies, in only a very few technologies does such research create anything resembling new processes and products, and in the economy as a whole industrial spending on R. & D. dwarfs that in universities.

2.1. *The Importance of Doing R. & D. In-House*

A central question that a theory of the firm must address therefore is this: why is it that industrial research and development laboratories—specialized organizations engaged in R. & D. closely tied to firms engaged in the production and sale of products—have become the dominant vehicle for industrial invention? Actually, there are two questions here. Why specialized laboratories? And why tied to individual firms?

The answer to the first question is that while during the first Industrial Revolution experience with the art generally was sufficient for effective inventive work, by the time of the second this was no longer the case. To do effective inventing required high-level schooling in the relevant sciences or engineering disciplines. Also, a good portion of inventing now required access to special equipment, or teamwork, or both. The differentiation of the people involved in R. & D. from those involved in production, and the requirement for dedicated equipment and organization, led naturally to the specialized laboratory committed to inventing and distance from on-line production.

The answer to the second question is that in most industries organizational distancing could not be too far or too sharp. Contrary to popular beliefs, even in the new era of science-based technologies, industrial inventions seldom represented simply the creative application of publicly known science to a particular practical objective. In the first place, industrial technologies--the way processes work, how products are put together—are often quite complex, and effective inventive work generally requires sophisticated understanding of the technology at least as much as it requires command of the relevant science. While part of the knowledge can be and is imparted in school, in many cases firms, not schools, are its repository. Secondly, in many instances inventive work is oriented towards improving or extending the product or process technology of particular firms. As I shall argue later, these tend to have an idiosyncratic element. Even where an inventive effort aims at something drastically new for the firm, effective implementation often requires fitting into or alongside old structures. Thus to be effective, industrial R. & D. generally needs good communication channels to the firm whose problems are being addressed, and who will in the end use the product of R. & D.

There are three aspects of these information channels that warrant attention. First, there are initial costs, often substantial, of setting them up and getting them to operate properly. There are major advantages for a firm of having a continuing relationship with a laboratory with which it knows how to interact. Similarly, a laboratory has a significant investment in firm-specific contacts and modes of communication. To change partners would require both parties to invest in set-up experience again. Second, since in many instances, if a firm is to reap significant returns from the R. & D. it sponsors, it must quickly get the resulting process on product innovation into use or production and sale, the firm needs to be closely appraised of developments so that it can assess and marshal what Teece (1986) has called necessary co-specialized assets. Third, much of what flows, both ways, in the channel is proprietary, in the sense that the firm would not like to see that information leak to other organizations, and the laboratory loses its value to its sponsor if the information leaks. Information about prevailing processes and products, and perceptions of market demand, are often trade secrets. New process and product designs, or solutions to problems, can give a sponsoring firm a proprietary edge if other firms are blocked from quick access, but if competitors quickly become privy to what is going on and respond, the sponsoring firm may gain little edge. Thus there are major advantages to the firm if the research and development organization with which it does business does not also do business with competitors.

Thus, while there are exceptions, where effective and sustained industrial R. & D. is important for competitive success, a firm is going to be forced to establish a long-term relationship with a research and development laboratory which, in turn, is committed to it at least in its particular line of business.[5] A firm cannot operate effectively if it must, as a general rule, try to contract for the R. & D. it needs on a project by project basis. Nor, because of the uncertainties involved regarding future needs, is a long-run contract an appropriate vehicle. One cannot specify far in advance just what one will want a research and development organization to do. And cost-plus contracts have proven very inefficient vehicles, for a variety of understandable reasons. All this strongly pulls towards having an in-house laboratory

The discussion above is not only rather abstract, but also undifferentiated. What I say is far more true about some technologies than others. It applies well to aircraft and semiconductors. Pharmaceuticals, on the other hand, can be found and tailored or constructed virtually exclusively in a laboratory, by scientists who know little about how pharmaceuticals are mass produced and marketed, although a lot of follow-on work may be needed on the production process that does require intimate knowledge of manufacturing technology. And new pharmaceuticals can be protected by patents. Some research projects can be and are farmed out. However, I propose as a strong first approximation that there are good reasons why a firm would want R. & D. connected with its core business and proprietary interests to be performed in a tightly connected R. & D. facility. And, again as a first approximation, laboratories that are basically free-standing and aim to sell their services or output to firms that manufacture and sell products, cannot always protect their wares from pilferage. They are not likely to be important sources of new product and process technologies in most industries.

In contrast with the need for intimate links between manufacturing laboratories, contacts and channels to the world of public science tend to be available to anyone with the relevant knowledge and credentials, at least where the relevant science is not at the cutting-edge of a field. Libraries are available, and current journals. More important, if one is in a field, one knows who is likely to know. Later I shall consider the situation where industrial R. & D. needs to be closely tied in to academic research, as well as to the firm it serves, but this seems to be rare.

2.2. The Outward Look of Industrial R. & D.

Discussion of the need of an industrial laboratory to be able to tap into the world of public science leads naturally to a more general matter. An industrial R. & D. laboratory needs to look two ways: towards the mother firm—for guidance on high-priority product and process innovations, and to deliver on these in privacy—but also outwards. As Cohen and Leventhal (1987), among others, have argued, R. & D. activities outside the firm provide both opportunities and challenges for the firm. They need to be monitored. Monitoring requires familiarity with the fields of R. & D. in question, so that the significance of new developments can be appreciated and implications for the firm understood. Close familiarity, and in particular capacity to respond appropriately without undue delay, is likely to require some continuing hands-on work in the field, which can also if needed serve as a solid basis for marshalling resources to more directed efforts.

Thus industrial R. & D. reflects the fact that technology in capitalist economies has both a private and a public aspect, and is also a major reason why this is the case. Industrial R. & D. laboratories create new proprietary technology, and at the same time their efforts to keep up with a field make it difficult for anyone to keep new technology fully proprietary for very long.

The nature of the R. & D. activities that an industrial R. & D. laboratory has to monitor differs from industry to industry. However, I propose that most industries fall into one or some combination of the cases described below.

There is persuasive evidence that in many industries technological advance is what Winter and I (Nelson and Winter, 1982) have called cumulative, in the sense that today's new technology not only provides enhanced operational capabilities but serves as a new starting-point for tomorrow's efforts to further advance technology. Science may be involved as well, but in most industries science seems to be tapped as a body of general knowledge relevant to problem-solving, with 'new' findings not playing a special role. Where new science is not particularly important, a steady flow of newly minted scientists and engineers suffices to keep the laboratory adequately up to date with the world of public science. The principal intelligence function of the laboratory is directed towards following technological developments in its own field. To a considerable extent this involves keeping up with what competitors are doing.[6]

Interestingly enough, while at first one would have thought that competitors would hold what they are doing and what they have accomplished close to their chest (this was the implication of the earlier discussion about the importance of confidentiality in industrial R. & D.), there is apparently a good deal of open communication about certain things. (See for example von Hippel, 1986.) Partly this is because of general recognition that sharing enhances the common good, but recognition alone cannot solve a 'prisoner's dilemma' problem. However, the following facts of life do go a long way toward explaining the phenomenon. First, in most industries a significant head start with respect to a particular product or process suffices enables a firm to reap returns if it can move quickly to market or use it. Second, generic or general knowledge of what has been done is bound to leak out or be ferreted out anyway. If the product is sold, it can be reverse-engineered by competitors. If it is patented, while the patent may block direct mimicking, it also may provide a lot of useful generic information. Scientists and engineers are mobile in many industries, and while trade-secrets laws limit what their new employers can do with what they can tell, they bring their knowledge with them when they move.

The discussion above is about industries where today's technical advance grows from yesterday's advance in the same field. In some technologies the stage for today's advances is set not so much from advances within the core technologies of the firms in the industry as from advances in components and materials that are created 'upstream'. Of course industries do not always belong simply to one type or the other; in many industries both sources are important.

Monitoring and anticipating upstream developments has something in common with monitoring work by competitors, but has important differences as well. What is the same is that active monitoring of upstream developments probably involves some active R. & D. in the relevant areas. A firm needs to be able to assess which developments are useful to it and which are probably not. In many cases, in order to make use of upstream developments special tailoring is required, or the firm needs to modify somewhat its own process and product technologies, or both. However, unlike competitors, upstream suppliers are likely to be active sources of help in this endeavour. It should also be noted that the flow of information on new developments from suppliers to users, which is in the suppliers' interests, means that a portion of such information is available to their competitors. Technical societies play an important role in moving information up and down stream. Willy-nilly, they also serve to enhance the ability of competitors to monitor each other.

I suggested earlier that industries where new public science is a principal input to technical advance seem to be uncommon. However, there is a number of such industries. The biotechnology industries are a prominent current example. In these industries a firm's R. & D. laboratory must monitor the relevant fields of science closely, which generally means both that they must pay attention to what is going on in universities where most of the science proceeds, and that they engage in some of this research themselves, so that they can understand the significance of what the academics are finding or doing. In Nelson and Winter (1982) these technologies we called 'science-based' in contrast with 'cumulative'.

Both detailed technological histories which trace the sources of important antecedent scientific events, and data collected through a questionnaire that we put together a few years ago (Levin *et al.*, 1987) indicate that where technical advance is strongly fed by new scientific developments, the relevant sciences are likely to be what might be called 'applied sciences'. These are scientific fields in which potential applications play a major role in focusing research effort. Included here are fields like oncology, plant genetics, computer science, metallurgy, and the various engineering disciplines that have research traditions. Thus in science-based industries, not only is it so that new technology must be seen as growing out of new science. New science is to a considerable extent engendered through attempts to understand prevailing technology and its limitations.

In turn, the applied sciences and engineering disciplines draw strength from the underlying basic sciences. The power of their research efforts to solve problems and illuminate options is greatly enhanced when basic understanding is enhanced. However, it would be a mistake to see these applied sciences as merely the mechanisms that translate basic scientific understanding into the potential for technical advance. It is apparent that these fields have strong research traditions of their own and represent bodies of knowledge with their own integrity.

In this discussion I have only scratched the surface of describing the mechanisms through which firms stay abreast of relevant technical and scientific developments originating elsewhere. Like Cohen and Levinthal (1987) I have stressed the role of internal R. & D. Joint or co-operative R. & D. projects are important in some industries. Sometimes these join complementary skills, as when a firm and an equipment supplier, or firms producing different components of a system, work together. Recently there have been cases of otherwise rival firms co-operating on pre-competitive research. Patents and publications in technical journals provide information, and engineers share information with colleagues in other firms through a variety of informal mechanisms. Universities and firms are connected through various channels—consulting agreements, industrial-liaison programmes, industry funding of university research and fellowships, etc. The intelligence mechanisms are complex and appear to vary across industries. They are much in need of more detailed empirical and theoretical analysis. However, I propose, as a strong first approximation generalization, that one cannot intelligently follow and rapidly respond to new developments in a field unless one is doing some work in that field too.

3. CHARACTERIZING TECHNICAL CAPABILITIES

I would like now to sharply differentiate the view of technological capabilities of firms latent in the discussion above from that contained in orthodox economic theory. I want to do this on two counts; first, concerning what is involved in technological mastery, and how to characterize technological capabilities of firms; second, regarding what aspects of mastery are private and what are public.

3.1. On Technological Mastery

Orthodox economics tends to define technological capabilities in terms of 'knowledge', or 'information', with a strong connotation that industrial technology is understood by particular individuals and is readily articulable and communicable from one individual to another with the requisite background training. Industrial technology is like a recipe. The recipe is written down in texts, articles, patents, and private documents, and a firm's engineering staff is custodian and master of the collection of recipes. The articulated part of a technology, like the recipe written up in a cookery book, necessarily takes certain operations as primitive, like 'beat for two minutes', or 'bake at 400 degrees'. However, it is presumed that workers with relevant training will know how to carry out these operations, with suitable instructions from on high. An implication of very major importance is that what one firm can do, other firms can do too, if they had access to the relevant documents.

Sidney Winter and I have argued that this point of view is at best half-true, and in many cases badly misleading. To continue to use the recipe metaphor, there is much more to making a good soufflé than is written down in a cookery book, as anyone with limited experience who has tried knows. Written recipes and ancillary written material in a cookery book provide a starting-place, but that is all. Making a good soufflé requires, indeed, knowing a good recipe, which may or may not be written down, but also experience, a light touch, and perhaps even a special twist. Even experienced chefs using the same nominal recipe will not produce the same soufflé. So too with 'technologies'. What is written down—the recipe, the textbook discussion, the patent—provides a start, but only in the sense that a recipe provides a start. Knowing how to produce a product, like a soufflé, or an aircraft, is as much experienced tacit skill as articulable knowledge. And, contrary to the implicit general theory, the tacit skills of one 'skilled in the art' are not interchangeable. Who works with the recipe makes a difference.

Moreover, modern production is a team activity, and team skills are involved in its undertaking. The capabilities of a football team are not decomposable into the capabilities of the coach and the individual players, but involve teamwork that must be co-ordinated and practised and effort that must be motivated. So too the production capabilities of a firm are in truth organizational capabilities. There are two important consequences. One is that while a chef or a stand-alone craftsman can work with a recipe in mind that is not actually written down, relying on memory of similar tasks and how they were accomplished, it is virtually impossible to proceed by team action without some articulation of the recipe. A basketball move may be dreamed up by the coach on the spur of the moment. However, for the team to know what he has in mind requires instructions expressly laid out in some way. While, as I noted above, an articulated recipe is 'only a start' on command of technology, for team production it is a necessary start. When a firm decides to produce a product it has not produced before, even though basic knowledge of how to do this is in its engineers' heads and handbooks, a considerable amount of production engineering is needed to lay out just how this firm will do it so that the various players can see their parts. Secondly, the organizational aspects of mastery of technology—the division of labour, work flow, who has the power to make on the spot decisions, monitoring mechanisms, even attitudes—cannot be clearly separated from the recipe aspects because the way the recipe is effected is determined by how work is organized.[7]

It may be objected that I am muddling together technology which is to some extent articulated, individual skills that most would agree are less so, and organizational capabilities which are something else again. I am well aware of the logical distinctions. My point is that I am not muddling the logical distinctions but arguing that practice does not separate like the conventional logic and therefore the conventional logic is not a useful basis for theory. If the question concerns what a firm can do, these three aspects overlap to a considerable degree. Winter and I were stressing this point long before the issue arose about the source of Japanese prowess in the making of semiconductors and motor-cars. As the evidence on this accumulates, it is increasingly clear that the Japanese advantage does not lie particularly in better design, or scientific and technical sophistication, but in a whole range of little things involving individual actions and organizational procedures. But these amount to better mastery of the technology.

This point is central to the issue of how to characterize the technological capabilities of firms. Traditional economic theory depicts firms' capabilities in terms of a clearly defined production set, the boundaries of which define a sharp production possibility frontier or a production function. The idea is that a firm, or an economy, either has and knows a recipe or it doesn't. Relatedly, technological advance is a sharp concept, involving the introduction of a recipe that did not exist before. Winter and I have argued that this is badly misleading.

Rather, the appropriate concept is something like a function defined over techniques giving the expected time and effort needed to get a technique under control, together with some measure of uncertainty about these matters, and about performance once it is under control. At any time there are a limited number of techniques a firm controls reliably, in the sense of having the capabilities to invoke these with clearly predictable inputs and outputs. There are others a firm could control with some set-up investment in production engineering and in experimentation and practice, and with some uncertainty regarding just how the process would settle down. Still others clearly require significant R. & D. for mastery, and it may not be clear just how much time and effort would be required, or exactly what the technique will be like when got into shape and under control. And so on.

This points to an interesting question. What determines the kinds of capabilities a firm decides to get under strong control, and how do these capabilities evolve over time? This is the topic addressed by Dosi, Winter, and Teece in Chapter 6, which is strongly complementary to mine, both in subject-matter and in general theoretical outlook.

3.2. *Private Capabilities and Public Knowledge*

I read the evidence as suggesting strongly that the principal things that distinguish the technological capabilities of individual firms from the public knowledge bearing on the technology in question are, first, a worked-out recipe tailored to the general capabilities of the firm, and second, the tacit individual and organizational skills acquired in the course of learning how to make that recipe work. Together, these requirements for mastery of a technique impose a significant entry price for a firm with a good grasp of public knowledge but without the set-up investment and actual experience with the particular technique in question.

I have not mentioned patents, or overt secrecy, as key protectors of proprietary technology. There clearly are certain industries where patents effectively bar competitors from the use of key recipes which without patent protection would be relatively easy to reconstruct and learn to implement, but these seem limited to products which are basically chemical compounds, and those which are clear designs. The Yale survey (Levin *et al.*, 1987) and other studies show that in most industries patents are not particularly effective—they can be invented around—and even where they are, they mostly add to imitation costs that in any case would be substantial (see also Teece, 1986). And, from another point of view, the details contained in patents become public knowledge. While the patent may protect the particular instance, the patent provides usable information about how things can be done more generally. Similarly, there are cases where overt secrecy has been important, but secrets are hard to keep bottled up for long from outsiders who are knowledgeable about the general state of the art.

The public knowledge consists of generic understanding, with much of all but the newest codified in texts and taught in courses, with new developments reasonably rapidly spread among the *cognoscenti*. Generic public knowledge provides the material for specific recipe generation. But the public knowledge includes more than just generic knowledge. It includes as well knowledge not necessarily of the specifics of the recipes actually employed, but of the broad characteristics of most of them. This is important knowledge. Generic knowledge contains clues as to the kinds of recipes that will work plus a basis for working out and testing recipes, but efforts that strive for something significantly different from the tried and true are risky. Knowledge that someone else has made something work indicates that a recipe of that general type can work, and this provides very useful guidance to others.

The picture I am trying to draw is one in which public knowledge provides the basis for the development of new private technological capabilities, but in which the existence of the latter as they come into being in turn becomes an important part of the public knowledge. The inducements to private R. & D. and the other costs and risks of attempted innovation are those associated with a head start. The head start may amount to a temporary monopoly only, or it may yield a durable lead in a race, but in either case what was originally private soon becomes public.

This is basically the Schumpeterian model, although I hope I have fleshed it out a bit, and made clearer the relationship of individual firms to the social enterprise. Technical advance proceeds by exploring not the completely unknown but the only partially known. The firms are the explorers. We all learn from what they find.

NOTES

1. This is very much the orientation, for example, of the collection of readings put together by Tushman and Moore, 1982, on managing innovation.
2. See e.g. Coombs, Saviotti, and Walsh, 1987; Kay, 1979, 1984; Cohen and Leventhal, 1985, 1987; Levin *et al.* 1987; Mowery, 1983; Teece, 1986; and Winter, 1987.
3. The following analysis draws extensively on Nelson, 1987.
4. The rise of industrial R. & D. has been discussed extensively by Mowery, 1983; Freeman, 1982; Reich, 1985; and Rosenberg, 1985, among others.
5. Basically the same argument has been developed by Teece, 1986, and in earlier works of his, and still earlier by Williamson, 1975.
6. See e.g. Price, 1965, 1969; and Allen, 1966, 1970.
7. Many different 'theories of the firm' (see my earlier discussion) bear on this proposition. Here it suffices to cite Winter, 1987; Dosi, Teece, and Winter, Chapter 6 of this volume, and Lazonick, 1986 and Chapter 4 of this volume.

REFERENCES

ALLEN, T. (1966), 'Studies of the Problem Solving Process in Engineering Design', *IEEE Transactions in Engineering Management* (June).

—— (1970), 'Roles in Technical Communications Networks', in C. Nelson and D. Pollock (eds.), *Communication among Scientists and Engineers* (Lexington, Mass.: Health Lexington Press).

COHEN, W., and LEVENTHAL, D. (1985), 'The Endogeneity of Appropriability and R and D Investment' (Carnegie Mellon University, mimeo).

—— and —— (1987), 'Innovation and Learning: The Two Faces of R and D' (Carnegie Mellon University, mimeo, Mar.).

COOMBS, R., SAVIOTTI, P., and WALSH, V. (1987), *Economics and Technological Change* (Totowa, NJ: Rowman & Littlefield).

FREEMAN, C. (1982), *The Economics of Industrial Innovation* (London: Frances Pinter).

KAY, N. (1979) *The Innovating Firm* (New York: St. Martin's Press).

—— (1984), *The Emergent Firm* (New York: St. Martin's Press).

LAZONICK, W. (1986) 'The Cotton Textile Industry', in B. Elbaum and W. Lazonick, *The Decline of the British Economy* (Oxford: Oxford University Press).

LEVIN, R. C., KLEVORICK, A. K., NELSON, R. R., and WINTER, S. G. (1987), 'Appropriating the Returns from Industrial R and D' (*Brookings Papers in Economic Activity*).

MOWERY, D. C. (1983), 'The Relationship Between the Contractual and Intra-firm Forms of Industrial Research in American Manufacturing 1900–1940', *Explorations in Economic History*.

NELSON, R. R. (1987), 'Capitalism as an Engine of Progress' (Columbia University, mimeo).

—— and WINTER, S. G. (1982), *An Evolutionary Theory of Economic Change* (Cambridge, Mass.: Harvard University Press).

PRICE, D. de S. (1965), 'Is Technology Historically Independent of Science? A Study in Historical Historiography', *Technology and Culture*.

—— (1969), 'The Structure of Publication in Science and Technology', in W. Gruber and D. Marquis (eds.), *Factors in the Transfer of Technology* (Cambridge, Mass.: MIT Press).

REICH, L. (1985), *The Making of American Industrial Research: Science and Business at G. E. and Bell 1876–1926* (New York: Cambridge University Press).

ROSENBERG, N. (1985), 'The Commercial Exploitation of Science by American Industry', in K. B. Clark, R. H. Hayes, and C. Lorenz (eds.), *The Uneasy Alliance: Managing the Productivity–Technology Dilemma* (Boston: Harvard Business School Press).

SCHUMPETER, J. A. (1950), *Capitalism, Socialism, and Democracy* (New York: Harper).

TEECE, D. J. (1986), 'Profiting from Technological Innovation', *Research Policy* (Dec.).

—— (1988), 'Technological Change and the Nature of the Firm', in G. Dosi, C. Freeman, R. Nelson, G. Silverberg (eds.), *Technical Change and Economic Theory* (London: Francis Pinter).

TUSHMAN, M. L., and MOORE, W. L. (1982), *Readings in the Management of Innovation* (Cambridge, Mass.: Ballinger Publishing Company).

VON HIPPEL, E. (1986), 'Informal Know-How Trading Among Competing U.S. Firms' (MIT, mimeo).

WILLIAMSON, O. E. (1975), *Markets and Hierarchies: Analysis and Antitrust Implications* (New York: Free Press).

WINTER, S. G. (1987), 'Knowledge and Competence as Strategic Assets', in D. Teece (ed.), *The Competitive Challenge* (Cambridge, Mass.: Ballinger Publishing Company).

Excerpt from *Technology and Enterprise in a Historical Perspective*, Giovanni Dosi, Rena Giannetti and Pier Angelo Toninelli (eds.)

[2]

7

Some Foundations for a Theory of the Large Innovating Firm

Keith Pavitt

1. INTRODUCTION

In what follows, I shall identify and discuss variables and processes that should be part of any theory of the large innovating firm. There are at least three reasons for doing this.

First is that large firms are a major source of technology and innovations. Although formal R. & D. activities exaggerate the share of firms with more than 10,000 employees, their contribution is at least in proportion to their sales and employment, and an earlier study suggested that it is greater and growing (Pavitt *et al.*, 1987). They make particularly big contributions in the chemical, electrical and electronic, aerospace, and motor-car sectors, where innovative activities in most countries are concentrated in relatively few firms (Patel and Pavitt, 1989). Differences amongst these firms in their technological dynamism can have a major impact on the technological activities and competitive performance of whole countries.

The second is that large innovating firms in the twentieth century have shown remarkable resilience and longevity, in spite of successive waves of radical innovations that have called into question their established skills and procedures (Mowery, 1983*a*). Such institutional continuity in the face of technological discontinuity cannot be explained simply by the rise and fall of either talented individual entrepreneurs, or groups with specific technical skills. The continuing ability to absorb and mobilize new skills and opportunities has to be explained in other terms.

The third is that we have no satisfactory theory of the large innovating firm. J. Schumpeter has had most influence, with his (contested) assertion that large firms are at an advantage mobilizing resources and taking risks, given their size and monopoly power.

He also observed that the locus of innovative activities was shifting from the talented individual entrepreneur to the organized R. & D. laboratory. As a consequence, he argued that innovation would become a routine and predictable activity. Unfortunately, he had little to say about the organizational and other characteristics of innovating firms that succeed in maintaining the existence over long periods by continuously changing their products, processes, market, and operating procedures.

In Section 2, I identify some of the essential characteristics of innovative activities, and in Sections 3–6 I discuss the four essential dimensions defining the large innovating firm: firm-specific competences, their organization, their development over time, and their implications for methods of allocating resources within the firm.

2. KEY CHARACTERISTICS OF INNOVATIVE ACTIVITY

Past empirical research has helped to delineate the following key characteristics of innovative activities in firms.

1. They are largely *firm specific* in nature, and show *cumulative* development over time. Most technological knowledge emerges from the development, testing, production and use of specific products within firms. Actual and desirable performance characteristics are usually multidimensional and complex, and cannot easily be reproduced from scratch. Tacit knowledge obtained through experience is therefore of central importance. Although firms can buy in some technology and skills from the outside, what they have been able to do in the past strongly conditions what they can hope to do in the future.

2. They are highly *differentiated*. The range of feasible choice open to a firm is limited strongly by the extent to which its accumulated technology skills are proximate to other technologies: for example, skills in the development of pharmaceutical products may be applicable in the closely related field of developing pesticides, but they are not much use in the design and building of motor-cars. Technological competences also vary in their proximity to other technological opportunities. For firms in chemicals and electronics, there are rich possibilities in closely related products and markets; for firms in textiles, there are not.

3. In large firms, innovative activities involve continuous and intensive collaboration amongst *professionally* and *functionally specialized groups*. Knowledge inputs for any specific innovation normally draw on a wide variety of professional skills within both science and engineering, and within management. In nearly all innovating firms with more than 10,000 employees, these skills are organized into product divisions and functional departments.

4. Contrary to Schumpeter's prediction, innovative activities have remained *highly uncertain* in relation to their commercial outcome (Freeman, 1982). Only about one in ten R. & D. projects turns out to be a commercial success, and about half industrial R. & D. firms find no profitable application. In addition, both practitioners and theorists still have great difficulties predicting the rate and direction of radical technical change (see, for example, recent experience of electronic mail compared to FAX; also, the continuing uncertainties about satellite and cable television).

In the light of these characteristics, it is plainly misleading to assume—as in standard production theory in economics—that technological strategies in firms consist of easily implemented choices from amongst a large and easily accessible range of process technologies, on the basis of clear signals about relative factor costs. Given the specific, differentiated, and cumulative nature of technological development, the range of possible choices about both product and process technologies depends on accumulated, firm-specific competence (Section 3). Given functional and professional specialization, the implementation of technological choices requires organization and orchestration across disciplinary, functional, and divisional boundaries (Section 4). Given cumulative development and uncertainty, the improvement of these competences requires continuous and collective learning (Section 5). And in the light of all these characteristics, it is necessary to develop appropriate systems for evaluating progress and allocating resources (Section 6).

3. FIRM-SPECIFIC COMPETENCES

Firms gain profitable innovative leads through firm-specific competences that take time or are costly to imitate. Innovating firms can discourage imitation through secrecy and patent protection. But their competitive advantage essentially results from the ability to do useful and difficult things better than their competitors. The nature of these competences, the rate and directions in which they can be developed, and their implications for management, vary greatly according to the firm's size and its core business activities. Table 7.1 summarizes an attempt to map out this variety.

Innovating small firms are typically specialized in their technological strategies, concentrating on product innovation in specific producers' goods, such as machine tools, scientific instruments, specialized chemicals, or software. Their key strategic strengths are in the ability to match technology with specific customer requirements. The key strategic tasks are finding and maintaining a stable product niche, and benefiting systematically from user experience.

Large innovating firms, on the other hand, typically have a broad front to their technological activities, and are divisionalized in their organization. The key technological strengths can be based on R. & D. laboratories (typically in chemicals and electrical-electronic products), or on the design and operation of complex production technology (typically in mass production and continuous-process industries), and—increasingly—on the design and operation of complex information-processing technology (typically in finance and retailing).

In technologies based on R. & D., the key strategic opportunities are horizontal diversification into new product markets. The key strategic problems are those of mobilizing complementary assets to enter new product markets (e.g. obtaining marketing knowledge when a pharmaceutical firm moves into pesticides), and continuous organizational redesign to exploit emerging technological opportunities (e.g. personal computers cutting across previous responsibilities in computers, office machinery, and even consumer electronics; see Section 4 below).

In production-based and information-based technologies, the key strategic opportunities are in the progressive integration of radical technological advances into products and production systems, and in diversification vertically upstream into potentially munificent production inputs (e.g. CAD/CAM, robots, and software). The key strategic tasks are ensuring diffusion of best-practice technology within the firm, and choices about the degree of appropriation (i.e. internalization) of production technology.

This pattern of innovating activities shows clearly that firms do

TABLE 7.1 *Basic technological trajectories*

Definition	Source of technology	Trajectory	Typical product groups	Strategic problems for management
Science-based	R. & D. laboratory	Synergetic new products,	Electronics, chemicals	Complementary assets, integration to exploit synergies, patent money
Scale-intensive	Production engineering and specialized suppliers	Efficient and complex production and related products	Basic materials, durable consumer goods	Balance and choice in production technology between *appropriation* (secrecy and patents), *vertical disintegration* (co-operation with supplier), and *profit centre* 'Fusion' with fast-moving technologies. Diffusion of production technology amongst divisions, exploiting product opportunities, patent money
Information intensive	Software/systems dept. and specialized suppliers	Efficient (and complex) information processing, and related products	Financial services, retailing	
Specialized suppliers	Small-firm design and large-scale users	Improved specialized producers, goods	Machinery, instruments, speciality chemicals, software	Matching technological opportunity with user, absorbing stable or new product 'niches'

not have completely free choice as to whether or not to operate on a broad front or to be specialized, or whether to be product- or process-oriented. In one sense, the pattern can be seen as the result of cumulative and differentiated diffusion amongst firms of four pervasive technological systems over a long period of time, each with its distinctive skills and sources of technological competence: (1) mechanical (design and production engineering); (2) chemical (R. & D. laboratories); (3) electrical–electronic (R. & D. laboratories); and (4) software (systems departments). These long-term processes are well described and analysed by US economic and business historians (in particular, Mowery, 1983*a*, 1983*b*; Rosenberg, 1976, 1985; Mowery and Rosenberg, 1989; Hounshell and Smith, 1989), and for the spread of software in services by Barras, 1986. More detailed analyses of the development of competences include the development of video-cassette recorders, particularly by Japanese firms (Rosenbloom and Cusamano, 1987), and Canon's strategy in optical recording technology (Doz *et al.*, 1989).

4. ORGANIZING FIRM-SPECIFIC COMPETENCES

Firm-specific competences are based on skills and knowledge that are functionally and organizationally specialized. An essential dimension of a firm's competence is the ability to combine these competences into unique and effective units for developing innovations.

Considerable progress has been made in understanding the importance for the successful implementation of innovation of interfunctional integration, stretching back to the pioneering work of Burns and Stalker in the early 1960s. Horizontal communications across functional boundaries, flexibility in the definition of tasks, links with outside sources of expertise and with users, and the authority and experience of responsible managers, are all factors that influence a successful implementation, in addition to the quality and competence of R. & D. and related technological activities (Rothwell *et al.*, 1975; Rothwell, 1977; Cooper, 1983; Maidique and Zirger, 1984).

Reviews of our state of understanding have been published by (amongst others) Shrivastava and Souder, 1987, and by Burgelman, 1985. However, the translation of these research findings into action remains a non-trivial task, especially in large and hierarchical organizations. The recent major reductions in product development times in the Japanese car industry appear to have resulted from organizational changes entirely consistent with the result of this earlier academic research (Clark *et al.*, 1987). But car firms in other parts of the world have apparently not yet been so successful.

However, whilst considerable progress has been made in our understanding of the problems that the functional form of organization poses for the management of innovation, we know much less about the problems posed by the divisional form of organization, which Chandler (1977) identifies as a key element in the managerial revolution and the emergence of large firms. At first sight, this may appear puzzling given that, as we have seen above, innovative activities—especially in science-based firms—are often a major source of entirely new ranges of products, and consequently of new product divisions. Perhaps this is because theories of the M-form organization (e.g. Williamson, 1975) have been essentially static rather than dynamic, concerned with the efficient management of what exists, rather than with the creation and management of the new. In any event, there are at least three problems that innovative activities pose for the 'pure' M-form organization, and that deserve further research.

The first is the existence of synergies across divisions or, alternatively, of core competences that pervade the technologies of all product divisions. As we can see in Table 7.1, these are of central importance in all types of large innovating firms. At the very least, they require central (i.e. corporate-wide) co-ordination, and inter-divisional exchanges of personnel and experience (Aoki, 1986). They may also require strong direction from central R. & D. or production engineering activities. We have little systematic evidence on the nature and determinants of inter-firm and inter-sectoral differences in this corporate-wide technical function. We would expect it to be strongest in sectors and firms with competences in technologies that have rich and pervasive opportunities.

Second, the development of technological opportunities and competences does not necessarily fit tidily into established divisional structures. Rigid definition of divisional markets and missions can result in missed opportunities that cut across existing organizational structures: for example, PABX (small telephone exchanges) probably requires competences from divisions in telecommunications, office machinery, computers, and components. In addition, existing firm-specific technological competences may not be sufficient to exploit new opportunities, and may require the purchase of what Teece (1986) has called 'co-specific assets': for example, marketing competence for pharmaceutical firms moving into pesticides, or for computing firms moving into telecommunications.

Thus, the strategic management and deployment of firm-specific technological competences is an intrinsically untidy process, so that tidy divisional structures can either stifle or miss opportunities. Together with Robson and Townsend (1989), I have shown elsewhere that a high proportion of significant innovations in the UK between 1945 and 1983 were not developed in designated product divisions. And in their more finely grained case-studies of the corporate development of competences, Doz *et al.* (1989) conclude that success depends (amongst other things) on not being locked into established organizational forms, and on flexibility and variety in exploring opportunities.

Once again, this organization and deployment of corporate competences requires, at the very least, central co-ordination of decisions to redefine divisional missions and competences in the light of emerging technological opportunities. It may also require some strong central involvement in technological and managerial activities, based on an established R. & D. laboratory. Again, we have little systematic evidence on inter-industry and inter-firm differences.

The third organizational problem is the inevitable tension and balance between corporate centralization and decentralization. On the one hand, the strategic exploitation of technology at the corporate level requires a strong technological and managerial input at the centre, in order to exploit synergies and to redesign organizational missions and competences in the light of emerging opportunities (for example, see Reader, 1975, on the emergence of plastics in Imperial Chemicals Industry). On the other hand, we have seen that effective transformation of technology into commercial advantage requires decentralization, with effective horizontal communications and rapid decision-taking. Thus, too much centralization is likely to result in ambitious, radical, and ill-conceived innovations. Too much decentralization is likely to result in incremental and safe innovations in established businesses, with consequent diminishing returns and decline.

The continuing debates and fashions about the appropriate

degree of centralization and decentralization of R. & D. and related technical activities suggest that what is 'too much' or 'too little' is not easy to define in practice. This is another challenging research problem for the future.

5. FIRM-SPECIFIC LEARNING

In an activity as complicated and unpredictable as innovation, it is misleading to represent innovative activities in large firms as once-for-all decisions on the content of strategies or policies which, when taken, are easily and predictably implemented. This is well recognized in the 'processual' school of industrial organization, some of whose members even argue that 'strategies' and 'decisions' are no more than resultant or outcomes of political battles amongst functional and professional groups within the firm (Pettigrew, 1985).

Whilst this element of corporate behaviour is certainly real, it cannot be the whole story. Corporations are purposive organizations. As a result, they develop what Nelson and Winter (1982) call 'routines', or rules of thumb, to help them cope with a murky, messy, and ever-changing world. In the terminology of this chapter, 'routines' embody 'firm-specific competences'; and their adaptation and change in the light of experience and further information is 'learning'. The processes of learning dissolve the sharp distinction between the 'content' and 'process' schools of strategy. As Fig. 7.1 shows, they allow continuing feedback between the content of strategy on the one hand, and knowledge of the context outside the firm, together with competences within the firm, on the other. Three other features also emerge from Fig. 7.1.

First, sources of learning are very diverse, and their relative importance will vary according to the nature of the core competences of the firm:

- 'learning by doing' (i.e. by producing) is of greatest importance in production- and information-intensive firms, coping with complex and interdependent process technologies; it is also likely to involve some 'learning from suppliers';
- 'learning by using' (i.e. by operating production equipment) is of greatest importance to specialized suppliers of capital and intermediate goods, the limitations and further potentialities of whose products emerge from experience in use;

- 'learning by failing' (i.e. by modifying failed product innovations) is important over a wide range of sectors, where market experience enables modifications to products innovations, and where so-called 'robust designs' can be extended in different dimensions to cover a range of market segments;
- 'learning by studying' (i.e. by experimenting and assessing in R. & D. laboratories) is important for exploring potential opportunities and uses of pervasive technologies and radical discoveries. It is likely to be most important in science-based firms, where eventual assimilation into firm-specific competences may involve 'strategic alliances', and will eventually require 'learning by hiring';
- 'learning from competitors' (i.e. evaluating competitors' performance) is important for all innovating firms, although the means of doing so will vary. Reverse engineering is the

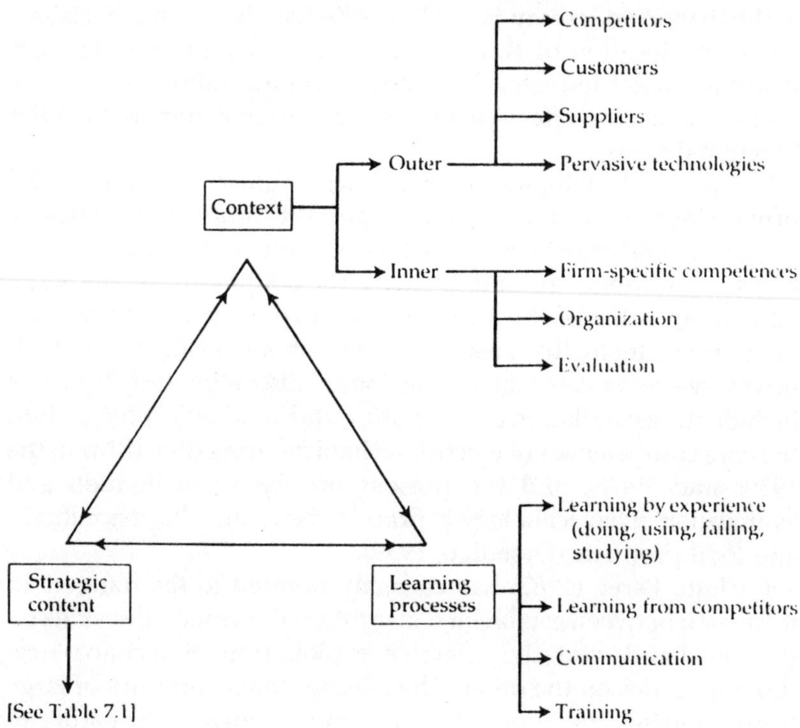

FIG 7.1 Learning processes in the innovative activities of large firms

main method of learning from competitors' product technologies. Equipment suppliers and informal exchanges among production engineers are channels for learning about process technologies. A number of firms are now beginning to use bibliometric methods to assess competitors' strategies.

Second, in the large firm, learning is a collective activity requiring frequent communication amongst specialists and functions. Since knowledge accumulated through experience is partly tacit, and the tasks to which such knowledge is applied are complex and loosely structured, personal contact and discussions are the most frequent and effective means of communication and learning. Policies for effective learning therefore go beyond training and organization to include those of geographical location. Tom Allen (1977) and other scholars have shown the importance of physical location in influencing patterns of communication, both within the technical function and between the technical and other functions in the firm. And Jeremy Howells (1990) has shown that decisions about the location of R. & D. laboratories by firms in the UK pharmaceutical industry have been strongly influenced by the requirements for effective internal communication with other functional areas.

Third, R. & D. laboratories and other technical functions in the firm are necessary and centrally important components in nearly all the dimensions of learning described above. This may help to explain the longevity and capacity for adaptation of the large innovating firms of the twentieth century. R. & D. laboratories have given them the means to learn about, evaluate, and—if necessary—assimilate major technological discontinuities. Examples include the assimilation of computing and solid-state physics into the core competences of electromechanical firms (like IBM) in the 1950s and 1960s; and the present process of evaluation and assimilation of biotechnology in firms in chemicals, pharmaceuticals, and food products (Orsenigo, 1989).

Carlotta Perez (1983) has correctly pointed to the dangers of mismatch between established institutional routines and skills on the one hand, and the effective exploitation of technological discontinuities on the other. The institutional continuity of large firms confirms the cumulative and multidimensional nature of firm-specific technological competences. Few (if any) technological discontinuities are now 'competence destroying' in that they make completely obsolete the whole range of existing competences in large innovating firms with strong technical functions (Tushman and Anderson, 1986). Cumulative assimilation and adaptation appear to be the norm.

6. ALLOCATING RESOURCES

The ultimate purpose of firm-specific competence is to create profitable investment opportunities that improve the present value of the firm. Put another way, the nature and extent of potentially profitable investment opportunities are defined in large part by accumulated firm-specific competences. The allocation of resources to create such competences poses unusual and difficult problems, since they involve processes with considerable uncertainties, which can take a long time, and whose outcomes (organization-embodied competences) are often intangible and difficult to measure. Given these properties, *ex ante* project assessments turn out to have wide and irreducable errors, and attempts to use purely quantitative 'figures of merit' for projects and programmes can be seen as part of processes of advocacy rather than as attempts to improve rigour and precision in resource allocation. Empirical research has shown that, given these uncertainties, methods of resource allocation tend to be incremental and closely coupled to the learning processes described in Section 5, with constant feedback from improved knowledge and experience to evaluation (Freeman, 1982).

None the less, the appropriate criteria and procedure for allocating resources to the creation of firm-specific competences still pose important and unresolved problems for theory and practice. These emerge from the growing debate (particularly in the UK and the USA) about the appropriateness of financial performance criteria as guides for the allocation of resources to R. & D., or even to commercialization of innovative products and production processes. Some practitioners argue that this leads to 'short-termism', reflected in under-investment in technological activities and undue concentration on safe and largely incremental modifications (EIRMA, 1986).

As Stewart Myers (1984) has pointed out, these problems result in part from the use of incorrect financial performance criteria.

Instead of assessing investments using discounted cash flow (DCF) as the basis of their contribution to the net present value of the firm, emphasis is often given to short payback periods, to internal rates of return, and to quarterly profit performance, all of which introduce bias against longer-term projects. However, Myers also points out that to use DCF is inappropriate when there are strong links between today's investment and tomorrow's opportunities, which is precisely the case for cumulative learning and firm-specific competences. Here, investment projects do not necessarily produce additions to the net present value of the firm now, but have an 'option value' for further investments at some time in the future. Since it cannot deal with option values, 'DCF is no help at all for pure research and development. The value of R&D is almost all option value. Intangible asset's value is usually option value' (Myers, 1984: 135).

There is also the more complex case when a firm builds on its own firm-specific competences to enter a new product market, or to adopt a radically new production technology. Such investments often combine both contributions to the net present value of the firm, and strong components of learning, with expectations of follow-on investments in the future. They normally take a long time before making a profit. Complete reliance on DCF and neglect of their 'option value' will presumably reduce the growth of the net present value of the firm.

As Myers points out, coping with option values is an important challenge to the developments in financial theory. In the mean time, large innovating firms have developed (more or less effective) 'routines' to deal with learning and option values, and these are often observable in the qualitative checklists commonly used by such firms to evaluate R. & D.-related programmes and projects. Again, we would expect characteristics of the technico-economic expertise at the centre of the corporation to be particularly important in the formation of corporate expectations about future developments in technology, and their implications for R. & D. programmes to open up future investment opportunities. Evidence shows that such expertise is stronger in Japanese firms than in those in Western Europe and the United States (Patel and Pavitt, 1987). We also know that expectations can sometimes turn out to be badly wrong: in the 1950s and 1960s, some very distinguished large electrical firms bet more heavily on nuclear energy than the semiconductor. We have little evidence on the nature and determinants of the formation of technological expectations in large innovating firms and their impact on the allocation of resources to R. & D. and on investment decisions.

7. CONCLUSIONS

We have identified and explored in this chapter four central characteristics of the large innovative firm:

- differentiated and firm-specific competences that dictate the direction and range of technological opportunities that firms are capable of exploiting;
- organizational forms that reconcile both the decentralization required for effective implementation, and the centralization required for the exploitation of core technologies and for the definition of appropriate divisional boundaries;
- processes of learning that enable firms to augment specific competences as a consequence of experience, monitoring of the external environment, and assimilation of radical technologies;
- methods of resource allocation that reconcile the requirements for profitable investment opportunities now, with the creation of profitable investment opportunities in the future.

These four characteristics are proposed as a common basis for the definition of the central tasks of the strategic management of technology, for the building of models at large innovating firms, and for the specification of hypothesis for empirical test. May it suffice at this stage to draw two conclusions.

The first concerns the role of R. & D. and other technical functions within the large firm. Our analysis shows that the technical function involves more than being a major actor in the implementation of innovations. It also involves contributions to the definition of the appropriate divisional boundaries, to the exploration of radical technologies and other forms of corporate learning, and to the formation of technological expectations that influence the corporate allocation of resources to R. & D. and to investment opportunities.

Second, although we now have considerable possibilities, through bibliometric techniques, for collecting systematic evidence on many dimensions of firm-specific technological competences, we still lack systematic knowledge of the organizational forms, learning processes, and methods of resource allocation that determine the development and exploitation of these competences. As has been made abundantly clear in this chapter, economic and business historians have a major role in enlightening us.

REFERENCES

ALLEN, T. (1977), *Managing the Flow of Technology* (Cambridge, Mass.: MIT Press).

AOKI, M. (1986), 'Horizontal vs. Vertical Information Structure of the Firm', *American Economic Review*, 76 (Dec.), 971–83.

BARRAS, R. (1986), 'Towards a Theory of Innovation in Services', *Research Policy*, 15/4: 161–73.

BURGELMAN, R. (1985), 'Managing the New Venture Division: Research Findings and Implications for Strategic Management', *Strategic Management Journal*, 6: 39–54.

CHANDLER, A. (1977), *The Visible Hand: The Managerial Revolution in American Business* (Cambridge, Mass.: Harvard University Press).

CLARK, K., CHEW, W., and FUJIMOTO, T. (1987), 'Product Development in the World Auto Industry', *Brookings Papers on Economic Activity*, No. 3.

COOPER, R. (1983), 'A Process Model For Industrial New Product Development', *IEEE Transactions in Engineering Management*, EM-30.

DOZ, Y., PRAHALD, C., and ANGELMAR, R. (1989), 'Managing the Scope of Innovation: A Dilemma for Top Management', in R. Rosenbloom and R. Burgelman (eds.), *Research on Technological Innovation Management and Policy*, iv (Greenwich, Conn.: JAI Press).

European Industrial Research Management Association (EIRMA) (1986), *Developing R. and D. Strategies*, Working Group Report No. 33 (Paris).

FREEMAN, C. (1982), *The Economics of Industrial Innovation* (London: Francis Pinter; 1st edn. published by Penguin in 1974).

HOUNSHELL, D., and SMITH, J. (1989), *Science and Corporate Strategy* (Cambridge: Cambridge University Press).

HOWELLS, J. (1990), 'The Location and Organisation of Research and Development: New Horizons', *Research Policy*, 19: 133–46.

MAIDIQUE, M., and ZIRGER, B. (1984), 'A Study of Success and Failure in Product Innovation: The Case of the US Electronics Industry', *IEEE Transactions on Engineering Management*, EM-31.

MOWERY, D. (1983*a*), 'The Industrial Research and Firm Size, Survival and Growth in American Manufacturing, 1921–1946: An Assessment', *Journal of Economic History*, 43: 953–80.

—— (1983*b*) 'The Relationships Between Intrafirm and Contractual Forms

of Industrial Research in American Manufacturing', *Explorations in Economic History*, No. 20.
—— and ROSENBERG, N. (1989), *Technology and the Pursuit of Economic Growth* (Cambridge: Cambridge University Press).
MYERS, S. (1984), 'Finance Theory and Finance Strategy', *Interfaces*, 14/1: 126–37.
NELSON, R., and WINTER, S. (1982), *An Evolutionary Theory of Economic Change* (Cambridge, Mass.: Belknap Press of Harvard University Press).
ORSENIGO, L. (1989), *The Emergence of Biotechnology* (London: Francis Pinter).
PATEL, P., and PAVITT, K. (1987), 'Is Western Europe Losing the Technological Race?', *Research Policy*, 16: 57–85.
—— and —— (1989), 'A Comparison of Technological Activities in FR Germany and the UK', *National Westminster Quarterly Review* (May), 27–42.
PAVITT, K., ROBSON, M., and TOWNSEND, J. (1987), 'The Size Distribution of Innovating Firms in the UK: 1945–83', *The Journal of Industrial Economics*, 35/3: 297–316.
—— —— and —— (1989), 'Technological Accumulation, Diversification and Organisation in UK Companies, 1945–1983', *Management Science*, 35/1: 81–99.
PEREZ, C. (1983), 'Structural Change and Assimilation of New Technologies in the Economic and Social Systems', *Futures*, 15: 257–375.
PETTIGREW, A. (1985), *The Awakening Giant: Continuity and Change in Imperial Chemical Industries* (Oxford: Blackwell).
READER, W. (1975), *Imperial Chemical Industries: A History* (Oxford: Oxford University Press).
ROSENBERG, N. (1976), *Perspectives on Technology* (Cambridge: Cambridge University Press).
—— (1985), 'The Commercial Exploitation of Science by American Industry', in K. Clark, R. Hayes, and C. Lorenz (eds.), *The Uneasy Alliance: Managing the Productivity/Technology Dilemma* (Cambridge, Mass.: Harvard Business School).
ROSENBLOOM, R., and CUSUMANO, M. (1987), 'Technological Pioneering and Competitive Advantage: the Birth of the VCR Industry', *California Management Review*, 29/4: 51–76.
ROTHWELL, R. (1977), 'The Characteristics of Successful Innovators and Technically Progressive Firms', *R&D Management*, 7/3: 191–206.
—— FREEMAN, C., JERVIS, P., ROBERTSON, A., and TOWNSEND, J. (1975), 'SAPPHO Updated—Project SAPPHO Phase II', *Research Policy*, 3/3: 258–91.
SHRIVASTAVA, P., and SOUDER, W. (1987), 'The Strategic Management of Technological Innovation: A Review and a Model', *Journal of*

Teece, D. (1986), 'Profiting from Technological Innovation: Implications for Integration, Elaboration, Licensing and Public Policy', *Research Policy*, 15/6: 285–305.

Tushman, M., and Anderson, M. (1986), 'Techriological Discontinuities and Organisation Environments', *Administrative Science Quarterly*, 31: 439–65.

Williamson, O. (1975), *Markets and Hierarchies: Analysis and Anti-Trust Implications* (New York: Free Press).

[3]

The Size of Firm and the Growth of Knowledge

Testimony of Jacob Schmookler before the Subcommittee on Antitrust and Monopoly of the Committee on the Judiciary, United States Senate, May 26, 1965.

The effect of the size of firm and market structure on economic progress has been a source of controversy for centuries. Much has been said but very little has been proved on both sides. If public policy is to reflect wisdom on this score, we shall have to replace plausible conjecture with hard fact.

This we are just beginning to do, but in my judgment many new facts will have to be put in evidence before abandonment of our antitrust laws can be justified. Quite the contrary, unless the evidence examined below can be explained away, it suggests that a more vigorous antitrust policy than our present one may be appropriate. Such an interpretation may indeed prove correct in the end. However, while I believe the evidence damages the case for bigness, it hardly seems solid enough to prove the case for smallness.

I shall confine my discussion to the effect of size of firm on the production of technological knowledge, because I know more about it. Of course, a balanced view of the relation of antitrust to economic progress would also need to take into account the effect of size of firm on the *use* of knowledge, and of market structure on both the production and use of technological knowledge, as well as other matters.

As soon as we ask, how does an increase in the size of firm affect the rate at which it produces new technological knowledge, a very embarrassing problem arises to cast a pall over the whole proceedings: how do you measure new technological knowledge? Theoretically it could be measured, at least for the purposes of the present problem, by the increase in output per unit of composite input that it makes possible. We have no such measure now, and until we do, any inferences based on other data must necessarily be provisional.

There are, however, two other kinds of indexes of output of knowledge available that are at least worth looking at: statistics of all patented inventions, and statistics of important inventions. While both kinds of data leave much to be desired, they point to identical conclusions with respect to the question at hand. For this reason if for no other, they merit serious attention.

Considered jointly, these two kinds of evidence suggest that beyond a certain not very large size, as a firm gets bigger it has to spend more to get an invention, the probability that it will use the inventions it makes declines, and so does the probability that its inventions will be significant.

Before discussing the implications of these conclusions, let me present the evidence on which they are based.

Table 1 shows the research and development expenditures per patent pending for firms in six different industries by size of firm. The data relate to 1953, because unfortunately the National Science Foundation Office of Special Studies stopped collecting data on patents in its subsequent surveys. Only firms having R & D programs are included.

Inspection of Table 1 reveals that in every industry, firms with 5,000 or more employees spent more on R&D per patent pending than did firms with less than 1,000 employees. Moreover, in every industry except chemicals, firms with 5,000 or more employees spent more on

Table 1. Research and Development Outlays per Patent Pending, 1953 (thousands of dollars)

Industry	Size of firm: Under 1,000 employees	Size of firm: 1,000–4,999 employees	Size of firm: 5,000 or more employees	Industry average (unweighted)
Machinery	$ 8.5	$14.2	$24.2	$15.6
Chemicals	11.2	24.4	23.6	19.7
Electrical equipment	15.7	12.6	25.6	18.0
Petroleum products and extraction	10.0	8.4	15.6	11.3
Instruments	15.8	14.4	37.5	22.6
All other industries	15.4	7.1	27.8	16.8
Average, all industries (unweighted)	$12.8	$13.4	$25.6	

Source:

National Science Foundation, *Science and Engineering in American Industry: Final Report on a 1953–54 Survey*, NSF 56–16, Appendix Tables A-34 and A-35.

R&D per patent pending than did firms in the next smaller size class. As shown by the last line in the table, for all industries combined, firms in the largest size class spent about twice as much as did firms in the two smaller size classes.

The greater R&D outlays per patent of large firms has several possible explanations. One is simply that large firms patent relatively less. This is probably true. However, well over three-fourths of the large firms participating in the N.S.F. survey reported in interviews that they patented "everything that is patentable or worthwhile".[1] In any case the difference in outlays per patent between large and small firms seems too great to be accounted for in this way.

Another possible explanation is that (*a*) the patents pending per firm reflect the efforts of both R&D and other technical and supervisory personnel, and (*b*) the inventive output of non-R&D personnel declines as the size of firm increases. The result is that small firms have a smaller R&D outlay per patent, because some of the relevant costs of inventing are borne by operating departments. I believe that a significant part of the result is to be explained this way, and I shall return to this point in another context later in the paper.

Still another possibility is that the inventions made differ according to size of firm, with big firms producing bigger, better, and more expensive inventions. Undoubtedly the inventions of big firms differ from those of small ones. They certainly seem to be more expensive. That they are, on the average, better in some significant sense seems very doubtful, however. Professor Edwin Mansfield found that holding R&D outlays constant, the number of significant inventions made by large firms in the chemical, petroleum, and steel industries declined as the size of firm increased. In Mansfield's words, "Thus, contrary to popular belief, the inventive output per dollar of R&D expenditure in most of these cases seems to be lower in the largest firms than in large and medium-sized firms."[2] Since the inventions made by large firms are more likely to receive publicity than those by small firms, and therefore were more likely to be included in Mansfield's study, I regard this result as especially significant since it emerges despite the presumptive bias in the opposite direction.

[1] National Science Foundation, *Science and Engineering in American Industry: Final Report on a 1953-54 Survey*, NSF 56-16, p. 37.

[2] Edwin Mansfield, "Industrial Research and Development Expenditures: Determinants, Prospects, and Relation to Size of Firm and Inventive Output," *Journal of Political Economy*, August 1964, 336.

39 Size of Firm and Growth of Knowledge

The assumption that the inventions of large firms tend to be inferior to those of small ones receives support from another direction. Two independent studies indicate that small firms use commercially a larger proportion of their patented inventions than do big firms. A very extensive survey, based on a 2-percent random sample of inventions patented in 1938, 1948, and 1952, conducted by the Patent Foundation of George Washington University, found that large firms used only 51 percent of their inventions commercially compared to 71 percent for small firms. "Large" firms were defined as those holding over 100 patents, or with some patents and over $100 million in assets. The group also included some firms with between 75 and 100 patents. "Small" firms consisted simply of those not defined as large.[3] The use rate for patents held by large firms, indeed, was not significantly larger than the 49-percent use rate for inventions made by independent inventors.[4] This fact is indeed striking, since independents often have to go into business for themselves before they can use their ideas.

An entirely different study, conducted at the Harvard Business School, of very large firms exclusively, and covering all patents held by the firms surveyed, found that firms with 1956 sales exceeding $500 million used only 51 percent of their patented inventions, compared to 56 percent for firms with sales of less than $500 million.[5]

In brief, existing comprehensive indexes of output of new technological knowledge suggest that beyond a certain not very large size, the bigger the firm the less efficient its knowledge-producing activities are likely to be. Evidently, as the size of firm increases, there is a decrease per dollar of R&D in (*a*) the number of patented inventions, (*b*) the percentage of patented inventions used commercially, and (*c*) the number of significant inventions.

These apparent disadvantages of bigness could be offset if R&D expenditures per firm rise as fast as R&D efficiency declines. While there is some tendency for this to happen, the tendency does not persist among the largest firms – the size group that is the natural focus of antitrust questions. It is true that the proportion of firms doing research and development increases as the size of firm increases. For

[3] Barkev S. Sanders, "Patterns of Commercial Exploitation of Patented Inventions by Large and Small Companies," *Patent, Copyright, and Trademark Journal*, spring 1964, Tables 1 and 3.

[4] *Patent, Copyright, and Trademark Journal*, conference supplement, 1958, Appendix Table 11.

[5] Frederic M. Scherer *et al.*, *Patents and the Corporation*, a report on industrial technology under changing public policy compiled by nine graduate students at Harvard Business School during the 1957-58 academic year, p. 112.

example, in 1958, about nine out of ten manufacturing firms with 5,000 or more employees had their own R&D programs, compared to only about six out of ten manufacturing firms with between 1,000 and 5,000 employees, and a much smaller fraction for firms with less than 1,000 employees. And it is also the case that in 1960, among manufacturing firms with R&D programs, firms with 5,000 or more employees spent 2.0 percent of the sales dollar on R&D compared to 1.4 percent for firms with 1,000 to 5,000 employees.[6]

However, impressions derived from such gross size classifications of firms are not fine enough for the present issue. No one today regards a firm with even 10,000 employees as big enough ordinarily to pose a potential antitrust problem. The real question is whether the very biggest firms in each industry engage in proportionally more research and development than firms somewhat smaller, thereby offsetting their presumptive R&D inefficiency. The answer is that, on the whole, they do not. Mansfield found that in petroleum, drugs, glass, and steel, over the period 1945 to 1959 the largest firms spent less on R&D as a percentage of sales than did somewhat smaller firms. Only in chemicals did the largest firms spend more on R&D per dollar of sales.[7]

This kind of result, together with the tendency for the number of significant inventions produced per dollar of R&D to fall as the size of firm increases, led Mansfield (in another article) to advance the very tentative estimate that the sixth largest firms in petroleum and coal mining, and an even smaller firm in steel, would have been of optimal size from the standpoint of maximizing the rate of innovation.[8] Because I believe that technological progress comes more from many small inventions that from a few large ones, and that smaller inventions are more likely to come from small firms than big firms, I am inclined to the view that the optimal size from this standpoint is even smaller than he estimated.

The impression derived from the evidence sketched above that, on the whole, very large firms contribute less to the advancement of technology in proportion to their size than do firms somewhat smaller

[6] National Science Foundation, *Research and Development in Industry, 1960: Final Report on a Survey of R&D Funds and R&D Scientists and Engineers*, NSF 63-7, Table 17, p. 41.

[7] Mansfield, note 2, 334. D. Hamberg reports substantially similar results in his article, "Size of Firm, Oligopoly and Research: The Evidence," *Canadian Journal of Economics and Political Science*, February 1964, 70-74.

[8] Edwin Mansfield, "Size of Firm, Market Structure, and Innovation," *Journal of Political Economy*, December 1963, 556-576.

is, of course, at variance with much contemporary opinion on the subject. To a considerable extent, the contrary view that the larger the firm, the greater its relative contribution to the progress of technological knowledge, rests on deductive logic. I do not wish to enter into that debate here. However, I should like to point out that deductive logic can contribute to our knowledge on the subject only to the extent that it begins from premises that approximate the salient facts. I submit that we do not yet know or understand all the salient facts, and that the logical briefs on behalf of the progressiveness of big firms ignore aspects of the problem which may be crucial.

On the factual side, the case for bigness appears to rest on two kinds of evidence: (*a*) the apparent progressiveness of a few big firms, and (*b*) the greater participation of large firms than small firms in formal R&D. Let me simply note, with respect to the first point, that the case for bigness has to be proved in terms of the record of big firms generally against that of small firms generally, because the performance of individual firms is far too variable for individual cases to prove anything.

The use of R&D data in the debate has been misleading on two counts. A study I did over a decade ago of inventions patented in 1953, when private industry was spending about $4 billion of its own money on R&D and the government was spending a good deal of its own, indicated that R&D men contributed about 40 percent of the inventions patented in that year, while supervisors, engineers, and scientists in the operating end of industry contributed about 20 percent, and independents contributed about as many patents as did the R&D men – roughly 40 percent.[9]

There is, of course, much that we would like to know, and do not, about the relative quality of the inventions involved. However, as indicated above, the inventions of independents are evidently commercialized about as frequently as are those of R&D men in big firms. Moreover, indirect evidence suggests that the inventions of operating personnel may be more often used than those of R&D men.[10]

[9] See my article, "Inventors Past and Present," *Review of Economics and Statistics*, August 1957, 321-333.

[10] Barkev S. Sanders, Joseph Rossman, and L. James Harris, in "Patent Acquisition by Corporations," *Patent, Trademark, and Copyright Journal*, fall 1959, Table 15, indicate that the corporately owned patents on inventions made by employees not under contract to assign patent rights are used commercially with greater frequency than are those on inventions by employees contractually obligated to assign patent rights.

Undoubtedly, the substantial growth of corporate research and development since the period covered by my study has increased its relative importance. Yet there is no question that the contributions of the hundreds of thousands of engineers, scientists, and supervisors in the operating end of business and of independents still remains very substantial.

What is more, even on their own terms and disregarding the serious questions of accuracy that can be raised about them, the R&D data relate entirely to expenditures and employment in R&D. They are thus measures of input alone. They tell us nothing whatever about output. There is no reason to suppose that input and output are perfectly correlated, and many reasons to suppose that they are not. Both indexes of output cited here – statistics of all patented inventions, and statistics of the number of important inventions – present a picture of the effect of bigness on the production of knowledge that differs radically from that derived from data on R&D inputs.

Indeed, if we tentatively accept the proposition that on the average a big firm has to spend more than a small firm to get an invention, or an invention it can use, or an important invention – as the evidence indicated above suggests – then the greater participation of big firms in the R&D process is explained not by the advantages of bigness but by its disadvantages: big firms have to run faster to stay in the same place. Actually, such an explanation is probably too simple. There are advantages as well as disadvantages of bigness, and what we observe (and that very inadequately) is the net effect of both. Hopefully, as research on the problem proceeds, the nature of the problem will become clearer.

In the balance of this paper I should like to focus on some aspects of the production of technological knowledge which seem to me to help explain the apparent disadvantages of excessive bigness suggested by the facts presented above. Producing a useful addition to technological knowledge requires many things. Among these are (*a*) an understanding of the problem to be solved, (*b*) a knowledge of whatever technological lore, science, or engineering bears on the problem, (*c*) access to the facilities needed to try out solutions, and (*d*) time.

With greater resources, big firms are better able than small ones to hire those knowledge specialists who can solve problems that others have indentified – when the solutions require formal training in science and engineering. They can give these men more time and better laboratory facilities. Yet, despite these clear advantages, the production of technological knowledge seems less efficient, on the average, in large than in small firms. Why? To answer this question it is helpful to consider the efforts of R&D personnel and operating personnel separately.

Professor Arnold C. Cooper of the Graduate School of Industrial Administration at Purdue University suggests several tentative conclusions based on a study of the problem which accord with my own impressions. After reporting that most R&D managers in both large and small firms agreed in interviews that a large company typically spends far more on R&D to develop a given product than does a small firm, Cooper suggested that the main reasons for this were that (*a*) "the average capabilities of technical people are higher in small firms than in large ones;" (*b*) R&D people in small companies "are more often concerned over how much a project costs than are their counterparts in larger companies;" and (*c*) "the problems of communication and coordination tend to be less acute in the smaller organizations."[11]

The suggestion that the technical people in small firms are likely to be, on the average, more competent than those in large firms may come as a shock to some, and possibly Cooper's finding is biased since his sample was drawn from the New England area – which is, after all, famous for the technological complex that radiates out from Cambridge. Nonetheless, the reasons he advances for his finding are cogent and bear reflection. Small, research-based firms, he argues, attract men who want to have a substantial effect on their environment and who want that effect recognized and appreciated; and such firms hire men of proven creativity rather than green graduates. By contrast, he argues, big firms tend to provide a haven for the mediocre in search of anonymity.

That the R&D men in small companies should be more sensitive to costs, and have smaller problems of communication and coordination, is obvious and I shall not pursue these points here, important though they are in explaining the phenomenon under discussion.

Instead, let me turn from R&D to the comparative contribution to knowledge of operating personnel in small and large firms. The big advantage of an operating man – whether he is a production engineer, a production supervisor, or a sales engineer – over an R&D man is that he

[11] Arnold C. Cooper, "R&D Is More Efficient in Small Companies," *Harvard Business Review*, May-June 1964, 78-81.

bumps up against technical problems and opportunities daily. He is therefore often in the best position to identify what needs bettering. However, I would suggest that there is a marked difference in both opportunity and incentive between an operating man in a small firm and one in a large firm. Even if this is not intuitively obvious, the fact was very apparent in the sample of inventions covered in my study cited above, for virtually all the inventions made by operating men in my sample were employees of small and medium-sized firms.

From the logic of the situation and from discussions with scientists, engineers, and executives in firms of all sizes, I would suggest several possible reasons why operating men in small firms are likely to contribute more new ideas than are those of large firms.

(*a*) A simple increase in size of the enterprise tends to make the productive process less comprehensible to the men who engage in it, simply because each man sees less of it.

(*b*) Big firms tend to be more than mere scaled-up versions of small ones: they cut the work up finer and narrow each man's responsibility, thereby further reducing his range of vision. Understanding less of what is going on, each man is less able to contribute to its improvement. Adam Smith was probably correct for his time when he suggested that inventiveness increases as the division of labor progresses, because the shops of the eighteenth century were tiny by our standards. However, the small factories of today would have been giants then. There seems little doubt that many of today's firms have subdivided labor so much that operating employees no longer understand the relation of their work to the rest of the enterprise well enough to make significant improvements, even when they have the native ability to do so.[12]

(*c*) The growth in the size of enterprise militates against invention by operating personnel in another way. The extreme division of labor and the larger numbers of individuals involved create a greater need for coordination and control by management, and of one tier of managers by another above it. Each man's influence is watered down, and his suggestions have less chance of acceptance. There are more managers, and therefore more potential vetoers. The channels of communication tend to become clogged if only because they are so long. Creative men in operating departments tend to find themselves frustrated by red tape and bureaucracy.

[12] This does not imply that such extreme division of labor is unjustified from the standpoint of current operating costs.

(*d*) As the organization becomes more formal, engineering and research men who are supposed to do the thinking tend to resent and discount suggestions by production and sales men who are supposed to do the doing. Whether warranted or not I do not know, but old-timers in the production line in any large-scale enterprise tell of good ideas that engineers blocked because they couldn't claim credit for them.

(*e*) Finally, there is the additional fact that independent inventors tend to steer clear of fields dominated by big firms. Indeed, even big firms in one industry tend to avoid fields dominated by established big firms. The reason in both cases is simply that outsiders are discouraged from entering such fields as producers, and generally are at a bargaining disadvantage if they try to sell their ideas to large potential customers for them. Hence, independently of any adverse effects of bigness on the *internal* rate of production of ideas, there is also an adverse effect on the rate at which ideas are produced outside the industry.

Let me now try to summarize my position. While we have much still to learn about what affects the rate of production of new technological knowledge, existing evidence suggests that there are significant adverse effects of bigness. On the average, very large firms seem to have to spend more than small ones to produce a patentable invention, to produce a useful invention, or to produce an important invention. These disadvantages are to some extent offset by the greater participation of large firms in the R&D process. However, very large firms generally do not participate in that process relatively more than do firms somewhat smaller.

The superior efficiency of smaller firms in the R&D process apparently reflects the superior quality of their technical personnel, greater cost consciousness, and better understanding of the problem to be solved resulting from closer contact with the firm's operations and better communications.

To the advantages of the small firm in R&D are added its more hospitable atmosphere for creative contributions by operating personnel, in consequence of the lesser division of labor and greater influence of individuals in such enterprises. There is the further fact that industries characterized by small firms provide a greater inducement for invention by outsiders.

By contrast, the case for bigness, while not wholly without merit, appears to rest on incomplete assumptions and misinterpreted evidence. The assumptions fail to take into account the effect of size of enterprise on individual incentives and opportunities to generate new ideas, while the evidence commonly adduced in support of bigness – statistics of R&D input or cost – presuppose a correspondence between input and output which apparently does not exist. While part of the greater participation of big firms in R&D – to the extent that it is indeed greater – probably does reflect some advantages of big firms, it also probably reflects the effort of large firms to compensate for their comparative inefficiency in the R&D process.

In the light of these considerations I conclude that it would be very imprudent to relax the antitrust laws in the hope of inducing thereby a more rapid rate of growth of the nation's technological knowledge.

Part II
From Coal Tar Dyes to IG Farbenindustrie

[4]

Coal Tar Dye Manufacture and the Origins of the Modern Industrial Research Laboratory

By John J. Beer *

TO the best of my knowledge, the industrial research laboratory, as we know it today, first appeared in the German dyestuffs industry during the second half of the nineteenth century. It is not surprising that this important scientific institution should have made its initial appearance in the chemical industry, for during the middle of the last century chemistry was further advanced in its technological applications than any other science, with the possible exception of that branch of physics we call mechanics. That it should have been the dyestuffs branch of chemical manufacture that first developed the industrial laboratory can in large part be explained in terms of the peculiarity of the dyestuffs trade, which, being an auxiliary of the dry goods and clothing business, was highly susceptible to the whimsical fluctuations of taste and the insatiable appetite for novelty characteristic of the fashion world. New colors enjoyed only brief popularity. For two or three years they could be sold at high profits, but thereafter the price plunged rapidly downward, often to a point considerably below the actual price of manufacture.[1]

To remain in business under such circumstances, the dye manufacturer (usually a chemist himself) was obliged to do two things. First, he had to improve the efficiency and quality of his production in order partially to offset the inevitable downward trend of dye prices; [2] and second, he had to find new colors to replace the old that were no longer profitable. At first he conducted his own scientific and technological investigations; but in the two decades that followed the founding of the coal tar color industry in 1857, as competition between the mushrooming color factories intensified, and as the chemistry of aromatic compounds grew ever more complex, he found that he alone could no longer master the diversified researches becoming indispensable to the survival of his business. He therefore began to hire help, and in this manner was born the industrial laboratory.

The establishment of commercially sponsored research, however, was not as simple as all that — as witnessed by the fact that both the French and the British, despite their initial leadership in the manufacture of aniline colors, were unable to set up a single industrial research laboratory worthy of the

* Eastern Illinois University.

[1] This point is well illustrated by the price drop of 100% pure alizarin in the following years:

1869	270 marks per kilogram
1873	120
1878	23
1881	17.50
1908	1.78

Franz Thissen, *Die Stellung der deutschen Teerfarbenindustrie in der Weltwirtschaft* (*vor, in und nach dem Kriege*) (a published thesis, Giessen, 1922), p. 25.

[2] In a similar price plunge, rosaniline crystals, valued at 250 to 300 francs in 1862, were worth only 25 to 30 francs in 1867. August W. Hofmann, George de Laire, Charles Girard, "Report on the Colouring Matters Derived from Coal Tar, Shown at the French Exhibition, 1867," in appendix of M. Reimann, *Aniline and its Derivatives* (London, 1868), pp. 104–105.

name during the entire period that preceded the First World War. They paid for this failure with the total collapse of their once thriving dyestuffs and fine chemical industry.[3] The Swiss, and more particularly the Germans, were more successful at introducing organized science into industry. They owed their success to several factors, the most important of which were their ample supply of academically trained chemists, their national tradition of painstaking work, and their pressing need to convert the few raw materials which they possessed into valuable products so as to be able to provide for their rapidly expanding populations.

Nor must it be assumed that the research laboratory, as we now think of it, came into being merely with the employment of one or more research assistants. It took a period of more than forty years, from 1868 to 1910, for that institution to mature. However, when it comes to making broad statements about the evolution of the industrial laboratory, it must be confessed that in the past we have not possessed a single case history upon which to base such generalizations. In order, therefore, to bring the history of corporate industrial research down to specific cases, most of the remainder of this paper will be devoted to a consideration of the growth of the laboratories of the Farbenfabriken vorm. Friedrich Bayer & Co., of Elberfeld in Westphalia. Since, however, this firm was not among the very first dye companies to gather a research staff, and as it has been selected for study on the sole ground that it is the only plant with an adequate record of its own past, I would, before launching into its history, briefly turn to a consideration of two other dye companies among the pioneers in this field. These firms, the Farbwerke vorm. Meister Lucius & Brüning of Höchst, near Frankfort-on-the-Main, and the Badische Anilin- und Soda-Fabrik of Ludwigshafen,[4] as early as 1868 began systemat-

[3] Failure on the part of the British and French to found industrial research laboratories of the type developed in Germany and Switzerland in the last three decades of the nineteenth century caused their organic and fine-chemicals industry to shrink to insignificance. So dependent did these two countries become on German dyestuffs, pharmaceuticals, and photographic supplies, that the outbreak of the First World War faced them with a serious crisis. Their failure to develop industrial research laboratories before 1914 must in large part be attributed to their national characters. In both lands the industrial scientist was looked upon as inferior to the academic scientist. For this reason the education of industrial scientists and engineers was neglected. Institutions of higher learning refused to take any part in training men to work in industry. It must also be said that, in general, the British and French were more likely to have a strong dislike for the monotonous drudgery and the restrictions on freedom of scientific inquiry which industrial research almost inevitably imposes. To these factors it should also be added that, in the period under consideration, both these countries were prosperous enough to be able temporarily to forego the benefits which industrial research brings.

[4] The Höchst works of Meister Lucius & Brüning was founded in 1863 by two university-trained chemists, Eugen Lucius and Adolf Brüning, and two businessmen, August Müller of Berlin and Carl Friedrich Wilhelm Meister, who had recently returned to Germany after many years of successful trading in Manchester, England. The company's first products, fuchsin and aniline blue, sold for barely enough to keep the company solvent, but still in the first year of operation Eugen Lucius made a discovery that spelled almost immediate prosperity for the little company. He found that the dye, aldehyde green, which had been discovered the year before by the Frenchman, Cherpin, could be precipitated from solution into a pasty mass that dyed cloth with greater evenness and intensity than was possible if the dye was dissolved in an alcohol dye bath, as had been the practice heretofore. Except for the fact that Eugen Lucius continued to direct the research of the Höchst plant and employed the help of a few academically trained chemists, we know practically nothing about the research activities of the firm during the next decade. *Farbwerke vorm. Meister Lucius & Brüning* (Höchst/M., 1913), pp. 3–5; and William Perkin, "The Origin of the Coal-Tar Industry and the Contributions of Hofmann and his Pupils," *Journal of the Chemical Society*, 1896, *69*: 623.

The Badische Anilin- und Soda-Fabrik was founded in 1865, but was actually a successor of the firm, Sonntag, Engelhorn & Clemm (1863), which in turn was a successor to the aniline dye company, located in Mannheim, by the name of Chemische Fabrik Dyckerhoff, Clemm & Co., (1861). The guiding spirits of these firms were the two brothers, August and Carl Clemm, both able chemists who had received their training under Liebig in Giessen. Associated with them was a Mannheim industrialist, Friedrich Engelhorn, a man of remarkable business genius and foresight. From the beginning, the Badische Anilin- und Soda-Fabrik was a large-scale undertaking which

ically to hire university trained chemists. It is regrettable that their company records shed little light on the specific purpose for which these men were employed. In all likelihood, the event which drove both plants into the rapid acquisition of a research staff was their eagerness to develop a commercially feasible reaction for the manufacture of alizarin, an old natural dye with an immense market, which the chemists, Carl Graebe and Carl Lieberman, had succeeded in synthesizing from anthracene a short time before in their Berlin laboratory.[5] Little is known about the operation of the laboratories in either of these two pioneering companies. We have no complete record of their staffs, of the problems they studied, or of the manner in which they coordinated the investigations of their chemists. The only item about which we do have some information relates to the close ties they maintained with such noted professors of organic chemistry as August Wilhelm Hofmann and Adolph Baeyer.[6] By leaning heavily on academic consultants and by employing the same rigorous, theoretical approach to scientific problems as practiced in the universities, these two factory laboratories provided their plants with such a steady flow of useful discoveries that they thoroughly convinced the management that the promotion of research was a profitable venture. Because of their progressive attitude toward science, both the Badische Anilin- und Soda-Fabrik and the Farbwerke of Höchst in the 1870's rose into first and second place respectively among the dye factories of the world.

If it is a pity that our knowledge of these two pioneering plants is so skimpy, we do, in now turning our attention to the aforementioned Bayer Company of Elberfeld, at least have this advantage, that we obtain an insight into what a smaller, less progressive dye factory was obliged to do in the line of scientific research in order to keep pace with the leading competitors in the field. We thus have the opportunity to observe the typical rather than the exceptional, and for that reason will be less likely to obtain a distorted picture of the rate at which the dye industry embraced science.

Like all the dozen or more dye factories that sprang up in Germany in the

---

distilled its own coal tar, made most of its own heavy chemicals, and, of all the German dyestuffs factories, manufactured the widest variety of coal tar dyes. Dr. August Clemm conducted the company's research (probably single-handed) till 1868, when two or three chemists were hired, among them the celebrated dye chemist, Heinrich Caro. Heinrich Caro, "Über die Entwicklung der chemischen Industrie von Mannheim-Ludwigshafen a. Rh.," *Zeitschrift für Angewandte Chemie*, 1904, *2*: 1343–1362.

[5] Carl Graebe and Carl Liebermann, *Berichte der Deutschen Chemischen Gesellschaft*, 1868, *1*: 1, 49, 104, 186; and *ibid.*, 1869, *2*: 2, 14, 332, 505. See also Louis F. Fieser, "The Discovery of Alizarin," *Journal of Chemical Education*, 1930, *7*: 2590–2633.

[6] Because of his contact with other chemists, Heinrich Caro, director of research at the Badische Anilin- und Soda-Fabrik, was able to provide his company with a steady stream of new dyes. In 1868, working in collaboration with Graebe and Liebermann, he developed a commercially feasible synthesis for alizarin. In 1874, collaborating with Adolf Baeyer then at Strasbourg, he discovered eosin. Two years later he traveled to England, where he secured from Otto N. Witt the secret of making chrysoidine. While on the island he also visited Peter Griess, then working for the Allsopp Brewery at Burton-on-Trent, and from him obtained samples of a whole series of colored azo compounds that Griess had made in his spare time and kept stored in a cigar box. One of those colored compounds proved to be a dye and was marketed by the Badische Company as helianthin. Meanwhile, Caro was also in touch with Emil and Otto Fischer, who were seeking to establish the structure of the triphenylmethane dyes. Caro's laboratory furnished these experimenters with some pure triphenylmethane compounds and with a guess as to what the structure might be. This guess proved to be correct. Caro received no new colors from this cooperative venture with the two Fischers; rather, this incident shows that his collaboration with other chemists was not always a one-sided affair in which only he benefited. In 1880, Caro's laboratory, working jointly with researchers from the Höchst dye works, initiated a program of close cooperation with Adolf Baeyer, now at Munich, trying to make the latter's indigo synthesis commercially feasible. This research project met one obstacle after another, and it took twenty years to bring it to a successful conclusion. A. Bernthsen, "Heinrich Caro," *Berichte der Deutschen Chemischen Gesellschaft*, 1912, *45*: 2007–2022.

early 1860's, the aniline company founded by the Westphalian businessman, Friedrich Bayer, copied its manufacturing processes from the British and the French, who had pioneered in the discovery and large-scale production of coal tar colors in the previous decade. The skill of Friedrich Bayer's business partner, Friedrich Weskott, a successful dyer by profession, was adequate enough to plan and operate a manufacturing process comparable in quality and efficiency to that of other dye firms. But after three years of operation, as competition between the plants intensified and greater efficiency was called for, Weskott and Bayer felt the need of the services of a university-trained chemist to help them improve the efficiency of the synthesis of their most important product, the dye fuchsin. The man they hired, a certain Dr. Schoenfeld, seems not to have come up to the owners' expectations, for he was shortly dismissed.[7] That he was not replaced by another academic chemist indicates that the factory solved its problem by other means. That same year the company employed a vocationally trained foreman, fresh from the newly founded textile trade school of Krefeld. This man, Eduard Siller by name, proved equal to the task of rationalizing the firm's fuchsin production, and the company immediately proceeded to add three more foremen of similar schooling to its payroll.[8]

For a full decade, from 1864 to 1874, these vocationally trained foremen were in sole charge of production and research at the Bayer plant. In this decade they adapted several new colors to large-scale production, most notably alizarin. So indispensable did two of these four men become, that the management took them into partnership lest they leave to found independent factories of their own.[9]

The type of experimenting done by these vocational chemists was confined almost exclusively to the improvement and adaptation to production of manufacturing processes about which they had heard or read. They did a little, but not much, in the line of trying to improve the methods by which dyes are applied to cloth. If ever they did make an effort to find new colors, they never succeeded. Though their method of research was not blindly empirical, it could not be termed theoretical or highly rational. They had been taught to reason by crude analogy, to think in such simple terms as oxidation or reduction and alkalinity, acidity, or neutrality; and they were acquainted with the reagents necessary to achieve these chemical conditions. Their "laboratory" consisted of a lead-covered table in a corner of the plant, upon which stood a few reagent bottles and some chemical glassware. Many experiments were made on a large scale while manufacturing was in progress; that is to say, the conditions or ingredients ordinarily used to make a batch of dyestuffs were altered slightly to see what effect the change would have on the yield or the quality of the product. To hide failures from the management, production figures were often falsified.[10]

About the year 1874, the Bayer Company again undertook to employ uni-

[7] Hermann Pinnow, *Werksgeschichte der Gefolgschaft der Werke Leverkusen, Elberfeld, und Dormagen zur Erinnerung an die 75. Wiederkehr des Gründungstages der Farbenfabriken vorm. Friedr. Bayer & Co.* (Munich, 1938), pp. 9–14, referred to hereafter as Pinnow, *Werksgeschichte.*

[8] The other three foremen were named Hempel, Fritz Rübel, and Eduard Tuft. *Ibid.*, pp. 19–22.

[9] The two men in question were Eduard Siller and Eduard Tuft; *ibid.*, pp. 34–35; Carl Duisberg, *Meine Lebenserinnerungen* (Leipzig, 1933), pp. 55–56.

[10] Carl Duisberg, "Selbsterlebtes und Schlussbetrachtungen," *Geschichte Entwicklung der Farbenfabriken vorm. Friedr. Bayer & Co., Elberfeld in den ersten 50 Jahren* (Munich, 1918, but never released to the public), pp. 601–603, referred to hereafter as Duisberg, "Selbsterlebtes," *Geschichte der Farbenfabriken.*

versity-trained chemists, for the foremen were becoming increasingly incapable of dealing with the growing complexity of higher requirements of dye manufacture. A determined effort was made to obtain men with some experience in the supervision of dyestuffs production. The company therefore resorted to luring chemists away from other factories. It was particularly skillful at enticing one fuchsin and two alizarin specialists away from Höchst. Yet, despite occasional successes, the policy of hiring chemists from competing firms proved unsatisfactory. Competitors bent every effort to keep their best men. Those who defected were usually second-rate scientists, their originality partly burned out, and they could never be fully trusted.[11] It was expected that the new chemists would spend the bulk of their time in the plant itself, occupying themselves with problems of cost, quality, and yield. In a comparatively short time their duties tended to become administrative and routine, with no time to search for new colors.

Two events in the year 1876 forced the Bayer Company to give serious attention to the problem of discovering new dyes. That year the German patent act was passed, replacing the hopeless patchwork of patent laws that had existed in Germany prior to that country's unification in 1871, with a single law applicable to the entire nation. Almost overnight the copying practices, used heretofore with telling success by the German dye firms, came to an end. Factories would now have to find their own new colors or be threatened with extinction.[12] A similar fate would await them if they failed to take advantage of the recent discoveries of Peter Griess, Otto Witt, and Z. Roussin, who in the year 1876 opened up the vast, immensely promising field of the azo dyes.[13] No plant could afford to neglect the azo dyes, least of all the Bayer Company, which for lack of new colors was rapidly approaching a business crisis.

[11] Dr. Oskar Gürke, from the Höchst factory, helped the Bayer company to set up an improved method for making fuchsin in which nitrobenzene replaced the poisonous and expensive arsenate as the oxidizing agent. The two alizarin experts, Doctors Hammerschlag and Heinrich Rose, brought the secret of oxidizing anthracene with saltpeter instead of the more costly potassium chlorate that had been used heretofore. *Ibid.*

[12] As long as German dye research lagged behind that of Britain and France, as it did from 1856 until about 1871, the chaotic condition of patent protection existing in the country prior to unification was actually an advantage, since it prevented foreigners from obtaining adequate safeguards for their inventions and thus permitted the Germans to copy at will. When, however, the German dye chemists overtook their British and French rivals, and copying was no longer profitable, then the German dye factories stood to lose more from the cou..try's ineffective patent system than they could gain. It was their good fortune that political unification came just at the right moment to make possible the passage of an effective, nationwi.le patent law, now that the country needed such a law to safeguard its growing number of discoveries. "Verein zur Wahrung der Interessen der chemischen Industrie Deutschlands E. V.," *Ausgewählte Kapitel aus der chemisch-industriellen Wirtschaftspolitik 1877–1927* (Berlin, 1927), 285.

[13] Peter Griess discovered the first diazonium compounds in 1858 while a student in Kolbe's laboratory at Magdebourg. Throughout the remainder of his life, most of which was spent in England, he continued to study this class of compounds and the phenomenon of azo coupling. In the course of these researches he came across several vividly colored compounds which convinced him that this class of chemicals had great possibilities as dyes. Despite the many colored azo compounds which he found, few were valuable as dyes. However, through his exploration of the azo coupling process, the results of which he finally published in 1876, he set the stage for the discovery of many valuable dyes in the decades that followed. August W. Hofmann, Emil Fischer, and Heinrich Caro, "Peter Griess," *Berichte der Deutschen Chemischen Gesellschaft*, 1891, *24*: xvii.

Otto N. Witt, another German chemist who worked in England during part of his career, served as chemist for the firm Williams, Thomas & Dower at Brentfort near London. Using the azo coupling principles of Griess, he discovered in 1876 (simultaneously with Heinrich Caro) the valuable azo color chrysoidine and the tropeolin family of dyes. Stephen Miall, *A History of the British Chemical Industry* (London, 1938), pp. 267, 270.

Z. Roussin was a French pharmacist, who in 1875 (simultaneously with Caro and Witt) discovered the first representative of the important sulfonated naphthalene branch of the azo dye family. This color was first introduced to the market in 1876 by the firm Poirier of Paris under the name "orange," and was a great commercial success. A. Wahl, "Le Problème de la Fabrication des Matières Colorantes Organiques en France," *Revue Scientifique*, 1916, *54*: 7.

At first, the management sought to meet the situation by employing more university graduates in order to give each chemist in the plant more time for research. This policy met with some success but apparently not enough to suit the management, for in the year 1881 it was decided to employ a young chemist named Seidler and to allow him to spend the bulk of his time in the laboratory searching for new colors. Unfortunately, Seidler, after two years of effort, failed to come through with any significant findings.[14]

At this point, Carl Rumpff, one of the firm's directors, hit upon a new scheme to provide the company with dye discoveries. He employed three young chemical doctors, newly graduated from the university; but instead of putting them to work in the Elberfeld plant, where there was always a danger that they would be drawn into routine production duties, he sent each of them to a different university, on what we today would call a post-graduate research grant, where they were expected to work on some project of interest to the company. It was a clever idea that not only provided the young chemists with laboratories far superior to what they would have had at the factory, but it was also an indirect and inexpensive way of employing the professor under whose direction the men were placed as a consultant in the firm's research program. This original program failed miserably, largely because of the unrealistic assignments given each of the young trio by inexperienced research directors in the home plant. Within less than a year they were told to report for work at Elberfeld.[15]

Now at last, and rather unexpectedly, fortune smiled on the firm's floundering research program. One of the three young men, Carl Duisberg, was an outstanding chemist. Within the span of two years (1884–1886) he discovered three important azo colors [16] which so increased the company's business that he was quickly detached from all factory responsibilities, placed on full-time research, and, as the months passed, given a growing number of research assistants.[17] These assistants, increasing from four in 1886 to twelve two years

[14] In 1882 Seidler found two violet azo dyes of slight commercial value. In the fall of 1883 he was sent to Burton-on-Trent in England to assist Peter Griess in finding a commercially feasible synthesis for a blue azo color (later called sulfonazurin) which the old brewery chemist had recently discovered. His mission, however, was a failure and shortly after his return to Elberfeld he left the company. Bernard Heymann, "Die erfinderische Tädigkeit der Farbenfabriken auf dem Gebiet der Teerfarben-Industrie," *Geschichte und Entwicklung der Farbenfabriken vorm. Friedr. Bayer & Co. Elberfeld in den ersten 50 Jahren* (Munich, 1918, but never released to the public), pp. 13–14.

[15] The threee chemists were named Carl Duisberg, Oscar Hinsberg, and Wilhelm Herzberg. Duisberg was sent to study under Fittig in Strasbourg, Hinsberg went to the University of Freiburg, and Herzberg to the Technische Hochschule of Munich. The company wanted Duisberg to find a commercial synthesis for indigo from isatin, but the process was too complicated and yielded an artificial indigo whose cost was far above that of the natural product. Duisberg's findings were purely academic. He clarified the reaction by isolating an intermediate compound which he called imisatin, whose presence had previously been unsuspected. The nature of Hinsberg's assignment has been lost, but Herzberg was charged with attempting to sulfonate in the cold, powdered, or coarsely ground chunks of beta-naphthol. Duisberg, "Selbsterlebtes," *Geschichte der Farbenfabriken*, p. 587; Duisberg, *Lebenserinnerungen*, pp. 26–27.

[16] Benzopurpurin, a vivid red especially useful for cotton; benzoazurin, a blue also used primarily for cotton; sulfonazurin, a blue for woolens which could compete with the natural dye of indigo.

[17] The first chemist to be engaged for the purpose of assisting Duisberg in his many tasks was Dr. Low, whom Adolph Baeyer had recommended as one of his most capable students. Low was soon joined by Dr. Emil Meyer, who was transferred from the alizarin division to Duisberg's laboratory in order to work out a commercial synthesis for dianisidine, an important base that Duisberg had recently discovered. That same year (1886) Dr. Myrtil Kahn, probably the first graduate from a *Technische Hochschule* to be hired for research by the Bayer firm, joined the staff, and he was soon followed by Dr. Moritz Ulrich (formerly of Höchst) who was transferred into the research unit from the fuchsin department. Then in 1887 two more men were hired, Doctors Israel and Ott, former assistants to Professors Geuther of Jena and Fittig of Strasbourg, respectively. Six new men were added during the next two years, Doctors Krekeler, Heymann, Nieme, Bonnhoefer, Lauch, and Pfitzinger. The number grew larger year by year, as illustrated, at least to some degree, by the following table showing the total number of chemists employed

later, were urgently needed to exploit Duisberg's discoveries to the limit and to fend off all attacks upon or circumventions of his precious patents. In this manner was born the company's research division — a separate unit concentrating on the search for new colors. Technological research continued to be part of the various production divisions such as the azo dye division, the alizarin division, etc.

In 1890 the directors of the plant voted to house the new research division in a laboratory costing 1½ million marks. This modern building incorporated some original features of interior arrangement, conceived by Duisberg, that have been widely copied by industrial laboratories ever since.[18] It was staffed with glass blowers, dyers, analysts, and a host of other technicians, as well as with young laboratory boys, all of whom relieved the chemist of routine work and permitted him to concentrate exclusively on study and thought. Through the purchase of the personal libraries of such recently deceased chemists as August Kekulé, Victor Meyer, and Henry Roscoe, the new laboratory was rapidly equipped with an adequate library.[19]

Quite early in the organization of his staff, in 1886, Duisberg instituted a bi-monthly staff conference obviously patterned closely after the university seminar. Through these conferences he hoped to keep his men abreast of current literature in the field, to keep them from growing overspecialized, and to allow them to profit from each other's scientific insights.[20]

Some three years later Duisberg and his colleagues conceived and put into effect a uniform program for training all newly employed chemists. The newcomers spent their first year in the various divisions of the plant and laboratories. Depending on their interests and aptitudes, they were then assigned either to production or research. A decade later two separate training schools were created, one for the chemists destined for the plant, the other for those expected to discover new products. The curriculum for this latter group involved training in the dyeing and printing of cloth, in the testing of samples for exact shading and fastness, in the synthesis of representative types of dyes, and finally in the analysis and resynthesis of dyes currently being sold by

---

by the company:

1881 — 15
1885 — 24
1890 — 58
1895 — 90
1896 — 104.

Unfortunately these figures do not differentiate between those chemists engaged in research and those employed in production. In 1885, roughly one-fifth of the company's chemists were engaged in the search for new dyes, and in all likelihood that proportion increased rather than diminished in succeeding years. Duisberg, *Lebenserinnerungen*, p. 59; Ferdinand Fischer, *Das Studium der technischen Hochschulen Deutschlands und das Chemiker-Examen* (Braunschweig, 1897), pp. 46.

[18] Duisberg's innovations pertained to the layout of the working space in the laboratory room. As a student, he had noticed the great advantage in terms of mutual stimulation which the Liebig type of laboratory afforded. It was simply a large room with rows of long tables. A segment of each bench was assigned to each student. However, Duisberg disliked the complete lack of privacy which such a layout made inevitable. He wished to combine the advantages of this kind of laboratory with those of the private laboratory, and he hit upon the idea of a box-like arrangement of the working space within a large room. Each of the twelve chemists in the laboratory was to have a U-shaped niche and was separated from his neighbor on either side by two long work benches topped with reagent shelves. These came to shoulder height, enabling the men to see and converse with each other. Within his box, each man had for his own exclusive use the apparatus and reagents he commonly needed. The advantages of the box arrangement were many. First of all, it was more economical to build and maintain than twelve private laboratories. It made for *esprit de corps*, for keener competition, and for ready intellectual cross-fertilization. Finally, it made it difficult for a chemist to hide important discoveries in the hope of leaving the company and selling them to someone else, as the Bayer chemist Böttiger had done with Congo red eight years before. Duisberg, "Selbsterlebtes," *Geschichte der Farbenfabriken*, p. 652; Duisberg, *Lebenserinnerungen*, pp. 69–70; Carl Duisberg, "Das wissenschaftlich-chemische Laboratorium der Farbenfabriken vorm. Friedr. Bayer & Co., zu Elberfeld," *Chemiker Zeitung*, 1893, *17*: 191–197.

[19] Duisberg, *Lebenserinnerungen*, p. 70.

[20] Duisberg, "Selbsterlebtes," *Geschichte der Farbenfabriken*, p. 609.

competing firms. During the first year the students were expected to learn as much as possible about the chemistry of dyes, but they were carefully kept away from company secrets. Those that survived the rather difficult program were then gradually initiated into one specialized area of research and for security reasons kept from learning too much about the investigations of colleagues working in areas different from their own. Contracts, issued on a yearly basis, were so written as to make betrayal of company secrets very difficult.[21]

Of the research work done in the Bayer laboratories, only a fraction was devoted to the search for dyes of radically new chemical composition. By far the largest amount of time was devoted to experiments of an unspectacular or routine nature in which the theoretical principle was well understood. Only on rare occasions in the course of this work did a reaction refuse to go as predicted and thus point up a new theoretical problem. The number of routine experiments that had to be conducted to find a single promising color was large. When such a color was discovered, it was sent to the dye-testing division, where it was subjected to a battery of tests to indicate whether and under what conditions it would tint any one of the common fibers, or such other items as wood, paper, leather, fur, or straw. Then each item successfully tinted was subjected to several agents of destruction to determine fastness. Of 2,378 colors produced and tested in the year 1896, only 37 reached the market.[22] This tedious, meticulous experimentation, in which a thousand little facts were wrenched from nature through coordinated massed assault, admirably illustrates the method and spirit introduced into scientific inquiry by the rising industrial laboratory of the late nineteenth century.

During the three decades that preceded the outbreak of the First World War, the Bayer research laboratory grew at such a phenomenal rate that changes in the organizational structure of the staff were frequent. New divisions were continuously created as dye chemistry became more ramified and the company diversified its production to include pharmaceuticals, heavy chemicals, photographic supplies, and synthetic polymers. Aside from knowing the dates when the new divisions were created, who worked in them, and what products they discovered, we are still very much in the dark about the manner in which they were organized and administered. So far, company records have failed to shed much light on how the research of individual staff members was coordinated, by whom and by what criteria research projects were selected and their success evaluated. For lack of adequate answers to these queries, we are not yet able to judge precisely to what extent the complex research team, as we know it today, had evolved by the outbreak of World War I.

From information gathered to date, it would appear that by 1904 the Bayer laboratory, which by that time had fully caught up with the leading industrial chemical laboratories in Germany, was aware of and had effectively moved to counter some of the potential evils inherent in the bureaucratization of science. With regard to the current question of what should be the ideal balance between long-range fundamental research and short-term inquiries of more immediate practicality, it would appear that this problem had not yet received its present formulation. The danger of applied research outdistancing the discovery of fundamental laws and concepts, though seen as a distinct possibility

[21] Carl Duisberg, "The Education of Chemists," *Journal of the Society of Chemical Industry*, 1896, *15*: 427–432; Sydney H. Higgins, *Dyeing in Germany and America with Notes on Colour Production* (Manchester, 2nd ed., 1916), p. 109.

[22] Pinnow, *Werksgeschichte*, p. 107.

in the future, had not yet become acute. The Bayer management (and in this it was quite typical) seems to have felt that fundamental research was the province of the academic institutions,[23] that it was to the long-range interest of the firm to cultivate close relations with outstanding professors, to help these with grants and rare chemicals wherever possible, and to obtain their services as consultants in return for very generous honoraria.[24] The primary function of the plant's laboratory, as the Bayer directors saw it, was to exploit with the utmost speed and thoroughness any scientific idea whose practical applications had come into view. To that end the research staff was given such a flexible organization that a crash program involving the close collaboration of several types of experts could be initiated at short notice.

Except for the fact that they relied more heavily on academic research and themselves did little in the line of purely speculative inquiry, it can be said in conclusion that the industrial chemical research laboratories of Germany in the decade prior to the First World War were already highly institutionalized and closely resembled the commercial research laboratories of our own day.

[23] Through such organizations as the *Deutsche Chemische Gesellschaft* and the *Verein zur Wahrung der Interessen der Chemischen Industrie* the German dyestuffs companies, working in cooperation with the entire chemical industry, brought heavy and steady pressure to bear upon the governments of the German states to improve and expand the chemical training offered at the universities and *Technischen Hochschulen*. Some directors of dye firms also took an intense personal interest in the educational program of Germany's institutions of higher learning. Of these Heinrich Caro and Carl Duisberg are probably most outstanding. All Duisberg's numerous official pronouncements on education are reprinted in: Carl Duisberg, *Abhandlungen, Vorträge und Reden aus den Jahren 1882–1921* (Berlin, 1923), pp. 117–240.

When it became evident about 1908 that the universities and *Technischen Hochschulen* were no longer the ideal places for the increasingly complex and time-consuming job of probing deeply into the unknown, the Bayer Company combined with the other leading chemical works of Germany to erect and finance the operation of an elaborate chemical research institute within the newly created *Kaiser Wilhelm-Gesellschaft*. Max Planck, *25 Jahre Kaiser Wilhelm-Gesellschaft zur Förderung der Wissenschaften* (3 vols., Berlin, 1936), I, pp. 1–6, 31–33, 67–71.

[24] Cooperation with university chemists was so actively sought by the various German companies that a veritable competitive struggle arose between them over the "control" of the most important academic laboratories; and in time each company managed to establish strong ties with certain schools to the exclusion of all other rivals. The Bayer firm had particularly close ties with Wislicenus at Würtzburg and with the University of Göttingen. Through the efforts of Carl Martius, the Agfa of Berlin had a virtual "monopoly" on Hofmann's findings and a first choice in the hiring of his students. Höchst and the Badische Anilin- und Soda-Fabrik competed, or sometimes cooperated, for the favor of Adolph Baeyer, who taught successively in Berlin, Strasbourg, and Munich, and of Baeyer's many gifted students, such as Graebe, Liebermann, Emil and Otto Fischer, and Ludwig Knorr. Duisberg, "Selbsterlebtes," *Geschichte der Farbenfabriken*, pp. 602–603.

[5]

The Industrialization of Invention: A Case Study From the German Chemical Industry

By Georg Meyer-Thurow*

IN THE LAST DECADES of the nineteenth century the relationship between science and industry changed in a decisive way. Invention was industrialized. Large company laboratories were set up. The consulting scientist and the scientific entrepreneur were replaced by the salaried industrial research worker. Applied science became a driving force for technical development and economic growth.[1]

The German chemical industry has often served as the outstanding example of an early science-based enterprise. In particular, it has become commonplace to stress the importance of research for the rise to world supremacy of German dye manufacturers. But while this truism has been readily accepted, the manner in which science was turned into a productive force has not been subject to detailed investigation. "Because it turns out, ex post," Paul M. Hohenberg wrote in 1967, "that the gamble was a good one for the chemical industry in 19th-century Europe, identifying the sources of the industry's growth reduces largely to finding the factors affecting the decisions to take the gamble." The prevalence of this attitude has meant that apart from the work of John J. Beer, little is known about industrial research in the German chemical industry up to 1914.[2]

The first small dyestuffs firms were founded in several western European countries in the late 1850s and early 1860s. Although the number of workers employed and the amount of capital invested always remained small as compared with older ventures such as textiles, coal, or iron, the rise of the dye industry was remarkable in at least two ways. First, dye manufacturing was the prototype of a new type of industry based on science. Second, Germany dominated the manufacture of dye-

*Department of History and Philosophy, University of Bielefeld, Federal Republic of Germany (FRG).

This article was first presented to a conference on "Obstacles and Innovations in the Applied Sciences" at the Centre Nationale de la Recherche Scientifique in Paris in May 1979. I wish to thank Edwin Layton and Arnold Thackray for their suggestions.

[1]See John D. Bernal, *Science and Industry in the 19th Century* (London: Routledge & Kegan Paul, 1953), p. 151. To Bernal I owe the phrase "industrialization of invention." More recently see David F. Noble, *America by Design: Science, Technology, and The Rise of Corporate Capitalism* (New York: Knopf, 1977).

[2]Paul M. Hohenberg, *Chemicals in Western Europe: 1850–1914* (Chicago: Rand McNally & Co., 1967), quoting p. 83; John J. Beer, "Coal Tar Dye Manufacture and the Origins of the Modern Industrial Research Laboratory," *Isis*, 1958, *49*:123–131; and John J. Beer, *The Emergence of the German Dye Industry* (Urbana: Univ. Illinois Press, 1959; Arno rpt. 1981), still the best work on this subject. The latest monograph to deal in some detail with the history of the German dyestuffs industry is Peter Borscheid, *Naturwissenschaft, Staat und Industrie in Baden (1848–1914)* (Stuttgart: Klett, 1976).

stuffs. This leap to hegemony, almost to world monopoly, constituted, as David Landes has pointed out, "Imperial Germany's greatest industrial achievement."[3] But these traits were not apparent at the outset. For many years science and industry were only loosely connected in dyestuffs manufacturing. Industrial research, often considered the most typical characteristic of German dyestuffs companies, was an evolutionary development. It was as much a result of the growth of these companies as a precondition for their further development. The turning point lay somewhere in the 1880s.

The present study examines the history of one company, the Bayer AG (founded in 1863 and incorporated in 1881 as Farbenfabriken vorm. Friedrich Bayer & Co.), during the years 1880–1914, when scientific research began to play a decisive role in the development of the dyestuffs industry. It concentrates on the main scientific laboratory at the Bayer Company and on the research staff working in this laboratory. After a short sketch of the early history of the dyestuffs industry in the 1860s and 1870s, the study deals with five aspects of industrial research at the Bayer Company from the 1880s to 1914 in some detail: the origins of the industrial research laboratory; the gradual expansion of research laboratories and the elaboration of an infrastructure for research; the organization of the research chemist's work; the transformations that the process of research underwent; and the effects of industrial research.[4]

THE DYESTUFFS INDUSTRY, CHEMISTS, AND CHEMISTRY IN THE 1860s AND 1870s

Up to the late 1870s the development of the dyestuffs business was divided into two distinct phases. First came the period of the aniline dyes, a handful of artificial colors of great beauty but little fastness, mainly used to dye silk. Then, after the synthesis of alizarin in 1868 by two German scientists, the production of alizarin dyes dominated the field. Their successful synthesis led to the replacement for the first time of a natural dye, here madder, which was next to indigo in importance and widely used for the dyeing of cotton.

The real growth in production of dyestuffs is hard to measure, especially for these early years, because rapidly falling prices characterized the industry. Even so, the total value of dyes produced seems to have risen rapidly (see Table 1).[5] Even though one has to rely on the estimates of contemporary chemists when trying to identify the proportional

Table 1. Estimated world production of dyestuffs

Year	Value (in millions of marks)
1862	7.5
1867	22.5
1872	30
1878	70
1883	92

[3]David S. Landes, *The Unbound Prometheus* (Cambridge: Cambridge Univ. Press, 1969), p. 276.

[4]For a fuller account of the history of the Bayer Company and its chemists see Georg Meyer-Thurow, "Industriechemiker und Chemische Industrie: Eine Fallstudie zur Symbiose von Wissenschaft und Industrie, 1857–1914," (Ph.D. diss., University of Bielefeld, FRG, 1982).

[5]Figures for Tables 1 and 2 are based on estimates made by contemporary academic chemists closely connected with the dyestuffs industry. For 1862–1872 see August W. Hofmann, "Introduction," *Gruppe III: Chemische Industrie: Wiener Weltausstellung. Amtlicher Katalog der Ausstellung des Deutschen Reiches* (Berlin: Von Decker, 1873), p. 108; the figures for 1872 do not include the value of the production of alizarin dyes which, according to the same source (p. 109), amounted to 12 million marks in 1873. For 1878 see M. Ch. Lauth, *Exposition Universelle Internationale de 1878 à Paris. Group V, Classe 47: Rapport sur les produits chimiques et pharmaceutique* (Paris: Lacroix, 1881), p. 115. For 1883 see Carl Engler, "Uber den heutigen Stand der Teerfarbenindustrie," *Zeitschrift des Vereins Deutscher Ingenieure*, 1884, *28*:937–940, 953–958, on p. 957.

Table 2. Germany's estimated contribution to the dyestuffs industry (in millions of marks)

Year	World	Germany	Percentage of total
1878	70	44	62.86
1883	92	60	65.22

contribution of different nations to the overall value of dyes produced, there is no doubt that the German manufacturers dominated the field from a very early date (see Table 2). A small group of corporations played a leading role in this development. While the total value of German dye production in 1883 was about 60 million marks, the production of four companies accounted for 60 percent of the total, with three firms far ahead of all the others: the Badische Anilin und Soda-Fabrik (BASF) at 14 million, Höchst at 10 million, and Bayer at 9 million marks.[6]

At the outset, the dyestuffs industry was only loosely connected with scientific research. In the early years chemist-entrepreneurs (trained chemists who started their own or became part owners of dye firms) bridged the gap between theory and practice. They kept contact with academic research and were in control of production. Academic science profited from the intensified interest in organic chemistry. In turn, academic chemists produced important inventions which had a profound influence upon the dyestuffs industry. Academic chemists also contributed to the development of production techniques. But science was not fully institutionalized in industry, as the development of industrial research at the Bayer Company demonstrates. The laboratories Bayer established in the 1860s were small and poorly equipped. The number of chemists employed remained small, their mobility was high, and they were generally restricted to quality control in production. As late as 1877, when the Bayer Company employed seven chemists, four were occupied with routine analysis and two with color tests. Moreover, these chemists were hampered because production remained under the control of foremen who anxiously guarded their manufacturing secrets. It was not until 1878 that the Bayer Company began to place the manufacture of dyestuffs under the management of chemists.[7]

[6]Figures from "Die Lage der deutschen Teerfarbenindustrie," *Frankfurter Zeitung*, No. 219, 7 Aug. 1883. The difference in size between Bayer and BASF or Höchst was greater than the sales figures seem to suggest; Bayer was behind in respect to diversification and vertical integration and considerably smaller in respect to manpower and financial strength. The respective share capital in 1883 was (in millions of marks) BASF, 16.5; Höchst, 12; Bayer, 7.5. See *Handbuch der Deutschen Aktiengesellschaften* for 1914/1915, Vol. I, pp. 1490–1491, 1493, 1495.

[7]The seventh chemist was engaged to reorganize the aniline dye department and left after his job was done. See Hugo Hassencamp, "Die bei meinem Eintritt in die Firma F. Bayer & Co. (1877) vorgefundenen und in den darauffolgenden Jahren beobachteten Verhaltnisse: Ein Beitrag zur Geschichte der Farbenfabriken," (1909), MS, Bayer archives, Leverkusen, FRG, 1/6.1, Vol. 2. For the early history of the Bayer Company see *Geschichte und Entwicklung der Farbenfabriken vorm. Friedrich Bayer & Co., Elberfeld, in den ersten 50 Jahren* (Munich, 1918) (printed but not published); Hermann Pinnow, *Werksgeschichte zur Erinnerung an die 75. Wiederkehr des Grundungstages der Farbenfabriken vorm. F. Bayer & Co.* (Munich: Bruckmann, 1938); and Beer, *Emergence*, pp. 72–93. For the early cooperation of dyestuffs firms with academic chemists see Beer, *Emergence*, pp. 61–69; Borscheid, *Naturwissenschaft*, pp. 111–157; and Curt Schuster, *Wissenschaft und Technik: Ihre Begegnung in der BASF wahrend der ersten Jahrzehnte der Unternehmensgeschichte* (Ludwigshafen: BASF AG Abt. Offentlichkeitsarbeit/Unternehmensarchiv, 1976). For an American example of academic cooperation with industry in research see John W. Servos, "The Industrial Relation of Science: Chemistry at MIT, 1900–1939," *Isis*, 1980, *71*:531–549.

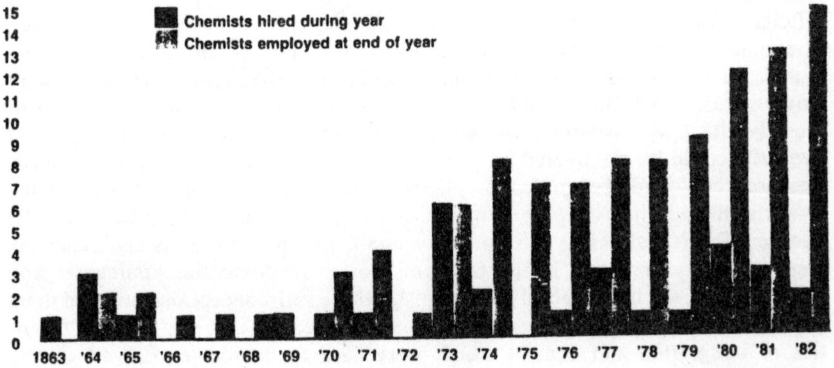

Figure 1. Recruitment of chemists at Bayer, 1863–1882. Out of thirty-one chemists engaged during this period, only eighteen remained with Bayer at the end of 1882; three of these entered into partnership (two in 1872, one in 1877) and no longer appear in the tally of chemists employed after that point. Turnover rates were high: six chemists left during their first year of service, two during their second, and four during their third.

Systematic and continuous industrial research at Bayer had to wait several more years. Until the 1880s progress was due to chance discoveries, which relied in part on empirical trial-and-error, in part on the results of academic research. Only once (in 1864) did Friedrich Bayer and Friedrich Weskott, founders of F. Bayer & Co., engage a chemist to devote his time in the laboratory to the "invention of new colors." About half a year later they ended this first unsuccessful experiment in industrial research and did not dare repeat it for almost two decades.[8] Producing industrially useful knowledge was not yet the main part of any chemist's job; what counted was technical experience, not scientific competence. For some time, therefore, Bayer's management preferred to recruit chemists from other companies, that is, from within the industry, rather than from the universities or *Technische Hochschulen*. So far as can be determined, six of the eight chemists hired during 1863–1871 and five of the six hired during 1874–1878 were recruited from other companies. No doubt this recruitment was one of the main routes of the transfer of knowledge between companies; it usually took place after the introduction of new products or new manufacturing processes. Figure 1 summarizes development at Bayer from 1863 to 1882. Peaks in the figure (based on a variety of sources in the Bayer archives) clearly reflect the effect on company recruiting policy of the wave of dyestuff inventions in 1863–1864, the start of the alizarin production in the early 1870s, and the introduction of new aniline and azo dyes in the late 1870s and early 1880s. Comparison of the number of chemists hired each year with the number employed also reflects the chemists' high mobility.

Knowledge of organic chemistry as well as technical expertise certainly played a role in the development of the dyestuffs industry. But the phenomenal success of the German chemical manufacturers—and the success of a small number of firms, among them Bayer, which quickly grew into big corporations—was due more to entrepreneurial strategies and favorable political and sociocultural factors than to

[8]See Friedrich Bayer's business correspondence, Bayer archives 16/1.2, 16/1.3, quoting Bayer to Philip Passavant, 12 May 1864.

research effort or scientific open-mindedness. The necessary preconditions for efficient industrial research were still missing: the chemical formulae and modes of action of successful dyes remained mysterious. For instance, though Kekulé elucidated a structure for the benzene molecule in 1865, organic chemists were slow to realize that this would elucidate coal-tar chemistry. The constitution of most dyestuffs was still unknown, and the possibility that commercially important dyestuffs could be discovered by systematic research was not as yet envisioned. Research could not be guided or planned, its success was uncertain, and the financial risks of investing in "human capital" were incalculable. Moreover, the number of artificial colors was still very small, and manufacturers and managers were primarily absorbed in the problems posed by production, purchase, and sales. Finally, the high mobility of chemists whose self-conceptions oriented them towards teaching, private consulting, or an entrepreneurial career raised difficulties of integration and control which were hard to overcome. But all of this changed in the late 1870s and the early 1880s.

THE ORIGINS OF THE INDUSTRIAL RESEARCH LABORATORY

The factors that precipitated the institutionalization of science in industry fall into three categories.

The importance of research for industrial progress was rapidly becoming apparent. A new interest in organic dye chemistry, beginning in the 1860s, led towards the end of the 1870s to a whole series of scientific and technical breakthroughs. Emil and Otto Fischer, then pupils of Adolf von Baeyer, professor of chemistry at the University of Munich, disclosed the structure of the first aniline dyes in a series of masterful experiments. Von Baeyer himself managed to synthesize for the first time the most important natural dye, indigo, after many years of investigation. The diazo reaction, a coupling reaction between diazonium compounds and amines or phenols, had been discovered by Peter Griess in the late 1850s and early 1860s. His reaction showed the way to countless possible dye products. But it was only in the second half of the 1870s that the pioneer work of one English and one French dyestuff firm raised the active interest of the dyestuffs industry in the azo colors, which in a very short time were to become the most numerous and most important class of dyestuff colors. The interest of organic chemists turned to the application of dyestuffs as well. The work of a German chemist engaged to an English dyestuff firm, Otto N. Witt (later professor of chemistry at the Technische Hochschule Berlin), provided the first steps toward a deeper understanding of the relationship between the constitution and the quality of dyes. And finally, in the early 1880s academic and industrial chemists realized that intermediates used for the production of dyestuffs could also be used for the manufacture of pharmaceutical products.

Economic factors encouraged institutional change in response to this rapid advance of scientific work in the field of dyestuff chemistry. A small group of big corporations, among them Bayer, now dominated the market and therefore possessed enough financial power to risk investing in the uncertain future of research. In addition, the crisis in the dyestuffs trade from 1875 to 1878 and after 1882 made these companies receptive to change: research offered a strategy that might reduce dependence on the ups and downs of the business cycle. Furthermore, the depressed economy, especially in the mid-1880s, deeply affected chemists by reducing their chances of employment and restricting their mobility, thus promoting their involuntary acceptance of the role of white collar workers.

Aside from scientific advances and ecomomic developments, the first German patent law encouraged the transplantation of research from academe to industry Enacted in 1877, this patent law had at least three remarkable effects. The publication of patent applications and granted patents provided information to all companies on advances in the state of the art. Since the patent protection was confined to new chemical processes, chemists were stimulated to find and patent all possible processes, thus blocking their competitors. Last but not least, the resulting endless patent litigation constituted a sort of legal warfare that intensified the connections between manufacturers and the academic scientists cailed upon to provide expert testimony. Thus cooperation between academic scientists and industry, characteristic of the early history of the dyestuffs industry, continued to be important. Yet when industrial research began on a larger scale, the relationships between industry and academe were decisively altered, as the history of the Bayer Company well demonstrates.

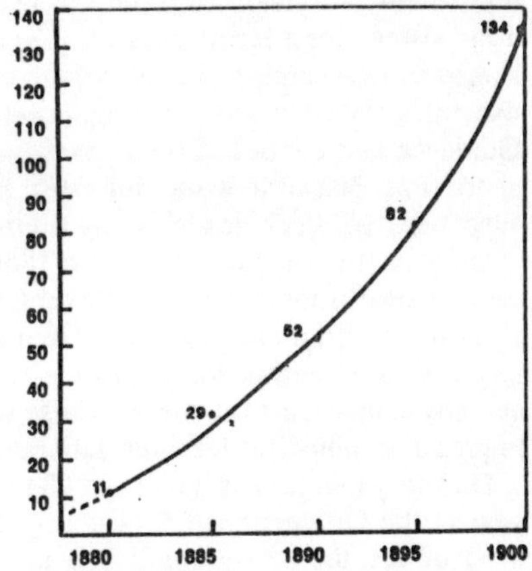

Figure 2. Number of chemists employed at Bayer, 1880–1900.

The necessity of investing in the production of knowledge, however, was only slowly accepted. Three important changes did occur early on in the internal structure of the Bayer Company in response to the scientific, economic, and legal developments of the late 1870s and early 1880s. From 1882 on the number of chemists employed, which started to rise continually only in 1878, increased at an accelerated rate, as indicated by Figure 2.[9] At the same time recruitment of chemists shifted more and more from rival firms to the universities, concentrating almost exclusively on Ph.D.s. Finally, chemists replaced foremen as the directors of the process of production. Scientific control and improvement of manufacturing processes were institutionalized. As a by-product, production personnel gradually began to engage in the development of new production methods and the discovery of new dyes.

The company also reorganized its relationship with academic research according to its growing need for innovations. The Bayer management followed three different strategies. It formed connections with several outstanding academic research chemists and formalized these connections by means of contracts securing every commercially important scientific result for the company. It engaged newly graduate Ph.D. chemists for one year and sent them to a university or *Technische Hochschule* laboratory, where they worked on research problems related to the company production program. In 1886 Bayer also began to support a laboratory,

[9]Figures (the number of chemists employed at the end of the given year) are based on the Bayer chemists whom I could identify. Discrepancies with data used by Beer and others (cf., e.g., *Emergence*, pp. 83–84) may be due to my inability to identify every chemist engaged by Bayer; to differences of definition, including apothecaries, physiologists, etc.; or to differences in the time the chemists were counted.

although the firm did not own it, and it was separated from the manufacturing plant by more than a hundred miles. At first one and later two chemists spent some time there doing research for Bayer.

These early attempts to make research serve the company's goals were less than a complete success and reflected a certain unwillingness to integrate research fully into company activities. The Bayer Company was not yet prepared to create its own research division, still preferring external laboratories. Instead of formulating a consistent long-term research program, the management handed a variety of short-term research topics to young chemists who never had been a part of the dyestuffs industry and were ignorant of the industry's possibilities and needs. Guidance and control of their investigations were left to academic teachers. For a short time research work done for Bayer continued in this transitional stage, suspended between academic institutions and industry.

Only in the second half of the 1880s did industrial research clearly begin to become institutionalized. The Bayer Company made a decisive step when it emancipated industrial chemists both from external academic research and from auxiliary activities within the company. The key figure in this change was a young chemist named Carl Duisberg, whose work led to the foundation of one of the most impressive industrial research laboratories known at that time.[10]

Duisberg had joined Bayer in 1884 after having done research for Bayer for one year at the University of Strassburg. He was appointed assistant chemist to the head of the then very small azo manufacturing plant. There he continued his research on azo direct dyes, a class of artificial organic colors introduced in 1884, which revolutionized dyeing practice because they could dye cotton directly, without prior treatment with a mordant.

Duisberg's research proved to be extremely successful. A series of discoveries followed, including dyes that turned out to be of the utmost commercial importance. Without any planning, a group of research assistants began to gather around this young inventor. These assistants were destined to follow up the lines of research initiated by Duisberg and to make sure that an invention was fully exploited: laboratory findings had to be tested with regard to their technical qualities and turned into production processes, and patent protection had to be gained and upheld against any possible infringement.

While Carl Duisberg became the personal nucleus of an inspired group of research-oriented young chemists, the group's cooperative effort decisively influenced the further development of Bayer. Duisberg's group performed many functions. Its members had to make rapid shifts among research, development, production, and sales. But their work meant an infusion of scientific modes of work, behavior, and thought into industry, and it demonstrated the value of a combined research effort. By about 1890 Bayer had recognized the distinctive role of the research chemist and made research a salaried, lifelong, specialized occupation.[11] Yet many difficulties remained. The laboratories were scattered all over the plant, located wherever there happened to be a free room. Their facilities were primitive and not always suitable for scientific research. And as the number of young research chemists increased, there was a growing need for mechanisms to organize and coordinate their work. These problems led to a momentous decision;

[10]See also Beer, *Emergence*, pp. 80–88; Hans-Joachim Flechtner, *Carl Duisberg: Vom Chemiker zum Wirtschaftsfuhrer* (Dusseldorf: Econ, 1961).

[11]Dr. Bernhard Heymann, who joined the Bayer Company in 1889 as a research chemist, was the first chemist to stay in research (until 1914).

the Bayer Company in 1889 decided to build a new main scientific laboratory, in which its full-time industrial research was to be concentrated. Bayer opened the new laboratory in August 1891.[12]

The arrangement of the laboratory reflected the special needs of industrial research; it was widely imitated throughout the industry in subsequent years. One large room was fitted up for twelve chemists with twelve box-like working places, each of which resembled a separate laboratory, although the whole room retained its unified character. This arrangement made it possible for each chemist to concentrate on his line of research without interruption and with a modicum of privacy, yet it permitted communication and consultation with fellow-chemists. Competition was stimulated, and the esprit de corps of a scientific community strengthened. Most importantly, the box arrangement facilitated the general supervision of the laboratory, allowing a strict control of the activities of each chemist if necessary. Moreover, the laboratory contained all necessary auxiliary services. In this respect Duisberg's laboratory surpassed every university laboratory then in existence.

These institutional innovations of the Bayer Company capped the process of separating research from other functions of chemists in industry. By 1891 research in the main laboratory had separated from the part-time research work done in the laboratories for the technical control of the manufacturing processes, so that research found an acknowledged place beside production and sales. The locus of invention had been transferred from external, mostly academic institutions to industry. But the institutionalization of research in this laboratory was but one aspect of the industrialization of invention.

CREATING AN INFRASTRUCTURE

The main scientific laboratory became the center of a developing research organization that continuously extended and improved until 1914. In the 1890s there were limitations. First, research in this laboratory remained concentrated on only some significant sectors of dye chemistry; it neglected, for instance, the chemistry of pharmaceuticals and of alizarin, both of which already were part of the production program. Second, the research chemist's work still included some analysis, some development of products, and, most importantly, a lot of routine patent work. Third, for a long time the latent function of the main scientific laboratory was to qualify and socialize chemists joining Bayer. Whenever the rapidly expanding works needed a production chemist, the management preferred to recruit him from the main laboratory staff. This practice led to a high rate of turnover, which interrupted the laboratory's research. In an attempt to correct these defects, several changes were made in the organizational structure of research. Three trends can be discerned: the expansion of research through the creation of new research laboratories, the further division of labor, and the elaboration of refined managerial techniques.

Table 3 shows the expansion in the number of research laboratories and research chemists at Bayer between 1897 and 1912. This table gives an approximate idea of the different fields of research and their relative importance.[13]

[12]For a detailed description of the laboratory see "The Appreciation of Science by German Manufacturers," *Nature*, 1893, *48*:29–34. Cf. Beer, *Emergence*, pp. 84–85.

[13]These four years have been chosen because they are representative for those Bayer research

Table 3. Number of chemists in research laboratories at the Bayer Company

	1897	1902	1906	1912
Main scientific laboratory (intermediates, aniline, azo and indigo)	15	15	16	16
Pharmaceutical laboratory	—	8	6	12
Bacteriological laboratory	—	1	—	—
Inorganic laboratory	—	—	4	7
Alizarin laboratory	—	—	4	8
Photographic and technical products laboratory	—	—	4	3
Artificial rubber laboratory	—	—	—	11
Total of research chemists, including heads of laboratories	15	24	34	57
Total of chemists	99	138	171	262
Percentage of research chemists	15	17	20	22

As the number of research laboratories increased, the research became more intensive, if one accepts the percentage of research chemists employed as a crude yardstick. While the main laboratory remained the largest laboratory until 1912, the organization of research became more and more decentralized as specialists in a given branch of chemistry were combined in separate laboratories. Diversification of the research program tended to lag slightly behind diversification of production. Not until 1912 did all production departments have a corresponding scientific laboratory. But the creation of an artificial rubber laboratory in 1909 reflects Bayer's deciding for the first time to invest heavily in research before undertaking production.[14]

These laboratories, however, constituted only the core of the research work done in industry. Apart from the "external differentiation" of research from production and the respective technical laboratories, an "internal differentiation" of research came about, which led to the creation of a complex research infrastructure.[15]

laboratories existing before 1914, and only for these years do the sources (all in Bayer archives) provide easy calculation of the overall distribution of Bayer's chemists according to place of work. For (May) 1897: "Statistik der Farbenfabriken vorm. F. Bayer & Co.," pp. 22 ff. (1/5.2); for 1902: "Bildalben fur F. Bayer zum 25 jahrigen Jubilaum als Teilhaber und Direktor der Farbenfabriken" (1902), 3 vols.; for 1906: "Tabellarische Ubersicht uber die Organisation des Gesamtgeschafts der Farbenfabriken vorm. F. Bayer & Co., Elberfeld: Anlagen zur diesbezuglichen Denkschrift" (10/3.1); for 1912: "Tabellarische Ubersichten uber die Organisation der Farbenfabriken vorm. F. Bayer & Co." (10/3.2). Again, these figures may differ slightly from those given in other studies.

[14]See Flechtner, *Duisberg*, pp. 227–230.

[15]For the terminology see William Kornhauser, *Scientists in Industry* (Berkeley: Univ. California Press, 1963).

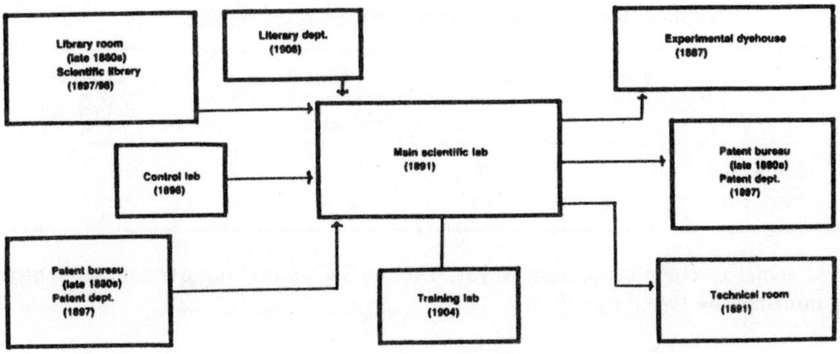

***Figure 3**. Research facilities at Bayer*

Figure 3 shows the highly developed network of facilities grouped around the scientific laboratory to support the research chemist. The collection of information had become a necessary adjunct to the work of the research chemist. The library helped the researcher keep up with the most recent scientific advances by stocking a large collection of chemical literature and a constantly growing number of scientific periodicals (190 in 1901, 360 in 1907, and 550 in 1912). By 1906 the sheer mass of articles published became so difficult to survey that a literary department was installed, which circulated abstracts of all scientifically or technically relevant articles. The patent bureau documented all German and foreign patent applications and granted patents, allowing a close and early watch of any technical and commercial progress made in the chemical industry. The control laboratory undertook the task of analyzing the constitution of any dyestuff, patented or not, that came onto the market.

Much of the success of industrial research depended on detecting the possible value of laboratory findings, on patenting research results, and on converting the discoveries of the laboratory into marketable products. The experimental dyehouse tested all dyestuff compounds delivered by the research laboratory and, if necessary, developed techniques for their application. The patent bureau had to ensure the legal protection for all inventions and to defend this protection against rival claims. The technical room was concerned with chemical engineering; it helped to transform laboratory findings into manufacturing processes.

In the end Bayer's management, under heavy pressure from Duisberg, who in 1900 joined the Board of Directors, also slowly accepted the need to set the research chemist free to engage full-time in research. Even after the establishment of the main scientific laboratory at Bayer in 1891, routine tasks, mechanical labors, and indirect service functions had distracted the research chemist from his research problems. Step by step these tasks were taken off his shoulders. The control laboratory was established in 1896 to carry out the routine task of analyzing the constitution of any dyestuff marketed by rival firms. Prior to 1897 the chemists in the scientific laboratory carried out much patent work. In 1897 the patent bureau was reorganized as the patent department so as to relieve them of many of the onerous tasks associated with patent litigations. And in 1904 a training laboratory was established, which ended the high rate of personnel turnover characteristic of the main scientific laboratory so long as it was also used to qualify

Table 4. Personnel turnover at the main scientific laboratory

	Positions	Entries	Exits due to		Death or Retirement
			Transfer	Notice	
1897–1904	15	28	23	4	—
1905–1912	16	10	8	—	2

and socialize chemists joining Bayer. Table 4 shows that turnover subsequently diminished by two thirds.[16]

ORGANIZING RESEARCH

The growing number of research laboratories and the intensified division of labor opened up new fields of activity and multiple career patterns for chemists. The differences between academic institutions and industrial research facilities were reduced, and this eased the transition from university to industry. But the personal contact between management and scientists diminished, confronting management with the problem of how to integrate the scientific expert within the organization and how to control the process of invention.

From the mid-1880s onward the difficult task of directing, coordinating, and controlling research activities kept the management of Bayer occupied. Labor contract regulations served to keep research and its results strictly secret, and the management gradually developed elaborate techniques with which to control and direct research processes.[17] Research became embedded in a highly developed and interwined complex of communication and supervision, incentives and controls. The Bayer Company employed four distinct managerial methods.

First, financial incentives served a double purpose. The board of directors offered prizes and premiums for the solution of specific research problems defined by management. Bayer also gave every research chemist a royalty consisting of a fixed percentage of the net profits accruing to the company from patented inventions; this was thought to stimulate the individual chemist's research and direct him to fields of commercial and industrial interest. The amount of money earned from these royalties was considerable and often surpassed the researchers' regular annual salary. Table 5 gives the figures (in marks) for all chemists employed in the main laboratory throughout the period.[18]

Second, the management introduced regular conferences of chemists in 1888. These conferences brought together all Bayer chemists to exchange information on the latest scientific and technical advances, on new patents, and on broad trends in industrial chemistry. Initially held weekly, these conferences were reduced to

[16]Figures from *Beamten-Verzeichniss*, p. 2 (Bayer archives); *Tabellarische Ubersichten* 1906 and 1912 (cit. n. 3); and the personnel files of the various chemists.

[17]These regulations provided that chemists employed by Bayer keep all business secrets strictly secret and abstain from all communications on methods of production at Bayer both during their engagement and for a period of up to three years after, and that upon leaving the Bayer Company they not join rival firms, also for up to three years. On the role of management in industrial enterprises see Alfred B. Chandler, *The Visible Hand: The Managerial Revolution in American Business* (Cambridge, Mass./London: Belknap Press of Harvard Univ. Press, 1977).

[18]Figures from the chemists' personnel files, Bayer archives. The top line belongs to the head of the department.

Table 5. Financial incentives for Bayer research chemists (in marks)

Chemist	1902		1906		1912	
	Annual salary	Royalties	Annual salary	Royalties	Annual salary	Royalties
1	7,500	19,041	9,500	28,685	20,000	51,727
2	5,400	14,084	6,000	22,095	6,750	39,878
3	4,800	3,993	6,250	7,655	7,750	13,214
4	4,800	744	6,000	1,058	7,750	3,785
5	4,000	651	4,600	5,766	6,800	9,227
6	3,600	559	5,500	462	8,000	3,050
7	3,000	—	4,300	575	6,600	1,377

biweekly in 1891 and later were held monthly. Patterned after the university seminar, the meetings were supposed to counteract trends toward overspecialization, isolation, or an all-too-academic orientation. They compensated for the strictly limited flow of communication with the outside professional community by encouraging informal coordination of work and exchange of ideas. Finally, they helped develop an esprit de corps.

A third method of controlling the chemist's research activities consisted of the periodic reports each chemist had to make on the progress and results of his work. When first introduced in 1887 the reports were required daily. This was changed to weekly in 1891. In 1906 the Bayer management asked the research chemists to supplement the weekly reports by a yearly report. The head of the appropriate department and the board of directors received these reports and thus were kept informed about the ongoing investigations. Apart from satisfying the need for information and control, these written reports also helped decide conflicting claims of priority between research chemists. Moreover, they could be used as legal evidence of priority against the patent claims of competitors.

Finally, the role of the research administrator was created. His duty was to coordinate and direct industrial research and to stimulate, discipline, and control the research chemist. His most important functions were overseeing all scientific and technical publications and checking, promoting, and controlling subjects and methods of research, using a cooperative style of supervision; maintaining contact with the patent department to secure patent rights for all inventions made; undertaking all administrative tasks related to the library, the stores, and the technical room; handling the qualification and final selection of chemists joining Bayer; and finally, heading the regular conferences of chemists.[19]

At first Carl Duisberg, because of his many successes as inventor and patent attorney, supplemented by his organizational ability, became the head of the growing research staff at Bayer. In 1888, after only four years with the company, he received the title of *Prokurist*, which underlined his outstanding position as first chemist of the company.[20] But the administration of research was not his only

[19]See Carl Duisberg's working instructions for the research department "Wissenschaftliche Abteilung der Farbenfabriken vorm. F. Bayer & Co., Elberfeld," Bayer archives 103/17.E.3.

[20]The title *Prokurist* empowered him to transact business in the name of the company. For most chemists the title was only honorary.

task. He was head of the patent bureau until 1897. He planned and organized the partial transfer of Bayer's manufacturing plant from Elberfeld to Leverkusen beginning in the early 1890s. And long before he was appointed a member of the board of directors in 1900, he had in fact become managing director of the works, sharing the responsibility for the scientific and technical aspects with Friedrich Bayer, the son of the company's founder.

Only with the general reorganization of Bayer's management structures at the department level in early 1896 was research administration formally institutionalized. Following Duisberg's proposals, Bayer's board of directors appointed a section manager with sole responsibility for administering research within the main scientific laboratory: Bernhard Heymann, who held a doctorate from the University of Munich and came recommended to Duisberg by Adolf von Baeyer in 1889. After seven years in the main scientific laboratory engaging in successful research and handling patent litigations competently, Heymann was an obvious choice for the new post as head of the research department. But not everyone shared this view.

For the trend towards bureaucratic control by management ran counter to the professional consciousness of the research chemists. The company's policy of promoting chemists into the managerial hierarchy no doubt served to lessen these tensions, but it did not eliminate them. On the contrary: while the elaborate system of conferences and work reports, to some a symptom of needless overorganization, usually met with only passive resistance from individual chemists, the appointment of Heymann almost produced a minor palace revolution. When informed about the forthcoming changes in late September 1895, a group of six research chemists submitted a petition to the board of directors dated 30 September 1895, demanding that this appointment be rescinded and that instead a renowned scientist be recruited from outside the company. But this protest met with no success. Just one day after their petition had been written, Duisberg, in the name of the board of directors, confronted these chemists with a simple alternative: submit or quit. Three chemists preferred to leave Bayer.[21]

This conflict between chemists and management was not simply a complaint about promotion procedures, heated up by the fact that Dr. Heymann was a Jew. It was the first and only collective protest by research chemists prior to 1914 and, as such, an obvious signal of rebellion against the introduction of hierarchical structures into the small and relatively homogeneous group of Bayer's research chemists. This protest can also be understood as a symptom of the research chemist's difficulties in adjusting to qualitative changes in the process of industrial research itself.

THE INDUSTRIALIZATION OF INVENTION

The German dyestuffs industry reacted to the challenges and promises which the scientific and technical developments of the late 1870s and early 1880s had offered by creating splendid opportunities for research and organizing the research process in ingenious ways. But dyestuffs corporations did not simply recognize

[21] Information about this protest comes from the fragments of the petition in the personnel file of Dr. R., and from the selections of an almost identical letter placed in the personnel files of Drs. Ko., N., S., and St. They had sent the letter, dated 1 or 2 Oct. 1895, to the board of directors to justify their petition.

and accept the importance of research for their own further expansion; their growing dependence on a continuous supply of inventions made them adapt science to their particular requirements. An irreversible industrialization of invention was the result. It involved two processes: industrial research became increasingly autonomous as it met industrial and commercial needs, and imitations, inventions, and innovations were increasingly reduced to routine.

Table 6 below conveys an idea of how far Bayer relied on the German system of higher education to supply its research chemists.[22] All the research chemists had a university Ph.D.; several had also studied at a *Technische Hochschule*, but the universities played the dominant role in their education. After passing their examination, the majority of these chemists had spent one or two years working as *Assistenten*, assisting the university or *Hochschule* professor in research or teaching. Only a small number of chemists were recruited from other chemical companies. Thus not only the Ph.D. but prolonged academic training in all branches of chemistry was regarded as more useful for research chemists than scientific and technical experience or knowledge of the dyestuffs industry. Reliance of industry on academia was not one-sided. After the mid-1880s industrial needs came to shape academic curricula, as dissatisfaction with chemical education in Germany led the German chemical industry to sponsor reform at universities and *Hochschulen*.[23] The goals of the reform movement were to improve the quality of training in chemistry, standardize curricula, and acquaint chemistry students with the type of

Table 6. Qualifications of the main laboratory staff

	Place of study		
	University only	University and *Technische Hochschule*	*Technische Hochschule* only
1897	9	6	—
1912	11	4	1

	Final examinations		
	University Ph.D. only	Ph.D. and *Technische Hochschule* diploma	*Technische Hochschule* diploma only
1897	13	2	—
1912	15	1	—

	Post-doctorate work at university or *Hochschule*		Industrial work experience	
	Yes	No	Yes	No
1897	12	3	3	12
1912	9	6	4	12

[22]See the personnel files of the chemists involved, Bayer archives.

[23]Lothar Burchardt, "Die Ausbildung des Chemikers im Kaiserreich," *Zeitschrift fur Unternehmensforschung*, 1978, 2:31–53. Cf. also Jeffrey A. Johnson, "The Chemical Reichsanstalt Association: Big Science in Imperial Germany," (Ph.D. diss., Princeton University, 1980), pp. 144–149.

work that the industry required; it gained momentum from the growing importance of scientific research for, and the changing character of that research in, the chemical industry.

After the late 1880s industrial research was more and more characterized by two interdependent factors: specialization and cooperation.[24] Specialization meant that the investigations of the chemist working in the main scientific laboratory were restricted to a very narrow field of dye chemistry. A training course introduced in the early 1890s, which lasted at least one year, tested the chemist's abilities and inclinations and gradually directed him into a special branch of color chemistry. This concentration on a tiny sector of chemistry seems to have been widely accepted as necessary for successful research. Nevertheless, Bayer's management, especially Duisberg, was well aware of the dangers of overspecialization and isolation from the general advances of chemistry. The system of conferences was considered one preventive. In addition, the working instructions of the research department explicitly warned the chemists against training as "one-sided specialists," stressing that every research chemist not only had the right to turn to other fields of chemistry temporarily, but had the duty to follow up all new developments in chemistry.[25] In practice, the research chemist remained committed primarily to his own special field of dye chemistry and to his principal task: inventions leading to innovations. And specialization did have the welcome—for management—secondary effect of reducing the chemist's mobility, by minimizing his value to other chemical firms, should he leave Bayer.[26]

The necessary complement to this carefully designed division of labor was cooperation. Apart from the informal consultation practiced in regular conferences, two types of interaction took place. A close and intensive personal cooperation existed between chemists in the main laboratory working on the same subfield of dye chemistry; they constituted a small work group that the head of the research department sometimes joined. An impersonal functional cooperation was required between the research chemist in the main scientific laboratory and the color chemist, pharmacist, physician, or other production employee who checked the research results for applicability and developed methods of application in the experimental dye-house, the physiological laboratory, and elsewhere. Table 7 below illustrates the ever-growing importance assumed by the screening of research results and the transformation of inventions into useful innovations, taking the dyestuffs sector as an example.[27]

Cooperative research in a bureaucratic structure involved problems. Chemists complained about having their ideas stolen, and conflicts arose over how to allot recognition among the different chemists working on a single project. Bayer had to shape its recruitment policy to the new needs; from the early 1890s the ability to cooperate became as important as scientific competence.

With discoveries and inventions, not to speak of practical innovations, becoming the result of cooperative efforts by highly specialized experts, the entire

[24]Compare Borscheid, "Die Industrie im Zeichen der Spezialisierung und Teamarbeit," in *Naturwissenschaft* (cit. n. 2), pp. 173–207.

[25]See Duisberg, "Wissenschaftliche Abteilung" (cit. n. 18).

[26]Cf. Beer, *Emergence*, pp. 87–88; see also labor contract regulations in n. 16 above.

[27]See Max Kitschelt, "Die Entwicklungsgeschichte der Farberei Abteilung," in *Geschichte der Farbenfabriken . . . Bayer* (cit. n. 7), p. 328. For comparable figures for the Höchst Company see Hermann Pinnow, *Zur Erinnerung an die 75. Wiederkehr des Grundungstages der Farbwerke vorm. Meister Lucius & Bruning* (Munich: Bruckmann, 1938), pp. 62, 65.

Table 7. Screening of dyes at Bayer

	New dyestuff compounds from laboratories	Dyes tested on a larger scale after a first screening	Dyes marketed*
1896	2,378	36	37
1906	2,656	60	36

*Includes new mixtures of old dyes and older dyes of competitors that had lost their patent protection.

process lost most of its individualized, art-like character. Empiricism, chance discoveries, or the heroic deeds of lone inventors gave way to the systematic production of knowledge, guided by what Otto Witt, then professor at the Technische Hochschule Berlin, in 1913 had called the "principle of purposive synthesis."[28] The azo dyestuffs, the largest group of dyes and economically the most important one, illustrate these changes well.

First, the countless possible products of the diazo reaction, presumably more than one hundred million, exceeded the research capacity of any one chemist. Individual research was superseded by the new "scientific mass work" (*wissenschaftliche Massenarbeit*). Second, because the mechanisms of the diazo reaction and the ways to make azo dye compounds were well known, inventive activity was turned into a routinized "construction bound to scientific rules" (*an wissenschaftliche Regeln gebanntes Konstruieren*).[29] This was true only when the constitution of chemical substances and the relation between their constitution and their qualities were known. Even then, individual originality and creativity and chance discoveries continued to have their place in industrial research. However, terms like "mass work" and "construction" show that "industrialization of invention" is not an artificial construct imposed on the past after the fact: these terms grew out of the experience of contemporary research chemists, who were all too familiar with their routine, frequently fruitless research.[30] Moreover, the terms characterize changes that produced two further negative effects.

First, in spite of the continuously growing prosperity of the company after the deep economic crisis of the mid-1880s, the chemist's share in the profits based on his inventions was reduced. Chemists who in the late 1870s and early 1880s had received five percent of the net profits accruing to the company from their patented inventions were paid only three percent from the late 1880s onward. The commercially valuable invention had lost its exceptional importance for the growth of the company as the number of research chemists rose and a stream of new discoveries and inventions got underway.

Second, the company became recognized as a sort of collective inventor. The emergence of the word *Etablissementerfindung* in the course of the heated public debate on the payment of salaried inventors in the early twentieth century reflects this perception. The company management provided the means, methods, and

[28]Otto N. Witt, "Wechselwirkungen zwischen der chemischen Forschung und der chemischen Technik," in *Die Kultur der Gegenwart*, Pt. 3, Div. 3, Vol. 2: *Chemie* (Leipzig: Teubner, 1913), p. 521.

[29]Quoting Heinrich Caro (BASF director of research, 1883–1890), *Uber die Entwicklung der Theerfarben-Industrie* (Berlin: Friedlander, 1893), p. 141; and Heinrich Caro to W. von Oechelhauser, 12 July 1906, Caro correspondence, Deutsche Museum, Munich; cf. Caro, *Uber die Entwicklung*, p. 137.

[30]See Table 7 and n. 26.

organization for research and directed its course, and the company became the repository for a body of collective knowledge and experience from which researchers drew in making new discoveries. The research chemist had become a small part in the total organization. Doubtless managers and entrepreneurs of the chemical industry used this term in order to repudiate the claims of research chemists who asked for a more appropriate reward for their inventions.[31] But all the same it affords a glimpse of the depersonalized character of the research of the time.

THE EFFECTS OF INDUSTRIAL RESEARCH

By the early twentieth century institutionalization of research was essentially complete at the Bayer Company. In the late 1870s and early 1880s scientific, economic, and legal developments had given the first impetus to a gradual and evolutionary change. The many successes of a young inventor, Carl Duisberg, accelerated this process and paved the way for the opening of the main scientific laboratory in 1891. Other research laboratories followed. Meanwhile a complex research infrastructure developed. Organizing research became the foremost managerial task, and the process of invention itself became industrialized, the routinized work of cooperating specialists following the imperatives of industrial needs in a bureaucratic and hierarchical setting. While admittedly tentative, the remarks that follow address the further effects of the new role of industrial research.

The status and prestige of the research department within the Bayer Company remained ambiguous. Remarkably, when Bayer's management structures were reorganized at the top level in 1906, the head of the research department became a member of the board of directors (without giving up research and research administration!). But research chemists were scarcely represented among the company's directors and *Prokuristen*, as Table 8 shows.[32]

Except for the position of head of the research department, until 1914 research was a dead-end career, since the only outlet for promotions was in management. Similarly, although Bayer's board of directors showed its appreciation of chemists' work by conferring the honorary title of *Prokurist* on them, the board passed over the research chemists with only one exception. It is therefore no surprise that most scientists attracted to industry did not go into research. Between 1897 and 1912 Bayer's management transferred thirty chemists from the main scientific laboratory, primarily to production, but only two chemists went the other way. One left again for a production post after a few years. The other stayed, but even he had vehemently opposed his transfer from production to research. Neither the generous patent royalties nor the superb working conditions succeeded in counterbalancing this tendency.

One reason for this low standing of research in industry may have been that its autonomy remained subordinated to industrial and commercial needs and that its range remained restricted. Truly basic research not directed towards industrially

[31]See Carl Duisberg, "Die Angestelltenerfindung in der chemischen Industrie," in *Abhandlungen, Vortrage und Reden aus den Jahren 1882–1933*, 2 vols. (Berlin/Leipzig: Verlag Chemie), Vol. II (1933), pp. 738–748. For the role of the *Etablissementerfindung* in the German jurisdiction see E. C. Rassbach, *Betrachtungen zur wirtschaftlichen Lage der technischen Privatangestellten in Deutschland* (Karlsruhe: Braun, 1916), pp. 136–147.

[32]For a list of the company's board of directors and Prokuristen see the appendix to Pinnow, *Werksgeschichte*.

Table 8. Chemists in management (1913)

	Board of directors	*Prokuristen*
Total	17	43
Chemists	10	25
Research chemists	1	1

and commercially valuable products was left to private or state research institutions. "We esteem highly every branch of pure science," Bayer's board of directors explained its research policies in 1910, "and we appreciate every activity on any field of theoretical chemistry or teaching, but from the chemists working for our company we have to demand industrially useful results of their work."[33] Moreover, Bayer's management rarely allowed research chemists to publish scientifically interesting results, even when company secrets were not involved. Consequently, the total number of publications by research chemists working in Bayer's main scientific laboratory was low; for the seven years from 1907 to 1914, it was exactly zero.[34] Quite a few chemists, therefore, left Bayer for an academic career.[35]

The dissatisfaction of individual chemists who resented having their freedom to pursue interesting problems curtailed was not the only limitation of industrial research. The dyestuffs industry knew well that, at least until 1914, it still depended upon the exploration of new fields undertaken at "free" research institutions by men like Emil Fischer, Paul Ehrlich, and Fritz Haber,[36] although academic research could no longer successfully compete with industrial research where routine applied science was concerned. That research program, following entrepreneurial strategies of diversification, was extremely effective at maintaining and extending the company's superiority wherever it had established itself in the market. Azo direct dyes, alizarin dyes, and pharmaceuticals are some well-known examples. But when Bayer tried to break into markets established by other companies or to break new technical and scientific ground, industrial research proved to be less effective. Bayer endeavored in vain to produce indigo. Research on perfumes and fertilizers came to nothing of importance. Neither the photographic division set up in 1902 nor the intensive research directed towards the production of artificial rubber turned out to be successful. The reasons for failure were manifold. Scientific and technical problems could be insurmountable. The barriers erected by patents could withstand any research attack. And last but not least, lack of experience in production or lack of knowledge of the market could prevent economic success, even if the research chemist had done his job. Thus

[33]Board of directors to Dr. J. (who left Bayer for an academic career), 11 Apr. 1910, personnel files, Bayer archives.

[34]This is the result of a search of the author index of the *Decennial Index to Chemical Abstracts;* Vol. 1–10, 1907–1916. For refusals to allow publication see the letter of Dr. B., 30 Jan. 1902, and the letter of Dr. Z., 23 Jan. 1913, personnel files, Bayer archives.

[35]One of the better known was Otto Dimroth, professor of chemistry at the universities of Greifswald (1913–1918) and Wurzburg (1918–1937).

[36]Cf. Beer's judgement: "The constant dependence of the German dye companies on academic science, even after their research laboratories had become mammoth undertakings, gives testimony to the limitations which secrecy and lack of intellectual freedom imposed on industrial research" (*Emergence*, p. 72). For industry's support of academic science see Johnson, "The Chemical Reichsanstalt Association" (cit. n. 2).

industrial research was not a master key to entrepreneurial growth. Its positive aspect was its ability to concentrate on a restricted range of products and cover that field with an unmatched intensity, reacting to every impulse from the market, be it industrial or scientific.

It is extremely difficult to measure exactly the contribution of industrial research to economic growth.[37] The facts enumerated here illustrate developments loosely related to industrial research. The number of products which Bayer produced rose steadily, from about 200 in 1886 to more than 1000 dyes and about 40 pharmaceutical products in 1900, and to about 2000 dyes and about 150 pharmaceutical products and photographical preparations at the end of 1913. Similarly, the number of German patents taken out by Bayer (in class 22) rose from 36 in the period 1877–1886, to 99 in the period 1887–1890, to 512 in the period 1891–1900. This rise reflects the growing intensity of research in the late 1880s and the stabilization of the research effort due to the foundation of the main scientific laboratory in 1891. Bayer held more than 2500 German and foreign patents in 1900, a quantity that again was more than trebled before 1914, to over 8000. The significance in the jump in patents held in the mid-1880s is confirmed by the statistics on Bayer's dyestuffs innovations from 1870 to 1899. While the number of innovations was 12 for the period 1870 to 1884, it was 10 in 1885 alone and 151 for the period 1885 to 1899. The annual reports of German Chambers of Trade in the late 1880s and early 1890s offer further confirmation in the form of complaints from smaller dyestuffs firms directed against the large joint-stock companies, which, with their splendid laboratories, their research staff, and their patents, endangered the smaller firms' existence.[38]

One of the indirect effects of industrial research was to make it harder for a new firm to enter the field. No new dyestuffs firms were successfully founded between the mid-1880s and 1914; the corporations already dominating the dyestuff business in the early 1880s, before industrial research made headway, were still in the top rank in 1914. Their previous economic success obviously was a prerequisite for their successful industrial research, and their ingenious management strategies of expansion, integration, and diversification surely remained a necessary complement to the successful industrialization of invention. Industrial research was not, as many contemporary chemists (and some historians) had it, the "prime factor" in the development of the coal-tar color industry prior to 1914,[39] but after 1880 it certainly played an important role in the development of firms like Bayer, which turned from a small dyestuffs company into a big chemical corporation. No doubt industrial research then contributed to Bayer's world-wide dominance on many a chemical market, and its accumulating stock of knowledge has helped Bayer survive as a giant chemical firm until today.

[37]As far as I could find out, figures for Bayer's financial expenditures in "research" are available for one year only, 1909, when the costs of *all* scientific laboratories were calculated at 827,128.29 marks (Bayer archives 15/11.1), about 1% of Bayer's sales for the year. A different approach would be to calculate from the royalties annually paid to the research chemists. Unfortunately, the material is incomplete, and the exact methods of calculation are unclear, but a very first assessment seems to indicate that for 1902, 1906, and 1912 about 20% of Bayer's net profits can be traced to inventions of chemists in the main scientific laboratory.

[38]See *Jahresberichte der Handelskammer von Barmen fur 1894* (Barmen, 1895), pp. 32–33; *Jahresbericht der Handels- und Gewerbekammer zu Leipzig 1893* (Leipzig, 1894), p. 157; *Handelskammer der freien Stadt Frankfurt: Jahresbericht fur 1889* (Frankfurt, 1890), p. 271.

[39]Quoting Raphael Meldola, "Tinctorial Chemistry, Ancient and Modern" (1910), in Walter M. Gardner, ed., *The British Coal-Tar Industry: Its Origin, Development, and Decline* (London: Williams & Norgate, 1915), p. 265.

[6]

BJHS, 1992, **25**, 91–111

The emergence of research laboratories in the dyestuffs industry, 1870–1900

ERNST HOMBURG*

INTRODUCTION

The focus of this paper is the emergence of the research laboratory as an organizational entity within the company structure of industrial firms. The thesis defended is that, after some groundwork by British and French firms, the managements of several of the larger German dye companies set up their own research organizations between the years 1877 and 1883. The analysis of the emergence of the industrial research laboratory in the dyestuffs industry presented here makes clear that both the older study on the subject by John J. Beer and a later paper by Georg Meyer-Thurow contain some serious defects. Beer, like so many other authors of the 1950s who studied the 'marriage' between science and industry during the 'Second Industrial Revolution', incorrectly correlates the engagement of university-educated chemists with the rise of industrial research. The appointment of academic chemists by BASF and Hoechst at the end of the 1860s, for instance, was described as 'the...acquisition of a research staff'.[1] This reveals a misunderstanding of the roles of chemists within the nineteenth-century chemical industry. University-trained personnel were, in fact, working in industry as early as the start of the nineteenth century. However, they were employed as managers, works chemists and analysts, and only exceptionally in research.

Meyer-Thurow's article, by contrast, suffers from the serious limitation that it presents an in-depth study of only one German dye firm – the Bayer company at Elberfeld.[2] Here it is shown that the Bayer company was actually *atypical* in several important respects.

* Department of Chemistry, Katholieke Universiteit Nijmegen, Toernooiveld 1, 6525 ED Nijmegen, The Netherlands.

An earlier version of this paper was presented to the autumn meeting of the Genootschap voor Geschiedenis der Geneeskunde, Wiskunde, Natuurwetenschappen en Techniek, The Hague, 25 October 1986. I am grateful to Henk van den Belt, Casper Hakfoort, Peter Morris, Tony Travis and two referees for helpful comments on earlier drafts of this paper. I would like to thank my colleague Willem Hornix for the generosity with which he placed several sources at my disposal, and the archivists of BAYER AG (Leverkusen), BASF AG (Ludwigshafen), and HOECHST AG (Hoechst a/M) for responding to my questions and for the access they provided to the documents under their care.

1 J. J. Beer, 'Coal tar dye manufacture and the origins of the modern industrial research laboratory', *Isis* (1958), **49**, 123–31, especially 124–5. For the historiography and theoretical implications of the concept of a 'Second Industrial Revolution', in relation to the rise of industrial research, see E. Homburg, 'De "Tweede Industriële Revolutie". Een problematisch historisch concept', *Theoretische Geschiedenis* (1986), **13**, 367–85.

2 G. Meyer-Thurow, 'The industrialization of invention: a case study from the German chemical industry', *Isis* (1982), **73**, 363–81.

Moreover, to generalize from the results of Meyer-Thurow's study to the emergence of industrial research laboratories in the German chemical industry as a whole would strongly underestimate the role played by the management of the German dye firms; it would also give an ill-balanced insight into the factors responsible for the institutionalization of industrial research. The present study will redress the balance by examining the creation of the research organizations of the seven largest German dye companies. By doing so, the main attention is not directed at the emergence of industrial research *per se*, but explicitly at the origin of the research laboratory as an integral part of the industrial organization.

An emphasis on the details of company structure is characteristic of other recent studies on the institutionalization of industrial research. Instead of using indicators on the aggregate level of an entire company (e.g. numbers of chemists, numbers of patent applications), as Beer and the other historians of the 1950s have done, it is necessary to open the black box of the firm, look inside, and seek out the answers to questions concerning its organizational structure. For example, in what departments were the chemists who made the patentable discoveries located and what exactly were the tasks of the chemists working in a certain company? These are the types of question that really matter if the relationship between science and industry is to be properly understood.

This approach, which draws on the work of Alfred Chandler, can, for instance, be found in recent works on the emergence of industrial research laboratories in the United States: Leonard Reich's book on General Electric and AT&T, and David Hounshell and John K. Smith Jr's study on Du Pont.[3] Meyer-Thurow, in his study of the establishment and growth of the research laboratory of the Bayer Company, also pays careful attention to the internal affairs of the company and to the specific tasks and positions of the appointed chemists.[4]

In this organizational approach, the historical search for the roots of the industrial research laboratory is clearly defined. It is necessary to investigate the evolution of the organizational structures of the companies involved. When, how and why did organizational units emerge with the specific tasks of performing scientific and technical

3 L. S. Reich, *The Making of American Industrial Research. Science and Business at GE and Bell, 1876–1926*, Cambridge, MA, 1985; D. A. Hounshell and J. K. Smith Jr, *Science and Corporate Strategy: Du Pont R&D, 1902–1980*, Cambridge, MA, 1988. Other publications on the emergence of industrial research laboratories in the US include: L. S. Reich, 'Industrial research and the pursuit of corporate security: the early years of Bell Labs', *Business History Review* (1980), **54**, 504–29; G. Wise, 'A new role for professional scientists in industry: industrial research at General Electric, 1900–1916', *Technology and Culture* (1980), **21**, 408–29; G. Wise, *Willis R. Whitney, General Electric, and the Origins of U.S. Industrial Research*, New York, 1985; L. Hoddeson, 'The emergence of basic research in the Bell Telephone System, 1875–1915', *Technology and Culture* (1981), **22**, 512–44; J. L. Sturchio, 'When industrial research was young. Chemistry and corporate strategy at Du Pont', *Research Management* (1984), **27**, 10–15 and 18. These studies are discussed in two essay reviews: M. A. Dennis, 'Accounting for research: new histories of corporate laboratories and the social history of American science', *Social Studies of Science* (1987), **17**, 479–518; and J. K. Smith Jr, 'The scientific tradition in American industrial research', *Technology and Culture* (1990), **31**, 121–31.

4 Meyer-Thurow, op. cit. (2). Other recent studies on industrial research include: H. Schubert, 'Emergence of industrial research', in *History of Solid State Physics* (ed. L. Hoddeson, E. Braun, J. Teichmann and S. Weart), New York, 1987; G. Bowker, 'Recherche industrielle et industrialisation de la recherche', *Culture technique* (1988), **18**, 146–53; and G. Bindon and D. P. Miller, '"Sweetness and Light": industrial research in the Colonial Sugar Refining Company, 1855–1900', in *Australian Science in the Making* (ed. R. W. Home), Cambridge, 1988, 170–94.

investigations that were distinctly separate from production or chemical analysis? Here I provide the answer for the case of the German dye industry.

First, I investigate the founding of the individual research laboratories in the seven largest German dye firms. Care has been taken to date the emergence of the first laboratories, and to analyse, in a few cases, the reasons for a company to set up a separate organizational unit for research. Secondly, I ask when and why the 'research laboratory' emerged as a more general organizational concept, independent of local definitions and circumstances, but available to the management of a company when discussing the structure of the firm. Before coming to these issues, it is useful to make a few general remarks on the way the work of chemists was organized in the German chemical companies.[5]

CHEMISTS IN INDUSTRY

In the early 1860s, when the German dye companies were small and of uncomplicated internal structure, there already existed a distinction between 'line' and 'staff' functions in an embryonic form. Accordingly, well before the first research laboratories were established, there were two well-defined positions for chemists:[6]

- works chemists (*Betriebschemiker*), and
- laboratory chemists (*Laboratoriumschemiker*).

Works chemists supervised production departments, and had both chemical and managerial tasks. Their position in the company hierarchy was clearly above that of the laboratory chemists. In many cases the works chemist also had a small laboratory, connected to the factory under his care, where he could do some chemical analysis or other tests and investigations.

Laboratory chemists, however, were, for reasons of secrecy, usually strictly forbidden from entering a production department.[7] It would be a great mistake to equate the 'laboratory chemist' with the 'research chemist'. The latter category forms only a subset of the former. More often the laboratory chemists were analysts, busy with testing the raw

5 Apart from Meyer-Thurow's paper, studies of the German dyestuffs industry that pay attention to the internal company organization are rare, and not very informative with respect to the emergence of the research laboratory. The few examples are: S. Kaku, 'The development and structure of the German coal-tar dyestuffs firms', in *Development and Diffusion of Technology, Electrical and Chemical Industries* (ed. A. Okochi and H. Uchida), Tokyo, 1980, 77–94; and J. Liebenau, 'Corporate structure and research and development', in *The Challenge of New Technology. Innovation in British Business Since 1850* (ed. J. Liebenau), Aldershot, 1988, 30–42.

6 An example is offered by the 1863–64 contract of the chemist E. Ullrich at Kalle, who could only work in the laboratory, and was not allowed to enter the production rooms. H. Voelcker, *75 Jahre Kalle. Ein Beitrag zur Nassauischen IndustrieGeschichte, 1863–1938*, Wiesbaden (1938), 70. For a more general account of the work of chemists in the German, French and British dye industries, see E. Homburg, 'De inschakeling van chemici in de kleurstofindustrie', in *De ontwikkeling van de kleurstofindustrie* (ed. H. van den Belt *et al.*), University of Nijmegen, 1984, 107–63.

7 Compare the very strict contracts that the laboratory chemists of the Oehler company had to sign in the early 1880s. 'Vor Vierzig Jahren', *Chemiker-Zeitung* (1922), 22, 525–6.

materials or final products, or performing analytical tasks in connection with control of the production processes.[8]

In relation to the company hierarchy, one can distinguish between four types of academically educated laboratory chemists (Figure 1):[9]

- the laboratory chemist (A) who was an assistant of one of the factory owners or directors;
- the laboratory chemist (B) who was an assistant of a works chemist;
- the laboratory chemist (C) who had both laboratory duties and some managerial power, and therefore had a more independent status in the company, in contrast to the laboratory chemist of type A; and
- the laboratory chemist (D) who was the assistant of a laboratory chemist of type C.

I have used this scheme as an analytical tool in my reconstruction of the founding dates of the first research laboratories.[10] Where possible, I have chosen the company's *own definition* to decide if, or when, a specific 'research laboratory' had in fact been founded. In those instances where the information from company histories or archives was inadequate, I have considered the following situation to be the decisive step towards the start of a research organization:

> a *research* chemist of type A or B, moving into a position C by obtaining his own, academically educated, *assistant*.

The rationale behind this *operational definition* is the understanding that the appointment of a single research chemist is a necessary but not a sufficient condition for the emergence of a separate organizational unit for research. Both of the words emphasized in the definition are essential. The head of the analytical laboratory, for instance, is a chemist of type C, but not a research chemist. A research chemist without an assistant, on the other hand, does not, generally speaking, represent a discrete part in the organization of a company.[11]

In this way the results of an extensive biographical research project have been used to extend the company histories of some smaller and less well researched firms like Cassella, Oehler and Kalle. And, by so doing, the history of the industrial research laboratory can, as I shall show, be placed upon a firmer basis.

8 Laboratories, as such, are an old phenomenon within the chemical industry. The earlier literature did not differentiate carefully enough between 'laboratories' and 'research laboratories'. For a good overview of the wide range of laboratory types inside and outside industry, see R. P. W. Visser and C. Hakfoort (eds.), *Werkplaatsen van wetenschap en techniek. Industriële en academische laboratoria in Nederland, 1860–1940*, Amsterdam, 1987 (= *Tijdschrift voor de geschiedenis der geneeskunde, natuurwetenschappen, wiskunde en techniek* (1986), 9(4)).

9 I restrict my analysis to university or polytechnic chemists, which is permissable in the German case. Next to them there were also assistants on lower levels of the hierarchy, for instance laboratory boys.

10 Apart from the published histories and some archival documents mentioned here, my analysis is to a great extent based on a large data base of biographies of several hundred German industrial chemists (see especially the chemists named in Tables 1 and 3). The membership lists of the *Deutsche Chemische Gesellschaft* and obituaries published in the *Berichte der deutschen chemischen Gesellschaft*, the *Chemiker-Zeitung* and the *Zeitschrift für angewandte Chemie* were my most important sources.

11 One could even argue, and the examples given seem to support this view, that not one, but perhaps three or four assistants were necessary before the companies used the term 'research laboratory'. In my opinion the creation of a hierarchy in the organization of research work is, however, the most crucial step. But care is required.

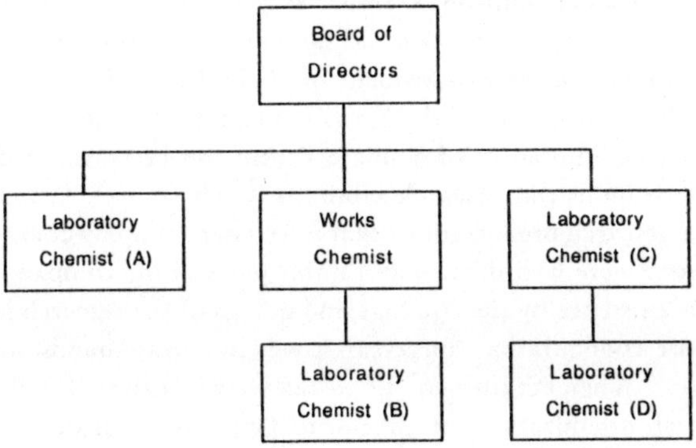

Figure 1. The organization of a chemical company: different functions for laboratory chemists.

THE FIRST INDUSTRIAL RESEARCH LABORATORIES

Prior to the emergence of regular industrial research laboratories, a key role in the field of technological innovation was, in several cases, played by the so-called inventor–entrepreneur. He was the man who had made the inventions that formed the technological and commercial basis of the company founded by him. Also, after the start of his enterprise, the typical inventor–entrepreneur was more than just a businessman. He remained active as a researcher and inventor, working on his own, or with the help of some assistants. Thomas Edison, who in 1876 created his 'inventions-laboratory' at Menlo Park, New Jersey, is undoubtedly the most famous example of this 'species'.[12]

In the domain of synthetic dyes, William H. Perkin, the inventor–discoverer of the first commercial aniline dye, is probably the best known example of the inventor–entrepreneur.[13] In the early 1860s Perkin set up a private laboratory northwest of London at Greenford Road, Greenford Green, close to his dye factory. Perkin used this laboratory not only for routine analyses, but also for several researches. In all probability Perkin's laboratory assistants were not on the payroll of Perkin & Sons. As far as can be deduced from the surviving records, the first of these private assistants was James Thomas Brown, engaged by Perkin in, or around, 1862. After Brown had advanced to the position of works chemist at Perkin & Sons, the vacant position was given to Robert Williamson, who seems to have assisted Perkin in his research on the derivatives of alizarin.[14] When the Greenford Green

12 Typically, this 'inventions-laboratory' was under Edison's direct control. It was not owned by the Edison Electric Light Company, formed in 1878, or its successor, the General Electric Company, formed in 1892. There was no direct historical connection between Edison's laboratory and the corporate research laboratory finally set up by General Electric in 1900. Wise (1980), op. cit. (3); G. Wise, 'The first corporate R & D organization', *Chemtech* (1987), 17, 394–6; Reich (1985), op. cit. (3), 42–5 and 60–73.

13 E. Homburg, 'The influence of demand on the emergence of the dye industry. The roles of chemists and colourists', *Journal of the Society of Dyers and Colourists* (hereafter *JSDC*) (1983), 99, 325–33; A. S. Travis, 'Perkin's mauve: ancestor of the organic chemical industry', *Technology and Culture* (1990), 31, 51–82.

14 From the biographies of Brown and Williamson it can be concluded that Williamson was 'a' successor of Brown (he was appointed to a vacant position). It is not certain whether he was Brown's direct successor. On Brown, see *Journal of the Chemical Society* (hereafter *JCS*) (1894), 65, 382–3; and on Williamson see *JCS* (1915), 107, 590–2.

works were sold to Brooke, Simpson & Spiller at the end of 1873, Williamson, unlike the two works chemists, Brown and Stocks, did not stay at Greenford Green under the new ownership. He continued in the employment of W. H. Perkin for another few years, and assisted the latter in his research work in a new private laboratory at nearby Sudbury.

In a way the private laboratory of William Perkin can be considered as an embryonic form of the modern industrial research laboratory. There was, however, an important difference with the research organizations that were later set up by companies like Hoechst and BASF. The latter were well-defined and stable parts of the company organization, not threatened in their existence by the comings and goings of the research leaders. In the case of Perkin, the entire configuration for research was in private hands, and not part of the company structure.[15] When Perkin's colour works were transferred to Brooke, Simpson & Spiller, the 'research organization' at Greenford Green was left out of the resulting sale agreement.

There are other examples of such 'embryonic' research laboratories in the early dye industry. In Britain, Brooke, Simpson & Spiller employed a 'research chemist', or 'scientific chemist' in their laboratory at the Atlas Dye Works, Hackney Wick, northeast of London, from at least 1874. The research was directed towards the investigation of the derivatives of aniline and its homologues, and, later, to the azo compounds.[16] Another British dyestuffs manufacturer that took out a considerable number of patents in the 1860s, and had several, mainly German, chemists in its employ, was the Lancashire firm of Roberts, Dale & Co. Although it has been claimed that several of the chemists engaged by this company carried out research work, in addition to the supervision of production and the performance of analytical and coloristic tasks, the historical sources do not permit any conclusion concerning the possible existence of a special 'research laboratory'. The company organization of Roberts, Dale & Co., and the positions of the chemists therein, are largely unknown.[17]

In France, the large company known as La Fuchsine possessed a *laboratoire des recherches chimiques* in Paris around 1865, headed by the dyestuffs chemist Charles Girard. He was a stockholder of the company, and – together with his colleague Georges de Laire – the inventor of several of its most remunerative patents.[18] In comparison to the other

15 A clear illustration of this situation is offered by the fact that William H. Perkin's (second) alizarin patent of November 1869 (British patent no. 3318), was licensed by him to his own company (Perkin & Sons). W. H. Cliffe, 'Pacta Conventa – the last days of Perkin & Sons', *JSDC* (1957), **73**, 318–22.

16 The first 'research chemists' at the Atlas Dye Works were R. J. Friswell (1874–77?), R. Meldola (1877–85) and A. G. Green (1885–94). In the 1860s, Simpson, Maule & Nicholson – the predecessor of Brooke, Simpson & Spiller – employed several chemists, who took out patents (e.g. D. S. Price, *c.* 1858–62 and F. Field, 1862–66), but very little is known about their place in the company organization. See 'Richard John Friswell', *JCS* (1909), **95**, 2204–6; 'Raphael Meldola', *JCS* (1917), **111**, 349–59; M. R. Fox, *Dye-Makers of Great Britain 1856–1976*, Imperial Chemical Industries, Manchester, 1987, 104–10 and 114–15.

17 Chemists at Roberts, Dale & Co. included H. Caro (1859–66), A. S. Leonhardt (*c.* 1860 and *c.* 1864–67), C. A. Martius (1863–65) and L. Schad (*c.* 1864–73) (all Germans). See: (on John Dale), *JCS* (1890) **57**, 446–8; Fox, op. cit. (16), 125–7, 131–2 and 249; and A. S. Travis, 'Heinrich Caro at Roberts, Dale & Co.', *Ambix* (forthcoming).

18 Reference to this research laboratory is made in the *Annuaire du commerce. Didot-Bottin*, Paris, 1865. It was situated at 18 rue des Marais, Saint-Germain. I thank Henk van den Belt for providing me with access to this source. To my knowledge, this constitutes the oldest example of the mention of the term '(industrial) research laboratory'!

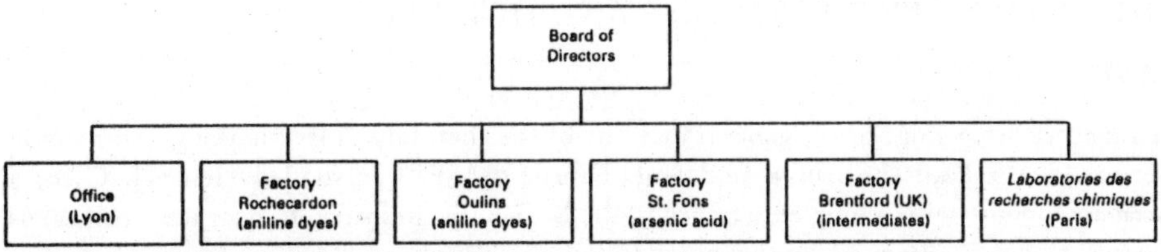

Figure 2. The company structure of 'La Fuchsine SA' in 1865/66.

dyestuffs companies then in existence, the organizational structure of La Fuchsine was rather complex. The company owned factories both at Lyon and near London (Figure 2).[19] La Fuchsine therefore is a good example of the type of firm where a need for coordination and for central research activities was considered essential. From this point of view one could argue that the laboratory of La Fuchsine was the first modern corporate laboratory established in the dye industry. On the other hand, Girard was one of the founders of the new corporation, although not a member of the board of directors. He can be considered as an inventor–entrepreneur in his own right. Girard's laboratory in Paris may be viewed as the transitional stage between the private laboratory of the inventor–entrepreneur and the corporate research laboratory, which was yet to come.

The foregoing examples show that the first steps towards the creation of industrial research laboratories took place in Britain and France. In those countries, the existence of comprehensive patent systems meant that in the field of synthetic dyes patent portfolios contributed in an essential way to the prosperity of the companies that owned them. Proprietors and chemists like Perkin and Girard had clearly discovered that a monopoly based on new inventions could considerably improve the economic vitality of their firm. It was in this legal setting that the first industrial research laboratories were born. Recent historiographical studies on the emergence of industrial research in the United States have shown that the most important explanatory factors for the founding of corporate laboratories were also related to patents and patent positions.[20] In Germany, by contrast, there was no comprehensive patent law before 1877. These circumstances make it particularly interesting to analyse the first steps towards the 'invention' of 'the' industrial research laboratory in Germany in the years before and after 1877.[21]

19 This diagram is based on a large number of sources, the most important of which are the biographies of several chemists; the *Guides Indicateurs* and *Annuaires* of Lyon and Paris (1855–70); J. Bouvier, *Le Crédit Lyonnais de 1863 à 1882: Les années de formation d'une banque de dépôts*, 2 vols., Paris, 1961; and P. Cayez, *Crises et croissance de l'industrie Lyonnaise, 1850–1900*, Paris, 1980. See also Henk van den Belt, 'Why monopoly failed: the rise and fall of Société La Fuchsine', in this issue.

20 See the literature cited in note 3.

21 For the history of the German patent law, and its relation to the chemical industry, see A. Fleischer, *Patentgesetzgebung und chemisch-pharmazeutische Industrie im deutschen Kaiserreich (1871–1918)*, Stuttgart, 1984; and H. van den Belt and A. Rip, 'The Nelson–Winter–Dosi model and synthetic dye chemistry', in *The Social Construction of Technological Systems. New Directions in the Sociology of Technology* (ed. W. E. Bijker, T. P. Hughes and T. J. Pinch), Cambridge, MA, 1987, 135–58, especially 149–55.

THE GERMAN PIONEERS: BASF AND HOECHST

BASF

In 1868 Friedrich Engelhorn, general director of the then largest dye-making company in Germany – the Badische Anilin und Soda-Fabrik (BASF) – persuaded Heinrich Caro, a German chemist with a long experience in the British dye industry, to accept a position as 'deputy technical director' (*mitleitender technischer Direktor*) in his company. Caro then had a private laboratory in Heidelberg and had worked from time to time as a consultant for BASF.[22] More information about his appointment can be obtained from his 1868 contract:

> Mr. H. Caro will preferably execute operations in the laboratory that are necessary from both a theoretical and a practical point of view. He is, however, obliged to manage the industrial production departments when this would be necessary in case of travels, illness or death.[23]

Apart from his fixed salary, Caro was to receive a number of royalties, including a 5 per cent royalty on 'apparent savings on the production accounts'.

Caro was, therefore, primarily a laboratory chemist who had to make himself available in the factory if he was needed. As deputy technical director, Caro had a relatively independent position in the company and was not a laboratory chemist of the then common type, that is, subordinate to a works manager and badly paid. Neither could his starting position at BASF be called a fully developed research function. Research was simply one of his regular tasks. No more than that, but also no less.

That the conduct of research was among Caro's duties can be concluded from the word 'theoretical' in his contract. The use of this term in an industrial context can be considered innovative in itself. Until this time, laboratory chemists were appointed to perform practical and analytical tasks. In Caro's case a new direction was taken. It is, however, important to note that this research was not primarily directed towards the discovery of new dyes – the type of innovation then prevalent in Britain and France and also on which the later fame of the German industrial dye laboratories would be based. Significantly there was no provision in Caro's contract for royalties on new products. He would receive his royalties on cost-saving process innovations alone. This is an interesting observation because it characterizes the situation of the German dye industry around 1870. Since a comprehensive patent law was absent in Germany, research was necessarily directed towards reduction in the cost of manufacturing and rationalization within the limits set by the ongoing production processes.

22 The life and work of Caro are well documented. His laboratory notebooks, both from his British years and his time at BASF, are preserved in the BASF Archives, Ludwigshafen, and at the Deutsches Museum, Munich. For biographies, see A. Bernthsen, 'Heinrich Caro', *Berichte der deutschen chemischen Gesellschaft* (1912), **45** (II), 1986–2042; C. Schuster, 'Heinrich Caro', in *Ludwigshafener Chemiker* (ed. K. Oberdorffer), Düsseldorf, 1960, ii, 45–83; C. Schuster, 'Heinrich Caro. Aus Anlass seines 50. Todestages am 10. September 1960', *Tradition, Zeitschrift für Firmengeschichte und Unternehmerbiographie* (1961), 6, 49–64.

23 Quoted from E. Wiedemann, 'Die Konstruktion der richtigen Formel. Strukturaufklärung und Synthese des Indigblau dargestellt an Hand der Briefwechsel Baeyer-Caro', Ph.D. diss. T.U. Munich 1978, ii, 292. See also, C. Schuster, *Wissenschaft und Technik. Ihre Begegnung in der BASF während der ersten Jahrzehnte der Unternehmensgeschichte*, Ludwigshafen, 1976 (= *Schriftenreihe des Unternehmensarchivs der BASF Aktiengesellschaft*, vol. 14), 36–7.

In his first years at BASF, Caro was indeed frequently active in the field of production. In cooperation with works chemists he developed the production of alizarin, and designed the manufacturing plant.[24] As far as research came into play at all, Caro's supervisor, Engelhorn, viewed science more as a lucrative gamble than a necessary element in company management. Carl Glaser, later a director of BASF, characterized the relationship between Engelhorn and Caro during the first half of the 1870s as that of a 'King [who] backed his goldmaker in alchemistic times'.[25]

At the end of 1873, Caro was, for the first time, provided with an academically educated assistant.[26] This assistant left after six months and was not succeeded by another chemist. At this time, however, Caro began his cooperation with Adolf Baeyer, who was a professor at Strasbourg.[27] From 1874 on, Caro's laboratory had its first successes in discovering new dyestuffs. Caro became less and less concerned with production. Research, which in 1868 was only one of his various duties, became more important. Slowly the idea of a 'research laboratory' took shape, not least through the discoveries of successful dyestuffs in and after 1874 (eosin, nitrosomethylaniline, methylene blue).

A second reason for bringing into existence a separate research function was the growth of the company. As more chemists were brought into the factories. Caro's assistance was less needed at the shop floor level. This assistance was, in any case, no longer greatly appreciated, owing to some serious conflicts within the management of BASF, which will be discussed later.

At the end of 1874, Caro for a second time obtained the services of an assistant, and from then on the post of assistant to Caro was a permanent one.[28] According to my criteria for the starting point of organized research, the relevant date in the case of BASF was therefore 1874. Formally, however, the 'research organization' was not created before 1877. In 1875 there were two academically educated assistants in Caro's laboratory, and in 1876 there were three. A research laboratory finally began to take shape. At the end of 1877 (significantly, the year in which the German patent law was enacted) Caro was formally appointed to the new position of full-time research director.[29] This was the decisive step in the institutionalization of research at BASF. By the beginning of the 1880s, five or six

24 Compare the correspondence between H. Caro and W. H. Perkin, 1871–73, Deutsches Museum, Munich (Sonder-Sammlung, hereafter SSDM), transcribed in E. Vaupel, 'Carl Graebe (1841–1927) – Leben, Werk und Wirken im Spiegel seines brieflichen Nachlasses', Ph.D. diss. Ludwig-Maximilians University, Munich, 1987, ii, 629–733; C. Glaser, 'Erlebnisse und Erinnerungen nach meinem Eintritt in die Badische Anilin- und Soda-Fabrik im Jahre 1869', MS written in 1921, BASF Archives, pp. 35–6, 38, 43, 51; and Schuster, op. cit. (23), 80.

25 Glaser, op. cit. (24), 14–34.

26 Richard Wienandt, a pupil of Hermann Kolbe. My analysis of the expansion of Caro's laboratory – and of the emergence of the other laboratories at BASF – is mainly based on the MS 'Chemiker der BASF, 1865–1896', BASF Archives. In this document – written in phases (and in different handwriting) between about 1884 and 1900 – some 166 chemists who entered the Ludwigshafen works between 1865 and 1896 are mentioned. With a few exceptions (for instance Félix Duprey, who in 1868 was the manager of the aniline dye department), this book seems to be almost complete. The chemists working in the Stuttgart factories of BASF are, however, not mentioned.

27 Wiedemann, op. cit. (23), passim.

28 Dr Friedrich Köhler, a 'type D' laboratory chemist, in my classification. The discussion that follows is based on my analysis of the MS 'Chemiker der BASF', op. cit. (26).

29 In a letter dated 18 November 1877, Peter Griess wrote to Heinrich Caro that he was very jealous of Caro, because of this new position, SSDM No. 1894. Cf. Schuster, op. cit. (23), 85–7.

academic chemists, including Caro, were working in this laboratory, then called the *Hauptlaboratorium* (main laboratory).[30] This term of course indicated the existence of other laboratories within BASF; indeed by 1887 there were eighteen.[31] Most of them were ordinary factory laboratories, but, significantly, a few research laboratories were set up as well.

From 1875 onwards, research laboratories were established in the alizarin factory, the aniline dye factory and the azo dye factory (see Table 1). This process began with the appointment on a limited scale of so-called *theoretischer Chemiker*. In the course of time, out of these functions grew organizationally clearly distinguishable research laboratories; for example the *wissenschaftliches Laboratorium der Alizarinfabrik* (scientific laboratory of the alizarin factory), about 1886.[32]

From such facts one would consider BASF to be a very progressive company, but the reality was more prosaic. Serious conflicts between Engelhorn and Caro and the director of the dyestuffs factories, August Clemm, stimulated the latter to found research laboratories on his own behalf.[33] Clemm disliked Caro and saw him as a threatening intruder in his dyestuffs realm. The communication between the *Hauptlaboratorium* and the factories was seriously strained. Not until Clemm had left the company in 1881, and Caro had changed his post from research director to head of the new patent laboratory, in 1887, was cooperation between the factory and the laboratory improved. Significantly, August Bernthsen, Caro's successor who had previously been professor at Heidelberg, had, in his first years as leader of the research department, to continue to work under the strict old rules that prohibited all direct communication between his laboratory and the production departments.[34]

In 1889 the *Hauptlaboratorium* was transferred to a new building, specially built for this purpose between 1887 and 1889. In Bernthsen's department, he and five other chemists were employed. In Caro's patent laboratory, in the same building, working space for some additional chemists was available. A few years later the *Hauptlaboratorium* was enlarged, so that it could accommodate more than twenty research chemists.[35]

Hoechst

The second largest German synthetic dyestuffs producer, the firm of Meister, Lucius & Brüning (Hoechst), at Höchst near Frankfurt, also set up a special laboratory organization at an early date. In 1870, Hoechst founded a separate *Analytisches Laboratorium*, headed

30 For the history of this laboratory, see C. Schuster, 'Die Geschichte des Hauptlaboratoriums der BASF von 1868–1938', typescript dated June 1958, BASF Archives.

31 Schuster, ibid., 8.

32 Schuster, ibid., gives a later date (1894). My date is based on the MS 'Chemiker der BASF', op. cit. (26).

33 Glaser, op. cit. (24), 14–34.

34 Glaser, ibid.; Schuster, op. cit. (30); A. Bernthsen, *Fünfzig Jahre Tätigkeit in chemischer Wissenschaft und Industrie*, Leipzig/Berlin, 1925, 39; P. Borscheid, *Naturwissenschaft, Staat und Industrie in Baden (1848–1914)*, Stuttgart, 1976, 174–7. An attempt to engage Emil Fischer failed in 1883. Cf. Schuster, op. cit. (23), 99–105.

35 Bernthsen, op. cit. (34), 35–7; Schuster, op. cit. (30); Schuster, op. cit. (23), 106. Borscheid's interpretation that the 'creation' of the *Hauptlaboratorium* in 1889 was the result of a growing cooperation between scientists and engineers is not based on any factual evidence. Borscheid, op. cit. (34), 177.

by Dr Hermann Strecker, a brother of the chemistry professor Adolph Strecker, who was a consultant to the firm. Hardly any research on new dyestuffs was done in this laboratory, but it was not just a routine analytical laboratory. New chemical methods for quality control of the raw materials, intermediates and final products of the alizarin synthesis were developed there.[36] Although not in any way a research laboratory, this central analytical laboratory is worth mentioning, because it was, like the laboratory of La Fuchsine, one of the first laboratories not connected to a specific production plant, but operating at the 'corporate level'.

During the early 1870s, research on new dyestuffs at Hoechst was not undertaken in the analytical laboratory, but in a laboratory directed by one of the company's owners, Dr Eugen Lucius. Accordingly, in those years, the situation at Hoechst was comparable to the situation at Perkin's firm.[37] Exact data on the situation of chemists at Hoechst are rare.[38] Lucius probably headed this laboratory until the end of the 1870s, when some new developments took place.

In 1878, Heinrich Baum, the manager of the azo dye factory at Hoechst, initiated extensive research in the field of azo dyes. From the same year there is the first indication that a research laboratory had been, or was going to be, founded: the chemist Theodor Diehl was appointed to a full time research function (according to his obituary). He was undertaking research in the *Labor der Farbenfabrik*, probably under the direction of Baum. Whether other chemists at Hoechst held similar positions at that time is not known.[39] However, a research organization was certainly in existence from around 1880, when a research group, under the direction of Dr Christian Rudolph, was investigating the synthesis of indigo.[40]

The fact that so few details are known about the emergence of a research laboratory at Hoechst makes it probable that the research organization evolved rather gradually out of

36 E. Luck, 'Methode der Antracenbestimmung', *Berichte der deutschen chemischen Gesellschaft* (1873), **6**, 1347–8. For an account of the work in the chemical laboratories of Meister, Lucius & Brüning, during the early period, see 'Aufzeichungen von Dr. Lucius und seine Mitarbeitern (1863–1874)', Hoechst Archives.

37 The assistants of Lucius were, according to my classification, laboratory chemists of type A. See 'Aufzeichnungen von Dr. Lucius', op. cit. (36); E. Fischer, 'Meister, Lucius und Brüning, die Gründer der Farbwerke Hoechst AG', *Tradition* (1958), **3**, 65–78, especially 66.

38 In a letter dated 12 October 1984, W. Metternich of the Hoechst Archives answered my queries concerning fifty-four chemists who worked at Hoechst between 1863 and 1890. In twelve cases no data at all could be found. In nearly all the other cases the personnel files did not contain any details on the kind of chemical work that a chemist carried out, or on his place in the organization. The recent book by E. Bäumler – *Farben, Formeln, Forscher, Hoechst und die Geschichte der industriellen Chemie in Deutschland*, Munich and Zurich, 1989 – gives no detail with respect to the organization of research.

39 In 1880, Hoechst employed twenty-five works and laboratory chemists. The laboratory chemists were assisted by twenty-five laboratory boys and other workers.

W. Grandhomme, 'Die Theerfarben-Fabriken der Herren Meister, Lucius & Brüning zu Höchst a. Main in sanitärer und sozialer Beziehung', *Vierteljahrsschrift für gerichtliche Medizin und öffentliches Sanitätswesen* (1880), n.s. **32**, 120–5, 280–320, and (1880), n.s. **33**, 78–126, especially **32**, 123; W. Grandhomme, 'Bericht über die Krankenbewegungen auf den "Farbwerken" vorm. Meister, Lucius & Brüning in Höchst a.M. pro 1880', *Correspondenzblatt des niederrheinischen Vereins für oeffentliche Gesungsheitpflege* (1881), **10**, 99–108, especially 100; *Farbwerke vorm. Meister, Lucius & Brüning 1863–1913*, Höchst, n.d., 18–19.

40 Biography of Eugen Fischer, *Berichte der deutschen chemischen Gesellschaft* (1918), **51**, 1–4; letters of A. Baeyer to H. Caro, dated 5, 12 and 14 July 1880, see Wiedemann, op. cit. (23), ii, 180–3.

a normal works laboratory (*Labor der Farbenfabrik*).[41] Only after the first reorganizations had taken place in 1883 did the research organization take on a more definite shape. More details on the research practice at Hoechst can be found in the company records and archives for the years after 1883. Nevertheless, the research activities going on earlier, should not be ignored.

In 1883, the board of Hoechst decided to enter the pharmaceutical field.[42] A production plant was erected and the management of the company set up a 'scientific laboratory' for the investigation of drugs. For this purpose, two university lecturers were recruited by Hoechst in 1883: Professor August Laubenheimer, from Giessen university, who later was elevated to membership of the board of directors of Hoechst, and *Privat-Dozent* Eduard von Gerichten, who became the leader of the new research group. Von Gerichten's laboratory soon received the title *Zentrallaboratorium*. Some dyestuffs research was later undertaken there as well.[43] Between 1889 and 1893, a new building for the *Zentrallaboratorium* was erected, called the *Löwenapotheke* by the staff of Hoechst, and in which the technical and analytical laboratories also found a home.[44]

It is important to point out that the founding of the Hoechst *Zentrallaboratorium* was not the result of an evolutionary process within the company. The laboratory was created as the result of a new corporate strategy, aimed at broadening the commercial and technical basis of the firm by entering a completely new product area. Research leaders in this new field were, as I have shown, attracted from outside the company, and a conscious research strategy of 'rational design' of drug molecules was introduced and practised. In a letter of 12 July 1884 to the Kaiserliche Reichs-Gesundheitsamt the Hoechst management argued that chemistry had now reached a state of development in which 'ganz bestimmte Körper,... mit einer, wie der Chemiker sagt, bestimmten Atom-Gruppirung,... [herausgesucht und] auf..., vorher erwartete, physiologische Wirkung untersucht [werden können]'.[45] When von Gerichten left Hoechst in 1893, another academic scientist was recruited by the company: Wilhelm Roser, previously professor of chemistry at Marburg.

THE EMERGENCE OF 'THE' INDUSTRIAL RESEARCH LABORATORY

BASF and Hoechst were the first German dyestuffs companies to establish laboratories specifically devoted to research. Developments in the other German dye firms will not be discussed in detail here. Table 1 presents the results of my investigation of the history of

41 See the biographies of T. Diehl, B. Homolka and A. Herrmann.

42 *Farbwerke vorm. Meister Lucius & Brüning*, op. cit. (39), 20.

43 Biographies of E. von Gerichten, H. Reissenegger and A. Laubenheimer; G. Ehrhardt, 'One hundred years of research', in *A Century of Chemistry* (ed. E. Bäumler), Düsseldorf, 1968, 273–337, especially 274–80; *Einführung der Serologie in Hoechst durch Prof. August Laubenheimer*, Höchst, 1968 (= *Dokumente aus Hoechster Archiven*, vol. 35), 7–9, 24–9, 80–1.

44 'Wissenschaftliches Laboratorium', drawing and ground-plan, n.d., Hoechst Archives; C. Graepler, 'Industriebauten der Hoechster Farbwerke im ersten Jahrhundert ihres Bestehens', *Jahrbuch Marburger Universitätsbund*, 1963, 137–52, especially 142–5.

45 E. Hickel, 'Die industrielle Arzneimittelforschung am Ende des 19. Jahrhunderts und die Durchsetzung einer reduktionistische Biologie', in *Materialistische Wissenschaftsgeschichte: Naturtheorie und Entwicklungsdenken*, Berlin, 1981 (= *Das Argument – Sonderband AS 54*), 132–54, especially 137–40.

Table 1. *Establishment of research laboratories in the German dye industry*

No.	Year	Company	Laboratory	Antecedents
1	1877	BASF	Caro's Laboratory (later) *Hauptlaboratorium*	1868–(1877) H. Caro
2	(1878)	Hoechst	'Labor der Farbenfabrik'	
3	1879	BASF	*Laboratorium für Azofarbstoffe*	
4	(1882)	Oehler	'Wissenschaftliches Laboratorium'	1879–(1882) E. Hepp
5	(1882)	Kalle	(later) *Hauptlaboratorium*	1879–(1882) R. Nietzki
6	1882	Agfa	*Versuchslaboratorium*	
7	(1883)	BASF	'(Wissenschaftliches) Laboratorium der Anilinfabrik'	(1877)–1882 F. Köhler 1882–(1883) C. Schraube
8	1883	Hoechst	*Zentrallaboratorium*	
9	(1885)	Cassella	'Wissenschaftliches Laboratorium'	1883–(1885) A. Weinberg
10	(1886)	BASF	*Wissenschaftliches Laboratorium der Alizarinfabrik*	1875–1877 E. Zickendraht 1879–1883 C. Meyer 1884–(1886) R. Bohn
11	(1886)	Bayer	'Laboratorium der Azofabrik', (later) *Wissenschaftliches Hauptlaboratorium*	(1881)–1884 P. Seidler 1884–(1885) C. Duisberg

Note:
Hauptlaboratorium = *official name.*
'Labor der Farbenfabrik' = unofficial name.
1882 = official founding date of the research laboratory.
(1878) = inferred from the appointment of assistants.
Sources: Biographies of the chemists mentioned, and those of H. Baum, T. Diehl, C. Rudolph, E. von Gerichten (Vongerichten), H. Reissenegger and A. Laubenheimer (Hoechst), V. Kallab, R. Henriques, P. Friedländer and B. Priebs (Oehler), A. Winther and E. Hepp (Kalle), O. Goll and F. Klingemann (Cassella) (see note 10); and Schuster (note 30); O. Horn, 'Research at Hoechst', *Chemistry & Industry* (1964), 1734–8; Meyer-Thurow (note 2); Dr Hagenböcker, 'Zur Geschichte des Werkes Offenbach', Offenbach, 1936 (typescript in Bayer Archives, 5/Fa.1), 30–56, 61; and Voelcker (note 6).

the research laboratories of the seven largest German dyestuffs companies. In the last column of this table ('antecedents') are those chemists known to have been appointed to a part-time or full-time research function.[46] As soon as assistants were appointed to these chemists (in the year shown in brackets), or when other historical information makes it

46 For instance the '*theoretische Chemiker*' at BASF.

clear that a research laboratory was set up (year not in brackets), I took that date as the first sign of the start of a research organization.

Patent law

Although the appointment of some research chemists predates the enactment of the 1877 Reichspatentgesetz (Patent law of the German Reich), the data presented in Table 1 show clearly that the establishment of almost all the research laboratories took place after that date. This is not a coincidence. As in the United States thirty years later, the screening of competitors' patents, and the building up of a strong company patent portfolio, were the main objectives of all the research laboratories (see below).

Company size

A second observation to be made in relation to Table 1 is that the chronological order in which the different laboratories were founded corresponds roughly to the size of the companies involved. Cassella and, especially, Bayer are the major exceptions to this rule. Table 2 gives some information on the numbers of workers employed in the seven largest German dye firms, and the leading position of BASF and Hoechst is clear. It is important to realize that Oehler and Kalle, two companies that had fallen behind by 1900 (and for that reason were largely forgotten by the historians) were in the late 1870s still relatively flourishing. Both companies were among the first to set up a research laboratory, but for them research was not a panacea for business success.

The Bayer company differed from the other German dye firms in many ways. Managed by a very able businessman, Carl Rumpff, it had by 1880 advanced to the third position in the German dyestuffs sector. It was, however, the only large German dye company without a university- or polytechnic-educated chemist on the board of directors! As contacts between teachers and their pupils frequently formed the basis of later connections between university and industry, the Bayer company lacked good contacts with the university world. Also, the company organization at Bayer was different from that found at the other market leaders. In the Bayer company, as Meyer-Thurow has shown, the foremen played a prominent role, at least until the 1880s. This certainly was not the usual practice in the German dye industry. At BASF, Hoechst, AGFA and the other companies, many – if not all – of the production departments were headed by academic chemists.[47]

I am inclined to think that the different type of composition of the board of directors, and the quite different organization of work in the Bayer company, were the main reasons why this company lagged behind in setting up a research laboratory. In any event, it is clear that Bayer was not typical of the German dye industry. It is therefore unfortunate – though

47 For BASF, see 'Chemiker der BASF', op. cit. (26); for Hoechst, see W. Grandhomme, *Die Theerfarben-Fabriken der Actien-Gesellschaft Farbwerke vorm. Meister Lucius & Brüning zu Höchst a.M. in sanitärer und sozialer Beziehung*, 2nd edn, Heidelberg, 1883; on AGFA a paper by the author is in preparation. On AGFA, Hoechst and the Swiss firm of Bindschedler & Busch (CIBA), see also the contemporary investigation by H. E. Roscoe published in the *Second Report of the Royal Commission on Technical Instruction*, London, 1884, i (= *Parliamentary Papers*, vol. XXIX), and (partly) in the *Journal of the Society of Chemical Industry* (1882), 1, 251–2.

Table 2. *Numbers of workers in the German dye companies*

Company	1870	1875	1880	1885	1890	1895	1900
BASF	500	850	1650	2366	3578	4444	6485
Hoechst	?	406	1088	1586	2243	2718	3555
Bayer	60	119	298	555	1264	2506	4515
Oehler	70 (1867)	?	130 (1878)	198	231	325	450
AGFA	—	?	100 (?)	476	690 (1892)	?	1790
Cassella	15	?	146	?	545	?	1800
Kalle	50	69	?	130	?	225	490

Note: Figures from different sources are frequently conflicting, often owing to unspecified decisions as regards the inclusion or exclusion of the staff in the total numbers of workers, the in- or exclusion of the workers of the foreign subsidiaries of the firm, and the day or month of the year to which the number refers. For the sake of comparison I have chosen, as far as possible, to present the numbers of workers (staff excluded) employed in the factories on German territory.

Sources: *Les produits chimiques et les matières colorantes. Le blanchiment, la teinture et l'impression des fibres textiles*, published by the *Revue Générale des Matières Colorantes*, Paris, 1900, 70–92; R. Sonnemann, *Zur Geschichte der Teerfarben-Industrie in Deutschland von ihren Anfängen bis zur Bildung der beiden Dreibunde (1905/07)*, Leuna-Merseburg, 1963, 39–40; 'Chronik der BASF – Der Mensch in der BASF', n.d., BASF Archives, 12–13; Grandhomme (note 47), p. 107; W. Grandhomme, *Die Fabriken der Aktien-Gesellschaft Farbwerke vorm. Meister, Lucius & Brüning, etc.*, 3rd edn, Frankfurt a.M., 1893, p. 75; idem. 4th edn, Frankfurt a.M., 1896, 77–9; H. Pohl, R. Schaumann and F. Schönert-Röhlk, *Die chemischen Industrie in den Rheinländen während der industriellen Revolution. Band 1: Die Farbenindustrie*, Wiesbaden, 1983, 122; F. Thissen, *Die Stellung der deutschen Teerfarbenindustrie in der Weltwirtschaft*, Ph.D. diss., Giessen, 1922, 81–2; *Chemiker-Zeitung* (1879), 3, 426; *Drei Postausgangsbücher der Chemischen Fabrik Karl Oehler, Offenbach (1852–1859)*, Hoechst a.M., 1967 (= *Dokumente aus Hoechster Archive*, 27), 19; R. Willstätter, 'Zur Geschichte der Agfa', *Berichte der deutschen chemischen Gesellschaft* (1931), 64, 140–9, especially 143; *Dreizehnter Geschäftsbericht erstattet vom Aufsichtsrath und der Direction der Actiengesellschaft für Anilinfabrikation*, Berlin, 1886, Bayer Archives, 5/E.a.40; *Amtlicher Bericht über die Weltausstellung in Chicago 1893*, Berlin, 1894, ii, 748; Voelcker (note 6), pp. 87, 95, 133; Kalle 1875 = mean of the figures for 1872 (76) and 1877 (62); for Kalle 1872 see *Amtlicher Katalog der Ausstellung des Deutschen Reiches auf die Wiener Weltausstellung. Gruppe III: Chemische Industrie*, Berlin, 1873, 126.

understandable because of the surviving company records – that this company has attracted most attention from historians.

The role of management

According to Meyer-Thurow, creation of the Bayer research laboratory was an evolutionary process, one not initiated by any direct action of the management. It grew out of the inventive activities of a group of young works chemists in the azo department, gathered around Carl Duisberg, and culminated in the move of this research group to a new laboratory building in August 1891.[48] Here it has been shown that the developments

48 Meyer-Thurow, op. cit. (2), especially 369–70. On the new laboratory, built between 1889 and 1891, see 'Das wissenschaftlich-chemische Laboratorium der Farben-fabriken vorm. Friedr. Bayer & Co. zu Elberfeld', *Chemiker-Zeitung* (1893), 17, 191–7; and H. E. Armstrong, 'The appreciation of science by German manufacturers', *Nature* (1893), 48, 29–34.

at BASF followed the same evolutionary pattern (although initiated by Friedrich Engelhorn's bold managerial decision to appoint a technical director who would spend part of his time in research). It would be wrong, however, to generalize and draw the conclusion that the creation of industrial research laboratories in all the dye companies was the result of such evolutionary developments. A few counterexamples will show that in certain instances it was the company directors who took the initiative to found laboratories that were true research laboratories right from the start.

The first time that this happened, as far as can be ascertained, was in the autumn of 1882 when the management of AGFA founded the *Versuchslaboratorium*.[49] Unlike Kalle and Oehler, who both appointed only one special research chemist in 1879, AGFA created, in 1882, a completely new research group and persuaded the Strasbourg *Privat-Dozent* Gustav Schultz to become its leader. At that time, Schultz was the most prominent authority in the field of 'applied' aromatic chemistry and had just finished a very comprehensive handbook on this subject.[50] At least one of the first chemists engaged in Schultz's laboratory at AGFA had previously been his assistant in Strasbourg.[51] Undoubtedly, AGFA had succeeded in recruiting one of the most knowledgeable academic aromatic chemists of the day, complete with a part of his research group.

The reasons for this remarkable initiative by the AGFA management are not well documented. From the yearly reports of Schultz's laboratory one gets the impression that the main cause must have been the time-consuming and laborious examination of the licences offered to the company and of the patent applications of the competitors. If one thing is clear from these reports, it is that there was no research programme at all. All activities were prompted by impulses coming from outside the company.[52] In the years before 1882, general director Carl A. Martius, or one of the works chemists, was given the laborious task of checking and evaluating the discoveries offered to the company by university professors and other chemists.[53] Other dye companies maintained close relations with one or more university chemists too, and encountered the same type of problem.[54]

49 Biography of Gustav Schultz, *Zeitschrift für angewandte Chemie* (1922), **35**, 172–5.

50 G. Schultz, *Die Chemie des Steinkohlentheers, mit besonderer Berücksichtigung der künstlichen organischen Farbstoffe*, Braunschweig, 1882.

51 Dr E. I. Erdmann. The names of two other early research assistants (1884) are known – Dr F. Streng and Dr A. Hagen – but their biographies are obscure.

52 G. Schultz, 'Arbeiten des Versuchslaboratorium in dem Jahre 1887', manuscript dated 21 March 1888; G. Schultz, 'Bericht über die im Versuchslaboratorium in dem Jahre 1888 ausgeführten Arbeiten', and the other yearly reports written by Schultz 1889–94. Bayer Archives, 5/E.a.26.

At the *Hauptlaboratorium* of BASF the situation was very similar. See Bernthsen, op. cit. (34), 39: 'Statt in den Betrieben Anregungen über die Bedürfnisse des Geschäfts zu erhalten, wurde ich für diese Anregungen auf das Studium der Patentanmeldungen der Konkurrenz verwiesen.'

53 Research for AGFA was done by Professor O. Wallach in Bonn (who had been a chemist in the factory in 1871) and by Professor C. Liebermann and the *Privat-Dozenten* O. Doebner and E. Jacobsen in Berlin. In the period 1877–83 the patents resulting from their discoveries were frequently taken out in Martius' name (and not in the name of the company). See '[Jahresbericht] an den Aufsichtsrach der Actiengesellschaft für Anilinfabrikation in Berlin [für 1881]', MS dated 18 March 1882, Bayer Archives, 5/E.a.13; and '[Jahresbericht] an den Aufsichtsrath der Actiengesellschaft für Anilin-Fabrikation [für 1885]', MS dated March 1886, Bayer Archives, 5/E.a.40.

54 I have not elaborated on this subject here. For general information, see Schuster, op. cit. (23); and Borscheid, op. cit. (34), 135–57.

The new division of labour initiated at AGFA in 1882 guaranteed that the works chemists and the directors could concentrate on their respective primary tasks.

The laboratory chemists in the research laboratory were, apart from the laboratory director Gustav Schultz, in the lowest ranks of the income hierarchy. The salaries of works chemists at AGFA varied in 1884 between 2300 marks and 6000 marks a year (managers and *Prokuristen* excluded), and those chemists received royalties as well. The three assistants of Schultz, however, earned only between 1800 and 2300 marks a year, without any rights to royalties! In contrast, payments to Schultz himself consisted in 1884 of a fixed amount of 9000 marks a year, supplemented with royalties on the patented discoveries from his laboratory.[55] These were the same feudal relations between a laboratory director and his assistants as those prevailing in German university life at that time. For a *Privat-Dozent* like Schultz, the prospect of earning a salary in industry well above that of a university professor must have made his decision to leave Strasbourg university relatively easy. Moreover, the laboratory facilities in industry were often superior to those at the universities. About 1889, Schultz and his research group moved to a new laboratory building which united all the 'corporate' laboratories at the Wiesenufer factory of AGFA in Berlin (i.e. analytical, patent, technical and '*Versuchs*' laboratories) under one roof.[56]

A second example of a deliberate action by a company management was the establishment of the *Zentrallaboratorium* at Hoechst, mentioned above. In this case the aim was clearly diversification into the field of pharmaceuticals. As at AGFA, highly qualified academics were head-hunted at the universities and appointed to high positions at Hoechst in order to direct the new research activities that the company management planned to undertake.

The 'research laboratory' as a general concept in industrial organization

These examples show that 'the industrial research laboratory' had become a recognizable organizational concept by 1882 or 1883. This is confirmed by a number of non-industrial sources. Already in 1878, Professor Heinrich Schwarz at the Graz Technical University wrote that there were only a few factories in Germany – for example the aniline dye factories at Berlin, Mannheim and Höchst – where a number of chemists had the sole task of making 'scientific' investigations. Although it is doubtful whether the label 'scientific' was justly applied to all the work done by the first industrial 'research chemists' (and the mention of Berlin was erroneous), we can infer from this statement that a few chemists outside industry were well aware that in some dye companies new professional possibilities for chemists had been created.[57]

55 For the salaries, see 'Gehalter-Kontierung 1884', Bayer Archives, 126/1; for the royalties, see 'Farben-Fabrikation und Production', vol. 1 (1867–1930), therein 'Notizen von Carl Duisberg', MS 31/12/1884, Bayer Archives, 111/2.

56 'Bau-Konzession Laborgebäude, 1889–1921', Bayer Archives, 5/E.a.21.

57 H. Schwarz, 'Der Studiengang für technische Chemiker', *Die Chemische Industrie* (1878), 1, 353–6 (354). Schwarz could only be referring to the companies of AGFA, BASF and Meister, Lucius & Brüning. The AGFA company, however, certainly had *no* research laboratory at that date (November 1878). It is doubtful if Schwarz had enough information on the internal affairs of these three dye companies. In my opinion, the most probable explanation is that Schwarz mentioned these companies because BASF, AGFA and Hoechst were the first Germany dye companies that obtained patents on coal-tar dyes as follows:

A few years later, an information booklet for chemistry students stated that '[those who want employment in a chemical factory, or in pharmacy ... must especially study chemical analysis]. An advanced study of synthetical processes is only advisable to those students who have the aspiration to enter one of those few companies which have an Experimental Station (*Versuchs-laboratorium*), for reasons of competition.'[58]

This quotation gives ample evidence that the concept '*Versuchslaboratorium*' had by that time been transformed from a local concept, with a company-specific meaning, to a more general concept of industrial organization.[59]

In summary, I have demonstrated that the first industrial research laboratories came into existence in an evolutionary way within the large British, French and German dye companies. In the period between 1877 and 1882, several German companies organized their research so that separate, relatively autonomous, organizational units for this purpose were formed. These five years can be called the formative period of 'the' industrial research laboratory. By 1883 the German dye industry had created an 'organizational model' which could be investigated, imitated or adapted, depending on local circumstances.

THE RECRUITMENT OF THE RESEARCH LEADERS

On the basis of the evidence presented so far, it is possible to conclude that the managing boards of the German dye firms played a much more active role in the emergence of the industrial research laboratory than was suggested by Meyer-Thurow in his 'industrialization of invention' paper. Further support for this claim is offered by the results of an investigation into the antecedents of the research directors and 'theoretical chemists' that were appointed by the German dye firms between 1868 and 1896 (Table 3).

In the period under consideration, twenty-five different appointments of research leaders and/or independent research chemists took place in the seven largest German dye companies.[60] For sixteen appointments, research chemists were recruited from outside the

DRP 1886	15-12-1877	BASF	Methylene blue
DRP 2096	16-12-1877	BASF	Acid magenta
DRP 4322	26-2-1878	AGFA	Malachite green
DRP 5411	12-3-1878	BASF	Fast red
DRP 3229	24-4-1878	Hoechst	Ponceau

The patent no. 4322 was not invented but only filed by AGFA. The inventor was the chemist Oscar Doebner, who worked in the laboratory of A. W. Hofmann at the University of Berlin.

58 O. Gumprecht, *Wie studirt man Chemie und die beschreibenden Naturwissenschaften? Mit Berücksichtigung der sächsischen, preussischen und baierischen Prüfungsordnungen*, Leipzig, 1885, 28–9.

59 See also G. Wunder, *Die Vorbereitung für den Eintritt in die chemische Technik. Eine Schrift zur Orientierung für künftige Techniker, nebst Beschreibung des neuen Laboratoriums der technischen Staatslehranstalten in Chemnitz*, Chemnitz, 1879, 11; and P. Friedlaender, *Fortschritte der Theerfarbenfabrikation und verwandter Industriezweige 1877–1887*, vol. 1, Berlin, 1888 (reprinted 1920), pp. iii–iv.

60 There were actually only twenty-four different persons, since Eduard Hepp was counted twice. Hepp started his industrial career as a research assistant to Heinrich Caro in the research laboratory of BASF. In 1879 he went to Oehler, where he was appointed a research chemist and later the leader of a research group. He succeeded A. Winther in 1884 as director of the research laboratory of the Kalle company. In 1895, he moved to the research laboratory of Hoechst, but not, this time, as the head of that department.

Table 3. *Antecedents of the research directors and 'theoretical chemists'*

Previous function outside the company					
Professor or *Privat-dozent*			Other		
1882	G. Schultz	6	1868	H. Caro	1
1883	A. Laubenheimer?	8	1875	E. Zieckendraht	10
1883	E. von Gerichten	8	1879	F. Fuchs	3
1884	P. Friedländer	4	1879	E. Hepp	4
1887	A. Bernthsen	1	1879	R. Nietzki	5
1893	W. Roser	8	1882	A. Winther	5
1894	F. Klingemann	9	1884	E. Hepp	5
			1884	R. Bohn	10
			1893	W. Herzberg	6
Previous function inside the company					
c. 1877	F. Köhler	7			
1882	C. Schraube	7			
1884	C. Duisberg	11			
1896	B. Heymann	11			
Details unknown					
1878	Th. Diehl	2			
1879	C. Meyer	10			
c. 1880	C. Rudolph	2			
c. 1881	P. Seidler	11			
1883	A. Weinberg	9			

Note: BASF = 1, 3, 7 and 10; Hoechst = 2 and 8; Bayer = 11; Oehler = 4; AGFA = 6; Cassella = 9; Kalle = 5. (These numbers correspond to the numbers in Table 1.)
Source: Biographies of the chemists mentioned.

company (80 per cent!); in four cases they came from a position inside the company, and in five cases the details of the initial appointment of these chemists are unknown. The number – seven – of former professors and *Privat-Dozenten* among them is striking. Some university assistants were appointed as well, and they often had considerable research experience (e.g. Nietzki, Herzberg).

From this analysis it is clear that in the majority of cases the evolution of the research laboratory did not follow the unplanned, haphazard route that Meyer-Thurow has described for Bayer. At that firm the research laboratory evolved 'bottom up'. Chemists in one production department, namely that of azo dyes, played a significant role. In the other companies, however, the management played an important part by appointing former university teachers, and research chemists who had worked for other companies (cf. Hepp), directly to specific research positions. After 1882, especially, the development pattern of corporate organization in the German dye industry was no longer evolutionary. The research laboratory was clearly recognized as an essential ingredient of a company's organizational structure, and management took the necessary steps – for instance by recruiting university professors – to ensure that this part of their organization would become a success.

Only at Bayer, and to a lesser extent at BASF, were chemists from inside the company promoted to the position of a research leader, or an 'independent research chemist'. This provides additional support for my claim that the Bayer company followed a pattern completely at variance with developments elsewhere.[61]

CONCLUSION

The genesis of the research laboratory as an organizational entity in the German dyestuffs firms was the result of a rapid process. Within a few years of the enactment of the Reichspatentgesetz of 1877, nearly all the larger German dye companies had founded special departments for research, so that by 1882 'the industrial research laboratory' had emerged as a clearly distinguishable organizational concept.

The first steps towards the creation of special units for research were somewhat unplanned and haphazard. The examples of BASF and Hoechst have shown that the seeds of their later research organizations were sown during contacts between academe and industry (Caro – Baeyer), and from the bench of a regular factory laboratory (Hoechst). Very soon, however, intentional managerial decisions laid the basis for a further expansion of the industrial research system in the German chemical industry. In the historical studies published so far (Beer and Meyer-Thurow), this role of management in the German firms was seriously underestimated. The main cause for this state of affairs has been the almost exclusive attention to developments taking place in the Bayer company. That the Bayer company differed in many respects from the other large German dye firms was insufficiently discerned.

Apart from the role played by the size of the firm, the answer to the question of why the industrial research laboratory came into being in the German chemical industry between 1877 and 1882 can be found in the interplay between two developments that affected both German industry and German science: (1) the new situation created by the patent law of 1877; and (2) the acceleration of discovery and invention in the field of organic chemistry and dye chemistry, as a result of the growing number of academic chemists and of the oligopolistic competition between the major firms. The Reichspatentgesetz of 1877 can be called the single most important *general* impetus for the systematic creation of research activities within the German dye firms, not only because it stimulated inventive activity of the dye firms themselves, but also because it created a 'market' whereby knowledge of organic chemistry and dye chemistry could be sold. Of course, there were also several *local* factors at work, which differed from company to company, as the examples of BASF, Hoechst, Bayer and Agfa have shown.

The German patent law strongly affected both the interaction between industry and the universities, and the mutual relations between the firms. A growing flow of patent applications by all the dye firms, and a growing number of patentable inventions offered to industry by researchers from the universities, could not be effectively checked and

61 At BASF, five out of seven theoretical chemists and/or research leaders were recruited from outside the company. The only two chemists who were promoted from another post within the company to the position of independent research chemist (Dr Friedrich Köhler and Dr Georg Conrad Schraube), had both previously been assistants in Caro's *Hauptlaboratorium*.

evaluated in the traditional works laboratories without disturbing the day-to-day duties related to dyestuffs production. It was this circumstance, as I have shown, that was a major cause for the emergence of research laboratories in the German dyestuffs industry. The German patent law, in combination with the oligopolistic market structure, also ensured that a successful organizational innovation in one company (BASF) would eventually be 'copied' by the others.

[7]

History and Technology, 1994, Vol. 12, pp. 23–77
Reprints available directly from the Publisher
Photocopying permitted by license only

STRATEGIES FOR SUCCESS: RESEARCH ORGANIZATION IN GERMAN CHEMICAL COMPANIES AND IG FARBEN UNTIL 1936*

ULRICH MARSCH

Deutsches Museum, Graduiertenkolleg, Museumsinsel 1, 80538 München, Germany.

Merging six leading chemical companies to form one single enterprise meant also finding adequate ways to coordinate more than 1000 research chemists in more than 50 laboratories scattered all over Germany. IG Farben was successful in organizing corporate research by using the experience and structures of the former independent companies. In the corporate innovative process, the applied works laboratories, the large Technical Stations and semi-works were of much greater importance than has so far been realized. Their existence goes well back into the late 19th century, but their role has been neglected. In analyzing the research structures of the core chemical companies in Germany until 1925, we find IG's organization anticipated. Though IG used most of the traditional instruments of the former companies, it pursued a much more active research and product policy than known today and developed its own concepts of organized research, which were in contrast to those of its competitors, DuPont in the USA and ICI in Great Britain.

Much attention has been paid to the origins of organised and institutionalized research in chemical companies in 19th century Germany. The focus has been on the establishment of the Central Research Laboratory, and most descriptions have been limited to this new corporate element and have ended well before the First World War. But if we want to know more about research structures and strategies in chemical firms, we have to broaden our view to all scientific and technical institutions within these firms, and we must observe the development and changes of the overall organization, structures and strategies over a longer period.

The Central Research Laboratory was a new and important element in the process of the 'incorporation of science'. But equally important in this strategy was the creation of large Technical Stations where the new discoveries were applied to mass production processes. These institutions were centralized, but decentralization of research occurred right from the start of the chemical firms:

Work on this project has been supported by the Volkswagen Foundation. My thanks to the staff of the archive of the Bayer AG, Leverkusen, Germany, especially to Mr Pohlenz and Mr Pogarell, and to the staff of the Hagley Museum and Library in Wilmington, Delaware, USA. I am indebted to Dr Peter Morris, Science Museum London, Dr Helmut Trischler, Deutsches Museum, Munich, Prof. Dr. Ulrich Wengenroth, Technische Universität München, Declan O'Reilly, Cambridge University, and David Hounshell, Carnegie Mellon University, Pittsburgh, PA, for their help and useful comments.

the works laboratories pursued scientific and applied investigations that were crucial for the development and improvement of existing products and contributed much to the success and growth of the firms. Here the large majority of the firms' chemists were employed, and here the close connection of research and production became fruitful. The cooperation of chemists, engineers, and technicians was the real "industrialization of invention". The distinction between research and development, which is still common in American and British firms, dissolved away, so that we have to state that for German chemical firms after 1900: development is part of research.

When IG Farben was founded in 1925, it consisted of six major companies with similar organizational patterns. For the following few years, IG continued with the decentralized structure of research inherited from its predecessors. However, it sought to change this after 1930 to gain cost reductions, and altered this structure and policy to adapt to contemporary needs in a different economic environment. The analysis will stop in 1936, when rearmament and the Four Year Plan strongly influenced IG's overall policy so that the period from 1936 to 1945 is not representative for the company's research policy. In this paper I want to show that IG Farben pursued a planned research and product policy. IG's management and structure was dominated by chemists who had developed a strong belief in the almost unlimited possibilities of research. With these mainly chemically trained managers, the technocrats[1], IG was in control of its scientific and technological affairs.

The works of Beer, Meyer-Thurow and Homburg[2] put the emphasis on the creation of a Central Research Laboratory, seeing this as the starting point of the process of the "industrialization of invention". For these historians the creation of a centralized Research Laboratory meant the beginning of using science as a corporate strategy for further growth. They thereby neglect the role of the engaged works-chemists, which were not working in centralized institutions such as the later research laboratory but carrying out their scientific investigations in the works-laboratories and technical stations. These chemists did research as well, which today would be regarded as "applied research", investigations mainly in the improvement of existing products, or of production processes and especially in gaining greater efficiency. Their role in an industrial entreprise has so far been neglected.

In his latest book, Alfred Chandler describes IG Farben, its structure and development, but his picture of IG's research structure and strategy is incomplete.[3]

Although Gottfried Plumpe recently dedicated a whole book to IG's history, technology and political events, he tells us very little about IG's corporate research, structure and policy. His, and also Hubert Sales', picture of IG Farben's research policy and structure is lacking in many respects.[4] This is especially annoying because David Hounshell and John Smith supply a detailed and informative analysis of DuPont's (IG's competitor on international markets) process of incorporating science and its corporate research policy during eighty years.[5]

This paper aims to fill this gap regarding long-term research policies at German chemical firms and particularly at IG Farben. Special attention will be paid to the works laboratories and organizational developments in the history of corporate industrial research.

I. THE COMPETITIVE PERIOD UNTIL 1914

1. *Emergence of the Chemical Industry*

The 1850s and early 1860s saw the foundation and rapid growth of chemical companies, mostly dyestuffs producers.[6] The more important of them were: "Oehler & Co" at Offenbach/Main (1850); "Chemische Fabrik Griesheim" uniting in 1898 with Chemische Fabrik Elektron to "Griesheim-Elektron"[7]; "Kalle & Co" (1862/63)[8]; the "Farbwerke Fried. Bayer & Co" (today Bayer AG)[9] and Meister Lucius & Brüning at Hoechst near Frankfurt (today Farbwerke Hoechst AG), both in 1863.[10] In 1865 followed "Badische Anilin- und Soda-Fabrik" (BASF)[11], and in 1867 the "Aktiengesellschaft für Anilinproduktion" (Agfa) in Berlin.[12]

The founders of these companies were often themselves chemists or had some chemical knowledge: Wilhelm Kalle held a PhD in chemistry, Friedrich Bayer received his business training in a chemical retail shop, two founders of BASF, Carl and August Clemm, were university trained chemists, and so were the founders of Agfa. In the firms' early lives, the entrepreneur was also the inventor: the owner secured with his inventive work the further existence of his firm. So it was not surprising that, along with the founding of these firms, laboratories were erected almost simultaneously. When Wilhelm Kalle founded his company in 1862/63, he included in his new buildings a laboratory for his and a hired chemist's scientific work.[13] Just two years after its foundation, BASF created its first works-laboratory in 1867 and was successful in hiring Heinrich Caro, a well known dye-chemist, in 1868.[14] When the first chemist at Bayer was employed in 1864, laboratories existed already within the works.[15] In Hoechst the first works-laboratory existed by 1863[16], and one additional chemist was hired in 1867 to work closely with the company's founder Dr Heinrich Brüning.[17] Similar to Kalle fifteen years earlier, Edmund ter Meer established a laboratory when he founded his company in 1877.[18] In general, the main task of these works laboratories was the monitoring of delivered raw-materials and testing new compounds, but investigations were also made to find new colours and substitutes for expensive natural dyes.

Chemical companies in Germany enjoyed rapid growth, as can be seen best from the employment figures in Table 1. Material on sales or dividends is more difficult to obtain. Most of the companies were transformed to share-issuing, public enterprises during the 1880s and then published their accounts. In the years around the turn of the century, the leading dyestuffs companies payed annual dividends of between 18 and 26%.[19] By the end of the 1870s, a number of dynamic companies had emerged, all of which had some kind of stations where chemical investigations were made alongside the basic testing of raw materials.

In the period after 1877, we find a change in the companies' organisation, which is closely connected with the introduction of the German Patent Law in 1877. The new law meant that companies could no longer simply copy a chemical or technical process to produce a certain colour or intermediate; differing from French Patent Law, only processes, but not chemical products, could be patented. Also, patents granted were published so that everyone could protect his invention from copying and competitors knew that they had to develop different processes if they wanted to place the same product on the market. All inventions in a patent application had to pass a preliminary test to make sure that they really were original and new. Following this new legal situation, an organisational change took

Table 1. Employment figures for German chemical companies.

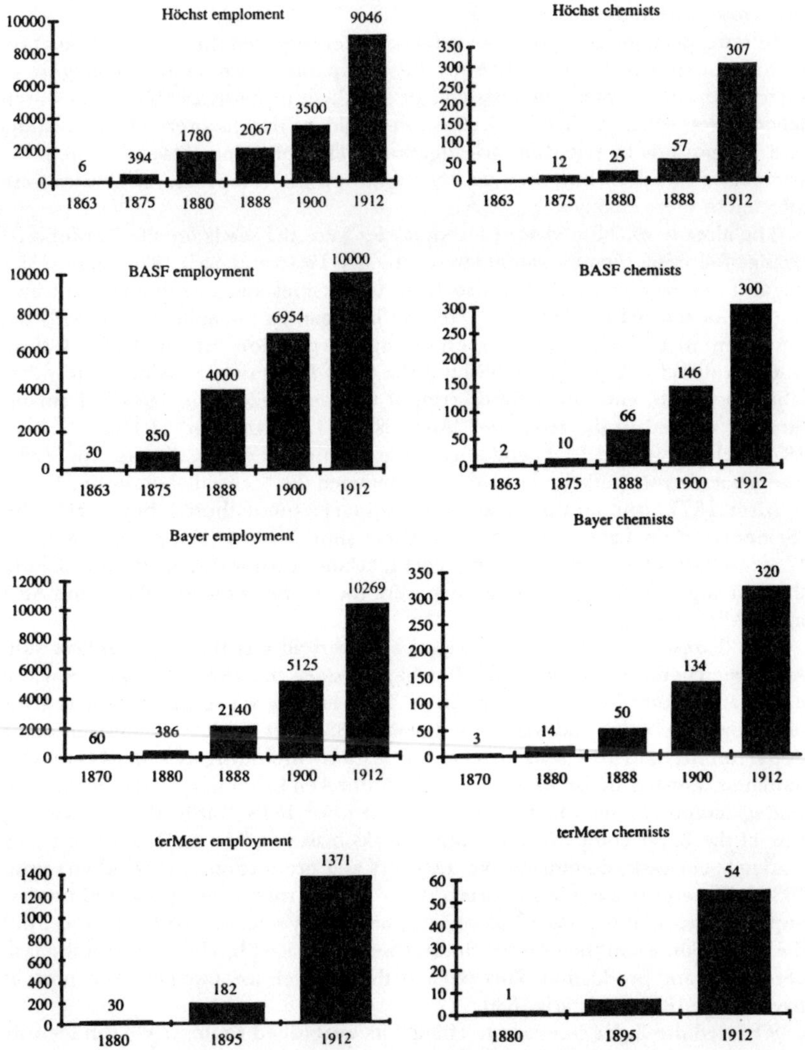

Sources: Hoechst 1863–1913, p. 19, 38: Pinnow, ***Hoechst 1863–1938***, p. 55; Ernst Bäumler, Ein Jahrhundert Chemie. (Düsseldorf 1963), p. 23; Meyer-Thurow, "Industrialization of Invention", pp. 366, 368; Homburg, "Dyestuffs industry research laboratories", pp. 105/106; Pinnow, *Bayer 1863–1938*, pp. 46, 108; *Geschichte des Werkes Uerdingen*, former Weiler-terMeer, pp. 37, 71; *Beiträge 100 Jahre Bayer*, pp. 35, 414; *Die Badische Anilin-& Soda-Fabrik 1900*, p. 4; *Die Badische Anilin-& Soda-Fabrik 1906*, p. 3; *Die BASF* 1922, p. 27; Curt Schuster, *Vom Farbenhandel zur Farbenindustrie, Die erste Fusion der BASF.* Schriftenreihe des Unternehmensarchivs der BASF, Vol. 11, 1973, pp. 10, 80; S. Kaku, "The Development and Structure of the German Coal-Tar Dyestuffs Firms". *Development and Diffusion of Technology. Electrical and Chemical Industries. Proceedings of the Fuji Conference.* Ed. by A. Okochi and H. Uchida. (Tokyo 1980), 77–95, here pp. 78, 80.

place in the chemical companies, which can be seen in Table 1 for the years between 1875 and 1888: the number of employed chemists within the expanding firms rose constantly.

In retrospect, industrialists and historians later regarded this new legal situation to have been a major incentive for the companies' owners and managers to increase the number of chemists and to put them in newly established research laboratories, where their only occupation would be the discovery of new colours and compounds by scientific investigation.[20] The following closer look at some companies will show how this was done and how the companies' structures were affected by these changes.

The already established works-laboratories were the seeds for the institutional change following the new patent law. In the late 1860s and early 1870s, companies began to engage research chemists from universities and to support them by a number of trained assistants. In 1868, BASF began its research strategy with the employment of Heinrich Caro, changing his position into that of full-time research director in 1877, establishing the Main Laboratory in addition to other laboratory units, and finally transferring it to a new building in 1889.[21] Hermann Strecker worked in the separated "Analytisches Laboratorium" of Hoechst from 1870, being followed by Theodor Diehl as full-time research director in 1878.[22] The laboratory was enlarged in 1883 and renamed the "Zentrallaboratorium".

After 1877, one company after another renamed their laboratories the "Scientific Main Laboratory" or gave them similar titles (Hauptlaboratorium, Wissenschaftliches Laboratorium, Versuchslaboratorium), sometimes placing them in larger buildings. Among them were the dye-firms Oehler, Kalle, and Agfa in 1882.[23]

The 'Farbwerke vorm. Fried. Bayer' at Elberfeld was the last to follow suit. Academic chemists had been hired at Bayer's since the 1870s, and we know the names and status of these chemists: in 1879, Dr Frank was engaged for the analysis of discovered colours and was assisted from 1883 by Dr Voeller; chemists in other departments had to check intermediates and incoming raw materials (Dr Sintenius, assisted by Dr Zarniko, in the Anthracen-Laboratory). Also in the dye testing station chemists had been employed since 1878. But in the outstanding case of the Bayer company, the trained works masters ("Meister"), assisted (!) by academic chemists, dominated the discovery and production of new colours until 1885. All chemists spent large parts of their time in routine analytical and process improving work: none was employed to pursue only scientific investigations until the mid-1880s. From then on the masters were displaced by chemists with the task of research and production. This explains the relatively low number of chemists at Bayer in the 1870s and early 1880.[24]

What led the Bayer company to change its established strategy? We can identify three main motivations that distinguish Bayer from the other companies. Firstly, it was the success the other companies enjoyed after having created in-house research facilities that brought new products and commercial success. This is clearly stated by Carl Duisberg himself, the later director of Bayer. Secondly, a number of years after the new patent law was introduced, new technologies and processes were needed to substitute for successfully selling, but aging, products. Intensified scientific work could help achieve this aim. Thirdly, the end of the Alizarin convention in 1885, which had guaranteed stable prices for five years, and the onset of a depression at about the same time stimulated Bayer to look for new colours if it wanted to stay in business.[25] Last of the major dye companies, Bayer established in 1885 a laboratory solely for the purpose of invention and acquired renown for its large and modern Main Laboratory in 1891, which had been planned and realized by Carl Duisberg and was regarded as the most modern company laboratory at that time.[26]

Between 1877 and 1886, all the major companies had established special units for scientific research (apart from the still existing analytical and works laboratories), whose tasks were not only to improve colours and the techniques to produce them, but also to do research in new and related fields to widen the product-range of the companies and thus to be ahead of competitors.

While the works-laboratories did their investigation mainly in one already established product line to improve existing colours and processes, the new research units had the clear order to search for new products, new syntheses, and advanced processes in new and related fields.

During the formation-period of the new research units (1877–1886), the companies' managements played different roles. At Hoechst and BASF, the new units evolved out of the contacts with academics, followed by the management's plan to enlarge the research stations after 1877; in the case of Hoechst, the management decided to enter the pharmaceutical field, accompanied by the creation of a laboratory to search for new drugs; Agfa in Berlin established a group headed by a chemist only to do research in the new "Experimental Laboratory" in 1882. At these companies, the management was the driving power behind the incorporation of science, additionally pushed by the new patent law. The Bayer company, clearly guided by the example of the other companies, reacted mainly to pressures from outside the firm in creating a centrally directed research structure.[27]

By 1886, scientific research was generally accepted as corporate strategy for the future well-being of the firm. The foregoing economic success (as exemplified by rising employment) was a necessary precondition to enable firms to invest in research. The 1877 Patent Law set the legal framework which forced companies to pursue original research as it stopped the copying of products and processes of other firms.[28] It also became clear to the companies' managements that in-house research units were needed as an entry ticket to understanding externally performed research, to evaluate discoveries and possible applications. The in-house research department also served as a tool to screen competitors' patents and to circumvent them.[29]

2. *Growth and Diversification*

The story of the German chemical companies is also a story of diversification and integration. From the 1880s onwards, the major companies integrated forward, firstly by establishing sales-organizations in foreign countries, and later, by investing there in production plants. Until 1914, the German chemical companies had subsidiaries in all important markets outside Germany: 33 were located in Great Britain, 28 in the US, 27 in France, 22 in Russia, followed by 10 in the Austro-Hungarian Empire.[30] Some companies, such as Bayer in France, first leased production facilities and enlarged them later on; others, as Hoechst in England at Ellesmere Port or Agfa and Weiler-terMeer in France, erected new plants from the outset[31]. Backward integration for these enterprises meant buying coal mines for securing energy and raw materials, and later the move to the production of inorganic intermediates for dyes, pharmaceuticals, photographic articles and plastics.[32]

By 1900, to a greater or lesser extent, all companies had diversified into special dyes, pharmaceuticals, photographic products, plastics, and artificial fibres. Bayer and AGFA were among the most diversified companies until World War I, but rival groups were also following the same track: Kalle in dyes and pharmaceuticals, Griesheim in the wide range of dyes for different materials, Hoechst in pharmaceuticals. BASF and Hoechst also integrated vertically by stepping up the production of inorganic chemicals and intermediates (sulphuric and hydrochloric acid in 1880).[33]

Research also provided the means for diversification: it was used as a tool to broaden the companies' basis. In most companies, diversification of products went simultaneously with the creation of new research facilities. When Griesheim-Elektron entered new fields in colours and inorganic chemicals, or erected new plants (Maintal, Elektron) in the 1890s, new research laboratories were immediately established.[34]

In 1882, Agfa employed a university chemist Dr Gustav Schulz to pursue research in a new laboratory. Additionally, another dye-chemist, Dr Momme Andressen, was engaged in 1887 to enlarge the scientific basis of the Agfa-company. He was also interested in photographic articles, mainly developers. Just two years later, he was successful in achieving a major technical breakthrough in this field, which after 1891 was sold under the name "Rodinal". Encouraged by this success, Agfa established in 1892 a Photographic Department, including a photochemical research laboratory, headed by Andressen.[35] In the case of Agfa, diversification of products and research went hand-in-hand.

In 1883, the management of Höchst decided to start production of pharmaceuticals and accordingly set up a Pharmaceutical Department, whose first head was the chemist Dr Eduard von Gerichten. On his instruction, a pharmaceutical research laboratory was built in the same year.[36] After 1892, Hoechst cooperated with Robert Koch and Paul Ehrlich in the new field of serum therapy and bacteriology. Shortly after the cooperation had been established, a new department was founded ("Bakteriologische Abteilung") and several laboratories in a newly erected and separated building were constructed.[37]

When BASF decided in 1891 to enlarge its applications of colours to paper, a new research station for this purpose was set up ("Papiertechnisches Laboratorium").[38]

In the case of Bayer, research in these new fields lagged somewhat behind the move into production. Although Bayer decided in 1888 to make pharmaceuticals, a Pharmaceutical Laboratory did not exist until 1902, and a Bacteriological Laboratory followed in the same year. With a time lag of several years and following their growing range of products, Bayer's research organization was continuously enlarged until by 1912 it consisted of a Pharmaceutical, Inorganic Chemicals, an Alizarin, and a Photographic and Technical Products Laboratory. The Artificial Rubber Laboratory (1906) was the only one to be set up before production started.[39] These few examples show that Bayer, where the diversification of research lagged behind the diversification of products, is again the exception. Other companies enlarged their range of products and of research almost simultaneously.

All major dye companies had created dye testing stations before 1880, in which the newly discovered colours were tested on different materials (cotton, wool, leather, plastics, paper, silk, rayon, feathers, inks, pelts, varnishes, later also metals). In the beginning, these stations worked on the "trial and error" principle, it was simply observed after what time period and under which conditions the colours on the tested materials would vanish. But in the 1890s, as the companies engaged more and more chemists and trained, non-academic specialists, they enlarged these stations to departments (comparable to the sales, patent or other technical departments) and renamed them, "Coloristische Abteilung", "Färbereiabteilung" or "Versuchsfärberei" (Dyeing Department, Experimental Dyeing Station). Accordingly, the investigative methods became scientific, and before the First World War these departments were seen as part of the scientific basis of the enterprises.[40] These testing departments were most important for the continuing high quality of the companies' products, which was decisive for their ongoing growth. The rigorous nature of this testing, reflecting the need for so many chemists, can be seen by some figures: in 1896, Bayer invented 2378 new colours, of which only 37 were actually sold. At the turn of the century, Höchst's laboratories synthesized 3500 new colours, but only 18 were placed on the market.[41]

In the middle 1890s, these dye testing laboratories had their original purpose transformed; with the increasing scale of the activities, they had become a kind of development department for the syntheses and processes that had been invented in the scientific laboratories. As a consequence of this, the Technical or Experimental Department was introduced at Bayer and BASF in 1895, which had its origins in the dye testing stations. At BASF, the "Technical Dyeing station" (as it was later renamed) moved to new, separate buildings, where all kinds of technical and dyeing applications were constructed and tested. An experimental station already existed at the Bayer works, but when plans were made for a new, large plant in Leverkusen, it was decided that the enlarged Experimental Station should be closely connected to the Scientific Main Laboratory.[42] These Experimental Stations were seen as additional institutions to the Dye Testing Stations, and leaders believed that they should apply the new discoveries to an industrial application by fostering co-operation between chemists and academically trained engineers.

At the same time, when new institutions were introduced, the old works laboratories still existed and provided continuity with the original research tradition. Besides the newly created centralized research laboratories, these works-laboratories continued their function after 1885, their number even growing due to the wide diversification of their firms. The chemists working here improved existing products or rationalized production. The number of chemists in these sites doing applied research and development was much larger than in the new centralized research units. So we find that after 1890 a maximum of 20% of all employed chemists were working in the centralized research institution with the other 80% in other laboratories, technical departments, or in the large sales agencies.

From 1875 onwards, BASF had created research laboratories congruent with its production lines, e.g. in the alizarin factory, the aniline dye factory, and the azo dye factory. In 1888, BASF employed approximately 57 chemists in 18 laboratories, but only six (including research director August Bernthsen) were doing research in the "Main Laboratory". Heinrich Caro in the company's "Patent Laboratory" was assisted by roughly the same number of chemists. Out of these 57 chemists, only about 10–12 were doing research in the centralized departments, but 70–80%

of the chemists at BASF were at the production-plant level.[43] Most probably, the two centralized laboratories had the largest number of chemists of all laboratories: the average plant laboratory only required two to three chemists, supervising production and analyzing materials. This numerical proportion continued well into the 20th century. In 1921, BASF had a total employment of 38,949, among them 284 academically educated chemists.[44] In the Main Laboratory, 30–40 chemists were doing research at the centralized level, or less than 14% of all employed chemists. The great majority of chemists worked in the scientific (!) laboratories of the departments, the works laboratories of each production plant, the dye-testing laboratories (a "Coloristische Abteilung" including a laboratory existed since 1895), and the technical-experimental stations in each plant.[45]

The situation at Bayer was broadly similar. Before Carl Duisberg's entry into the company in 1884, Bayer employed 10–14 chemists, who were working in the so-called "laboratories", being in reality badly equipped niches in the plants. In 1885, about 29 chemists were engaged at Bayer, among them five in the newly created Main Laboratory.[46] Out of the 99 chemists in the company in 1897, 15 were in the Main Laboratory (15%), the rest were spread in the works laboratories and technical stations of the different departments. Until 1912, the number of chemists in the centralized Main Laboratory remained fairly stable (15–16), while new scientific laboratories of the new departments (bacteriological, inorganic, alizarin, photographic, artificial rubber), the work laboratories, the dye-testing laboratory and the technical stations employed more and more chemists. Out of 262 chemists at Bayer in 1912, 57 were clearly identified as research chemists (= 22%), but only 16 of them worked in the Main Laboratory, representing about 6% of all the company's chemists.[47] We find very similar conditions in the companies Hoechst, Agfa and Griesheim-Elektron.[48]

From these examples we see that companies decentralized research as part of their growing diversification strategy. Comparing the location of chemists within the companies we find that research at BASF was to a higher degree centralized than that at Bayer.

By 1914, striking characteristics in the organisation and structure of German chemical companies had developed that separated them from their British and American competitors. Research had been incorporated to such an extent that laboratories and technical stations were automatically constructed whenever a new chemical field was entered by the companies or a new plant was erected. A discussion on how to organize research best inside the firm and, in particular, whether it should be centralized or decentralized, had not taken place. The structures evolved in forty years, proving effective and useful. Besides the central scientific main laboratory, a number of scientific and technical departmental research stations were also erected. These institutions did less risky research than the Main Laboratory, and their efforts were clearly product oriented. Thus, the companies pursued a dual strategy of research, having one centralized and several decentralized investigative scientific and engineering facilities.[49]

By 1914, this decentralized dual strategy of research was applied at the major German chemical companies under review here. But from 1904 on, the structure of the chemical market in Germany changed its character in a more oligopolistic direction. During the First World War especially, new organizational arrangements for the companies in a changing environment had to be found. In the next section we will look at the problems and ideologies regarding the restructuring efforts.

II. DUISBERG'S IDEAS ON ORGANIZATION, COMPETITION AND RESEARCH

Carl Duisberg (1861-1935) was one of the most influential German industrialists before World War II, and the formation of IG Farben owed much to his initiative.[50] Soon after finishing his chemical dissertation, Carl Duisberg entered the "Farbenfabriken Bayer" in 1884, and after a successful career as research chemist, he was quickly promoted to the management, being responsible for the construction of the new Bayer plant in Leverkusen in the mid-1890s. By 1899, he was already one of five directors of the Bayer company, and after the retirement of Friedrich Bayer (the son of the company's founder) in 1912, he became "Generaldirektor", the most powerful manager within the company. It is thus important to look at Duisberg's ideas on how to best organize a large company and his views on research in differently structured markets: in competitive markets such as Germany until 1904, and in oligopolistic markets especially after 1905.

1. The 1895 Memorandum

In 1895, Duisberg drew up an organizational plan for the new Bayer plant in Leverkusen.[51] The striking characteristic of this plan was departmentalization. Seven departments were to be set up: five of them for production, each of which should run its business independently and decentralized from the Main Administration; the sixth department supplied energy, workshops, packaging etc; finally the central administration was located in the last department, including the Scientific Main Laboratory, the large library, the patent office, the Dye Testing Station and the newly introduced Technical and Experimental Station. Also, in this department, specialized laboratories were to work besides the Main Laboratory: one for analytical, one for physiological, and one for electrolytical investigations. As each of the five production departments should work independently in different chemical fields, each production department would also have its own scientific laboratory dealing with the field of chemistry, a technical experimental laboratory (consisting of academic engineers and chemists), and small libraries. Each department would have its own scientific institutions, just as a small company, and would thus be working without assistance from the Main Laboratory. By 1913, these plans had been realized: a decentralized company with independently operating units had emerged. The new organization can be roughly observed in Table 2.[52]

But the Bayer company was not the only enterprise to develop such an organization, combining production and research, centralized and departmental laboratories. The Farbwerke Hoechst had grown along a similar pattern. Hoechst had established a Main Laboratory in 1878, but also many smaller laboratories besides:

> ... for almost a hundred years, Hoechst did not possess a homogeneous research department. The company developed too quickly for this to happen. Whenever a new plant was built, corresponding research laboratories sprung up alongside it... [53]

A departmental organization with a close connection of production and research, and with centralized and departmental laboratories can also be found at BASF and Agfa.[54] As we have seen in the previous section, generating new knowledge and its application were decisive conditions for a company's success in the competitive

Table 2. Organizational Scheme of Farbenfabriken Bayer & Co 1913

Farbenfabriken vorm. Friedr. Bayer & Co		
	1. Department (Inorganic Chemicals)	- Production - Administration - Scientific Lab. - Technical Lab.
	2. Department (Organic Intermediates)	- Production - Administration - Scientific Lab. - Technical Lab.
	3. Department (Alizarin Dyes)	- Production - Administration - Scientific Lab. - Technical Lab.
	4. Department (Aniline Dyes)	- Production - Administration - Scientific Lab. - Technical Lab.
	5. Department (Pharmaceuticals)	- Production - Administration - Scientific Lab. - Technical Lab.
	6. Department	packaging, energy, etc
	7. Administration	Acquisitions, Sales, Main Laboratory, Experimental Stations,

German market after the introduction of the new Patent Law in 1877. Scientific and applied research quickly became an important means to keep the company in the race and offered new opportunities to diversify the firm's business. Competition among the manufacturers spurred on investment in new fields of chemical research and therefore paved the way for new intrafirm scientific institutions, leading to the diversified and integrated modern company. Regarding the development of the German chemical industry and its large firms, Duisberg himself pointed out in 1904:

> This admirable diversity of their activities evolved only out of competition, out of the force of circumstance and the necessity to avail oneself of all helpful means in order to be successful in the struggle of one company against the other..... Those great scientific laboratories, those patent departments with the best legally trained and chemically educated staff, they would never have emerged if one company had not driven forward the other in the race, the struggle for priority that has been created by patent legislation in different countries.... [55]

The organizational structure of the Bayer firm and the other dye producers must be seen in this connection. Internal competition among the Main Laboratory, single specialized laboratories and the departmental laboratories might have resulted in overlapping and duplication of work, but it also offered the important possibility that one of these institutions would create new industrially applicable discoveries or would find ways to produce at the lowest costs by improving processes.

In his 1904 memorandum, it becomes clear that Duisberg appreciated inter-firm competition only as long as it secured technical progress, economic welfare, and market-domination for the German chemical companies in domestic and foreign markets. Returning home from a visit to the United States and being impressed by the merger movement there, after 1904 Duisberg promoted industry-wide trusts based on the US model.

2. *The 1904 Memorandum*

By 1925 Duisberg had written three memoranda in which he proposed a merger of the largest chemical companies: in 1904, in 1915, and in 1923. All organizational aspects of the proposed trust were already described in 1904 in great detail. Duisberg anticipated the threats to the chemical industry stemming from foreign rivals, tariffs in foreign markets, and over-capacities of German producers. This prompted him to write his 1904 memorandum, proposing a "Fusion", as he called it himself, of the 16 leading chemical and pharmaceutical companies to form a chemical trust.[56] He proposed rationalization of production and combined purchase of raw materials as the most important means to cut down costs, and therefore to remain competitive. Acquisitions, sales, distribution, and legal matters were to be centralized, while production and research should still be pursued in many different locations.

On the domestic market, real competition between firms was no longer wanted. Competition should only exist with chemical industries of other nations in their markets. This planned monopoly should eliminate any foreign competitor in Germany and secure German domination of the world's markets in dyes and chemicals:

> ... The present reign of the German chemical industry, especially of the dye industry over the whole world, would in my opinion then [i.e. by means of inside competition] be secured for the immediate future...[57]

However being fully aware that the severe competition in the German market had been a precondition for the earlier companies' success abroad, Duisberg analyzed the possible consequences of eliminating all competition:

> ... abolition of competition [and] impeding the natural development, i.e. paralyzing these two effective forces so conducive to progress will bring to an end the powerful growth of this great and mighty industry if it cannot be achieved by means of a suitable form of organization to eliminate outside competition on the one hand but on the other hand to maintain inside competition by not only conserving the individuality of the single enterprises and their institutions but also to elaborate them further.... Eliminating competition soon leads to a stagnation of progress, to a sit-and-wait attitude, and to lax manners...[58]

Clearly Duisberg wanted to avoid stagnation after the proposed merger. He sought to construct an enterprise that combined the advantages of trusts with those of necessary competition which he regarded as so important for the long-time profitability of the new company. This he called "Idealkonkurrenz" (ideal competition) and thus reduced it to intra-company competition. By this he meant that the new company should have two production works or departments for the same product and also two laboratories working in the same chemical field. Duisberg was most afraid of stagnation, and by having two competing teams of workers and of scientists in the same sector, he hoped to have a continuing stimulus for improvement of production and creation of new scientific and technical knowledge. This would bring down prices, keep the intra-firm peace as no company department would be arbitrarily dissolved, and would maintain a competitive German performance on international markets. Competing production and research sites *within* the trust should guarantee the beneficial effects that were formerly achieved by competition of independently acting enterprises. This internal system of "Idealkonkurrenz" (ideal competition) between plants and laboratories should, he argued, give a constant incentive to improvement and innovation.[59]

> ... the main organizational structures of the factories themselves should remain the same ones as those already in effect in the well-run factories.... in such a way that at least 2 of such plants maintain their existences at different locations so that the chemists and engineers at work there (who should participate as much as possible in the net profit of the production plants run by them) do not feel themselves to be without competitors, but rather work in constant mutual competition...[60]

Duisberg's 1904 memorandum was sent to all major chemical companies, but his ideas were only partially realized: not one single trust was created. But by 1905, two blocks of cooperating firms evolved: Bayer, BASF and Agfa on the one hand and Hoechst, Cassella and Kalle on the other. These "communities of interest" (Interessengemeinschaften) were loosely knitted inter-firm agreements, in which each company remained independent and just the financial gains and losses were spread among the members.[61] The period after 1905 was a time of limited competition: the number of dye producing firms fell. The "core" companies prospered, even grew, and were aiming to reduce competition still further. When World War I broke out, an oligopolistic market structure in Germany for chemical products already existed. But a form of even closer cooperation was yet to come.

3. *The 1915 memorandum*

It was World War I that pushed forward Duisberg's efforts to build a trust of the largest German chemical companies. As early as 1915, Duisberg foresaw that the world's markets would not be as open for German products after World War I as they had been before. He noticed that rival countries, such as the US and Great Britain, were directing their efforts to close the industrial gap in chemical products by supporting their own industries and building up new chemical corporations. Consequently he once again pursued the idea of a great merger to resist the economic problems of the chemical companies which he anticipated when the war was over. Duisberg rewrote his 1904 memorandum, adjusting it to contemporary conditions.[62] In two examples he showed the difficulties the German chemical companies would have to struggle against after the war. First, by describing the war-time efforts of each of the Allied states, he gave a pessimistic picture of the export potential of the chemical companies to these countries. Secondly he argued that the costs for the laboratories, the technical stations, and the semi-works had become so immense that pooling of these activities was necessary to save precious financial resources.[63] He again wanted the companies to merge to form one single enterprise with a centralized administration for accounting, book-keeping, sales and storage, but keeping the former production and research facilities decentralized and competing. Such an organization would have the advantage of concentrating all the activities of the participating companies and thus reduce costs for sales, purchasing, and production. Rationalizing research by closing down laboratories or technical stations was not mentioned. In his eyes, only a large, multi-divisional company of this kind would allow German firms to compete successfully with newly established competitors in foreign countries.

Once more, these plans were not fully realized: the two blocks of Bayer, BASF, Agfa and Hoechst, Cassella, Kalle united in 1916 to form one "Interessengemeinschaft", but without merging to form a single company. We again find the loosely knitted cooperation.

The years immediately following World War I confirmed Duisberg's forecasts : although Germany remained the largest single producer of chemicals and pharmaceuticals, its share of the world trade in this field had diminished. Nearly all countries had introduced tariffs on imported goods.[64] As long as the German currency was weak and, especially between 1922 and 1923, when rapid inflation led to low production costs in Germany, German chemical products were still competitive. After 1923, this short period of inflationary growth came to an end when the Mark was stabilized by a currency reform. Also, it became clear that the "Interessengemeinschaft" of 1916 did not work as well as its founders had intended. From the end of 1923, Duisberg was pondering the weaknesses and defects of the 1916 agreement in practice. He was distributing his ideas among the directors of the associated enterprises, and finally in 1925, after the third memorandum, they voted for a merger of the six leading German companies to form the public enterprise "Interessengemeinschaft Farbenindustrie AG". The new joint stock company that came into being in December 1925 was one of the world's largest enterprises and the biggest chemical trust to date.[65]

III. FORMAL ORGANIZATION OF IG FARBEN

In this section, we will take a closer look at the ideas for organizing IG Farben regarding the production and research facilities rather than giving a detailed account of its history.[66]

1. *IG's Organization*

Legally, the firms of Bayer, Hoechst, Agfa, Griesheim-Elektron, and Weiler-terMeer were taken over by BASF, which at the same time changed its name into "IG Farbenindustrie AG" and eventually moved its headquarters to Frankfurt/Main. The production plants of these six firms were scattered all over Germany and their range of products ran from dyes to hydrogenation products, from pharmaceuticals to films and from plastics to fibres. The participating companies each had some major plants, which are listed in Appendix 1 and also a number of smaller plants.

A loose regional organization between neighbouring plants was first introduced in 1926.[67] These four operating plant communities ("Betriebsgemeinschaften", Appendix 2) more or less resembled the former firms. In joint conferences and sessions, the plant directors of each regional community exchanged data on production, sales and research and informed each other about new production and research projects as well as about patents or ways to gain higher yields from new processes. A flow of information also took place in conferences between the plant directors of all regional communities. The plant directors were on these levels also informed about decisions of IG Farben's senior management: the "Vorstand" (Managing Board), which dealt with day-to-day matters, and the "Aufsichtsrat" (Supervisory Board), representing the major stockholders and deciding on general business policies.

The basis for this decentralized structure were Carl Duisberg's and Carl Bosch's (director of BASF) desire to maintain generally the production and research sites of the traditional companies, and thus to make full use of the infrastructure of the former independent firms. Not only the larger, but also the smaller works remained small complete firms with facilities for production, research, libraries, patent and engineering departments, stores and sales offices. In keeping the former structure of the traditional companies to such a large extent, Duisberg and Bosch deliberately followed a different organizational path to that of some large American chemical companies.[68] This structure they called "decentralized centralization".

By 1929, it was clear that this mainly regional organization was inefficient. IG Farben, facing excessive costs for production and research compared to its falling sales, looked for ways to rationalize these two areas. The solution seemed to be an administrative reorganization along a vertical line. The operating units were divided into three technical branches or divisions ("Sparten") organized by product type. The aim was to rationalize by avoiding overlapping production at several plants. Similar products and processes were combined in single plants, instead of having two or more plants for the same product (e.g. fertilizers, dyes, or expensive projects in hydrogenation). Production within the branches or divisions was for the first time concentrated in the most efficient plant, and the old structure of "decentralized centralization" was therefore abandoned.[69]

Divisions of IG Farben from 1929[70]

I. Division (Carl Krauch)	*II. Division (Fritz terMeer)*	*III. Division (Fritz Gajewski)*
nitrogen	heavy chemicals	photo articles
methanol	dyes	cellulose
synthetic fuels	magnesium	artificial silk
synthetic oils	organic intermediates	films
lubricants	solvents	vulcanized fibre
metal carbonyls	pharmaceuticals	plastics processing
nickel	pesticides	
lignite	synthetic tans	
	softening agents	
	plastics	
	synthetic rubber	
	detergents	

2. Planning and supervising research: who is in charge ?

In their decision-making process, the Managing Board ("Vorstand") and the Supervisory Board ("Aufsichtsrat") could draw on information and data supplied by two assisting committees: the Technical Committee (TEA) and the Commercial Committee (KA), which were both founded in 1926. Hierarchically, they stood above the divisions, but below the top-level Managing Board and the Advisory Council.

The Technical Committee TEA

The TEA consisted of the technical members of the Managing Board (Vorstand), some managers of the larger plants, technical engineers of the divisions, the heads of the regional plant communities, and later also of the heads of the divisions ("Sparten").[71] The TEA met the day before the meetings of the Managing Board, and had to deal with all technical questions of production and research, and expansion of IG's business. Major decisions in these fields were prepared by the TEA before being submitted to the Managing Board. The TEA discussed investment proposals, which were initiated by directors of the plants, the four regional plant communities, and the heads of the divisions. Until 1929, a further sub-division of the TEA existed along the nine main product lines, which also submitted project plans via the plant directors. Many sub-committees, ultimately more than thirty, grew up to cope with the many special problems which could not be dealt with satisfactorily in the full Technical Committee.[72]

The TEA-Büro[73]

In fulfilling these tasks, the TEA's work was co-ordinated and assisted by its office, the TEA-Büro. This too was set up in 1926, but had its great influence after 1929, when IG Farben was organized along technical lines. During the entire existence of IG Farben, the TEA-Büro was under the supervision of Dr Ernst Struss, an academically trained chemist. Originally intended only to prepare and to organize the TEA meetings, the Büro eventually had to deal with a number of technical matters centrally for all of IG's activities and was therefore able to enlarge its influence over time. It had to gather financial and technical data, and then to present them, in such areas as:

- Programme for new construction and installation
- Budget for new construction and installation
- Expenditure on new construction and installation
- Research costs (expenditure for laboratories, special experiments and new lines)
- Statistics on technical personnel, in particular chemists
- Reports on sales of all groups for the TEA meetings...

Particular attention was given to programmes for new construction and installation which were examined from the technical point of view and cleared so far that the TEA could decide upon them. Furthermore the TEA-Büro had to co-operate and to co-ordinate all measures of technical rationalization and to deal with similar problems.

In addition to the above, the TEA-Būro also acted as a central office for Group II (Sparte II, Dyestuffs, Chemicals, Pharmaceuticals) and as such it dealt with statistics on progress of new products.[74]

Struss' duties of gathering information were enhanced by a special privilege:

In order to keep informed about all technical developments the writer [Dr Struss] had the right to be present at all meetings of scientific, technical or application committees of I.G. and also of the special meetings of Group I and III which handled all technical problems within their competence.[75]

In 1932, the TEA-Būro introduced its own meetings to improve contact with the leading chemists of the divisions ("Sparten"). Later, one or two directors of the major plants were also invited to discuss the latest scientific developments and technical improvements. As these meetings usually took place on a Monday, they were soon called "Monday meetings".[76]

No other corporate office at IG Farben was able to make use of such a wealth of statistical and technical data: through this, the TEA-Būro became increasingly influential. Decisions of the TEA or the Managing Board to enlarge a plant or to abandon a technology were based on the data presented by the TEA-Būro. If plant directors or research chemists, being responsible for large budgets, wanted to ensure future investments or increasing expenditures, they had to make sure that their proposals were approved by the Monday Meetings. Subsequently such proposals found their way up to the TEA before being placed on the agenda of the Managing Board. Approval by the Managing Board was then a mere formality, as terMeer recounted in 1953: votes were never taken, discussions on investment were like "small talk" between old friends, the agenda of investment and research projects usually found unanimous approval.[77]

The case of marketing new dyes provides an example. A new dye is synthesized in a laboratory and then presented to a commission, consisting of chemists and colourists from IG's dye plants and laboratories. This commission had control over pilot plant expenditure and decided whether semi-works operation should be carried out on the new product. If this new dye seemed promising, the decision was made to manufacture and if the semi-works tests could confirm the laboratory results, then the final product was given to the Sales Department within one of IG's divisions ("Sparten"). As DuPont's director of the Chemical Department, Elmer Bolton, noted in 1936, the difference between IG's and DuPont's development policy was striking:

> In the IG the decision to place a new dye on the market is made by the chemists and not by the salesmen as is the practice in our Company.[78]

If further investment was needed for additional production capacity or completely new installations, then the plant director, in whose plant the research laboratory and the semi-works were located, could request financial resources at the Monday Meetings of the TEA-Būro. His request then made its way up to the TEA and the Managing Board.

The TEA-Būro can be identified as the technological heart of IG Farben. Scientifically trained managers, directors, and research chemists could bring here their ideas and plans, which were then realized after being approved by the TEA and the Managing Board. Using this concept, Peter Morris' description of the "technocrats" can be confirmed:

> The technological policy of I.G. was entirely under the control of the technical

Table 3. Route of technical and investment decisions at IG Farben 1932.

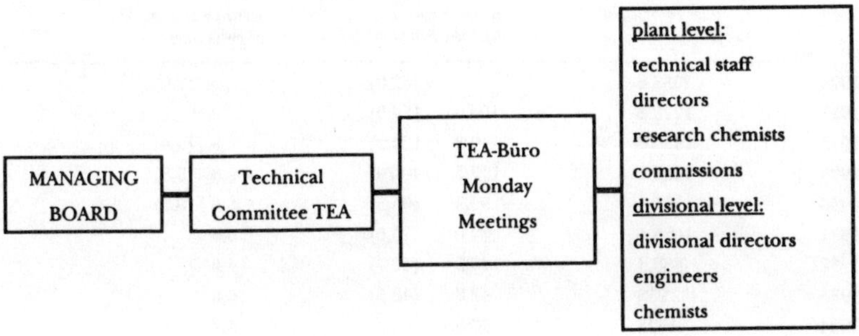

Table 4. Employment of chemists at IG Farben

Year	*tot. employment at 31 Dec. each year*	*chemists (approx.)*	*research chemists*	*research intensity+*
1926	81,601	ca. 2,000	958	24,5 (10)
1927	108,034	ca. 2,000	1,000	18,5 (9,2)
1928	114,185	2,055	1,050	18 (9,1)
1929	97,750	1,760	1,100	18 (11,3)
1930	79,963	1,760	1,100	22 (13,7)
1931	68,022	1,632	1,050	24 (15,4)
1932	66,508	1,729	1,000	26 (15)
1933	76,748	1,995	1,000	26 (13)
1934	92,492		1,020	(11)
1935	98,051		1,060	(10)
1936	109,957	2,223*	1,100	(10)
1937	122,679		1,150	
1938	133,462		1,200	
1939	127,595		1,250	
1940	146,414	2,611*	1,290	
1941	158,307		1,250	
1942	171,394	2,762*	1,250	
1943	190,884		1,100	

Sources:
total employment: 1926–36: terMeer, *IG Farben*, 1953, p. 80;
1937–43: Plumpe, *IG Farben*, p. 614;
total number of chemists: re-calculated with Plumpe, *IG Farben*, table at p. 461;
number of research chemists in: Table on expenditure on research at IG Farben, 9 July 1945, Bayer Archive 13/13 IG Ausschüsse, Kommissionen; see also Nuremberg War Crimes Trial, Microfilm 95, Defense terMeer, Document 59.
*figures from list 3, prepared on 7 July 1945, Bayer Archive 13/13, IG Ausschüsse, Kommissionen;
+ research intensity as number of chemists to 1000 employees;
in brackets ( ): number of research chemists to 1000 employees

Table 5. IG Farben's expenditure on research

Year	*sales in Million RM*	*research cost in Million RM*		*research costs as % of gross sales*	
1926	1038,8		(82,0)	–	(7,9)
1927	1279,3	154,0	(154,0)	12	
1928	1430,8	140,3	(134,9)	9,8	(9,4)
1929	1435,2	137,2	(140,79)	9,6	(9,8)
1930	1165,9	99,9	(95,5)	8,6	(8,2)
1931	1025,6	70,0	(70,0)	6,8	
1932	883,4	40,5	(42,7)	4,6	
1933	933	42,8	(42,5)	4,6	
1934	1034	43,8		4,2	
1935	1153	57,1		5,0	
1936	1371	68		5,0	
1937	1610	82,2		5,1	
1938	1756	93,3		5,3	
1939	2138	100,5		4,7	
1940	2420	105,9		4,4	
1941	2861	109,2		3,8	
1942	3315	105,3		3,2	

Sources:
sales 1926–1932: Plumpe, *IG Farben*, p. 434;
1932–1942: list 4, prepared on 22 August 1945, Bayer Archive 13/13 IG Ausschüsse, Kommissionen
research costs 1927–1942: list 1, prepared on 9 July 1945, Bayer Archive 13/13 IG Ausschüsse, Kommissionen;
research costs 1926–1933 in brackets: Plumpe, *IG Farben*, table at p. 473;
percentage 1926–1942: author's calculation
percentage 1932–42: list 4, prepared on 22 August 1945, Bayer Archive 13/13 IG Ausschüsse, Kommissionen;

men within the organisation and this also gave them considerable influence over general corporate policy... In more general terms the power within I.G. lay in the hands of scientifically qualified managers. This combination of a technologist who was also a manager can be loosely termed a technocrat... A typical technocrat sprang from the professional ranks of the middle class, took a doctorate in science or engineering, spent around five or six years in developement work, learnt business methods within the company and in his middle thirties took command of one of I.G.'s many units....[79]

IV. THE SCALE OF RESEARCH

Speaking about IG Farben's research efforts, we have to keep in mind that Germany's second largest company had a total employment of 81,601 after the merger in 1925/26, increasing to more than 108,000 at the end of 1927 and reaching 114,200 in 1928.[80] Approximately 2000 chemists were engaged, among them 1000 doing scientific research.[81]

As we have seen, production and research were traditionally closely tied together in the German chemical companies, and this basic principle was maintained at IG Farben. Thus in more than 40 plants scattered all over Germany, we also find a high number of research institutions. In a report drawn up by IG Farben in 1926/27, no less than 65 research institutions are listed.[82] This figure does not include the large dye testing stations, employing large numbers of chemists, and the laboratories in the nitrogen works at Oppau. The number of research chemists identified in the 1927 report is 567. If we add the research chemists in the dye testing laboratories and in the Oppau plants, the total number goes up to 958.[83] Table 4 shows that the number of employed research chemists remained fairly stable (around 1000) until 1936.

IG could afford to employ this large number of chemists because of relatively low wages. In 1936, a PhD-chemist earned approximately $960 a year. However taking into account the cost of living in Germany at that time, it was the equivalent of only $600 a year. At the same time, ICI in England paid its chemists $1500, and DuPont in the US $2500 a year.[84]

V. INTERNAL RESEARCH COORDINATION 1926–30

In this and the following sections it will be shown how IG Farben managed more than 1000 research chemists pursuing scientific work in nearly all chemical fields in more than 50 laboratories including 10 large Main Laboratories, how the internal research policy changed over the years (1927 to 1936), and how IG's research policy differed from that of its predecessors.

Right from the start, IG's general organization was polarized between two extremes. On the one hand there was Duisberg's idea of "Idealkonkurrenz" of competing plants and decentralized laboratories working in the same field. On the other there was the need for rationalization of production and sales offices and standardization of products, meaning generally more concentration. During the depression, IG's organization was changed to achieve these latter aims regarding sales and production. In research things were different. The research strategy, involving the distribution of funds to promising fields and the fostering of fundamental or applied research, did change after 1930 and will be looked at separately. The research structure, being many different laboratories and development stations, did in general not change over the years. This is why in the following section archival material from the 1920s and 1930s is equally suitable.[85]

IG Farben had no single Central or Main Laboratory. But we find Main Laboratories in each of the merging firms, and additionally each large plant included a set of all kinds of laboratories and Technical Stations. Table 6 gives an overview of IG's general research structure. On top of the hierarchy of all laboratories stood the Centralized Main Laboratory (being 10 in 1926) in each of the largest plants of the IG's constituent companies.

1. *Centralized Main Laboratories*

All of the six former companies had traditionally centralized main laboratories. Some companies had grown organically by merger (Griesheim-Elektron or Weiler-terMeer) and had kept the former independent production and research sites. We find that IG Farben had no less than ten Main Laboratories, employing 296 chemists in 1926. Some of them were very large, giving work to 50–100 chemists, such as the laboratories at Oppau, Elberfeld, and Leverkusen, but others were smaller with an employment of 16–26 chemists, e.g. in Mainkur, Griesheim, and Offenbach.[86] Research in these institutions was specifically linked to certain products and it sought to find new products and chemical reactions in planned and coordinated team-work of a scientific rather than an empirical nature. But general topics in dyes, inorganic, catalytic and molecular organic chemistry and pharmaceuticals were also pursued. This research work could today be regarded as fundamental research, because these investigations were not immediately product oriented. As one of IG's research directors pointed out in 1935, only 5–10% of the staff in the Main Laboratories was occupied with such pure fundamental work. This was seen by scientists and managers as a kind of "life insurance" for the long-term well-being of the enterprise.[87] Research on "new lines" (meaning activities in product sectors that IG had not yet entered) was mainly done in these old established laboratories, e.g. rubber and macro-molecular research in Leverkusen, high pressure technology in Ludwigshafen and Oppau, bacteriological and pharmaceutical investigations in Hoechst, and plastics in Ürdingen.

The laboratory directors were free to organize their laboratories, to define interesting projects, and to distribute money to promising fields. Table 7 shows the typical organization of one of IG Farben's Main Laboratories.[88] The company's administration, being the "Vorstand" (Managing Board) or the Technical Committee, appointed the research director and granted the research expenditure. The Scientific Literature Office or Patent Department screened patents in all important industrialized countries, analyzed scientific journals from all over the world in search of new ideas or promising technologies, and then distributed abstracts to all research institutions within IG Farben. The research director supplied the group leaders of his Main Laboratory with these abstracts. The group leaders supervised small teams of scientists, differing in size, but having a maximum of 8 researchers. Experience had taught the research directors that some scientists were most fruitful when working alone on certain problems. This was accepted and these researchers were "one-man-teams", but they were still incorporated into the overall structure and supervision of the Main Laboratory and its director.[89] Choosing the right man for the position as research director was decisive, but IG Farben seemed to have been successful in its recruitment policy, as Elmer Bolton from DuPont noted in 1936:

> ...Dr [Otto] Bayer of the Central Laboratory of the Leverkusen plant does not employ any project system, but is budgeted a certain amount of money each year which he allocates to the different lines of activity under his direction. On the other hand, Dr [Hans G.] Grimm of the High Pressure Laboratory at Oppau, is employing a project system somewhat similar to our own. Dr Bayer stated that from time to time a review of the research results of his laboratory had been made in order to determine if the results justify the cost of the research, but in every case the management has been satisfied with the progress of the laboratory. In all the laboratories of I.G. which were visited it was very apparent that the Director took a very active part in the guidance of research work and delegated the administrative duties to others. As a result, the Laboratory Directors we met were outstanding men in some particular field of research, and their activities were

Table 6. IG Farben Research Structure – Organizational Scheme

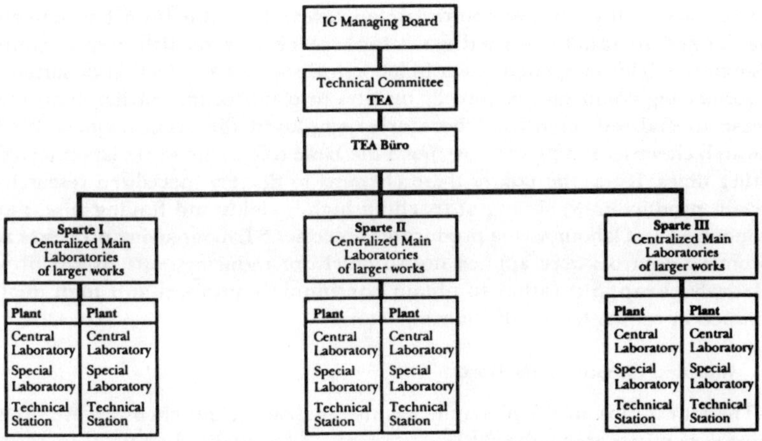

Table 7. Laboratory Organization of IG Farben's Main Laboratories

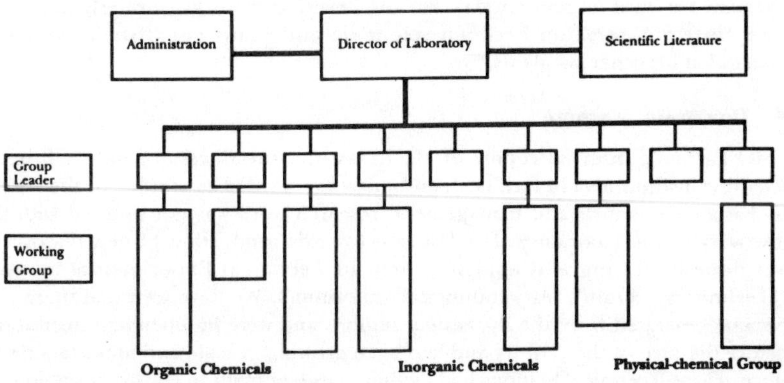

confined very largely to intimate guidance of the various researches of the laboratory...[90]

Pursuing research this way was very expensive: a calculation from 1927 shows that nearly RM 30.000 per chemist were spent in these Main Laboratories per year, including assistants, apparatus and chemicals. The scientists were aware that these expenses would only pay off after years or even decades. But scientists and managers shared the broad conviction that such a long-term policy was absolutely necessary for the enterprise.[91]

2. Departmental Laboratories

Contrary to the relatively unrestricted research in the Main Laboratories, specialized product-oriented work took place in scientific departmental laboratories. This happened either in laboratories of the production departments of former big companies or directly on sites in plants of the smaller companies. These specialized scientific laboratories employed the great majority of the research chemists and played now much the same role as the works laboratories of earlier times. It was the task of these chemists to do very specialized research in certain product-lines, aiming at reaching higher yields and finding new, more economical and labour saving production processes.[92] Labour-saving processes and automatic controls were applied not so much for reducing costs, as labour was relatively cheap, but rather to obtain continuously uniform and high quality products by relying less on the human factor.[93]

3. Analytical Chemists in the Works

The lowest rank in the hierarchy was the analytical plant chemists. We do not know very much about the duties and work of the plant chemists, but certain institutions must have continued to exist in which the plant chemists worked on the best ways to supervise the plants' operations and to improve the plants' performance. Also, they tested incoming raw materials and processes under operation. Clear borderlines between the work of the scientific and the plant chemists are hard to define today. For the Leverkusen works (formerly Bayer) we find close cooperation between scientific and plant chemists, as was also demanded for other IG plants.[94]

4. Development is research

IG Farben's internal report of 1927 classifies the development facilities as scientific institutions (10 facilities), and this was regarded as sensible in the eyes of IG Farben's scientists and management; research work was not finished with the discovery in the laboratory. The last step towards production of new discoveries was done by testing and applying them in Technical Experimental Stations ("Technischer Raum", "Anwendungslaboratorium"). We have seen that these sites became separated from the dye testing stations and were independent institutions before the turn of the century and had been growing in scale and importance ever since. These Technical Stations had facilities large enough to construct semi-works to test the industrial application of new processes. Since the late 1880s, development work was carried out by the chemists who had discovered a new process, assisted by additional chemists and academically trained engineers.[95] These chemists were willing to change position within the company, from the Main Laboratory to the testing stations or to plants as heads of departments. Here the Group Leader in the Main Laboratory was crucial: he decided when a discovery or an invention was taken to the Technical Experimental Station by the inventing scientist, and the Group Leader continued his supervision until all problems had been solved and production ran efficiently.[96] Application to industrial production was done step by step, as the results of chemical reactions often changed with increasing amounts of chemicals being used. Then, bigger and bigger apparatus was constructed, followed by semi-works, and finally a new plant was built. We find these large Technical Experimental Stations at Hoechst, Ludwigshafen (former BASF), Ürdingen (terMeer), Mainkur (Griesheim), and at Leverkusen (Bayer), which was constructed in 1923 and was one of the largest.[97] The duties of these sites were not just to do product development in the sense of less ingenious and more practical work. Basic engineering happened here as well as chemical research by a co-operative strategy of research and plant chemists, leading to the construction of demonstration apparatus by engineers and chemists.[98] Some chemical products were produced only in these testing sites, as the demand remained too low to justify its own plant, or the apparatus could not be built on a larger scale. Inventive and innovative action did not take place only in the Main Laboratories or Scientific Departmental Laboratories. An IG internal report of 1936 shows that leaving the testing stations out as mediator between laboratory and large scale production proved disastrous for a number of products.[99] In 1936, these semi-works or experimental stations were described as being something outstanding:

> ...The most striking thing about the semi-works was the wealth of all kinds of equipment and the large size of apparatus... The semi-works plants that were visited were very large and complete in every detail. There seemed to be an extravagance in the amount of space and equipment in these plants...[100]

No clear borderline was drawn between research and development, neither by the management nor by the scientists themselves. Development was seen as part of the scientific process, included in the calculations of general research expenditure and seen as necessary in the science-based chemical industry by IG Farben's and other companies' management.[101]

5. *Flow of Information*

The flow of information between 1000 research chemists was secured by meetings of scientific commissions, which had been established along the line of products or research interests, and consisted of chemists of all laboratories and plants working in this field. The Patent Department distributed abstracts of interesting developments from all over the world to all researching institutions within IG Farben. When a new chemical field was entered, the Technische Ausschuß (TEA) decided to create new commissions to deal with this subject: a scientific and a technical-engineering commission. In these commissions the research work was coordinated, results discussed and exchanged, and further research efforts distributed among the participating laboratories. In 1925, when IG Farben came into being as a single company, there were some 26 commissions and more than 40 sub-commissions, and these numbers increased to 30 full commissions and many more sub-commissions in 1930.[102]

A fundamental difference in research management and strategy between IG Farben and its predecessors in 1929–1930 cannot really be seen. IG Farben still used the same research strategy and structure of its predecessor companies and, despite the expense, was seemingly attached to Duisberg's ideas of "Idealkonkurrenz" and deliberate duplication of research efforts. But the onset of the Great Depression, with falling sales and shrinking markets, forced IG Farben to re-think its general production and research policy, and most important: to rationalize production.

VI. RATIONALIZATION OF PRODUCTION

In the course of the First World War, the German chemical companies had lost 50% of their dye-sales, mainly in foreign markets. The challenge in 1918 was twofold: to find new products to replace lost exports, and to re-arrange dye-production in Germany, because oversupply led to falling prices and diminishing profits. One possible answer to these needs was the 1925 merger to form IG, in order to adapt then the company to the changed situation: by closing down plants and rationalizing production.[103] By pursing this grand conception, IG's managers aimed finally at achieving an enterprise that had modernized its dye-production and had thus set resources free for the development and production of new goods. The latter aspect was very important in the eyes of IG's managers, as IG's main interests between 1914 and 1925 had shifted from dyes to inorganic products (nitrogen, fertilizers, fibres, intermediates), and the cessation of dye-production in some factories was to be accompanied simultaneously by the production of newly developed chemicals in them. IG's managers realized that it would take years to achieve this, but went ahead nevertheless, wanting the transition to be as smooth as possible to avoid conflicts in the artificial and elaborate construction of IG Farben.[104]

Although important Directors in IG's Managing Board, Dr Karl Krekeler (Bayer) and Dr Paul Duden (Höchst), generally accepted the need to close down some of IG's numerous dye-factories, they differed substantially about maintaining Duisberg's concept of "Idealkonkurrenz" of competing plants. They agreed that sites which produced less than 10% of all IG's articles in one sector, were to be shut down.[105] Karl Krekeler wanted to continue production of great volumes in two or even three locations, but felt that old established products in which further improvements could not be hoped for should be concentrated in one single plant. Paul Duden pleaded for maintaining Duisberg's concept, arguing that even in old product lines improvements had been found to which all plants had contributed. He thought the concept of "Idealkonkurrenz" still important and that it could help to improve IG's performance over time.[106]

IG's rationalization process started in 1927/28, when 33,000 dyes, sold under more then 50,000 brand names, were revised and by 1932 reduced to 6,000. Most attention was lavished on the standardization of dyes and their production in larger quantities by single plants rather than in lower volumes on many different sites. Detailed plans were drawn up as to which site should give up dye production completely in favour of other plants. The most important aspects for choosing production sites were capacity, efficiency, employment, overall costs, different chemical processes, transport and energy supply.[107]

In the process of rationalization, Höchst was the big loser. Höchst lost azo dyes and its intermediates entirely, capacities in alizarin and sulphuric dyes (which had already lost their importance and profitability), and had to give pharmaceuticals almost entirely to Bayer. Höchst's fifty years' tradition in this sector, stemming from the cooperation with Robert Koch, Paul Ehrlich and Emil von Behring, was over, and even Höchst's products were sold under the "Bayer Cross" trademark. Only production and research of a few drugs (Pyramidon, anti-pyrene) remained at Höchst.[108] The Bayer works at Leverkusen and Elberfeld became IG's main centre for dye and pharmaceutical production, and the Leverkusen plants continued research and production of pesticides. The BASF plants gained some additional capacity in vat dyes.[109]

Other chemicals and articles were also looked at and production was concentrated mainly at one single plant in the years up to 1936. Höchst enlarged its activities in inorganic chemicals by stepping into acetylene chemistry (polyvinyl acetate), and fulfilled all of IG's requirements for aceto acetic acid.[110] The former BASF plants in Ludwigshafen kept on producing intermediates and organic chemicals. High pressure technology for synthetic ammonia and nitrogenous fertilizers was continued at the Oppau plants of BASF. The Kalle works at Bibrich concentrated on one new product, cellophane. The Weiler-terMeer part of IG at Ürdingen mainly took over intermediate chemicals and chrome production, losing sulphuric and azo dyes. Metals, such as magnesium and aluminium, became the focus of the Griesheim plants at Bitterfeld in Central Germany. AGFA and Wolfen continued producing photographic articles (films, cameras) and artificial silk (Viscose), AGFA at Lichtenberg also acetate silk and spun rayon. Bayer at Dormagen and Premnitz (former Köln-Rottweil) specialized in different lines of artificial fibres as those of AGFA and Wolfen: Dormagen in copper silk and Cuprama, Premintz in Vistra.[111] Apart from the changes in dyes, re-organization in the other sectors was less drastic and done quickly, as the products had been introduced by the constituent companies and simply remained there.

The number of sales organizations was also reduced from 91 to 45. The products were marketed by sales offices in the production plants and sold in Germany by local agencies. Wherever possible, IG attempted to continue to make use of the existing infrastructure of its constituent companies.[112]

Rationalizing production also meant reducing employment at IG: from 114,000 in 1928 to 66,500 in 1932 (see Table 4). It was mainly workers who were made redundant, while the employment of chemists did not change so dramatically. Simply, positions of retiring chemists remained empty.[113]

Although a few products continued to be produced in two sites in the hope of making improvements, Duisberg's original grand conception of "Idealkonkurrenz" in all production sectors to maintain the company's dynamic through internal competition was quietly dropped during IG's process of rationalization.[114]

In an international comparison, IG resembled very much the United States Steel Corporation. Both enterprises were amalgamations of a number of companies, and both trusts did not rationalize production or close down plants immediately after the merger. Both companies had their backs to the wall in the Great Depression before they reorganized the overall corporate structure, rationalized production by closing down plants and re-defined the remaining plants' duties.[115] Quite differently, IG's American competitor DuPont combined acquisition of smaller companies and immediate rationalization. For example, DuPont acquired the Grasselli company in 1928, and only one year later DuPont concentrated all of its dispersed activities in research and production of white pigments at the former Grasselli works, therefore abandoning its established plants.[116]

VII. CHANGES IN THE RESEARCH ORGANIZATION UP TO 1936

We have seen that research and production were closely tied together at IG Farben. Rationalization and re-organization of manufacturing had therefore some influence on IG's traditional combination of research and production. Extremely important as well were the changes in IG's management until 1936, because research strategies and organizational ideas were often integrally connected to certain persons of IG's management.

1. Centralizing efforts

Carl Duisberg's influence at IG Farben's daily business was not very great in the years from 1926 to his death in 1935. Carl Bosch had succeeded Duisberg as head of the "Aufsichtsrat" (Supervisory Board) in 1935 and Hermann Schmitz (previously Chief Financial Officer) became head of the much more important "Vorstand" (Managing Board), succeeding Carl Bosch. One of the most influential positions, immediately below Bosch and Schmitz, was held by Fritz terMeer, who was head of Division II from 1929 to 1945 and head of the Technical Committee TEA from 1932–45. He was thus responsible for the coordination of all technical decisions and scientific developments, in effect for the scientific and technological future of IG Farben. He was assisted by Dr Struss, head of the TEA-Būro because Duisberg's influence was weakened by being too far away from day to day decision making and Bosch possessed a distaste for general organizational matters, Fritz terMeer was able to restructure IG Farben's research policy from 1929 on until this process was finished in 1936. TerMeer's notion of research structure was considerably opposed to Duisberg's and Bosch's ideas and practice.[117]

Bosch and Duisberg had always agreed on the necessity of having competition between plants regarding research and production in the same sector to secure the best results in cost and quality. TerMeer did not see this as necessary. In 1936, Bolton summarized terMeers's opinion on internal competition:

> ...[this scheme] has not worked out very satisfactorily inasmuch as production on many items has not been large enough to warrant manufacture in two separate plants. Since 1929 Dr terMeer has been rationalising manufacture of many of their products and having the entire quantity produced in the plant most advantageous to the corporation. He does not seem to be impressed by the idea that it is necessary to have competition between plants in order to arrive at the lowest cost. The long experience in the manufacture of many of the products gives them a measure as to good performance.[118]

This statement about terMeer makes clear that he had deliberately abolished the established idea and practice of having competing plants, and therefore concentrated production at the most advantageous works. The era of uncentralized and uncoordinated research at all existing 50 laboratories was over. By 1936, the coordinating work for IG's future research and the allocation of financial resources to the laboratories and fields of investigation were taken over by the Technical Committee (TEA), headed by terMeer and assisted by Dr Struss from the TEA-Būro. Numerous commissions helped avoid duplication of research between different laboratories working in the same fields of production.[119] When Bolton visited IG Farben's research sites in 1936, he already saw the results of terMeer's efforts to re-organize IG Farben's overall structure. According to the concentration of production, the laboratories at Leverkusen (formerly Bayer) were now mainly engaged in research on dyes, pharmaceuticals, and insecticides, and also on synthetic rubber and polymerisation. Hōchst was the most important location for research in inorganic chemicals (acetylene), and for some pharmaceutical research.[120] Oppau (previously belonging to BASF) concentrated on high pressure technology (fertilizers, ammonia, synthetic gasoline), Ludwigshafen (formerly BASF) on synthetic intermediates (styrene and ethylene), polymerisation, and high pressure technology. Elberfeld (the original Bayer works) now worked mainly on dyes and pharmaceuticals, Bitterfeld on inorganic research, and Berlin and Wolfen (Agfa) on photographic articles and different kinds of artificial fibres.[121] Concentrating research efforts meant that the once desired duplication of research in the same product field had generally come to an end, although some overlapping still occurred. However the close connection between research and development remained unchanged. Semi-works continued to exist in all major works next to the central laboratories of the plants. They adjusted their work according to the new fields that were under investigation in the major laboratories after the concentration of production and research had ended in 1936. The established research sites, consisting of central laboratories in the major plants, specialized laboratories in all production departments, and large semi-works, were not reduced nor did employment there decrease. Even in the long and bitter German depression 1929–1933, not a single scientist was discharged:

> the directors of the IG consider their traditional research organisation as a very valuable asset, and... it has taken many years to develop the organisation and personnel. They feel that had they reduced their research organisations it would have taken many years to build them up again.[122]

2. *Centralized versus Decentralized Research*

By 1926, it became clear that the scale and diversity of IG's research activities made a new coordinating policy necessary. Carl Bosch, then head of the "Vorstand", proposed to create new large scientific institutions in which long-term research on interesting problems should be pursued.[123] Also, further recommendations were made in the 1927 report.[124] The existing system of commissions should be kept and even be enlarged. Further, special laboratories for exclusive research in important fields were to be established, similar to the already existing departmental institutions. Vat dyes, indigoid vat dyes and naphtol AS were seen as promising topics for these specialized laboratories, of which the total number should be limited to avoid excessive cost. Although the creation of one centralized main laboratory for all of IG's fundamental research was not considered, the need for central planning for the different laboratories' work was recognized. For example, research on intermediates for synthetic rubber, cellulose and emulsions could only be successful if several scientists in different locations cooperated. Central planning for research could happen through new, relatively unspecialized scientific commissions, consisting of chemists in all research areas, meeting twice a year. Thus they could discuss their work and come to know each other. This model was proposed after it already had proven effective in the commissions for Alizarin and Azo dyes, cellulose, rubber and resins. The proposal to introduce this enlarged commission system was seen as the most important way to avoid overlapping and doubling research work. However despite these plans to concentrate research efforts, little was actually done to bring this effectively into being.

Although IG attempted to reduce the employment of researchers due to high costs and too much overlapping work, and despite the re-organizing efforts from

1930 on, which resulted in the closure of some production plants and some reduction of IG's employment of workers and chemists, the decentralized structure of research and development was deliberately continued. The urgent need for several Main Laboratories as the most important institutions for basic, rather than product-oriented research was clearly recognized and accepted. The idea of having only one centralized large Main Laboratory was heavily critizised by research directors: IG was too large and its activities were much too dispersed for one single institution to cope with the challenges of finding new products in so many different chemical fields. Having ten different institutions for this purpose was seen as the necessary strategy. In the existing memoranda, reports, minutes, and publications we find no hint that a general re-structuring of IG's scientific institutions in favour of only one R&D Department should have taken place. Nothing like this was actually planned. Because IG's managers insisted on the close connection between research and production and because this structure had proven valuable during so many years, IG's managers saw no reason to change it.[125]

VIII. PRODUCT POLICY – RESEARCH POLICY 1926–36

Unlike its predecessors, IG pursued a research policy by allocating research expenditure to promising fields.

The years from 1913 to 1929 saw a shift of sales in the German chemical firms from mainly dyes to high pressure technology, and substitutes for natural materials. While dyes accounted for 63% of the sales of the later IG firms in 1913, this had diminished to 28.6% in 1929.[126]

Product sales deriving from high pressure technology, such as nitrogen, had surmounted the sales of dyes, accounting for 11% of sales in 1917, reaching a peak of 42.4% in 1926, and then falling like all commodity prices to 20.6% in 1932.[127] The other product sector of high pressure technology, synthetic oils and gasoline, was immensly expensive in its development but never achieved a substantial part of IG's sales.[128] Appendices 12 and 13 list product sales as percentage of total IG sales, and research costs regarding products as percentage of total IG's research expenditure.[129] These two appendices lead to the following conclusions:

Nitrogen by high pressure technology had high sales and yields. Once research on nitrogen production was nearly over, the high sales guaranteed good returns. However, due to a worldwide nitrogen oversupply beginning in the late 1920s, sales had dropped by almost a third by 1936. Research costs had been low when sales were high, but increased when sales fell. That is, as sales in nitrogen halved between 1926 and 1932, research costs more than tripled in the same period. The sharp movement out of synthetic gasoline research set resources free for further nitrogen research (and therefore: new fertilizers and agrochemical products).

Research on synthetic gasoline at first promised good returns as the scientific breakthrough at the laboratory scale had occurred by the mid-1920s. Unfortunately adaption to large-scale production proved to be more expensive than anticipated, swallowing more than 30% of IG's research expenditures in 1927. Although these outlays were cut back to 7.5% by 1930, they increased again to nearly 10% in 1932 and to more than 12% after the introduction of the "autarky-programme" of the National Socialist government in 1936. Despite this enormous investment, sales never became really important until 1936, rising from

1% in 1929 to 5% in 1935 of all of IG's sales. And even 7.9% of all of IG's sales in 1938 was not an impressive figure. Combining nitrogen and synthetic gasoline as products of Sparte I (Division I) "High Pressure Technology", we see that sales fell (from over 40% in 1926 to around 20% in 1934), while these products absorbed between 32% and 20% of IG's research expenditures. Though this seems reasonable, it was not: sales fell despite increasing research, and already in the early 1930s, it was clear to IG's management that the hydrogenation project was fundamentally uneconomic. In 1936, terMeer stated that the profits made by hydrogenation of coal were small and mainly possible on a very large scale only because of the interest of the government.[130]

The story of dyes is somewhat different. Dyes had already lost their lead in sales to nitrogen and nitrogenous products before 1920. Though falling slightly from 33% in 1926 to 28.6% in 1929, the overall trend of dyes sales was stable, counting for roughly a third of all of IG's business until 1936. In 1930, dyes once more surmounted nitrogen in sales, and this did not change until IG's demise in 1945. Also, dyes were high value products, they generated between 38% to 74% of IG's profits.[131] Two developments were decisive for this. First, the international dyestuffs cartel provided secure market agreements, so that research efforts were not as risky as on free markets.[132] Second, the diffusion of high pressure technology to other industrial countries had come to an end by the mid-1920s, and the technological superiority of IG Farben dwindled as other countries supplied former German export markets. Because IG maintained technological leadership in dyes, these products became once again highly profitable and led to a new concentration of IG's business in dyes. The renewed importance of dyes can also be seen in the research expenditure as part of IG's total research: dyestuffs research grew from only 9% of total expenditure in 1926 to more than 15% in 1929. The expected 43% in 1932 is certainly an exception due to the collapse in the company's other lines of products, but the years until 1942 saw a substantial research effort of around 20–25%.

The development of IG's business in pharmaceuticals, pesticides, photographic articles and artificial silk had similar characteristics. These fields counted for the smallest part of IG's sales until 1929. But as IG enjoyed scientific and technological leadership in these sectors, sales grew vastly between 1926 and 1936. Even the general drop of IG's business in the depression of 1929–32 could not harm these special sectors: sales in pharmaceuticals and pesticides grew by 80% between 1926 and 1931 (Appendix 8), in photographic articles by more than 100% (Appendix 10), and by more than 800% in different lines of artificial fibres (acetate silk, Vistra, Cuprama) in the same period (Appendix 11). This development was recognized by IG's management, and increasing sales were rewarded by a growing research expenditure to secure the competitive leadership in these fields. A statement of terMeer in 1936 makes clear that this was a deliberate strategy:

> A number of years ago very large profits were realised from the manufacture of ammonia and fertilisers and the principal income of the I.G. was obtained from these fields. The management at that time had a feeling that the smaller fields, such as dyes, pharmaceuticals, and miscellaneous chemicals, had lost their importance from an income standpoint, so that there was a very definite tendency to suppress research in these fields and to favour expansion in the ammonia and fertiliser fields. Inasmuch as these plants are now only operating at 25% capacity,

Table 8. Distribution of research chemists in major fields

Major research groups	*chemists in 1926*	*chemists in 1936*	*chemists in 1938*
dyes and dye testing stations	200 research 91 testing	250 research 150 testing	275 research
pharmaceuticals, pesticides	64	100	110
inorganic, misc. organic research, intermediates	included in rubber	200	210
rubber and plastics	384	–	125
fibres and photographics	29	100	100
Oppau high pressure	100	100	100
other hydrogenation	–	100	170
inorganic chemicals (only 1926)	90	see above	
total number of research chemists	958	1000	1090

Sources: 1926: Meyer, Report on the Scientific laboratories at IG Farben, Bayer Archive 103/17.05.5; 1936: Bolton to Crane, June 11th, 1936, p. 7, DuPont Company Papers. Series II, part 2, Box 1038, Hagley Museum and Library;
1938: terMeer, *IG Farben*, 1953, p. 110.

the smaller fields have assumed a greater importance.[133]

The shift of IG's business interests can also be seen from the distribution of research chemists in different fields. (Table 6) The growing importance of the dye business is represented by an increase in the number of research chemists from 200 in 1926 to 275 in 1938. In pharmaceuticals and pesticides, the number of scientific chemists nearly doubled (from 64 to 110) between 1926 and 1938, and it almost tripled in artificial fibres and photographic articles. More difficult is the situation in rubber, intermediates, and inorganic chemicals: these areas engaged 474 chemists in 1926. This number declined until 1938, when rubber and plastics needed 125 researchers, inorganic and miscellaneous chemicals 210. The high-pressure laboratory in Oppau continued its stable level of employment with 100 chemists. The increase in the total number of research chemists from 958 in 1926 to 1090 in 1938 can then be explained by additional efforts in hydrogenation (supported by the National Socialist government), and in dyes pharmaceuticals, and fibres.

New product lines

Research intended to create new products in various chemical fields occurred mainly in the large centralized Main Laboratories of the constituent companies. One is tempted to call this "fundamental research" because of the nature of these investigations, which were done in unexplored fields. But the borderline is difficult to draw; fundamental investigations had to be made before a given task could be identified and applied programmes carried out.

Sometimes the field of activity was defined by a decision of the TEA or the Managing Board ("Vorstand") as was the case artificial fibres. In late 1925, the TEA discussed the possibilities of substituting natural fibres such as cotton, wool, linen and silk.[134] Different methods of production were presented as were the application of each single type of fibre and its profitability. Before making a decision, a fibre's possibilities on the German and international markets were evaluated and the existence of competitors was checked. IG then decided that artificial fibres should be produced on the basis of already known processes, but that investigations of new types should also be undertaken. In many of IG's laboratories research started, undertaking basic explorations in polymers and macro-molecular chemistry, and IG was successful in 1936 in synthesizing polyamide 6 or nylon 6, shortly after the discovery of nylon 66 by Carothers at DuPont.[135]

In other fields, research was pursued without any specific new order. That is, because Hoechst and Bayer had stepped into pharmaceuticals in the 1890s, their laboratories simply continued research in this field and were doing fundamental research in narcotics, hormones, vaccines, vitamins, and cancer (Dr Gerhard Domagk). The High Pressure Laboratory at Oppau also continued its work on all kinds of high pressure technology (e.g. the action of atomic hydrogen on petroleum). It was estimated that in no other of IG's Central Laboratories was so much fundamental research under way as in the Oppau laboratory.

In its first years, IG spent large sums on this basic research in new fields, but this expenditure decreased sharply during the Depression and then remained well below its pre-1929 level (see Table below)

Table 9. Research costs of IG Farben in new lines 1928–42

	Sparte I in Mill RM	*Sparte II in Mill RM*	*Sparte III in Mill RM*	*tot. costs of new lines in Mill RM*	*tot. costs of IG research in Mill RM*	*new lines as % of tot. IG research*
1928	66.7	3.9	–	70.6	134.9	52.3%
1929	53.1	4.8	0.4	58.3	140.7	41.4%
1930	35.6	3.7	0.8	40.1	95.5	42.0%
1931	–	–	–	–	70.0	–
1932	1.9	1.7	0.7	4.3	42.7	10.1%
1933	1.5	1.7	0.8	4.0	42.5	9.4%
1934	–	2.6	–	2.6	46.8	5.6%
1935	–	3.9	–	3.9	53.3	7.3%
1936	5.9	6.0	2.2	14.1	69.4	20.3%
1937	5.5	3.4	5.3	14.1	82.1	17.2%
1938	7.9	4.1	1.5	13.5	93.5	14.4%
1939	12.4	4.2	2.1	18.6	101.0	18.4%
1940	11.8	2.2	2.3	16.2	105.9	15.3%
1941	4.8	1.7	0.7	7.1	108.7	6.5%
1942	5.0	3.0	0.7	8.7	105.1	8.3%

Source: Umsätze und Forschungskosten der I.G. Farbenindustrie 1926–43. Direktionsabteilung T, Dr. Struss, December 11, 1953, Finanzwesen Bayer-Forschung, Forschungskosten allgemeine, Bayer Archiv 15/9.2.
The figures for IG's total research expenditure presented in the 1953 calculation conflict with list 1 of 9 July 1945, Bayer Archive 13/13 IG Ausschüsse, Kommissionen. Though the differences could not be explained and therefore remain, the trend of IG's research expenditure throughout the years is correct.

The immense costs of research in hydrogenous fields (Sparte I) accounted for the largest part of IG's research in new lines. But when IG's total sales fell from over 1400 Million RM in 1929 to 883 Million RM in 1932, it was clear that IG could not maintain such an expensive programme. Also, as hopes for future sales in this area and for large returns dwindled, further efforts were sharply cut, even coming to a short halt in late 1931. We see a return to work in hydrogenation after 1933, largely payed for by the state.[136]

Sparte II represents fundamental research work in rubber, intermediate chemicals, dyes, plastics, and inorganic chemicals, and also the basic investigations in macro-molecular structure for artificial fibres and silk (although other research in this sector and production was done by Sparte III). Expenditures in these sectors were reduced by 50% during the depression but rose quickly with the improvement of the economic situation after 1933. Intermediate chemicals, pharmaceuticals, and Buna-rubber accounted for the major part of the expenditure on new lines in this Sparte. This was reasonable as they were seen to be promising fields for future business, and research could be undertaken without the immense costs found in hydrogenation technology.

The situation is similar in photographic articles and artificial fibres, the main activities of Sparte III. Fundamental research in these areas experienced a general upward trend, disrupted only shortly by the depression in 1931. These sectors accounted for 9–15% of IG's total sales, increasing over time, and therefore renewed research activity was justified.

The heavy investment in hydrogenation distorts the picture of IG's research in new lines because of its dominant position. Therefore, when IG stopped hydrogenation research to save money, overall research expenditures in new lines fell as a part of IG's total research. However, we cannot argue that IG neglected promising fields. Nor can we state that the overall decrease of research expenditure in new lines meant an abandonment of explorations in new fields.[137] In general we see a decline of research expenditure in new areas, but in Sparte II and III, such efforts increased. They even grew fourfold in Sparte III between 1929 and 1936. IG kept paying considerable attention to new, promising fields, as the growing importance of general research and work in these new lines in Sparte II and III clearly demonstrates.

IX CONCLUSION

The corporate structure of IG Farben and its research organisation had its origins well before the 20th century. This is the reason why the analysis of corporate research started in these early stages of the major German players in the chemical industry. Much early research was done by the entrepreneurial inventor or the inventing entrepreneur. His work contributed greatly to the economic success of his young firm and this success gave the financial background to enlarge the scientific basis of the firm, and later to diversify its business by scientific discoveries. While the introduction of the new German Patent Law in 1877 can explain the establishment of a new institution in all major companies, i.e. the in-house scientific laboratory, the role and tasks of the works or departmental laboratories were not changed by this new legal situation. Well before 1900, a new corporate element in the system of chemical industrial research was introduced: the Technical Station, in which chemists, technical chemists and engineers worked cooperatively on the application of new knowledge. By 1900, the organizational structure for industrial research in the German chemical industry had evolved: centralized and decentralized scientific institutions. The centralized Main Laboratory conducted unspecialized research in new fields. Young chemists, having just left universities, found their first industrial employment here and kept the Laboratory staff and senior researchers up to date with the latest developments at the universities. Next in the hierarchy and nearby was a large Technical Station, where new processes were applied to mass production in those chemical fields that did not yet fit into the production departments. Research in the production departments mirrored this structure, but we also find works laboratories in addition to the scientific and technical sites. In the departments, three different institutions took care of the scientific and technical well-being of the firm. This structure indicates the strategy of the companies: doubling the research effort by doubling the institutions and having overlapping work and internal competition.[138] The close local connection of research and production guaranteed that there was a constant exchange of data and knowledge between scientific and production chemists, and the works were always under the control of a scientific chemist. The dual strategy of research and close physical location of research and production are striking characteristics in the development of the German chemical companies until World War I.

Pursuing research this way was extremely expensive. Duisberg's memoranda of 1904, 1915, and 1925, which aimed to unite the leading chemical firms, must also be seen as a means to pool the research efforts and the scientific and technological experiences of these firms.[139] He not only anticipated the administrative and productive advantages of his proposed merger but also forsaw the future high costs of chemical research. This would only be possible in very large companies: none of the leading German firms alone could realize these ambitions.

When IG Farben was founded in late 1925, the organization of research was not a major issue. Decentralization of research had been practiced by all constituent companies for more than forty years was thus simply continued. Only Carl Bosch proposed to create additionally large laboratories in different chemical fields (hydrogenation was not among them), but this should not be seen as a plea for centralizing research. The addition of the research institutions of six companies brought difficulties for planning and managing research and made it hard to define something like a research policy. IG's policy was dominated by hydrogenation, and it was four to five years after IG's foundation until some kind of control of research or a research policy became visible.

Looking at the distribution of the research expenditures, the periodisation of IG's research policy by the Hack's can be accepted.[140] In the period until 1930, hydrogenation accounted for the largest part of IG's research, and the newly founded IG can also be seen as a tool to continue this expensive research project. But only four years after its foundation, IG changed its policy under the impact of the depression, the discoveries of new oil fields in the US and unanticipated technological problems.

The years 1930 to 1936 saw the need for rationalization in both production and research. This also meant abandoning research in hydrogenation, or at least most parts of it, and it set resources free for research on new products that could immediately be placed on the market (dyes, pharmaceuticals, and plastics). Rationalization was achieved by concentration, by which the tasks of works and laboratories were re-defined. The company's interest shifted, leading away from hydrogenation to other, already established product lines. When IG had finished this re-structuring, the last period of research policy emerged.

The "autarky-programme" of the National Socialist government, begun in 1936 to prepare for war, and leading to World War II, limited IG's possibilities to conduct research freely and made an orientation on the needs of the international markets impossible. I believe that this latter period is unrepresentative for a research policy and have neglected it to some extent in this paper.

In the second period, we clearly find a research policy conducted by the technical leaders of IG's management. This policy was dominated by Fritz terMeer, head of Sparte II (chemicals, dyes, pharmaceuticals). Hydrogenation, represented by Carl Bosch and Carl Krauch, had to step back. As head of the important Technical Committee TEA and its office, the TEA Büro, terMeer used these two coordinating bodies masterly to restructure IG's research sites and to re-define IG's interests. Until 1936, these interests were not in hydrogenation. In 1936, high value added products (dyes and pharmaceuticals) and new products (artificial fibres, photographic articles, and plastics) accounted for a substantial part of IG's research efforts, unlike eight or ten years earlier.

The change of IG's senior management from Duisberg and Bosch to terMeer (and Schmitz) was more than just a change in personnel. It also meant a shift in the company's structure and strategy. Rationalization paved the way for concentration: the era of decentralization to the greatest possible extent was over, and hydrogenation gave way to other, more promising and better selling products. After Duisberg's idea of internal competition in production had been dropped by 1930, internal competition in research and overlapping scientific work also became suspect and had come to an end by 1936.

IG was under control of the technical men: members of the Managing Board and the key-Technical Committee. A decisive element in the technical structure of IG became the TEA-Büro, though it was never meant to play that role. Here the scientific and technical directors of the works and laboratories met and proposed via the Technical Committee, TEA, new projects or investment. We have also seen that the decision to place a product on the market was made by scientifically trained middle managers (heads of works laboratories, and technical commissions) after the successful completion of tests and application processes. Unlike DuPont, where salesmen were more involved in future products, at IG the scientists were in charge.

There are even more differences between IG and DuPont that are worth noting. As it was put recently, IG had a much "leaner" management with fewer hierarchies between production and administration, and especially a closer contact between researchers, research directors and the top management.[141] The project and budget system, and the absence of a rigid controlling system of the laboratories' expenses exemplify the relative autonomy of the scientific men inside IG Farben. Also, the existence of independently working individual scientists within the organized research work of the large laboratories and the close contact between those scientists and the research directors shows the importance of the individual scientist right into the 20th century, and casts doubts on the absolute priority of team work stressed today. On the other hand, in terms of product policy, DuPont seemed to have been the more modern enterprise. Since the 1920s, DuPont followed deliberately an expensive programme of fundamental research to find new products and to replace old ones. IG paid less attention to this replacement-policy (e.g. only 5 % of researchers in the Central Laboratories), but rather stressed its fifty years old way of finding new products: substituting natural resources. As Carl Bosch said to DuPont's Jasper Crane in 1930:

"... We suppressed natural madder, then natural indigo, now we are suppressing Chile nitrate, also the wood distillation industry, and we will soon supplant petroleum by hydrogenated coal..."[142]

Crane recognized IG's successes so far and realized to what extent science had been incorporated there:

"...IG's management has faith that their chemical research will develop every few years a new process of outstanding importance, a big income producer. Their faith has indeed been justified by the impressive list of achievements to their credit, and examination of their affairs shows that without the developments of their own research they would be very badly off financially. We cannot help but be impressed, therefore, by the IG's vision of the possibilities of research..."[143]

For Crane, the lack of modernity in this old strategy and the lack of ingenuity in chemical work at IG was substituted by the mass attack of researchers on chemical problems, represented by the dual strategy and the stress on development work. He was not convinced that these were the right answers in times of fierce competition between IG, ICI, and DuPont. For the company he represented, he was quite convident:

"...Yet I do not have the feeling or the fear that we will be outdistanced, but my view is, rather, that we ourselves are awake to the importance of research and that in the duPont Company we can and will produce results at least on a par with those of any other organisation.[144]

It has been argued here that IG attempted to change its research policy, to shift its attention to new fields. Crane's statement was right in 1930, but the situation looked different in 1936. IG Farben was created to cope with the problems after World War I: lost markets, oversupply, new competitors and technological diffusion. The new company also had to face special German developments: hyper-inflation, the very long and hard depression, and the NS-regime. IG's history shows how the company attempted to deal with these problems. IG and its predecessors traditionally manufactured producer-oriented chemicals and IG, under the leadership of terMeer, was about to change this towards consumer-oriented articles when terMeer's strategy came to a halt with the NS autarky programme. The increase in sales and research expenditure in artificial fibres, photographic articles and pharmaceuticals attest to this new, consumer-oriented strategy that emerged between 1930 and 1936. However, the German market developed differently than IG had anticipated and IG could also not exploit the chances of consumer-oriented products such as fibres on the much smaller German market, compared to the US or the British Empire. The bottom of the depression was reached in winter 1932/33 (and unemployment decreased). But available income and the standard of living did not rise significantly. Between 1913 and 1938, real wages rose by 9% (c.f., US 53%; GB 33%), income per head in 1939 was only 46% of the comparable US figure and 66% of the British figure. The consumption of durable consumer goods fell until 1939, while the cost of living rose.[145] For example, the living standard in 1935/36 was threatened by sharply rising prices. An estimated 25% decline in the living standard could only be offset by working more hours.[146] This means that unlike the US or Great Britain, Germany did not develop into a "consumer society" until well after World War II. While it was reasonable for DuPont to find new products for privately consuming markets (e.g. cellophane and nylon), it was also reasonable for IG to do the same between 1930 and 1936 when IG anticipated a similar development in Germany. But this was unrealistic after the introduction of the Four Year Plan and rearmament in 1936, when overall prospects were gloomy and private consumption was deliberately restricted by the regime. There simply was no market for new goods in Germany, and IG's main income continued to come from producer oriented chemicals. So the changes initiated in product and research strategy could not be maintained.

The argument that the absence of a centralized Research and Development Department at IG caused strategical disadvantages is somewhat misleading: all major German chemical firms before 1925 had grown that way. Putting the main focus only on the Main Laboratories means telling only a small part of the whole story. It was the close connection between research and production and the parallel existence of centralized scientific laboratories, scientific works laboratories, product works laboratories and many technical stations that built the basis for the success of the German chemical companies for more than forty years. IG continued this tradition successfully. The innovative process happened in many locations, and the more research and application sites there were, the more likely was invention and innovation. IG Farben did not grow at the same rate as its predecessors for many reasons: new competitors, tariff barriers, technological diffusion of once profitable processes, and the special development of Germany during the depression and the National Socialist era.

References

1. Peter Morris, *The Development of Acetylene Chemistry and Synthetic Rubber by IG Farbenindustrie Aktiengesellschaft 1926–1945.* PhD Dissertation Oxford 1982, pp. 43–45.
2. John J. Beer, *The Emergence of the German Dye Industry.* (Urbana, Illinois 1959); John J. Beer, "Coal Tar Dye Manufacture and the Origins of the Modern Industrial Research Laboratory", *Isis* 49 (1958), 123–131; Georg Meyer-Thurow, "The Industrialization of Invention: A Case Study from the German Chemical Industry", *Isis,* 73 (1982), pp. 363–381, p. 367; Ernst Homburg, "The Emergence of Research Laboratories in the Dyestuffs Industry, 1870–1900", *British Journal for the History of Science,* 25 (1992), pp. 91–111, here pp. 98–102.
3. Alfred D. Chandler Jr., *Scale and Scope. The Dynamics of Industrial Capitalism.* (Cambridge Mass., 1990), pp. 564–584.
4. Gottfried Plumpe, *Die IG Farbenindustrie A.G. Wirtschaft, Technik und Politik 1904–1945.* (Berlin, 1990), esp. pp. 471–475; Hubert Sales, "La Recherche-Dévelopement dans la Stratégie des grandes Enterprises Chimiques de l'Entre-Deux-Guerres: DuPont de Nemours, Imperial Chemical Industries, I.G. Farben", *Revue d'Économie Politique,* 94 (1984), 446–464.
5. David Hounshell and John Kenly Smith, *Science and Corporate Strategy. DuPont R&D, 1902–1980.* (Cambridge Mass., 1988).
6. Walter Wetzel, *Naturwissenschaften und Chemische Industrie in Deutschland. Voraussetzungen und Mechanismen ihres Aufstiegs im 19. Jahrhundert.* (Stuttgart, 1991), pp. 50–85; for a broader international comparison see the still excellent two volumes of L.F. Haber, *The Chemical Industry during the Nineteenth Century. A Study of the Economic Aspect of Applied Chemistry in Europe and North America.* (Oxford, 1958); and *The Chemical Industry 1900–1930. International Growth and Technological Change.* (Oxford, 1971); Peter Borscheid, *Naturwissenschaft, Staat und Industrie in Baden 1848–1914,* (Stuttgart, 1976).
7. Gustav Pistor, *Hundert Jahre Griesheim 1856–1956. Ein Beitrag zur Geschichte der chemischen Industrie.* (Tegernsee, 1958), pp. 27–51.
8. Heinrich Voelcker, *75 Jahre Kalle. Ein Beitrag zur Nassauischen Industriegeschichte.* (Wiesbaden, 1938), p. 57.

9. *Geschichte und Entwicklung der Farbenfabriken vorm. Fried. Bayer & Co Elberfeld in den ersten 50 Jahren 1863–1918.* (München, 1918); Hermann Pinnow, *Werksgeschichte Farbenfabriken vorm. Fried. Bayer 1863–1938.* (München, 1938); *Beiträge zur 100jährigen Firmengeschichte 1863–1963.* (Köln, 1963); *Meilensteine. 125 Jahre Bayer.* Ed. by Gottfried Plumpe and Heinz Schultheis (Leverkusen, 1988).
10. *Farbwerke Meister Lucius & Brüning zu Höchst am Main 1863–1913.* (Höchst, 1913); Hermann Pinnow, *75 Jahre Werksgeschichte Höchst. Zur Erinnerung an die 75. Wiederkehr des Gründungstages der Farbwerke vorm. Meister Lucius & Bruening* (München, 1938).
11. *Die Badische Anilin-& Soda-Fabrik.* (Ludwigshafen, 1922); Curt Schuster, *Badische Anilin-& Soda-Fabrik Ag. Ein Beitrag zur Geschichte der chemischen Technik.* (Ludwigshafen, 1961); Wilhelm Roggersdorf, *Im Reiche der Chemie. 100 Jahre BASF.* (Düsseldorf, 1965).
12. Richard Willstätter, "Zur Geschichte der AGFA", *Berichte der Deutschen Chemischen Gesellschaft* 64(1931), 140–149.
13. Voelcker, *75 Jahre Kalle*, pp. 57, 95.
14. Schuster, *BASF*, p.5 and Homburg, "Dyestuffs industry research laboratories", p. 98.
15. *Geschichte und Entwicklung der Farbenfabriken vorm. Fr. Bayer in der ersten 50 Jahren*, pp. 14, 15; Carl Duisberg, Selbsterlebtes und Schlussbetrachtung, in *Geschichte und Entwicklung der Farbenfabriken vorm. Fr. Bayer in der ersten 50 Jahren*, p. 590; *Beiträge 100 Jahre Bayer*, p. 7 and Meyer-Thurow, "The Industrailzation of Invention", p. 365.
16. *Höchst 1863–1913*, p. 11.
17. Pinnow, *75 Jahre Höchst*, p. 26.
18. *Geschichte des Werkes Uerdingen der Farbenfabriken Bayer Aktiengesellschaft* (former Weiler-ter Meer & Co), (Leverkusen, 1956), pp. 26–28.
19. Carl Duisberg in his 1904 Memorandum. See Carl Duisberg, "Denkschrift über die Vereinigung der deutschen Farbenfabriken 1904. Carl Duisberg, *Abhandlungen, Vorträge und Reden aus den Jahren 1882–1921.* (Berlin, 1923), pp. 343–369, here p. 347.
20. Carl Duisberg, "Selbsterlebtes und Schlußbetrachtung", in *Geschichte und Entwicklung der Farbenfabriken vorm. Fr. Bayer in der ersten 50 Jahren*, (1918) p. 590; Wetzel, *Naturwissenschaften und Chemische Industrie in Deutschland*, pp. 249 passim and 299; Arndt Fleischer, *Patentgesetzgebung und chemisch-pharmazeutische Industrie im deutschen Kaiserreich 1871–1918.* (Stuttgart, 1984), esp. pp. 329–339; Eberhard Schmauderer, "Der Einfluß der Chemie auf die Entwicklung des Patentwesens in der zweiten Halfte des 19. Jahrhunderts", *Tradition*, 3/4 (1971), 144–176; Homburg, "Dyestuffs industry research laboratories", pp. 110, 111; Jonathan Liebenau, "Patents and the chemical industry: tools of business strategy", in *The Challenge of New Technology. Innovation in British Business since 1850.*, edited. by Jonathan Liebenau, (Aldershot, 1988), pp. 135–151.
21. Homburg, "Dyestuffs industry research laboratories", p. 100.
22. Ibid., pp. 100, 101.
23. Ibid., pp. 103–109.
24. Chemists were not even allowed to enter the production facilities; see Duisberg, " Selbsterlebtes und Schlußbetrachtung". pp. 590, 593, 601, 605.
25. Duisberg, "Selbsterlebtes und Schlußbetrachtung", p. 602; Wetzel, *Naturwissenschaften und Chemische Industrie in Deutschland*, pp. 172, 173; H. Rose, "Alizarinrot oder die Entwicklung der Alizarinfabrikation", *Geschichte und Entwicklung des Farbenfabriken vorm. Fr. Bayer 1863–1913*, pp. 289–296. here p. 293;
26. *Geschichte und Entwicklung des Farbenfabriken vorm. Fr. Bayer 1863–1913*, p. 16; "Das wissenschaftlich-chemische Laboratorium der Farbenfabriken vorm. Fried. Bayer & Co zu Elberfeld". *Chemiker-Zeitung* 1893, p. 191; reprinted in Carl Dusiberg, *Abhandlungen, Vorträge und Reden aus den Jahren 1882–1921*, pp. 103–114; "The Appreciation of Science by German Manufacturers". *Nature* 48 (1893), pp. 29–34.
27. Homburg, "Dyestuffs industry research laboratories", pp. 102, 110–111; *Höchst 1863–1913.* p. 20; Willstätter, "Geschichte der Agfa", p. 143; *Geschichte und Entwicklung der Farbenfabriken vorm. Fr. Bayer 1863–1913*, p. 16.
28. Homburg, "Dyestuffs industry research laboratories", p. 110; Fleischer, *Patentgesetzgebung und chemisch-pharmazeutische Industrie 1871–1918.*, esp. pp. 329–339; Liebenau, "Patents and the chemical industry: tools of business strategy." in: *The Challenge of New Technology*, 1988. pp. 135–151.
29. David Mowery and Nathan Rosenberg, *Technology and the Pursuit of Economic Growth.* (Cambridge Mass., 1989), p. 290 passim.
30. Harm G. Schröter, "Die Auslandsinvestitionen der deutschen chemischen Industrie 1870 bis 1930." *Zeitschrift für Unternehmensgeschichte* 35 (1990), 1–23, here Table 1, p. 4.
31. Plumpe, *Die I.G. Farben*, pp. 58, 59; for Höchst's foreign investments see also Friedrich Spoun, *Höchst in England 1901–1914. Zur Geschichte des Werkes Ellesmere Port der Meister Lucius & Brüning Limited.* Dokumente aus Höchster Archiven 45, Höchst 1971; Freidrich Spoun, *Höchst in Frankreich 1881–1914. Zur Geschichte des Werkes Creil der 'Compagnie Parisienne des Coleurs d'Aniline'.* Dokumente aus Höchster Archiven 44, Höchst 1970. Friedrich Spoun, *Aktiengesellschaft der Moskauer Chemischen Fabrik Farbwerke vorm. Meister Lucius & Brüning.* Dokumente aus Höchster Archiven 43, (Höchst,

1970).

32. Haber. *The Chemical Industry 1900–1930,* p. 16; *Höchst 1863–1913,* p. 19; for AGFA and the connection to coal mines near Bitterfeld see Willstätter, 'Geschichte der AGFA", p. 146.
33. In the case of Höchst, the rapidly increasing sales of Alizarin in the late 1870s led to a growing need for sulphuric acid for the production of Alizarin; because of this and profitable forecasts, the management decided in 1880 to start producing the needed intermediates. See M. Simon, *Schwefelsäure Höchst. Vom Kammerverfahren zum Kontraktprozeß 1880–1914.* Dokumente aus Höchster Archiven 47, (Höchst, 1975), p. 9. *Höchst 1863–1913,* pp. 19, 20; Chandler, *Scale and Scope,* p. 478; Willstätter, 'Zur Geschichte der AGFA", p. 147; *BASF* 1922, p. 20; Plumpe, *I.G. Farben,* p. 43; Hermann Raschen. *75 Jahre Chemische Farbrik Griesheim-Elektron.* (Frankfurt am Main, 1938), p.33
34. Raschen. *75 Jahre Griesheim-Elektron.* p.33.
35. Erich Stenger, *100 Jahre Photographie und die Agfa 1839–1939.* (München, 1939), pp. 23, 24.
36. *Wie die ersten Heilmittel nach Höchst kamen.* Dokumente aus Höchster Archiven 8. (Höchst. 1965), pp. 8, 21.
37. *Höchst 1863–1913,* pp. 28/29; H. W. Flemming, *Die Zusammenarbeit Behring-Höchst 1892–1904.* Dokumente aus Höchster Archiven 37. (Höchst 1968), pp. 7–9; see also Jonathan Liebenau, "Paul Ehrlich as commercial scientist and research administrator", *Medical History,* 34 (1990), 65–78; Johathan Liebenau, "Industrial R&D in pharmaceutical firms in the early twentieth century", *Business History,* 26 (1984) 329–347; Jonathan Liebenau, "Ethical business: the formation of the pharmaceutical industry in Britain, Germany, and the United States before 1914", *Business History,* 30 (1988), 116–130.
38. *75 Jahre Papiertechnisches Laboratorium der BASF.* Ludwigshafen 1966, p. 5.
39. Horst-Bernd Dünschede, *Tropenmedizinische Forschung bei Bayer.* (Düsseldorf, 1971), p. 6; Meyer-Thurow, "Industrialisation of Invention", p. 371.
40. *Geschichte und Entwicklung der Farbwerke vorm. Fried. Bayer 1863–1918,* p. 321; Pinnow. *Bayer 1863–1938,* p. 107; *Höchst 1863–1913,* p. 36; *BASF 1922,* p. 53; Raschen. *75 Jahre Griesheim-Elektron,* p. 33;
41. Pinnow. Bayer 1863–1938, p. 107; Pinnow, *75 Jahre Höchst,* pp. 64, 65; Wetzel, *Naturwissenschaften und chemische Industrie in Deutschland,* p. 214.
42. *BASF Laboratoriumsbauten der Anwendungstechnik.* (Detmold, 1963), p. 9; Carl Duisberg, "Denkschrift über den Aufban und die Organisation der Farbenfabriken zu Leverkusen" (1895). *Abhandlungen, Reden und Vorträge 1882–1921,* pp. 387–409, here p. 391; also Chandler, *Scale and Scope,* pp. 476, 477.
43. Homburg, "Dyestuffs industry research laboratories", p. 100; Michael Hörskens, "Fortschrittsformel Forschung. 125 Jahre BASF". *Chemische Industrie* No 5 1990, p. 78.
44. *BASF 1922,* p. 26, 27
45. Ibid., p. 49. BASF, *Ein Beitrag zur Geschichte der chemischen Technik.* (Ludwigshafen, 1961). p. 13, 14.
46. *Geschichte und Entwicklung der Farbenfabriken vorm. Fried. Bayer & Co 1863–1913,* p. 17.
47. Meyer-Thurow, "Industrialisation of Invention", p. 371. v. Beckrath, *Beiträge 100 Jahre Bayer.* p. 414, gives 320 chemists for the year 1912. In this case,only 17,8% of all chemists would then have worked as research chemists.
48. Pinnow, *75 Jahre Höchst,* pp. 158, 159; Willstätter, "Geschichte der Agfa", p. 147; Pistor, *Hundert Jahre Griesheim,* pp. 50 passim.
49. For a description of the general organisation of the German chemical works see *The Chemical Industries of German Rhineland.* A Summary of the British Chemical Mission on Chemical Factories in the Occupied Area of Germany Ed. by the Association of British Chemical Manufacturers. (London, 1920). Although this report gives a picture of the 1919 conditions, it is also true for the pre-1914 years as the war changed the scale of the works but not their general organization.
50. See Hans-Joachim Flechtner, *Carl Duisberg. Vom Chemiker zum Wirtschaftsführer.* (Düsseldorf, 1959).
51. Carl Duisberg, "Denkschrift über den Aufban und die Organisation der Farbenfabriken zu Leverkusen" (1895). *Abhandlungen, Reden und Vorträge 1882–1921,* pp. 387–409.
52. This table supplies only a stylistic drawing of Bayer's organization. Exact drawings for the years 1913, 1921, 1924 can be found in the Bayer Archive Leverkusen, BAL 10/3.3 – 10/3.4. The organizational scheme after 1913 did not change apart from the expansion of the company and the integration of new parts in the existing organization.
53. Ernst Bäumler, *Milestones of Progress in the Years 1863–1966.* (Frankfurt am Main, 1967). p. 7.
54. *BASF 1922,* pp. 183, 207; Willstätter, "Geschichte der Agfa", pp. 144–147.
55. "... Erst durch die Konkurrenz, durch den Zwang und die Notwendigkeit, sich alle ihr dienlichen Mittel und Wege nutzbar zu machen, um im Kampf der einen Firma gegen die andere Erfolge zu erzielen, ist jene bewundernswerte Vielgestaltigkeit ihrer Tätigkeit ausgebildet worden....Jene großen wissenschaftlichen Laboratorien, jene mit besten juristisch geschulten und chemisch vorgebildeten Kräften besetzten Patentbureaus, sie wäen nicht entstanden, wenn nicht in dem Wettrennen, in dem Kampf um die Priorität, wie ihn die Patentgestzgebung in den verschiedenen Ländern verursacht, einer den anderen vorwärts getrieben hätte..." Carl Duisberg, "Denkschrift über die Vereinigung der deutschen Farbenfabriken (1904)" [Memorandum on the Fusion of the

German Dye Manufacturers] *Abhandlungen, Reden und Vorträge 1882–1921,* pp. 343–369, here p. 354/355.

56. Bayer, BASF, Höchst, Farbwerk Mühlheim A. Leonhardt & Co, Chemische Fabriken Weiler terMeer, Cassella, Oehler, Kalle, Dahl, Chemische Fabrik von Heyden, Schering, Merck, Böhringer, Chininfabriken Zimmer & Co, Knoll, Riedel.
57. "...Die jetzt bestehende Herrschaft der deutschen chemischen Industrie, speziell der Farbenindustrie über die ganze Welt, würde dann meines Erachtens für die nächste Zukunft gesichert sein..." Duisberg, Memorandum of 1904, p. 361.
58. "...Mit der Beseitigung der Konkurrenz, mit der Hemmung der natürlichen Entwickung, mit der Lahmlegung dieser beiden wirksamen und den Fortschritt fördernden Kröfte hört aber auch die Erstarkung der groß und mächtig gewordenen Industrie auf, wenn es nicht gelingt, durch geeignete Organisation die Konkurrenz zwar nach außen zu beseitigen, sie aber im Inneren durch Erhaltung der Individualität der einzelnen Unternehmungen und ihrer Einrichtungen nicht nur bestehen zu lassen, sondern noch weiter auszubilden....Mit der Aufhebung der Konkurrenz tritt leicht eine Stagnation des Fortschritts, ein Ausruhen und sich Gehenlassen ein..." Duisberg, Memorandum of 1904, p. 355.
59. Duisberg, Memorandum of 1904, pp. 351, 365.
60. "...Die Fabriken selbst wären in ihren großen Organisationszügen so zu belassen, wie sie in den gut geleiteten Fabriken bereits bestehen...., daß immer mindestens zwei solcher Betriebe an verschiedenen Orten erhalten bleiben, damit die darin tätigen Chemiker und Ingenieure, die in weitestgehendem Maße an dem Reingewinn der ihnen unterstellten Fabrikationsbetriebe zu beteiligen sind, sich nicht konkurrenzlos fühlen, sondern dauernd im gegenseitigen Wettbewerb miteinander arbeiten..." Duisberg, Memorandum of 1904, p. 365.
61. This story has often been told, and will not be repeated here at length again. See F. Steckel, "Cartelization of the German Chemical Industry 1918–1925", *Journal of European Economic History* 19(1990), 329–351, here p. 331; Haber, *The Chemical Industry 1900–1930,* pp. 247–279; Chandler, *Scale and Scope,* pp. 506, 563–569; Plumpe, *I.G. Farbenindustrie AG,* pp. 45–50 and 131.
62. Carl Duisberg, "Die Vereinigung der deutschen Farbenfabriken" (1915), introduced by Wilhelm Treue, *Tradition* 8 (1963), 193–227.
63. Ibid., pp. 202, 209.
64. *The Chemical Industry.* Documentation of the International Economic Conference Geneva May 1927. Publications of the League of Nations, II, Economic and Financial Section, 1927. II. 4. For Germany's economic position see Tables on pp. 23, 25, 27, 28, 42; a list of import duties is supplied at pp. 55–59.
65. Plumpe. *IG Farbenindustrie AG,* pp. 131 passim, 143, 183; Francis C. Steckel, "Cartelizataion of the German Chemical Industry 1918–1925", p. 348.
66. This is done by Plumpe, *IG Farben;* Chandler, *Scale and Scope,* pp. 564–584; H. Tammen. *Die IG Farbenindustrie AG 1925–1933. Ein Chemiekonzern in der Weimarer Republik.* Berlin 1978; Peter Hayes, *Industry and Ideology. IG Farben in the Nazi era.* (London, 1987); for a detailed description of IG Farben's administration and managerial hierarchies see Wilfried Feldenkirchen. "Big Business in Interwar Germany: Organizational Innovation at Vereinigte Stahlwerke, IG Farben, and Siemens." *Business History Review* 61 (1987), 417–451.
67. The most accurate and best descriptionis given by Peter Morris, *The Development of Acetylene Chemistry and Synthetic Rubber by IG Farbenindustrie Aktiengesellschaft 1926–1945.* PhD Dissertation Oxford 1982, pp. 33–49.
68. Fritz terMeer, *Die I.G. Farben. Ihre Entstehung, Entwicklung und Bedeutung,* (Düsseldorf, 1953), p. 30; Wolfram Fischer, "Dezentralisation oder Zentralisation-kollegiale oder autoritäre Führung? Die Auseinandersetzung um die Leitungsstruktur bei der Entstehung I.G. Farben-Konzerns" in *Recht und Entwicklung der Großunternehmen im 19. und frühen 20, Jahrhundert,* edited by Norbert Horn and Jürgen Kocka. (Göttingen. 1979), 476–487.
69. Plumpe, *IG Farben,* pp. 147–149.
70. Feldenkirchen, "Big Business in Interwar Germany", p. 437; Plumpe, *IG Farben,* p. 148.
71. Geschichte des Technischen Ausschusses der IG Farben, p. 3. Bayer Archive 13/13 IG Ausschüsse, Kommissionen.
72. Minutes of the meetings of the TEA, 14 October 1925, Bayer Archive 13/15 vol 1. See also Records of the U.S. Nuremberg War Crimes Trials, Military Tribunal no. 6 case VI, microfilm no. 95, Prosecution Basic Information on IG Farbenindustrie, 20 August 1947, p. 17; Karl Winnacker, *Nie den Mut verlieren.* Erinnerungen an Schickalsjahre der deutschen Chemie (Düsseldorf, 1974), p. 74.
73. I am much indebted to Dr Peter Morris, Science Museum London, who brought this point and its importance to my attention.
74. Report of Dr Struss, Functions of the TEA-Büro, p. 2, 12 September 1945, Bayer Archive 13/13 IG Ausschüsse, Kommissionen
75. Ibid..
76. Dr Struss, Comprehensive History of the TEA-Büro, pp. 6–7, 4 March 1947 [Ausführliche

Geschichte des TEA-Büros] Bayer Archive 13/13 IG Ausschüsse, Kommissionen.

77. terMeer, *IG Farben*, 1953, pp. 54/55; also Hayes, *Industry and Ideology*, pp. 23–25; Chandler, *Scale and Scope*, pp. 579–580.
78. E.K. Bolton, head of the DuPont Chemical Department, to Jasper Crane, DuPont Vice President, 11 June 1936, Report on a visit to IG Farben's major research laboratories, p. 4, DuPont Company papers, Series II, part 2, Box 1038, Hagley Museum and Library, Wilmington, Delaware, USA. See also Winnacker, *Nie den Mut verlieren*, pp. 80–81.
79. Morris, *Development of Acetylene Chemistry*, 1982, pp. 43–45
80. Statement Dr Bosch's at the Committee on the Conditions of Production and Sales of German Goods. Subcommittee III: The Chemical Industry. [Ausführungen des Sachverständingen Dr. Bosch, in: *Ausschuß zur Untersuchung der Erzeugungs- und Absatzbedingungen der deutschen Wirtschaft.*. Enquete-Ausschuß, III. Unterausschuß: Die deutsche Chemische Industrie.] (Berlin 1930)p. 114; terMeer, *IG Farben Industrie*, 1953, p. 80; Plumpe, *IG Farben*, p. 459.
81. Records of the U.S. Nuremberg War Crimes Trial, Microfilm Roll 95, Defense terMeer, Document 59: Dr Struss, Wissenschaftliche Forschung bei der IG, 11 Feb. 1948; terMeer, *IG Farben*, 1953, p.111. This figure is also confirmed by the report of 1927 mentioned below
82. Report on the scientific laboratories at IG Farben, by K.H. Meyer, head of Ludwigshafen Main Laboratory, June 1927, submitted to the TEA, 28 Oct. 1926. [Bericht über die wissenschaftlichen Laboratorien der Farbenfabriken von K.H. Meyer, Juni 1927, TEA-Beschluß vom 28. Oktober 1926] Bayer Archive 103/17.05.5.
83. An explanation of why this difference was made could not be found.
84. Bolton to Crane, Report on a visit to IG's major research laboratories, p. 23, 11 June 1936, DuPont company papers, Series II, part 2, Box 1038, Hagley Museum and Library;
85. See Report on the scientific laboratories at IG Farben, 1927, Bayer Archive 103/17.05.5; Bolton to Crane. Report on visit to IG's major research laboratories, 11 June 1936, DuPont company papers, Series II, part 2, Box 1038, Hagley Museum and Library; Hans Georg Grimm. "Organisation der Forschung in der chemischen Industrie". *Stahl und Eisen*, 55(1935), 349–352;
86. Report on the scientific laboratories at IG Farben, 1927, Bayer Archive 103/17.05.5, Appendix.
87. Grimm, "Organisation der Forschung in der chemischen Industrie", 1935, p. 350. Dr Grimm was director of the Oppau High Pressure Laboratory.
88. Taken from Grimm, "Organisation der Forschung in der chemischen Industrie", 1935, p. 350. See also Bernd Dornseifer, "Zur Bürokratisierung deutscher Unternehmen im späten 19. und frühen 20. Jahrhundert." *Jahrbuch für Wirtschaftsgeschichte* 1993/2, 69–91, p. 84–86.
89. For the importance of single scientists even in corporate research see John Jewkes, David Sawers, Richard Stillerman, *The Sources of Invention.* (Edinburgh, 1969).
90. See Bolton to Crane, Report on visit to IG's major research laboratories, p. 8, 11 June 1936, DuPont company papers, Series II, part 2, Box 1038, Hagley Museum and Library.
91. Report on the scientific laboratories at IG Farben, 1927, p. 10, Bayer Archive 103/17.05.5; Grimm, "Organisation der Forschung in der chemischen Industrie", 1935, p. 350.
92. Report Prof. Lechner on the reduction of employment of IG Farben's scientific chemists, 1929, p. 12 [Prof. Lechner, Über die Herabsetzung der Zahl unserer wissenschaftlich arbeitenden Chemiker, 29. November 1929] Bayer Archive 12/13 Technische Direktionskonferenz; Grimm, "Organisation der Forschung in der chemischen Industrie", 1935, p. 349.
93. Committee on the Conditions of Production and Sales of German Goods, Subcommittee III: The Chemical Industry. [*Ausschuß zur Untersuchung der Erzeugungs- und Absatzbedingungen der deutschen Wirtschaft.* Enquete-Ausschuß, III. Unterausschuß: Die deutsche Chemische Industrie.] (Berlin 1930), pp. 32–33.
94. Report on the scientific laboratories at IG Farben, 1927, pp. 5, 10, Bayer Archive 103/17.05.5.
95. Originally, this was the policy of only the Griesheim company. But the development of the Haber-Bosch process shows that other companies used similar strategies. See Report on the scientific laboratories at IG Farben, 1927, p. 7, Bayer Archive 103/17.05.5.
96. Grimm, "Organisation der Forschung in der chemischen Industrie", 1935, p. 351.
97. See 1927 Report on Scientific Laboratories of IG Farben and Josef Mersch, *Geschichte des Technischen Raumes Wissenschaft Leverkusen 1923–58*, Bayer Archive 1.6.6.28.
98. Committee on the Conditions of Production and Sales of German Goods, Subcommittee III: The Chemical Industry. *Ausschuß zur Untersuchung der Erzeugungs- und Absatzbedingungen der deutschen Wirtschaft.*, Die chemische Industrie, 1930, p. 24.
99. See Mersch, *Technischer Raum Wissenschaft*, p. 16, Bayer Archive 1.6.6.28.
100. Bolton to Crane, pp. 4, 24, 11 June 1936, DuPont Company Papers, Series II, part 2, Box 1038, Hagley Museum and Library.
101. Committee on the Conditions of Production and Sales of German Goods, Subcommittee III: The Chemical Industry. [*Ausschuß zur Untersuchung der Erzeugungs- und Absatzbedingungen der deutschen Wirtschaft.* Enquete-Ausschuß, III. Unterausschuß: Die deutsche Chemische Industrie.] (Berlin 1930) p. 24.
102. Fritz terMeer, "Die I.G. Ihre Entstehung, Entwicklung und Bedeutung." *Chemische Industrie* 1952,

777–805, here p. 783. Also H. Flemming, *Gründung der wissenschaftlichen Kunststoffkommission (Kuko) im Jahre 1930.* Dokumente aus Höchster Archiven 16 (Höchst, 1966).

103. Statement Dr Bosch's at the Committee on the Conditions of Production and Sales of German Goods. Subcommittee III: The Chemical Industry. [*Ausschuß zur Untersuchung der Erzeugungs- und Absatzbedingungen der deutschen Wirtschaft.* Enquete-Ausschuß, III. Unterausschuß: Die deutsche Chemische Industrie.] (Berlin 1930) p. 112–113; see also Chandler, *Scale and Scope,* pp. 569–573.
104. Dr Duden (Höchst) to the Members of the TEA, 7 June 1926: Zusammenlegung der Fabrikation, [On the Concentration of Production.], pp. 1–2, Bayer Archive 13/15 vol 1, Niederschriften des TEA.
105. Dr Krekeler (Bayer), Zusammenlegung der Fabrikation, [On the Concentration of Production], 11 May 1926, Bayer Archive 13/15 vol 1 Niederschriften des TEA.
106. See Krekeler's and Duden's reports on the Concentration of Production, Bayer Archive 13/15 vol 1 Niederschriften des TEA.
107. See Krekeler's and Duden's reports on the Concentration of Production, Bayer Archive 13/15 vol 1 Niederschriften des TEA and Chandler, *Scale and Scope,* p. 571.
108. Winnacker, *Nie den Mut verlieren,* pp. 80/81, 103/104; Georg Kränzelin, "Deutsche Teerfarbstoffe", *Chemiker-Zeitung* 59, 1935, p. 3; Bolton to Crane,p. 10, 11 June 1936, DuPont Company Papers, Series II, part 2, Box 1038, Hagley Museum and Library.
109. Statement Dr Bosch's at the Committee on the Conditions of Production and Sales of German Goods. subcommittee III: The Chemical Industry. [*Ausschuß zur Untersuchung der Erzeugungs- und Absatzbedingungen der deutschen Wirtschaft.* Enquete-Ausschuß, III. Unterausschuß: Die deutsche Chemische Industrie.] (Berlin 1930) p. 113.
110. Bolton to Crane, 11 June 1936, DuPont Company Papers, Series II, part 2, Box 1038, Hagley Museum and Library.
111. See list of Plants of Main Group 3, September 22nd 1945, Bayer Archive 13/13 IG Ausschüsse, Kommissionen; Chandler, *Scale and Scope,* p. 571; *Werksgeschichte Dormagen,* 1953, Bayer Archive 1/6.6.1; Erich E. Beyer, *Chemiefasern bei der Bayer AG und deren Rechtsvorgängern.* (Leverkusen, 1982), pp. 16/17.
112. Statement Dr Bosch's at the Committee on the Conditions of Production and Sales of German Goods. subcommittee III: The Chemical Industry. [*Ausschuß zur Untersuchung der Erzeugungs- und Absatzbedingungen der deutschen Wirtschaft.* Enquete-Ausschuß, III. Unterausschuß: Die deutsche Chemische Industrie.] (Berlin 1930) p. 112.
113. Richard Willstätter, "Über den Chemikerbedarf der deutschen Industrie", *Chemiker-Zeitung* 54, 1930. 365-366; Richard Willstätter, "Betrachtungen über Chemikerentlassung und Anstellungssperren", *Chemiker-Zeitung* 55, 1931, 1–2.
114. Flechtner, *Carl Duisberg,* pp. 350–351.
115. William T. Hogan, *Economic History of the Iron and Steel Industry in the United States.* (Lexington, Mass., 1971) Volume 3, pp. 1193–1204.
116. Hounshell and Smith, *Science and Corporate Strategy,* pp. 210–219.
117. The following is based on Bolton to Crane, 11 June 1936, DuPont Company Papers, Series II, part 2, Box 1038, Hagley Museum and Library.
118. Ibid., p. 20.
119. Ibid., p. 7.
120. Georg Kränzlein, "Neue Kunststoffe", *Chemiker-Zeitung 59,* 1935, 33–34.
121. Bolton to Crane, 11 June 1936, DuPont Company Papers, Series II, part 2, Box 1038, pp. 2–19, Hagley Museum and Library.
122. Ibid., p. 20.
123. Minutes of the Meeting of the TEA, 6 May 1926 [Sitzung des TEA] Bayer Archive 13/15 vol. 1.
124. Report on the scientific laboratories at IG Farben, 1927, p. 12/13, Bayer Archive 103/17.05.5.
125. Report Prof. Lechner on the Reduction of Employment of IG Farben's Scientific Chemists, November 29, 1929, p. 12, Bayer Archive 12/13 Technische Direktionskonferenz; terMeer, *IG Farben,* 1953, p. 30; Committee on the Conditions of Production and Sales of German Goods. Subcommittee III: The Chemical Industry. [*Ausschuß zur Untersuchung der Erzeugungs- und Absatzbedingungen der deutschen Wirtschaft.* Enquete-Ausschuß, III. Unterausschuß: Die deutsche Chemische Industrie.] (Berlin 1930) p. 24/25; Grimm, "Organisation der Forschung in der chemischen Industrie", 1935, p. 349.
126. Teltschick, *Geschichte der deutschen Großchemie,* table 2 at p. 47; Appendix 12 at the end of this paper.
127. Ibid..
128. For the history of high pressure technology see in general Plumpe, *IG Farben,* pp. 203–318; Carl Krauch, "Technische und wirtschaftliche Betrachtungen über Kohleveredelung unter besonderer Berücksichtigung der Hochdruckverfahren", *Stahl und Eisen,* 47 (1927), 1118–1128.
129. We have to note here that nearly no contemporary figures could be found. Where possible they have been used. The earliest available figures comparing products derive from autumn 1945, the most precise table from December 1953. Though they might differ in absolute terms by 0.5 – 2

Millionen Reichsmark, and by 0.3 – 2 per cent in relative terms, the general development remains unchanged.

130. Bolton to Crane, 11 June 1936, p. 20, Hagley Museum, DuPont Company Papers.
131. Plumpe, *IG Farben*, table at p. 440.
132. Harm G. Schröter, "Cartels as a form of concentration in industry: the example of the international dyestuffs cartel from 1927 to 1939." *German Yearbook on Business History*, 1988, 113–144.
133. Bolton to Crane, 11 June 1936, p. 21, Hagley Museum, DuPont Company Papers. Bolton is once more summarizing his conversation with terMeer.
134. Akten-Notiz über die Sitzung des TEA am 10.12.1925 Kunstseide betreffend. Bayer Archive 13/15 vol. 1
135. Plumpe, *IG Farben*, pp. 318–325.
136. See also Thomas P. Hughes, "Technological Momentum in History: Hydrogenation in Germany 1898–1933", *Past and Present*, 44 (1969), 106–132.
137. This is contrary to Plumpe, *IG Farben*, p. 472.
138. William Lever of Unilever also followed the idea of having internal competition among his subsidiaries. See Mowery and Rosenberg, *Technology and Pursuit of Economic Growth*, p. 104.
139. See Irmgard Hack, Lothar Hack, *Die Wirklichkeit die Wissenschafft, Zum wechselseitigen Begründungsverhältnis von 'Verwissenschaftlichung der Industrie' und 'Industrialisierung der Wissenschaft'.* (Franfurt am Main, 1985), pp. 148–158.
140. Ibid., p. 149.
141. Bernd Dornseifer, "Zur Bürokratisierung deutscher Unternehmen im späten 19. und frühen 20. *Jahrhundert." Jahrbuch für Wirtschaftsgeschichte* 1993, 69–91.
142. J E Crane, "One way of Looking at Research", talk presented at the chemical directors' meeting, May 26, 1930, p. 2 Accession 1416, Box 7, Hagley Museum and Library.
143. Ibid., p. 2.
144. Ibid., p. 4.
145. Richard J. Overy, "Blitzkriegswirtschaft? Finanzpolitik, Lebensstandard und Arbeitseinsatz in Deutschland 1939–1942." *Vierteljahrshefte für Zeitgeschichte*, 36 (1988), 379–437, here p. 384.
146. So Walter Carpenter's impression in 1937. Report on Germany, 19th April, 1937, p. 3. W S Carpenter Papers, Series II, Part 2, Box 823, Hagley Museum and Library.

Appendix 1, Main plants of the merging six companies

BASF plants: Ludwigshafen, Oppau, Merseburg, Niedersachswerfen;
Farbenfabriken vorm. Fried. Bayer & Co plants: Elberfeld, Leverkusen (near Cologne), Dormagen;
Farbwerke vorm. Meister Lucius & Brüning, Hoechst/Main plants: Hoechst near Frankfurt/Main, Gersthofen (Southern Germany), Biebrich/Rhine;
Agfa plants: Berlin-Treptow, Berlin-Rummelsburg, Wolfen near Bitterfeld (Central Germany);
Chemische Fabrik Griesheim-Elektron plants: Frankfurt am Main, Offenbach/Main (former K. Oehler's firm), Bitterfeld. Rheinfelden(Baden, South-West Germany);
Chemische Fabriken vorm. Weiler-terMeer plants: Cologne, Uerdingen near Cologne, Müngersdorf, Riehl.
Several plants were operating in different chemical fields on the same location (e.g. Bitterfeld), so that the number of IG Farben's production facilities is not identical with its number of production sites. IG counted 37 production plants in 1925 (but only 36 actually operating), and 46 in 1929 (43 operating), according to the economic situation and having enlarged existing sites and added newly acquired ones.

Sources: Der Farben-Konzern. Die I.G. Farbenindustrie AG. Ihre Tochtergesellschaften und Beteiligungen, Aufbau, Statistik, Finanzen. *Das Spezial-Archiv der deutschen Wirtschaft* (Berlin, 1927) pp. 8–11.
Statement of Dr Carl Bosch, Director of the Central Committee of IG Farben at the Commission on the Conditions of Production and Sales of German goods in 1930, III Subcommittee: The Chemical Industry. [Ausführungen des Sachverständingen Dr. Bosch, in: *Ausschuß zur Untersuchung der Erzeugungs- und Absatzbedingungen der deutschen Wirtschaft.* Enquete-Ausschuß: Die deutsche Chemische Industrie, Berline 1930, p. 111]

Appendix 2. IG Farben's regional plant communities in 1926:

Plant community	plant location	products
UPPER RHINE	1. Ludwigshaen	dyes, intermediates, soda and chlorine-products
	2. Oppau	synthetic ammonia, fertilizers
	3. Leuna/Merseburg	synthetic ammonia, nitrogen, synthetic fuels
	4. Niedersachswerfen	plaster
MID-RHINE	5. Hoechst/Main	dyes, intermediates, acids, chlorine products, pharmaceuticals, pesticides
	6. Gersthofen	indigo-products
	7. Biebrich/Rhine	chemicals, pharmaceuticals
	8. Knapsack	lime, nitrogen
LOWER-RHINE	9. Leverkusen	dyes, inorganic products, acids, chlorine, pharmaceuticals, pesticides, photographic chemicals
	10. Elberfeld	ditto
	11. Dormagen	ditto
	12. Barmen	ditto
	13. Heggen in Westphalia	lime
	14. Munich	photographical products
	15. Ürdingen (terMeer)	dyes, intermediates, acids
CENTRAL GERMANY	16. Wolfen-Bitterfeld	dyes, fertilizers, chemicals, perfumes
	17. Wolfen	photographic films, artificial silk
	18. Grieheim-Elektron in Bitterfeld	dyes, chlorine and phosphorous products
	19. Teutschenthal	hydrogen, oxygen
	20. Berlin-Treptow	intermediates, photographic products
	21. Doeberitz	ditto
	22. Delvendahl, Werder	perfumes
	23. Mainkur/Frankfurt	dyes, auxiliaries, varnishes, inorganic products
	24. Griesheim/Frankfurt	ditto and autogenous welding tools
	25. Offenbach/Frankfurt	ditto and Naphtol-AS dyes
	26. Mühlheim	ditto
	27. Rheinfelden	ditto
PLANTS COLOGNE-ROTTWEIL AG	28. Kunigunde	explosives
	29. Roehnsahl	ditto
	30. Hamm-Siege	ditto
	31. Bobingen	artificial silk
	32. Premnitz/Havel	textile fibres "Vistra", artificial silk,
	33. Rottweil	artificial silk, explosives
	34. Düneberg/Hamburg	vulcanized fibre, Triolin
	35. Zell/Wiesenthal	explosives

Source: Taken and shortened from: *Farben Konzern*. Spezial Archiv der deutschen Wirtschaft 1927, pp. 11–13

Appendix 3. Sales and Research in Nitrogen

	Sales in Mill RM	*as % of tot. IG sales*	*research costs in Mill RM*	*as % of tot. IG research*	*research costs as % of product sales*
1926	440	42.4%	2.6	3.2%	0.6%
1927	504	39.4%	–	–	–
1928	540	37.7%	3.9	2.8%	0.7%
1929	502	35.0%	12.8	9.3%	2.5%
1930	330	28.3%	12.6	12.6%	3.8%
1931	234	22.8%	–	–	–
1932	182	20.6%	4.8	11.9%	2.6%
1933	170	18.2%	–	–	–
1934	173	16.7%	–	–	–
1935	184	16.0%	–	–	–
1936	224	16.3%	9.3	13.7%	4.1%
1937	232	14.1%	10.1	12.3%	4.3%
1938	266	15.1%	11.7	12.5%	4.4%
1939	311	14.5%	11.5	11.4%	3.7%
1940	313	13.0%	13.0	12.3%	4.1%
1941	291	10.2%	13.0	12.0%	4.4%
1942	305	9.2%	11.4	10.8%	3.7%

Source: Umsätze und Forschungskosten der I.G. Farbenindustrie 1926–43. Direktionsabteilung T, Dr. Struss, December 11, 1953, Finanzwesen Bayer-Forschung, Forschungskosten allgemeine, Bayer Archiv 15/9.2.
and Plumpe, *IG Farben*, table at p. 258; the nitrogen reasearch costs for the years 1926–1931 are derived by subtracting the figures for synthetic gasoline from the research budget for Spart I.

Appendix 4. Sales and Research in Synthetic Gasoline

	Sales in Mill RM	*as % of tot. IG sales*	*research costs in Mill RM*	*as % of tot. IG research*	*research costs as % of product sales*
1926	–	–	–	–	–
1927	–	–	48.7	31.6%	–
1928	6	0.4%	20.3	14.5%	38.3%
1929	14	1.0%	15.9	11.6%	113.6%
1930	16	1.4%	7.5	7.5%	46.9%
1931	19	1.9%	6.4	9.1%	33.7%
1932	27	2.8%	4.0	9.9%	14.8%
1933	25	2.7%	–	–	–
1934	35	3.4%	–	–	–
1935	59	5.1%	–	–	–
1936	98	7.1%	8.6	12.6%	8.7%
1937	119	7.4%	10.3	12.5%	8.6%
1938	138	7.9%	12.6	13.5%	9.1%
1939	162	7.6%	13.2	13.0%	8.1%
1940	192	7.9%	14.5	13.6%	7.6%
1941	229	8.0%	16.2	14.8%	7.1%
1942	340	10.3%	14.0	13.3%	4.1%

Source: Umsätze und Forschungskosten der I.G. Farbenindustrie 1926–43. Direktionsabteilung T, Dr. Struss, December 11, 1953, Finanzwesen Bayer-Forschung, Forschungskosten allgemeine, Bayer Archiv 15/9.2; Forschungskosten der IG, list 2, compiled July 7, 1945, Bayer Archive 13/13 IG Ausschüsse; Walter Teltschik, *Geschichte der deutschen Großchemie,* Entwichklung und Einfluß in Staat und Gesellschaft (Weinheim, 1992) Table 5 p. 112; Plumpe, *IG Farben,* tables at p. 258, 294.

Appendix 5. Sales and Research in Dyes

	Sales in Mill RM	*as % of tot. IG sales*	*research costs in Mill RM*	*as % of tot. IG research*	*research costs as % of product sales*
1926	346	33.3%	7.4	9.0%	2.1%
1927	405	31.6%	–	–	–
1928	436	30.5%	–	–	–
1929	411	28.6%	21	15.3%	5.1%
1930	366	33.4%		–	–
1931	354	34.5%	–	–	–
1932	317	35.9%	17.1	43.7%	4.5%
1933	329	35.3%	–	–	–
1934	339	32.8%	–	–	–
1935	341	29.6%	–	–	–
1936	349	25.4%	17.9	26.2%	5.1%
1937	385	24.0%	18.6	22.6%	4.8%
1938	364	20.7%	17.9	19.2%	4.9%
1939	398	18.6%	17.8	17.7%	4.4%
1940	354	14.6%	17.7	16.7%	4.9%
1941	398	–	–	–	–
1942	387	–	–	–	–

Source: Umsätze und Forschungskosten der I.G. Farbenindustrie 1926–43. Direktionsabteilung T, Dr. Struss, December 11, 1953, Finanzwesen Bayer-Forschung, Forschungskosten allgemeine, Bayer Archive 15/9.2; Forschungskosten der IG, list 2, compiled July 9, 1945, Bayer Archive 13/13 IG Ausschüsse; separates dyes from auxiliaries, if added, the figures are closely to the ones provided by Dr. Struss in 1953. The overall trend of the dye business remains unchanged.

Appendix 6. Sales and Research in Chemicals
(inorganic, solvents, smelling products, metals)

	Sales in Mill RM	*as % of tot. IG sales*	*research costs in Mill RM*	*as % of tot. IG research*	*research costs as % of product sales*
1926	143	13.8%	–	–	–
1927	213	16.6%	–	–	–
1928	233	16.3%	–	–	–
1929	266	18.5%	21	15.3%	7.9%
1930	224	19.2%	–	–	–
1931	186	18.1%	–	–	–
1932	151	17.1%	4.2	10.4%	2.8%
1933	171	18.3%	–	–	–
1934	206	19.9%	–	–	–
1935	269	23.3%	–	–	–
1936	332	24.2%	8.3	12.2%	2.5%
1937	415	25.8%	17.4	21.2%	4.2%
1938	474	27.0%	24.2	25.9%	5.1%
1939	637	29.8%	26.7	26.6%	4.2%
1940	756	31.2%	31.4	29.6%	4.2%
1941	973	34.0%			
1942	1162	35.0%			

Source: Umsätze und Forschungskosten der I.G. Farbenindustrie 1926-43. Direktionsabteilung T, Dr. Struss, December 11, 1953, Finanzwesen Bayer-Forschung, Forschungskosten allgemeine, Bayer Archive 15/9.2; Plumpe, *IG Farben*, tables at pp. 330, 338;

Appendix 7, Sales and Research in Plastics

	Sales in Mill RM	*as % of tot. IG sales*	*research costs in Mill RM*	*as % of tot. IG research*	*research costs as % of product sales*
1926	–	–	–	–	–
1927	–	–	–	–	–
1928	–	–	–	–	–
1929	–	–	–	–	–
1930	–	–	1.1	1.1%	–
1931	–	–	1.2	1.7%	–
1932	–	–	0.8	2.0%	–
1933	–	–	0.6	1.4%	–
1934	–	–	1.0	2.3%	–
1935	1.6	0.1%	0.8	1.4%	50.0%
1936	1.6	0.1%	1.1	1.6%	68.8%
1937	5.0	0.3%	1.3	1.6%	26.0%
1938	11.4	0.6%	1.6	1.7%	14.0%
1939	28.3	1.3%	1.4	1.4%	5.0%
1940	62.7	2.6%	0.4	0.4%	0.6%
1941	92.6	3.2%	0.01	–	–
1942	114.2	3.4%	0.2	0.2%	0.2%

Source: Plumpe, *IG Farben,* tables at p. 330, 338.

Appendix 8. Sales and Research in Pharmaceuticals and Insecticides

	Sales in Mill RM	*as % of tot. IG sales*	*research costs in Mill RM*	*as % of tot. IG research*	*research costs as % of product sales*
1926	53	5.1%	1.6	2.0%	1.5%
1927	65	5.1%	–	–	–
1928	85.1	–	–	–	–
1929	92.8	6.5%	4.8	3.5%	5.2%
1930	93.9	8.1%	–	–	–
1931	92.8	9.0%	–	–	–
1932	82.1	9.3%	4.6	11.4%	5.6%
1933	85.6	7.4%	–	–	–
1934	88	8.5%	–	–	–
1935	96	8.3%	–	–	–
1936	103	7.5%	5.6	8.2%	5.4%
1937	121	7.5%	5.6	6.8%	4.6%
1938	134	7.6%	6.2	6.6%	4.6%
1939	152	7.1%	6.5	6.5%	4.3%
1940	156	6.4%	5.8	5.5%	3.7%
1941	185	6.5%	–	–	–
1942	212	6.4%	–	–	–

Source: Umsätze und Forschungskosten der I.G. Farbenindustrie 1926–43. Direktionsabteilung T, Dr. Struss, December 11, 1953, Finanzwesen Bayer-Forschung, Forschungskosten allgemeine, Bayer Archives 15/9.2; Forschungskosten der IG, list 2, compiled July 9, 1945, Bayer Archive 13/13 IG Ausschüsse; Plumpe, *IG Farben,* table at p. 447.

Appendix 9. Sales and Research in Synthetic Rubber BUNA

	Sales in Mill RM	*as % of tot. IG sales*	*research costs in Mill RM*	*as % of tot. IG research*	*research costs as % of product sales*
1926	–	–	–	–	–
1927	–	–	0.1	0.08%	–
1928	–	–	2.7	1.9%	–
1929	–	–	4.1	3.0%	–
1930	–	–	2.6	2.6%	–
1931	–	–	0.7	1.0%	–
1932	–	–	0.3	0.8%	–
1933	–	–	0.4	0.9%	–
1934	–	–	0.1	2.4%	–
1935	–	–	2.0	3.6%	–
1936	–	–	–	–	–
1937	12.8	0.8%	–	–	–
1938	20.6	1.2%	–	–	–
1939	62.3	2.9%	–	–	–
1940	102.9	4.3%	–	–	–
1941	167.8	5.9%	–	–	–
1942	340.8	7.3%	–	–	

Source: Plumpe, *IG Farben*, tables at p. 354, 388; Peter Morris, *The Development of Acetylene Chemistry and Synthetic Rubber by IG Farbenindustrie Aktiengesellschaft 1926–1945* PhD Dissertation Oxford 1982, p. 158.

Appendix 10. Sales and Research in Photographic Articles

	Sales in Mill RM	*as % of tot. IG sales*	*research costs in Mill RM*	*as % of tot. IG research*	*research costs as % of product sales*
1926	40	3.9%	–	–	–
1927	62	4.8%	–	–	–
1928	85	5.9%	–	–	–
1929	100	7.0%	–	–	–
1930	84	7.2%	–	–	–
1931	81	7.9%	–	–	–
1932	72	8.2%	1.6	4.0%	2.2%
1933	65	7.0%	–	–	–
1934	66	6.4%	–	–	–
1935	76	6.6%	–	–	–
1936	83	6.1%	2.4	3.5%	2.9%
1937	99	6.1%	1.9	2.3%	1.9%
1938	109	6.2%	2.4	2.6%	2.2%
1939	137	6.4%	3.1	3.1%	2.2%
1940	150	6.2%	4.2	4.0%	2.8%
1941	171	6.0%	3.3	3.0%	1.9%
1942	175	5.3%	2.7	2.6%	1.6%

Source: Umsätze und Forschungskosten der I.G. Farbenindustrie 1926–43. Direktionsabteilung T, Dr. Struss, December 11, 1953, Finanzwesen Bayer-Forschung, Forschungskosten allgemeine, Bayer Archive 15/9.2; Forschungskosten der IG, list 2, compiled July 7, 1945, Bayer Archive 13/13 IG Ausschüsse.

Appendix 11. Sales and Research in Artificial Silk and Spun Rayon

	Sales in Mill RM	*as % of tot. IG sales*	*research costs in Mill RM*	*as % of tot. IG research*	*research costs as % of product sales*
1926	5	2.0%	0.6	0.7%	12%
1927	27	2.1%	–	–	–
1928	41	2.9%	–	–	–
1929	56	3.9%	–	–	–
1930	53	4.5%	–	–	–
1931	43	4.2%	–	–	–
1932	41	4.6%	1.4	3.5%	3.3%
1933	46	4.9%	–	–	–
1934	76	7.4%	–	–	–
1935	66	5.7%	–	–	–
1936	106	7.7%	2.2	3.2%	2.0%
1937	139	8.6%	1.8	2.2%	1.3%
1938	151	8.6%	2.3	2.5%	1.5%
1939	164	7.7%	2.7	2.7%	1.9%
1940	173	7.1%	2.0	1.8%	1.1%
1941	200	7.0%	3.9	3.6%	1.9%
1942	209	6.3%	5.1	4.8%	2.5%

Source: Umsätze und Forschungskosten der I.G. Farbenindustrie 1926–43. Direktionsabteilung T, Dr. Struss, December 11, 1953, Finanzwesen Bayer-Forschung, Forschungskosten allgemeine, Bayer Archive 15/9.2; Forschungskosten der IG, list 2, compiled July 7, 1945, Bayer Archive 13/13 IG Ausschüsse; 1927 report on IG Farben scientific research laboratories.

Appendix 12. Farben Sales Structure 1926–1942

	nitrogen	*synthetic gasoline*	*dyes*	*chemicals*	*plastics*	*Pharma., pesticides*	*BUNA*	*photo. articles*	*artificial silk*
1926	42.4%	–	33.3%	13.8%	–	5.1%	–	3.9%	2.0%
1927	39.4%	–	31.6%	16.6%	–	5.1%	–	4.8%	2.1%
1928	37.7%	0.4%	30.5%	16.3%	–	–	–	5.9%	2.9%
1929	35.0%	1.0%	28.6%	18.5%	–	6.5%	–	7.0%	3.9%
1930	28.3%	1.4%	33.4%	19.2%	–	8.1%	–	7.2%	4.5%
1931	22.8%	1.9%	34.5%	18.1%	–	9.0%	–	7.9%	4.2%
1932	20.6%	2.8%	35.9%	17.1%	–	9.3%	–	8.2%	4.6%
1933	18.2%	2.7%	35.3%	18.3%	–	7.4%	–	7.0%	4.9%
1934	16.7%	3.4%	32.8%	19.9%	–	8.5%	–	6.4%	7.4%
1935	16.0%	5.1%	29.6%	23.3%	0.1%	8.3%	–	6.6%	5.7%
1936	16.3%	7.1%	25.4%	24.2%	0.1%	7.5%	–	6.1%	7.7%
1937	14.1%	7.4%	24.0%	25.8%	0.3%	7.5%	0.8%	6.1%	8.6%
1938	15.1%	7.9%	20.7%	27.0%	0.6%	7.6%	1.2%	6.2%	8.6%
1939	14.5%	7.6%	18.6%	29.8%	1.3%	7.1%	2.9%	6.4%	7.7%
1940	13.0%	7.9%	14.6%	31.2%	2.6%	6.4%	4.3%	6.2%	7.1%
1941	10.2%	8.0%	–	34.0%	3.2%	6.5%	5.9%	6.0%	7.0%
1942	9.2%	10.3%	–	35.0%	3.4%	6.4%	7.3%	5.3%	6.3%

Appendix 13, Distribution of research costs as % of total IG research expenditure 1926–1942

	nitrogen	*synthetic gasoline*	*dyes*	*chemicals*	*plastics*	*Phama., pesticides*	*BUNA*	*photo. articles*	*artificial silk*
1926	3.2%	–	9.0%	–	–	2.0%	–	–	0.7%
1927	–	31.6%	–	–	–	–	0.08%	–	–
1928	2.8%	14.5%	–	–	–	–	1.9%	–	–
1929	9.3%	11.6%	15.3%	15.3%	–	3.5%	3.0%	–	–
1930	12.6%	7.5%	–	–	1.1%	–	2.6%	–	–
1931	–	9.1%	–	–	1.7%	–	1.0%	–	–
1932	11.9%	9.9%	43.7%	10.4%	2.0%	11.4%	0.8%	4.0%	3.5%
1933	–	–	–	–	1.4%	–	0.9%	–	–
1934	–	–	–	–	2.3%	–	2.4%	–	–
1935	–	–	–	–	1.4%	–	3.6%	–	–
1936	13.7%	12.6%	26.2%	12.2%	1.6%	8.2%	–	3.5%	3.2%
1937	12.3%	12.5%	22.6%	21.2%	1.6%	6.8%	–	2.3%	2.2%
1938	12.5%	13.5%	19.2%	25.9%	1.7%	6.6%	–	2.6%	2.5%
1939	11.4%	13.0%	17.7%	26.6%	1.4%	6.5%	–	3.1%	2.7%
1940	12.3%	13.6%	16.7%	29.6%	0.4%	5.5%	–	4.0%	1.8%
1941	12.0%	14.8%	–	–	–	–	–	3.0%	3.6%
1942	10.8%	13.3%	–	–	0.2%	–	–	2.6%	4.8%

References Cited

1. *Ausschuß zur Untersuchung der Erzeugungs- und Absatzbedingungen der deutschen Wirtschaft*. Enquete-Ausschuß, III. Unterausschuß: Die deutsche Chemische Industrie. (Berlin 1930).
2. *Die Badische Anilin-& Soda-Fabrik*. (Ludwigshafen, 1922).
3. BASF, *Ein Beitrag zur Geschichte der chemischen Technik*. (Ludwigshafen,1961).
4. *BASF Laboratoriumsbauten der Anwendungstechnik*. (Detmold, 1963).
5. Ernst Bäumler, *Milestones of Progress in the Years 1863–1966*. (Frankfurt am Main, 1967).
6. Bayer AG (Ed.), *Beiträge zur 100jährigen Firmengeschichte 1863–1963*. (Köln, 1963).
7. John J. Beer, "Coal Tar Dye Manufacture and the Origins of the Modern Industrial Research Laboratory", *Isis* 49 (1958), 123–131.
8. John J. Beer, *The Emergence of the German Dye Industry*. (Urbana, Illinois 1959).
9. Erich E. Beyer, *Chemiefasern bei der Bayer AG und deren Rechtsvorgängern*. (Leverkusen, 1982).
10. Peter Borscheid, *Naturwissenschaft, Staat und Industrie in Baden 1848–1914*. (Stuttgart, 1976).
11. *The Challenge of New Technology. Innovation in British Business since 1850*. Edited by Jonathan Liebenau, (Aldershot, 1988).
12. Alfred D. Chandler, Jr., *Scale and Scope. The Dynamics of Industrial Capitalism*. (Cambridge Mass., 1990).
13. *The Chemical Industry*. Documentation of the International Economic Conference Geneva May 1927. Publications of the League of Nations, II, Economic and Financial Section, 1927. II. 4.
14. *The Chemical Industries of German Rhineland*. A Summary of the British Chemical Mission on Chemical Factories in the Occupied Area of Germany. Edited by the Association of British Chemical Manufacturers. (London, 1920).
15. Bernd Dornseifer, "Zur Bürokratisierung deutscher Unternehmen im späten 19. und frühen 20. Jahrhundert. "*Jahrbuch für Wirtschaftsgeschichte* 1993/2, 69–91.
16. Horst–Bernd Dünschede, *Tropenmedizinische Forschung bei Bayer*. (Düsseldorf 1971).
17. Carl Duisberg, "Die Vereinigung der deutschen Farbenfabriken" (1915). introduced by Wilhelm Treue, *Tradition* 8 (1963), 193–227.
18. Carl Duisberg, *Abhandlungen, Vorträge und Reden aus den Jahren 1882–1921*. (Berlin, 1923).
19. Der Farben-Konzern. Die I.G. Farbenindustrie AG. Ihre Tochtergesellschaften und Beteiligungen. Aufbau, Statistik, Finanzen. *Das Spezial–Archiv der deutschen Wirtschaft* (Berlin 1927).
20. *Farbwerke Meister Lucius & Brüning zu Höchst am Main 1863–1913*. (Höchst, 1913).
21. Wilfried Feldenkirchen, "Big Business in Interwar Germany: Organizational Innovation at Vereinigte Stahlwerke, IG Farben, and Siemens." *Business History Review* 61 (1987), 417–451.
22. *75 Jahre Papiertechnisches Laboratorium der BASF*. (Ludwigshafen 1966).
23. Wolfram Fischer, "Dezentralisation oder Zentralisation-kollegiale oder autoritäre Führung? Die Auseinandersetzung um die Leitungsstruktur bei der Entstehung I.G. Farben-Konzerns". *Recht und Entwicklung der Großunternehmen im 19. und frühen 20. Jahrhundert*. Edited by Norbert Horn and Jürgen Kocka. (Göttingen, 1979), 476–487.
24. Hans–Joachim Flechtner, *Carl Duisberg. Vom Chemiker zum Wirtschaftsführer*. (Düsseldorf, 1959).
25. Arndt Fleischer, *Patentgesetzgebung und chemisch-pharmazeutische Industrie im deutschen Kaiserreich 1871–1918*. (Stuttgart, 1984).
26. *H. Flemming, Gründung der wissenschaftlichen Kunststoffkommission (Kuko) im Jahre 1930*. Dokumente aus Höchster Archiven 16 (Höchst, 1966).
27. H.W. Flemming, *Die Zusammenarbeit Behring-Höchst 1892–1904*. Dokumente aus Höchster Archiven 37. (Höchst. 1968).
28. *Geschichte des Werkes Uerdingen der Farbenfabriken Bayer Aktiengesellschaft* (former Weiler-ter Meer & Co), (Leverkusen, 1956)
29. *Geschichte und Entwicklung der Farbenfabriken vorm. Fried. Bayer & Co Elberfeld in den ersten 50 Jahren 1863–1918*. (München, 1918).
30. Hans Georg Grimm. "Organisation der Forschung in der chemischen Industrie". *Stahl und Eisen*, 55 (1935), 349–352.
31. L.F. Haber, *The Chemical Industry during the Nineteenth Century. A Study of the Economic Aspect of Applied Chemistry in Europe and North Americal*. (Oxford, 1958).
32. L.F. Haber, *The Chemical Industry 1900–1930. International Growth and Technological Change*. (Oxford, 1971).
33. Irmgard Hack, Lothar Hack, *Die Wirklichkeit die Wissen schafft. Zum wechselseitigen Begründungsverhältnis von 'Verwissenschaftlichung der Industrie' und 'Industralisierung der Wissenschaft'*. (Frankfurt/Main, 1985).
34. Peter Hayes, *Industry and Ideology. IG Farben in the Nazi era*. (London, 1987).
35. Michael Hörskens, "Fortschrittsformel Forschung. 125 Jahre BASF". *Chemische Industrie* 1990, No 5, 78.
36. William T. Hogan, *Economic History of the Iron and Steel Industry in the United States*. (Lexington, Mass., 1971).

37. Ernst Homburg, "The Emergence of Research Laboratories in the Dyestuffs Industry, 1870–1900", *British Journal for the History of Science*, 25 (1992), 91–111.
38. David Hounshell/John Kenly Smith, *Science and Corporate Strategy. DuPont R&D, 1902–1980.* (Cambridge Mass., 1988).
39. Thomas P. Hughes, "Technological Momentum in History: Hydro-genation in Germany 1898–1933", *Past and Present*, 44 (1969), 106–132.
40. John Jewkes, David Sawers, Richard Stillerman, *The Sources of Invention*. (Edinburgh, 1969).
41. S. Kaku, "The Development and Structure of the German Coal-Tar Dyestuffs Firms". *Development and Diffusion of Technology. Electrical and Chemical Industries.* Proceedings of the Fuji Conference. Edited by A. Okochi and H. Uchida. (Tokyo, 1980), 77–95.
42. Georg Kränzlein, "Deutsche Teerfarbstoffe", *Chemiker-Zeitung* 59, 1935, p. 3.
43. Georg Kränzlein, "Neue Kunststoffe", *Chemiker-Zeitung* 59, 1935, 33–34.
44. Carl Krauch, "Technische und wirtschaftliche Betrachtungen über Kohleveredelung unter besonderer Berücksichtigung der Hochdruckverfahren", *Stahl und Eisen*, 47 (1927), 1118–1128.
45. Jonathan Liebenau, "Paul Ehrlich as commerical scientist and research administrator", *Medical History*, 34 (1990), 65–78.
46. Jonathan Liebenau, "Industrial R&D in pharmaceutical firms in the early twentieth century", *Business History*, 26 (1984), 329–347.
47. Jonathan Liebenau, "Ethical business: the formation of the pharmaceutical industry in Britain, Germany, and the United States before 1914", *Business History*, 30 (1988), 116–130.
48. Fritz terMeer, "Die I.G. Ihre Entstehung, Entwickllung und Bedeutung." *Chemische Industrie* 1952, 777–805.
49. Fritz terMeer, *Die I.G. Farben. Ihre Entstehung, Entwicklung und Bedeutung*, (Düsseldorf, 1953).
50. *Meilensteine. 125 Jahre Bayer.* Edited by Gottfried Plumpe and Heinz Schultheis (Leverkusen, 1988).
51. Georg Meyer-Thurow, "The Industrialization of Invention: A Case Study from the German Chemical Industry". *Isis*. 73 (1982), 363–381.
52. Peter Morris, *The Development of Acetylene Chemistry and Synthetic Rubber by IG Farbenindustrie Aktiengesellschaft 1926–1945.* PhD Dissertation Oxford 1982.
53. David Mowery/Nathan Rosenberg, *Technology and the Pursuit of Economic Growth*. (Cambridge Mass., 1989).
54. Richard J. Overy, "Blitzkriegswirtschaft ? Finanzpolitik, Lebensstandard und Arbeitseinsatz in Deutschland 1939–1942. *"Vierteljahrshfte für Zeitgeschichte*, 36 (1988), 379–437.
55. Hermann Pinnow, *Werksgeschichte Farbenfabriken vorm. Fried. Bayer 1863–1938.* (München 1938).
56. Hermann Pinnow, *75 Jahre Werksgeschichte Höchst. Zur Erinnerung an die 75. Wiederkehr des Gründungstages der Farbwerke vorm. Meister Lucius & Bruening.* (München, 1938).
57. Gustav Pistor, *Hundert Jahre Griesheim 1856–1956. Ein Beitrag zur Geschichte der chemischen Industrie.* (Tegernsee, 1958).
58. Gottfried Plumpe, *Die I.G. Farbenindustrie A.G. Wirtschaft, Technik und Politik 1904–1945.* (Berlin, 1990).
59. Hermann Raschen, *75 Jahre Chemische Farbrik Griesheim-Elektron.* (Frankfurt/Main, 1938).
60. William J. Reader, *Imperial Chemical Industries. A History.* Vol. II, The First Quarter-Century 1926–1952, (Oxford, 1975).
61. Wilhelm Roggersdorf, *Im Reiche der Chemie. 100 Jahre BASF.* (Düsseldorf, 1965).
62. Hubert Sales, "La Recherche-Dévelopement dans la Stratégie des grandes Enterprises Chimiques de l'Entre-Deux-Guerres: DuPont de Nemours, Imperial Chemical Indsutries, I.G. Farben", *Revue d'Économie Politique*, 94 (1984), 446–464.
63. M. Simon. *Schwefelsäure Höchst. Vom Kammerverfahren zum Kontaktropzeß 1880–1914.* Dokumente aus Höchster Archiven 47, (Höchst, 1975).
64. Eberhard Schmauderer, "Der Einfluß der Chemie auf die Entwicklung des Patentwesens in der zweiten Halfte des 19. Jahrhunderts". *Tradition*, 3/4 (1971), 144–176.
65. Harm G. Schröter, "Cartels as a form of concentration in industry: the example of the international dyestuffs cartel from 1927 to 1939." *German Yearbook on Business History*, 1988, 113–144.
66. Harm G. Schröter, "Die Auslandsinvestitionen der deutschen chemischen Industrie 1870 bis 1930." *Zeitschrift für Unternehmensgeschichte* 35 (1990), 1–23.
67. Curt Schuster, *Badische Anilin-& Soda-Fabrik AG. Ein Beitrag zur Geschichte der chemischen Technik.* (Ludwigshafen, 1961).
68. Friedrich Spoun, *Höchst in England 1901–1914. Zur Geschichte des Werkes Ellesmere Port der Meister Lucius & Brüning Limited.* Dokumente aus Höchster Archiven 45, Höchst (1971).
69. Friedrich Spoun, *Höchst in Frankreich 1881–1914. Zur Geschichte des Werkes Creil der 'Compagnie Parisienne des Coleurs d'Aniline'.* Dokumente aus Höchster Archiven 44 (Höchst 1970).
70. Friedrich Spoun, *Aktiengesellschaft der Moskauer Chemischen Fabrik Farbwerke vorm. Meister Lucius & Brüning.* Dokumente aus Höchster Archiven 43, (Höchst, 1970).
71. F. Steckel, "Cartelization of the German Chemical Industry 1918–1925", *Journal of European Economic History* 19 (1990), 329–351.

72. Erich Stenger, *100 Jahre Photographie und die Agfa 1839–1939.* (München, 1939).
73. H. Tammen, *Die IG Farbenindustrie AG 1925–1933.* Ein Chemiekonzern in der Weimarer Republik. Berlin 1978.
74. Walter Teltschik, *Geschichte der deutschen Großchemie.* Entwicklung und Einfluß in Staat und Gesellschaft (Weinheim, 1992).
75. Heinrich Voelcker, *75 Jahre Kalle. Ein Beitrag zur Nassauischen Industriegeschichte.* (Wiesbaden, 1938).
76. Walter Wetzel, *Naturwissenschaften und Chemische Industrie in Deutschland. Voraussetzungen und Mechanismen ihres Aufstiegs im 19. Jahrhundert.* (Stuttgart, 1991).
77. *Wie die ersten Heilmittel nach Höchst kamen.* Dokumente aus Höchster Archiven 8. (Höchst, 1965).
78. Richard Willstätter, "Über den Chemikerbedarf der deutschen Industrie", *Chemiker-Zeitung* 54, 1930, 365–366.
79. Richard Willstätter, "Betrachtungen über Chemikerentlassung und Anstellungssperren", *Chemiker-Zeitung* 55, 1931, 1–2.
80. Richard Willstätter, 'Zur Geschichte der AGFA', *Berichte der Deutschen Chemischen Gesellschaft* 64 (1931), 140–149.
81. Karl Winnacker, *Nie den Mut verlieren.* Erinnerungen an Schickalsjahre der deutschen Chemie (Düsseldorf 1974).

Part III
The Research Laboratory and the US Corporate Economy

[8]

By Leonard S. Reich
ASSISTANT PROFESSOR OF HISTORY
RENSSELAER POLYTECHNIC INSTITUTE

Research, Patents, and the Struggle to Control Radio: A Study of Big Business and the Uses of Industrial Research.*

¶ *The advent of high-technology industries around the turn of the century created the modern industrial research department and placed a new emphasis upon the search for patentable innovations. While some of this research led to advances in basic scientific knowledge, and much of it produced product or process improvements that were directly applicable to a firm's business, a great deal was undertaken to enhance firms' bargaining powers with each other in order to preserve monopoly positions. In the early years of radio, the structure of the industry changed repeatedly with every innovation in apparatus or circuitry, a situation that led, as Professor Reich shows, to heavy investment in "non-productive" research.*

What is the relationship of scientific research and technological advance to industrial innovation? Over the past twenty-five to thirty years a substantial beginning has been made toward a better understanding of this problem.[1] A number of historical works have shown that, while many factors are involved in the evolution and ultimate control of markets, the use of scientific knowledge and technical advance, often well-protected by patents, has taken on considerable importance during the last century. Changing technology has played a substantial role in businessmen's attempts to control their markets and thus to secure their own corporate destinies. Therefore, to comprehend corporate relationships to markets it has become necessary in many cases to know how technical advances were affected and how they were used commercially.

Business History Review, Vol. LI, No. 2 (Summer, 1977).

* I am indebted to Professor Louis Galambos for much helpful criticism during the writing of this article.

[1] See, W. Rupert Maclaurin, *Invention and Innovation in the Radio Industry* (New York, 1949); Harold Passer, *The Electrical Manufacturers, 1875–1900* (Cambridge, Mass., 1953); John Enos, *Petroleum Progress and Profits* (Cambridge, Mass., 1962); Thomas Hughes, *Elmer Sperry, Inventor and Engineer* (Baltimore, 1971); and Reese Jenkins, *Images and Enterprise* (Baltimore, 1975).

This article seeks to understand this process by considering how research and development work at several industrial laboratories was employed in the struggle to control the emerging radio market from World War I to 1926. Corporate strategies and conceptions of markets were important factors in the competition, to be sure, but most often, holding the right patents at the right time turned out to be the key to success. I will show that several companies were able to use their research-generated patents to control markets by often circuitous routes, such as trading of particular patent rights to potential competitors. Such findings would indicate that industrial research has been useful in ways heretofore unemphasized if not unrecognized, and that, in order to understand the place of science and technology in industry, we would do well to look more closely at the direction of industrial research and the relationships of the laboratories to their corporate sponsors.

Historical works on industrial science and technological innovation generally take a more narrow perspective on the place of patents in the corporate armory. In particular, they consider the uses of patents to be straightforward: the preclusion of competitors from chosen markets by control of crucial patent rights or by maintenance of prohibitively dense patent structures. Such methods have indeed been prevalent in the cases of one- or main-product firms, especially at a time when they were building their initial market positions. However, once primary markets had been secured, many corporations realized that survival required extensive and somewhat nondirective research efforts for the purpose of gaining patents, some of which were only peripherally related to their contemporary areas of commercial interest. As Willis R. Whitney, first director of research at General Electric, explained: "Our research laboratory was a development of the idea that large industrial organizations have both an opportunity and a responsibility for their own life insurance. New discovery can provide it." [2]

This type of "life insurance" protected the corporate "policy holder" in two ways. First, it opened up new areas to exploit commercially, providing opportunity for growth and ensuring that the company would have a strong hand in directing important new markets. Secondly, it offered the company protection from encroachment on its primary markets by placing it in the technological forefront, and, just as importantly, by putting in its hands patents that could be used to check potential competitors by attacking them in *their* primary markets. Thus, the use of patent rights could take on both

[2] Quoted in John Broderick, *Willis Rodney Whitney* (Albany, N.Y., 1945), 188.

an offensive and a defensive character, and the defense could be indirect.

While considerable research and development work on radio systems had been completed before industrial research laboratories took it up in 1912, within five years they transformed radio communication from a weak, distorted, and erratic medium to one that was strong, clear, and reliable. In so doing, the laboratories built patent structures for their corporate sponsors to use in the attempt to dominate the newly-profitable radio market of the 1920s. By studying how these patent rights were exerted we should acquire an appreciation of the complexities and subtleties of the commercial situation and an understanding of what types of commercial considerations provided direction to research efforts in the laboratories.

This is by no means the first effort to understand the types of research undertaken by industrial laboratories in the development of radio technology. For example, Kendall Birr's *Pioneering in Industrial Research: The Story of the General Electric Research Laboratory* (Washington, D.C., 1957) discusses vacuum tube and radio research in some detail, and G. L. Archer's *History of Radio to 1926* (New York, 1938) and *Big Business and Radio* (New York, 1939) look at the external factors that affected the development of the industry. What these authors share with numerous others who have written on the history of radio is a certain naïveté concerning the effect of commercial exigencies on the shaping of research. The standard interpretation is that the various companies maintained extensive programs exclusively for the purpose of putting themselves at the forefront of product innovation – an advantage to be used naturally in a competitive marketplace. But it is also true that these companies pursued research for monopoly control of markets, not for shares of the market, and that their eagerness for patents provided both direction and constraint to their research efforts. Let us begin by recounting the development of the radio art and by considering the attempts of several companies to dominate the new-born radio industry, before proceeding to examine the role of research and patents in attaining corporate goals, taking examples from the commercial history of radio.

Pre-World War Developments

The development of radio goes back to the nineteenth century. Following Heinrich Hertz's first successful transmission of "ether waves" in Germany in 1887, a young Italian named Guglielmo

Marconi developed a functional system of "wireless telegraphy" and by 1897 succeeded in organizing a British company to produce and lease the devices. This "wireless" used what is known as a spark-gap oscillator to generate waves: a single pulse of electricity in a specially shaped antenna produced one series of waves that corresponded to a dot or dash (according to duration) at the receiver, and subsequent amplification was minimal. In 1899 Marconi launched an American subsidiary and set out to lease equipment for use on ocean-going vessels. The U.S. Navy became his best customer.

Among Marconi's consultants in Britain was Ambrose Fleming. Familiar with incandescent lamp phenomena from having worked with the Edison Electric Light Company of London, Fleming in 1904 patented a device that resembled an electric lamp but that had certain electrical characteristics that made it a superior "detector" of received radio waves. Fleming called this two-element device a "diode," and assigned the patent rights to the Marconi Company. In 1907 the American inventor Lee deForest made a major improvement to the diode when he inserted a third element or "grid," into the tube. The grid of the "audion," as deForest called his improved device, made it a much more sensitive detector, and after its powers of amplification became known, the triode tube eventually became the heart of both transmitting and receiving apparatus. Within a few years, deForest had constructed a functional radio transmitter and receiver, though operation of the system proved to be erratic.

For more than a decade, meanwhile, the American Telephone and Telegraph Company (AT&T) had been interested in the commercial possibilities of long-distance telephone communication. Since signals attenuate with distance, the problem was to provide an amplifier or "repeater" as AT&T preferred to call it. Thus, when Professor Michael Pupin of Columbia University obtained a patent on the "loading coil," which reduced the need for amplification, AT&T immediately purchased the patent rights in 1900. AT&T had organized its own research department in 1881, but it was concerned, almost exclusively with the engineering development of existing apparatus. In 1906, the AT&T chief engineer reported to the president: "Every effort in the Department is being exerted toward perfecting the engineering methods; no one is employed who, as an inventor, is capable of originating new apparatus of novel design. In consequence of this it will be necessary in many cases to depend upon the acquisition of inventions of outside men, leaving the adaptation of them to our engineers and to the Western Electric Company [an AT&T subsidiary]."[3]

The Panic of 1907 brought a change of management to the company and a new chief engineer to the Research Department. Both the new president, Theodore N. Vail, who returned to the company from retirement, and the new chief engineer, J. J. Carty, were concerned with applications of science to technology because they realized that science-fostered technological advance could have great impact on the Bell System. Carty was aware that radio in particular presented a threat to the wire-dependent Bell patent position. Thus, while one of Carty's first goals was to solve the long-distance telephone problem by development of a suitable "repeater," he kept an eye on the radio situation. In 1909, for example, he based a plea for expanded facilities on the anticipated value of work in what came to be called "electronics": "One additional argument making for vigorous work upon the development of a more powerful repeater I call to your particular attention. . . . Whoever can *supply and control* the necessary telephone repeater will exert a dominating influence in the art of wireless telephony [radio] when it is developed. . . . A successful telephone repeater, therefore, would not only react most favorably upon our service where wires are used, but might put us in a position of control with respect to the art of wireless telephony should it turn out to be a factor of importance."[4]

By 1911, radio work was in full swing at AT&T, and the annual report noted that, "to make adequate progress with this work, it was decided to organize a branch of the engineering department which would include in its personnel the best talent available and in its equipment the best facilities possible for the highest grade laboratory research. . . . A number of highly trained and experienced physicists have been employed."[5] This represented a complete turnabout from the corporate policy maintained as late as 1907.

The realization in 1912 that deForest's audion could amplify as well as detect prompted the Western Electric researchers to take a close look at it, and they very much liked what they saw. How-

[3] *Federal Communications Commission Investigation Pursuant to Public Resolution #8*, 74th Congress, Docket #1, Exhibit #2112, p. 105. (This is an FCC investigation of the Bell System undertaken in 1934 and published in seventy-seven volumes. Copies are available at the FCC Library at 1919 M St., NW, Washington, D.C. A condensed, one-volume report was issued in 1938, then revised and reissued in 1939; see the following two footnotes.)

[4] Federal Communications Commission, *Investigation of the Telephone Industry in the United States* (Washington, D.C., 1939), 189–190. Italics added.

[5] Federal Communications Commission, *Proposed Report Telephone Investigation* (Washington, D.C., 1938), 212-213.

ever, their early work confirmed that deForest's patent pre-empted virtually all possibilities for a better repeater. That is, while the Telephone researchers could design vacuum tubes far superior to those of deForest, his basic patent on the concept of inserting another element (or elements) between the diode's cathode and anode lay at the heart of every one of them and was protected by patent. Therefore, AT&T had to purchase all patent rights to the deForest tube in order to develop it. This was accomplished in 1913 and 1914 when the company attempted to obtain exclusive rights to all of deForest's inventions, a policy pursued wherever possible. However, deForest insisted on retaining rights under the patents for his own use.

The "triode" which the AT&T researchers took over in 1913 was very weak. It choked and distorted when they applied a loud speech current and it did not consistently give the same response to the same stimulus. The researchers mounted a massive attack on its theory, its mathematics, its construction, and its behavior.[6] First, they increased its vacuum, resulting in essentially pure electron emission that gave the tube more stable operating characteristics. They achieved other advances quickly, including a new type of filament that emitted more electrons and lasted longer, and improved the design of all three elements – cathode, anode, and grid. By 1915, AT&T had made the triode a reasonably reliable, stable device, and when the first transcontinental telephone line opened in that year, triodes performed the necessary amplification. During these years, AT&T had built up a substantial research organization, increasing the number of research personnel from 192 to 959 and the research budget from less than $500,000 to over $1,500,000.[7] As Carty explained to the stockholders: "This condition has been brought about by new demands for research in the fundamentals of the science of telephony; together with larger and very important activities in new branches of the telephone, telegraph, and wireless arts." [8]

The AT&T researchers undertook their first extensive wireless work in 1914, when a program was initiated with the objective of building an apparatus that could bridge the Atlantic with continuous-wave, voice-carrying signals. They accomplished this in short order. First tested between Montauk, Long Island and Wilmington, Delaware, an AT&T radio system was successfully transmitting to Paris by October 1915, though there were still great

[6] Maclaurin, *Invention and Innovation in the Radio Industry*, 89, 90.
[7] FCC, *Investigation of the Telephone Industry*, 196.
[8] *Ibid.*

problems of fidelity and especially of static to be solved. While not alone in radio-related research, AT&T was the first company to launch an organized effort aimed at perfecting a functional system of transmission and reception.

AT&T pursued these activities in order to learn more of the wireless art and to explore the properties and behavior of vacuum tubes, circuits, and radio waves, not to prepare for overseas commercial message service, which it was content to leave to the cable and Marconi companies. Rather, the company claimed that it intended to develop a two-way radio system for short-range use. As Carty noted, "the results of long distance tests show clearly that the function of the wireless telephone is primarily to reach inaccessible places where wires cannot be strung. It will act mainly as an extension of the wire system and a feeder to it."[9] As a "feeder" to the wire system, the AT&T directors saw two-way radio telephony as the natural province of the Telephone Company, and they intended to dominate the field. This orientation brings out the essential conservatism of the AT&T management; for, while a potentially profitable market for the new device loomed over the horizon, their first thought was securely fixed on protection of the company's wire interests.

When America entered World War I, the government decided to promote production of wireless apparatus for military purposes. This was accomplished by removing restrictions on the use of all devices and methods covered by patents, with the Government assuming financial responsibility for infringements. Thus, several companies, including AT&T, were able to spend the war years in the further development of radio systems.[10]

Although improvements certainly were made in circuit and tube design, the advance of greatest consequence effected during the war was the achievement of mass production of standardized tubes and sets. Previously, vacuum tubes and radios had generally been made one-at-a-time and were often one-of-a-kind (at least in their individual characteristics, if not their individual designs). Now,

[9] *Electrical World*, October 9, 1915, p. 790; About such pronouncements, Lloyd Espenschied, an engineer in the AT&T laboratory and a participant in the long-distance radio tests, later noted: "[T]he company was fearful that its own success [would] lead the public [to] believe that wires were about to be supplanted by radio, whereby they might sell their telephone stock! So, credit was given to no one, save a blanket commendation of its own workers, and the company began to preach the limitations of radio – the words spread all over creation, were not secret, were subject to interference, only a limited number of stations could operate in the common medium. . . . So much for the great radio events of 1915 – the company had a bear by the tail!" Quoted from a manuscript titled "Advent of Electronic Telephony," Box 8, *Lloyd Espenschied Papers*, Division of Electricity and Nuclear Energy, Smithsonian Institution.

[10] Lloyd Espenschied, "The Origin and Development of Radiotelephony," Institute of Radio Engineers *Proceedings*, vol. 25 (September, 1937), 1109.

the manufacturers produced thousands of identical devices, all with similar operating characteristics. Just as important for the nascent industry, thousands of military-trained radio operators were dispersed around the country after the war. These two factors – an advanced technology of mass production and a pool of potential employees, customers, and servicemen – provided an important spur to the commercial development of radio.

In 1919, AT&T resumed its former program by authorizing an expenditure of $500,000 for wireless development, including $360,000 for a marine transmitting and receiving station.[11] Through this program and through its early development of the deForest triode and various types of circuits, the Telephone Company by 1920 had acquired an important stake in radio technology. Although the company itself could not dominate the radio industry, it maintained a strong enough position to prevent any other company from doing so.[12]

And, of course, others would try. General Electric had been at work on the development of various aspects of vacuum tubes and radio since 1903. At that time the company had built a high-frequency alternator under contract to an inventor who wanted a carrier-current source for continuous-wave broadcast experiments. Further orders and further improvements followed and by 1907 the company possessed an alternator, designed and built by E. F. W. Alexanderson, that was an acceptable high-frequency current source for transmission. Alexanderson produced other important inventions for GE: an improved microphone in 1912, a better transmission antenna in 1916, and a circuit that helped eliminate static in 1918.

GE also undertook research that was initially directed at the incandescent lamp and x-ray tubes for medical purposes, but which bore on vacuum tube problems. First, they developed a new, more effective filament made from tungsten. Then, research on thermionic currents independently led the GE researchers to the realization that high-vacuum, pure electron discharge would make the deForest triode a more stable device. GE's Irving Langmuir filed for a patent on this use of high-vacuum; however, the application quickly became involved in a suit charging interference with an AT&T-sponsored patent application by H. D. Arnold on essentially the same advance. The interference continued for many years.

When the war brought the nascent radio industry to life, GE and

[11] Maclaurin, *Invention and Innovation in the Radio Industry*, 93.
[12] *Ibid.*

AT&T supplied the government with many radio parts. Westinghouse did so as well, although it had no patents of its own. Of course, during the war it needed none. The apparatus these companies produced in such quantity during the course of scarcely two years forcefully demonstrated the capabilities of the triode as the heart of a radio system. For example, as an oscillating source of broadcast waves, it proved to be far more versatile than the Alexanderson Alternator, which was not easily portable and limited to the longer waves (3000 meters and up). Thus it became apparent to GE, Westinghouse, Marconi, and other companies considering the manufacture of radio systems that in order to do so they had to acquire rights to the triode, which AT&T controlled so closely.

Maneuvering for rights to the triode began as early as 1915, when the Marconi Company, which held the patent rights to the diode, brought suit against deForest, claiming that the triode infringed the diode patent. DeForest brought countersuit, claiming that the Marconi Company was violating his patent by its use of a third element in some of its tubes. In September 1916, the court ruled in favor of and against both parties: (1) upholding Marconi, it declared the triode an infringement of the diode, and (2) that ruling notwithstanding, it decided that Marconi had, in fact, infringed the deForest patent. Neither of them could manufacture the triode! [13] The two companies later tried to work out a scheme whereby deForest produced the tubes and Marconi distributed them, but that arrangement was already falling apart when the exigencies of war made it unnecessary.

Thus, as the war drew to a close and patent rights were about to be reasserted, the stalemate had to be resolved, along with many other interferences over tubes and circuits. Each manufacturer, pushing its researches as far as possible, repeatedly encountered some device or circuit that was already patented by another.[14] Obviously, such a situation could not long endure.

Post-World War Developments

With the end of the war, the Marconi Company decided to convert its wireless telegraph services to voice transmission, and began negotiations with GE for the purchase of several large alternators. GE was eager to make some profit from the alternator, as it was in no position itself to exploit the device, having rights to neither the

[13] *Ibid.*, 85.
[14] N. R. Danielian, *AT&T: The Story of Industrial Conquest* (New York, 1939), 109–110.

diode nor triode. But the Marconi policy of refusing to sell patented equipment outright aroused great antagonism in its largest customer, the U.S. Navy, which found itself dependent on a foreign company. Thus, when news of the impending agreement between GE and Marconi reached the higher Navy staff levels, there was strong sentiment against its consummation. With the assurance of a lucrative contract as one incentive, a high-ranking officer proposed to the GE directors that rather than selling alternator rights to the Marconi Company, they should instead purchase the American Marconi Company, and thus acquire all the Marconi patent rights for the United States. For reasons that are not entirely clear, GE agreed.

The British Marconi Company, aware of U.S. Government opposition to its control of American radio and cognizant of the obstacles the Government could place in its path, shortly came to terms. In October 1919, GE bought the British company's stock in American Marconi (about half the total) and the American stockholders agreed to a conversion of their holdings to securities of the new Radio Corporation of America (RCA), which GE formed to take over the Marconi assets. GE contributed over $3,000,000 to the venture, receiving only RCA stock in return. The plan from the outset was that RCA would not manufacture radio sets or tubes; these would be supplied by GE and sold through RCA. The two companies exchanged all patent rights, including those for inventions made up to 1945. This still left deForest's indispensable triode outside the new combine, so as one of its first acts, RCA in July 1920 negotiated a cross-licensing agreement with AT&T, which bought RCA stock valued at $2,500,000.

The companies wrote into the contract certain interesting restrictions concerning patent rights. The restrictions did not cover particular patents, but rather defined the fields of exploitation open to each company. AT&T received exclusive licenses under everyone's patents in wire telephony and telegraphy and certain specified rights to radio in conjunction with the wire telephone network. RCA and GE similarly acquired rights to use all parties' patents in wireless telegraphy, in international two-way radio communication, and "to make, use, lease, and sell all wireless telephone apparatus for amateur purposes".[15]

Not to be left out, Westinghouse had been busy since early 1920 building a formidable patent position by the purchase of strategic

[15] "License Agreement, General Electric Company and American Telephone and Telegraph Company, July 1, 1920," Article V, Paragraph 4, Section d,3. In Federal Trade Commission, *Radio Industry* (Washington, D.C., 1923), 134.

rights. Westinghouse concentrated on circuits rather than on tubes, and by October of that year controlled the patent rights to two of the most sophisticated detection systems, the heterodyne and superheterodyne circuits. There is some question as to what use Westinghouse would have been able to make of them. GE, for example, immediately contested the superheterodyne patent when Westinghouse acquired it, claiming that it infringed an earlier GE patent. Still, these and other patent rights that Westinghouse had bought were quite valuable in producing a receiver of advanced design, so GE invited the Pittsburgh electrical manufacturer to join the cross-licensing agreement. On June 30, 1921, the GE-RCA agreement was amended so that RCA purchased 60 per cent of its radio apparatus from GE and 40 per cent from Westinghouse.

Thus, by the middle of 1921 a formal structure had been carefully built in an industry that had been chaotic less than two years before. Primarily responsible for this was Owen D. Young, who, as legal counsel for GE, had handled the purchase of American Marconi and the establishment of RCA and who, as head of the new Corporation, had successfully brought AT&T into the fold. To Young goes the credit for bringing the different companies with different interests into an agreement that would probably have survived to everyone's satisfaction, had the market not changed significantly soon thereafter.

Based on its goals and its assets, each party to the agreement got what it desired and what it thought it deserved. We can see this best by considering these goals in relation to the expected market. To do this, we must divide the companies into two main groupings: GE, RCA, and Westinghouse on the one hand (hereafter often referred to as the "RCA group") and AT&T, with its manufacturing subsidiary, Western Electric, on the other. The first group intended to concentrate on wireless communication, the latter on wired or closed-circuit transmission. While the original goals of RCA were rather modest by industrial standards, they covered most areas of radio expected to be profitable. When it took over from American Marconi, RCA was engaged solely in wireless telegraph service, both transoceanic and ship-to-shore. Almost immediately, however, the Corporation began to sell apparatus to "amateurs" – those who experimented with radio as tinkerers, serious inventors, or scientists and those who broadcast and/or received as a hobby. When Westinghouse entered the agreement in 1921, it was to take a portion of this market. Since the sales of amateur apparatus had grown considerably in the years after the war and had prospects of continuing to do so, GE's concession of 40 per cent of RCA's sales to Westinghouse was preferable to an entangled legal situation; for, while the giants engaged in their internecine struggle, others might sneak in to grab a substantial share of the market.

One other advantage brought to the RCA group by the agreement is so obvious as scarcely to need mention but so important as to require emphasis: protection from the most potent of competitors. As recently as April 1919, AT&T had been seriously considering a foray into the sphere of transoceanic and ship-to-shore communications.[16] Though AT&T lacked certain important patent rights held by the others, there was no assurance that the rapid pace of scientific and technological advance would not bring it other equally or more valuable rights in the near future. Thus, by keeping the Telephone Company out of wireless communications, the RCA group protected itself from what in 1920 and 1921 appeared to be its only substantial rival. Expansion into the radio apparatus market, however, was natural for companies like GE and Westinghouse. Capable of manufacturing products of high technological sophistication, they were already diverse; so that adding a new product line proved to be relatively easy. Vacuum tubes, for example, could be fabricated using machinery similar to that for making light bulbs. Furthermore, the companies were committed to this type of expansion. To maintain corporate integrity in the long term, they needed to secure new lucrative markets to compensate for those that would ultimately deteriorate for technological or competitive reasons. With research, production, and distribution organizations already established, GE and Westinghouse had an advantage from the start, and with corporate security at stake, they had a reason to enter and to try to dominate the market for radio apparatus.

For its part, AT&T gained security in pursuit of its outstanding corporate goal, monopoly of public-service, two-way telephone communications. AT&T had been fighting to maintain that monopoly ever since the original Bell patents ran out in the 1890s. First, it had protected itself by acquiring patents on telephone exchanges, which were necessary for the building of telephone networks. Then, it sought to dominate long-distance telephony by its acquisition of the Pupin and deForest patents. With absolute control of long-distance lines, it could refuse to connect to competing local systems, smothering them in isolation. Finally, in obtaining an exclusive license under the patents of potential competitors to "make, use, lease and sell wireless telephone apparatus connected to or operated

[16] *FCC Investigation*, Exhibit #2112, pp. 27–28.

as part of a public service telephone communication system, whether wire or wireless," [17] AT&T stymied the competition of two-way radio with its extensive wire network. While it had no need, desire, or intention to further develop two-way radio except as an adjunct to the wire system, by conceding minor profits to the RCA group in the form of ship-to-shore and transoceanic revenues, it blocked the "perennial gale" of competition before the first telephone line was blown over.

It should be emphasized that research leading to patents in radio helped to construct this new shield. In a memorandum after the 1920 agreement, Frank Jewett, AT&T's new chief engineer, told vice president J. J. Carty: "As I look back on it, it seems to me that this enlarged and enhanced position [in radio research] played no small part in enabling us to reach our present satisfactory understanding with the General Electric Company and the Radio Corporation of America and that if we never derive any other benefit from our work than that which follows the safe-guarding of our wire interests we can look upon the time and money as having been returned to us many times over." [18]

The Market — Destruction and Reformation

The careful work that went into building the industrial structure came to naught within a few years. As quickly as it had been created, the structure was destroyed by developments unforeseen as late as mid-1921. While technological innovation contributed to the crisis, it did not directly precipitate it. Rather, new technology brought new commercial possibilities, and the sudden realization of these possibilities initiated the collapse.

Usually credited with operating the first radio "station," Westinghouse began regular broadcasts in Pittsburgh in November 1920. The company expected station KDKA to stimulate sales of Westinghouse apparatus, and it intended to maintain the broadcasting operation, which was only on the air for a few hours each evening, out of profits from radio-related sales. With these sales growing rapidly in Pittsburgh, Westinghouse decided to open stations in the Boston and New York areas in the fall of 1921. WBZ, Springfield, Mass., went into operation in September, and WJZ, Northern New Jersey and New York, began broadcasting in October of that year. RCA actually operated a station in competition with WJZ for a few

[17] "License Agreement, July 1, 1920," Article V, Paragraph 4, Section e,1. In FTC, *Radio Industry*, 135.

[18] FCC, *Investigation of the Telephone Industry*, 209.

months in the winter of 1921–1922, but decided very shortly to join forces with Westinghouse and pay half of WJZ's operating expenses when the directors perceived that their interests coincided.[19]

RCA and its partners found themselves ill-prepared for the surge of orders for receivers and for transmitters from new groups who wished to become broadcasters that followed the opening of these stations. Under the generally accepted definition of the post-war period, the new-found radio audience consisted of "amateurs," and according to the agreement, the RCA group could "make, use, lease and sell" radio apparatus for them. Thus, GE and Westinghouse responded by beginning production of radio apparatus and by speeding up work on the development of improved receivers with an eye to getting them into production as quickly as possible, for in that early phase of the business the company that could supply the demand would gain a priceless advantage.[20] The agreement, however, prohibited the RCA group from selling transmitting apparatus to others.[21]

As long as the amount of apparatus to be sold remained relatively small, and as long as broadcasting appeared to be on an expense rather than a profitable undertaking, all parties to the contract were content with this arrangement. But by late 1921, it had become apparent that this was no longer the case, and AT&T had the most reason for concern. While it had secured its wire telephone plant, AT&T had passed up a share in new profits to be made in the manufacture and sale of radio sets, an art to which its engineers had made significant contributions. True, it had sole right under the contract to sell broadcast equipment; but the number of transmitters to be supplied numbered at best in the thousands, whereas the potential market for receivers was many millions. If it wanted to realize substantial profits, AT&T had two choices, either to make broadcasting pay or to force its way into the receiver market. It decided to try the first, though it kept its options open on the second by continuing research and development work on all facets of the radio art.

AT&T's attempt to dominate broadcasting presents the fascinating case of a company using its monopoly in one field as leverage to

[19] Gleason L. Archer, *Big Business and Radio* (New York, 1939), 245.

[20] *Electrical World*, vol. 79 (March 4, 1922), 419. An historian of radio has written: "The least expensive vacuum-tube receiver [1922] was a one-tube set manufactured by Westinghouse and marketed by the Radio Corporation . . . for $79.50. General Electric receivers employed three tubes and sold for $250. And even at these prices for simple sets, it was a year before suppliers were able to catch up with orders on hand." Quoted from Donald McNicol, *Radio's Conquest of Space* (New York, 1946), 341–342.

[21] "License Agreement, July 1, 1920," Article V, Paragraph 4, Section 2. In FTC, *Radio Industry*, 134.

force others out of an adjacent market area. Since commercial broadcasting had not been anticipated at the time of the 1920 agreement, the agreement contained no provision specifically granting rights for the manufacture and sale of broadcasting equipment to any party, though the RCA group was expressly prohibited from doing so. This was a strange arrangement, to be sure. The prohibition had been inserted "for the protection of the Telephone Company," apparently without full consideration as to the possible consequences. Under the circumstances, AT&T held the right to lease or sell broadcasting apparatus only under its own patents; but by 1922, these were insufficient for the purpose.

By late 1921, the AT&T directors realized the importance of blocking the proliferation and influence of stations established by rivals, especially those of the RCA group. Thus, their first act was defensive in nature. Using both their patent rights under the agreement to all activity related to wire telephony and their position as a monopoly of that service, they prohibited the RCA group from using telephone circuits to send signals from remote pick-ups, such as sporting events and public lectures, to the broadcasting studio.[22] Since most stations at the time considered broadcasting of this type to be an important function, AT&T's action represented a substantial threat. The Telephone Company continued this practice, denying pick-ups to the RCA group and even to its own licensees (those who bought its equipment) if an AT&T station opened in the vicinity. While Westinghouse, GE, and RCA could build transmitters for their own use under the contract, others had to rely on AT&T for equipment. None too anxious to outfit a troop of competitors, the company refused most requests for transmitters,[23] and those it did supply were granted only under a license stipulating that they could not be operated for hire – no advertising! As further dissuasion, AT&T set the cost of a transmitter and license very high: $9000 for a low power model (100 watts), $12,500 for a larger one (500 watts).[24]

By mid-1922, AT&T had decided on a course of action that would make broadcasting pay. Beginning in August, the company opened its own stations, intending ultimately to link them together as a network. Revenues were to be generated by selling air time, or broadcasting for toll, much like telephone service. With station WEAF about to be opened in New York, AT&T made the following an-

[22] *FCC Investigation*, Exhibit #289, p. 78.
[23] William P. Banning, *Commercial Broadcasting Pioneer: The WEAF Experiment, 1922–1926* (Cambridge, Mass., 1946), 55.
[24] Internal company memo, "Radio Telephone Broadcasting", quoted in *Ibid.*, 74.

nouncement: "Anyone desiring to use these facilities for radio broadcasting should make arrangements with Mr. Drake, general commercial manager. . . . Mr. Drake can advise fully with reference to all particulars concerning the use of the station including information as to the periods of operation and charges thereof. He is also in a position to give helpful suggestions with reference to the arrangement of programs and the kind of subject matter which it is thought will be most acceptable to the radio audience."[25] But no one applied!

The company expected to produce its own broadcasts for a few days until, it was believed, others would pay to take over. Finally, twelve days after opening, WEAF sold its first air-time, but only as an advertisement, not as an entire show.[26] Thus, a misconception of the public demand for broadcasting time put the company in a position that it had roundly condemned as undignified and inappropriate when it had refused to let others broadcast for hire with AT&T equipment. That winter, the company decided on another policy, best explained in the announcement sent out to the Associated Bell Companies:

> In each locality an important group of people will get together and form a broadcasting association. In that group of people should be the type that the community looks to as being the leaders of that community. In it I would expect to see the chamber of commerce, the important news papers [sic], the department stores. . . . For that association we would erect, own, and operate a broadcasting station; they to provide all the programs; they to give the public what the public desires but we to have the latest facilities known to the art and all of the things that go with them including remote-control lines and speech-input equipment. That station is to be operated by the Bell System under definite guarantees from the association as to expenses plus a reasonable return.[27]

By May 1924, AT&T had opened two other stations, one in Philadelphia and one in Washington, D.C., and operated them occasionally as a network. However, the idea of community stations never caught on, the AT&T management having been unable to organize local broadcasting associations.

AT&T was at a loss as to how to operate a broadcasting network at a profit, but its management harbored no doubts that such a monopoly was wise. In the same bulletin to the Associated Companies quoted above, an assistant vice president wrote:

[25] Archer, *Big Business and Radio*, 54.
[26] Gleason L. Archer, *History of Radio to 1926* (New York, 1938), 276.
[27] FCC, *Investigation of the Telephone Industry*, 388.

> We have been very careful, up to the present time, not to state to the public in any way, through the press or in any of our talks, the idea that the Bell System desires to monopolize broadcasting; but the fact remains that it is a telephone job, that we are telephone people, that we can do it better than anyone else, and it seems to me that the clear, logical conclusion that must be reached is that, sooner or later, in one form or another, we have got to do the job. . . .
>
> I may state to you that I have talked this idea over with Messr. Thayer [president], Gifford and Bloom [vice presidents] and each of them think [sic] it is a proper set-up.[28]

However, AT&T's aggressive tactics engendered strong opposition. Frequent newspaper and magazine editorials condemned the nascent monopoly.[29] Rumblings came even from the highest levels of government. Secretary of Commerce Hoover was clearly referring to AT&T when he wrote in 1924: "I can state emphatically that it would be most unfortunate for the people of this country to whom broadcasting has become an important incident of life if its control should come into the hands of any single corporation, individual, or combination."[30] Such opposition to a broadcasting monopoly posed a further threat, as it could create poor relations with AT&T customers and perhaps ultimately lead to opposition to an AT&T *telephone* monoply. As one apologist for the company put it: "Positive action to prevent other broadcasting operations . . . of course, 'very much involved public relations'. For one thing, such could easily be construed as an arbitrary and selfish effort to deprive the public of entertainment on which it was relying. And any such local feeling would necessarily react on telephone service itself which depends, for general usefulness and for growth, upon cooperative and understanding relations between company and customer."[31]

Had AT&T succeeded in its early schemes to organize and finance broadcasting, the public would have been presented with a *fait accompli*, a new monopoly. However, constrained by its directors' social and business attitudes, the company failed to consolidate its position before a groundswell of opinion made such an eventuality impossible. By 1924, it had become apparent that broadcasting would not be dominated by any one concern. AT&T had missed its chance for profit in this market, its first choice; so it turned to its second.

The greatest profits in the new radio industry came, of course, in

[28] *Ibid.*, 389.
[29] For example, see *Radio Dealer*, vol. 1 (April, 1922), 30.
[30] Maclaurin, *Invention and Innovation in the Radio Industry*, 114.
[31] Banning, *Commercial Broadcasting Pioneer*, 136.

the sale of radio receiving sets to the public.[32] The 1920 agreement placed this market in the hands of the RCA group, providing that the Telephone Company had "no license under this agreement to make, lease or sell wireless telephone apparatus *except* as part of or for direct use in connection with transmitting apparatus made by it." (The exception here provided for two-way radio used as part of the Bell communications system.) However, AT&T had built up a momentum in radio and vacuum tube research and development that by early 1924 placed a superior receiver in the hands of the company management. Seizing on the exception, AT&T began to sell sets with fixed frequencies of reception tuned to AT&T broadcasting stations. AT&T even went so far as to have one installed in the White House, an action which caused great consternation at RCA.[33] In addition, the Telephone Company began to claim that once the signal had come in the antenna and passed through the detector tube (or tuner) it was no longer wireless but wire telephony and thus open to them under the contract. An RCA internal memorandum of the period noted:

> The Telephone Group is actually manufacturing and selling loudspeakers, amplifiers, vacuum tubes, head telephones and other accessories, with full knowledge that such items are being used for the purpose of radio broadcast reception (our exclusive field). The Telephone Group contends that the items enumerated constitute a loud speaker device which it had been selling in the past for wire telephone purposes and therefore feels at liberty to continue to sell even though such devices are attached to and become part of radio broadcast receivers.
>
> The practical meaning of this activity on the part of the Telephone Company is that its subsidiary, the Western Electric Company, is selling radio devices or parts which constitute about 80 to 90% of a complete broadcast receiver.[34]

Because such a situation was intolerable to the RCA group, it initiated official arbitration as provided for by the 1920 AT&T-GE agreement.

Meanwhile, competition from other quarters continually dogged the RCA group in its drive to control the receiver market. Others'

[32] Sales of sets grew rapidly. The gross sales figures for RCA in its first three years in the receiving set market were: (1921) $1,469,000; (1922) $11,286,000; (1923) $22,465,000. Figures from Archer, *Big Business and Radio*, 139.

[33] The *1922 Annual Report*, Engineering Department, Western Electric Company (copy at Bell Telephone Laboratories, Murray Hill, N.J.), p. 44, stated: "During the year development work was carried out on receiving equipment and this resulted in the design and manufacture of a number of different types of receiving equipment. The types of receiving sets which have been designed cover the whole field efficiently and we are in a position to meet any reasonable commercial demand." On the set in the White House, see letters from D. Sarnoff to F. P. Guthrie (April 30, 1924) and to J. G. Harbord (May 19, 1924), quoted in Archer, *Big Business and Radio*, 143, 145.

[34] Archer, *Big Business and Radio*, 78.

sets had appeared on the market as early as 1922, many with components free from RCA patent control. This was possible because when deForest sold the triode patent rights to AT&T in 1914, he had insisted on retaining a license for himself, and when the Fleming diode patent held by RCA ran out in 1922, deForest was free to produce triodes for sale in competition with RCA. While the RCA group controlled most of the important circuit patents for receivers, some others were workable, and one in particular, the neutrodyne, gave very good results with the deForest Company's tubes. The first of these sets appeared in the spring of 1923, and in all, fourteen companies were soon manufacturing neurodyne-equipped receivers.

Despite the impressive research departments at GE and Westinghouse, the RCA group had difficulty getting up-to-date sets on the market. The problem generally occurred at the "development" stage, where engineers designed new receivers and prepared them for full production. Since two essentially independent companies produced apparatus for sale under a common label it became necessary to standardize products and, in certain instances, to coordinate production. As one commentator remarked, "RCA . . . found itself under the necessity of harmonizing the views of three research and engineering departments for design and construction and, at the same time, of harmonizing those views with the public demand as determined by the sales department." A GE radio researcher noted, "RCA required that the two companies standardize their designs so that the casual purchaser could not tell the difference. Sometimes this 'standardization' took as long and cost as much as the original development." [35] The inter-company administrative structure worked so slowly that even after RCA approved final designs, several months usually passed before the first product deliveries. Since the research needed to produce an acceptable battery-powered receiver had been largely accomplished by 1922, the competition that grew up in the first few years consisted of rather limited technological advance or development. Science-oriented research departments such as those at GE or AT&T could produce break-throughs that destroyed the structure of an industry; but for the latest-and-best-model competition within a temporarily stable structure, the far less sophisticated development staffs of any number of manufacturing enterprises proved to be just as effective.

Thus, in the early 1920s the RCA group received stiff competi-

[35] *Ibid.*, 25; William C. White, "The Story of Electronics Development at the General Electric Company," unpublished MS, c.1955 (copy at Division of Electricity, Smithsonian Institution), p. x-c-2.

tion from companies capable of usurping its lead in product design. Unable to launch an effective offensive, RCA decided to fall back on its patents for protection. It challenged the neutrodyne circuit patent, claiming in 1924 that certain aspects of the circuit had been anticipated by GE and Telephone Company workers as early as 1913. RCA contended, not that the neutrodyne patent was invalid, but that like the triode as an extension of the diode, the neutrodyne built on circuits to which RCA held patent rights, and that those wishing to use the neutrodyne circuit had to procure a license from RCA as well. RCA lost the initial decision in district court but pressed the case and won in the Court of Appeals in 1927.[36] Thereafter, RCA controlled all of the major circuit patents for radio receivers, and any company wishing to produce sets needed an RCA license. RCA patent royalties increased from $136,000 in 1926 to $3,300,000 in 1927 as a result of the decision.[37] Some smaller companies went out of business because of the 7.5 per cent royalty rate, which forced up their prices and took away their competitive edge. Others did not even get a chance, for RCA restricted the granting of licenses to "$100,000 customers." [38] To maintain its position as the dominant force in the field, RCA further required that all licensees give it an option to acquire any radio-related patents they might secure. There was no choice but to comply.

An invention that came out of the Westinghouse laboratory put a lock on the RCA patent position in 1928. Up until that time it had been necessary to operate all radio sets with direct current, which was usually supplied by batteries, although a special "battery-eliminator" could be used to rectify alternating house current, which, applied directly to vacuum tube filaments, created a hum in the receiver. The invention of vacuum tubes with indirectly heated cathodes and special circuits for their use permitted radios to be plugged directly into the home electrical system, a saving in the cost and inconvenience of batteries and battery-eliminators. This innovation led to a new round of radio set sales beginning in 1928, and the $388,000,000 retail sales of that year swelled to $592,000,000 in 1929.[39]

Such figures show that great profits could be made in the sale of receivers. Although the method of market control had not become clear by 1923, both RCA and AT&T were then aware that

[36] Maclaurin, *Invention and Innovation in the Radio Industry*, 129.
[37] *FCC Investigation*, Exhibit #2112, p. 57.
[38] Maclaurin, *Invention and Innovation in the Radio Industry*, 136.
[39] John F. Rider, "Broadcast Receiver Equipment Then and Now: History of Receiver Design", *Radio News* (May, 1931), 1072; T. T. Eoyang, *An Economic Study of the Radio Industry* (Doctoral dissertation, Columbia University, 1936), 75.

patent rights could be used to great effect, and that their laboratories would supply them with new weapons in the years to come. While the arbitration first begun in 1923 aimed at interpreting the 1920 license agreement and thus clarifying the rights of the parties under it, AT&T wanted an entirely new contract that would redefine the market and restructure the industry. Since the 1920 agreement the RCA group had continued to expand its broadcasting interests, violating the spirit if not the letter of that contract. RCA and its partners claimed that the provisions permitting them to operate their own stations implicitly conferred the right to do so profitably if possible, while the agreement envisioned a more restricted role. The RCA group maintained extensive research and development programs in radio transmission as well, taking out patents that, under the contract, did not revert to AT&T for that purpose, and thus threatening the Telephone Company's position.

The details of moves and counter-moves made during the course of arbitration, negotiation, and settlement of the dispute are voluminously recorded in G. L. Archer's *Big Business and Radio*. The initial instigator of contractual scrutiny was actually RCA, who, perturbed by AT&Ts refusal to supply pick-up wires or to meet its other demands concerning broadcasting, hired an outside attorney to make an "impartial interpretation" of the agreement in mid-1922. The attorney's reading of the often ambiguous document left neither side satisfied. In fact, it led to the realization by both parties that areas each had assumed to be closed off might be open under one or another interpretation of the contract, which led to emboldened thrusts into the market such as that by AT&T described above.

Article XIII of the 1920 GE-AT&T agreement stated that either party could call for formal arbitration if "any difference . . . shall arise which the parties are unable to adjust between themselves."[40] In May 1924, by mutual consent, they undertook formal arbitration. The main points in dispute were (1) whether the RCA group could use AT&T pick-up lines for broadcasting (that is, whether these lines were "wire telephony" and thus strictly in the Telephone province, or whether they were an adjunct to the RCA group's right to maintain its own stations) and (2) whether AT&T could manufacture radio sets pre-tuned to receive only its own stations. Since AT&T hoped to eventually monopolize broadcasting, a favorable decision on the latter question would have opened the way into the entire receiver market.

In a preliminary "Draft Decision" dated November 13, 1924, the

[40] "License Agreement, July 1, 1920", Article XIII. In FTC, *Radio Industry*, 138.

arbitration referee found virtually all disputed points in favor of RCA. Momentarily stunned, the Telephone Company quickly regained its composure, and in March 1925, produced a legal opinion that obviated the need to put the draft into final form. The crux of the new argument was that, as construed by the referee, the agreement violated the Sherman Antitrust Act because it constrained the Telephone Company from selling radio sets even though it could do so entirely under its own patents. Much to its dismay, the RCA group could find no counterpoint, and negotiations entered a new stage. Through the summer of 1925, AT&T held the upper hand in bargaining, because the dispute no longer concerned whether AT&T could enter the radio market, but what share of that market it could have.[41]

Then in the fall a stunning reversal occurred. The previously mentioned Arnold-Langmuir interference suit on the high-vacuum tube, which had been in litigation since before the World War, was decided by the U.S. Court of Appeals in favor of Langmuir and RCA. Since virtually all phases of the radio art required high-vacuum tubes, a new situation arose: blocked in the use of the Arnold patent, AT&T could no longer produce receivers completely under its own patents, which nullified the antitrust argument. Nor, it should be noted, could AT&T any longer sell broadcast equipment![42] Appeal was made to the Supreme Court as a matter of course, but the possibility of a final adverse decision now loomed large in the dispute.[43]

In the summer of 1926 the two groups finally came to terms. Repulsed in its battle to enter radio sales and unsuccessful for a number of reasons in its attempt to monopolize broadcasting, the Telephone Company retreated to its main line of defense, maintenance of its telephone monopoly. In the new license agreement, AT&T received the telephone monopoly reassurance and acquired control of all two-way communication systems, including ship-to-shore and transoceanic radio telephony, a market area which had gone to GE in 1920. The agreement specifically stated that members of the RCA group had the right to use telephone lines for station pick-up and for network transmission; in fact, it obligated them to do so (as opposed to stringing their own or using Western Union's), thus

[41] Archer, *Big Business and Radio*, 207–208.

[42] *DeForest Radio Co. v. General Electric Co.*, U.S. Circuit Court of Appeals for the Third Circuit, 44 F. (2nd) 931 (1925). This point seems to have been missed by Archer and by those who, like Maclaurin, used him as a source.

[43] In fact, the Supreme Court ruled in 1931 that high-vacuum was not a patentable invention and therefore all patents relating to it were invalid. *DeForest Radio Co. v. General Electric Co.*, 283 U.S. 664.

ensuring AT&T of what was to become a substantial source of revenue.[44] For its part, the RCA group accomplished what it had wanted all along: domination of the radio receiver market and the right to use telephone lines as an adjunct to broadcasting for hire. AT&T sold its stations to RCA and agreed to stay out of broadcasting. The two groups agreed to compete in the least financially rewarding market area, the sale of broadcast equipment.

Thus, in mid-1926 a new structure had been established in the radio industry. The preamble of the new agreement stated that because, in 1920, "the art in certain fields dealt with in said [1920] agreement had not progressed to a point at which it was possible fully to comprehend the problems involved, disputes have arisen between the parties as to the meaning of various provisions of said agreement," and as a result "certain provisions of said agreement are not, as a practical matter, workable in the present state of the art." Unless the parties settled these disputes and made the agreement workable in practice, "progress in the fields dealt with in said agreement will be greatly hampered and delayed." This really meant that the companies were making a new attempt to define more clearly the boundaries of the exclusive domain of each party so that none would overstep the others' territory.[45]

Research, Patents, and Competition

Industrial research consists of many activities. Of particular importance are (1) research leading to the discovery of laws of nature, often in conjunction with the creation of new or improved products or processes, and (2) development work to better utilize already-known principles for the modification of existing devices or to reduce the costs of production. Each of these activities has its place in the research scheme, and each is important in its turn. The "end product" in both cases is often a patent, which means that to comprehend the directions of industrial research we must understand the corporate value of these patents. Our case study has already revealed extensive use of patent rights to gain control of a market, and if we look more closely at the motives of the companies that acquired and held these rights we should be able to understand why they undertook extensive research operations.

Patent rights were of primary importance to the Telephone Company, which used them quite effectively to extend its control of

[44] *FCC Investigation*, Exhibit #2112, p. 30. The AT&T revenues for this service amounted to over $28,000,000 between 1926 and 1935.
[45] Danielian, *AT&T*, 127.

telephone service. It first relied on the original Bell patents, then on exchange patents, then on those for long-distance telephony. In 1907, a transition year for research at AT&T, the president of the company testified in court: "One of the first things that was fully developed in our minds was the necessity of *occupying the field*; not only that but of surrounding ourselves with everything that would protect the business. . . . Just as soon as we started into the district exchange system we found out that it would develop [sic] a thousand and one little patents and inventions with which to do the business which was necessary, and that is what we wanted to control and get possession of."[46] Before 1907, "getting possession" of patents usually entailed buying them, but after that date the company began to rely on in-house production and acquired only those patents vital to its interests, such as the deForest triode and feedback circuit. "A thousand and one little patents" meant just that; AT&T took out patents on every minor modification or advance which qualified under the patent laws as invention. The company realized that such patents might become useful later.

Pursuit of patents to achieve corporate goals exerted a major directive force on the work of the research departments. If a researcher had a particular objective in mind – for example, the improvement of a tuning or amplifying circuit – he often continued work on a design that he would have preferred to abandon simply because the company wanted patents not only on better circuits but on all circuits that qualified under the patent laws.[47] This, of course, would keep them out of the hands of competitors. A 1912 exchange of letters between J. J. Carty, an AT&T vice president, and C. E. Scribner, the chief engineer of its subsidiary, Western Electric, makes this policy explicit:

> [Directive:] Naturally, I would not wish to delay the practical development of the repeater [deForest triode] but I should hope that there would be some way whereby alternative methods might be thought out and patented. I think it is important that you go as far in this direction as you can. . . . I think that the patenting of the alternatives is a very important feature of our work and I suggest that the routine covering such matters be looked carefully over and strengthened if necessary, the idea being that as far as possible on all new developments of substantial importance, we patent the alternative methods.
>
> [Response:] I sometime ago pointed out . . . the desirability of doing

[46] *FCC Investigation*, Exhibit #2112, p. 20. Italics added.

[47] For example, R. A. Heising of AT&T applied for three patents on modulating circuits within a year, one of which he acknowledged to be a minor modification to that of another AT&T researcher. See R. A. Heising patents 1,137,315; 1,199,180; and 1,343,562. Heising's statement in "Modulation in Radiotelephony", Institute of Radio Engineers *Proceedings*, vol. 9 (1921), 316.

> what you propose and, after discussion with Mr. Thayer [AT&T President], it was decided that we should adopt the policy of developing alternate methods to the more important lines of development, and carry on this alternative to a point where patents might be secured. . . . The plan is to keep the actual physical tests down to the point where simply enough information is gained to justify the prosecution of a patent application.[48]

Even though, as the Western Electric chief engineer pointed out, the laboratory kept the "actual tests" to a minimum, the process, including patent application and attorney's fees, had to be expensive. Why, we may ask, was it worthwhile to pursue patents that the company did not intend to use in production?

To answer this question we must look at how the companies actually exerted their various patent rights. There are essentially three ways to use patents for purposes other than production. The first is the most obvious: *to prevent competition.* AT&T's "thousand and one little patents" on the district exchange system kept competitors out of the market by making its product (the telephone tied into an extensive exchange system) distinctly superior to what the others could offer; for, while there are many ways to build an exchange, AT&T held patent rights on some essential aspect of all of them. When its exchange patents ran out, AT&T sought to similarly control long-distance telephony so that any potential competitor's district exchange system would be kept isolated.

Secondly, the holding of important patents can be used *to prevent competitors from acquiring a strong patent position through their own research.* RCA accomplished this by inserting in its license contracts an option to acquire rights under all radio-related patents developed by its licensees. This, together with the wide patent coverage of the RCA group, destroyed the licensees' incentive for original research through the 1930s.[49] Other ways to use patents to stymie competitors' patent positions were more subtle but no less important. Sometimes, inventions which originally appeared to have no strategic value became quite useful later. An excellent example brought out in the narrative is RCA's use of pre-World War patents (which had been of no consequence for ten years) in a successful move to regulate the use of the neutrodyne circuit in 1924. Frank Jewett, chief engineer of AT&T after the World War and head of Bell Telephone Laboratories from its founding in 1925, discussed this type of strategy in 1932:

[48] *FCC Investigation*, Exhibit #2112, p. 2.

[49] Maclaurin, *Invention and Innovation in the Radio Industry*, 155. The license agreement had a further advantage for RCA: it assured that the licensees recognized the validity of RCA patents.

While it is obvious that the basic inventions which control a large new field are not made very often, one can never tell where or when they will crop up. . . . When . . . [they] do, possession of a strong and unmortgaged patent position on the part of an industry needing rights is frequently the most powerful and sometimes the only available weapon for securing those rights. *Ability to stop the owner of a fundamental and controlling patent from realizing the full fruits of his patent by the ownership of necessary secondary patents may easily put one in the position to trade where money alone might be of little value.*[50]

This statement brings out the third "non-productive" use of patents: *trading*. The story of the creation and re-creation of structures in the radio industry has of course, been essentially one of trading, although the companies traded market area rights to all patents rather than all rights to particular patents. Jewett made clear just how worthwhile he believed AT&T's research work in radio to have been – even though most of it eventually wound up in the hands of the RCA group – when he testified in 1936: "I told you earlier . . . that I considered this expenditure in radio one of the most profitable expenditures the American Company [AT&T] had ever made, and I still do. . . . It has created a patent situation that has protected the Associated Companies, either directly or through the American Company from the attacks of people who would have otherwise obtained patents or inventions which relate not only to wireless and radio, but to all the wire plant." [51]

Perhaps nowhere has the big business attitude towards patents, research, and the trading of patent rights been more fully or forcefully expressed than in an AT&T internal memorandum written in 1927 and revealed in the Federal Communications Commission's investigation of the telephone industry:

The regulation of the relationship between two such large interests as the American Telephone and Telegraph Company and the General Electric Company and the prevention of invasion of their respective fields is accomplished by mutual adjustment within "no man's land" where the offensive of the parties is recognized as a natural defense against invasion of the major fields. Licenses, rights, opportunities, and privileges in connection with these competitive activities are traded off against each other and interchanged in such a manner as to create a proper balance and satisfactory relationship between the parties in the major fields. . . .

It seems obvious that the best defense is to continue activities in "no man's land" and to maintain such strong engineering, patent and commercial situation [sic] in connection with these competitive activities as to always have something to trade against the accomplishment of other parties.

[50] *FCC Investigation*, Exhibit #1946, pp. 15–16. Italics added.
[51] *FCC Investigation*, Exhibit #2110, p. 99.

> If the American Telephone and Telegraph Company abandons its activities in the commercial competitive field and other potentially competitive interests continue their activities, it means that they will carry their offensive right to the wall of our defense and our trading must be in our major field against activities in their outlying commercial fields. The nearer the trading can be carried to the major field of our competitors the more advantageous trading position we are in.[53]

Thus, based on the philosophy that the best defense was a strong offense, the rational course of action was to develop a strong patent position as close as possible to a competitor's main field of commercial interest. And this is exactly what AT&T did in the six years between license agreements. It only failed to acquire a large share of the receiver market and control of broadcasting because of an adverse legal decision on crucial patent rights in the former case and because of inept handling of an ill-conceived market in the latter. But, it is important to note that, although AT&T had failed dismally, it had only to retreat as far as the lines of its monopoly on telephone service, and to an extended monopoly at that. The RCA group for its part, maintained strong research and patent positions relating strictly to broadcasting even while the first agreement was in force – when it had few rights to exploit broadcasting commercially. The effect of this policy was to keep pressure on AT&T and to provide material should trading have become necessary or desirable.

Conclusion

Research-generated patents were used during the course of the early commercial development of radio for the purpose of gaining market control, just as they had similarly been used in petroleum refining and electrical equipment. Companies used them both offensively and defensively: either to gain concessions from competitors, or to short-circuit new inventions that might have had disruptive possibilities. In almost every case, research became a more important factor in competition for monopoly control than in competition for shares of the market.

With the arrival of high-technology industries came a new conception of the uses of industrial research. But, although few have argued that industry undertook research from a sense of corporate altruism, the fact is that industrial research has most frequently been perceived in terms of "progress." Standard works such as H. R. Bart-

[53] FCC, *Investigation of the Telephone Industry*, 210.

lett's "The Development of Industrial Research in the United States" treat its growth in simple quantitative terms, as if twice as much money and twice as much manpower has meant twice as much scientific and technological advance.[53] However, while some large laboratories have given relatively free reign to a few of their best scientists, it is important to realize that the majority of men and resources has been devoted to pursuit of those types of patents useful for market control. In a significant sense, then, science has been compromised to the extent that research funds and researchers have been sacrified to the essentially unproductive work needed to gain or maintain monopoly position, and pursuit of patents has been at the heart of this process.

[53] Howard R. Bartlett, "The Development of Industrial Research in the United States", National Resources Planning Board, *Research – A National Resource* (Washington, D.C., 1940), vol. II.

[9]

Suppressing Innovation: Bell Laboratories and Magnetic Recording

MARK CLARK

From 1930 to 1941, scientists and engineers at Bell Laboratories worked to perfect a practical magnetic recorder. They produced several prototypes, including one that was arguably the best in the world in the mid-1930s. Recorders designed by Bell Labs engineers were in regular service by the late 1930s in telephone central offices, and one engineering group even had a fully functioning magnetic recording answering machine attached to its telephone line for six months in 1935. Despite this early work, and the contracts for magnetic recorders that Bell and Western Electric (the manufacturing subsidiary of Bell's parent company, American Telephone and Telegraph [AT&T]) filled for the United States government during World War II, Bell fell behind other companies working on magnetic recording after 1940. The Bell system did not produce a recorder comparable in quality to those of other American and German companies, and it played little role in the exploitation of American and German magnetic recording technology by American companies after the war.

The work Bell engineers did in the 1930s is almost unknown today.[1]

DR. CLARK is a graduate of the doctoral program in the history of technology at the University of Delaware, where he was a Hagley Fellow. He is currently writing the centennial history of the university's College of Engineering. He wishes to acknowledge the assistance of David Hounshell, Sheldon Hochheiser, and Kenneth Lipartito, as well as the *Technology and Culture* referees, all of whom made many valuable suggestions. Work on the article was supported by fellowships from the Institute of Electrical and Electronics Engineers Life Members Fund for Electrical History and the Charles Babbage Institute's Tomash Fellowship in the History of Information Processing.

[1]No published comprehensive scholarly history of early magnetic recording based on archival materials exists at present. See Mark Clark, "The Magnetic Recording Industry, 1878–1960: A Business and Technological History" (Ph.D. diss., University of Delaware, 1992). A number of popular articles and unpublished works have treated various aspects of the overall picture. Most mention the work done at Bell Laboratories only in passing, if at all. One exception to this is the work of William Lafferty. In his "The Early Development of Magnetic Sound Recording in Broadcasting and Motion Pictures, 1928–1950" (Ph.D. diss., Northwestern University, 1981), Lafferty describes

0040-165X/93/3403-0007$01.00

Why? As this article will show, the impressive technical successes of Bell Labs scientists and engineers were hidden by the upper management of both Bell Labs and AT&T. Although AT&T used magnetic recorders for central office applications, it refused to develop magnetic recording for consumer use and actively discouraged its development and use by others. This suppression stemmed not from technical criteria but from policy considerations that went to the heart of AT&T's corporate culture.[2] The following is a history of efforts at Bell Laboratories to build a practical magnetic recorder in the 1930s and of why those efforts are now largely forgotten.

The Early History of Magnetic Recording

Magnetic recording was not a new technology when Bell engineers first started to work with it during World War I. Oberlin Smith, inventor and owner of the Ferracute Machine Company, had first conceived of using electromagnetism to record sound in 1878, only a few months after he visited Thomas Edison's Menlo Park laboratory and saw the newly invented phonograph. Smith believed that the phonograph had a fundamental flaw: both the recording and the reproduction of sound required physical contact. The phonograph needle moving in its groove inevitably produced noise along with the sound recorded, noise that reduced the quality of the reproduction. Smith's conceptual leap was to record sound by subjecting a recording medium to magnetic rather than physical vibrations. A magnetic material exposed to a magnetic field retains a portion of the magnetism to which it has been subjected, and this leftover, or "remnant,"

---

a number of the recorders developed by Bell Laboratories prior to 1940. However, his account is based solely on articles published by Bell scientists and engineers, not on archival sources, and as a result is incomplete and misleading about the overall thrust and level of activity at the laboratories. Published histories of Bell Laboratories contain even less information about the work done there on magnetic recording. For example, the monumental Bell Telephone Laboratories, *A History of Engineering and Science in the Bell System*, 2 vols. (1975, 1978), contains only a single reference to magnetic recording research done in the early 1950s and makes no mention of any developments prior to World War II.

[2]An extensive body of scholarship pertaining to the early history of the Bell system and the development of a distinct corporate culture centered on centralized control of the entire telephone system exists. See, e.g., Claude S. Fischer, " 'Touch Someone': The Telephone Industry Discovers Sociability," *Technology and Culture* 29 (January 1988): 32–61; Robert W. Garnet, *The Telephone Enterprise: The Evolution of the Bell System's Horizontal Structure, 1876–1909* (Baltimore, 1985); George David Smith, *The Anatomy of a Business Strategy: Bell, Western Electric, and the Origins of the American Telephone Industry* (Baltimore, 1985); and Leonard S. Reich, *The Making of American Industrial Research: Science and Business at GE and Bell, 1876–1926* (New York, 1985).

magnetism can be detected and played back as sound. Since magnetic fields can be transmitted without physical contact, Smith thought that the mechanical noise that plagued the phonograph could be avoided with a magnetic recorder.[3]

Smith performed a number of experiments in an attempt to prove his concept, but, because of financial pressure and the demands of his other business interests, he was unable to develop a working device.[4] He eventually published his ideas in 1888, hoping someone else might be able to perfect his concept.[5] Several other inventors patented recorders based on electromagnetic principles at about the same time, but none of them was able to build a working machine either.

A functional magnetic recorder was eventually built in 1898 by a Danish inventor, Valdemar Poulsen, whose work appears to have been independent of Smith's. His machine used a thin metal wire as a recording medium (today's familiar magnetic tape was not invented until 1929 in Germany and did not come into common use in the United States until after World War II). Poulsen demonstrated his invention, which he called the "telegraphone," at the Paris International Exhibition in 1900, garnering a gold medal and a great deal of coverage in the international technical press. On the strength of this success, Poulsen set up manufacturing firms in both Denmark and the United States. Initially, an investment syndicate provided Poulsen with the capital for the American venture. The syndicate included American Bell Telephone, parent company of AT&T. Executives at American Bell had become aware of the telegraphone as a result of an article published in *Scientific American* in 1900 on the Paris exhibition. Bell joined the syndicate in 1902 but withdrew the next year before any money changed hands.[6] As a result, Bell had no involvement in the subsequent manufacture and sale of magnetic recorders by the American Telegraphone Company, the firm set up to purchase Poulsen's patents in the United States. Because of poor management,

[3]Smith's idea, though conceptually correct, is not entirely accurate for real-world magnetic recording. Stray magnetic fields lead to noise just as physical contact does, so all magnetic recorders have some degree of noise in their background. Additionally, early magnetic recorders (roughly those built before 1940) operated at such high speeds that they produced a fair amount of mechanical noise as well.

[4]James W. Gandy, "Bridgeton: The Birthplace of Magnetic Recording," *South Jersey Magazine* 18 (Summer 1989): 8–12.

[5]Oberlin Smith, "Some Possible Forms of Phonograph," *Electrical World* 12 (September 8, 1888): 116. An examination of Smith's other papers indicates that, although this article was published in 1888, all of the research was done prior to 1879.

[6]W. H. Martin, "Development History in Field of Automatic Telephone Answering Service and Message Recording," September 20, 1950, Case 22070-80, AT&T Archives, Warren, N.J.

American Telegraphone produced fewer than a thousand recorders over the next twenty years, fell into receivership in 1920, and was finally dissolved in 1944.[7] After American Telegraphone entered receivership, several large corporations, including Westinghouse and General Electric, purchased telegraphones and subsequently patented inventions related to magnetic recording.[8] However, no commercial products resulted from these inventions.

The Bell system initiated research into magnetic recording in 1916 or 1917, after engineers at Western Electric obtained a telegraphone for experimental work.[9] This led to four patents relating to signal multiplexing and to sound delay for public address systems in large auditoriums.[10] But after the last of the four patents was filed in December 1920, there is no evidence of any further work on magnetic recording at Bell until January 1930, nor is there any evidence of any commercial products resulting from these patents.

During the mid-1920s, there was relatively little research activity related to magnetic recording anywhere in the United States or in Europe. Poulsen had abandoned magnetic recording for the more profitable area of radio by 1904, and no telegraphones were built in Europe after 1913 at the latest.[11] The large American firms discontinued their research by 1922, and only a few individual inventors persisted, devoting their efforts unsuccessfully to developing a magnetic recording-based system for sound motion pictures.[12]

The only active commercial ventures involving magnetic recording in the 1920s were European. Curt Stille, a German independent inventor and entrepreneur, had first become involved with the telegraphone in 1903, and he continued to experiment with magnetic recording throughout the 1920s. In 1928 he was able to obtain funding from both British and German sources, resulting in the development of two machines: the Blatnerphone in England, a modified form of which was eventually used by the British Broadcasting Corporation for broadcast

[7]For a complete history of American Telegraphone, see Clark, chap. 3.

[8]U.S. Patents 1,664,243 and 1,883,907.

[9]It is possible, though not likely, that the device was left over from tests done in 1911–12 to determine if an improved device built by the American Telegraphone Company was suitable for telephone use. See Martin, pp. 1–4.

[10]U.S. Patents 1,358,053; 1,365,470; 1,465,732; and 1,624,596.

[11]After 1914, customers in Europe were supplied by the American Telegraphone Company. For a description of Poulsen's career in radio and a history of the development of the Poulsen arc transmitter, see Hugh G. J. Aitken, *The Continuous Wave: Technology and American Radio, 1900–1932* (Princeton, N.J., 1985).

[12]See, e.g., Harry E. Chipman, U.S. Patents 1,480,992; 1,612,359; 1,832,097; 1,883,559; 1,883,560; 1,883,561; 1,883,562; or Edward Everett Cothran, U.S. Patent 1,588,706.

purposes in the 1930s, and the Dailygraph, a German machine intended for dictation. At the same time, Fritz Pfleumer, an Austrian consulting chemist working in Berlin, patented a process for coating paper with magnetic powder, which could then be cut into strips and used for recording. Pfleumer's coated tape was potentially much cheaper than competing recording media, solid tapes, or wires made from expensive magnetic steel. Pfleumer's idea was purchased by the German electrical firm Allgemeine Elektrizitäts Gesellschaft (AEG) in 1932, which then asked the chemical firm I. G. Farben to manufacture tape based on Pfleumer's concept. I. G. Farben researchers eventually developed plastic recording tape in the late 1930s, the direct ancestor of present-day recording tape.[13] Despite some commercial success in Germany, machines based on Stille's and Pfleumer's ideas were unsuccessful in the American market.

Bell Laboratories and Magnetic Recording

When Bell Telephone Laboratories took up the question of magnetic recording in 1930, the motive was not pressing outside competition. American Telegraphone was out of business, other American companies had abandoned the magnetic recording research programs they had initiated in the early 1920s, and German firms had had no success in the United States. Rather, Bell's motives had to do with new research strategies at AT&T. After Theodore Vail took over leadership of AT&T in 1907 and articulated his ideals of universal service and a single integrated system, the company switched its strategy from buying inventions from outsiders to developing new technology from within.[14] The success of this new strategy led to the creation of Bell Telephone Laboratories in 1925. This organization centralized existing staff and expanded the range of AT&T's research efforts. Increasingly, Bell Labs scientists and engineers were allowed to pursue projects that, while they might have potential applications, did not relate directly to AT&T's immediate needs.

Researchers at Bell Labs undertook work on magnetic recording in 1930 for two reasons. First, they were already involved in sound-recording research. Through its predecessor organization, the Western Electric Engineering Department, Bell Labs had worked on

[13] The first plastic tapes were made with cellit, an I. G. Farben cellulose acetate. In 1943, I. G. Farben introduced tape made with polyvinyl chloride, the basis for the subsequent postwar development of magnetic tape. See Lafferty (n. 1 above), pp. 128–30.

[14] Reich (n. 2 above), pp. 152–53.

sound recording as early as 1913.[15] Research on sound and sound reproduction was central to the telephone business, and AT&T exploited the fundamental work done by the scientists and engineers of Bell Laboratories for uses both inside and outside the telephone network. A notable example of this was Bell's development of the Vitaphone system in 1926, the first commercially successful sound motion-picture process.[16]

Clearly, Bell Laboratories and AT&T had an active interest in sound research all through the 1920s. This is not enough to explain the renewed interest in magnetic recording, however, since Bell had worked with and abandoned magnetic recording in the late 1910s. It was pressure from outside firms trying to interest AT&T in some form of telephone answering device that served as the immediate inducement for Bell Labs to undertake further work on magnetic recording. A number of different companies contacted AT&T during the 1920s, and almost all of their products were referred to Bell Laboratories for evaluation. For example, between 1921 and 1924 commercial agents attempted to sell AT&T patent rights to the Telegraphon (not to be confused with the Poulsen Telegraphone), a wax-cylinder phonographic device that could be attached to telephone lines to record conversations or answer the phone when the owner was absent. AT&T rejected these advances.[17] In 1925, A. V. Bodine, vice president of the Dictaphone Corporation, wrote to Bell Laboratories, pointing out that his company had seen "a rather insistent demand" for a device that would allow a standard dictating machine to record a telephone conversation. He went on to describe a device developed by a private inventor that allowed such a connection. The Pennsylvania Power and Light Company was already using this device on its private phone lines. Bodine believed a definite market existed for such devices and wanted to know whether Bell was interested in manufacturing and selling them to Dictaphone.[18]

Despite Bell Laboratories' past commercial association with Dictaphone, no agreement was reached at that time.[19] But Dictaphone

[15]Sheldon Hochheiser, "AT&T and the Development of Sound Motion-Picture Technology," in *The Dawn of Sound,* ed. Mary Lea Bandy (New York, 1989), p. 25. Photographic evidence from the AT&T Archives, such as photograph 41269, documents sound on film devices from as early as 1900, but it is unclear if these were made at Bell at that time or were acquired from outside sources later.

[16]Hochheiser, pp. 23–33.

[17]Lyman Morehouse to E. B. Craft, September 4, 1924, Case 33251, AT&T Archives.

[18]A. V. Bodine to E. B. Craft, June 1, 1925, folder 74-07-03-03, AT&T Archives.

[19]In the early 1920s Bell Laboratories had developed electrical recording and reproducing equipment to be supplied to Dictaphone for use with its dictating

(along with its competitor Ediphone) continued to try and interest Bell Laboratories and AT&T in the idea of a telephone recorder.[20] Although the president of Dictaphone was still trying to get Bell Laboratories to build and supply telephone answering machines to his company as late as 1933, AT&T and Bell did not begin to furnish equipment that allowed the connection of recorders to public telephone lines until 1948.[21] Indeed, AT&T prohibited the connection of recorders to public telephone lines prior to 1948, and it was 1951 before AT&T offered a telephone answering machine to its customers.

By then, the Bell Labs research program related to telephone recorders had a long history. During the late 1920s Bell engineers had continued to examine devices submitted to AT&T by outside companies.[22] The majority of those systems were based on phonographic recorders, but Bell also tested a number of telegraphone (i.e., magnetic) recorders. At first, Bell engineers rejected all of these devices, citing their poor performance as justification. In a 1929 memorandum, Bell engineers explained that the primary problem was an inability to deal with the variation in apparent volume between calls from nearby and distant stations.[23]

However, they also suggested that Bell Laboratories' recent innovations in phonograph and theater sound systems could be applied to the problem of telephone recording. As a result, in late 1929 the director of Bell Labs, Frank B. Jewett, concluded that technical reasons alone no longer sufficed for rejecting telephone recorders.[24]

machines. See E. B. Craft to E. H. Colpitts, June 3, 1925, folder 74-07-03-03, AT&T Archives.

[20]"Telephone Speech Recording Equipment—Case 33251," file memorandum, November 2, 1929, Case 33251, AT&T Archives.

[21]"Memorandum for File," April 3, 1933, folder 74-07-03-03, AT&T Archives; "Plans Set for Installing First Connector Units for Voice Recorders," *Bell Laboratories Record* 26, no. 8 (August 1948): 351–52.

[22]For example, see the correspondence between AT&T and the Dictaphone Corporation in folder 74-07-03-03, AT&T Archives, which contains a number of references to examination of the equipment of Dictaphone and others by Bell engineers.

[23]"Telephone Speech Recording Equipment—Case 33251," November 2, 1929, Case 33251, AT&T Archives.

[24]F. B. Jewett, "Development of Telephone Speech Recording Equipment," August 9, 1929, folder 74-07-03-04, AT&T Archives. Jewett was the head of Bell Laboratories from its founding in 1925 until 1940. After receiving his doctorate in physics from the University of Chicago in 1902, he worked as an instructor at Massachusetts Institute of Technology until 1904, when he joined Bell. He pioneered the development of long-distance telephone lines at AT&T and was instrumental in turning the De Forest Audion into a practical electronic amplifier. As a member and later president of the National Academy of Sciences, he was a major advocate for the benefits of industrial

He decided to pursue two separate avenues of research. First, he hired Clarence Hickman and assigned him a small budget to investigate magnetic recording in general.[25] Although there is no mention of telephone recording applications in Hickman's initial instructions, it appears that the decision to renew work on magnetic recording was due to the exposure of Bell Labs to the magnetic recorders that had been offered to AT&T for telephone recording use.[26]

Second, Jewett dramatically increased funding for the case under which telephone recording research fell, no. 33251.[27] Funding in 1929 was less than $14,000. Jewett planned to spend $77,200 in 1930, but actual spending exceeded $100,000.[28] This represents an increase in the share of the total Bell Labs budget from .06 percent to .42 percent, a rather large share for a project that was peripheral to the primary interests of the labs at the time, namely, switching systems and long-distance transmission.[29] Clearly, Jewett had made a commitment to explore in earnest the development of a marketable telephone recorder with a focus on phonograph technology.

Bell engineers used the increased funding to build and test a telephone recorder based on the phonograph, a technology with which they were already familiar because of Bell's work on the Vitaphone system and electrical recording. As early as 1924, Bell engineers had constructed a simple telephone recorder using a

research. See *Current Bibliography 1946*, pp. 283–85; Daniel J. Kevles, *The Physicists* (Boston, 1987), pp. 98–100, 170, 252, 265.

[25]Hickman, trained as a physicist (Ph.D., Clark University, 1920), had a rather interesting career prior to joining Bell Labs. He worked with Robert Goddard on rocket development in 1918. After receiving his Ph.D., he went on to do research for the Bureau of Standards and the U.S. Navy Yard in Washington. When he joined Bell Labs, he had been doing acoustics research since 1925 at the American Piano Company, a manufacturer of player pianos. During World War II, he headed the National Defense Research Council Rocket Development committee, as a result of his prior work with Goddard. See Winfield Scott Downs and Edward N. Dodge, *Who's Who in Engineering* (New York, 1959), p. 1108; press release from Bell Telephone Laboratories, January 1946, Box 96-06-16, AT&T Archives.

[26]Additional evidence for this explanation is the fact that the first practical applications for magnetic recording that Hickman worked on after his initial general development work were telephone answering systems, as discussed later in this article.

[27]Each research project at Bell Laboratories was referred to as a "case" and was assigned a number.

[28]Case Survey Report for 1929, January 1, 1929; Case Survey Report for 1930, January 1, 1930; and "Revised Authorization for Work," October 14, 1930, all from Case 33251, AT&T Archives.

[29]Calculation based on figures from "Bell System Research and Development Costs through 1971," May 1972, Box 175-11-03, AT&T Archives. Hickman's budget for this period is not known but was definitely much smaller.

wax-cylinder phonograph.[30] In 1927 they built an improved machine and supplied it to AT&T for field tests.[31] These efforts had been fairly crude because of limited funding, but the increased support in 1930 enabled the design and construction of a series of sophisticated, purpose-built phonographic recorders derived from those early machines.

The key innovation was a volume compressor circuit, which, by limiting the range of input to the recording device, solved the problem that had plagued all preceding devices: the wide difference in signal strength between a local and a long-distance call. Engineers judged that the playback of recorded calls was "substantially equal to that of the original telephone conversation . . . [and] may in some long line connections be somewhat superior." Recorded on a flat paper disk coated with wax, the messages were also cheap to produce and easy to file. Although each disk held only two minutes of conversation, the provision of two turntables that worked alternately allowed continuous recording. According to a 1930 year-end report, the problem of recording quality had been solved; all that remained was to make the device less complex, less bulky, and, above all, less expensive.[32]

However, a variety of technical problems, as well as Depression-related funding cutbacks, inhibited progress toward a commercially practicable recorder between 1931 and 1934. In April 1934, W. H. Martin, director of Local Transmission Development, one of the divisions involved with recorder development, noted that Dictaphone was marketing a new device superior in most respects to Bell's and that new magnetic recording techniques discovered by others at Bell were much more promising. He recommended that funding for further research into phonograph-based telephone recorders be greatly reduced.[33] By 1935, all such research had ceased.[34]

Clarence Hickman and Magnetic Recording at Bell Laboratories

In contrast to the phonograph project, Hickman's work on magnetic recording was successful. His success was due to an incremental approach to improving the quality of the recording process, and this steady improvement was instrumental in management's decision in

[30] E. H. Colpitts, "Telephone Conversation Recorder for Dictaphone Corporation," June 15, 1925, folder 74-07-03-03, AT&T Archives, p. 1.

[31] "Telephone Speech Recording Equipment—Case 33251."

[32] "Revised Authorization for Work," November 14, 1930, Case 33251, AT&T Archives.

[33] W. H. Martin, Memorandum for File, April 23, 1934, Case 33251, AT&T Archives.

[34] "Case Closing Notice," December 31, 1934, Case 33251, AT&T Archives.

1934 to focus on magnetic tape as a means to record telephone conversations. Magnetic recording for telephone calls had been considered by AT&T before, but the poor quality of available recorders led to its rejection. At least twice during the 1920s outside companies had offered AT&T the chance to examine and acquire interests in machines based on Poulsen's designs. In 1929, for example, a representative of Stille's Dailygraph Recorder Company attempted unsuccessfully to sell AT&T rights to develop telephone recording applications for his company's device.[35] Despite their rejection of the machine, Bell researchers did order the translation of two articles about the Dailygraph (along with several other articles on home recorders) in late 1930, and the recorder probably was demonstrated at Bell Laboratories in January 1931.[36]

Hickman's work inspired others at Bell Labs to think about the possible applications of magnetic recording. As early as 1931, a Bell engineer, Rudy F. Mallina, proposed an answering machine based on a magnetic recorder. Drawing on ideas developed in the course of building the phonograph-based recorder, Mallina's design incorporated a number of sophisticated automatic features; it operated, for example, only if the caller stayed on the line after the machine played its message. Although no actual working model of this design was produced at the time, the proposal was circulated to those involved in the telephone recorder project, making them aware of the possibility of magnetic recording.[37] Mallina cited the "progress made on magnetic tape recording during the last year."[38] Taken in the context of the designs shown in the drawings attached to the memo, he was undoubtably referring to Hickman's efforts. That work, already showing fruit after a year, led in 1934 to the opening of a new case at Bell Laboratories that funded magnetic recording research in general and the building of a magnetic telephone recorder in particular.

Unfortunately, the case file documenting Hickman's work on magnetic recording from 1930 to 1934 is not available. But Hickman's laboratory notebook for this period does survive, and it tells us a great

[35]"Telephone Station Recording Devices," July 12, 1929, folder 74-07-03-03, AT&T Archives. The Dailygraph company was later purchased by the German company Lorenz, which updated the machine and continued to produce it until the early 1940s (exact details are unknown as company records were lost in World War II).

[36]"Competitive Recording and Reproducing Equipment—Case 33251," December 30, 1930, Case 33251, AT&T Archives.

[37]"Automatic Telephone Message Recorder (Magnetic Tape Type)—Case 33251," Memorandum for File, March 30, 1931, Case 33251, AT&T Archives.

[38]Ibid., p. 7.

deal about Hickman and his research.[39] In the first few months of 1930, he made daily entries, recording in considerable detail his initial work. As he became more involved in research, his entries became less frequent, until by the mid-1930s as much as six months might pass between notations. Partly this was because Hickman was committing his ideas and designs to paper in the form of memoranda to the file rather than in his notebook.[40] The primary reason for this change, however, was the increasing tempo of the work. He only wrote brief summaries of his work in later notebook entries, suggesting that, as his experiments started to show valuable results, he spent most of his time doing experiments rather than writing in his notebook.

Hickman's daily entries for January and February 1930 allow us to get a clear picture of the state of magnetic recording research at Bell as he began his work. In his initial orientation, Hickman was informed by Jewett that his primary research area was to be "the telegraphone," meaning in this case magnetic recording in general.[41] Jewett was interested in developing four devices: a sound-delay system, a tone synthesizer, a prerecorded announcing system, and a musical instrument based on the tone synthesizer.[42] Although Jewett did not mention telephone answering machines, the prerecorded announcing system would have been a potential part of such a device.

His orientation completed, Hickman started his research on January 4, 1930. He obtained a telegraphone from another department of the laboratory, eventually devoting over two months to research in attempts to improve its function. Descriptions in his notebook suggest that the machine was a standard reel-to-reel wire recorder made by the American Telegraphone Company, intended for recording office dictation. After a systematic examination of the machine and its associated patents, Hickman read a number of published technical articles. Examination of the articles he lists in his notebook reveals a number of significant points. First, Hickman relied almost exclusively on articles published in the 1900–1903 period, which described Poulsen's initial invention of magnetic recording and the first commercial models produced in Denmark, including both wire and

[39]Laboratory notebook of Clarence N. Hickman, notebook no. 10935, AT&T Archives (hereafter Notebook).

[40]Hickman refers to several such memoranda in the notebook, for example, on p. 64 to a memo on magnetic recording and reproducing, the page dated January 14, 1931. These memoranda are not available.

[41]Notebook, p. 1.

[42]Although AT&T engineers had worked on a magnetic recorder-based sound-delay system in the late 1910s, there is no evidence that Jewett made Hickman aware of those efforts.

disk-type recorders. With one very minor exception, he ignored all of the literature published on magnetic recording between 1903 and 1929.[43] Second, Hickman did not read, at least not at this time, any articles in the German-language literature that dealt with the physics of the recording process. This seems to explain the course of his early research, which consisted of varying the components of the telegraphone apparatus on an empirical basis to improve recording quality, seemingly unguided by theoretical considerations.[44]

Hickman followed two major strategies to improve the quality of recording. First, he varied the location and electrical connections of the recording heads. After several weeks, he settled on an arrangement of two coils on opposite sides of the wire, directly aligned with each other. Second, he experimented with variations in the recording medium. He increased the size of the wire with some success and also tried a flat disk made out of Bakelite resin mixed with powdered steel.[45] But he abandoned both the disk and wire recording in favor of solid metal tape after only one month of research, as a result largely of his conviction that the twisting of the wire between the recording and the playback process was responsible for considerable distortion.[46] Tape does not twist and so provides a far more consistent orientation between playings.

Hickman also tried varying both the metal used in making tapes and the manufacturing process itself. He found that by alloying the steel used to make his tape with various other metals he could markedly improve its magnetic response. Similarly, heat treatment of the tape after it was rolled out also improved its performance. Hickman's heat-treated alloy-steel tape was markedly superior in its magnetic properties to the wire used in the telegraphone.

No single invention was responsible for the improvement Hickman made during 1930. Rather, trial and error with a series of different

[43]At least fifteen articles appeared between 1904 and 1929 on magnetic recording, most of them in English. The only article from this period Hickman consulted, which contains no technical information, pertained to the use of magnetic recording in railroad applications.

[44]Hickman's failure to use these articles does not appear to be the result of an inability to read German; several of the articles he listed in his notebook are in that language (Notebook, pp. 5–7). He did draw on his knowledge of magnetic phenomena, however. One of his first observations in testing was that the wire appeared to be oversaturated by the recording process (Notebook, p. 2). His response, however, was rather empirical in nature—to increase the wire size and change the type of microphones used.

[45]Notebook, pp. 10–16.

[46]The tape Hickman used was made from solid alloy steel and should not be confused with present-day magnetic tape, which is made from plastic coated with metallic or metal-oxide particles.

materials and processes led eventually to the dramatic increase in sound quality. These improvements in metal-tape performance seem to have made the alternative recording materials, wire and metal-impregnated plastic, less attractive and resulted in a focus on metal tape for recording at Bell Laboratories that was to last until after World War II.

By the end of 1930, Hickman had considerably improved the recording clarity of his machines. The combination of alloy-steel tape and a new recording head resulted in markedly better sound.[47] Hickman demonstrated a prototype to lab personnel and several patent attorneys on December 31, 1930, and those who listened "seemed to be much impressed."[48] Subsequently, Hickman claimed that the background noise for this machine was less than for a phonograph, a claim that, if true, would make his magnetic recorder the best in the world at the time.[49] The next day Hickman witnessed a demonstration of a German-made wire recorder. Though he did not enter the name of the inventor or the machine in his notebook, it was almost certainly a Dailygraph.[50] Hickman noted that the sound quality of the German machine was inferior to his own and that it operated at much higher speed with greater mechanical noise.

This comparison encouraged Hickman, and his notebook over the next few weeks of January 1931 records a number of additional experiments. He tested several different recording heads and types of tape, determining that thinner tape would function as well as thicker and that heat-treating the tape for hardness resulted in a better signal.[51] By mid-1931, work had progressed to the point that managers were considering the first commercial application: a device for recording telephone messages for hotels.[52] Hickman and others completed the machine later that year, and tests indicated that the noise levels for this recorder were lower than for the recorder that used wax records. Although work on the wax record-based system continued for several more years, this was the first indication that

[47]Notebook, pp. 54, 57.

[48]Ibid., p. 59.

[49]Ibid., p. 61. To my knowledge, no other magnetic recorder maker from this period claimed performance approaching that of a standard phonograph.

[50]Ibid., p. 60.

[51]Ibid., pp. 62–65.

[52]Ibid., pp. 69–70. The recorder was described in a memorandum for file, unavailable at the time of this writing, so little appears in the notebook about it beyond a brief description of its performance. A later patent granted to Hickman that describes a magnetic recorder attached to a telephone switchboard appears to match the notebook description (U.S. Patent 2,006,255, filed March 31, 1932, granted July 2, 1935).

magnetic recording would eventually bring an end to research on wax recording.

Because his experiments were recorded in memoranda for his case file, Hickman's notebook is much less detailed over the next four years. He alludes to several projects, such as a portable tape recorder, a delay tape machine, an endless recording loop, and a machine used in connection with a stock exchange, but he apparently described them more fully somewhere else.[53] Although he and his coworkers continued to focus on improvements in recording heads and tape manufacture, their research is best characterized as refinement and cost reduction rather than fundamental discovery. Hickman eventually received four patents based on his work prior to 1934. One was for the hotel message system mentioned above; the other three were related to minor improvements in tape manufacture and mechanism design.[54] By 1934, when a new research case file (no. 20872) for magnetic recording was opened, magnetic recording had become a practical method for sound reproduction, one which had a number of potential commercial applications.

Over the next two years, Hickman continued to work on projects related to magnetic recording. He was involved in the design of a number of experimental systems, and his work led to four more patents.[55] Eventually, as the project turned away from research and development and more toward the design and building of prototypes for field testing, Hickman was reassigned to other projects.[56] Prior to his departure from the magnetic recording project team, however, he was intimately involved in the design and construction of the Bell Labs prototype magnetic recording telephone answering machine. One of the first reports in the new case file is a description of that machine.[57] Built in early 1934, this apparatus functioned substantially along the lines outlined in Mallina's 1931 memorandum—and in a fashion identical to that of a present-day answering machine, responding to a telephone call with a prerecorded message and then recording the message left by the caller. A rather large and complicated device, the machine stood over 6 feet tall and used a plethora of relay switches. Employing the volume-limiter circuitry developed for

[53]Notebook (n. 39 above), pp. 75, 79, 86, and 109.

[54]U.S. Patents 1,944,238; 1,982,810; 2,003,968; 2,006,455.

[55]U.S. Patents 2,086,130; 2,106,350; 2,144,844; 2,185,300.

[56]Notebook, p. 137; U.S. Patent 2,086,130. Hickman's last patent filings and his last notebook entries related to a magnetic recording date from mid-1936. He continued to work at Bell Labs until 1950, but never again on magnetic recording.

[57]R. F. Mallina, "Automatic Telephone Message Recorder—Case 20872-1," February 27, 1934, Case 20872, AT&T Archives.

the wax phonograph telephone recorder, the machine's recordings were clear and well defined. Although obviously an experimental device, it worked well (Hickman's coworkers left it connected for almost six months) and had definite potential.

Suppressing Magnetic Recording

Despite Bell Laboratories' interest in sound recording, production of a working answering machine, and a demonstrated desire to commercialize the products of laboratory research, the Bell system delayed in offering an answering machine to its customers until the early 1950s, almost twenty years after the production of a successful prototype. This delay was not due to a faulty perception of demand on the part of Bell and AT&T executives; memoranda from the 1920s and 1930s refer constantly to customer requests for recorders and outside companies trying to interest Bell in their own devices.[58] In addition, a number of electrical utilities and railroad companies had installed recorders on their private phone systems and were requesting that they be allowed to use them on AT&T's public lines.[59]

Neither was the delay in production due to the quality of the equipment. Bell Laboratories' telephone recorder met all reasonable engineering requirements for performance, and similar equipment was in fact used successfully in a series of field tests from 1935 to 1940. In 1936, for example, Bell Labs built and supplied to four news agencies on a trial basis recorders to enable reporters to call in and leave stories for later transcription. After overcoming problems in providing satisfactory tape, the experiment seems to have succeeded.[60] In 1937, Bell installed a system in Hightstown, New Jersey, that allowed farmers to call in to the local office of the State Department of Agriculture and hear prerecorded reports on potato prices.[61] In this application, the machines functioned flawlessly.[62] Bell Laborato-

[58]See, e.g., "Memorandum for Mr. E. H. Colpitts, Assistant Vice President," August 8, 1929, and E. H. Colpitts to H. Nagatomi, July 25, 1932, both in Case 35825-173, AT&T Archives.

[59]See, e.g., letter to E. B. Craft, vice president, Bell Laboratories, from A. V. Bodine, vice president, Dictaphone Corporation, June 1, 1925, and other correspondence in folder 74-07-03-03 and 74-07-03-04, AT&T Archives.

[60]Case Survey Report, January 1, 1937, Case 20872, AT&T Archives.

[61]"Market News and Magnetic Tape Voice Recording," Memorandum for Mr. S. L. Andrew, chief statistician, January 11, 1938, Jewett Collection, AT&T Archives. This system of price reporting had been anticipated in a memorandum written in 1930 on providing services to rural areas; magnetic recording made the idea practical. See Memorandum, May 26, 1930, Box 38-02-02, AT&T Archives.

[62]The experiment as a whole was not successful because of other factors. Bell executives offered a number of explanations for the failure. According to one report,

ries, in conjunction with AT&T, developed and used a number of other message machines in the prewar period, including a weather-announcing system tested by the New York Telephone Company in 1939.[63] This machine is of particular interest because it used a closed-loop system designed by Hickman in 1934.[64]

These field trials show that Bell had developed magnetic recording to the point of practical application. In fact, evidence suggests that Bell had what was possibly the best magnetic-recording technology in the world through most of the 1930s, until the development of machines that used AC-bias after 1938.[65] The German recorder Bell engineers examined in 1931 was inferior to Hickman's experimental model, and later tests of Bell equipment gave results as good as phonograph recording. Moreover, during a 1935 visit to Germany Mallina examined commercial magnetic recorders made by Lorenz and found them to be inferior to Bell's efforts.[66]

---

a lack of sufficiently local price news for the smaller growers who made up most of the potential users of the service, combined with a too-short trial period, led to lack of use ("Market News," Jewett Collection, AT&T Archives, p. 1). Another report attributed the failure to the fact that the bottom dropped out of the potato market just as the test started, resulting in farmers being uninterested in the uniformly bad prices (letter to N. R. Powley, president, the Pacific Telephone and Telegraph Company, November 15, 1937, folder 74-03-03-04, Jewett Collection, AT&T Archives).

[63]W. Bennett, "Weather by Telephone," *Bell Laboratories Record* 28, no. 3 (November 1939): 66–72.

[64]Compare Notebook (n. 39 above), p. 116, and ibid., p. 71, fig. 1.

[65]Essentially all magnetic recorders (except for the first few built by Poulsen) use some form of bias in recording sound. Bias is defined as adding an additional electrical signal to the input current prior to its being used to record on the magnetic medium. Poulsen and his fellow researcher Peder Pedersen determined in 1901 that the quality of recordings could be substantially improved if a constant voltage was added to the input signal before it was recorded. This technique is known as "DC-bias." The AC-bias technique adds a very high frequency alternating signal (above the limits of human hearing) to the input signal in the same fashion as DC-bias. It reduces noise to a much greater extent, and is used in all but very specialized modern magnetic recorders. Researchers at the Washington Navy Yard discovered and patented AC-bias in the early 1920s, but, with the decline in interest in magnetic recording in the late 1920s in the United States, their work was forgotten. The AC-bias technique was independently rediscovered in Japan, the United States, and Germany in the late 1930s and was used in military recorders during World War II in the last two countries. It was central to the commercial success of magnetic recording after World War II, as it dramatically improved the quality of sound reproduction. See Clark (n. 1 above), chap. 7.

[66]R. F. Mallina to C. N. Hickman, May 16, 1935, Case 20872, AT&T Archives. Mallina went on to indicate that he planned to visit AEG to see its recorder, but no record of that visit, if it took place, is in the case file. Other sources from the period indicate that the AEG recorder at this time was considerably inferior to those of Lorenz.

Given Bell's successful experiments and the evidence of its superior technology, why was there little or no attempt to commercialize the developments of Bell's engineers? Lack of consumer demand is not the explanation; as noted above, the Bell system was constantly being approached during the 1920s and 1930s by individuals and corporations who were interested in magnetic recorders or in telephone answering machines. Even if AT&T did not wish to exploit magnetic recording itself, the company could have licensed its technology to other companies as it had done earlier with its phonograph developments.[67] That it did not indicates that there must have been other motives involved.

One possible explanation relates to Bell's choice of recording medium. All of Bell Laboratories' machines used steel recording tape, and the high quality of the recordings stemmed in large part from the quality of the steel used and the care taken in rolling it out. Bell Laboratories devoted most of its research in magnetic recording after 1936 to improving the quality and consistency of tape through the development of better rolling and heat-treatment methods.[68] Even so, the cost per linear foot for metal tape was high; Bell engineers calculated that even modest lengths of tape would cost more than the recorder itself. However, given the commercial success of a number of magnetic recorders in Europe in the early 1930s, which had similar machine-to-recording-medium cost ratios, the high cost of tape cannot be considered the overriding determinant.[69]

Another potential explanation is that executives feared that magnetic recorder development would interfere with profits from other parts of AT&T's business. In contrast with the broadcast radio stations of Europe, which by the latter half of the 1930s often used prerecorded programming (much of it recorded magnetically), American network programming was exclusively live.[70] All of those network

[67]In 1925 the Victor and Columbia phonograph companies licensed the electrical recording system developed by Western Electric/Bell Telephone Laboratories. This system later became the basis for prerecorded radio broadcast systems used by non-network stations in the United States. See Michael Jay Biel, "The Making and Use of Recordings in Broadcasting before 1936" (Ph.D. diss., Northwestern University, 1977), esp. pp. 120–28.

[68]See Case 20872, AT&T Archives, especially documents for the 1937–38 period related to magnetic recording.

[69]See the sales brochure attached to "Competitive Recording and Reproducing Equipment—Case 33251," Memorandum for File, December 30, 1930, Case 33251, AT&T Archives.

[70]Non-network radio stations did make use of prerecorded programming, distributed on phonograph disks, but this was seen by the radio industry as an inferior method of programming. Bell Labs designed and Western Electric marketed a 16-inch phono-

programs were distributed by AT&T via its long-distance network, a very lucrative arrangement for the company. The introduction of magnetic recorders similar to the ones used in Europe might have meant the end of this arrangement, as programs could have been distributed as recordings rather than live via AT&T. Though it is plausible on the surface, there is no documentary evidence to support this thesis. In all of the available files examined at the AT&T archives, there is no mention of the possibility of magnetic recorders being used in the broadcast industry. Additionally, the networks showed no interest in adopting new recording technologies during this period. The magnetic recorders being built in Europe in the 1930s for broadcast purposes were not imported to the United States, and this suggests that AT&T did not have to be concerned with the networks switching to prerecorded programming.[71] Though it is certainly possible, in the absence of positive evidence it appears that the threat of the broadcast use of magnetic recording played little if any role in AT&T's failure to commercialize magnetic recording.

Rather, through concerted effort, upper-level executives at AT&T sought to suppress the commercial exploitation of magnetic recording for ideological reasons stemming from the corporate culture of the Bell system. Internal memoranda reveal that AT&T's management worked to dampen outside demand for recorders and to limit the use of magnetic recording within the Bell system to selected areas. Magnetic recorders that were used by the Bell system in its own offices by its own employees were permitted; all other applications were not. Thus, devices like answering machines or personal recording devices were unacceptable.

---

graph disk system for recording and distributing radio programs to non-network stations in the 1930s. Biel, pp. 11–18 passim.

[71]The American radio broadcast networks eventually adopted magnetic recording for programming in the late 1940s, as a result largely of competitive pressure from the American Broadcasting Company (ABC) network. Newly formed after World War II as a result of the antitrust-related breakup of the National Broadcasting Company, ABC used recorded programming to give itself a competitive advantage. The most notable example of this strategy is ABC's signing of singer Bing Crosby in 1946. Crosby had left his highly rated radio show in 1944 because he no longer wished to perform live on a regular schedule. He was allowed by ABC to prerecord his programs, first on phonograph and later on magnetic tape, freeing him from a regular schedule and so making him willing to perform again. Crosby was so pleased at the freedom magnetic recording gave him that he played a major role in the founding of the magnetic recorder manufacturer Ampex. ABC's and Crosby's involvement with Ampex legitimized the network use of magnetic recording, and by 1950 all of the networks used magnetic recording extensively. See Clark, chap. 9.

Why? The reason was very simple: Management feared that the availability of a recording device would make customers less willing to use the telephone system and so undermine the concept of universal service. This fear is explicit in the memos that were written discussing the possibility of commercializing magnetic recording. The fear took two forms. First, if conversations became matters of record in the same way as letters or other contracts, managers felt that customers would abandon the telephone for critical negotiations and return to the mails, where a slip of the tongue would not prove fatal.[72] In the words of one manager, "If at any time there was a reasonable probability that such a device [i.e., a magnetic recorder] was connected at one end or the other, it would change the whole nature of telephone conversations and would in our opinion render the telephone much less satisfactory and useful in the vast majority of cases in which it is employed. It would greatly restrict the use of the telephone."[73] Such a decline in use was in direct opposition to the goal of universal service.

Second, if conversations could be recorded, matters of an illegal or immoral nature, which some executives estimated made up as much as one third of all calls, would no longer be discussed by phone.[74] The net result of this perceived loss of privacy would be a great reduction in the number of calls and a reduction in the trust individuals placed in the phone system, meaning a loss in both revenue and prestige for AT&T. Although one might expect that managers would stress the economic half of this equation, in fact they paid far more attention to the question of trust and image. During this period AT&T had constant public relations problems, largely the result of antitrust investigations. Thus, the fact that magnetic recording might create a major upset with only a minor gain for a few customers who could afford to have their calls answered automatically was simply not acceptable for managers concerned with preserving AT&T's good name.

The fears of AT&T's executives were probably not entirely irrational, given that questions about the privacy of telephone conversations had been a matter of public concern almost since the invention of the telephone. For example, the party line with its great potential for

[72]B. Gherardi (File Copy of Circular Letter), November 26, 1930, folder 74-07-03-03, AT&T Archives. This argument is repeated in a number of other documents in this file over the next ten years.

[73]Ibid., p. 2.

[74]Hearing before the Committee on Patents, United States Senate, 72d Cong., 1st sess., on S. 1301, *A Bill to Renew and Extend Certain Letters Patent, March 10, 1932* (Washington, D.C., 1932), p. 30.

overhearing secrets was a staple of turn-of-the-century fiction.[75] Of more immediate concern to AT&T officials in the 1930s was a growing awareness of the problem of wiretapping. The Supreme Court's finding in the 1928 case *Olmstead v. United States* that wiretapping was legal led to considerable public outcry, and several bills were introduced in Congress to ban the practice.[76] In 1934 the Radio Act of 1927 was amended to make intercepting electronic communication illegal, a change the Supreme Court later ruled applied to wiretaps.[77] Moreover, Congress created the Federal Communications Commission (FCC) in 1934, and charged it with regulating telephony as well as radio and television.[78] Given this context of concerns about telephone privacy, it is not surprising that AT&T officials were worried about the impact of magnetic recorders and moved to prevent their attachment to the telephone network.

This desire to control the nature of the telephone system and to prevent the attachment of devices like telephone recorders that were outside system control was an integral part of corporate culture at AT&T. As Leonard Reich has pointed out, the focus at the Bell system since 1907 had been on increasing the control of the central authority.[79] To provide universal service at a superior level of quality became AT&T's formula for corporate prosperity. Such quality service, AT&T management believed, could only be insured if AT&T had control over every part of the telephone network and everything attached to it. Any technical innovation that might affect service was treated by AT&T with suspicion; magnetic recording was no exception. It had consistently fought against the attachment to its circuits of devices it did not control. Magnetic recorders, which could be easily connected to telephones without AT&T's permission, were anathema to Bell's executives.

Moreover, AT&T operated during the 1930s under increasing scrutiny by the federal government. In the words of Peter Temin, "cooperation gave way to conflict" during this period, largely as a result of a general suspicion about the activities of big corporations in relation to antitrust.[80] During this period AT&T was forced to

[75]John Brooks, "The First and Only Century of Telephone Literature," in *The Social Impact of the Telephone*, ed. Ithiel de Sola Pool (Cambridge, Mass., 1977), p. 213.

[76]"Federal Government Information Technology: Electronic Surveillance and Civil Liberties," Office of Technology Assessment, OTA-CIT-293 (Washington, D.C., 1985), p. 31.

[77]Ibid., p. 32.

[78]Peter Temin, *The Fall of the Bell System: A Study in Prices and Politics* (New York, 1987), p. 11.

[79]Reich (n. 2 above), p. 142.

[80]Temin, p. 11.

withdraw from business activities not directly related to the telephone, such as radio, motion pictures, and television.[81] The company was also subjected to unfriendly investigations, most notably the FCC's Special Telephone Investigation in 1936.[82] Such increasing governmental hostility toward AT&T's participation in non-telephone-related business no doubt made it easier for executives to reject the idea of building and marketing magnetic recorders.

Thus, Bell Laboratories and AT&T, although developing in the laboratory sophisticated magnetic recording systems, effectively suppressed commercial development that went against the firm's own interests. Although the company did publish articles in its technical journals about its research, these articles reveal a consistent pattern: those applications of magnetic recording useful to AT&T's central offices were shown; those useful to individuals were not. Magnetic recorders were exhibited by AT&T at both the 1933 Chicago Century of Progress Exhibition and the 1939 New York World's Fair, but requests by AT&T subsidiaries to exhibit those machines elsewhere were denied, and individuals who wrote to Bell Labs asking about the machines were told that AT&T had no plans for further commercial exploitation, even though in actuality research was constantly going on at Bell into various applications.

The suppression was so effective that historians who have relied on the published record have rendered a highly incomplete picture of Bell Laboratories' activities. Bell's 1935 telephone answering machine with magnetic recording is largely unknown, and the extensive research and testing that Bell Laboratories performed in the 1930s is unmentioned in the histories of early magnetic recording that have appeared to date. This paucity of information reflects the success of the policy of suppression; those articles that did appear in Bell's own publications covered only central office applications, such as price reporting and weather announcing, and made no mention of applications that might hint at home recording, such as the experiment with news office recording or the telephone answering machine built in 1934. Publication outside of Bell's own journals was similarly limited by official policy.[83] As a

[81]Ibid., p. 15; and Alan Stone, *Wrong Number: The Breakup of AT&T* (New York, 1989), pp. 60, 66.

[82]Stone, p. 62.

[83]R. W. King, Memorandum for Mr. F. A. Cunningham, distributor contract specialist, Western Electric Company, November 4, 1935, folder 74-07-03-05, AT&T Archives.

result, the true nature of Bell Laboratories' work has remained largely unknown.[84]

Overall, the history of magnetic recording research at Bell Laboratories in the 1930s is consistent with previous accounts of AT&T's behavior. Hickman's initial research program was general in scope, but it was soon focused on the solution of problems of interest to the Bell system. This confirms the essentially practical orientation of the laboratory's work in this period, though the freedom Hickman had in doing his early work anticipates the less structured research style the laboratory was to have in the period after World War II. The laboratory was also careful to protect Hickman's innovations with patents, while at the same time limiting publication about his work, a pattern consistent with earlier work at AT&T.[85] Finally, the determination of how to exploit Hickman's work was not based on strictly economic criteria. The goal of AT&T in researching magnetic recording was to anticipate possible market developments and to enhance the functioning of the Bell system as a whole. Magnetic recorder applications that fit with the Bell system goals of universal service, as defined by Bell, were permitted. Commercial applications that would cause what executives perceived as problems were not. The course of magnetic recording research in this period is a clear demonstration of the values of AT&T in the 1930s.

Aftermath

The implications of AT&T's policy for World War II and postwar development of magnetic recording in the United States are less clear. Bell engineers who had worked on magnetic recording went on to serve in wartime in research and development posts related to magnetic recording, and Bell Laboratories manufactured a number of recorders for the military during the war. Yet the bulk of wartime work on magnetic recording was done by other organizations, most notably the Armour Research Foundation and the Brush Development Company. The technological lead Bell Laboratories and AT&T had developed in

[84]This policy of secrecy is consistent with previously documented behavior by AT&T. It initially concealed its development work related to Lee de Forest's Audion (ancestor of the vacuum tube) and various radio techniques (such as the concept of sidebands), as well as other research results, from its competitors in the 1920s. See Reich, pp. 186–88. Of course, these developments were made public once Bell had gained a competitive advantage from the secrecy and could cite these inventions for publicity purposes. Magnetic recording research did not lead to an equivalent competitive advantage for AT&T, perhaps accounting for the historical obscurity of Hickman's work at Bell Labs.

[85]For example, AT&T engaged in radio research and took out patents to protect the competitive position of its long-lines telephone connections.

the 1930s and the suppression of magnetic recording they had engineered were shattered by the events of the 1940s.

The thousands of recorders Bell and others built for the American military in World War II led directly to industry and consumer demand for magnetic recorders and the manufacture of hundreds of thousands of magnetic recorders in the years just after the war. However, those recorders were not built using technology developed by Bell Labs. Wartime research funded by the United States government and conducted by Armour and Brush resulted in the movement of a wide variety of new technologies out of the laboratory into production. These innovations included ferrite tape, AC-bias, and improved magnetic materials. After the war, these developments were combined with captured German technology to form the basis for the postwar boom in magnetic recording.[86] By 1948, consumer pressure had forced AT&T to allow the attachment of recorders to telephone lines, and by 1951 the Bell system was finally offering answering machines to its customers, over twenty-five years after Bell Labs first began to work on the problem. The irony is that those recorders were not built by AT&T; they came from an outside contractor.[87]

[86]My stress on the importance of wartime American research is a somewhat controversial view of the origins of the postwar boom. Previous writers on this topic have focused on the importance of captured German technology to American developments and have largely ignored American wartime research. That focus is partially due to the paucity of source material on American companies from this period, most of which are now out of business. However, I have been able to locate some records for the two most important American firms, Armour Research Foundation and the Brush Development Company (both of whose principal engineers are still alive and who I have had the pleasure of interviewing), and it is clear that there was a well-developed American magnetic recording industry well before German technology became available for use after the war. See Clark (n. 1 above), chap. 6–8.

[87]The first answering machine offered by AT&T to its customers was a device called the Peatrophone, manufactured by the Telephone Answering and Recording Corporation. It was not a magnetic recorder; it used phonograph disks to record both the incoming and outgoing message. See J. J. Gilliod, "Field Trial—Automatic Telephone Answering Service," August 30, 1950, and "Automatic Telephone Answering and Recording," March 12, 1951, both in Case File 22070-80, AT&T Archives. The first magnetic recording telephone answering machine was not offered by AT&T until 1953. See C. R. Keith, "The New Telephone Answering Set," *Bell Laboratories Record*, November 1953, pp. 439–43.

Excerpt from *Technology and Enterprise in a Historical Perspective*, Giovanni Dosi, Rena Giannetti and Pier Angelo Toninelli (eds.)

[10]

8

Continuity and Change in the Management of Industrial Research: The Du Pont Company, 1902–1980

David A. Hounshell

INTRODUCTION

The often used expression, 'the more things change, the more they stay the same', perfectly fits the history of the Du Pont Company's research and development programmes from their formal beginning in 1902 until 1980. Managing research during this period was always problematical. Despite changes in the size of the company and the nature of its business, despite new developments in the worlds of science and technology, despite changes in the scale and sophistication of the company's research efforts, Du Pont's executives have from the very outset faced a certain set of perennial issues in the management of research: what role should research play in the company's strategy? Should the company conduct research in a centralized organization or along decentralized, departmental lines? How much money should the company spend on research, and in what areas? What kind of time horizons should the company's research programmes have—short or long range? Should researchers be closely managed or pretty much left alone? This chapter is an attempt to highlight briefly the major areas of change and continuity in Du Pont's twentieth-century research programmes while summarizing very briefly the rich history of the company's research experience.

If I were forced to identify a single, dominant element of continuity in Du Pont's R. & D. history during the first eight decades of this century, I would say that it was faith in and commitment to R. & D. Despite changes in overall corporate

This chapter draws from Hounshell and Smith (1989), to which we refer for a detailed bibliography.

strategy, the company's leaders have maintained a steadfast faith in the value of research and have consistently outspent—both in absolute figures and as a percentage of sales—its major competitors.

For the sake of analysis, I have divided the history of Du Pont's R. & D. programmes into six periods: founding (1902–10), centralization and diversification (1911–21), decentralization, acquisition, and rationalization (1921–35), nylon and the 'new nylons' mentality (1936–60), maturation and response (1960–70), and rethinking and restructuring (1970–80). I present also some charts (see Appendix) to provide a quick guide to the major events of the period, the internal and external forces that affected the company's R. & D. programmes, and the major fruits of those programmes in terms of products and processes.

FOUNDING, 1902–1911

In 1902 two signal events occurred in the Du Pont Company. First, the hundred-year-old firm passed into the hands of three younger du Pont cousins, T. Coleman, Pierre S., and Alfred I. The older generation of du Ponts was tired, leaderless, and without any real idea of the assets of their company. In the new generation, Coleman and Pierre du Pont in particular saw their acquisition as a major opportunity to consolidate the entire explosives industry of the United States. By 1905 they had succeeded, bringing three-quarters of the industry under the control of a central management, the Executive Committee of E. I. du Pont de Nemours Powder Company. The consolidated Du Pont Company manufactured three types of explosives: black powder (an old, fragmented, and declining business), dynamite (a newer, concentrated, growing, and highly profitable product), and smokeless powder (a product aimed mainly at the military market and one that was just becoming profitable).

The second major event of 1902 occurred when the Eastern Dynamite Company, a subsidiary of Du Pont soon to be consolidated into the larger company, established the Eastern Laboratory, a facility divorced from any immediate manufacturing responsibilities and with a specific mission to improve the company's high explosives products and processes through scientific research and development. From the time Du Pont interests had first entered the high explosives business, its managers had been aware of the economic benefits that accrued from technical improvements. A small increase in chemical yields meant important savings in materials that could easily be measured in dollars and cents. But not until 1902 did the managers of the dynamite business decide to build a laboratory removed from the day-to-day operations of the plant. The wisdom of this decision was quickly demonstrated. Under the direction of the German-educated Dr Charles L. Reese, Eastern Laboratory made a big showing in only a few years. In 1911, for example, Reese calculated the annual profits stemming from only four among many of Eastern Laboratory's developments at a million dollars; the total cost of research at Eastern was but a third of these profits.

Eastern Laboratory's success contrasted sharply with the experience of the Experimental Station, the second Du Pont R. & D. laboratory, established in 1903 north of Wilmington, Delaware, on the Brandywine River immediately downstream from the company's original hundred-year-old black-powder works. Shortly after Coleman du Pont had organized the executive committee to run the newly acquired company, Francis I. du Pont (another cousin) and Alfred—both members of the Executive Committee—decided not to establish laboratories similar to Eastern for their respective businesses, smokeless powder and black powder. Rather, they agreed with A. J. Moxham, the executive Coleman had appointed to watch over the newly organized development department, that the company should organize a 'general experimental laboratory' whose mission would be to support all three of the company's existing businesses and to do pioneering research for the company along broad lines such as the synthesis of natural raw materials used in explosives manufacture. Hence the Experimental Station, which today must be one of the largest corporate research complexes in the world.

When Charles Reese's counterpart at the Experimental Station was compelled in 1911 to prepare a statement of the Station's contributions to profits, he was unable to demonstrate any such profits. In fact, he was hard pressed to show that the laboratory operated on a break-even basis. He argued that several of the Station's projects were of enormous proportions and, though of great cost and risk, might ultimately pay handsome dividends. Executives in 1911—at least most of them—accepted this argument because it was consistent with the policy established in 1907 by Pierre S. du Pont while he was briefly in charge of supervising the director of the Experimental Station. Pierre had maintained that:

> In our Experimental [Station] Laboratory we should at all time endeavor to have some investigations in which the reward of success would be very great, but which may have a correspondingly great cost of development, calling for an extended research of possibly several years, and the employment of a considerable force. (Pierre S. du Pont, 1908)

Pierre du Pont's statement captures the very essence of the Du Pont Company's philosophy of research for much of this century—pursue big, high-risk, high-reward projects that demand extensive commitment of research personnel and financial resources over a long period of time.

The director of the Experimental Station stressed that his laboratory's performance had to be measured by means other than its immediate contribution to profits. Specifically, he alluded to the qualitative success of the Station in that the United States army and navy ordnance officials continued to view Du Pont as a scientifically advanced manufacturer of smokeless powder in spite of the growing public criticism by Progressives of Du Pont as an evil 'trust'. The army and the navy continued to purchase smokeless powder from Du Pont even though some members of Congress wanted to prohibit these purchases. Congress actually passed the 1908 Naval Appropriations Bill with a specific clause banning the Navy from buying powder from a 'trust'.

Du Pont's success in being viewed by military officials as a progressive manufacturer of smokeless powder was not accidental. In 1903, Francis I. du Pont had written what essentially became the charter of the Experimental Station and had stressed the necessity of Du Pont's remaining well out in front of the government in the development of military propellants. Both the army and the 'new' navy had established their own R. & D. laboratories for smokeless powder and had also built or planned to build small-scale experimental manufacturing plants. The full weight of Du Pont's success in staying ahead of the army and navy in the development of military propellants was not felt until after Du Pont lost its first big anti-trust case. In 1912 the company was convicted of violating the Sherman Act and was forced to divest a big chunk of its business. But the army and navy interceded on Du

Pont's behalf and convinced the court to keep intact all of the company's smokeless powder operations, largely because of Du Pont's R. & D. prowess in this area. The consent decree went into effect in 1913, so Du Pont was perfectly positioned to reap several hundred million dollars in profits from sales of smokeless powder that mushroomed when World War I broke out in Europe in 1914.

CENTRALIZATION AND DIVERSIFICATION, 1911–1921

In 1911, however, this vast fortune from smokeless powder manufacture was not evident. A minority of members of the executive committee were not convinced that the pursuit of research in a general or central laboratory like the Station was the best allocation of resources. That year, for reasons that had nothing to do with R. & D., T. Coleman and Pierre du Pont reorganized the managerial structure of the company. As Alfred Chandler has written in his classic *Strategy and Structure* (Chandler, 1962:63), the net effect of this reorganization was the centralization of the company under one general manager, Hamilton Barksdale. The selection of Barksdale had an important impact on the company's R. & D. He had been the cofounder of the highly successful Eastern Laboratory and the staunchest opponent of the Experimental Station as a general laboratory. Seeking increased funding for the expansion of Eastern Laboratory in 1904, Barksdale had led a campaign to convince the Executive Committee to choose once and for all how it would manage its R. & D. (in decentralized departmental laboratories like Eastern or in a centralized general laboratory like the Experimental Station). Barksdale believed that the Station's fuzzy mission, its duplication of Eastern's high explosives research, and its location away from manufacturing plants resulted in inefficient and ineffective research. He strongly advocated the establishment of separate R. & D. laboratories for black powder and smokeless powder, to be located at appropriate plant sites and managed by the respective industrial divisions.

Barksdale's efforts led to extensive debates within the full executive committee about how best to manage industrial R. & D. As was its custom, the committee formed a subcommittee to study the issue, which issued two majority reports accompanied by two minority reports. When the full committee took up the issue, it too was unable to resolve the issue in 1904 and decided instead that there was no 'one best way' to manage research. Because of the company's adequate resources, the committee decided to continue to pursue both centralized and decentralized approaches to research.

When Barksdale assumed the position of general manager in 1911—only a month after the directors of Eastern Laboratory and the Experimental Station had issued their reports about their respective laboratories' contributions to profits—he quickly seized the opportunity he had been waiting for since 1904. He determined to put the unwieldy Experimental Station under the control of Charles L. Reese, the founding director of the Eastern Laboratory, who by this time was known as the Chemical Director of the High Explosives Operating Department. In that position, Reese was in charge of not only the Eastern Laboratory but also the chemical control of all of the department's plants. That is, he was responsible for monitoring the actual plant yields as compared to the theoretical yields derived by the Eastern Laboratory. Barksdale created a corporate-wide position of chemical director and elevated Reese to this position. Barksdale obviously hoped that Reese would do for the Experimental Station what he had done for the Eastern Laboratory—that is, to align research closely with production problems and product opportunities.

Paradoxically, Barksdale's move resulted in the greatest centralization of research in the company's history. In the short run, Reese did exert considerable pressure on the Station to look more closely at the company's businesses, but for reasons discussed below, the Station took increasing responsibility for the diversification of the company. This meant that the bulk of its research moved away from immediate commercial needs in smokeless powder and black powder. Moreover, Reese liked power, and over the next five or six years, he built a large, centrally controlled organization that for some in the company seemed to exist for itself.

This increasing centralization in the management of research and development occurred at precisely the time that the company's top managers had determined to diversify the company. Talk of diversification began when Progressive agitation against the 'powder trust' threatened Du Pont's sales to the armed services. In 1908, when Congress prohibited the navy from buying any powder from 'trusts', Du Pont's executives were forced to close down much of the company's smokeless powder manufacturing plant and to begin searching for products to make with this idle plant. Smokeless powder is nitrocellulose, so Du Pont concentrated on diversifying into other nitrocellulose products, such as celluloid plastics and nitrocellulose-coated fabrics. Even at this early date, Du Pont had developed a noticeable hubris about its R. & D. capabilities, and managers initially thought they could easily develop products and processes in these areas. But after doing enough work to learn the true complexity of these businesses, Du Pont bought its way into both the 'artificial leather' (nitrocellulose-coated fabrics) business and the celluloid industry.

Although the Congressional attack on Du Pont abated and thus took immediate pressure off the company to find ways to utilize its excess smokeless powder plant capacity, World War I suddenly amplified this concern to unprecedented proportions. With the company building more and more capacity, its executives worried more and more about overcapacity in the post-war era. In 1916, after two years of anxiety, Du Pont's executives realized that their strategy of finding uses for the company's smokeless powder plants did not utilize Du Pont's greatest resources, which were people, know-how, and plenty of cash.

In place of the old plan, Du Pont's Development Department fashioned a new diversification strategy that was so bold that it was expected to take a generation to execute fully. The goal was to move Du Pont away from its core business in explosives and turn the firm into a well-diversified chemical giant. Formulated largely by Fin Sparre, the director of the Experimental Station who was on loan to the Development Department and who would play a key role in Du Pont's diversification until his retirement in 1944, the new strategy including entry into vegetable oils, paints, pigments, and organic chemicals, including dyestuffs, pharmaceuticals, and photographic chemicals.

Du Pont bought its way into paints and pigments but stayed out of vegetable oils. With organic chemicals, however, the company ventured into dyestuffs on its own largely because there was no American dye firm to buy; the Germans totally dominated the industry, and after 1914 the United States was cut off from German supplies. Reese quickly brought all of the company's technical resources to bear on the dyestuffs venture, and he began an intensive campaign to hire organic chemists who knew something about the dye industry—an impossible mission. Du Pont quickly learned that there was far more to the synthetic dye industry than its chemists knew and that it would be difficult and expensive to do in a few years what it had taken the Germans fifty years to achieve. The company soon worked out an agreement with the British dye house of Levenstein, which, excepting synthetic indigo, knew little more about the business than Du Pont did. Reese had had organic chemists at both the Station and the Eastern Laboratory working on dyestuffs, but in 1917, he consolidated these efforts in the new Jackson Laboratory, located at the site of the company's new dye works. Soon well over half of the company's thousand-strong R. & D. staff were working at Jackson Laboratory and its sister institution, the Technical Laboratory, the end-use research laboratory also located at the dye works.

DECENTRALIZATION, ACQUISITION, AND RATIONALIZATION, 1921–1935

By 1920, Du Pont was rapidly losing money on its dyestuffs business and all of its other non-explosives businesses. This trend continued into mid-1921 and reached such proportions that after extended and hotly debated discussions, the executive committee took a dramatic step and decentralized the company's organization into separate industrial departments, each with its own manufacturing and sales divisions. The committee believed that if each business had better control over its functions it would perform better than in the centrally managed firm.

This decentralization radically affected Du Pont's R. & D. programme because the men whom the executive committee put in charge of the various businesses—general managers, as they were called—had been generally unsatisfied with the chemical department's performance. They believed that the department served its own ends rather than theirs. Soon after assuming their new duties, the general managers banded together and demanded that R. & D. functions also be decentralized—that the chemical department be split up. Although the president of the company, Irenee du Pont, did not like the idea, he was unwilling to vote against the general managers. The chemical department was 'disembodied', as one research manager observed, but not totally eliminated. Eastern Laboratory became the responsibility of the explosives department; the new dyestuffs department took charge of the Jackson Laboratory; and so on. (By this time, the chemical department was managing five geographically separated laboratories.) A mere vestige of the once glorious chemical department survived; eighteen researchers remained at the Experimental Station under the nominal direction of Charles Reese, who retired a few years later.

Looking back on the decade from 1911 to 1921, the most important development—at least as far as the future of Du Pont was concerned—was the decision to diversify the company. That decision led eventually to the decentralization of the company's organization, which in turn facilitated future diversification activities. The decentralization of the company in 1921 forced the company's executives to debate again the merits of centralized versus decentralized management of R. & D. The resulting decision to split up the once powerful central chemical department reaffirmed the executive committee's decision of 1904 to support both approaches to R. & D. Since 1921 Du Pont has not materially departed from this basic structure for managing R. & D.

Du Pont's experience with diversification in the half decade after 1916 led not only to the decentralization of the corporation but also to some important conclusions by executives about how future diversification would be undertaken. In 1921 executives knew that they were in well over their heads in dyestuffs. The dye works had proven to be essentially a sink-hole for Du Pont's cash. Du Pont had invested perhaps $20 million in the business and faced investments of another $20 million before profits would become apparent. No doubt with some regrets, executives had accepted the fact that Du Pont did not have all of the necessary know-how in dye chemistry and dye manufacturing, so they had approved and helped carry out a somewhat clandestine operation to recruit a number of German dye chemists to provide this critical know-how. The lesson these executives drew from the dye experience was that diversification by acquisition was easier than diversification through the internal generation of science and technology. Over the next decade and a half, du Pont built a chemicals empire through acquisitions, not through internal means.

Between 1921 and 1933, Du Pont's executives pursued a growth strategy predicated on diversification through acquisition of companies or technologies. The decentralized structure of the corporation suited this strategy well because when Du Pont brought a company or a major technology, the acquisition often became a new industrial department. In terms of R. & D., Du Pont typically built a research organization around the acquisition and then began the very important process of rationalizing the acquired technology, pushing it to its limits, finding new markets for its products, and developing new products as well. In this way Du Pont moved into the production of rayon fibres (French technology purchase), cellophane films (French technology purchase), synthetic ammonia and other high-pressure reaction products (French technology purchase), tetraethyllead (TEL) gasoline antiknock (Du Pont manufacturing process of compound discovered by GM), and Freon refrigerants (Du Pont manufacturing process of compound discovered by GM), titanium-dioxide pigments (initial process purchased and the chloride process developed later), and electrochemistry (purchase of Roessler and Hasslacher Chemical Company). The last major acquisition Du Pont made during this period was the Remington Arms Company in 1933, which was well outside its chemical industry realm and therefore did not really fit the diversification strategy developed by Fin Sparre in 1916. Indeed, not until very recently did Du Pont purchase the balance of Remington's stock and fully merge the company into the larger corporation.

The growth of the Du Pont Company during the period from 1921 to 1935 was clearly world class. After their formation, two other firms were in the same league, I. G. Farben and Imperial Chemical Industries. Du Pont had had international market-sharing agreements with predecessors of both firms and clearly wanted those agreements, which were restricted to certain explosive products, to be put in force again and to be greatly expanded to cover other products. Wanting to capture part of the large American market, I. G. Farben was not anxious to divide world markets. ICI and Du Pont, however, reached an agreement in 1929 called the Patents and Processes Agreement, which was clearly designed to divide world markets between the two firms. Du Pont's Sparre believed that by licensing one another's patents and processes, Du Pont and ICI could divide world markets without violating the Sherman Act.

Although clearly a market-sharing agreement, this device affected Du Pont's R. & D. programmes because the presidents of the two chemical companies also reached a gentlemen's agreement that their R. & D. laboratories would exchange not only information on existing products and processes but also on all research being done in the two companies (that is, on research from its inception). Until anti-trust action ended this arrangement in the late 1940s, the ICI agreement significantly broadened Du Pont's research base. For example, in the mid to late 1940s when Du Pont was trying to develop dyes for its new, hydrophobic Orlon acrylic and Dacron polyester fibres, research managers relied heavily upon the dye research unit at ICI for fundamental work. When the agreement was cancelled, Du Pont's research managers estimated that Du Pont would have to hire thirty first-class researchers to make up for the loss of ICI information. Moreover, since ICI did many things differently from Du Pont, considerable cross-fertilization of ideas occurred when the hundreds of research delegations from the two companies visited each other's laboratories and plants and when researchers from both companies read the tens of thousands of research reports that were exchanged during the lifetime of the agreement.

During this period from 1921 well into the 1930s Du Pont had access to much of the world's chemical research. The company's strategy called for R. & D. to play a supporting role of assimilating, improving, and extending acquired technology and knowledge. Du Pont did not see its R. & D. as the principal means of generating new technology and new business. But within a very short period, owing to both internal and external factors, the company completely changed its strategy. Du Pont moved into an explosive new era.

NYLON AND THE 'NEW NYLONS' MENTALITY, 1936–1960

In the 1920s, while the industrial departments were laying the foundations of the Du Pont chemical empire, the remnant of the central chemical department struggled to find a role in the company. Under the leadership of Reese's successors, Charles M. A. Stine in the 1920s and Elmer K. Bolton in the 1930s, the chemical department rose to prominence primarily because of the discoveries of neoprene synthetic rubber and nylon by Wallace H. Carothers's fundamental research group. Until this time, the chemical department had survived by doing research for the industrial departments essentially on a contract basis. But Stine sought a more central and independent role for his department. In 1926 he asked the executive committee for more money for general investigations and proposed what he called a 'radical departure', the establishment of a 'pure research' programme. The executive committee was uneasy about the concept of pure research and asked Stine to reformulate his proposal.

When Stine came back to the committee, his ideas had developed significantly. Now he proposed a 'fundamental research' programme, not 'pure research'. He convinced the executive committee that Du Pont's seemingly disparate businesses really possessed common scientific bases. If these common scientific bases were better understood, Du Pont's businesses would inevitably benefit. But because of the shorter-range focus of each of the industrial department research divisions, such fundamental work was not apt to get done; only the central chemical department could engage in such important work on a long-term basis and thus it could now fulfil its logical role as a central laboratory in the now well-diversified Du Pont Company.

Stine identified several areas in which his department could contribute to the welfare of the company. One item on his list was polymers. He argued that there was virtually no fundamental understanding of polymer chemistry, yet many of Du Pont's businesses were polymeric in nature. These included paints, plastics, rayon, and cellophane film. Du Pont could not rely on universities to generate knowledge of polymer chemistry because universities simply were not doing such chemistry. The same was true, argued Stine, in such areas as catalysis, high-pressure reactions, physico-chemical phenomena, and chemical engineering. Stine stressed that his plan would not only lead inevitably to improvements in Du Pont's products and processes but more importantly would serve as a major piece of public relations, thereby allowing Du Pont to recruit higher-calibre researchers and improve the company's relationship with colleges and universities. Thoroughly convinced by Stine, who incidentally was a great, almost evangelical salesman of research, the executive committee granted Stine a much larger budget for 'general research' and generous funds for his fundamental research programme.

Soon the chemical department was building a new laboratory at the Experimental Station to hold the researchers Stine envisioned recruiting for his programme. Other researchers at the Station promptly and somewhat facetiously dubbed this laboratory 'Purity Hall'. But even with a new laboratory, very high salaries, generous research support, and a supposedly 'unlimited' publication policy, Stine found it impossible to recruit the top-tier chemists away from the universities. These men—and they were all men in those days—were very suspicious of industrial research and reckoned it industrial slavery. Rebuffed by established university professors, Stine had to turn to young academics who could be more easily lured by the money and promising opportunities.

One of these more vulnerable chemists was Wallace H. Carothers, an instructor at Harvard, who agreed to join Du Pont only after repeated assurances that he could pursue 'pure science'. Carothers and Stine agreed that he would investigate polymers and would have a group of several scientists with fresh Ph.D.s helping him. Within three years, Carothers had conducted his classic research on the nature and formation of polymers that propelled him to the forefront of the new discipline, thereby accomplishing Stine's original goal of building Du Pont's image as a high-calibre scientific organization. But this academic era would not last much longer.

In April 1930, two of Carothers's assistants made unexpected discoveries, neoprene and the first truly synthetic fibre. Carothers was pleased by these discoveries but quickly moved on to other areas of polymer chemistry that interested him more, leaving the organic-chemicals department to develop neoprene and dropping the synthetic fibre work altogether after publishing several papers on the subject. A few months after the initial discoveries, Stine was promoted to the executive committee and was replaced by Elmer Bolton, who had directed Du Pont's dyestuff research since the early 1920s and who had voiced objections to Stine's fundamental research programme when he first heard about it. Whereas Stine had symbolized the expansive visions of the 1920s, Bolton represented the practical, tough-minded Depression decade. He believed that Du Pont could not afford to allow its élite group of scientists in Purity Hall to pursue only scientific goals, and he accordingly started to bring fundamental research down from its ethereal realm. After constant badgering by Bolton and his assistant Ernest Benger, who had come from the Rayon Department's research division, Carothers agreed in 1934 to renew work on fibres, which had languished for about three years after the initial exciting results. He soon discovered nylon.

This discovery resulted in the chemical department being turned inside out to develop nylon into a commercial product. While nylon was being developed, Carothers essentially went into a tail-spin; his bouts of depression, which had plagued him since early manhood, grew worse and worse until he committed suicide in 1937. By this time Bolton had essentially done away with Stine's original fundamental research programme (that is, a group of élite researchers pursuing their own agenda as long as it seemed to be related to Du Pont's businesses). Most of those that had been in the fundamental research programme found themselves working in one way or another on the development of nylon. Bolton made fundamental research something that all of the chemical department's groups did some of when it was appropriate.

The virtually instantaneous success of nylon came at roughly the same time as the more radically oriented members of President Roosevelt's 'brain trust' had gained the upper hand in diagnosing the United States' economic problems. They laid the Depression at the feet of big business in America and launched the late New Deal's attack on large corporations. Under Thurman Arnold, a sharp critic of Du Pont, the Anti-trust Division of the Justice Department expanded rapidly between 1938 and 1942 and instituted some 180 cases—50 per cent of all those that had been pursued since the Division was created in 1903! Arnold and his successors vigorously went after Du Pont, bringing some eight anti-trust cases against the firm by the end of the 1940s. To Du Pont executives in the late 1930s and early 1940s, this new wave of attacks on the firm yielded one very solid conclusion: Du Pont's future growth would have to stem from some means other than acquisition.

Nylon provided the way. It became the new archetype for Du Pont's strategy: nylon (and neoprene) had emerged from fundamental research. If Du Pont were to expand its fundamental research efforts, not only in the central chemical department but also by carrying it out in the industrial research divisions, then its laboratories would become cornucopias of 'new nylons'. Executives derived this new formula for success between 1941 and 1945 as the United States was engaged in World War II. Du Pont's highly successful experience in turning a piece of exotic physics into the massive Hanford Plutonium Works, a critical part of the Manhattan Project, reinforced the power of this formula.

Between 1946 and 1954, Du Pont dramatically expanded the firm's R. & D. facilities, personnel, and expenditures. In particular, executives convinced the general managers of most of the industrial departments (who were very independent in those days) to build new laboratories on the grounds of the Experimental Station and to devote these facilities to basic and pioneering research. The Station rapidly took on the appearance of a university campus, and a similar atmosphere, with all of its polar attributes of interchange and insularity, also followed. The model of fundamental science inducing corporate growth became the foundation of Du Pont's corporate culture. Science was now firmly at the centre of Du Pont's corporate strategy.

During this period, the industrial departments became more autonomous than before. Many of them carried out their own fundamental research programmes on polymers, the area of chemistry that now formed Du Pont's core technology. Ironically, the chemical department—soon to be renamed the central research department—was instructed to leave polymer research to the industrial departments, a move contrary to the central research philosophy that Stine had articulated in 1927 when he won approval to establish the company's first fundamental research programme. In developing a mentality of 'new nylons', executives and research managers alike had forgotten why the company had so easily and swiftly developed nylon. The chemical department's pioneering work had fitted neatly into the company's existing businesses, technologies, and levels of expertise. With central research's programmes being pushed away from the company's commercial interests, the nylon model became skewed.

The role of central research in the company therefore became highly problematic. The department was urged simply to explore new areas rather than to provide the fundamental knowledge base for Du Pont's technologies. Executives still believed, however, that central research would discover new nylons. As the department moved into physics and biology, it began to lose contact with many of the industrial departments and consequently took on the trappings of a high-powered scientific research establishment divorced from any commercial objectives. Eventually, many in the company began to criticize central research for pursuing science that was out of step with corporate strategy.

These sentiments began to grow when it became apparent that

Du Pont had not come up with another nylon. This is not to say that research had not produced anything, but rather that no new products were developed during this period that approached nylon's impact on the company's growth and earnings. During the 1950s Du Pont developed important insecticides and herbicides and new plastics with outstanding properties. The textile fibers department (successor to the rayon department) had produced the most significant products with its Orlon acrylic and Dacron polyester fibres. Moreover, intensive R. & D. on nylon itself had yielded new products and processes that continued to make nylon a major money-maker. Commanding the lion's share of total company R. & D. expenditures, textile fibers provided almost half of Du Pont's earnings on one-third of the company's sales in 1960.

Du Pont and its R. & D. programmes were affected by other trends external to the company during the 1940s and 1950s. The most important was the growth of competition in the chemical industry. World War II had rapidly accelerated the development of Du Pont's competitors. The wartime emergency had led many of them to engage in R. & D. projects and manufacturing operations that were extraordinary for them at the time but that would become commonplace after the war. In this sense, the war was a great leveller in terms of scientific and technical talent in the chemical industry. Du Pont's executives perhaps underrated the abilities of their competitors; certainly most of Du Pont's R. & D. personnel believed they were head and shoulders above their competition. This hubris sometimes had dire consequences for Du Pont's products.

MATURATION AND RESPONSE, 1960–1970

By the late 1950s, Du Pont's rate of growth and earnings began to decline, a sign that the chemical industry had begun to mature. Although the company's research programmes had been expanded significantly, the number of important new products introduced annually had not. The increasing size of the company required R. & D. to turn out a larger number of new products each year if Du Pont were to achieve its performance goals. After assessing the situation, the executive committee concluded that Du Pont research was more productive than ever but that the results were not getting beyond the laboratory, particularly those discoveries that would lead Du Pont into new businesses. Therefore, in 1960 the committee called for a corporate-wide push to develop products from Du Pont's accumulated riches of research. Executives likened their new venture programme to Du Pont's first diversification efforts when the company successfully moved away from being purely an explosives manufacturer. Rather than acquisition of new technologies and small companies, however, the new venture programme would be based on Du Pont's own R. & D.

Du Pont had learned in the 1910s how difficult it was to diversify through R. & D. alone. The anti-trust climate of the post-war era, however, and the company's self-definition as a chemical company (a definition stemming from the post-war planning conducted by the executive committee during World War II) limited Du Pont's alternatives. Two brief examples illustrate these factors. Had Du Pont had the legal go-ahead and the intention to purchase a large pharmaceuticals company, the company's life sciences programme would probably be far ahead of where it is today. Second, Du Pont pioneered the production of pure silicon, but not wishing to compete with its primary customer, Texas Instruments, and seeing itself as a chemical firm, Du Pont did not integrate forwards into doping and eventually lost out when TI integrated backwards to produce its own silicon.

Not discouraged by the setbacks of the late 1950s, all of Du Pont's technical organizations responded enthusiastically to the executive committee's call for new products. The company had money to spend on new ventures—about $300 million, the executive committee estimated. In the ensuing decade of the 1960s, Du Pont doubled its rate of new product introduction from 2.5 to 5 per year but at an extraordinary cost—$2 billion.

The new venture programme led to the revitalization of the once-powerful development department, which had atrophied following Fin Sparre's retirement in 1944. The Executive Committee gave the department responsibility for developments that did not fall logically into the province of any industrial department or that the industrial departments did not want to pursue. This marked something of a departure from the company's post-war decentralized research strategy, which placed complete responsibility for diversification on the industrial departments. The Development Department launched seven major new ventures and invested in a like number of embryonic companies (i.e. became minority shareholder in what the development department hoped would become emerging technology-driven growth companies). Five of the new ventures, which included the successful Automatic Clinical Analyzer, were commercialized, but Du Pont eventually divested itself of its holdings in the small companies.

Although the development department was beefed up, the Executive Committee's actions fell well short of what the manager of the new venture programme, Edwin Gee, wanted. Gee advocated the establishment of a diversification department, but the committee did not see the need and reasserted that the industrial departments were still expected to be the principal seat of innovation within the firm. Instead of moving the company in new directions, the new venture programme of the 1960s resulted in a deepening of commitment to Du Pont's existing lines of business—to more of the same, in other words. Many of the largest and most expensive ventures were, in fact, in the textile fibers department. There is irony in this because much of the rationale for launching the entire new venture programme was to lessen the company's dependence on textile fiber earnings, which had dominated the company and which could not last forever.

Without a programme for reallocating corporate resources, textile fibers continued to dominate the company. When that department's earnings suffered a significant drop in 1966, just six years after the beginning of the new venture programme, doubts began to arise about the entire programme. The new ventures that had been launched were not yet capable of contributing much to the company's profits. In fact, most were still losing money. Moreover, textile fibers' profits dived just when Du Pont needed unprecedented amounts of cash to finance its new venture programme. Sizing up the situation in 1967, the executive committee concluded that a major cause of the deterioration of the company's position was poor investment decisions in new ventures, and estimated that about one-third of them would have to be written off as losses.

In the late 1960s, Du Pont attempted to deal with this situation by terminating some ventures and scrutinizing more carefully departmental requests for additional new venture funds. Mostly, though, the company rode out the storm initiated by the burst of innovation that had marked the first part of the decade. By the end of the 1960s, the executive committee declared the entire new venture programme a failure; the company had spent $2 billion to develop over forty new products. But many of these were facing extended periods of operating losses, and some appeared to be beyond hope.

Du Pont's predicament called into question the company's entire post-war strategy. Large-scale projects had become so expensive and risky that the company could afford only a limited number of them—a far different situation from that which had previously prevailed, when the company funded any project that had sound technical merits. Nevertheless, the executive committee remained reluctant to play a stronger role in the direction of research and development activities until capital shortages in the mid-1970s forced it to select particular projects for development and commercialization. Only gradually did the committee assume a larger role in the management of Du Pont research.

RETHINKING AND RESTRUCTURING, 1970–1980

When the executive committee declared the entire new venture programme a failure, thereby also condemning the policy of long-term, high-risk, high-investment, high-projected-earnings innovation first enunciated by Pierre S. du Pont in 1907, it was truly at a loss about what to do. Therefore, during much of the 1970s, a decade marked by extreme economic turmoil, the company used its research organizations to keep its existing businesses healthy and to try to salvage the more promising of the new ventures of the 1960s. In the early part of the decade, the committee was unwilling or unable to revise or restructure the way in which the company managed its R. & D. function. Simple cutting back on expenditures was the easiest course to follow, and that is what the company did. But the Arab oil embargo of 1974 hit the company very hard in all of its polymer areas; textile fibers was particularly vulnerable. Du Pont faced unprecedented cash shortages. The situation called for radical action.

The election of Edwin A. Gee to the executive committee in 1970 provided the critical catalyst for change. As noted earlier, Gee had been the one selected by president Crawford Greenewalt to head the development department's new venture efforts of the 1960s. He had called for the creation of a diversification department charged with the responsibility of moving the company into new businesses and had lost that battle. Both as the development department's new ventures manager and later as head of the development department Gee had also lost all of his attempts to wrest primary responsibility for diversification away from the industrial departments. He had consistently argued that the industrial departments were too busy protecting their businesses to think radically about the future and that until Du Pont had some means to reallocate its research resources, it would be business as usual. Later, as general manager of the fledgling photo-products department, Gee well knew how much he could have done in developing new businesses for the company if he had only had some of textile fibers' huge R. & D. funds to work with. Nevertheless, by taking over the development department's Automatic Clinical Analyzer venture and purchasing Berg Electronics, Gee's photo products became a star performer. The department demonstrated that Du Pont could build high-return businesses in entirely new areas without vast outlays of R. & D. expenses and capital.

Gee brought to the executive committee an agenda for the restructuring of Du Pont R. & D. When Irving Shapiro, an attorney without any intellectual or emotional commitments to R. & D., became the chief executive officer, Gee was able to achieve most of his goals before he left the company to head International Paper. Gee's agenda contained several items. First, he proposed to cut fundamental research and concentrate on areas in which Du Pont had an existing competitive advantage. He argued that central research's areas of excellence should follow, not precede, the company's areas of business and corporate commitment. Gee also maintained that Du Pont should be more wary of launching high-risk and high-investment ventures. After a major venture was initiated, the executive committee had to pay more attention to it than in the past, because the venture could become a significant drain on the company's earnings. Gee also advocated cutting R. & D. expenditures on products that had fallen to commodity status, and he called for more emphasis on the purchase and sale of technology. Much of Du Pont's research, Gee believed, should be aimed at lowering production costs as a means of gaining competitive advantage. Gee urged his colleagues on the executive committee to find a means by which the committee could assume a greater role in the allocation of the company's research resources; that is, Gee wanted to establish a mechanism for the reallocation of research resources in the company. Finally—and this was a bitter pill for those in the research ranks to swallow—Gee advocated cutting R. & D. spending by 10 to 15 per cent in the early 1970s in addition to reductions caused by inflation.

Except for the final recommendation, the Executive Committee did not act immediately on Gee's recommendation. But by the end of the decade, the committee had seized control of the research process along the lines advocated by Gee. Short-range research on existing businesses was emphasized under both Gee's leadership as research liaison on the committee and that of his successor in that role, Richard Heckert, the current chief executive officer at Du Pont. Central research was forced to take a more careful look at how it could aid the industrial departments. The vehicle Heckert arrived at to achieve this was to establish a goal for CRD's funding; within a few years Heckert wanted 50 per cent of CRD's budget to come from industrial department research contracts. (As late as 1950, CRD had derived half of its expenditures from the departments, but by the 1970, this figure was less than 10 per cent.)

In 1975, the executive committee split up the development department. A small piece of the department was grafted on to central research, and that department became known as central research and development. The new department was given primary responsibility for diversification, and the industrial departments were encouraged to stick close to the businesses that they were already in.

The other segment of the old development department was renamed the corporate plans department and given the important charter of evaluating overall corporate strategy, helping chart the company's future strategy, and recommending to the executive committee how it should allocate its resources to implement that strategy.

To provide tighter control over the company's extensive R. & D. activities the executive committee decided in 1978 to have Central Research and Development report directly to one of its members. The committee also charged this research executive with the responsibility for overseeing the work of the research directors of the industrial departments, although these research directors continued to report formally to their respective industrial department heads. In Du Pont parlance, the research executive had 'dotted line' responsibility over all departmentally controlled R. & D.

Thus, these 1978 changes signalled yet another step back towards the centralization of research that had prevailed at Du Pont in the 1910s.

At the same time that the strategic functions of research management were being centralized, there was also a countercurrent to this development: the proliferation of profit centres throughout the company. Profit centres were created to make particular segments of Du Pont's businesses more responsive to the market. The thinking was along much the same lines as the original decentralization of 1921: industrial departments had become too large and unwieldy to be responsive to the market. By decentralizing the departments through the creation of profit centres, whose managers were given responsibility for research, manufacturing, and marketing functions, these businesses would be more competitive.

While achieving their goals, profit centres have created definite problems in the management of research at Du Pont. Foremost among these problems is the process of promotion. With relatively small research groups associated with these profit centres, little opportunity exists for advancement except by leaving research and going into general management. Reassignment of research personnel has also proved problematical. Because of these problems, Du Pont has not reached a set formula for the creation or management of profit centres, and it has been done case by case, with each centre tailoring its research to its own particular needs.

When the executive committee determined to have central research and development report directly to one of its members, Edward G. Jefferson was chosen to assume that expanded role. Working with both the head of the corporate plans department and the director of central research and development, Jefferson determined to move the company into an expanded research and development programme in the life sciences. The rapid development of recombinant DNA technology and related developments in the life sciences in the later 1970s appeared to be opening major growth opportunities for Du Pont if it correctly positioned its R. & D. Jefferson saw particular possibilities for the new technology in crop protection, pharmaceuticals, and medical diagnostics, areas in which Du Pont had been trying to become leader. Largely through Jefferson's persuasion, Du Pont announced in 1980 that it would be dramatically expanding its life science research and would build a major new laboratory at the Experimental Station to house this research. This facility, dedicated in 1984 as the Greenewalt Laboratory, is located at the Experimental Station not far from a greatly expanded agricultural chemical research laboratory. The company has also invested in other laboratory space at other locations devoted to life sciences research in a number of areas. The life sciences venture represents the most dramatic expansion of Du Pont R. & D. since the late 1940s and perhaps the most significant departure from its once strictly tailored self-image as a chemical company.

Jefferson's subsequent purchase of Conoco in 1981 definitely ended that self-image. But more importantly, for reasons that are beyond the scope of this chapter, it has allowed the top management at Du Pont to centralize more fully control over Du Pont's industrial departments, thereby reinforcing the slightly earlier centralization of the company's overall research and development programme.

CONCLUSION

It is difficult to summarize in a few words eight decades of research and development at Du Pont. What impresses me, however, is the continuity in the company's faith in R. & D. throughout this century. Except in the turbulent decade of the 1970s, Du Pont did not question its faith in R. & D. In that decade, some executives recognized that the company's zealousness about R. & D. had led to sloppy management of R. & D. and excessive allocation of resources. Therefore they determined not to destroy R. & D. but rather to cut back on it a little and, more importantly, to give it sharper focus—a development that was probably long overdue. While being fully faithful to the value of R. & D. to the company, executives at Du Pont, from members of the first executive committee to those on the committee in the 1980s, have not really found a permanent solution to the problem of finding the best way to manage R. & D. Du Pont's management of R. & D. has varied significantly over the last eight decades, moving back and forth between decentralized and centralized structures. This is but one of the perennial problems of managing research in a large modern corporation.

APPENDIX
DEVELOPMENT OF DU PONT R. & D.

1. *Founding, 1902–11*

Company focus: explosives manufacture; consolidation and rationalization of the industry.

Major institutional development:
- consolidation of two-thirds to three-fourths of explosives industry;
- establishment of Eastern Laboratory for high explosives research, 1902;
- estabishment of Experimental Station for general research, 1903;
- decision to pursue both centralized and decentralized research simultaneously, 1904;
- beginning of anti-trust action against Du Pont, 1907;
- initial thoughts on need for diversification.

Major R. & D. projects:
- Eastern Laboratory:
  non-freezing dynamite;
  safety explosives;
  improved manufacturing processes.
- Experimental Station:
  Stabillite military powder (failure);
  continuous black powder manufacturing process (failure);
  pulp powder keg (failure);
  glycerine synthesis (failure);
  stabilization of smokeless powder with diphenylamine.

Company sales and earnings ($ millions)

	1904	1906	1907	1911
Gross receipts	26.1	30.8	31.7	33.4
Net earnings	4.4	5.3	3.9	6.5
Total assets	57.2	66.6	70.9	83.2

R. & D. expenditures

	1904	1906	1907	1911
Total expenditure ($ 000)	n.a.	150	219	309
% of sale	n.a.	0.5	0.7	0.9
% of earnings	n.a.	2.8	5.6	4.7
R. & D. staff (no.)	n.a.	n.a.	n.a.	111

Note: n.a. = not available

2. *Centralization and Diversification, 1911–21*

Company focus: movement away from being purely an explosives manufacturer, diversification into chemicals, paints, plastics, nitrocellulose-coated sheet structures.

Major institutional developments:

- centralization of company under one general manager, 1911;
- centralization of research, 1911;
- tighter centralization from reorganizations in 1914;
- anti-trust conviction, 1912;
- forced divestment of part of black powder and high explosives businesses (formation of Hercules and Atlas), 1913;
- initial diversification efforts and approval by Executive Committee of Development Department's master plan for diversification:
  purchase of Fabrikoid Company, 1910,
  purchase of Arlington Company, 1915,
  purchase of Harrison Brothers Paint Company, 1917,
  other minor acquisitions, 1917–20;
- establishment of Jackson Laboratory (dyestuffs), 1917;
- establishment of Delta Laboratory (Celluloid), 1917;
- establishment of Redpath Laboratory (photographic film), 1920;
- World War I (vast expansion of smokeless powder capacity, vast profits, vast problems of utilizing capacity after war).

Major R. & D. projects:

- TNT;
- TNX;
- tetryl;
- synthesis of amyl acetate;
- synthesis of glycerine (failure);
- synthesis of diphenylamine;
- synthesis and manufacture of dyestuffs.

Company sales and earnings ($ millions)

	1911	1914	1916	1918	1919	1921
Gross receipts	33.4	25.2	318.8	329.1	105.4	55.3
Net earnings	6.5	5.6	82.1	43.1	17.7	7.6
Total assets	83.2	83.4	217.9	303.3	241.0	252.2

R. & D. expenditures

	1911	1914	1916	1918	1919	1921
Total ($ 000)	309	250	639	2,748	3,447	1,734
% of sales	0.9	1.0	0.2	0.8	3.3	3.1
% of earnings	4.7	4.5	0.8	6.4	19.5	22.8
R. & D. staff (no.)	111	95	200	475	652	135

3. *Decentralization, Acquisition and Rationalization, 1921–35*

Company focus: diversified chemical company, achieved through acquisition and rationalization of technologies and companies.

Major institutional developments:

- decentralization of management (including R. & D.), 1921;
- acquisition of French rayon technology, 1920–1;
- acquisition of French cellophane technology, 1923;
- acquisition of French (Claude) high-pressure ammonia process, 1924;
- joint venture with French in photographic film, 1924;
- acquisition of French cellulose acetate technology, 1928;
- other acquisitions: Viscoloid (1925), National Ammonia (1926), Grasselli (1928), Krebs Pigment (1929), Roessler and Hasslacher Chemical (1930), Commercial Pigments (1931), Newport Chemical (1931), Remington (1933);
- establishment of élite fundamental research programme in central Chemical Department, 1927;
- Patents and Processes Agreement with ICI, 1929.

Major R. & D. projects:

- TEL process;
- Freon process;
- straightening out Claude process;
- Duco fast-drying lacquer;
- discovery and development of neoprene, 1930 onwards;
- discovery of first truly synthetic fibre, 1930;
- discovery and development of nylon, 1934–40;
- major fundamental research work in polymer chemistry under Wallace H. Carothers.

Company sales and earnings ($ millions)

	1921	1926	1931	1936
Sales	55.3	90.4	163.5	250.1
Operating income	7.3	90.4	19.0	44.3
Ave. operating investment	84.7	140.8	365.4	449.5
Ave. no. employees	n.a.	15,228	31,041	45,938

Note: n. a. = not available.

R. & D. expenditures

	1921	1926	1931	1936
Total exp. ($ 000)	1,734	2,224	5,400	7,652
% of sales	3.1	2.5	3.3	3.0
% of earnings	23.8	23.9	28.4	17.3
R. & D. staff (no.)	135	241	755	912

4. *Nylon and the 'New Nylons' Mentality, 1936–60*

Company focus: 'chemical' company growing through internal generation of technology based on fundamental research.

Major institutional developments:

- nylon a smash hit, grows like Topsy;
- late New Deal attacks on Big Business, resulting in eight anti-trust cases against Du Pont, 1944–52;
- Nye Committee hearings, 1934–46; Du Pont labelled 'Merchant of Death';
- World War II (Du Pont not anxious to be involved in armament);
- Manhattan Project (Du Pont designs, builds, and operates Hanford plutonium works; designs and builds semiworks pile and separation facilities at Oak Ridge, Tenn.);
- total 'new nylons' strategy worked out by 1944–6;
- dramatic expansion of Du Pont R. & D., particularly at Experimental Station;
- rapid growth of world-wide competition in post-war period.

Company sales and earnings ($ millions)

	1936	1941	1946	1951	1956	1961
Sales	258.1	480.1	648.7	1,531.1	1,888.4	2,191
Operating income	44.3	57.5	84.0	139.4	254.8	258
Ave. operating investment	449.5	628.4	891.8	1,553.2	2,251.7	3,121
Ave. no. employees	45,938	60,029	72,002	86,878	89,449	87,057

R. & D. expenditures

	1936	1941	1946	1951	1956	1961
Total ($ millions)	7.65	12.4	26.0	71.9	140.3	160.8
% of sales	3.0	2.6	4.0	4.7	7.5	7.3
% of earnings[a]	17.3	21.6	31.0	51.6	55.1	62.3
R. & D. staff (no.)	912	1,341	1,800	3,376	4,582	3,787

[a]These figures are 30–40% lower if calculated as exp./exp. + income.

5. *Maturation and Response, 1960–70*

Company focus: diversification through new ventures based on storehouse of Du Pont knowledge.

Major institutional developments:
- court-ordered General Motors stock divestiture;
- reactivation of the Development Department;
- establishment of formal new venture programme, 1960;
- capital shortages begin to be experienced for first time in almost sixty years.

Major R. & D. projects:
- during this decade Du Pont commercialized forty-one new products. See Table 22.1 in Hounshell and Smith, 1988: 528.

Company sales and earnings ($ millions)

	1961	1965	1970
Sales	2,191.0	2,999.3	3,618.4
Earnings	258.0	384.9	328.7
Ave. operating investment	3,120.6	4,267.0	6,505.0
Ave. no. employees	87,057	106,013	115,126

R. & D. expenditures

	1961	1965	1970
Total expenditures ($ millions)	160.8	238.0	270.0
% of sales	7.3	8.0	7.5
% of earnings[a]	62.3	61.9	82.2
R. & D. staff (no.)	3,787	4,295	4,362

[a] Expenditures/income.

6. *Rethinking and Restructuring, 1970–80*

Company focus: fuzzy—prop up existing businesses, minimize launching of high-cost new products, contend with wide, energy-related swings in the economy, cash flow out of existing businesses, coping with apparent collapse of synthetic fibres business.

Major institutional developments:

- tightening reins on industrial department heads;
- increasing centralization of research units of company under one . Executive Committee member;
- break-up of Development Department:
  addition of one piece to form the Central Research and Development Department,
  other piece becomes the Corporate Plans Department;
- increasing willingness of Executive Committee to reallocate research resources along consciously developed strategic lines;
- purchase of Continental Oil Company (Conoco), 1981.

Major R. & D. projects:

- Kevlar development;
- process improvement in a number of key intermediates;
- life sciences research launched.

Company sales and earnings ($ million)

	1970	1975	1980
Sales	3,618.4	7,221.5	13,652.0
Net operating income	328.7	271.8	716.0
Ave. operating investment	6,585.0	11,418.0	17,448.0
Ave. no. employees	115,126	132,235	135,900

R. & D. expenditures

Total expenditures ($ millions)	270.0	336.0	484.0
% of sales	7.5	4.7	3.6
% of earnings[a]	82.2	123.6	67.6
R. & D. staff (no.)	4,362	4,593	4,050

[a]Expenditures/income.

REFERENCES

CHANDLER, ALFRED D., Jr. (1962), *Strategy and Structure* (Cambridge, Mass.: MIT Press).

DU PONT, PIERRE S. (1908), Letter to C. M. Barton of 17 Aug., Papers of E. I. Du Pont de Nemours & Co. (ser. II, pt 2, box 167, Hagley Museum and Library, Wilmington, Del.).

E.I. Du Pont de Nemours & Co. (n.d.), *The Story of the Eastern Laboratory, 1902–52* (Pamphlet Collection, Hagley Museum and Library, Wilmington, Del.).

HOUNSHELL, D. A., and SMITH, J. K. (1988), *Science and Corporate Strategy: Du Pont R & D, 1902–80* (New York: Cambridge University Press).

[11]

DISCUSSION PAPER

• *ABSTRACT*

Industrial research is acclaimed as the motor of contemporary American economic growth. This paper discusses several recent histories of corporate laboratories and industrial research by explicating common themes and assumptions. Alfred D. Chandler's work on business organization in late nineteenth-century America is the interpretative framework used in two recent and important histories – Leonard Reich's The Making of American Industrial Research: Science and Business at GE and Bell, 1876–1926, *and George Wise's* Willis R. Whitney, General Electric, and the Origins of US Industrial Research. *Both works provide rich accounts concerning the reasons why firms invested in research laboratories. Both also reveal the problems inherent in using Robert Merton's sociology of scientists in discussions of researchers working outside of the discipline-oriented university. The remainder of the paper replaces the normative sociological framework found in these recent histories with an analysis that locates the first corporate laboratories within the history of late-nineteenth- and early twentieth-century American science.*

Accounting for Research: New Histories of Corporate Laboratories and the Social History of American Science

Michael Aaron Dennis

Move over capital and labour! Make way for research and development. The traditional engines of economic theory are not yet antiquated, but contemporary economists, businessmen and government bureaucrats agree that American industry must invest heavily in scientific research, as well as in the development of new technologies, to counter the competitive threats posed by 'high tech' European and Asian firms. As trade deficits mount and economic recovery slows, research and development are seen as the foundations of a revitalized America. Supporters of President Reagan's expensive Strategic Defense Initiative (SDI) justify

Social Studies of Science (SAGE, London, Newbury Park, Beverly Hills and New Delhi), Vol. 17 (1987), 479–518

the massive expenditures with promises of profitable, advanced technology 'spinoffs'. Even America's trading partners compete for SDI contracts; no nation wants American corporations to become the sole beneficiaries of government largesse. The equation underlying such behaviour is simple: research generates new ideas and new technologies, both of which yield profits. Few today would disagree with the claim of a previous generation's students: '[i]ndustrial research is today a major source of new production techniques and new or improved products or services'.[1] All attempts to discover the exact relationship between research and profitable technologies – projects TRACES and HINDSIGHT – yield, at best, ambiguous results.[2] Nevertheless, the belief that research pays big dividends is widespread, and it is an important aspect of current discussions on the state of the American economy.

Contemporary American observers assume the indispensable character of a company's research and development laboratory, yet such laboratories are of recent vintage. General Electric founded the nation's first corporate research laboratory in 1900. Although several other firms (most notably AT&T and Kodak) also founded laboratories in the years between 1900 and 1914, the majority of such labs date from the period between the world wars. Clearly the assumption that research and development are corporate necessities is a twentieth-century belief. Until recently, corporate laboratories were seldom examined by historians of American science and technology. Now there is a growing body of literature on the topic. Two recent works – Leonard Reich's study of the labs at GE and AT&T and George Wise's biography of Willis Whitney, the GE Research Laboratory's first director – testify to the growing interest in the relations between science and industry.[3] Combined with earlier works – like David Noble's on science, technology and corporate capitalism; Stuart Leslie's biography of Charles Kettering, the first director of the General Motors Research Corporation; studies on research and development at Du Pont by Jeffrey Sturchio, John Smith and David Hounshell; and Reese Jenkins on the founding of the Kodak Research Laboratory – we can present an interesting pointillistic picture of the development of relations between science and industry in late nineteenth- and early twentieth-century America.[4]

The first part of this paper summarizes this literature, paying particular attention to those works dealing with laboratories founded before World War I, as well as focusing on the reasons why firms established research laboratories. The need to understand and respond to new technologies threatening existing markets was the major reason for the creation of corporate research laboratories in the years between 1900 and 1914. Patents were the main products of the new laboratories; corporations used patents to protect existing markets as well as define new ones. The most detailed and extensive work discussed in this section is Leonard Reich's *The Making of American Industrial Research*. Although other works (especially those by Wise and Jenkins) are used to supplement Reich's argument, no other work integrates the organizational and intellectual aspects of corporate laboratory life so successfully. Reich's book incorporates the types of arguments made in Wise's biography of Whitney and Jenkins's discussions of the Kodak lab.

Part one of this paper, then, summarizes histories of corporate laboratories; part two exploits the methodological quandary at the centre of each work – the use of an ahistorical sociology of science in accounts of an important historical change – to develop a more historical understanding of the emergence of the first corporate laboratories. This part locates these new institutions within the existing framework of the sciences in late nineteenth-century America by exploring two issues. First, the question of 'pure science' and its relation to the new laboratories is discussed. What did scientists, especially the often-quoted Henry Rowland, author of 'A Plea for Pure Science', mean when they used the phrase 'pure science'? If we recognize the complex moral and political dimensions of 'pure science', can we write an interesting supplement to the new organizational histories of company labs? Second, we address the question of the massive explosion in the absolute number of corporate laboratories and the widespread use of the phrase 'industrial research'. How are we to account for the proliferation of industrial research laboratories? In 1920 there were 300 company labs, by 1927 almost 1000 self-proclaimed industrial research laboratories.[5] Reich and Wise succeed in showing why individual firms established laboratories, but their focus obscures the truly innovative character of these new institutions within the history of the sciences. The founding of corporate laboratories, coupled with the establishment of other *research* institutions in the early part of the twentieth century, marked the beginnings of a new set of relations between American scientists and their patrons.

Science *and* Industry

Scientists and industry were not strangers before the establishment of the General Electric Research Laboratory in 1900. On the contrary, a small number of scientists found full-time employment in corporate testing and quality-control laboratories or as part-time consultants. Both practices revealed a company's willingness to *use* science, but an unwillingness to *do* science. Prior to the creation of corporate research laboratories, science and technology were products firms purchased, not processes in which companies invested. In testing laboratories, which were often simply benches in the manufacturing plant, scientists, especially chemists, confirmed that raw materials met specifications, monitored production quality, and made existing production processes more efficient. For example, in 1875, the Pennsylvania Railroad hired Charles Dudley, a Yale PhD, to manage a testing laboratory for the analysis of raw materials, to understand the cause of boiler scale, and to examine the relationship between steel's chemical and physical properties.[6] By the time he retired, Dudley's staff included nearly three-dozen chemists. The standardization and calibration of electrical measuring devices, as well as the setting of company-wide values for important electrical constants (for example, the ohm) were the tasks of workers in the General Electric Standardizing Laboratory established in 1896.[7]

With the advent of testing 'labs', businesses purchased the scientist's analytical skills and ability to make improvements in *existing* processes and equipment. Invention, the development of patentable devices and processes capable of altering a firm's competitive ability, remained the task of individuals outside the firm. Patents were the foundation of the science- and technology-based chemical and electrical firms – Edison General Electric, Thomson-Houston, AT&T, Eastman Kodak – developing in the 1880s and 1890s. Patents were also one reason behind the wave of mergers that swept American business in the 1890s. Take the case of General Electric, formed from the union of Edison General Electric and Thomson-Houston in 1892. By 1890 each firm found its advancement in several markets – electric traction, lighting and power – stymied by the possibility of infringement suits on patents held by each company. Merger solved the legal problems while the integration of the two firms by the Thomson-Houston management created a seemingly more powerful corporation.[8] Merger alone did not change pre-merger policies of either buying patents that threatened existing markets, or of temporarily hiring independent inventors or part-time academic consultants to develop new technologies or to solve problems.

Consulting opened a profitable role for academics and independent inventors. In 1864, Benjamin Silliman, Yale's chemistry professor, earned $50,000 for work done on behalf of California oil interests. In 1898 two MIT chemistry professors, Arthur A. Noyes and Willis Whitney, developed a process to reclaim valuable solvents from the waste generated by the American Aristotype Company's photographic paper production process. While Aristotype cut costs and effectively competed with industry giants like George Eastman and Leo Baekeland, the two professors earned royalties of $1000 a month.[9] The success of Arthur D. Little's scientific consulting firm, founded in 1886, demonstrated the widespread character of 'hiring out' for invention and development.

The use of outside consultants and the purchase of menacing patents were corporate strategies for dealing with technological change. These practices reveal two important corporate assumptions concerning invention and technology. First, many executives apparently believed that technologies were relatively stable and subject to only minor changes. One cause for such a belief was that firms were still labouring in the 17-year monopoly period granted by the original patent. Engineers in manufacturing plants could, and did, make patentable improvements and developments in products whose underlying technologies were relatively unchanging.[10] Such patents often extended the lifetime of the original patent's monopoly. Unfortunately, once a dramatic technological change occurred, as in the switch from direct to alternating polyphase current, engineers experienced with only the older technology were at a disadvantage. Under these conditions a firm could either buy the patent(s) to the invention or employ individuals familiar with the new technology to design a similar device while avoiding the new patent(s). For example, Westinghouse, GE's main competitor, held the rights to Nicola Tesla's AC motor patents. General Electric hired two engineers to develop an AC motor circumventing Tesla's patents. Although successful, the two engineers were not long in GE's employ.[11] As late as the 1890s, GE executives still believed that inventions came from outside the firm.

Second, these two strategies for dealing with change were supported and justified by a psychological portrait of the inventor as an individual often unsuited to working in the engineering department of a large corporation.[12] Nicola Tesla, the flamboyant Yugoslavia-born inventor of an alternating current motor, undoubtedly acted as the example proving this psychology's accuracy.[13] Rather than employ and thereby constrain such individuals, firms like GE and AT&T put proven inventors, like Elmer Sperry and Thomas Edison, under contracts securing the company's rights to patents in particular fields. Contrary to 'pop' psychology, inventive individuals did work in the large companies. Charles Proteus Steinmetz, GE's master of alternating current, designed AC motors at GE's Schenectady works from 1893 until his death in 1923.

Corporate assumptions concerning science, technology and invention generated strategies – patent purchasing and the use of hired hands. Because each approach relied upon resources beyond the firm's control, such strategies were intrinsically uncertain. There was no guarantee that a firm could purchase the necessary patents, or hire individuals with the requisite skill whenever a need arose. Given this past, what forces lay behind the creation of research laboratories at GE, Western Electric and Eastman Kodak in the period between 1900 and 1912?

Science *in* Industry

Before this new work on corporate research labs, there were two views about the origins of labs at GE and other companies. The first emphasized the convergence of corporate need and the development of the sciences. As the products of the science-based electrical and chemical industries became more sophisticated, only individuals with advanced training in the relevant sciences could make useful developments. The second approach made the new labs appear as 'universities-in-exile' because original work of university quality sometimes emerged from behind corporate walls. Historians confused the work of individuals with the work of the laboratories as a whole.[14] Rather than focus upon individual researchers or intellectual developments, the new histories examine the corporate contexts in which company labs originated. Large corporations invested in research because they were faced with the possibility of losing valuable markets to competitors with more advanced technologies. The new laboratories replaced older, more uncertain methods of dealing with technological change by making the firm itself the source of patentable products and processes.

But corporate need alone did not bring the new institutions into existence. There were also important 'supply-side' prerequisites. Only large, profitable corporations with secure market positions, firms like GE, AT&T and Kodak, possessed the capital to invest in laboratories, assets that might not yield immediate dividends. General Motors only founded the GM Research Corporation in 1920, after Alfred P. Sloan transformed a loose-knit amalgamation of different companies into a single, decentralized firm with a set of secure market shares.[15] Equally important was the existence of an educational infrastructure. By 1900 American universities were producing BAs and PhDs in the physical sciences as well as the new engineering disciplines, capable of solving problems of relevance to industry. Although the authors of the new histories discuss these same supply-side prerequisites, each argues that these were necessary, but insufficient conditions for the establishment of corporate research labs. Only after individuals within a firm recognized that technological changes threatened to undermine the company's market position did companies found laboratories 'separate from production facilities, staffed by people trained in science and advanced engineering who worked towards a deeper understanding of corporate related science and technology'.[16]

Let us now briefly examine the reasons why three companies – GE, AT&T and Kodak – established research laboratories in the years between 1900 and 1912.

At General Electric, German and American developments in incandescent lighting technologies had the potential of making GE's major revenue source, the carbon filament light bulb, obsolete. In Germany, Walter Nernst invented a lamp generating 50 percent more light and much less heat than GE's bulb. In the United States, Westinghouse, GE's domestic rival in the bulb business, financed the work of inventor Peter Cooper Hewitt in the development of mercury vapour lamps. GE did nearly no lamp development work. Charles P. Steinmetz, GE's resident scientist-engineer, recognized the importance of these outside developments; in July 1897 he suggested to his superiors that GE establish an 'electrochemical laboratory' to pursue developments in the lighting field. Even after enlisting the support of inventor Elihu Thomson, a GE consultant and founder of one of GE's corporate predecessors, Steinmetz failed to convince GE's fiscally conservative management to expend money on a lab not directly connected to production. Only after Steinmetz enrolled GE's chief patent attorney Albert Davis, did the proposal make headway. Davis made the appropriate argument to GE's cost-conscious management: 'If someone gets ahead of us on this development [referring here to mercury vapor lamp technology] we will have to spend large sums in buying patents or patent rights, whereas if we do the work ourselves this necessity will be avoided'.[17] With Davis's and Thomas's support, GE's management finally accepted Steinmetz's proposal, but only after making it clear that the lab needed to produce useful results and not waste money. As the Research Laboratory's first director, Willis Whitney of MIT, realized, GE hired him for a purpose; the lab was 'not primarily an asylum for indigent chemists'.[18]

Reich and Jenkins tell similar stories about the founding of laboratories at AT&T in 1909 and Kodak in 1912.[19] The installation of new management at Bell, following J. P. Morgan's 1907 acquisition of the company, started a series of events leading to the founding of the Western Electric Research Branch in 1909. Even though the new management operated under severe budget constraints, company president Theodore N. Vail and chief engineer J. J. Carty were dedicated to the goal of wiring the country and the ideal of 'one system, one policy, universal service'. To this end, the company engaged in the purchase of local telephone companies as well as the development of the technology (an electronic long-distance repeater) needed to provide coast-to-coast telephone service. Through the invention of a long-distance repeater AT&T provided long-distance service, while securing an important set of patents related to continuous wave radio transmission.[20] The technologies needed for long-distance wireless and wired telephony were nearly identical; by acquiring patents in both fields AT&T protected itself from unexpected technological changes in radio that might undermine its corporate investment in a wired national system. Unfortunately, the task of developing the devices for long-distance telephone service proved difficult; a year's effort by the staff of Western Electric's Engineering Department yielded little progress. Frank Jewett, a Chicago physics PhD (1902) and one of Western Electric's Assistant Chief Engineers, realized that contemporary work in the physical sciences, especially 'molecular physics' had important implications for communications technologies. Jewett outlined a plan in which scientists and engineers working 'with full coordination and under proper direction' explored possible repeater technologies drawn from relevant areas of physics. To do this Western Electric needed to hire 'skilled physicists'. J. J. Carty, Jewett's superior and AT&T's Chief Engineer, endorsed Jewett's plan to the AT&T Board of Directors, who approved the necessary expenditure.

Although we know far less about the immediate circumstances surrounding the founding of the Kodak research laboratory in 1912, the existing discussion in Jenkins's analysis of the photographic industry suggests a similar set of circumstances.[21] George Eastman, Kodak's entrepreneurial founder, feared that another firm might develop a colour film for the large and growing amateur photography market. To avoid such a possibility, Eastman hired Charles K. Mees, a University College London PhD who specialized in photographic chemistry, to organize and manage a research laboratory. Mees also had considerable business experience; he served as a managing director of a British dry plate company, Wratten & Wainwright, which Eastman purchased to acquire Mees's talents. Once again, markets and possible technological changes furnished the impetus for establishing a research laboratory.

The new accounts of the establishment of corporate research labs emphasize the importance of the corporate recognition of technological change and its effects on existing markets. In Reich and Wise, we are reading the influence of Alfred D. Chandler's work on business organization in nineteenth- and twentieth-century America.[22] Chandler argued that centralized, or vertically integrated firms developed in America in response to technological changes in both the production and distribution of goods. Centralization and administration began in the railroads where technological necessity made it important to centralize functions concerning scheduling, maintenance and finance. The railroad and telegraph lines created an infrastructure capable of supporting nation-wide markets. As technologies of mass production (continuous flow, assembly line, and so on) came into existence, salaried managers began linking their production facilities to supplies of raw materials as well as networks of distribution. The integrated firm put everything but the customer under the control of a single group of managers. To use Chandler's metaphor, the visible hand of corporate management replaced the invisible hand of market forces in allocating resources and distributing finished goods. According to the accounts summarized above, the founding of corporate research laboratories was another step in the development of the centralized firm. Just as United States Steel integrated backwards into the sources of its raw materials, so did firms like GE, AT&T and Kodak integrate backwards into another of their raw materials – knowledge. Chandler suggested as much in *The Visible Hand*, but Reich and Wise clearly demonstrate that, by financing research laboratories, companies partially insulated themselves from the problem of external technological change.[23] Research was a risk-reducing investment in two ways. We have already noted the first, and the second was similar to benefits reaped by other integrated firms – heightened barriers to entry. That is, any firm wanting to compete against GE or AT&T now had to invest in research as well. Finally, we should note that companies garnered a public relations benefit by appearing in a direct relationship with prestigious sciences like chemistry and physics.

If we now have a convincing organizational argument explaining why companies invested in research, we still lack an explanation as to the origins of the idea to invest in research. Charles Steinmetz championed the laboratory idea at GE before enrolling powerful supporters; Frank Jewett developed the plans for the Western Electric Research Branch and sold them to J. J. Carty. Steinmetz and Jewett were PhDs, one trained in Germany, the other educated in America. Within each corporation each occupied a role as the individual who linked the solution of company technological problems to the establishment of laboratory-based research facilities. Why? Both historians of GE assume that Steinmetz was well aware of the laboratories established by the German chemical, pharmaceutical and electrical firms in the 1870s and 1880s. In Jewett's case, Reich assumes both the influence of German examples as well as the GE lab's 1909 success in developing a ductile tungsten filament. In neither case is there any textual or archival material cited to substantiate these influences. If Steinmetz thought the German labs important, why did he not mention them in his proposal?[24] Jewett's brief appointment at MIT appears more important than either the success of GE's research lab or German influences in his plans for Bell. If GE's success was influential, how could Jewett fail to notice that it took Willis Whitney's lab nine years to develop the new, market-dominating filament? Given Louis Galambos's argument that each firm literally reinvented the processes of managerial integration in the period between 1900 and 1914, why should we assume AT&T learned anything from GE on establishing a corporate research laboratory?[25] The irony here is simple – in the midst of a wave of organizational histories we find that individuals made an important difference. If we are to understand the origins of the ideas about research in industry, we might profit from knowing more about men like Steinmetz and Jewett. We are dealing with the distant corporate cousins of a well-known kind of man, the inventor-entrepreneur, who by virtue of his own labour invented and patented devices to fill particular needs while raising the capital to manufacture and market his inventions. Working in another context, Steinmetz and Jewett were recognizing the technological challenges faced by their firms and enrolling the interests of superiors in pursuing laboratory research as a strategy to meet and conquer those challenges. The two were selling an institutional arrangement capable of meeting immediate demands, as well as long-term goals of either corporate diversification or the protection of existing markets. If men like Jewett and Steinmetz are related in their actions to the independent inventor-entrepreneurs who came before them, we should remember one crucial difference. Men like Elmer Sperry and Elihu Thomson originally worked for themselves.[26] Through their inventions, they sought public recognition and profit. Steinmetz and Jewett were employees; they defined themselves with respect to corporations that existed before their arrival and would continue after their departure. Their plans benefited the firm as they secured their own corporate positions. If the inventor-entrepreneur acted in the public world at large, corporation men acted in a new 'public' realm – the large integrated firms that developed in late nineteenth-century America.

Establishing laboratories did not guarantee their continuing existence. Corporate laboratories only gained institutional security after completing the project for which they were established. At GE, Willis Whitney, the Research Laboratory's first director, realized that the development of a new filament would take time; therefore, he solicited tasks from the manufacturing division in an attempt to make the laboratory an indispensable part of the firm. Whitney succeeded in improving GE's carbon filament lamp, as well as solving manufacturing problems, but neither he, nor his growing staff were able to beat European firms in the race to invent metallic filaments capable of rendering the improved carbon lamp obsolete. In 1907, several foreign firms developed processes for the manufacture of brighter and more powerful metal (tungsten) filaments; GE purchased the American rights to these patents. GE created the lab to avoid just this situation, yet the laboratory failed to provide GE with an in-house process. The penalty for failure came as the firm responded to the 1907 panic; GE management did not shield the lab from company-wide cutbacks as it had done in the past. Budget cuts forced Whitney to fire fourteen of his forty-four-member professional staff. Institutional security appeared in 1909, after William D. Coolidge, one of the remaining PhDs, invented a process to manufacture ductile tungsten filaments. In the aftermath of Coolidge's success the GE laboratory worked closely with the manufacturing division to improve the lamps while continuing research on other possible filaments.

Breakthroughs like Coolidge's marked only the beginning of a research agenda. Work on alternative filaments and related lamp technologies was a pre-emptive strike on potential competition. By patenting around the tungsten filament process, GE secured its market position for several patent lifetimes. Practices like these raised the cost of entering the lamp trade. Any firm wanting to compete with GE in the bulb manufacturing business had to avoid infringing upon GE patents; if the lab did its job well, the cost of developing an as yet unpatented filament technology would discourage all but the most determined and wealthy of competitors. The defensive use of patents also underlay the Western Electric Research Branch's radio research project.[27] Bell had no intentions of entering the radio field. On the contrary, the patents generated by the Research Branch made it impossible for anyone to use radio as a technological substitute for the wired telephone system. Corporate laboratories at GE and Bell were patent production centres, and patents were the instruments capable of preserving existing markets and defining new ones.

A Central Paradox

Although the new histories of industrial research discuss many facets of each laboratory, none are as interesting as Leonard Reich's attempt to understand the work done in the GE and AT&T labs. Discussions of corporate laboratory research are generally confused from the start, because authors feel a need to distinguish science from technology, or fundamental from applied science. Reich argues persuasively that trying to distinguish between science and technology in the new labs is an impossible task. Take the case of Irving Langmuir's studies of electron emission at GE. Such work came about as Langmuir sought an understanding of how the tungsten filament light bulb worked. Using standard off-the-shelf bulbs, Langmuir made observations and measurements yielding a sophisticated (and Nobel Prize-winning) account of how the bulb worked, as well as ways to improve the bulb, including changing the shape of the filament and introducing an inert gas into the bulb to prevent blackening. In an earlier article, Reich called Langmuir's method the 'natural study of the artificial', an interesting and revealing appellation.[28] Langmuir transferred simple laboratory techniques learned in the university for the study of objects 'found' in nature to the corporate lab for the study of manufactured goods. In Langmuir's work, Reich identifies the workings of the physical sciences in general since the mid-nineteenth century, not simply the character of 'industrial research'. To study electrons and the other aspects of atomic theory, one needed a device capable of emitting electrons. Langmuir used mass-produced light bulbs; researchers in university labs either purchased or made their own discharge tubes. University scientists were doing nothing different from Langmuir. They too were developing an understanding of the technologies that generated electrons. In their papers academic researchers might write about infinite, ideal radiators, but such discussions were grounded in an understanding of laboratory apparatus. Both Langmuir's work and that of his university colleagues hinged on an unspoken assumption: the 'natural' world worked like the devices that men used to study and to represent the 'natural' world.[29] Increasingly since the nineteenth century, if not earlier, the most esoteric modern science has been developed through the study of phenomena generated by equally esoteric instruments or technologies; it is pointless to argue over where science ends and technology begins, or vice versa.

Not everyone in the corporate labs was a Nobel Prize winner, but everyone in the labs at GE and Bell worked to manufacture the laboratory's common product – patents. In an analysis of Bell's radio work, Reich shows that such research was fundamental, fundamental to technology and to patent development.[30] In the lab, Jewett's skilled physicists studied Lee de Forest's audion (precursor to the vacuum tube) or any other technology of corporate interest, and developed 'technological theories', often mathematical in character, to explain the device's workings. Within the framework of these theories, laboratory researchers devised 'design methodologies', often graphical or mathematical statements, expressing acceptable parameters for components of different kinds of vacuum tubes. Using the design methodologies, other laboratory workers developed vacuum tubes for different purposes; many of them were patentable. At this juncture laboratory workers cooperated closely with patent attorneys, whose importance grows in these new histories. At GE, patent attorneys sometimes spotted inventions where researchers saw only their laboratory apparatus.[31] With the issuance of a patent, a researcher might publish his results or simply allow the patent to serve as publication. Patents emerge as an alternative literary form for the corporate scientists, one that secured priority for both the researcher and the firm within the sphere of corporate security and the marketplace. Technological theory found its material expression in two forms – the patent or published paper, and the device itself. Given this framework, even Langmuir's work is an example of technological theory – one leading to improvements in an existing device (the light bulb), as well as to a deeper understanding of electron emission.

Reich's extensive discussions and analyses of work in the GE and Bell labs are of more than passing historical interest. He is challenging those historians and philosophers of science who believe that the research done in a corporate lab is inherently inferior to the work done in a university lab. In his analysis of Langmuir's Nobel Prize-winning (and hence, unquestionably scientific) research, Reich identifies the technological basis of modern science. By arguing that it is impossible to differentiate between science and technology in the new labs, Reich upsets traditional hierarchical views of the science–technology relationship. Both moves confound those seeking to distinguish the work of corporate from university labs strictly on the basis of the content of each institution's research. Through his understanding of the work of the corporate researchers, Reich legitimizes these scientists as the intellectual equivalents of their academic colleagues and equally worthy of study by historians of science. Instead of arguing that there are distinct intellectual differences between the research done in corporate and university laboratories, Reich argues that the difference lies in the organization and goals of corporate research as opposed to university research. Reich and Wise both demonstrate how corporate goals clearly influenced the publication practices of company scientists and engineers. Firms only allowed their research staff to publish after patents were granted on the researchers' laboratory work.[32] Patent attorneys and management, rather than the individual researcher, decided when the scientist might publicly present his results. Such practices made it clear that the scientist was now a corporate employee, a white collar wage-earner. We are now on the shoals where all our historical accounts run aground; implicit in these discussions of publication practices are assumptions about the 'nature' of the scientific enterprise, assumptions that owe more to Robert Merton's famous sociology of scientists than to the history of American science.

From his famous study of seventeenth-century natural philosophers as well as from the writings of contemporary scientists, Merton extracted four norms – communism, universalism, disinterestedness, and organized scepticism – that accounted for the behaviour of scientists.[33] Communism, the recognition that scientific knowledge was communal or public property, was most important. Publication or the presentation of papers at conferences was the means through which scientific knowledge became public property, as well as providing the materials upon which the reward system of science, Merton's greatest interest, acted. Because of the restrictions upon their publications, corporate scientists are forever second-class citizens in the ideal community defined by Mertonian sociology. Merton also assumed that the scientist's aim was the 'extension of certified knowledge'. The phrase 'pure science' represents Merton's 'certified knowledge' in this historical literature, while supposedly describing the goals of élite university scientists. Corporate researchers clearly advanced the bounds of 'certified knowledge', but only to the extent that such advancement intersected corporate needs. Once again, by virtue of their employment, the corporate scientist is caught outside the Mertonian fence.[34]

We are confronting the paradox at the centre of recent works on corporate laboratories. On the one hand, the authors, especially Reich, demonstrate that the work done in the labs was of a sophisticated scientific and/or technological character worthy of study. On the other hand, the sociological view of the scientist implicit in these studies denies that corporate researchers were genuine scientists. We can easily rid ourselves (and these texts) of this paradox by abandoning Merton's timeless norms in favour of a more historical framework, one offered by recent examples of disciplinary history. To do so implies some revisions of the stories in the new histories of corporate laboratories, especially in their treatment of both the academic disciplines and the activities of university scientists.

Both Reich and Wise observe that disciplinary research projects did not survive in corporate settings, but they define academic disciplines in largely intellectual terms, emphasizing abstract theory over practice. Such an approach serves an important purpose in the works under discussion by denying the claim that corporate laboratories were universities-in-exile, but it does so at a cost – the misrepresentation of university science. The phrase 'pure science' becomes a label for either the work or the goals of university scientists, but the authors define 'pure science' as the study of causes, the search for 'law-like' regularities, or knowledge without direct commercial application. There is a kernel of truth in this definition, but not an appreciation of the complex social, political and moral relations embedded in the concept of 'pure science'. When George Wise describes the end-product of an experiment in the GE lab as a 'negative discovery in pure science', his claim derives more from his *a priori* definition of 'pure science' than from the actual meaning of the phrase in the texts under discussion.[35] Using such definitions is part of the problem of employing the Mertonian sociology. Science, according to both Merton and the historians of corporate laboratories, possesses its own 'internal dynamic', an essence manifesting itself in various institutional contexts. The underlying model here is one of 'social heredity'; the genotype, science, expresses itself in a variety of institutional phenotypes: academic science, corporate laboratories, government bureaus, and so on. After reading the new histories of corporate labs, we should abandon such a model. The genotype does not exist. Instead, these new histories suggest that there are only institutional phenotypes which are later linked together to forge the image of an underlying unity. In the act of forging the apparent unity, the past is selectively erased. Such an erasure allows Reich and Wise to claim that their works are on the beginnings or origins of 'industrial research', a phrase that the individuals in their works never use until after the founding of the corporate laboratories. It is to this problem that we must now turn in order to revise the understanding of academic science present in these works.

The University, the Laboratory and the Marketplace

When addressing the question of the character of late nineteenth- and early twentieth-century American academic science, historians tend to cite the same two pieces of evidence. Upon hearing that his student, Frank Jewett, wished to pursue a career at Western Electric, Albert A. Michelson, the University of Chicago's Nobel Prize-winning physicist, declared that a corporate career involved the prostitution of the physicist's training and ideals. Jewett's 1939 testimony before the Temporary National Economic Committee is the source of this colourful and often-quoted comment.[36] Even if we trust Jewett's memory, how are we to interpret this comment? Was corporate employment the intellectual equivalent of the 'social evil', or was it, as Reich notes, the presence of patents that Michelson found so distasteful? The other highly-cited source for an understanding of the university scientist is Henry Rowland's 1883 address to the American Association for the Advancement of Science, entitled 'A Plea for Pure Science'. Historians of American science seize upon Rowland's address as a clear statement of the undeveloped state of late nineteenth-century American science as well as a 'definition' of 'pure science' as the search for causes. Following the interpretation Daniel Kevles developed in his influential work, *The Physicists*, historians read Rowland's talk as an example of 'best science' élitism. That is, Rowland is read as a discipline builder, arguing for the 'construction of an institutional pyramid, oriented toward the advancement of abstract physics, open to the talents at the bottom, and commanded at the heights by a best-science élite'.[37] No doubt Rowland was a discipline builder, but we possess only fragments of Rowland's conception of a discipline or of 'pure science'. We seem to forget that the university's primary role was pedagogical; all discussions of 'pure science' were conducted within a framework in which teaching remained the university's primary role.

The importance historians of science attach to Rowland's speech is directly related to the speaker's institutional affiliation. As the first physics professor at the Johns Hopkins University, Rowland should epitomize the mythology surrounding that institution's origins. In his seminal history of the American university, Lawrence Veysey argued that Johns Hopkins, with its emphasis on laboratory and seminar instruction, was the American version of the German research university, *wissenschaft* in Baltimore. Unlike the pre-Civil War college, whose goal was the production of good citizens through a well-defined regime of intellectual studies and social activities, the new university's goal was the generation of knowledge through research in the laboratory and the seminar.[38] Research was important at Hopkins, but the university's first president, Daniel Coit Gilman, constantly remarked that teaching remained the university's first task. From its inception, Hopkins offered both advanced or graduate instruction, as well as a college or undergraduate education. As Owen Hannaway, and more recently Larry Owens, observed, Gilman replaced the traditional forms of college discipline with the discipline of the laboratory and seminar while retaining the goal of making worthy citizens.[39] The citizen-building institutions of the college were identical to those training the next generation's professors at the graduate level. There was, however, a crucial difference between the citizen-building, or disciplinary apparatus, of the pre-Civil War college and the disciplines of the post-Civil War college and university. The former prepared individuals to work and live as individuals in local communities. The latter prepared individuals to work and live as members of organized groups in the developing national society. Laboratories and seminars were pedagogical innovations emphasizing cooperation. Through their work with each other and their teacher, students experienced a model of 'social harmony'.[40] The value of science lay in the *methods* of producing scientific knowledge, not the content of such knowledge. Writing in his president's report of 1883, Gilman declared that

> I can never rid myself of the belief that the essential value of the university does not depend upon the discoveries it makes, or the knowledge it accumulates and imparts, but in the character which it develops.... In the hunt for truth, we are not first hunters and then men; we are first and always men, then hunters.[41]

Gilman and his faculty did not separate the intellectual growth of the disciplines from the development of moral character in their students. Research and teaching were bound together, the latter more important than the former, even at the university where historians most want to locate the origins of the research spirit.

The emergence of the disciplines as the organizational framework of the new university reflected the already-mentioned emphasis on method. At the institutional level, disciplines defined areas of 'academic territory' most visible in university departments. The word 'discipline' also referred to the beneficial effect of study in the laboratory and the seminar. Practical skills, not abstract theory, lay at the heart of the dual meanings of 'discipline' in the academic world. Exposed to the methods of science through their laboratory experiences, the college student emerged prepared for life in a dynamic and turbulent world.[42] Graduate students mastered the skills of working with scientific instruments in the laboratory, while preparing for a professorial career.

Instruments and techniques were the intellectual foundations of the natural sciences in the nineteenth century, but such skills were developed and transmitted within a particular institution, the research school. With its laboratory and charismatic leader, the research school was almost always associated with a university. Research schools were the functional units of the academic disciplines that emerged during the nineteenth century; they were grounded in the simple belief that although geniuses discovered, well-trained or disciplined students 'transformed discoveries into disciplines'.[43] In the school's laboratory, students became disciplined research workers capable of teaching as well as continuing the work being done in the school. Teachers, whether for the German *Gymnasia* or the American college, and their theses were the products of the research school. The success of a school hinged upon the ability of the school's leader to place his students in other professorial positions where the former students would spread the teacher's techniques. Control of a publication also helped make a research school successful by providing a forum in which students could publish and acquire reputations that reflected back upon the leader. Research remained part of pedagogical practice, not a separate activity pursued for its own sake. At Giessen, Justus Liebig pioneered the study of organic compounds through his 'combustion apparatus' while campaigning for the establishment of chemistry as a discipline within the philosophical faculty of the Prussian university system, and securing his own position by training pharmacy students seeking greater scientific legitimacy.[44] Students from Europe and America left Liebig's lab with a sophisticated knowledge of a technique which they could use for pedagogical, as well as investigative, purposes.

Liebig's laboratory was among the most famous, but the emphasis on technique appeared in other fields. At Cambridge, the physiologist Michael Foster founded a school dedicated to solving the problem of the mammalian heartbeat. Developing methods for the isolation of the heart as the object of study became crucial in Foster's project. Henry Newell Martin, a Foster student and the first physiology professor at Hopkins, emphasized isolation techniques in his work and that of his students.[45] Carl Ludwig's students in Leipzig acquired a knowledge of physiology grounded in the manipulation of instruments Ludwig developed. Before returning to America, Ludwig's students purchased the instruments needed to continue their teacher's pedagogical and research practices.[46] The instruments generated graphical records which became the focus of enquiry; Reich's notion of the 'natural study of the artificial' applies equally well in the biomedical disciplines.[47] In Ira Remsen's chemistry laboratory at Hopkins, graduate students acquired the tools to work with the various organic chemical structures, work Remsen began in Germany. Eighty-four of Remsen's students went on to hold full professorships in colleges and universities.[48] Henry Rowland's students worked in fields, like spectroscopy, closely related to their teacher's mastery of experimental practices.[49] Forty-five students took doctorates with Rowland during his twenty-two-year tenure (1879–1901); another 165 worked in his laboratory as advanced students during the same period. Rowland's students went on to head departments at MIT, Harvard, Yale, Princeton, Columbia, Wisconsin and the University of California. Although our list of research schools is international, we should note that there were important national differences. In Germany, especially Prussia, government ministers played important and direct roles in disciplinary innovation through their power to appoint professors within an established university system.[50] In the United States, government actions, like the Morrill Land Grant College Act of 1862, indirectly affected disciplinary innovation by expanding the academic market place and creating a demand for teachers to fill the new positions.[51]

Given the primacy of research schools, their associated techniques and instruments, and the inseparability of teaching and research, how are we to read Henry Rowland's 'Plea for Pure Science'? Rather than read the speech as a diatribe against applied science, which was clearly not its intent, we should read the 'Plea' as an oration on the relations between the university and the marketplace in a young and growing commercial society. Rowland used a classical framework, one relying heavily upon republican conceptions of the marketplace, to explicate the relations between the student, the professor and the university within commercial society. Put simply, Rowland articulated a political economy of science linking the future of American science to economic prosperity and national pride, while justifying such claims on moral and political grounds.

In Rowland's 'Plea', applied science did not threaten pure science, nor was the former the opposite of the latter. On the contrary, in a new nation where 'the curse of Adam' fell upon everyone, applied science produced the necessities of life.[52] The application of science yielded things; electric lights and the telegraph were among Rowland's favourite examples of the applications of science. 'Pure science', however, was not the converse; 'pure science' was neither an object nor a form of knowledge. 'Pure science' was a 'form of life' found only in research schools. It is a definition best understood by noting that Rowland reserved his most virulent salvo for those

> who have every facility for the pursuit of science, who have an ample salary and every appliance for work, yet who devote themselves to commercial work, to testifying in courts of law, and to any other work to increase their already present large income.[53]

These men might only regain Rowland's respect by renouncing the title professor and assuming one more appropriate to their actual work – consulting chemist or physicist. Such men failed to set the proper example for their students, 'young men looking forward into the world for something to do'. Instead of seeing a 'high and noble life . . . more honorable than the accumulation of wealth', students saw only the familiar desire for riches. 'Pure science' was a pedagogical ideal in which the professor, through his teaching and research, inspired his students to devote their lives to similar pursuits. As teachers, men involved entirely in commercial science simply did not set the proper moral example. 'If our aim in life is wealth', wrote Rowland, 'let us honestly engage in commercial pursuits, and compete with others for its possession.'[54] University positions insulated commercially-oriented professors from fully competing with others. Nevertheless, equal opportunity in the market place was not Rowland's greatest concern. When professors sold their skills, or the products of their skills, they transformed knowledge into just another commodity whose value was subjectively determined in the marketplace. Knowledge as commodity threatened the pedagogical basis of 'pure science', undermining the explicit goal of offering young men an alternative to the pursuit of wealth. Students, like men in general, were 'gregarious' animals dependent upon the 'sympathy' of those around them for their happiness.[55] Few were capable of pursuing an independent path; fewer students could resist the allure of wealth, especially when the professor eagerly engaged in the marketplace. Once in the marketplace, both knowledge and its producers were reduced to a level no different from that of any other manufacturer and his wares. Subjectivity and the problem of an individual's identity were Rowland's main concerns here. In commercial society, economic man generated his sense of self from the 'mirror gaze of other men'; an individual's appetites and desires were fuelled by the passions of others.[56] Man's gregarious nature conspired against him; new technologies like the telegraph and the railroad only accelerated the market's levelling effects. The discipline of the research schools offered an alternative, one in which the student's identity derived from his work or research in the laboratory. In the marketplace, men not only encountered each other, but the endless play of exchange-values, and opinion; in the laboratory, individual opinion encountered an unchanging and unyielding 'Nature', whose absolute qualities forced a recognition of opinion's errors. If the laboratory produced a test for opinion, such a test acted upon men who were no different from those engaged in the pursuit of wealth. Just as the market place played upon man's dependence upon 'the sympathy of those around him' for his happiness, so did the research school's laboratory. The student's 'natural' desire to resemble those around him moulded the student in the image of the school's leader and his fellow students while transmitting the professor's dedication to teaching, as well as to the further understanding of 'nature' through the research school's techniques. The laboratory acted as an antidote to the 'distorting' power of the marketplace, but only to the extent that the university and its research schools remained insulated from the marketplace.

Knowledge as commodity also affected the sources of support for the university, the only institution capable of supporting the research schools of 'pure science'. If professors with largely commercial interests represented science in the local press, it was the 'death-blow for science in that region'. The problem lay in the workings of commercial society itself, and the sources of support for science. Establishing research schools cost far more than the sum of student tuition; government subsidies were 'out of the question', since public monies opened the recipient to 'political trickery'.[57] Only universities with large and growing endowments could support 'pure science', and the only means of acquiring money was through private patronage. Unfortunately, individuals reading about the commercial success of university scientists would believe the sciences capable of supporting themselves and direct their charity elsewhere. Commercial success and private patronage were mutually exclusive; the former corrupted the young men of the university, while dragging the university into the marketplace; this made private patronage impossible. With a large endowment, universities could afford to equip research schools and pay professors enough to make commercial work unnecessary, while removing the university from the source of corruption, the marketplace. Freed from fortune's financial vicissitudes, research schools and pure science would flourish.

There was nothing novel in either Rowland's fear of the marketplace and the dominance of exchange-values, or in his recognition of the debilitating effects of commerce upon young men. Such concepts were the foundation of republican political thought from the eighteenth century onwards, yet Rowland's use of these ideas differed dramatically from earlier uses. In previous generations, republican theorists contrasted the marketplace and the republic (*res publica*) while struggling to find a way in which the classical ideals of citizenship and *virtu* might survive in commercial society. As property, the traditional basis of citizenship, became subject to exchange in the market, the classical ideals were steadily eroded; economic man began to replace man, the citizen – the political animal. How civic society might survive in commercial society was the topic of countless works. Perhaps the most famous was, and remains, Adam Smith's *Theory of Moral Sentiments*, especially the 1790 edition.[58] Even Smith, the market's most vigorous defender, advocated an educational system to counter the market's debilitating power.[59] What separates Rowland from these earlier discourses on citizenship and the marketplace, in addition to time, is that he located the problem in the university, not society-at-large. The university was the polity threatened by commerce and its citizens were the professors and students. Most students simply acquired college degrees before proceeding into the growing American economy. For these individuals, the laboratory demonstrated a standard of truth independent of the market; discipline acted as a shield against the distortions of commercial society. For the student who chose to stay and work towards an advanced degree, Rowland wanted only the most secure circumstances – well-endowed universities capable of supporting faculty and students so that each group could proceed with the serious work of teaching and learning. Financial autarky removed economic circumstance from the lives of those living within the university. Put simply, the discipline-oriented university was a new *res publica* whose citizens – the professors and students in the research schools – found their fulfilment through teaching and research, the twin motors of the disciplines. Each activity propelled the university and the disciplines into the future; the very word, '*re-search*' emphasized repetition and time.

In *The Visible Hand*, Alfred D. Chandler discussed how the new, large integrated firms of the late nineteenth century provided the successors of the founding managers with an organization possessing both a history and a future, independent of the movements of individuals. Successful integrated firms were bastions of permanence in an otherwise rapidly changing world. Managerial capitalism also enhanced its image of permanence through remarkably stable oligopolistic market conditions. Following the interpretation developed in Robert Wiebe's *The Search for Order*,[60] Chandler suggested that the large firms were one way in which individuals imposed order upon rapid and dislocating social change.[61] Other historians, especially those of the organizational school – Louis Galambos, Samuel P. Hays, Ellis Hawley, to name only a few – have made similar arguments about the emergence of the 'organizational society' in the period between 1880 and 1920.[62] Americans used a variety of institutional forms – including trade unions, trade associations, the professions and integrated firms – to manage and respond to the possibilities opened by new transportation and communication technologies in the wake of a violent and devastating Civil War. In the language of contemporary social theorists, the new organizations were the 'secondary groups' to which individuals needed to belong if society were to avoid chaos and divisive social unrest.[63] The university and the disciplines were integral parts of the 'organizational society', training men to live in the new institutions, but in Rowland's address we also see something that all aspects of the organizational society share:

a recognition of the marketplace's corrosive, levelling power. Integrated firms, the professions, trade associations and trade unions were all attempts to escape the negative effects of the marketplace. The origins of this common recognition lay in the transformation of a political language, republicanism, into an organizational language – a way of discussing the relationship between individuals, private institutions and a rapidly changing society. Republican or civic thought offered a powerful understanding of the relations between the polity, the marketplace and the individual. Rowland, among others, especially social scientists, translated the basis of the civic tradition, man as citizen deriving his sense of self through citizenship, into the discourses of professionalization, trade unionism and syndicalism. All three processes were related to one another in late nineteenth-century social thought; each activity imparted a distinctive sense of identity to participants. Individuals no longer found their fulfilment or identity in the duties of citizenship, but in their actions within their firm, profession or union. In the university (and, more importantly, in the research schools of the disciplines), individuals created a world whose very activities – teaching and research – guaranteed its permanence while offering each person the opportunity to participate in a new kind of polity. Insulated from the direct effects of the marketplace, individuals developed selves within their 'secondary groups', private versions of the once public republic.

An intellectual framework originally developed to discuss political problems related to the effects of commercial society upon the individual became part of the framework for understanding private institutions and the marketplace. How and why such a move took place are questions for both debate and study. If the move did occur, as I have suggested, we are reading in Rowland and his colleagues part of what Hannah Arendt and Richard Sennet called, for different reasons, 'the decline of public man'.[64] Was the possibility of the nation as *res publica* shattered by the Civil War? Or was the movement of a republican framework into the private world itself an optimistic move? In Rowland's texts, the university affected commercial society while remaining free of the marketplace. With its endowment the university supported the research schools of 'pure science'; professors need no longer 'degrade' themselves by entering the marketplace. Yet in its laboratories and seminars, society's young men were prepared for life in a rapidly changing world. Minds formed in the laboratory were society's future, as well as the university's; they were minds 'destined to govern the world in the future and to solve problems pertaining to politics and humanity as well as to inanimate nature'.[65] Ultimately, the vision of *res publica* embodied in the university would return to society at large, but only when all problems were viewed from within the research school's laboratory.

The Market for Money

Private patronage was the key to Rowland's vision of 'pure science'. Americans were a generous people, but funds would only appear if professors lived lives of 'pure devotion' to their sciences, not lives of 'indolent ease'.[66] Even with good behaviour, 'pure science' was a hard sell; individuals who supported the sciences gave money for things. For example, Americans generated little astronomical knowledge relative to the nation's 'many excellent [astronomical] observatories'. Wealthy people gave money to build or purchase telescopes and domes, but donors made no provision for the instruments' long-term use.[67] For private patrons, the sciences were goods to purchase and display in public, a form of 'conspicuous consumption'. Rather than have patrons supply instruments, Rowland wanted donations for university endowments controlled by the university, not the patron. Securing patronage was difficult; in the frenzy for funds only the most prestigious title – university – attracted private money. Names that once possessed certain meanings – university, college, school professor, teacher – became meaningless as colleges proclaimed themselves universities and schools called themselves colleges. According to the 1880 report of the commissioner of education, the United States harboured over 400 self-proclaimed universities. Only in a 'democratic country' could such a confusing state of affairs exist, but the levelling and misrepresentations Rowland observed were the effect of the marketplace and the scramble to acquire funds. Furthermore, because 'higher education' could never 'pay for itself', fund-raising was never-ending. Institutions, like individuals, were reduced to a level in commercial society.

Once again, the problem was one of identity and subjectivity. Save for living a brutally ascetic life, Rowland offered only one suggestion on how to secure support for 'pure science'. The disciplinary research schools of the university needed a way to distinguish themselves from others who appropriated their title, if not their practices. Rowland articulated a set of institutional definitions through which he sought to escape the marketplace. Only schools with endowments of three to four million dollars or more were 'genuine' universities. If patrons accepted Rowland's criteria, only eight centres of higher education remained from the government list of 400.[68] Only these eight could support and maintain 'pure science'; those patrons seeking to support Rowland's devoted workers need only hand their money over to the endowments of the universities in question. Patrons would first supply the needs of the universities, then the colleges, and finally the schools. Students trained in the university's discipline-based research schools for a teaching career in the colleges' pedagogical laboratories, setting the proper moral example for their students.

Rowland's definitions were élitist and representative of a 'trickle-down' approach to education, but the definitions were also an attempt to preserve differences erased in the philanthropic marketplace. It was an impossible scheme, if only for reasons that Rowland well understood. Every college aspired to university status; none need accept a predefined position, least of all one defined by a university scientist. Why acquiesce to second-place status when the existing system allowed otherwise? More importantly, who would enforce the distinctions Rowland sought to preserve? Once unleashed, marketplace mentalities were impossible to harness. No wonder Rowland offered no other methods of securing patronage besides patience and devotion; the alternatives were unacceptable. Scientists could engage in entrepreneurial practices, attract patrons and secure funds or enter the commercial world as consultants and earn the dollars needed to practise 'pure science'. A. A. Noyes, an MIT chemist, used the royalties earned ($1000 a month) from a successful consulting contract to establish a laboratory of physical chemistry at MIT. In either case, knowledge became a commodity. Ironically, the very practices which Rowland sought to counter through laboratory experience were the practices necessary to found and support the laboratories.

Reading Rowland reveals that 'pure science' was neither the study of causes nor the antithesis of applied science. Equally important, the 'pure science' ideal did not separate research from teaching; research itself was a pedagogical activity. More revealing is the flexibility of Rowland's political economy. As the Johns Hopkins University endowment stopped yielding dividends in the late 1880s, Rowland entered into the business of consulting and training electrical engineers as well as physicists. According to Rowland, the pursuit of applied science by electrical engineers might one day yield the wealth needed to practise only 'pure science'.[69] Just as the relations between virtue and commerce were determined historically, so were the relations between 'pure science' and commerce – the new poles of Rowland's political economy of science. Circumstances forced a partial retreat from the ideals of 'pure science', but the interpretation of Rowland developed here is useful for our purposes.

Corporate Research Laboratories Revisited

If discipline-based research schools and technique were at the centre of late nineteenth-century science, how is this related to the emergence of corporate research laboratories? According to the new histories of company labs, the threat of new technologies to existing markets lay behind the corporate decision to invest in research. As I observed earlier, a common event in each history was the recognition by an individual – Charles Steinmetz at GE and Frank Jewett at AT&T – that the firm needed a *research laboratory* to improve existing technologies and develop related devices. Neither one spoke or wrote of founding an *industrial* research laboratory. In their proposals Steinmetz and Jewett, the two PhDs, appropriated the research laboratory, the very heart of the research school, and moved it from the university into the integrated firm. Instead of teachers and theses, the corporate laboratories manufactured patents and devices. Research, severed from its pedagogical foundations, became a separate activity dedicated to the solution of particular problems of corporate interest. Although Reich and Wise comment upon the relatively 'narrow' focus of corporate work, each is implicitly comparing the work of a single laboratory with the diversity of work done in an academic field or discipline, itself a set of research schools. Clearly, the appropriate comparison is between a research school and a corporate lab. An important point made in recent works, especially Reich's, concerns the technological basis of corporate research. That is, the laboratories took existing technologies – the audion at Bell and the carbon filament lamp at GE – and developed a sophisticated understanding of each device before improving or transforming the product in question. At GE, such work first led to the GEM (GE Metallic) lamp and later a metal (ductile tungsten) filament. From their understanding of Lee de Forest's audion, the Bell researchers developed a variety of successful vacuum tubes.

Research schools were just as 'narrow' as corporate labs; both focused on a particular instrument or problem. The schools fell along a spectrum ranging from those oriented towards the solution of a particular problem through the development of instruments and techniques (for example, Michael Foster's physiology school at Cambridge) to those which applied a given instrument or technique to a set of closely related problems (for example, Liebig's use of the combustion apparatus to study any organic compound). Just as the corporate lab covered the field with patents, so did the research school cover the field with students and theses. Steinmetz and Jewett shared an insight: the practices of the research schools, practices with which each man was familiar, were capable of being directed towards non-pedagogical ends. Obviously, corporations used laboratories for purposes different from the research schools, yet Steinmetz and Jewett shared part of Rowland's understanding. In the laboratory's discipline, Rowland found an antidote for the problems produced by commercial society, as well as a partial escape from the marketplace's subjectivity. In the research laboratory Steinmetz and Jewett found the instrument with which to control the marketplace and conquer competition. What Rowland sought to escape, Steinmetz and Jewett sought to dominate. The research school's engine, the laboratory, was portable.

Separating research from teaching was the crucial innovation of the new corporate labs. We take the separation of the two for granted, but it was a genuine accomplishment in the first decade of the twentieth century. Clearly the ideal of 'pure science', as articulated by the often-quoted Rowland, offered no such severance. The difficulty of disconnecting research and teaching is most apparent in the early history of the Carnegie Institution of Washington (CIW); a new organizational species – the private research institution – developed during the first decade of the century.[70] Dedicated to the advancement of knowledge and the support of the 'exceptional individual', it was endowed with ten million dollars' worth of United States Steel Corporation bonds; the Carnegie is, perhaps, the most important and least-studied institution in the history of American science. Academics believed that the new organization would simply dispense funds to universities; others, claiming the title 'exceptional individual', simply wanted the Carnegie's cash. Robert S. Woodward (the institution's second president) and the Board of Trustees engaged in a near-decade long debate on how best to fulfil the new organization's mandate. The trustees wanted to support largely university-based researchers through a minor grants programme which Woodward sought to eliminate. According to Woodward, prolonged CIW support of university work created a damaging dependence as local funding sources atrophied due to the (over) use of CIW money. Instead of supporting the research of others, Woodward wanted to create a research institution – a set of autonomous departments engaging in work outside the scope or means of the university. To that end, the Carnegie departments were generally large-scale data-gathering operations in fields as diverse as terrestrial magnetism, human embryology, and economic history.[71] On one occasion, Woodward defined the CIW as a 'university in which there are no students' – an informative, if not contradictory, description. After all, a university without students is not a university.[72]

Yet Woodward's definition revealed the problem of discussing research separate from the university's pedagogical function. The Carnegie trustees, like the second president of the Johns Hopkins University, Ira Remsen, saw little merit in separating the two activities.[73] If the impoverished character of Woodward's language, as well as the CIW trustees' beliefs, betray the residue of 'pure science', they also highlight the innovation involved in deciding that a corporation needed a research laboratory.

'Industrial research' is a phrase appearing in the titles of recent works by George Wise and Leonard Reich, but it was not a phrase used by Steinmetz and Jewett before GE and AT&T established laboratories. In fact, the phrase only begins to appear with any frequency during and after World War I. In turn, we should remember that only a handful of company labs existed before the war; the majority date from the interwar period. Our understanding of corporate labs founded during the interwar period is extremely limited. To date, historians have studied labs founded after World War I at only two decentralized firms, Du Pont and General Motors; these firms were fundamentally different organizations from large, integrated companies like GE and AT&T.[74] Changing technologies undoubtedly forced companies to invest in research after World War I, but there was an important supply-side presence which did not exist before the war – the National Research Council (NRC) and its Division of Industrial Research and Engineering. Before the war, the idea of creating a corporate laboratory had been the insight of individuals; after the war, it was the insight of an organization.

Historians of American science have long understood the importance of World War I in shaping the sciences.[75] Experience with cooperative research – that is, groups of scientists working together to solve a problem – is generally cited as the great benefit of the wartime work, but one might also argue that the war's great effect lay in showing everyone that research might flourish when separated from its pedagogical foundations. In both Germany and America, scientists worked on problems of importance to victory on the battlefield – ranging from the synthesis of nitrates for explosives to submarine-detection devices. Beliefs about the power of science to transform society existed long before the war, but the war offered scientists an opportunity to demonstrate their value to American society. War also produced an institutional framework for the mobilization of the sciences (the NRC) which was not discarded at the armistice.[76] The importance of science for the war effort translated into appeals for the private support of the sciences after hostilities ended. According to the leaders of the NRC – scientist-entrepreneurs like George Ellery Hale, Robert A. Millikan, Robert M. Yerkes and Vernon Kellogg – post-war reconstruction and future economic prosperity demanded that American society invest heavily in scientific research. The first issue of the NRC's *Bulletin* contained a collection of articles under the title, 'The National Importance of Scientific and Industrial Research'.[77] Authors included George Eastman of Kodak, Theodore N. Vail, president of AT&T, Elihu Root, chairman of the CIW Board of Trustees, as well as Hale, the NRC's honorary chairman. All vouched for the power and importance of the sciences, as well as for the preservation of the wartime organization, the NRC, during the peace.

As in other sectors of the 'associative state', private initiative and support were valued above government aid.[78] Organized along disciplinary lines, the NRC acted as a trade association for the sciences. Under the direction of psychologist Robert M. Yerkes, the NRC's Research Information Service created a roster of all American scientists while routinely gathering statistics on the employment of researchers. The Division of Industrial Research and Engineering, under the direction of AT&T's J. J. Carty, provided corporation leaders with tours of existing company labs and encouraged firms to establish research laboratories. Apparently, a visit to the AT&T labs helped persuade Du Pont executives to invest in research; however, it is unclear if the visit took place under NRC auspices.[79] All these activities were valuable, but none so important as the relationships between the NRC leaders and the new private patrons of the sciences, the executives of the Rockefeller Foundation (George Vincent, Wycliffe Rose and Beardsley Ruml), the Carnegie Corporation (James T. Angell) and private industry.

Private patronage created and maintained the sciences in America during the interwar period.[80] The NRC, and later the Social Science Research Council (SSRC), became the funnels through which the foundations and industry poured money into the academic disciplines. Fellowships for promising young PhDs were the most popular funding instrument. For the foundations, such fellowships avoided the problem of direct aid to universities with its potential for turning the university into a permanent foundation-dependent. Fears first voiced by CIW president Robert Woodward before the war persisted in the post-war period. Neither the foundations nor private industry supported academic science for its own sake; foundation executives and industrialists wanted to establish strong academic science departments to produce scientists. As Robert Kohler recently observed, foundations wanted scientists as 'agents of "rational social change"', and industry wanted a steady supply of scientists for the rapidly growing number of research and development labs.[81] Corporations endowed fellowships while also contracting with university scientists for research of mutual interest.[82]

Foundation and industrial patronage, as well as the work of the NRC itself, transformed both scientists and their knowledge into commodities. The dual meanings of 'discipline' encompassed in Henry Rowland's 'pure science' were effectively erased. The discipline, not the university, became the institutional framework in which the scientist viewed himself. Institutions like the Carnegie were colonized by the academic disciplines in the post-war period. New markets for physical scientists, especially in corporate labs, created both new career paths and new vocabularies. Research became the common denominator in the making of scientific careers, while the emergence of research positions outside the university had important effects upon university science. For example, what one observer called a 'fever of commercialized science' within the university during the 1920s was, in part, a result of training men to work in commercial laboratories.[83]

Robert Kohler's 'political economy of science' offers us a framework within which we can understand the disciplines during the interwar period. In his study of the development of biochemistry as a discipline, Kohler argued that American biochemists were able to create strong biochemistry departments in medical schools because of the movement to reform American medical education and the biochemists' ability to provide valuable pedagogical services in the pre-clinical curriculum of the reformed university medical school. In turn, American biochemical research developed along largely clinical lines as physicians turned to biochemists for diagnostic techniques as well as scientific prestige.[84] Biochemists were not the only biomedical scientists to recognize the opportunity opened by the science-worshipping reformers of medical education. Physiologists successfully invented their discipline in America at roughly the same time; anatomists also used the medical education reform movement to transform anatomy into an 'experimental science' grounded in the study of human embryology.[85] In each case, biomedical scientists justified their role in the medical school by claiming that the knowledge produced within their respective disciplines would ultimately yield improved clinical practice. The 'discipline-building window' accompanying the reform of medical education adumbrated the events following World War I as the physical and social sciences forged relationships with the foundations and industry.[86] Contrary to popular belief, the biomedical disciplines, rather than the physical sciences (physics, chemistry, mathematics, astronomy, and the like), were the paradigms for discipline building in America. 'A fever of commercialized science' was simply one effect of having new markets for graduates of the physical sciences, but I am not arguing that industrial needs completely directed academic science. The only way to answer the question of problem choice is to examine individual research schools, a task as yet undone. However, it is not far-fetched to suggest that the push to produce physicists, mathematically-trained electrical engineers, and other scientists aided in the creation of positions for theoretical physics and quantum mechanics in the physics departments of American universities.[87]

The market relationship between university scientists and the contexts in which their students worked affected the identity of the university scientists, as well as their choice of problems. Although university scientists received private monies to train scientists, they began to see themselves as researchers first and teachers second. Yet the majority of physics PhDs still went into academia.[88] We can see signs of the change in the new vocabulary employed by scientists, especially those working with the NRC. New phrases like 'pure' or 'fundamental research' and 'industrial research' suggested that the main activity of the scientist was now research; the adjectives modified a now independent noun and activity. According to J. J. Carty, motive distinguished the two kinds of research. Scientists conducted 'pure scientific research' for a 'philosophic purpose, for the discovery of truth', whereas industrial research, done by industrial scientists and engineers, always had the goal of 'accomplishing some utilitarian end'.[89] These definitions still appear in many contexts, but the identity change is clearest when we recognize that during the interwar period, 'pure science' and 'pure research' were used interchangeably, with none of the meanings implied by Rowland. Industrial research, with its emphasis on the corporate context of scientific investigation, had arrived.

The 1920s and 1930s saw the development of a disciplinary economy, or system of scientific knowledge, in the United States. It also saw the beginnings of a particular history and sociology of science, one familiar to all historians of science. Internalist history, now a pejorative label, best describes this genre. Much of this history was written by scientists who extended the histories of their disciplines backwards until encountering a famous seventeenth-century figure; ideas and errors were the major actors in these accounts. Such 'internalist' histories legitimized a relatively new organization of the sciences, one that redefined the sciences as engines of research while erasing a pedagogical past. More important for our purposes was the emergence of a sociology of science that drew its raw materials from a study of seventeenth-century experimental philosophers, and treated knowledge as simply another commodity.

Merton's sociology of scientists found the norms governing the scientist's behaviour in the rules and rhetoric of research for an unchanging scientific community. In fact, Merton's sociology only described the behaviour of scientists within their disciplines.[90]

Conclusions

We have come full circle from the recognition of a particular sociological view implicit in recent histories of corporate laboratories to a suggestion that the sociology in question was itself a product of the period in which the disciplines achieved their current form. In the second part of this paper, I have attempted to understand the development of corporate laboratories and industrial research from a perspective other than the organizational focus employed by Leonard Reich and George Wise, authors of two recent and important histories. By focusing upon the academic texts Reich and Wise used in writing about 'pure science', I have argued that there were important linkages between the history of university science and the corporate laboratories. These relations, as well as the complex meaning of 'pure science', were obscured as long as we absorbed Rowland's language into our own. The ideas and assertations of this second part only supplement the historical works summarized in the first half of this paper. Detailed case studies, like Reich's, are the books upon which graduate students cut their hermeneutic teeth; they are also the type of books needed if we are to understand the complex organization of the sciences in the twentieth century. However, if the second part of this paper has any value, it lies in showing that a fusion of the organizational approach with the contextual reading of intellectual and political history might yield rich rewards. Ostensibly, this is a Discussion Paper; therefore, I conclude with some suggestions for further research.

After reading histories of corporate labs, one cannot doubt the laboratory's economic value. What we need to know is: how are laboratory ideas and devices translated into finished goods? What are the relations between researchers, manufacturers, marketers and distributors? For example, at GE, Willis Whitney used laboratory space to set up pilot production lines so that laboratory workers could solve production problems. On other occasions laboratory scientists worked in the manufacturing plant. In his recent work on the diffusion of scientific knowledge, Bruno Latour argued that knowledge never left the laboratory.[91] Instead, he suggested that scientists transformed non-laboratory settings into labs (or lab-equivalents) through the use of standard weights, measures and practices. Scientific knowledge worked in seemingly non-laboratory contexts because those contexts were, in fact, like laboratories. To what extent, if any, did the successful production of the fruits of corporate research hinge upon the transformation of the manufacturing plant into a laboratory? Contemporary 'high tech' manufacturing plants resemble the popular, antiseptic image of the laboratory, but what kind of representation is our image of the laboratory? What did such a transformation imply for the workers in the plant? How did it affect management's approach towards its employees?

Closer examinations of the actual work done in corporate (as well as in university) labs might yield a better understanding of the labour processes of the various disciplines, especially the changing relations between professionals, technicians, and administrators. The questions suggested above mine the ground disturbed by the books in question. We might also ask: how is the interwar period and the emergence of a discipline-based framework for the sciences related to the present? What has changed since the military became the towering patron of the sciences? Here we might draw upon another historical tradition of the 1930s found in the work of Lewis Mumford, especially his book *Technics and Civilization*. As David Hounshell observed in his history of mass production in America, it was Mumford who realized that the mass-production economy of the 1920s required the mass consumption only war could provide.[92] Organized destruction is the most efficient form of consumption. With the proliferation of corporate laboratories, science became part of the mass-production/mass-consumption culture. Could we not suggest that the policy of 'mass producing' scientists between the wars had the unexpected effect of creating a supply that only the military could use? After all, the mobilization of the sciences for World War II never ended; the sciences benefit from a continuing cold war and arms race. Research laboratories moved from industrial and academic contexts to the front lines of defence; the history of these events remains unwritten.

• *NOTES*

A National Science Foundation Pre-Doctoral Fellowship, a Smithsonian Institution Pre-Doctoral Fellowship, and the generous support of my parents, all made this paper possible. I am grateful to Mary Fissell, David DeVorkin, Pamela Smith and Bill Leslie for important criticisms of the ideas in this paper, as well as to the staff of the National Air and Space

Museum's Space Science and Exploration Department, for their patience as I worked on it. I am also grateful to Robert Kargon, Louis Galambos and Robert Kohler, each of whom introduced me to different aspects of the history of American science.

1. George Perazich and Philip M. Field, *Industrial Research and Changing Technology* (Philadelphia, PA: Works Progress Administration, National Research Project on Reemployment Opportunities and Recent Changes in Industrial Techniques, Report No. M-4, 1940), 1; a more recent popular discussion appeared in David Sanger, 'US Research Sponsorhip Soars at Universities', *New York Times* (8 September 1986).

2. See George Wise, 'Science and Technology', *Osiris*, Vol. 1 (1985), 229–46.

3. Leonard Reich, *The Making of American Industrial Research: Science and Business at GE and Bell, 1876–1926* (Cambridge & New York: Cambridge University Press, 1985); and George Wise, *Willis R. Whitney, General Electric, and the Origins of US Industrial Research* (New York: Columbia University Press, 1985).

4. David Noble, *America by Design: Science, Technology, and the Rise of Corporate Capitalism* (Oxford & New York: Oxford University Press, 1977); Stuart Leslie, *Boss Kettering: Wizard of General Motors* (New York: Columbia University Press, 1983); John K. Smith, 'The Ten Year Invention: Neoprene and Du Pont Research, 1930–39', *Technology and Culture*, Vol. 26 (1985), 34–55; Smith and David A. Hounshell, 'Wallace H. Carothers and Fundamental Research at Du Pont', *Science*, Vol. 229 (2 August 1985), 436–42; Jeffrey Sturchio, *Chemists in Industry: Studies in the Historical Application of Science Indicators* (unpublished PhD dissertation, University of Pennsylvania, 1981), Chapter 4; and Reese Jenkins, *Images and Enterprise: Technology and the American Photographic Industry, 1839–1925* (Baltimore, MD & London: Johns Hopkins University Press, 1979).

5. Perazich & Field, op. cit. note 1, 65.

6. See Sturchio, op. cit. note 4, 80–81.

7. Reich, op. cit. note 3, 58.

8. Ibid., 42–59.

9. George Wise, 'Ionists in Industry: Physical Chemistry at General Electric, 1900–1915', *Isis*, Vol. 74 (1983), 7–21, at 11; Wise, op. cit. note 3.

10. On the importance of manufacturing engineers in invention and development, especially in the electrical industry, see Thomas Parke Hughes, *Networks of Power: Electrification in Western Society, 1880–1930* (Baltimore, MD & London: Johns Hopkins University Press, 1983), 160–64.

11. Reich, op. cit. note 3, 60–61.

12. Reich notes this particular psychology at GE: ibid., 61.

13. Tesla's notoriety extends to the present. According to the historian of the Naval Research Laboratory, Bruce Hevly, individuals still believe that as yet unpublished Tesla papers in the Navy's possession contain secrets of unbelievable value to the national defence of the United States. A more modest discussion of the inventor is found in Hughes, op. cit. note 10, 112–21.

14. Kendall Birr's studies of General Electric's lab are an example of this interpretation: see Birr, 'Industrial Research Laboratories', in Nathan Reingold (ed.), *The Sciences in the American Context: New Perspectives* (Washington, DC: Smithsonian Institution Press, 1979), 193–209. A quick tour of the various interpretations is found in George Wise, 'A New Role for Professional Scientists in Industry: Industrial Research at General Electric, 1900–1916', *Technology and Culture*, Vol. 21 (1981), 408–29, esp. 409–10.

15. See Leslie, op. cit. note 4, 90–95. Louis Galambos argues that only after firms completed integration and controlled their markets could they move into research. The

majority of firms were still working out the problems of centralization in the years before World War I; few were therefore organized enough to engage in research before 1914. See L. Galambos, 'The American Economy and the Reorganization of the Sources of Knowledge', in A. Oleson and J. Voss (eds), *The Organization of Knowledge in Modern America, 1880–1920* (Baltimore, MD & London: Johns Hopkins University Press, 1979), 269–82, at 275–81.

16. This is part of Reich's definition of industrial research: Reich, op. cit. note 3, 3.

17. Quote from Reich, op. cit. note 3, 66; on GE's decision, see ibid., 62–69.

18. Quoted in Wise, op. cit. note 3, 91.

19. This discussion draws upon Reich, op. cit. note 3, 151–84, esp. 158–59; on Kodak, see Jenkins, op. cit. note 4, 300–18.

20. On the development of continuous wave transmission technologies in general, see Hugh G. J. Aitken, *The Continuous Wave Technology and American Radio, 1900–1932* (Princeton, NJ: Princeton University Press, 1985).

21. See Jenkins, op. cit. note 4, 300–18.

22. Alfred D. Chandler, *The Visible Hand: The Managerial Revolution in American Business* (Cambridge, MA: The Belknap Press of Harvard University Press, 1977).

23. Ibid., 375.

24. If Steinmetz did mention the labs of the German firms, why did neither Reich nor Wise cite this comment? For a discussion of a German corporate research laboratory, see George Meyer-Thurow, 'The Industrialization of Invention: A Case Study from the German Chemical Industry', *Isis*, Vol. 73 (1982), 363–81.

25. We should note that C. E. K. Mees, the first director of the Kodak Research Laboratory, did visit GE's lab in 1912: see Reich, op. cit. note 3, 254.

26. See Thomas Parke Hughes, *Elmer Sperry: Inventor and Engineer* (Baltimore, MD & London: Johns Hopkins University Press, 1971); and W. Bernard Carlson, *Invention, Science and Business: The Professional Career of Elihu Thomson, 1870–1900* (unpublished PhD dissertation, University of Pennsylvania, 1984).

27. Leonard Reich first made this point: see Reich, 'Research, Patents, and the Struggle to Control Radio: A Study of Big Business and the Uses of Industrial Research', *Business History Review*, Vol. 51 (1977), 208–35.

28. See Leonard Reich, 'Irving Langmuir and the Pursuit of Science and Technology in the Corporate Environment', *Technology and Culture*, Vol 24 (1983), 199–221; and Reich, op. cit. note 3, 120–27.

29. On this point in general, see Trevor Pinch, 'Towards an Analysis of Scientific Observation: The Externality and Evidential Significance of Observational Reports in Physics', *Social Studies of Science*, Vol. 15 (1985), 3–36; H. M. Collins, *Changing Order: Replication and Induction in Scientific Practice* (London & Beverly Hills, CA: Sage Publications, 1985); and Steven Fuller, 'A French Science (with English Subtitles)', *Philosophy and Literature*, Vol. 7 (1983), 1–14.

30. Reich, op. cit. note 3, 205–17.

31. See Reich, ibid., 103.

32. Although corporate needs obviously came first, GE and Bell did encourage their researchers to publish, for several reasons. First, publications brought the lab free advertising, and could serve as a recruiting device. Second, publication allowed corporate scientists to participate in the larger scientific community. The company benefited if such participation brought its staff into contact with new and potentially profitable ideas and techniques. Third, even if the corporation derived no profit from the research staff's actions in disciplinary affairs, like professional society meetings, the firm still benefited. Allowing

staff to publish or participate in the activities of professional societies was part of the management of professionals who were corporate employees, yet who still derived aspects of their identities from non-corporate sources.

33. Robert K. Merton, *The Sociology of Science* (Chicago, IL: The University of Chicago Press, 1973), 267–78; the particualr essay in question, 'The Normative Structure of Science', originally appeared in 1942. See also Merton, *Science, Technology, and Society in Seventeenth Century England* (New York: Harper & Row, 1970), originally published in *Osiris* (1938).

34. Barry Barnes makes a similar point about sociological studies of contemporary industrial researchers: see S. B. Barnes, 'Making Out in Industrial Research', *Science Studies*, Vol. 1 (1971), 157–75; also important as a critique of Merton's sociology is Ian I. Mitroff, *The Subjective Side of Science: A Philosophical Inquiry into the Psychology of the Apollo Moon Scientists* (Amsterdam: Elsevier, 1974).

35. Wise, op. cit. note 9, 15.

36. For some uses, see Wise, op. cit. note 3, 169, 399n; Reich, op. cit. note 3, 264n; and Spencer Weart, 'The Rise of "Prostituted" Physics', *Nature*, Vol. 262 (1 July 1976), 13–17.

37. Daniel J. Kevles, *The Physicists: The History of a Scientific Community in Modern America* (New York: Knopf, 1979), 42–44. A different reading of Rowland's text, written from the perspective of the history of technology, is found in David Hounshell, 'Edison and the Pure Science Ideal in 19th-Century America', *Science*, Vol. 207 (8 February 1980), 612–17 (I owe this reference to Robert Friedman).

38. Lawrence Veysey, *The Emergence of the American University* (Chicago, IL: The University of Chicago Press, 1965), 121–79.

39. Owen Hannaway, 'The German Model of Chemical Education in America: Ira Remsen at Johns Hopkins (1876–1913)', *Ambix*, Vol 23 (1976), 145–64, esp. 151–53; and Larry Owens, 'Pure and Sound Government: Laboratories, Gymnasia, and Playing Fields in Nineteenth Century America', *Isis*, Vol. 76 (1985), 182–94.

40. See Owens, op. cit. note 39; and David Hollinger, 'Inquiry and Uplift: Late Nineteenth-Century American Academics and the Moral Efficacy of Scientific Practice', in Thomas Haskell (ed.), *The Authority of Experts* (Bloomington, IN: Indiana University Press, 1984), 142–56.

41. Daniel Coit Gilman, *Eighth Annual Report of the President* (Baltimore, 1883), 64.

42. Owens, op. cit. note 39, 182–94; George Frederickson, *Inner Civil War: Northern Intellectuals and the Crisis of the Union* (New York: Harper & Row, 1965); and Peter Hall, *The Organization of American Culture, 1700–1900* (New York: New York University Press, 1982).

43. On the historical importance of research schools, see Simon Schaffer, 'Scientific Discoveries and the End of Natural Philosophy', *Social Studies of Science*, Vol. 16 (1986), 387–420, esp. 406–13, quote on 412; Gerald Geison, 'Scientific Change, Emerging Specialties, and Research Schools', *History of Science*, Vol.19 (1981), 20–40; and J. B. Morrell, 'The Chemist Breeders: The Research Schools of Liebig and Thomas Thomson', *Ambix*, Vol. 19 (1972), 1–45. On a research school not associated with a university, see James A. Secord, 'The Geological Survey of Great Britain as a Research School', *History of Science*, Vol. 24 (1986), 223–75.

44. On Liebig, see Morrell, op. cit. note 43; R. S. Turner, 'Justus Liebig versus Prussian Chemistry: Reflections on Early Institute-Building in Germany', *Historical Studies in the Physical Sciences*, Vol. 13 (1982), 129–62; and W. H. Brock, 'Liebigiana: Old and New Perspectives', *History of Science*, Vol. 19 (1981), 201–18.

45. Gerald Geison, *Michael Foster and the Cambridge School of Physiology* (Princeton, NJ: Princeton University Press, 1978); Owens, op. cit. note 39, 192 and passim.

46. See George Rosen, 'Carl Ludwig and his American Students', *Bulletin of the History of Medicine*, Vol. 4 (1936), 609–50; W. Bruce Fye, 'Why a Physiologist? The Case of Henry P. Bowditch', ibid., Vol. 56 (1982), 18–29.

47. On the role of instruments in a nineteenth-century biomedical science, see Timothy Lenoir, 'Models and Instruments in the Development of Electro-Physiology, 1845–1912', *Historical Studies in the Physical Sciences*, Vol. 17 (1986), 1–54; on the more general role of instruments that appear to author their own text, see Bruno Latour and Steve Woolgar, *Laboratory Life: The Social Construction of Scientific Facts* (Beverly Hills, CA & London: Sage Publications, 1979), 43–90, and Pinch, op. cit. note 29.

48. Hannaway, op. cit. note 39, 145.

49. Robert H. Kargon, 'Henry Rowland and the Physics Discipline in America', *Vistas in Astronomy*, Vol. 29 (1986), 131–36.

50. R. S. Turner, 'The Growth of Professorial Research in Prussia, 1818–1848: Causes and Contexts', *Historical Studies in the Physical Sciences*, Vol. 3 (1971), 137–82.

51. See Veysey, op. cit. note 38; on the effects of the Morrill Act upon American agricultural science in particular, see the essays in Charles Rosenberg, *No Other Gods: On Science and American Social Thought* (Baltimore, MD & London: Johns Hopkins University Press, 1976).

52. Henry Rowland, 'A Plea for Pure Science', *The Papers of Henry Rowland* (Baltimore, MD: The Johns Hopkins Press, 1903), 593–613, at 597.

53. Ibid., 598.

54. Ibid., 598–99.

55. Ibid., 595–96.

56. Quotation from Michael Ignatieff, *The Needs of Strangers* (New York: Viking Penguin, 1986), 94.

57. Rowland, op. cit. note 52, 604.

58. On the importance of the 1790 edition, see Laurence Dickey, 'Historicizing the "Adam Smith Problem": Conceptual, Historiographical, and Textual Issues', *Journal of Modern History*, Vol. 58 (1986), 579–609.

59. This understanding of the civic or republican tradition owes its origins to the work of J. G. A. Pocock: see his *The Machiavellian Moment: Florentine Political Thought and the Atlantic Republican Tradition* (Princeton, NJ: Princeton University Press, 1975), 432–500, and *Virtue, Commerce, and History: Essays on Political Thought and History, Chiefly in the Eighteenth Century* (Cambridge & New York: Cambridge University Press, 1985). Also important are: Ignatieff, op. cit. note 56, 105–32; Nicholas Phillipson, 'Adam Smith as Civic Moralist', in Istvan Hont and Michael Ignatieff (eds), *Wealth and Virtue: The Shaping of Political Economy in the Scottish Enlightenment* (Cambridge & New York: Cambridge University Press, 1983), 179–203. Other essays in the Hont & Ignatieff volume influenced the hypothesis offered here, especially Peter Jones, 'The Scottish Professoriate and the Polite Academy, 1720–46', 89–118, and John Robertson, 'The Scottish Enlightenment and the Civic Tradition', 137–78. A useful discussion of republicanism is found in Joyce Appleby (ed.), 'Republicanism in the History and Historiography of the United States' (Special Issue), *American Quarterly*, Vol. 37 (1985). Also important are Drew McCoy, *The Elusive Republic: Political Economy in Jeffersonian America* (Chapel Hill, NC: University of North Carolina Press, 1980); Lance Banning, 'Jeffersonian Ideology Revisited: Liberal and Classical Ideas in the New American Republic', *William and Mary Quarterly*, Vol. 43 (1986), 3–19; and Daniel Walker Howe, *The Political Culture of the American Whigs*

(Chicago, IL: The University of Chicago Press, 1979), 299–305. An intriguing comment on *virtu* and discipline is found in Charles Taylor, 'Foucault on Freedom and Truth', in his *Philosophy and the Human Sciences, Philosophical Papers*, Vol. 2 (Cambridge & New York: Cambridge University Press, 1985), 152–84, at 165, and passim.

60. Robert H. Wiebe, *The Search for Order, 1877–1920* (New York: Hill & Wang, 1966).

61. Chandler, op. cit. note 22, 6–12.

62. Louis Galambos, 'The Organizational Synthesis in American History', *Business History Review*, Vol. 44 (1970), 279–90; Galambos, 'Technology, Labor, and Professionalization', ibid., Vol. 57 (1983), 471–93; Samuel P. Hays, *The Response to Industrialism* (Chicago, IL: The University of Chicago Press, 1957); Hays, 'The New Organizational Society', in his *American Political History as Social Analysis* (Knoxville, TN: University of Tennessee Press, 1980), 244–63; and Ellis Hawley, *The Great War and the Search for a Modern Order* (New York: St Martin's Press, 1979).

63. On this aspect of late nineteenth-century social thought, see Henrika Kuklick, 'Professionalization and the Moral Order: American Social Scientific Thought at the Turn of the Century' (unpublished ms., Department of History & Sociology of Science, University of Pennsylvania, 1985).

64. Hannah Arendt, *The Human Condition* (Chicago, IL: The University of Chicago Press, 1958); Richard Sennett, *The Fall of Public Man* (New York: Knopf, 1978).

65. Henry Rowland, 'The Physical Laboratory in Modern Education', in *Papers*, op. cit. note 52, 614–18, at 618; originally published in *Johns Hopkins University Circular*, No. 59 (1886), 103–05.

66. Rowland, op. cit. note 52, 605.

67. Ibid., 604.

68. Ibid., 599–603.

69. On Rowland's training of electrical engineers, see Robert Rosenberg, 'American Physics and the Origins of Electrical Engineering', *Physics Today*, Vol. 38 (1983), 48–54; and Rowland, *Papers*, op. cit. note 52, 624.

70. The following discussion draws upon Nathan Reingold, 'National Science Policy in a Private Foundation: The Carnegie Institution of Washington', in Oleson & Voss (eds), op. cit. note 15, 313–41; and Howard S. Miller, *Dollars for Research: Science and its Patrons in Nineteenth-Century America* (Seattle, WA & London: University of Washington Press, 1971).

71. An interesting exception was the Carnegie geophysical laboratory: see John Servos, 'To Explore the Borderland: The Foundation of the Geophysical Laboratory of the Carnegie Institution of Washington', *Historical Studies in the Physical Sciences*, Vol. 14 (1984), 147–85.

72. Robert S. Woodward, 'President's Report', *Annual Report of the Carnegie Institution of Washington*, No. 4 (1905), 32.

73. On Remsen and private research institutions, see Hannaway, op. cit. note 39, 159.

74. See the works by Smith, Sturchio and Leslie listed in note 4.

75. On the sciences and World War I, see Kevles, op. cit. note 37, 102–38; D. J. Kevles, 'George Ellery Hale, the First World War, and the advancement of Science in America', *Isis*, Vol. 59 (1968), 427–37; Daniel P. Jones, 'Chemical Warfare Research During World War I', in J. Parascandola and J. Whorton (eds), *Chemistry and Modern Society* (Washington, DC: American Chemical Society, 1983), 165–85; and to a lesser extent, Carol S. Gruber, *Mars and Minerva: World War One and the Uses of the Higher Learning in America* (Baton Rouge, LO: Louisiana State University Press, 1975).

76. On the importance of the NRC, see Robert Kargon, 'The New Era: Science and

American Individualism in the 1920s', in Kargon (ed.), *The Maturing of American Science* (Washington, DC: American Association for the Advancement of Science, 1974), 1–29; and Nathan Reingold, 'The Case of the Disappearing Laboratory', *American Quarterly*, Vol. 29 (1977), 79–101.

77. See George Ellery Hale, 'The Purpose of the National Research Council', Elihu Root, 'The Need for Organization in Scientific Research', and Theodore N. Vail, 'Relations of Science to Industry', all in the *Bulletin of the National Research Council*, No. 1 (1919–21), 1–44, at 1–6, 7–10, 12–14. Robert A. Millikan made an eloquent speech for private patronage: see Millikan, 'The New Opportunity in Science', *Science*, Vol. 50 (26 September 1919), 285–97.

78. Hawley, op. cit. note 62; and Ellis Hawley, 'Herbert Hoover, the Commerce Secretariat, and the Vision of an "Associative State"', *Journal of American History*, Vol. 61 (1974), 116–40.

79. On Du Pont and AT&T, see Smith, op. cit. note 4, 39.

80. The following discussion draws upon Reingold, op. cit. note 76; Kargon, op. cit. note 76; Robert Kohler, 'Science and Philanthropy: Wycliffe Rose and the International Education Board', *Minerva*, Vol. 23 (1985), 75–95; and Stanley Coben, 'American Foundations as Patrons of Science: The Commitment to Individual Research', in Reingold (ed.), op. cit. note 14, 229–48. On the raw magnitude of support, see Roger Geiger, *To Advance Knowledge: The Growth of American Research Universities, 1900–1940* (Oxford, London & New York: Oxford University Press, 1986), 140–90.

81. Kohler, op. cit. note 80, 75.

82. On academic industrial relations, see John Servos, 'The Industrial Relations of Science: Chemistry at MIT, 1900–1939', *Isis*, Vol. 71 (1980), 531–49; and John Swann, *The Emergence of Cooperative Research Between American Universities and the Pharmaceutical Industry, 1920–1940* (unpublished PhD dissertation, University of Wisconsin, Madison, 1985).

83. The quotation is from Spencer Weart, 'The Physics Business in America, 1919–1940: A Statistical Reconnaissance', in Reingold (ed.), op. cit. note 14, 295–358, at 301–02.

84. Robert Kohler, *From Medical Chemistry to Biochemistry: The Making of a Biomedical Discipline* (Cambridge & New York: Cambridge University Press, 1982), 1–8, 121–57.

85. See Franklin P. Mall, 'On the Flexner Report', *Anatomical Record*, Vol. 4 (1910), 278–80, esp. 280.

86. Given the role of the foundations in the development of the biomedical disciplines, we should investigate the extent to which foundation officers saw a relationship between their earlier work with the reformers of medical education, and the post-World War I period.

87. On quantum theory in America, see Kevles, op. cit. note 37, 155–69; and Stanley Coben, 'The Scientific Establishment and the Transmission of Quantum Mechanics to the United States, 1919–32', *American Historical Review*, Vol. 76 (1971), 442–66.

88. For example, see the statistics for physics PhDs and their employment in Weart, op. cit. note 83, 331.

89. J. J. Carty, 'The Relation of Pure Science to Industrial Research', *Annual Report, Smithsonian Institution, 1916* (Washington, DC: Smithsonian Institution, 1917), 523–31, at 524.

90. On history of science in America, see Arnold Thackray and Robert K. Merton, 'On Discipline Building: The Paradox of George Sarton', *Isis*, Vol. 63 (1972), 473–95; Thackray, 'The Pre-History of an Academic Discipline: The Study of the History of Science in the United States, 1891–1941', *Minerva*, Vol. 28 (1980), 448–73; and Nathan Reingold, 'History of Science Today, 1. Uniformity as Hidden Diversity: History of Science in the

United States, 1920–1940', *British Journal for the History of Science*, Vol. 19 (1986), 243–62. Also of interest on the history of the history of science in America are the articles in the Sarton Centennial Issue of *Isis*, Vol. 75 (1984), by Thackray, I. B. Cohen, John T. Edsall, Thomas S. Kuhn and Victor L. Hilts. On Merton's treatment of scientific knowledge as a commodity, see M. D. King, 'Reason, Tradition, and the Progressiveness of Science', *History and Theory*, Vol. 10 (1971), 3–32. An interesting reading of Merton 'in-context' is found in David Hollinger, 'The Defense of Democracy and Robert K. Merton's Formulation of the Scientific Ethos', *Knowledge and Society*, Vol. 4 (1983), 1–16.

91. Bruno Latour, 'Give Me a Laboratory and I Will Raise the World', in Michael Mulkay and Karin Knorr-Cetina (eds), *Science Observed: Perspectives on the Social Study of Science* (London & Beverly Hills, CA: Sage Publications, 1982), 141–70.

92. Lewis Mumford, *Technics and Civilization* (New York: Harcourt, Brace and Company, 1934); and David Hounshell, *From the American System to Mass Production, 1800–1932* (Baltimore, MD & London: Johns Hopkins University Press, 1984), 315.

Michael Aaron Dennis is a graduate student in the history of science at the Johns Hopkins University.
Author's address: Department of the History of Science, The Johns Hopkins University, Baltimore, Maryland 21218, USA.

[12]

Technological Knowledge without Science: The Innovation of Flush Riveting in American Airplanes, ca. 1930–ca. 1950

WALTER G. VINCENTI

In the early 1930s most metal airplanes of American origin were held together by rivets with dome-shaped heads protruding beyond the external surface of the aircraft. A decade later almost all such airplanes had rivets flush with the surface. To observers outside the airplane industry this change, if they noticed it at all, must have seemed straightforward and even trivial. In fact, since flush riveting had never before been used in plates as thin as those in aircraft, a great deal of learning was necessary. We may therefore ask, Why and how did the change to flush riveting take place? And what knowledge did the learning process generate that was not available before?

In earlier papers I have explored two examples of technological knowledge close to the interface between engineering and science, one concerned with experiment and one with theory.[1] My purpose here is to move to the other end of the spectrum and take up a case far

MR. VINCENTI, a professor emeritus of aeronautical engineering in the Stanford University School of Engineering, continues teaching part-time in Stanford's Program in Values, Technology, Science, and Society. He wishes to acknowledge his special indebtedness to Howard Wolko of the National Air and Space Museum; James Hansen, historian at the Langley Research Center; George Rechton and Eugene Speakman of the Douglas Aircraft Company; David Richardson and Charles Wordsworth of the Lockheed-California Company; Edwin Layton of the University of Minnesota; and Todd Becker, Robert McGinn, Stephen Kline, and Nathan Rosenberg of Stanford University.

[1]"The Air-Propeller Tests of W. F. Durand and E. P. Lesley: A Case Study in Technological Methodology," *Technology and Culture* 20 (October 1979): 712–51; "Control-Volume Analysis: A Difference in Thinking between Engineering and Physics," *Technology and Culture* 23 (April 1982): 145–74. For a representative list of sources treating technology as knowledge, see n. 1 of the second of these. Two valuable references concerning knowledge specifically in the aircraft industry are E. W. Constant, *The Origins of the Turbojet Revolution* (Baltimore, 1980), and J. V. Becker, *The High-Speed Frontier* (Washington, D.C., 1980). Constant, pp. 6–8 and passim, also includes some pertinent and illuminating remarks about technological knowledge generally.

0040-165X/84/2503-0002$01.00

removed from science, on the assumption that some features of the cognitive structure of technology may be more visible there. To be most significant, this example should come from a modern, high-technology industry that is usually thought of as science based. Flush riveting in airplanes meets these requirements. Although it was a far from negligible development in an industry that was depending more and more on engineering science,[2] it did not itself involve anything that could remotely be described as science in the modern sense. Moreover, it did not even require a major technological invention or conceptual breakthrough. As airplane performance improved to the point that the aerodynamic drag of protruding rivets was no longer economically and militarily acceptable, anyone could see at once what had to be done: the rivets had to be made flush. What needed to be learned was "simply" how to accomplish the change in routine design and production, and this required the effort of a large number of people over a considerable period of time. Such humble productive activity in the context of science-based industry has received little attention from historians. It is, I suspect, more prevalent than is generally recognized.[3]

The study of flush riveting does in fact prove intellectually rewarding. Detailed examination of the learning process—and the essence of the learning can be found only in the details—reveals different examples of knowledge with different characteristics. Reflection on these differences then suggests a framework for thinking about technological knowledge that may have interest beyond the present example.[4] Except for some discussion by Sidney Winter in relation to economic

[2]See, e.g., P. A. Hanle, *Bringing Aerodynamics to America* (Cambridge, Mass., 1982).

[3]Another problem requiring such activity in the aircraft industry is the sealing of pressurized airplane cabins to prevent leakage at high altitude. Possible examples from other industries are the production of round passages with very tight tolerances, as occur in hydraulic cylinders and carburetors for automobiles, and the early changeover from wiring to (nonminiaturized) printed circuits in television receivers. For discussion of the importance and difficulties of the "lower" forms of technological knowledge involved in such problems, see N. Rosenberg, "Problems in the Economist's Conceptualization of Technological Innovation," in his *Perspectives on Technology* (Cambridge, 1976), pp. 61–84, esp. pp. 77–79.

[4]In anticipation of the distinctions involved in such a framework, I have resisted the temptation to describe everything about flush riveting as "know-how," the common catch-all term for a wide (and usually undefined) range of practical knowledge and/or ability; see, e.g., D. Sahal, *Patterns of Technological Innovation* (Reading, Mass., 1981), p. 42 and passim. Ingvar Svennilson's careful definition of know-how as "the capacity to use technological knowledge," though meaningful, preempts the term from describing the knowledge itself; "Technical Assistance: The Transfer of Industrial Know-How to Non-industrialized Countries," in *Economic Development with Special Reference to East Asia*, ed. K. Berrill (New York, 1964), pp. 406–9.

production theory, the epistemology of industrial production has been largely *terra incognita*.[5]

Flush riveting also tells us something—an unexpected bonus—about the nature of industrial innovation. The fundamental decision to remove the rivet heads from the airstream was a design decision based on aerodynamic considerations. Once it had been made, however, implementation would have been impossible without successful changes in the methods of production. Some knowledge did have to be generated for detail design, but the pivotal developments were in production, and it was there that the greater part of the innovative activity took place. Flush riveting can therefore be called a *production-centered innovation*. It thus differs from innovation derived from formal research and development, which has dominated the scholarship on technological change in mid-20th-century industry.[6] Comparable investigations of "*non*-R&D" innovation of whatever sort, such as Samuel Hollander's economic inquiry into plant-level process innovations in the rayon industry, or some cases in a policy-oriented study by Arthur D. Little, Inc., are relatively scarce and not much concerned with technical detail.[7] When examined at the level of such detail, the innovative activity of flush riveting appears widespread and simultaneous throughout the entire aircraft industry. This pattern is very different from the model of innovation by diffusion from an initiating source that has occupied the attention of economists and historians.

[5]S. G. Winter, "An Essay on the Theory of Production," in *Economics and the World around It*, ed. S. H. Hynan (Ann Arbor, Mich., 1982), pp. 55–93.

[6]Flush riveting was a tangible innovation, even if a secondary one, in a product, the airplane; at the same time, it depended critically on innovation in a process, that of airframe production. It thus does not fit neatly into either of the economist's standard classifications of product or process innovation. Classification of an innovation according to where it is embodied may not always be possible. Of course, classification according to where the crucial innovative activity is centered may be expected to have its problems too.

[7]S. Hollander, *The Sources of Increased Efficiency: A Study of Du Pont Rayon Plants* (Cambridge, Mass., 1965); A. D. Little, Inc., *Patterns and Problems of Technical Innovation in American Industry*, report to National Science Foundation (September 1963). Considerable related scholarship exists concerning production in 19th-century industry, notably with regard to the American system of manufactures; see, e.g., M. R. Smith, *Harpers Ferry Armory and the New Technology* (Ithaca, N.Y., 1977), chaps. 4, 8; P. Uselding, "Measuring Techniques and Manufacturing Practice," and D. A. Hounshell, "The *System*: Theory and Practice," both in *Yankee Enterprise: The Rise of the American System of Manufactures*, ed. O. Mayr and R. C. Post (Washington, D.C., 1981), pp. 103–52. In its concern for details of production, the present paper also relates to the vast literature on technique, most of it descriptive, amassed by archaeologists, industrial archaeologists, and historians of technology; e.g., *Dictionnaire archéologique des techniques* (Paris, 1963), and classification 10 in any annual "Current Bibliography in the History of Technology," *Technology and Culture*, April issues.

Origins and Nature of the Problem

Flush riveting appeared in American aircraft as part of the development of the modern aluminum stressed-skin airplane in the 1930s.[8] Significantly, however, the revolutionary airplanes that led this development in the first half of the decade did not incorporate flush rivets to any appreciable extent. The Northrop Alpha and Boeing Monomail, the single-engine commercial airplanes generally credited with starting the development in 1930, both utilized rivets that protruded into the airstream. So also did succeeding Northrop designs, as well as the Boeing and Martin twin-engine military and commercial planes that helped inspire the definitive Douglas DC-1, -2, and -3 of 1933–36. These twin-engine Douglas transports, especially the almost legendary DC-3, set the pattern for multi-engine commercial landplanes until the appearance of the jets in the 1950s. Besides stressed-skin structure, the Douglas designs incorporated all the major new innovations of retractable landing gear, wing flaps, and controllable-pitch propeller. These and other improvements raised the maximum speed of the DC-3 to 212 MPH, compared with the 170 MPH of the Northrop Alpha. Even the Douglas airplanes, however, continued to use protruding-head rivets.[9]

Though at first apart from the mainstream, flush riveting appeared early in the development of stressed-skin airplanes.[10] Anyone with an elementary knowledge of aerodynamics could see that protruding rivets caused undesirable air resistance; that this resistance could be eliminated by making the rivets flush would have occurred to numer-

[8]Riveting, of whatever sort, is used in aluminum airplanes, in preference to the welding that has replaced it in industries that employ other metals, because the aluminum alloys needed in aircraft tend to be affected adversely in their material properties and structural behavior by the high temperatures of the welding process.

[9]R. Miller and D. Sawers, *The Technical Development of Modern Aviation* (New York, 1970), pp. 18–20, 47–50, 63–65; C. G. Grey and L. Bridgman, eds., *Jane's All the World's Aircraft 1931* (London, 1931), pp. 303c–4c, and *Jane's . . . 1936*, pp. 276c–77c. For a recent discussion of the role in this development of long-range commercial flying boats, largely neglected by Miller and Sawers and others, see R. K. Smith, "The Intercontinental Airliner and the Essence of Airplane Performance, 1929–1939," *Technology and Culture* 24 (July 1983): 428–49.

[10]In Europe flush riveting appeared at least as early as 1922 on the Bernard C.1, a racing plane that was the hit of the Paris Air Show of that year; T. G. Foxworth, *The Speed Seekers* (New York, 1976), pp. 83–84. I have made no attempt to trace the details of the application in European commercial and military aircraft, but a cursory examination suggests it was more or less simultaneous with that in the United States. The innovation appears to have taken place essentially independently on the two sides of the Atlantic, until flush riveting had become well established in the 1940s. This fact further emphasizes the simultaneous, widespread nature of the development as mentioned in the introduction and elaborated below.

ous people almost without thinking. As often happens with obvious ideas, some of these people would be expected to employ such rivets on principle, even before data existed to prove them worth the extra trouble and expense.

In these circumstances—and because details such as flush riveting often go unrecorded—assessing priority is difficult, and pinpointing all the early uses in aircraft is impossible. The first proposal I find in the United States was in a patent granted to Charles Ward Hall in 1926 which covered numerous features of metal airplanes including flush riveting. By 1929 Hall had equipment for such riveting in operation at his Hall-Aluminum Aircraft Corporation in Buffalo, and he probably employed flush rivets on the Hall PH-1 of 1932, a twin-engine patrol flying boat for the U.S. Navy. That same year the Boeing Airplane Company used flush rivets on its P-26 single-engine fighter for the Army Air Corps.[11] Over the next few years an increasing number of aircraft apparently incorporated such riveting. A considerable fraction of these were flying boats or amphibians, perhaps because of early awareness by seaplane designers that any protuberances on a planing surface have an especially detrimental effect on takeoff from water. Many pioneering applications were undoubtedly partial; according to a news item of 1934, "some manufacturers" (unidentified) were "compromising the gain and expense" by limiting flush rivets to the upper surface of the wing near the leading edge or to other locations where protruding rivets are especially harmful.[12]

[11]U.S. Patent 1,609,468, C. W. Hall, *Metallic Construction for Aircraft and the Like*, December 7, 1926, pp. 3, 7. For Hall's equipment, see Aluminum Company of America, *The Riveting of Aluminum* (Pittsburgh, 1929), pp. 21–22, and C. W. Hall, "Weight Saving by Structural Efficiency," *SAE Journal* 28 (1931): 77–83, esp. 79, 80. A photograph in the files at the National Air and Space Museum (information from Howard Wolko) shows that the XFH-1, a Navy single-engine fighter and Hall's first airplane (1929), had protruding rivets. Photographs of the PH-1 are not conclusive but strongly suggest flush riveting. Boeing information is from personal correspondence with Edward C. Wells, an engineer and leading engineering executive with that company after 1931.

[12]Airplanes I have been able to identify as using flush rivets extensively in these years (though there were undoubtedly more) are the mammoth Hall XP2H-1 four-engine military flying boat of 1934 ("Navy's Giant," *Aviation* 33 [1934]: 317–19), the Sikorsky multi-engine commercial flying boats and amphibians of 1934 and 1935 (*Jane's . . . 1934*, pp. 311c–12c, *Jane's . . . 1935*, p. 335c; the S-42 of 1934 is an exception to the earlier statement about the lack of extensive flush riveting in the revolutionary aircraft of the first half of the 1930s), the Hughes H-1 single-engine racer of 1935 in which Howard Hughes set a landplane speed record of 352 MPH (*Jane's . . . 1935*, p. 311c), and the Boeing 299 of 1935, prototype of the four-engine B-17 Flying Fortress (Wells correspondence). The news item is "Flush Riveting," *U.S. Air Services* 19 (1934): 32. An instance of partial use is the Boeing 247D commercial transport of 1933, a key airplane in the mainstream of development, which had flush rivets limited to the engine cowling (see example in the National Air and Space Museum).

Concurrent with these applications, wind-tunnel tests of a specially built wing, reported in 1933 by the Langley Laboratory of the National Advisory Committee for Aeronautics (NACA), confirmed the aerodynamic penalty incurred by protruding rivets. For a hypothetical single-engine airplane, for example, eliminating the rivet-head drag was estimated from these tests to increase the top speed from 200 to 205 MPH. This firm evidence was presumably becoming known in the industry.[13]

Still, the fact remains: flush riveting in the first half of the 1930s was peripheral and lagged behind the major innovations. This lag was not due to problems in the riveting itself. Flush riveting did not depend critically on any prior technology, and its problems could have been solved—indeed, already had been in part—whenever people like Hall set their minds to it. The designers of the DC-3 and its predecessors simply had their eyes on bigger game. Only when the performance improvements caused by retractable gear, flaps, and the like had been realized did it become generally attractive to pursue the lesser gains offered by recessed rivet heads. However, as the speeds of commercial airplanes went up as a result of the other innovations, the cost of dragging exposed rivet heads through the air increased, whereas the extra cost of flush riveting stayed constant. And in military aircraft, as other methods of increasing top speed reached the point of diminishing returns, the smaller gain from getting the rivet heads out of the airstream became more inviting. By the mid-1930s the time was ripe for flush riveting throughout the industry. Only after this time do engineering reports and articles of any depth begin to appear.

To understand the problems of flush riveting in aircraft, some terminology and prior history are helpful. Riveting is an ancient art, practiced by at least the 2d century B.C., for joining together overlapping pieces of metal or other material by means of a malleable metal pin.[14] In modern times the pin—or *rivet*—as it comes from the manufacturer usually has a cylindrical *shank* with a preformed head, the *manufactured head*, at one end. Riveting is accomplished by putting the shank through matching holes in the pieces to be joined—usually metal plates or sheets of one kind or another—and hammering or otherwise upsetting a second head, the *upset head*, on the end opposite the manufactured head. In structural applications, riveting, besides joining the pieces together, transmits load from one structural element to another.

[13]C. Dearborn, "The Effect of Rivet Heads on the Characteristics of a 6 by 36 Foot Clark Y Metal Airfoil," *Technical Note No. 461*, National Advisory Committee for Aeronautics (Washington, D.C., May 1933), p. 5. This agency will hereafter be given its common abbreviation, NACA.

[14]H. Hodges, *Technology in the Ancient World* (New York, 1970), pp. 143–44, 187.

Both the manufactured and upset heads of rivets may be of various shapes (round, pointed, cylindrical), most of which protrude above the surface of the plates. To meet special needs, however, one (or occasionally both) of the heads may be made flush. This is accomplished by putting the head in a *conical recess* at the entrance to the hole in the plate (fig. 1). Most often nowadays the manufactured head is the flush head; in this case the head is preformed with a conical shape the same as that of the recess, as with an ordinary flathead screw. On occasion, however, the upset head is made the flush head; to accomplish this the shank of the rivet is hammered or pressed into the recess, and any excess material is cut off. Either way, the included angle of the conical head is called the *head angle*. Whichever method is used, flush riveting entails extra expense and is employed only when circumstances make it necessary or desirable. Such circumstances led to limited use, beginning at least in the 1830s, in iron and steel ships (to reduce water resistance on the hull or obtain a smooth surface on the deck plating) and in boilers, bridges, and other structural work (in situations where room for the rivet head is restricted by other parts of the equipment or structure). Here the conical recess was made by machine countersinking, that is, by cutting away metal with a rotating conically shaped tool. More often than not in this usage, the upset head was made the flush head.[15]

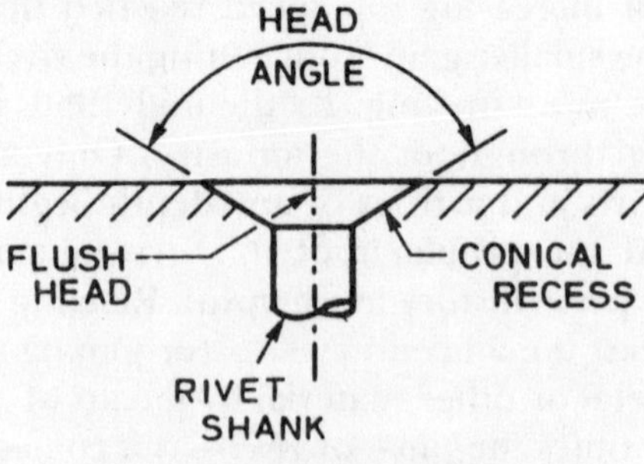

FIG. 1.—Typical flush rivet head

[15]The first comprehensive experiments on the strength of riveted joints, carried out by Eaton Hodgkinson in 1838 under the direction of William Fairbairn, included flush rivets; W. Fairbairn, "An Experimental Inquiry into the Strength of Wrought-Iron Plates and Their Riveted Joints as Applied to Shipbuilding and Vessels Exposed to Severe Strains," *Philosophical Transactions of the Royal Society of London* 140 (1850): 677–725, esp. 698. For use in ships and boilers, see, e.g., E. J. Reed, *Shipbuilding in Iron and Steel* (London, 1869), pp. 196–97, 329–331, 340, and C. H. Peabody and E. F. Miller, *Steam-Boilers* (New York, 1908), p. 200. For a complete history and bibliography of the (mainly nonaeronautical) literature up to 1944 on the strength of riveted joints generally, see A. E. R. de Jonge, *Riveted Joints: A Critical Review of the Literature Covering Their Development* (New York, 1945). The bibliography contains 1,209 items, each with an abstract carefully prepared by de Jonge—an impressive monument to technological knowledge with little or no science.

When flush riveting came to be applied to aircraft, a new problem arose. To save weight, airplane skins tend to be much thinner than the plates in ships, boilers, and bridges. In the small airplanes of the 1930s, the skins were often so thin relative to the size of rivet needed for strength that the machine countersink required by the head would have had to penetrate through one sheet of skin and into the next. This situation could be detrimental to the strength and rigidity of the riveted joint, and some other way to form the conical recess had to be found. Again, the solution was easy to perceive in principle: deform each sheet by pressing a conical dimple around the rivet hole and accommodate the flush head in the outermost of the nested dimples. This method was, in fact, employed by Hall and other early users; their experience showed that it worked in practice. With dimpling, all the basic ideas required for flush riveting in aircraft were in hand.

The task that remained for the aircraft industry at the middle of the 1930s was to learn to implement these ideas in various structural situations in volume production. The history of this learning process breaks down conveniently into two major parts, one concerned with production and the other with design. The first has to do primarily with the question of how to fabricate flush-riveted joints economically and reliably; the second deals with the proportioning of such joints to carry the structural loads. In dividing the material in this way, however, we must not lose sight of the fact that the two activities were in reality intimately related. The production process had to produce riveted joints that satisfied the requirements of aerodynamic and structural design. Conversely, experience about what was feasible in production generated information needed by the designer for deciding which type of flush riveting—dimpled, machine countersunk, or some combination thereof—to use in a given situation. Producibility also defined the types to be tested for strength by the structural engineer. Thus, while the innovation was production centered in the sense explained earlier, generation of knowledge for production and for design went forward together.[16] Knowledge of flush riveting from ships, boilers, and buildings was of little use for this purpose, both because aluminum alloys were so different from iron and steel and because dimpled riveting had not been used before. Nowhere in the aeronautical literature do I find appeal to the prior experience.

As we shall see, the aircraft industry solved its flush-riveting problems more or less simultaneously throughout the industry. Only a relatively small portion of this activity, however, was reported in detail

[16]The question of materials was central in both production and design. Considerations in this regard, however, were mostly no different for flush than for protruding riveting. I shall mention them only when something arises peculiar to flush riveting.

in technical publications or even company reports—shopworkers and design and production engineers have more urgent things to do than record their experiences. Even though a fair amount of material *can* be found, one soon develops the uncomfortable feeling of seeing only the tip of a very large iceberg. Some help can be had, of course, by interviewing participants who are still living. Additional insight comes from observing how continuing developments in riveting go on today, on the reasonable assumption that attitudes and activities in the riveting community have not changed greatly since our period. I also remember from my own experience as an aeronautical engineer how things took place in related areas in the 1940s and 1950s. I have drawn from all these sources in piecing together the historical account.

Production of Flush-riveted Joints

Although relatively unnoticed, riveting has played a sizable role in airplane manufacture. The airplanes of World War II had as many as 160,000 rivets in a medium bomber and 400,000 in a large bomber. In 1944, design, fabrication, and installation of riveted joints reportedly constituted 40 percent of the cost of a typical airframe.[17] Because of the large number of rivets, airplane riveting is a mass-production operation, even though airplane production itself is not. As with all mass-production operations, small savings per rivet mount up.

Aircraft riveting is also a subject of great complexity. Besides the ever-important matter of cost, the aircraft engineer must pay attention to the often conflicting requirements of weight, production quality, structural reliability, corrosion resistance, service maintenance, and appearance.[18] Even for only *flush* riveting, to grasp this complexity fully from a historical paper is impossible—one must go through a number of the original technical articles in their entirety. The reader will have to imagine the detailed and painfully complicated descriptions of tools and processes, the contrary assessments of the same process by different people, and the frequently unresolved arguments about the relative merits of different types of flush rivets or riveted joints. Of such is the fabric of complex technological activity.

The widespread, simultaneous nature of flush-riveting development is unmistakable. By the end of the basic stage of development, which occupied the second half of the 1930s, at least fifteen manufacturers were using the new type of riveting. That they worked largely indepen-

[17] *Aircraft Fabrication. Aircraft Riveting*, Part 1 (Scranton, Pa., 1943), p. 1; E. S. Jenkins, "Rational Design of Fastenings," *SAE Transactions* 52 (1944): 421–29.

[18] For some of the considerations involved prior to the shift to flush riveting, see "Riveting of Aluminum and Its Alloys," *Aero Digest* 30 (1937): 44–46.

dently is indicated by the wide range of head angles from one company to another (78–130 degrees) and the diversity of head dimensions even among companies using the same angle.[19] The need for flush riveting was apparently felt more or less simultaneously everywhere, and, once it was, work on such riveting welled up from below, so to speak, throughout the industry. This widespread simultaneity is the most striking characteristic of the activity we are examining. I will come to the reasons for it later, but it needs to be kept in mind throughout.

In the basic stage, manufacturers put their effort into learning to produce the various kinds of flush riveting and finding out which kind was best used where. Attention quickly settled on three types, depending on the thickness of the sheet being riveted (which is dictated by the load-carrying requirements of the structure). Where the thickness of the outer sheet was sufficiently greater than the height of the conical rivet head, old-fashioned machine countersinking proved suitable (fig. 2*a*). The limitation here was the desire to avoid the sharp, feather edge that occurs when the countersink extends to or beyond the thickness of the sheet. For the ⅛-inch-diameter rivets that predominated in aircraft construction, for example, the thinnest standard sheet that could meet this requirement for most of the various head proportions was 0.051 inches.[20] Some manufacturers tried countersinking thinner sheet by letting the countersink extend into the inner skin, but this was discarded as "poor design practice" when such joints proved weak or the resulting feather edges developed cracks in fabrication or service. For such thinner sheets, recourse was had to dimpling, as described earlier (fig. 2*b*). Here production engineers learned from experience that there were minimum and maximum sheet thicknesses beyond which the method could not satisfactorily be used—for example, 0.020 and 0.081 inches, respectively, for ⅛-inch rivets. If the thickness exceeded the maximum, the dimple tended to crack during forming; if it was less than the minimum, the sheet between adjacent rivets might buckle

[19] The number 15 comes from experiments on the strength of flush-riveted joints made by the National Bureau of Standards from 1941 to 1943 using specimens supplied by the manufacturers; see, e.g., W. C. Brueggeman, "Progress Summary No. 1. Mechanical Properties of Flush-riveted Joints Submitted by Five Airplane Manufacturers," *Wartime Report W-79*, originally *Restricted Bulletin* (unnumbered), NACA (Washington, D.C., February 1942). I find no record of the overall summary this report promised, but file cards, of individual progress reports, in the library of the Langley Research Center, NASA, reveal the head angles used by twelve of the manufacturers as follows: 78°—Curtiss-Wright, Grumman; 100°—Lockheed, Douglas, Vultee, Vought-Sikorsky; 110°—Brewster; 115°—Martin, Republic, Consolidated; 120°—Bell; 130°—Boeing. For differences in dimensions for the same head angle, see E. B. Lear and J. E. Dillon, *Aircraft Riveting* (New York, 1942), p. 72.

[20] The specified diameter of a rivet is always the shank diameter.

under load. These limits, however, were set at considerably different values by different manufacturers. When the inner sheet alone exceeded the maximum, the problem was solved by dimpling the outer sheet and machine countersinking the inner (fig. 2*c*). Various limitations also emerged for this combination. For each type of riveting the limits on sheet thickness varied with rivet diameter, and additional considerations entered when more than two overlapping sheets had to be riveted. Learning about these matters was a complicated and protracted business, much more so than a summary description can convey. Knowledge of the various limits, which could come only from manufacturing experience, was essential for the designer.[21]

Since dimpled riveting was new, development concentrated more on this than on countersinking. The problem was not an easy one. In the words of a tool-design engineer for Bell Aircraft, "At first glance,

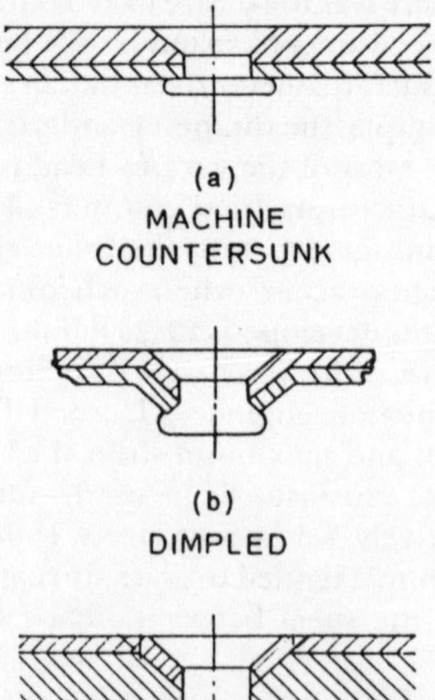

FIG. 2.—Three basic types of flush riveting. From G. Rechton, *Aircraft Riveting Manual. Addendum I. Riveting Methods*, Douglas Aircraft Co. (Santa Monica, Calif., 1942), pp. 7, 21, 26; by permission.

[21] *Aircraft Fabrication. Aircraft Riveting* (n. 17 above), p. 10; Part 2, pp. 7–12; A. H. Nisita, *Aircraft Riveting* (New York, 1942), pp. 83–89; interview (November 4, 1980) with Edward Harpoothian, from 1939 to 1962 chief of Structures Section, Douglas Aircraft Co., Santa Monica, Calif.

nothing seems simpler than putting a dimple into a sheet of metal, yet it has proved to be no simple problem for the best engineering talent in the industry." Even allowing for self-aggrandizement, this statement is probably not far from the truth. A description of later developments at Northrop Aircraft included the following description of the requirements on a dimple: "It is essential that the dimple does not crack during forming or after assembly of the sheet. The dimple must be dimensionally accurate to insure fastener flushness and to nest correctly in either machined countersunk recesses or in other dimpled recesses, regardless of sheet thickness. Furthermore, the dimple must provide maximum joint strength under all conditions, at the same time developing the very minimum of warpage or distortion in the sheet. Productionwise, the dimpling method must be rapid and accurate, and at the same time require simple, inexpensive tools and set ups."[22] Since there existed at the time no theory for calculating the deformation of the sheet in the press-forming process, the problem had to be solved entirely empirically. The nature and diversity of the work on this and other problems can best be described in terms of three aircraft companies for which written accounts exist. Statements in the literature indicate that parallel developments went on elsewhere.

The Douglas Aircraft Company in Santa Monica took up flush riveting for the DC-4E, an experimental commercial transport that first flew in June 1938. The -4E was Douglas's first venture into four-engine airplanes and had a maximum speed of 245 MPH, 33 MPH greater than the DC-3. Douglas aerodynamicists had insisted at the inception of design in 1936 that the time had come to get the rivet heads out of the airstream, and Arthur Raymond, director of engineering, agreed. Development of a suitable method of flush riveting was assigned to Vladimir Pavlecka, a Czech-born engineer in charge of structural research. Working under Pavlecka's direction between 1936 and 1938, a small group of engineers and production men devised what they called "the Douglas system of flush riveting."[23]

The Douglas system had two distinguishing features. First, the

[22] A. A. Schwartz, "Flush Rivets for Speedy Airplanes," *American Machinist* 85 (1941): 443–48, quotation from p. 446; G. C. Close, "Flow-Form Dimpling at Northrop," *Machinery* (N.Y.) 60 (1954): 191–93, quotation from p. 191.

[23] Miller and Sawers (n. 9 above), pp. 131–32; *Jane's . . . 1939*, p. 245c; Harpoothian interview; interview (February 5, 1981) with John Buckwalter, longtime engineer with Douglas and project engineer on the DC-4E. For a variety of reasons, the -4E did not prove to the liking of the airlines, and only one was built. It is not to be confused with the later (1942), somewhat smaller, and highly successful DC-4 (C-54 in World War II military designation). P. W. Brooks, *The Modern Airliner—Its Origins and Development* (London, 1961), p. 94, states that the DC-4E was "the first American airliner with flush-riveted skin." This statement is hardly correct in view of the use of flush rivets in the Sikorsky commercial aircraft of 1934–35 (n. 12 above).

manufactured head of the rivet itself provided the tool for press forming the dimples. (In our account, the manufactured head is always the flush head unless specified otherwise.) In this scheme the manufactured head impressed the dimples into both sheets in a female die in a single operation (fig. 3*a*, *b*); a second operation then upset the rivet shank to form the upset head. The Douglas method of "rivet dimpling" thus differed from the more usual "pre-dimpling," where the dimples were made in the two sheets, usually separately, by pressing between male and female dies (fig. 3*c*); the sheets were then assembled, the rivet inserted, and the upset head formed. Pavlecka argued that rivet dimpling would reduce the number of individual operations and hence the time and cost compared with pre-dimpling. He also anticipated a closer fit of the various parts and thus a tighter joint. Rivet dimpling was not new, having been used in Europe before 1934. There is no indication, however, that Pavlecka was influenced by the European work. The idea again was one that could be expected to occur to anyone who thought seriously about dimpled riveting.[24]

The second feature of the Douglas system was the use of a rivet with a head angle of 100°. This was a departure from the 78° that had long been used in boilers and pressure vessels and had been taken over by

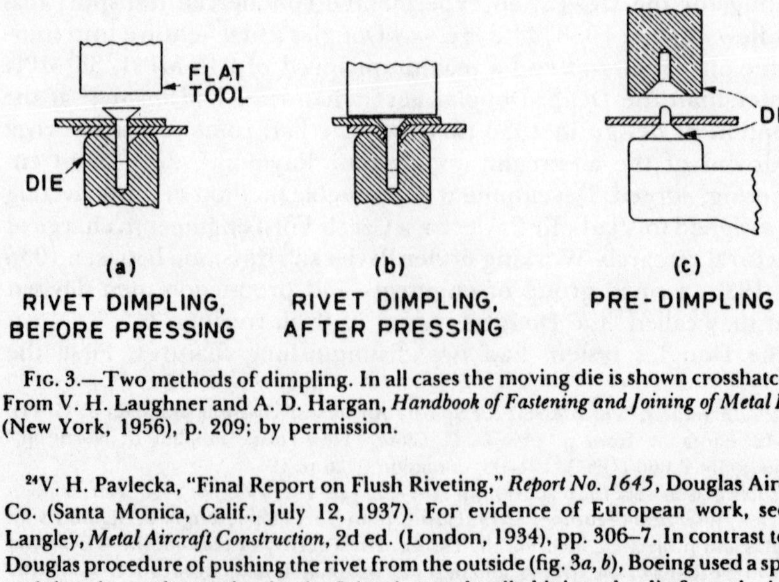

FIG. 3.—Two methods of dimpling. In all cases the moving die is shown crosshatched. From V. H. Laughner and A. D. Hargan, *Handbook of Fastening and Joining of Metal Parts* (New York, 1956), p. 209; by permission.

[24]V. H. Pavlecka, "Final Report on Flush Riveting," *Report No. 1645*, Douglas Aircraft Co. (Santa Monica, Calif., July 12, 1937). For evidence of European work, see M. Langley, *Metal Aircraft Construction*, 2d ed. (London, 1934), pp. 306–7. In contrast to the Douglas procedure of pushing the rivet from the outside (fig. 3*a*, *b*), Boeing used a special tool that clamped onto the shank of the rivet and pulled it into the die from the inside (Wells correspondence). This presumably was done to avoid possible scarring of the skin from the flat dimpling tool required by the Douglas system. I have seen no record of the Boeing scheme being used elsewhere.

the army and navy as standard for the flush rivets already in limited use in aircraft.[25] Pavlecka made the change after attempts to form dimples by means of 78° heads led to cracking of the sheet and to excessive deformation of the rivet head in the dimpling and upsetting operations. The shallower dimple required by the 100° head (the outside head diameter and rivet diameter being unchanged) eliminated these difficulties. The Douglas rivet also incorporated a thin constant-diameter addition on the top of the head. This was said to eliminate possible cracking of the edge of the head in dimpling and upsetting as well as damage to the otherwise sharp edge in the shipping of batches of rivets.[26]

At the Curtiss-Wright Corporation in Buffalo developments took a different course. Don Berlin, chief engineer, and Peter Rossman, production design engineer, in a paper read to the Society of Automotive Engineers in March 1939, reported an extensive study begun in 1937 in connection with a large order of military fighter planes. (The SAE concerns itself with airplanes as well as automobiles.) In contrast to procedures at Douglas, Curtiss-Wright opted for pre-dimpling and, for a number of reasons, preferred the 78° head angle. Berlin and Rossman, like the Douglas team, encountered cracking problems when they tried to pre-dimple to accommodate the standard 78° head. Like Douglas, they solved the problem by going to a shallower dimple. Unlike Douglas, however, they did this by keeping the 78° angle but reducing the head height to less than that of the standard head (for a given rivet diameter). They also gave careful attention to the precise shape of the dimpling dies and to details of the pre-dimpling operation. Also in contrast to Douglas, they specifically rejected the constant-diameter addition to the top of the head on the grounds that it led to an undesirable gap between the edge of the flush head and the adjacent sheet. In the discussion following the paper, Pavlecka explained and defended the Douglas system, and B. C. Boulton of the Glenn L. Martin Company told why his company, "after exhaustive tests," had

[25] G. B. Haven and G. W. Swett, *The Design of Steam Boilers and Pressure Vessels* (New York, 1915), p. 87. I am unable to discover why the seemingly idiosyncratic value of 78° was originally adopted. Perhaps a prominent early manufacturer of boilers with flush rivets happened to have a conical cutter with that angle available for his countersinking (which, of course, only displaces the mystery), and everyone afterward simply followed suit. Things sometimes happen in technology for no better reason.

[26] Douglas patented its riveting system in 1941; U.S. Patent 2,233,820, V. H. Pavlecka, assignor to Douglas Aircraft Co., *Method of Riveting*, March 4, 1941. Shortly thereafter Douglas released the patent to the aircraft industry in connection with the national defense program; "Flush Riveting in Aircraft," *Aviation Industries* 84 (1941): 518. I find no evidence, however, that the Douglas system was widely imitated except for the rivet head, as we shall see later.

adopted a 115° rivet. The Curtiss-Wright people, however, had clearly solved the problem to their own satisfaction. As in many technical matters, there was no single road to success.[27]

Bell Aircraft Corporation in Buffalo entered the flush-riveting field in 1939, about the time of the Berlin-Rossman paper. According to an article by Arthur Schwartz, the tool-design engineer quoted earlier, Bell at once "decided that a thorough study was in order before considering the adoption of any of the [diversity of] existing methods." This study concluded that rivet dimpling led to undesirable sunken areas surrounding each rivet head and made excessive demands (as I shall explain later) on the skill of the worker. Also, according to Schwartz, "Exhaustive experiments showed a decided superiority of [a] . . . rivet head angle in the neighborhood of 120 deg." Bell therefore opted for pre-dimpling at this angle. With carefully proportioned dimpling dies, this process was said to avoid the difficulties that Schwartz saw in other methods based on smaller angles. Bell also rejected the constant-diameter addition to the rivet head for the same reason as Curtiss-Wright.[28]

Douglas, Curtiss-Wright, and Bell all struggled with matters besides dimpling method and rivet shape. Both Curtiss-Wright and Bell put a great deal of effort into solving the production problems of machine countersinking as well as dimpling. All three companies found themselves forced to develop special tools for the shop processes required by the different types of riveting. Since obtaining the desired smoothness in a flush-riveted surface in routine production is difficult, Bell gave careful attention to the dimensional tolerances required on rivets, countersinks, and dimples; Bell also developed special gauges to check that the tolerances were attained in the shop. Curtiss-Wright, faced with a large production order, made detailed study of the time required to do each step in the various types of flush riveting; on this basis they arrived at estimates of production costs and how these might be expected to decrease with increased shop experience. And to satisfy themselves that flush joints could provide the necessary strength, Douglas and Curtiss-Wright (and presumably Bell, though this was not stated) made structural tests of a large number of joints under both static and vibrational loads.

Learning to solve these kinds of problems required minute attention to detail. To exhibit the detail, one problem may serve as typical: the shaping of dies for pre-dimpling. Although Curtiss-Wright, Bell, and

[27]D. R. Berlin and P. F. Rossman, "Flush Riveting Considerations for Quantity Production," *SAE Transactions* 34 (1939): 325–34.

[28]Schwartz (n. 22 above), quotation from p. 443.

Douglas all addressed this crucial problem (Douglas also went over to pre-dimpling when rivet dimpling proved unsuitable for large sheet thicknesses), the most detailed account came from R. L. Templin and J. W. Fogwell of the Aluminum Company of America. Alcoa, the main supplier of aluminum to the aircraft industry, dealt with such problems to assist and encourage its customers in the use of its product. Templin and Fogwell, working with 100° rivets, took it as their task to produce a set of dimpling dies that would (1) leave the sheet flat around the dimple after the dimpling operation, (2) form dimples that would nest tightly into one another, and (3) provide a cylindrically shaped hole so that no undesirably sharp corners would dig into the rivet shank. These goals they attained by trial-and-error variation of the geometrical parameters of the dies. They satisfied the flatness requirement by having the dies extend outside the dimple in such a way as to strain the surrounding sheet slightly in an angular direction opposite to that of the dimple; with the removal of die pressure, elastic springback of the metal then caused the sheet to lie flat. The precise amount of strain proved important, and a difference of ½° could be crucial. They solved the nesting problem by choice of a suitable die pressure and by careful selection of different maximum diameters for the dimpling cones on the male and female dies. The cylindrical hole was obtained by drilling the initial hole in the sheet to a diameter slightly smaller than that of the mandrel on the male die (which served to center the die in the hole, as in fig. 3*c*); the suitably tapered mandrel then forged the hole to the desired shape in the dimpling operation. The hole could not be made too small, however, lest radial cracks appear around its edge. To guide their work, the Alcoa engineers used microphotographs of sectioned specimens, as in figure 4*a* and *b*, which shows dimples with sharp-cornered and nearly cylindrical holes, respectively. Other developers of flush riveting used trial-and-error methods and microphotographs in much the same way. People in different companies, however, did not always agree on the best solution to a given problem.[29]

Other aspects of both dimpling and upsetting required attention throughout the industry. Forming the upset head on the shank of flush rivets presented no great problem—the methods already in use for protruding rivets could be taken over with only minor change. In the most usual method a first worker, using a pneumatic hammer with a

[29]R. L. Templin and J. W. Fogwell, "Design of Tools for Press-countersinking and Dimpling of 0.040-Inch-Thick 24S-T Sheet," *Technical Note No. 854*, NACA (Washington, D.C., August 1942); interview (September 2, 1981) with George D. Rechton, process engineer at Douglas, Santa Monica, Calif., during the period in question. Although the Templin-Fogwell work was done in the early 1940s, it may be taken as representative of earlier activities elsewhere.

concave fitting, applied a rapid succession of blows to the dome-shaped protruding head; at the same time a second worker held a heavy metal "bucking bar" against the end of the rivet's shank. The impact of the hammer working against the inertia of the bucking bar thus deformed the shank to produce the upset head. On sufficiently small subassemblies, a single worker upset the rivet by squeezing it between the ends of

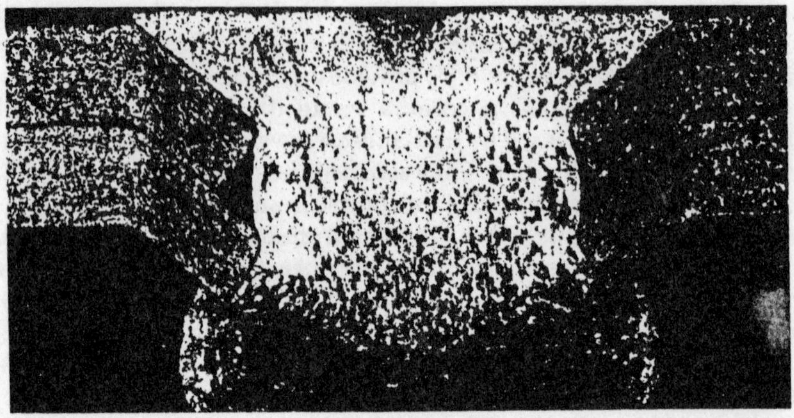

(a)

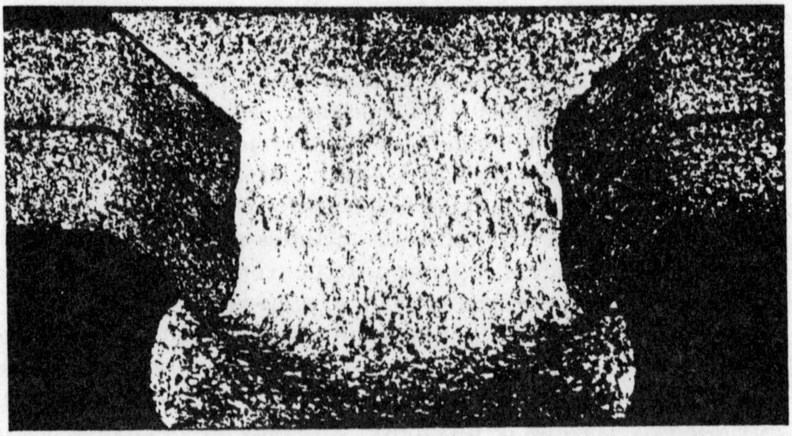

(b)

FIG. 4.—Microphotographs of sections through dimpled holes, actual rivet diameter ⅛-inch. (*a*) Drilled hole same diameter as mandrel of male die; note sharp corners digging into rivet. (*b*) Drilled hole 15 percent smaller in diameter than mandrel; note more nearly cylindrical hole without sharp corners. From R. L. Templin and J. W. Fogwell, "Design of Tools for Press-countersinking and Dimpling of 0.040–Inch-Thick 24S-T Sheet," *Technical Note No. 854*, NACA (Washington, D.C., August 1942), figs. 2 and 4.

a C-shaped tool that (depending on the depth of its jaw) could fit around the sheets and reach both ends of the rivet. The only special problem in upsetting *flush* rivets was that of applying the necessary force to the flush manufactured head without marring the adjacent sheet. For this purpose workers used slightly convex fittings on the hammer or squeezer, sometimes in conjunction with a swiveling device that could adjust itself to the surface of the sheet even if the tool was not oriented precisely.[30]

Where dimpling was needed, manufacturers found they could supply the required die pressure with the same hammers and squeezers used for upsetting by simply replacing the upsetting fittings with appropriate dimpling dies. In rivet dimpling by a two-worker team, a problem arose in that the female die (fig. 3*a*, *b*) had to be hand-held carefully at right angles to the sheet for the head to seat properly. This was the difficulty that helped discourage Bell from using rivet dimpling. Douglas, however, claimed to have overcome the problem by using a special swivel-mounted die. Douglas also found that rivet dimpling was done more satisfactorily by one or two heavy blows than by a rapid succession of weaker ones; they developed and patented a special pneumatic "one-shot" hammer for this purpose. (Considerable argument went on about the relative merits of rapid hammers, one-shot hammers, and squeezers for different operations and different dimensions of sheet and rivet.) Douglas also patented an air-operated two-position bucking tool useful for both the rivet-dimpling and upsetting operations. Such specialized tool development took place throughout the industry for both dimpling and machine countersinking.[31]

In such fashion—widespread, detailed, and entirely empirical—the aircraft industry had learned by the end of the 1930s how to produce flush-riveted joints. At the same time, additional aerodynamic tests plus articles about them in technical magazines had emphasized the gains from such riveting.[32] With this knowledge of how and why, no designer would any longer consider using protruding rivets in a high-

[30]*Aircraft Fabrication. Aircraft Riveting* (n. 17 above), pp. 18–46.

[31]*Aircraft Fabrication. Aircraft Riveting*, Part 2, pp. 1–15; Pavlecka, "Final Report . . . ," pp. 10–12; U.S. Patent 2,257,267, A. T. Lundgren, A. B. Rogers, and V. H. Pavlecka, assignors to Douglas Aircraft Co., *Percussive Tool*, September 30, 1941; U.S. Patent 2,274,091, V. H. Pavlecka, G. D. Rechton, and C. C. Misfeldt, assignors to Douglas Aircraft Co., *Bucking Tool*, February 24, 1942.

[32]M. J. Hood, "The Effects of Some Common Surface Irregularities on Wing Drag," *Technical Note No. 695*, NACA (Washington, D.C., March 1939), and "Surface Roughness and Wing Drag," *Aircraft Engineering* 11 (1939): 342–44; C. P. Autry, "Drag of Riveted Wings," *Aviation*, May 11, 1941, pp. 53–54.

performance airplane. The learning process, however, did not stop there. Three additional developments between the early 1940s and the first years of the 1950s require mention.

1. Although flush riveting had come a long way, it still left something to be desired regarding tightness of joint and smoothness of riveted surface. If the manufactured head did not seat firmly in the recess, the joint would yield under load; if the top of the head were too high or too low, owing to the recess being made too shallow or too deep, the surface would not be aerodynamically smooth. About 1941 researchers at NACA's Langley Laboratory had the problem of building a wing with the extreme degree of smoothness required by the new laminar-flow airfoils. After much consideration, they realized they could attain their goal by making the top of the manufactured head deliberately too high and carefully milling away the excess after the upsetting operation. At the same time, tightness was improved because the upsetting hammer, with its impact now concentrated entirely on the top of the head, forced the head firmly into the recess. They also learned with obvious pleasure that tightness proved even better if a round-head rivet was inserted from the back side and the upset head formed by pressing the shank into the recess (after which it was made smooth by milling). The last arrangement, reported in 1942, came to be known as the "NACA rivet." To perform the milling operation the Langley people developed and patented a power-driven, hand-held milling tool that they said would not mar the surrounding surface of the skin. The basic idea of the NACA rivet—of having the upset head be the flush head—had long been used, of course, in ships and boilers. Whether the NACA people got the idea from the earlier work is impossible to say, but it seems unlikely. Again, the concept is one that could easily have been rediscovered.[33]

Despite its increased difficulty and cost, the NACA rivet found application where extreme smoothness and tightness were paramount. More important, the different but related practice of installing the top

[33]E. E. Lundquist and R. Gottlieb, "A Study of the Tightness and Flushness of Machine Countersunk Rivets for Aircraft," *Wartime Report L-294*, originally *Restricted Bulletin* (unnumbered), NACA (Washington, D.C., June 1942); R. Gottlieb, "A Flush-Rivet Milling Tool," *Restricted Bulletin* (unnumbered), NACA (Washington, D.C., June 1942); R. Gottlieb and M. W. Mandell, "An Improved Flush-Rivet Milling Tool," *Restricted Bulletin No. 3E18*, NACA (Washington, D.C., May 1943); U.S. Patent 2,393,463, R. Gottlieb, assignor of one-half to A. Sherman, *Milling Tool*, January 22, 1946. For a popular account, see G. W. Gray, *Frontiers of Flight* (New York, 1948), pp. 197–202. The idea of having the upset head be the flush head had also appeared in the European aircraft industry before 1934 in the "De Bergue rivet" used in fuel tanks (Langley [n. 24 above], p. 267), and in a U.S. patent on a riveting procedure granted to a German in 1939 (U.S. Patent 2,147,763, E. Becker, *Flush or Countersunk Riveting*, February 21, 1939).

of a *manufactured* head deliberately too high and milling away the excess—so-called flush-to-high installation—became widespread. Two of the NACA workers obtained a patent on this procedure, which did not exist in earlier practice. Problems of milling the head without marring the skin, however, continued to trouble the industry in routine production.[34]

2. The dimpling methods developed up to the early 1940s worked satisfactorily with the aluminum-plus-copper alloys then in use. The new higher-strength aluminum-plus-zinc alloys of the mid-1940s, however, were less ductile, and the problem of cracking of the sheet reappeared.[35] Various aircraft companies tried different approaches with differing degrees of success. In the long run, the old solution of modifying the rivet head or the dimpling dies did not work. Heating the sheet around the hole and thus causing a local and temporary increase in ductility finally proved successful. Engineers tried various methods of heating, including resistance heating by passing an electric current through the sheet, induction heating by means of a high-frequency magnetic field, and conduction heating from electrically heated dies. Conduction won out, and aircraft tooling companies developed integral heaters and associated automatic controls for both squeezer and hammer-plus-bucking-bar applications. As in earlier aspects of dimpling, extensive trial-and-error and parametric tests were necessary to establish die temperature, pressure, and contact time as they depended on sheet thickness and rivet size.[36]

"Spin dimpling" also achieved some success with the higher-strength alloys. In this method a special male die, shaped so that only two small radial sectors rested on the (cold) sheet, was spun rapidly around its longitudinal axis during dimpling. Compared with hot dimpling, spin dimpling found less favor in the United States than in England, perhaps because the 100° rivet adopted as standard in this country in the early 1940s (as explained later) required greater deformation of

[34]U.S. Patent 2,395,348, A. Sherman and R. Gottlieb, *Flush-Riveting Procedure*, February 19, 1946; interviews with Eugene Speakman (August 31, 1981), employed at Douglas Aircraft Co., Long Beach, Calif., since 1947 and currently senior structural engineer specializing in fasteners, and with Neil Williams (September 2, 1981), materials and process engineer with the same company.

[35]Alcoa 24S-T, common in the early 1940s, had 4.5% copper, together with 1.5% magnesium and 0.6% manganese. The higher-strength 75S-T introduced in the mid-1940s had 5.6% zinc, 2.5% magnesium, 1.6% copper, and 0.3% chromium. Aluminum Company of America, *Alcoa Aluminum Handbook* (Pittsburgh, 1956), p. 34.

[36]K. F. Thornton, "Standard Dimpling Methods Adapted for 75S-T Aluminum Sheet," *American Machinist* 89 (1945): 106–8; "Hot Dimpling Widens Metal Use," *Aviation Week*, April 2, 1951, pp. 21–22; T. A. Dickinson, "How to Hot-Form a Dimple," *Steel Processing* 38 (1952): 172–74; Close (n. 22 above), pp. 191–93.

the sheet than did the British standard of 120°. The experience with the improved alloys illustrates the close relationship that exists among design, material properties, and manufacturing processes.[37]

3. In the first half of the 1940s, the need for increased production for World War II, coupled with the lack of skilled workers, caused a trend toward riveting by machine. The highly skilled two-person teams of the 1930s were replaced increasingly by larger and larger squeeze riveters operated by one worker. Stationary pedestal-type versions provided jaws as deep as 5 feet, and some were capable of upsetting a number of rivets simultaneously. At the same time structural engineers produced designs incorporating more and larger subassemblies to take advantage of the new machines.

In the second half of the decade, in response to the ever-present desire to reduce costs, the flush-riveting process became increasingly automated. The Engineering and Research Corporation, in cooperation with various of the aircraft companies, developed the Erco Automatic Riveter, which could punch and dimple the rivet hole on subassemblies and insert and upset the rivet, all in rapid automatic sequence. The General Drivmatic Riveter provided the equivalent for machine-countersunk rivets. The latter machine allowed North American Aviation to reduce the riveting time for the complete wing panel of an F-86 fighter from one day to two hours. At the same time, only 1 percent of the rivets had to be rejected as imperfect and redone, compared with 12 percent by previous hammer or squeeze riveting. Moreover, the machine was so precise in operation that rivets could be installed to the desired flushness without head milling. Thus what had been largely a hand process in the early 1940s became a highly mechanized procedure ten years later, with improvement in precision as well as economy.[38]

* * *

In our concern for specifics, we must not lose sight of the fact that the entire development from the second half of the 1930s to the early

[37]Thornton, pp. 106–7; "Spin-Dimpling," *Aircraft Production* 13 (1951): 3–7; interview (August 12, 1982) with David Richardson, currently and for the past twenty years an engineering specialist on fasteners with the Lockheed-California Company, Burbank, Calif.

[38]J. E. Cooper, "Production Riveting by Machine," *Western Machinery and Steel World* 36 (1945): 494–97; E. O. Baumgarten, "Riveting as Applied to Aircraft Production," *Automotive Industries* 102 (1950): 176, 265–66, 268, 270, 272; "Automatic Riveter Speeds Plane Assembly," *Aviation Week*, December 24, 1951, pp. 34–35; "Subassembly Riveting," *Aircraft Production* 13 (1951): 12–17. For remarks and references on the tendency toward mechanization generally, see R. R. Nelson and S. G. Winter, *An Evolutionary Theory of Economic Change* (Cambridge, Mass., 1982), p. 260.

1950s was a continuous and integral process. Experience in the shop and in service was a stern teacher that required numerous changes, mostly in close relation. From observation of what did and did not work in practice, for example, the limits on the different types of flush riveting became more sharply defined; as a result, the choice of which type to use in a given design situation became more apparent. When rivet dimpling was found lacking in tightness (contrary to original Douglas expectations), such dimpling tended to be avoided in structurally critical locations. As time went on, shopworkers and production engineers learned, frequently to their dismay, the idiosyncrasies and intractabilities of different materials in different situations. At the same time, agreement grew, though it was never complete, on which kind of tool best suited a particular operation. Most important, increasingly sophisticated dimpling methods made stricter tolerances feasible on the size and shape of dimples. Such improvement was prompted by growing realization that poor nesting of dimples could bring about sizable decreases in strength and that small gaps around rivet heads or a consistent departure of the heads from exact flushness by as little as 0.005 inch could cause significant increases in aerodynamic drag. North American made an important improvement in this regard in the late 1940s with so-called coin dimpling. In this method, the female die for pre-dimpling was made in two concentric parts that exerted different pressures on the sheet, thus forging a more sharply defined and precise dimple. All this affords an example of how increasing experience leads to increased knowledge for both production and design.[39]

As usual, generation of knowledge depended in part on its interchange. Although the various companies operated essentially independently in the beginning (and maintained their individual programs later), communication grew as time went on. As with the Berlin-Rossman paper to the SAE, presentations and discussion at technical meetings provided a useful if somewhat delayed channel. So also did technical journals and trade magazines of the sort cited in the footnotes. More immediate exchange took place by personal contact; according to an engineer at North American, "When I got in a jam, I'd call my college friends at other companies." As production activities increased and became internally specialized, an identifiable fastener community appeared. (Rivets are only one of many kinds of fasteners needed in airplanes.) By 1952 and probably earlier, a Panel on Fasten-

[39] "Spin-Dimpling," p. 3; S. F. Hoerner, *Fluid-Dynamic Drag* (Brick Town, N.J., 1965), pp. 5–8; "Coin-Dimpling. Part I. Basic Principles and Governing Factors," *Aircraft Production* 12 (1950): 181–83. For remarks on the evolutionary nature of innovative activity and references to other cases, see Sahal (n. 4 above), pp. 36–38.

ing and Mechanical Joining was holding sessions at the annual meeting of the SAE. Here experts from four or five companies—Convair, Douglas, Northrop, Lockheed, and Ryan in 1952, for example—presented short papers on many matters, including flush riveting, and answered questions from the audience and each other. News also spread via engineering representatives of Alcoa and of tool suppliers such as Erco, who traveled about the industry promoting their company's products. Employees at one aircraft firm also resigned and moved to another, taking their growing knowledge with them. Numerous means thus existed to cross-fertilize the learning process.[40]

Means also existed to pass the knowledge on to routine users. One of these was the "process specifications" that aircraft manufacturers issue for guidance of their detail designers and production engineers. These proprietary documents, jealously guarded by the companies, contain (among many things) specification of the company's standards for flush riveting. These appear as tables of the limits the company considers necessary in the various types of such riveting and statements about the fabrication processes and precautions to be specified on the drawings and followed in the shop. In such process specifications, which are continually revised on the basis of experience, the knowledge of flush riveting was packed down for everyday use.

In 1941–43, because of the need to train numbers of detail designers and shopworkers for wartime production, a spate of books on aircraft riveting also appeared.[41] These included information on the various types of flush riveting and detailed explanations of the tools and processes for their manufacture. Since much of the knowledge in any shop process is difficult or impossible to put into words, these books included many pictures and drawings. An especially impressive effort

[40]Interview (September 3, 1981) with William Barker, engineer with North American in the 1940s and 1950s and for some years leader of the SAE panel; Buckwalter interview; R. L. Hand, "Notes on the Use of Rivets and Other Fasteners." *SAE Journal*, June 1953, pp. 41–43; E. Baumgarten and C. A. Sieber, "Fasteners—Conventional and Special," *Report SP-317, SAE Production Forum*, Society of Automotive Engineers (New York, 1956), pp. 62–66. The importance of the fastener community is indicated by the fact that it has its own technical magazine, *Assembly Engineering*, formerly *Assembly and Fastener Engineering*, begun in 1958. For a useful general discussion of the role of communities of practitioners in technology, see Constant (n. 1 above), pp. 8–10. For a review essay on the rather unsatisfactory state of research concerning the relationship between generation and diffusion in technological innovation, see C. Ganz, "Linkages between Knowledge, Diffusion, and Utilization," *Knowledge—Creation, Diffusion, Utilization* 1 (1980): 591–612.

[41]Lear and Dillon (n. 19 above); Nisita (n. 21 above); A. M. Robson, *Airplane Metal Work*, vol. 4, *Airplane Pneumatic Riveting* (New York, 1942); *Riveting and Drilling*, Consolidated Vultee Aircraft Corp. (San Diego, 1943); *Aircraft Fabrication. Aircraft Riveting*, Parts 1 and 2 (Scranton, Pa., 1943).

was the *Aircraft Riveting Manual*, compiled in 1941–42 by George Rechton, a process engineer at Douglas, for internal use in that company. This manual and its addendum contained 226 pages of detailed photographs and drawings of tools and processes, many of them on flush riveting. Only a small fraction of the material was written text.[42]

But even a pictorial work cannot convey all the knowledge involved in a demanding industrial process. Some of it can be imparted only by personal demonstration. In 1942 newly hired riveters, at the Douglas plant built hurriedly in Chicago to increase wartime production of C-54 transports, were having trouble getting their flush riveting right. John Buckwalter, head of the plant, made an appeal to Santa Monica for Rechton, who came and was able to show the Chicago workers how to do it.[43] In the words of the North American engineer quoted earlier, "You can't build an airplane from the drawings." By the nature of things, some of what had been learned about flush riveting could only remain in the neuromuscular skills of the workers and the intuitive judgment of the engineers.[44]

Design of Flush-riveted Joints

Before a riveted joint can be manufactured, it must of course be designed. For a flush-riveted joint, this calls first for a choice (perhaps tentative) of rivet diameter and a decision on the type of riveting—dimpled, machine countersunk, NACA. The rivet diameter is conditioned by the general level of loading, and the type of riveting depends on skin thickness, smoothness requirements, and production considerations, as we have seen. Both decisions usually involve engineering judgment based on knowledge of current and past practice; the choice of type of riveting must also conform with the tables of limits in the company's process specifications. Completion of the design then requires calculation of the number of rivets needed per lineal foot of joint. On the assumption that the load on the joint is shared equally by the rivets, this calculation comes down simply to dividing the load per lineal foot (known from the overall structural design) by the load that

[42]G. Rechton, *Aircraft Riveting Manual* and *Addendum I. Riveting Methods*, Douglas Aircraft Co. (Santa Monica, Calif., 1941 and 1942). For related nonverbal elements in technology, see E. S. Ferguson, "The Mind's Eye: Nonverbal Thought in Technology," *Science* 197 (1977): 827–36, and B. Hindle, *Emulation and Invention* (New York, 1981), pp. 133–38.

[43]Buckwalter interview.

[44]The production and design engineers interviewed for this paper show a distinctly different temperament and mind-set from academic and research engineers who function close to the interface between engineering and science (cf. first two references of n. 1 above).

can be carried by a single rivet (taken from empirical tables obtained as described below).[45] If the number of rivets per foot turns out to be unsuitable for some practical reason, the design may have to be repeated with a different choice of rivet diameter. (Finally the layout pattern of the rivets must be decided, but that need not concern us here.) The key piece of knowledge in the design of a riveted joint is thus the load that can be carried by a single rivet, specified as a function of rivet diameter. The engineer calls this the rivet's "allowable strength."

The important loading on a riveted joint in an airplane is a tension (or compression) in the plane of the skin (fig. 5). With a protruding rivet the loading is counteracted by bearing of the sheet on the rivet shank and by resistance of the shank to shearing. Engineers calculate the maximum sustainable value of these forces, and hence the rivet's allowable strength, by simple theoretical considerations involving the dimensions and material properties of the rivet and sheet. The theory neglects added forces resulting from friction between the overlapping sheets and work hardening of rivet and sheet materials caused by upsetting the rivets. These oversimplifications are clearly conservative, however, and the theory provides a useful approximation. For flush rivets, even this crude theory fails. In the machine-countersunk joint (fig. 2*a*), wedging action between sheet and rivet head seriously alters the mechanics of the situation; in the dimpled joint (fig. 2*b*), bearing between the nested dimples complicates things further. As a result, when protruding rivets had to be discarded for use on aircraft surfaces, so also did the theory. Knowledge of the strength of flush riveting came entirely from experiment.

Such knowledge grew in parallel to that of production. In 1933–34, when flush riveting was in its infancy, tests by the Air Corps Matériel Division showed that machine-countersunk rivets were somewhat weaker than protruding-head rivets and should therefore be used with care. The tests were rough and fragmentary, however, and far from conclusive. After the mid-1930s, the many companies working on

FIG. 5.—Protruding rivet with tension loading in plane of sheet. (In a flush-riveted joint, measures are also taken to eliminate the step from one sheet to the next, but that does not concern us here.)

[45]The assumption is not in fact true. Attempts have been made to allow for unequal loading (see, e.g., Jenkins [n. 17 above], pp. 421–29), but they are complicated and do not appear to have caught on in routine design.

production problems also conducted strength tests to support their designs. Company test departments established allowable strengths by gradually increasing the tension load on specially constructed joints having a small number of rivets and noting the load per rivet at which the joint failed. Each company used the rivet-head angle it favored and covered a limited range of rivet diameter, sheet thickness, and material pertinent to its own design needs. In 1940 the National Bureau of Standards, working at the request of the NACA, reported tests of 865 specimens supplied by seven manufacturers, the navy, and itself. Considerable diversity existed in type and details of riveting, and even for a given type the load-carrying ability varied considerably from one manufacturer to another. Microphotographs revealed that "in the stronger [joints] the sheets conform more closely to each other and to the rivet, with little or no void between them and a greater portion of the manufactured head in contact with the sheets." The industry was learning of the close relation between strength and manufacturing technique.[46]

The variation in strength among different manufacturers caused problems for the government agencies responsible for certifying airplane designs. In early 1942 the Navy Bureau of Aeronautics therefore issued a specification of allowable strengths "for use in the design of naval aircraft," reasonable values being arrived at by analysis of test results from Boeing, Martin, the U.S. Army Air Corps, and the Bureau of Standards plus "other pertinent test data." A great deal of interpolation and what the engineer calls "educated guessing" were undoubtedly involved. Experience had also shown by this time that loads less than those for failure could cause flush-riveted joints to become unfit for service as the result of excessive permanent yielding. This possibility was taken account of in the values by a method too complicated to go into here. The final result was a tabulation of allowable strength as a function of rivet diameter, sheet thickness, and rivet and sheet material for each of three rivet-head angles (78, 100, and 115 degrees). Values were given separately for machine-countersunk and dimpled rivets, with the former turning out generally weaker and the latter stronger than the corresponding protruding rivets. Later in the year the same

[46] S. R. Carpenter, "Study to Determine the Suitability of Countersunk Rivets in Duralumin Airplane Construction," *Engineering Section Memorandum Report No. AD-51-114*, Air Corps Matériel Division (Dayton, Ohio, April 18, 1933); D. M. Warner, "Investigation of Efficiency of Countersunk Rivets with 110 Degree Heads," *Engineering Section Memorandum Report No. M-56-2601*, Air Corps Matériel Division (Dayton, Ohio, January 4, 1934). A typical company report is H. B. Crockett, "LS 1105 Flush Rivet Tests," *Report No. 2377*, Lockheed Aircraft Corp. (Burbank, Calif., August 4, 1941). The NBS report is W. C. Brueggeman and F. C. Roop, "Mechanical Properties of Flush-Riveted Joints," *Report No. 701*, NACA (Washington, D.C., 1940).

tables appeared in *ANC-5*, the handbook of structural requirements issued jointly by the Army Air Corps, the Navy Bureau of Aeronautics, and the Civil Aeronautics Administration. Companies had to see to it that their fabricated joints came up at least to the specified values.[47]

Knowledge of allowable strengths, however, was still far from complete. More refined testing by aircraft companies, the Bureau of Standards, the NACA, Alcoa, and several universities had by 1944 thrown doubt on the values in *ANC-5*. In July the Engineering Committee of the Aircraft War Production Council formed a special subcommittee with members from seven aircraft companies to study the problem, and an extensive experimental program ensued. Over the next 18 months, nine firms cooperated in testing 3,500 specimens of machine-countersunk joints and 3,600 of dimpled joints. The tests covered three rivet materials and six sheet materials (all aluminum alloy), in addition to the necessary range of rivet diameter and sheet thickness. Most specimens used only the 100° rivet head that was by then standard. A successor to the original subcommittee, under the chairmanship of Milton Miner of Douglas, analyzed the mountain of data. Its initial findings appeared in a first report in April 1946, with final reanalyzed values in an appendix in May 1948. The latter were incorporated in a revised edition of *ANC-5* dated May 1949.[48]

The *ANC-5* of 1949 showed clearly the growth and refinement of knowledge since the edition of 1942. This was especially evident in the allowable strengths for dimpled rivets, which now exhibited a complex interplay of rivet diameter and rivet and sheet material. The values for failure and for permanent yielding were now listed separately, giving

[47]R. S. Hatcher, "Strength of Flush Riveted Joints," *Structures Bulletin No. 68*, Bureau of Aeronautics, Navy Department (Washington, D.C., January 30, 1942); Army-Navy-Civil Committee on Aircraft Design Criteria, *ANC-5: Strength of Aircraft Elements*, rev. ed. (Washington, D.C., December 1942), pp. 5–27 through 5–29.

[48]ARC Rivet and Screw Allowables Subcommittee (Airworthiness Project 12), *Report on Flush Riveted Joint Strength*, Aircraft Requirements Committee, Aircraft Industries Association of America (Washington, D.C., original dated April 12, 1946, revisions June 24, 1946, September 20, 1946, December 3, 1946, and February 26, 1947, Appendix F added May 25, 1948) (the Aircraft Industries Assn. took the activity over from the Aircraft War Production Council at the end of World War II); Subcommittee on Air Force-Navy-Civil Aircraft Design Criteria, *ANC-5a: Strength of Metal Aircraft Elements*, rev. ed. (Washington, D.C., May 1949), pp. 76–78; interview (September 3, 1981) with Milton Miner. The number of specimens could have been greatly reduced and the analysis facilitated if the committee had examined their results in terms of a system of generalized parameters proposed in the mid-1940s at the Martin Company; G. E. Holbeck, "Structural Fastenings in Aircraft," *Detail Aircraft Structural Analysis. Supplemental Pamphlet No. 8.*, Glenn L. Martin Co. (Baltimore, undated but probably 1945–46 on basis of listed references). This system was not generally adopted, however, until some time in the 1950s; Richardson interview.

the designer a basis for more sophisticated judgments. In general, the allowable strengths for 1949 (now for standard 100° rivets only) were somewhat higher than for 1942; this increase may have come from more thorough and precise testing or from better manufacturing techniques—the record does not say. The tables, however, carried a cautionary note that any deviations from " 'good' manufacturing" practice could lead to significant reductions in strength. The subcommittee's report also contained the admonition that "individual manufacturers should ascertain that their methods are adequate to substantiate the values of [the tables] before using the data for design purposes."[49] The lesson about the relation between strength and manufacturing technique had been well learned.

Thus, as with production, the aircraft industry was well on top of the strength problems of flush-riveted joints by the early 1950s. How this was accomplished is an impressive, if undramatic, example of successful industrial response to a mundane but crucial general problem. Airline passengers who glance out at the upper surface of the wing probably rarely notice the flush riveting. Accurate knowledge of allowable strengths, however, is essential for their safety.

Standardization of Flush Rivets

As mentioned, the development of flush riveting included adoption of a standard head angle of 100°. How this happened deserves a closer look, since standardization is characteristic of items used throughout an industry.

The early diversity of head angles—from 78° to 130°—is understandable in view of the complexity of considerations in production and design. It could not be tolerated for long, however. The combinations of rivet length and diameter that had to be stocked by the military for field replacement of protruding-head rivets were already burdensome and expensive; to compound the problem by a variety of head proportions of flush rivets was unacceptable, especially in wartime. Within industry, the variety also presented costly difficulties for subcontractors, who were compelled to maintain different sets of countersinking and dimpling tools to do business with different prime contractors employing different head angles. To alleviate these problems, the aircraft companies formed intercompany standards committees at their own initiative in the eastern and western parts of the United States, and by the end of 1939 the western committee had settled on a standard head with the 100° angle. (I find no record of action by the

[49] ARC Subcommittee, p. 01.18.

eastern committee.) Apparently, however, different companies did not eagerly give up their separate ways (see n. 19 above). Finally, to coordinate activities throughout the country, the industry-wide Aeronautical Chamber of Commerce, on November 7, 1940, established the National Aircraft Standards Committee. As part of a broad program of standardization, the committee in April 1941 also approved a 100° rivet head. In December the Aeronautical Board of the Army and Navy made the same head mandatory in all aircraft supplied to the military services. The 100° angle then spread rapidly throughout the industry.[50]

The 100° head adopted by the Standards Committee was essentially the one developed by Pavlecka's team at Douglas. I have not been able to unearth the reasons for the committee's choice, but a reasonable conjecture can be made based on what had been learned. Experience in tests had indicated that too large a head angle produced a wedging action by virtue of the sheet's pushing on the head under load; this caused prying up of the head and roughening of the surface. Large angles were also known, according to the navy specification of 1942, to have lower allowable strengths, perhaps from the same wedging action. Too small an angle, on the other hand, caused the cracking problems we have witnessed in the dimpling process. These findings would have pushed the committee toward some intermediate angle, and, since 100° was already in use by Douglas and Lockheed, two of the largest manufacturers, that angle would have seemed a reasonable and expeditious compromise. With the angle settled, the head height and other dimensions of the Douglas rivet may have seemed desirable because of the large number of airplanes that company was already putting into the field.[51]

Observations and Reflections

Getting the rivet heads out of the airstream thus took a good deal of doing—more, I dare say, than people outside the aircraft industry would suspect. And all of it despite the fact that the essential ideas were

[50]D. G. Lingle and G. A. Seitz, "Army-Navy Aeronautical Standardization: The Record So Far," *SAE Journal* 50 (1942): 32–33, 66–80; H. Mingos, ed., *The Aircraft Year Book for 1942* (New York, 1942), pp. 253–56; Templin and Fogwell (n. 29 above), p. 2; National Aircraft Standards Committee, "Rivet—Countersunk Head 100°," *NAS 1*, Aircraft Industries Association of America (Washington, D.C., 1945) (the Aircraft Industries Assn. succeeded the Aeronautical Chamber of Commerce under which the Standards Committee was set up); personal letter from Captain John Ross, Directorate of Equipment Engineering, U. S. Air Force, October 19, 1981.

[51]This reasoning is my own, though I have had valuable assistance from Howard Wolko and David Richardson.

already readily available. Moreover, the development did not stop with the early 1950s but has continued to the present. Already in the 1940s structural engineers were concerned about the failure of flush-riveted aircraft joints from fatigue, that is, from a reduction in strength caused by frequently repeated (as compared with static) loading. Despite more than thirty years of testing and study, the question of the effect of fatigue on allowable strengths still remains a matter of design judgment. Over the years, increasing size and wing loading of aircraft and consequent increases in skin thickness have resulted in less emphasis on dimpling and more on machine countersinking. This shift was accelerated by flight experience in the 1950s showing that the bend in a dimple was particularly subject to fatigue cracks in the higher-strength aluminum-plus-zinc alloys. The detailed shape and tolerances of rivet heads have also undergone refinement, with an eye especially toward obtaining smooth installation without head milling. And finally, increased replacement of aluminum alloy in the 1970s by stainless steel, monel metal (nickel-and-copper alloy), and titanium has required changes in design considerations and riveting techniques. Flush riveting today is still not a closed book. But our story must conclude somewhere, and by the early 1950s the original innovation of flush riveting was completed in the sense that design and production had been reduced to routine, relatively trouble-free procedures. Both engineers and shopworkers now had a wealth of knowledge with which to carry out their tasks.[52]

Science, it should be evident, played no role in our account. No scientific knowledge was involved (unless one counts such commonplace things as the concept of force and the awareness of electric-resistance heating), and no enabling scientific discovery was necessary. The development of production techniques and the determination of allowable strengths took place entirely empirically by trial and error or parameter variation of a commonsense engineering kind.[53] No scientific theories were called for, and few mathematical equations appeared in the articles and reports, and then only for elementary engineering calculations. A good deal of analytical thinking *was* evi-

[52]Speakman, Richardson, and Baker interviews. For the fatigue problem at different stages, see, e.g., L. R. Jackson, H. J. Grover, and R. C. McMaster, "Advisory Report on Fatigue Properties of Aircraft Materials and Structures," *OSRD No. 6600, Serial No. M-653*, Office of Scientific Research and Development (Washington, D.C., March 1, 1946); J. Schijve, "The Fatigue Strength of Riveted Joints and Lugs," *Technical Memorandum No. 1395*, NACA (Washington, D.C., 1956); R. B. Heywood, *Designing against Fatigue of Metals* (New York, 1962), pp. 230–42.

[53]For definition and discussion of these methods in a different context, see "Air-Propeller Tests" (n. 1 above), pp. 714, 746, and passim.

dent, but such thinking is not solely a province of science. One looks in vain in the story of flush riveting for anything that could seriously be identified as scientific activity. This is neither to commend nor to disparage the methods the industry found necessary—it is simply a statement of fact.

A second set of observations has to do with the widespread and simultaneous character of the learning process. At the beginning—and, curiously enough, even after communication became established in the fastener community—each major aircraft company pursued its own innovation program. Reasons are not difficult to find. The first has to do with necessity. Although much could be learned about forming dimples or upsetting flush rivets from books or conversations, complete mastery required hands-on experience. All the acquired knowledge in matters of this kind is not susceptible of codification or communication. Each company knew it had to obtain its own firsthand knowledge no matter how much it might pick up from others. A great deal of duplication resulted, of course, but it was unavoidable. Situations of this kind appear frequently in production technology. A second reason is what some engineers call "the not-invented-here syndrome." Capable engineers, like any skilled professionals, take pride in their ability and consider that they—or at least their company—can do things better than the competition. They also have their share of human obstinacy. Desperation might rule in a jam, but otherwise "why should we go to *them*?" Necessity and pride are both reflected in a passage from "the typical day of [the] fastener engineer" as described in the report of an SAE panel: "He is sometimes influenced by innovations reportedly in use by competitors, but hard headedness obtained through experience prevails, enabling him to uphold his company's reputation and satisfy the customer."[54]

But reasons are one thing, the ability to act on them another. That such widespread, simultaneous innovation could occur when aerodynamic requirements prompted it depended on two circumstances. (1) As I have emphasized, all the fundamental ideas about flush riveting were either already known or conceptually obvious. (2) The aptitudes and training needed for their implementation were available throughout the industry. The latter circumstance was connected in part with the nonscientific nature of the activity. Just as no new or nonobvious ideas were required, so also no training in engineering science was needed to carry them out. Neither the creative insight of a James Watt nor the scientific training of a Rudolph Diesel was essential. A great deal of knowledge and resourcefulness about production and design

[54]Buckwalter interview; quotation from Baumgarten and Sieber (n. 40 above), p. 62.

was required, but it was of a kind that could be obtained from experience on the factory floor or in the engineering office. Such knowledge and resourcefulness were already available in every aircraft company. Every company thus could and did get into flush riveting as soon as the aerodynamic need was felt throughout the industry.

The pattern here—what I described earlier as a welling up from below simultaneously throughout the industry—is very different from the familiar one of diffusion downward and outward from one or more initiating sources. This latter, well-studied pattern has appeared in my own accounts of technological knowledge close to science, one dealing with experimental design data and the other with a theoretical method of engineering analysis. In both cases the basic ideas were not obvious, and the protagonists, who necessarily had to be versed in engineering science, were academics at one or two universities. The circumstances, as well as the pattern, were thus very different from those seen here.[55] The diffusion pattern has also appeared in countless instances *far* from science, an obvious case being Eli Whitney's cotton gin. Although the basic notion of the device was simple and no great training was required, the idea again was new and far from obvious before Whitney's creative insight brought it to light. To the extent that such examples are representative, the circumstances of the diffusion pattern, whether close to or far from science, appear to be different from those of the simultaneous pattern. The differences appear to lie in the nature of the required ideas and the range of availability of the necessary aptitudes and expertise. It would be interesting to know whether the simultaneous pattern exists elsewhere and whether it is more prevalent in production-centered than in other types of innovation.[56]

Flush riveting also exhibits interesting differences regarding a number of aspects of technological knowledge. Since science was not involved, these differences pertain unambiguously to technological knowledge alone. Table 1 lists the differences without elaboration for various aspects of knowledge, not mutually exclusive, according to whether the purpose was production or design. Knowledge for design must be further subdivided depending on whether it derived from

[55]Note 1 above. To borrow expressions current in social historiography, the studies close to science constituted an examination of *high* technological culture; the present study examines *popular* technological culture.

[56]The divorce from science, though a historical fact here, is not essential in principle for widespread innovative activity. If a high degree of scientific aptitude and training were present throughout an industry, widespread science-based innovation could occur when needed provided the essential ideas about the way to proceed were well known or obvious. Perhaps this may already have happened in the microelectronics industry.

TABLE 1

DIFFERENCES IN KNOWLEDGE ACCORDING TO PURPOSE

Aspect of Knowledge	Knowledge for Production	Knowledge for Design: From Production Requirements	Knowledge for Design: From Strength Requirements
1. Use	To translate design drawings into reality	To supply design drawings of riveted joints	
2. Character	Qualitative and quantitative	Mostly quantitative	Almost entirely quantitative
3a. Generation: actors	Production engineers and shopworkers		Structural and design engineers
3b. Generation: locale	Almost entirely in industry; government laboratory (NACA) in one instance		Industry, government laboratories, and universities; final resolution in industry
3c. Generation: organization	Teams within companies; no formal industry-wide organization		Formal industry-wide committees
3d. Generation: communication	Decriptive articles in technical journals and trade magazines; interchange via personal acquaintance and technical meetings		Evaluative technical reports; committee meetings and correspondence
4. Codification	Process and other specifications, differing from company to company		Military-commercial requirements, binding throughout industry
5. Published literature	Technical journals, trade magazines, books, descriptive manuals		Government bulletins
6. Example	Shapes and dimensions of dimpling dies	Limits on sheet thickness for different types of riveting	Allowable strengths

requirements of airframe production or structural strength. (Specific examples appear in row 6.) The table concerns only knowledge that can be set down explicitly in words, tables, diagrams, and pictures. The implicit knowledge involved in skill and judgment is temporarily ignored.

Table 1 shows clearly that knowledge generated for production and knowledge generated for design but derived from the needs of production have a great deal in common. As we saw, the latter also fits naturally into the historical section on production methods. The common elements are not surprising; they reflect the fact that both sorts of knowledge, despite their difference in purpose, were grounded in productive activity. They cause one to wonder, however, whether the first two examples of knowledge in row 6 may not differ epistemologically in some shared way from the third. Reflection shows that this is indeed the case.

The difference lies in the distinction between *descriptive* and *prescriptive* knowledge. Descriptive knowledge, as the term suggests, describes things as they are. Prescriptive knowledge, by contrast, prescribes how things should be to attain a desired end. Descriptive knowledge is thus knowledge of truth or fact; it is judged in terms of veracity or correctness. Prescriptive knowledge is knowledge of procedure or operation; it is judged in terms of effectiveness, of degree of success or failure. It follows that descriptive knowledge, while it may be more or less precise, is not subject to willful adjustment by technologists to serve their needs. Prescriptive knowledge can be altered at will to be more or less effective.[57]

Of the examples of row 6, the allowable strengths obviously qualify as descriptive. Although they serve a practical purpose in design, they are statements of fact. They say that a rivet of a certain type and diameter combined with sheets of a certain thickness will fail at a certain load. The accepted values of the failing loads may become more precise as experimental techniques improve, but they cannot be altered purposively by the research engineer or designer. As we have seen, they are the same for all companies.[58] The limits on sheet thickness for

[57]Mario Bunge ("Technology as Applied Science," *Technology and Culture* 7 [Summer 1966]: 329–47, esp. 338–41) contrasts the narrower categories of "scientific law" and "technological rule" in somewhat similar fashion. This, however, oversimplifies the distinction into one between science and technology and overlooks the possibility, evident below, of having to deal with statements of fact that would be hard to describe as "scientific."

[58]These statements require minor qualification. In the case of failing loads (though not of yielding loads), the published allowable strengths were obtained by reducing the original experimental values through division by a *judgmental* factor of 1.15 to allow for

the different types of riveting, on the other hand, are clearly prescriptive. They tell, for a given type of riveting and a given rivet diameter, what thicknesses should be avoided to make a producible and effective joint. The prescribed limits differ from company to company depending on the judgment of the production engineers. The designer in a given company must follow them, but the production engineer may modify them on the basis of changing judgment or increasing experience. The shapes and dimensions of the dimpling dies for a given rivet and sheet are likewise prescriptive knowledge. They differ from company to company and can be modified by the production engineer to make a more or less smooth (and hence aerodynamically more or less effective) riveted joint. In both the limits on sheet thickness and the shapes and dimensions of dimpling dies, engineers are free within bounds to make a tradeoff between technical effectiveness and other considerations such as cost. They have no such freedom with regard to allowable strengths. The limits on thickness and the contours of dies are thus both examples of prescriptive as against descriptive knowledge. The common elements between the two center columns in table 1 are thus paralleled by a commonality in the associated technological knowledge (though this does not imply that *all* knowledge used for or derived from production is necessarily prescriptive—see below). These observations suggest that, for epistemological discussion, classification of technological knowledge according to its nature as descriptive or prescriptive may be more fundamental than according to its purpose for production or design.

"Descriptive" and "prescriptive" denote varieties of explicit technological knowledge. To these we must add the implicit, wordless, pictureless knowledge essential to engineering judgment and workers' skills. Such *tacit knowledge* was evident here, for example, in the ability of workers to form dimples and upset rivets by hand. It appeared also in the attempt of production engineers to forecast the decrease in riveting cost with time and in the "educated guessing" by which structural engineers arrived at allowable strengths in the face of early incomplete data. Words, diagrams, and pictures can help suggest and promote tacit knowledge (as in the photographs in Rechton's *Riveting Manual*). The knowledge itself can come in the end, however, only from individual practice and experience. The fact that tacit knowledge is inexpressible does not mean that it is any the less knowledge, a fact

---

dispersion caused by experimental inaccuracies and other imponderables. This had the effect of introducing a secondary prescriptive element into the published figures. In this case, the original experimental values constitute the descriptive knowledge.

we recognize when we say such things as "Mary knows how to ride a bicycle" or "those workers know how to drive a good rivet."[59]

Tacit knowledge and prescriptive knowledge are closely related in practice in that they have to do with procedures. They can thus both be described as *procedural knowledge.* With this addition, our schema of technological knowledge can be summarized thus:

explicit knowledge

descriptive knowledge	prescriptive knowledge	tacit knowledge

procedural knowledge

The distinctions here are not meant to be sharp and discontinuous. In general, we can expect gradations of knowledge and cases that do not fit neatly into just one category. What we have is more in the nature of a framework for thinking about the substantive structure of technological knowledge. The extent to which it is general and useful must remain to be seen.[60]

Learning to produce flush-riveted joints required generation of a great deal of procedural knowledge of both the tacit and prescriptive varieties. It also drew upon descriptive knowledge with regard to material properties, though this was mostly already known or became known for other purposes. Learning to design flush-riveted joints required both descriptive and prescriptive knowledge that did not previously exist; tacit knowledge for design was minimal, though it did enter in several ways, such as the "educated guessing" mentioned above. In technology generally, proportions of knowledge may be expected to vary more or less similarly. Production by its nature is likely to rely heavily on both varieties of procedural knowledge, though the tacit component will tend to decrease as automation increases; descriptive knowledge is also likely to enter, especially through the physical properties of materials. Design of the detailed sort exemplified by flush riveting would be expected to depend mostly on the two kinds of explicit knowledge. Conceptual design of new or complex devices may

[59]The term "tacit knowledge" is taken from the writing of Michael Polanyi, *Personal Knowledge* (Chicago, 1962), pp. 69–245. Polanyi, however, uses "tacit" more broadly to describe the unspoken element indispensable to all knowing. For discussion of the relation of skills and tacit knowledge, see Nelson and Winter (n. 38 above), pp. 76–82.

[60]The ideas and terminology of the last four paragraphs have benefited crucially from discussion with Todd Becker.

in addition call for a great deal of tacit knowledge derived from experience.[61]

Technological knowledge and the activities that generate it are rich and complex. Such knowledge comes as much from production and design as it does from research and invention. Edwin Layton, writing about "Technology as Knowledge," viewed technology "as a spectrum, with ideas at one end and techniques and things at the other, with design as a middle term."[62] He went on to recommend to historians the value of focusing on design, a recommendation my present experience strongly supports. Since "things," the sine qua non of technology, must be made as well as designed, however, the techniques of production deserve equal time.

[61]Several historians have recently called attention (though in different words) to the tacit element in mechanical design; see, e.g., A. F. C. Wallace, *Rockdale* (New York, 1978), pp. 237–39, and B. Hindle (n. 42 above).

[62]*Technology and Culture* 15 (January 1974): 31–41, quotation from 37–38.

[13]

Ideology and Technical Choice: The Decline of the Wooden Airplane in the United States, 1920–1945

ERIC SCHATZBERG

In 1989 the "Ripley's Believe It or Not!" Sunday comic strip featured the British Mosquito combat airplane, which "during World War II . . . was one of the fastest planes in existence. The photographic reconnaissance version of this aircraft . . . was able to fly non-stop over Europe so high it was neither seen nor heard. *It was constructed entirely of wood.*" Believe it or not.[1]

"Ripley's" claim is accurate and even understates the Mosquito's success as a bomber and fighter in combat against metal aircraft.[2] "Ripley's" does not seek to provide historical instruction, but rather to evoke surprise and disbelief. Why should a successful airplane with a wood structure evoke surprise and disbelief? The reason lies in the symbolic meanings that our modern technological culture associates with different materials. Wood symbolizes preindustrial technologies and craft traditions, while metal represents the industrial age, technical progress, and the primacy of science. The airplane is one of the defining technologies of the 20th century, the age of science-based industry. The wooden airplane is thus a symbolic contradiction, representing both science and craft, modernity and tradition. These symbols not only shape our current perception of the wooden airplane, they also played a crucial role in its demise.

Between 1919 and 1939, metal replaced wood as the dominant material in the structures of American airplanes. The decline of wood

DR. SCHATZBERG is an assistant professor in the Department of the History of Science at the University of Wisconsin—Madison. He is finishing a book on the shift from wood to metal airplanes. He thanks Walter Vincenti, Mi Gyung Kim, Michal McMahon, and Ken Lipartito for helpful comments. Research for this article was supported in part by a postdoctoral fellowship at the Center for the History of Electrical Engineering, a NASA/AHA Aerospace History Fellowship, and a Dean's Graduate Fellowship from the University of Pennsylvania.

[1]"Ripley's Believe It or Not!" *Washington Post,* May 21, 1989 (emphasis in original).

[2]For a detailed account of Mosquito operations during World War II, see C. Martin Sharp and Michael J. F. Bowyer, *Mosquito* (London, 1967), pp. 117–371.

0040-165X/94/3501-0005$01.00

began when the American aeronautical community enthusiastically embraced the development of metal airplanes shortly after World War I. Despite the nearly universal belief among aviation engineers in the superiority of metal, wood remained an essential material for airplane structures into the early 1930s. The persistence of wood was the result, I argue, of the indeterminacy of the technical choice between wood and metal. In the 1920s, the technical evidence favored neither wood nor metal overall. Technical criteria thus cannot explain the aviation community's enthusiastic support for metal construction. In addition to technical arguments, supporters of metal invoked a nontechnical rhetoric that linked metal with progress and wood with stasis.[3] Using this rhetoric, aviation engineers expressed their belief in the inevitable triumph of the metal airplane, a belief I term the progress ideology of metal. This ideology insured that research and development resources went overwhelmingly to improving metal airplanes.

My interpretation differs fundamentally from the standard technical histories of the airplane, which accept at face value the arguments for the inherent superiority of metal and portray the shift from wood to metal as an essential step in the technical progress of aviation. These accounts are classic exercises in Whig history, judging the past in terms of its contribution to the present. Their heroes are the pioneers and prophets of the victorious path, the path leading to the all-metal stressed-skin airliners developed in the United States during the early 1930s. The standard histories do little more than codify the aviation community's own mythology and thus cannot reflect critically on the basic assumptions of that community.[4] My approach, in

[3]The boundary between the technical and nontechnical is, like all linguistic categories, subject to negotiation and dependent on the particular problem at hand. See Michel Callon, "Pour une sociologie des controverses technologiques," *Fundamenta Scientiae* 2 (1981): 390–93. I am using the distinction according to current convention, particularly as the terms are deployed in debates over the social shaping of technical change, where the nontechnical is typically equated with the social. See, e.g., Ron Westrum, *Technologies and Society: The Shaping of Peoples and Things* (Belmont, Calif., 1991), pp. 4–12.

[4]In the Anglo-American literature, the dominant view originated with Peter Brooks in the late 1950s. Brooks describes the origins of the fully cantilevered all-metal stressed-skin airliner, culminating in the first "modern airliners," the Boeing 247 and the Douglas DC-1, which first flew in 1933. Peter W. Brooks, *The Modern Airliner: Its Origins and Development* (London, 1961; reprint, Manhattan, Kans., 1982), esp. chap. 3. Other commonly cited technical histories follow and elaborate on Brooks's interpretation. Charles Gibbs-Smith, *Aviation: An Historical Survey from Its Origins to the End of World War II* (London, 1970), pp. 200–201; John B. Rae, "The Airframe Revolution," chap. 4 in *Climb to Greatness* (Cambridge, Mass., 1968); Ronald Miller and David Sawers, *Technical Development of Modern Aviation* (London, 1968), pp. 53–71. Because my interpretation diverges so fundamentally from these works, I make no attempt to

contrast, makes a critical assessment possible by adopting a "symmetrical" perspective, one that makes no presumption about the relative merit of the failed and successful technologies.[5]

American Enthusiasm for Metal Airplanes

American enthusiasm for metal airplanes took shape in 1920, largely as a response to German all-metal designs. This enthusiasm received its first programmatic statement in the 1920 *Annual Report* of the National Advisory Committee for Aeronautics (NACA). The NACA was the principal federal agency for aviation research, and an authoritative source of technical information for the American aviation community. The NACA report addressed the issue of metal airplanes in sober, technical language:

> All-metal construction of airplanes has received the careful attention of airplane manufacturers in Europe, with the result that apparently successful models have been constructed. The war was fought with machines constructed of wood, which from many standpoints is most unsatisfactory. . . . Wood has a nonhomogeneous structure, is uncertain in strength and weight, warps and cracks, and weakens rapidly when exposed to moisture. The advantages of using metal construction for airplanes are apparent, as the metal does not splinter, is more homogeneous, and the properties of the material are much better known and can be relied upon. Metal also can be produced in large quantities, and it is felt that in the future all large airplanes must necessarily be constructed of metal.[6]

The report then described a research program in light alloys to support the development of metal airplanes.

This excerpt from the NACA report seems quite sensible, an example of the uncreative yet practical rationality we expect from engineers. The praise of metal and criticism of wood do not seem strange, since the comparison accords with present perceptions of the relative merits of these materials. Yet it is precisely this familiarity that hinders historical understanding. The NACA report is a typical

engage them directly. Nonetheless, I do not dispute that the all-metal stressed-skin aircraft of the early 1930s were an important turning point in aviation history.

[5]See Trevor Pinch, "Understanding Technology: Some Possible Implications of Work in the Sociology of Science," in *Technology and Social Process*, ed. Brian Elliot (Edinburgh, 1988), pp. 75–76. My argument is also influenced by David Noble's work on alternative technical paths in *Forces of Production: A Social History of Industrial Automation* (New York, 1984).

[6]National Advisory Committee for Aeronautics, *Sixth Annual Report* (1920), pp. 52–53.

example of engineering rhetoric, but behind the apparent objectivity of the technical language there lurks a powerful prejudice in favor of metal construction. The NACA's resounding endorsement of metal was no sober assessment of the technical evidence, but rather an enthusiastic response to the metal airplanes developed in Germany during and immediately after World War I.

Roughly 170,000 airplanes were built during World War I, and the overwhelming majority had fabric-covered wooden structures. Although engineers in various countries experimented with metal structures during the war, only the Germans succeeded in developing serviceable metal airplanes. Of greatest practical significance was the welded steel-tube fuselage developed by Anthony Fokker, a Dutch entrepreneur working in Germany. German manufacturers built over 1,000 warplanes using the Fokker fuselage and wood wings.[7] Much more potent symbolically, however, were the "all-metal" designs of Hugo Junkers. All-metal airplane structures consisted entirely of metal, including the wing coverings, as distinct from fabric-covered metal frameworks or composite wood-and-metal designs. Junkers at first used sheet iron *(Eisenblech),* but then switched to duralumin, a high-strength aluminum alloy developed shortly before the war. Although a few of these duralumin airplanes saw combat late in the war, they had no significant military impact. After the Armistice, Junkers and other German designers turned to civil aviation and developed several all-metal passenger transports.[8]

Knowledge of German metal aircraft spread slowly after the Armistice. The American aviation community got its first detailed introduction to the new German airplanes in spring of 1920, when John M. Larsen began demonstrating an imported Junkers all-metal passenger airplane, the JL-6. The JL-6 generated great excitement in the aviation community, especially among army officers. In a typical response, one army pilot pronounced the JL-6 "the airplane of the future." General Charles T. Menoher, chief of the Army Air Service,

[7]John H. Morrow, Jr., *German Air Power in World War I* (Lincoln, Nebr., 1982), pp. 190–91; N. J. Hoff, "A Short History of the Development of Airplane Structures," *American Scientist* 34 (1946): 221; Henri Hegener, *Fokker—the Man and the Aircraft* (Letchworth, Hertfordshire, 1961), pp. 200–202, 207. For pre–World War I interest in metal airplane structures, see Tom Crouch, *A Dream of Wings: Americans and the Airplane, 1875–1905* (Washington, D.C., 1989), p. 70; Brooks (n. 4 above), p. 70.

[8]Hugo Junkers, "Metal Aeroplane Construction," *Journal of the Royal Aeronautical Society* 28 (1923): 428–29, 432, 436–37; Morrow, pp. 162–64; "Germany and Aviation," *Aviation* 12 (1922): 219. See also Richard Blunck, *Hugo Junkers: Ein Leben für Technik und Luftfahrt* (Düsseldorf, 1951); and Eric Schatzberg, "Ideology and Technical Change: The Choice of Materials in American Aircraft Design between the World Wars" (Ph.D. diss., University of Pennsylvania, 1990), pp. 27–41, 53–60.

wrote Larsen that "there can be no question that the all-metal plane is here and it behooves the rest of us to get busy in the near future if we hope not to be left entirely behind in the race."[9] An article in the *New York Times* reflected this enthusiasm: "Aircraft design and construction will have to be completely revolutionized as the result of the success of an all-metal airplane, the product of German genius, in the opinion of prominent American airplane manufacturers and Army Air Service officials. . . . It was said on good authority that one American company was going out of business, realizing the futility of continuing to manufacture planes along the present line of construction."[10] The JL-6 drove no American company out of business, and did not revolutionize American aviation. However, the favorable publicity did pay off for Larsen, who soon sold eight Junkers to the U.S. Air Mail at $25,000 each, quite a high price for the time, and six more to the army and navy.[11]

The excitement generated by the JL-6 led directly to the NACA's endorsement of metal in its 1920 *Annual Report.* Members of the NACA began discussing German metal airplanes in March 1920, but these discussions produced no concrete results. The army's enthusiastic embrace of the JL-6 in June roused the NACA to action. In July senior NACA officials proposed an ambitious program of research in all-metal aircraft, with hopes of funding from the military. Cuts in the postwar military budget apparently prevented the NACA from implementing most of the program, but its enthusiasm carried through into the endorsement of metal in the 1920 *Annual Report.* This endorsement presented the superiority of metal as an established technical fact, yet this "fact" depended on the very research that the NACA was proposing to undertake.[12]

The NACA's endorsement of metal helped stimulate a burst of activity in the design and construction of metal airplanes. The Army Air Service and the Navy Bureau of Aeronautics quickly established major metal aircraft programs. The navy developed duralumin fabrication techniques at the Naval Aircraft Factory, and then actively

[9]Charles T. Menoher to John M. Larsen, June 4, 1920, quoted in William M. Leary, *Aerial Pioneers: The U.S. Air Mail Service, 1918–1927* (Washington, D.C., 1985), p. 119.

[10]"All-Metal Plane Stirs Flyers Here," *New York Times*, June 20, 1920, sec. 2, p. 9.

[11]Leary (n. 9 above), p. 119.

[12]"Minutes of Meeting of Committee on Materials for Aircraft," March 22, 1920, p. 4; Leigh M. Griffith to NACA (Attn: Executive Officer [George W. Lewis]), July 12, 1920, box 217, file 42-6B; "A Program of Research Necessary for the Development of All-Metal Aircraft," August 12, 1920, box 218, file 42-6C, National Archives and Records Administration, Suitland Reference Branch, R.G. 255, Records of the National Aeronautics and Space Administration, National Advisory Committee for Aeronautics General Correspondence, 1915–42 (hereafter NACA Numeric File).

transferred these techniques to airplane manufacturers. In 1924 William Stout produced the first American all-metal commercial airplane. As early as 1923, some advocates of metal construction felt enough confidence to conclude that the "problem today is not the choice between wood and metal but rather how best to design and fabricate metal parts."[13]

This confidence proved premature. Although Fokker's welded steel-tube fuselage became standard by the mid-1920s, all-metal aircraft remained the exception. Beginning in 1925, Henry Ford put his skills and money behind Stout's all-metal airplanes, but few imitators followed.[14] Statistics of U.S. commercial aircraft demonstrate the continued dominance of wood wing structures. In 1930, there were 130 types of commercial aircraft certified by the federal government for use in interstate commerce. Of these 130, 107 used wood wing spars while fifteen used metal spars. (The spar is the main structural member of a wing.) Only seven of these fifteen also used metal covering, and four of these metal-covered airplanes were Fords. Thus in 1930, all-metal construction accounted for only 5 percent of the types of commercial aircraft in production.[15]

The persistence of wood wing spars into the early 1930s seems surprising in light of the NACA's unequivocal statement in 1920 on the advantages of metal. However, the aeronautical community's endorsement of metal was not the result of a careful consideration of its practical advantages. These advantages would arise only at the end

[13]Adm. William A. Moffett, "The Navy's Record in Aeronautics," *Aviation* 12 (June 19, 1922): 720–22; William F. Trimble, *Wings for the Navy: A History of the Naval Aircraft Factory, 1917–1956* (Annapolis, Md., 1990), pp. 55–59, 85–89; "Development of Metal Aircraft," *Aviation* 12 (May 29, 1922): 636; "The Stout Air Pullman: America's First All-Metal Commercial Plane, Built in Detroit, Passes Successful Flying Tests," *Aviation* 16 (May 19, 1924): 533–34; quotation from Roy G. Miller and F. E. Seiler, Jr., "The Design of Metal Airplanes: Outstanding Features of Metal Construction as Illustrated by Its Principal Exponents," *Aviation* 14 (February 19, 1923): 210. At the urging of the navy, the Aluminum Company of America began commercial production of duralumin in the early 1920s to supply the first American-built rigid airship (Margaret G. W. Graham and Bettye H. Pruitt, *R&D for Industry: A Century of Technical Innovation at Alcoa* [Cambridge, Mass., 1990], pp. 157–69).

[14]John T. Nevill, "Ford Motor Company and American Aeronautic Development," *Aviation* 27 (1929): 229; Henry Ladd Smith, *Airways: A History of Commercial Aviation in the United States* (New York, 1942), pp. 336–37.

[15]"Specifications of American Commercial Airplanes," *Aviation* 28 (March 22, 1930): 608–9 (eight types provide no data on spar construction). The preponderance of wood was even greater in terms of the number of aircraft produced since smaller airplanes, which invariably used wood wings, were produced in much greater numbers than larger airplanes. On federal certification of commercial aircraft, see Nick A. Komons, *Bonfires to Beacons: Federal Civil Aviation Policy under the Air Commerce Act, 1926–1938* (Washington, D.C., 1978), pp. 98–99.

of a long period of development, and even then the relative merits of wood and metal remained open to debate. During the 1920s, the choice between wood and metal remained highly indeterminate.

Technical Indeterminacy: Experience versus Rhetoric

Dominant technologies often obscure past alternatives, making the chosen path seem the only possible outcome. The rejected alternatives are thought to have failed due to objective technical or economic factors, factors that make the victory of the successful technology appear inevitable. In general, one can counteract this presentist bias through a close examination of technical controversies, that is, arguments over the alternative paths.[16] But such a strategy is not sufficient in the case of the wooden airplane. Although there was considerable discussion of the relative merits of wood and metal, there was very little debate, if one defines debate as the presentation of opposing points of view. In most technical discussions of aviation materials, the participants simply posited the superiority of metal.[17] Since I cannot rely on the participants themselves for a balanced assessment, I need to reconstruct the technical comparison of wood and metal, relying only on evidence available to the aviation community at the time. This evidence demonstrates that the choice of airplane materials remained indeterminate during the interwar period; in other words, neither theory nor experience could demonstrate the superiority of metal.[18]

Advocates of metal airplanes did not see any ambiguity in the choice, and they expressed their opinions in numerous strongly worded articles in the early 1920s. Whether published in Germany, France, Great Britain, or the United States, these articles advanced a fairly uniform set of arguments, claiming a multitude of advantages for metal. These advantages fall into four main areas: fire safety, weight efficiency, manufacturing costs, and durability. In each of these four areas, when the supposed advantages of metal were put to the test in the 1920s, the claims for the superiority of metal proved equivocal.

[16]The best argument for the importance of studying technical controversies remains Callon (n. 3 above).

[17]For evidence on the absence of debate, see below under "Progress Ideology and the Neglect of Wood."

[18]I am emphatically not arguing that wood was superior to metal, but only that the choice was indeterminate, i.e., the case for metal was equivocal. Wooden airplanes also had serious technical problems, as my discussion of glues reveals (see below). If at times my argument seems like a legal brief against metal, this impression is an unfortunate by-product of my need to compensate for present-day as well as historical prejudices.

Fires were a major cause of airplane accidents during World War I and the 1920s. Advocates of metal frequently claimed fire safety as a major benefit. In an early discussion of the Junkers metal airplanes, Samuel W. Stratton, head of the Bureau of Standards, declared them "incombustible." An army pilot who examined the Junkers JL-6 in June 1920 told the *New York Times* that "fireproof" metal construction had a "big advantage" over wood. Larsen himself argued that "these metal machines eliminate the aviator's greatest fear—fire."[19]

Yet when the fire safety of the JL-6 was put to the test, it proved to be tragically exaggerated. In airplanes of the interwar period, the chief fire danger came from the fuel rather than the structural materials. The Air Mail's experience with the JL-6 strikingly demonstrated this problem and the inability of aluminum alloys to withstand fuel fires. The U.S. Air Mail had purchased eight JL-6s from John Larsen in the summer of 1920 for $200,000, and began flying them in August. Unfortunately, the JL-6 had serious defects in its fuel system. Less than one month after the Air Mail began using the JL-6, one pilot narrowly escaped death when his feet were suddenly enveloped in flames that had burned through the metal floor. The next day two pilots were killed when their Junkers caught fire in flight. The Junkers were grounded but soon returned to service. On September 14, just two weeks after the first fatal accident, another JL-6 burst into flames while flying over Ohio, killing its two crew members. Despite extensive changes to the fuel system, another Junkers was destroyed in a fatal crash in February, probably as the result of a fire. After this incident, the Post Office sold the four remaining JL-6s for a mere $6,044.[20]

As the Air Mail's experience with the JL-6 demonstrates, combustible fuels posed the greatest fire danger for airplanes in the 1920s. Thin sheets of aluminum provided little protection against fuel fires. Aluminum alloys melt at about 1,000°F, about half the melting point of steel. Additional experience in the 1920s confirmed that metal offered little protection against airplane fires.[21]

However questionable the initial claims for metal's superiority in fire resistance, this issue was minor when compared with metal's greatest liability: weight, or more precisely, weight in relation to

[19] "Says Metal 'Plane' Opens New Era," *New York Times*, June 15, 1920, p. 14; Larsen to Menoher, June 3, 1920, quoted in Leary (n. 9 above), p. 118. Samuel Stratton quoted in "Minutes of Meeting of Committee on Materials for Aircraft," March 22, 1920, p. 4, box 217, file 42-6B, NACA Numeric File.

[20] Leary, pp. 121–27, 138–40.

[21] Joseph S. Newell, comment to H. V. Thaden, "Metallizing the Airplane," *ASME Transactions, Aeronautics* 52 (1930): 171.

strength. As designers began accumulating experience in metal wing construction during the early 1920s, they soon discovered that it was very difficult to build a metal wing as light as a wood structure. Neither theoretical comparisons, laboratory tests, nor practical experience could demonstrate an unequivocal advantage in weight efficiency for either wood or metal.

The history of aircraft design remains incomprehensible without understanding the role of weight. "Without doubt," argued T. P. Wright, a prominent airplane designer, "weight and weight distribution, or balance, are of more importance in airplane design than in any other branch of engineering." To be useful, an airplane had to carry, in addition to its own weight, a "useful load" consisting of passengers, freight, gas, oil, and crew. The weight of the airplane itself was termed "weight empty." The sum of weight empty and useful load, the total weight carried in flight, was called "gross weight."[22]

Within the boundaries of gross weight, variations in weight empty and useful load are a zero-sum process. Every ounce eliminated from weight empty becomes an addition to useful load, and conversely every item added to weight empty reduces useful load by an equal amount. These principles gave designers of aircraft structures a clear goal—to create structures of minimum weight while maintaining adequate strength as specified by government safety regulations. Airplane designers therefore had to negotiate a narrow path between two sources of failure, excess weight and inadequate strength.[23]

In the early postwar years, many airplane designers hoped that metal structures would prove lighter than wood. A few well-publicized cases seemed to confirm these hopes, most notably a set of metal wings designed for the navy in 1922 by Charles Ward Hall, an experienced civil engineer turned airplane designer. However, Hall's success remained exceptional.[24] More typical was the army's unsuc-

[22]T. P. Wright, "Aircraft Engineering," *Annals of the American Academy of Political and Social Science* 131 (May 1927): 30; Richard K. Smith, "The Weight Envelope: An Airplane's Fourth Dimension . . . Aviation's Bottom Line," *Aerospace Historian* 33 (March 1986): 30–33, "The Intercontinental Airliner and the Essence of Airplane Performance, 1929–1939," *Technology and Culture* 24 (1983): 428–31; *Airworthiness Requirements of Air Commerce Regulations*, Aeronautics Bulletin no. 7-A (Washington, D.C., 1929), p. 11. The modern term for gross weight is "maximum takeoff weight" (MTOW).

[23]John E. Younger, *Structural Design of Metal Airplanes* (New York, 1935), p. 3; R. K. Smith, "The Weight Envelope" (n. 22 above), pp. 30–31; Wright (n. 22 above), p. 30.

[24]Albert P. Thurston, "Metal Construction of Aircraft," *Aeronautical Journal* 23 (1919): 473; "Metal Airplanes," *Scientific American* 121 (September 20, 1919): 276; John D. North, "The Case for Metal Construction," *Journal of the Royal Aeronautical Society* 28 (1923): 5. On Hall's designs, see "Successful Design of Light Weight Metal Wings,"

cessful program to develop metal aircraft in the early 1920s. As part of this program, the Army Air Service in late 1920 contracted with the Gallaudet Aircraft Company for a monoplane bomber with metal-framework wings and an all-metal fuselage, named the DB-1. The company promised the army an airplane with a weight empty of 3,800 pounds and a useful load of 3,250 pounds, including bombs. As delivered in late 1921, the prototype DB-1 exceeded the estimated weight empty by more than a ton, which reduced the useful load to about 1,000 pounds. After deducting the weight of fuel and crew, this useful load was barely enough for a hand grenade. The Gallaudet DB-1 was just one of the army's three metal airplane projects during the early 1920s, all of which ended as expensive failures.[25] The Air Service wrapped these failures in military secrecy, and they have all but disappeared from American aviation history.

Not every metal airplane prototype was as overweight as the DB-1, but by the mid-1920s even supporters of metal construction admitted to weight problems, especially in wing structures. "Metal wings are undoubtedly heavier than those of wood and fabric," reported an army spokesman in a 1925 article hailing the advantages of metal. In a 1929 textbook, Alexander Klemin, a leading aeronautical engineer, estimated that metal wings weighed on average from 25 to 36 percent more than wood wings.[26]

Aviation 14 (January 8, 1923): 38, 41; Charles J. McCarthy, "Notes on Metal Wing Construction," *U.S. Air Services* 10 (March 1925): 10–11, 13–16. McCarthy's article compared wood and metal airplanes, but McCarthy could not demonstrate weight savings for metal wings, with the exception of two designs by Hall. The airplanes compared were mainly American but included some German all-metal types. McCarthy's position as a lieutenant in the Navy Bureau of Aeronautics gave him access to reliable weight data for many models, data that manufacturers rarely released.

[25] "Expenditures of Government with Aircraft Industry," *Aviation* 18 (January 26, 1925): 102; E. W. Dichman, "Resume of the Development of the Gallaudet DB-1B with Recommendations for the Future," *Air Corps Technical Report*, no. 2369 (May 13, 1924), pp. 2–3; D. B. Weaver, "Static Test of the Gallaudet DB-1 Day Bombardment Airplane," *Air Corps Technical Report*, no. 1957 (June 19, 1922), p. 4. On the army's other unsuccessful metal airplane projects, see Schatzberg (n. 8 above), pp. 177–89.

[26] Corley McDarment, "Will the Future Airplane Be of Metal?" *Iron Age* 115 (1925): 21; Alexander Klemin, *Airplane Stress Analysis: An Introductory Treatise* (New York, 1929), p. 116. See also William B. Stout, "The Modern Airplane and All-Metal Construction," *Journal of the Society of Automotive Engineers* 11 (1922): 499. French designers reached similar conclusions. M. E. DeWoitine, a leading French builder of metal airplanes, noted that designers "have met with quite considerable difficulties in the realization of an all-metal wing within compatible [comparable] limits of weight and performance" (DeWoitine, "The Metal Construction of Airplanes—Its Advantages—Its Present State—Its Future," U.S. NACA *Technical Memorandum*, no. 349 [February 1926], pp. 25–26).

One might think it relatively easy for engineers to determine which material would give the lightest structure: they could simply choose the material with the greatest ratio of strength to density. Such simple comparisons are inadequate, however, because no single measure of strength suffices to describe the behavior of a specific material. Materials exhibit different strength properties when subjected to compression, tension, shear, torsion, and bending. Some materials, such as wood, are highly anisotropic, meaning that their properties vary with direction.[27] A material might prove superior according to one strength criterion but inferior according to another. The appropriate measure of strength depends on the practical problem at hand.

Despite these difficulties, engineers did publish theoretical and empirical analyses of the relative weight efficiencies of aircraft materials. Sometimes these studies showed a definite advantage for metal. For example, when the army tested short blocks in compression, metal proved superior. For blocks of equal strength, aluminum alloy weighed 13 percent less than spruce, and heat-treated alloy steel weighed 24 percent less.[28] But metal fared less well according to another criterion, compressive buckling strength. The nontechnical reader can understand buckling by experimenting with a sheet of paper. If one grasps opposite edges of the sheet and pulls, the paper will resist a moderate amount of force without tearing. If the edges are pushed toward each other, however, the paper provides almost no resistance, usually bending in the middle of the sheet. This bending is compressive buckling.[29] Whenever a structure relies on thin sheets or long slender parts in compression, the possibility of buckling exists. Metal aircraft structures were especially susceptible to buckling because of the high density of metal, which necessitated the use of thin cross-sections. For example, steel is about one-fourteenth as thick as a plywood sheet of equal weight, while duralumin is one-fifth as thick.[30]

[27] On the technical properties of wood, see J. E. Gordon, *The New Science of Strong Materials, or Why You Don't Fall through the Floor,* 2d ed. (Princeton, N.J., 1976), pp. 129–53.

[28] J. A. Roché, "Selection of Materials for Aircraft Structures," *SAE Journal* 21 (November 1927): 494–95.

[29] The paper analogy was not uncommon. See Brian L. Martin, "Steel Spars," U.S. NACA *Technical Memorandum,* no. 458 (April 1928) (from the *Gloster,* September/December 1927), p. 4; Alfred S. Niles and Joseph S. Newell, *Airplane Structures,* 1st ed. (New York, 1929), pp. 153–54; Hoff (n. 7 above), p. 223.

[30] Paul Brenner, "Problems Involved in the Choice and Use of Materials in Airplane Construction," U.S. NACA *Technical Memorandum,* no. 658 (February 1932), p. 9 (translation of "Baustoffragen bei der Konstruktion von Flugzeugen," *Zeitschrift für Flugtechnik und Motorluftschiffahrt,* vol. 22 [1931]). Density data from George W. Trayer, *Wood in Aircraft Construction* (Washington, D.C., 1930), p. 63, and John E. Younger, *Mechanics of Aircraft Structures* (New York, 1942), pp. 64, 66.

A 1942 textbook by John Younger, a professor of aeronautical engineering at the University of Maryland, showed just how great a handicap metal had to overcome in structures limited by buckling strength. Younger estimated the relative weights of wings in terms of the buckling strength of a flat plate.[31] According to his calculations, given a plywood wing weighing 100 pounds, an aluminum wing of equal buckling strength would weigh 255 pounds and a steel wing 500 pounds. These calculations show why buckling was the most serious problem faced by metal airplane designers and one of the best arguments in favor of wood. Younger's analysis also showed why proponents of metal usually preferred aluminum to steel.[32]

An army project to develop metal wing spars clearly illustrates the difficulties that compressive buckling posed for designers of metal structures. In 1925 the Army Air Corps requested bids for metal wing spars designed for identical 10-ton loads, accepting bids for thirty different spars. The Air Corps also tested some standard wood spars for comparison. The initial results, published in 1927, showed that no metal spar performed as well as the best wood spars. According to the report, the principal cause of failure was "the liability of [metal] spars to fail by lateral buckling of the compression flange. The wood spars did not show the slightest tendency to buckle."[33]

Many metal advocates, aware of the buckling problem, believed metal better suited to large airplanes. Buckling became less serious as airplanes increased in size, because buckling strength increased with the cube of thickness. Because of this relationship, many proponents of metal construction were also advocates of large airplanes. However,

[31]Rather curiously, I have found no quantitative comparisons of the buckling strength of wood and metal before the late 1930s, even though the calculations in Younger's analysis are quite simple. In particular, the relationship between the buckling load of a flat plate and the cube of its thickness had been well understood since the 19th century. Since thickness is inversely proportional to density for flat plates of equal weight, buckling strength varies inversely with the cube of density, giving less dense materials like plywood a great advantage. F. C. Marschner, "Structural Considerations Favoring Plastics in Aircraft Structures," *Modern Plastics* 17 (September 1939): 41–42 and passim; Nathan Rosenberg and Walter G. Vincenti, *The Britannia Bridge: The Generation and Diffusion of Technological Knowledge* (Cambridge, Mass., 1978), p. 29; Stephen P. Timoshenko, *History of Strength of Materials* (1953, reprint, New York, 1983), pp. 299, 413–15.

[32]Younger (n. 30 above), pp. 62–73. Structures can often continue to support increasing loads even after certain small areas have experienced local buckling. For example, buckling in reinforced thin webs in shear was considered acceptable under certain conditions since the structures could continue to carry a load after the onset of buckling (Hoff [n. 7 above], pp. 374–79).

[33]A. S. Niles and E. C. Friel, "Progress Reports on Experimental Spars," *Air Corps Information Circular*, no. 590 (August 26, 1927), p. 1.

no one in the 1920s could predict how big airplanes had to become before metal became preferable to wood. Even in 1930, the largest American passenger airplanes revealed no clear advantage for metal in terms of weight efficiency.[34] The historical record presents very little evidence that airplane designers turned to metal to meet future requirements for large airplanes.[35]

After 1930 another design trend, the increasing use of stressed-skin structures, more than offset the reduction in buckling that larger airplanes had promised to provide. In a stressed-skin structure, the covering contributes a large part of the structure's strength, whereas in framework structures strength depends primarily on a skeleton of structural members. Stressed-skin construction solved two problems at once, providing a streamlined external surface for the airplane as well as a load-bearing structure. On the other hand, stressed-skin structures made buckling failures more likely. In a framework structure, most of the material is concentrated in a few major members with relatively thick cross-sections, whereas a stressed-skin structure spreads its material over a large area, resulting in relatively thin cross-sections. When stressed-skin structures were combined with metal, preventing buckling failures became the structural designer's principal problem.[36]

[34] "The Battleship of the Air," *Aviation* 3 (August 15, 1917): 93; F. H. Norton, "The Possibility of the Large Airplane," *Aviation* 10 (January 10, 1921): 48; Junkers (n. 8 above), pp. 416–17; DeWoitine (n. 26 above), pp. 7, 26. On the relationship between buckling strength and size, see J. E. Gordon, *Structures, or Why Things Don't Fall Down* (New York, 1978), pp. 310–11. The analysis of weight efficiency for American airplanes in 1930 is based on my calculation of the ratio of useful load to gross weight for the eight passenger landplanes in 1930 with gross weights greater than 10,000 pounds. See "Specifications of American Commercial Airplanes" (n. 15 above), pp. 606–9.

[35] The disadvantage of metal in buckling strength also decreases with increasing speeds, insofar as higher speeds are achieved through higher wing loadings (Miller and Sawers [n. 4 above], pp. 55–56). Miller and Sawers quote a 1944 source that shows clear awareness of this relationship; however, I have found no evidence that aviation engineers in the 1920s and early 1930s were aware of the connection between wing loading, buckling strength, and relative weight efficiencies. In any case, wood structures remained competitive with metal in weight efficiency even at wing loadings typical of World War II, as demonstrated by the de Havilland Mosquito Mk. 35 with its wing loading of 49 pounds/square feet. These wing loadings are several times higher than loadings typical of the early 1930s, when metal established its dominance (Mosquito data from Sharp and Bowyer [n. 2 above], p. 409).

[36] Gordon, *Structures* (n. 34 above), pp. 293, 311–13. Aviation engineers in the late 1920s and early 1930s seemed almost completely oblivious to the advantages of wood in stressed-skin construction. See, e.g., William Nelson, "The Monocoque Fuselage," *Aviation Engineering* 6 (April 1932): 32. I have found only one source that recognizes the advantage of wood for these structures: L.-L. Kahn, "Stressed Coverings in Naval and Aeronautic Construction," U.S. NACA *Technical Memorandum*, no. 447 (1928), p. 42

In practice, the structural efficiency of airplane materials fell somewhere between the values predicted by buckling strength and other criteria. In some applications steel would show clear superiority, while in others spruce might demonstrate a marked advantage. In most cases the efficiency of a complete structure depended as much on the skill of the designer as on the choice of material.[37]

Advocates of metal had always assumed that ingenuity would solve the buckling problem, and by the late 1920s this assumption proved justified. By the end of the decade, several firms were building metal-winged airplanes with weight efficiencies comparable to wooden-winged airplanes.[38] However, solving the buckling problem led to a new problem: metal structures were much more expensive to produce than equivalent wood structures.

Designers learned to prevent wing coverings from buckling through intricate systems of reinforcement, and they insured the stability of metal spars by using complex curved shapes.[39] These designs required a massive amount of riveting and a large number of different parts, which greatly increased manufacturing costs, particularly in stressed-skin structures. Initial costs are typically high in the early stages of most innovations. But even as manufacturers gained experience with all-metal construction, it still remained far more expensive than the composite airplane built with wood wings and a steel-tube fuselage.

Proponents of metal admitted that wood had some advantages for experimental designs and small quantities, but they insisted that these advantages would disappear with quantity production. Hugo Junkers argued that metal was essential for "modern methods of manufacture, such as . . . interchangeability, standardisation, [and] wide application of machine work." William Stout asserted, with characteristic hyperbole, that small metal airplanes could be produced at even lower cost than cars or trucks, given an equally large market.[40]

Despite these claims, high costs plagued the production of American metal airplanes in the early 1920s. In the United States these costs

---

(translation of "Les bordés travaillants en construction navale et aéronautique," *Bulletin technique du Bureau Veritas* [June 1927]).

[37]Edward P. Warner, discussion comment to "Symposium on Metal Aircraft Construction," *SAE Journal* 22 (April 1928): 433.

[38]The most important firms were Ford, Sikorsky, Curtiss (the B-2 and Condor models), and Boeing (its model 80).

[39]Hoff (n. 7 above), pp. 225, 374, 378–81.

[40]Junkers (n. 8 above), pp. 417–18; Stout, "The Modern Airplane and All-Metal Construction" (n. 26 above), p. 503. See also McDarment (n. 26 above), p. 21; DeWoitine (n. 26 above), pp. 8, 12–14.

were borne mainly by the army and navy, since private buyers could not afford the extremely high prices of metal prototypes. Yet even the military found the costs excessive and often refused to reimburse manufacturers for the losses they suffered under fixed-price development contracts. For example, the Gallaudet company lost $45,000 on its $103,000 contract for the prototype DB-1. The Air Service paid Gallaudet another $150,000 to redesign the DB-1, but the new model also proved unsatisfactory. The DB-1 was among the most costly development projects undertaken by the Air Service in the early 1920s. The navy had similar problems producing the intricate designs of Charles Ward Hall.[41]

The widespread faith that metal airplanes would ultimately prove cheaper to build than wood airplanes undoubtedly helped sustain support for metal construction during the 1920s. Yet subsequent experience proved this faith unfounded. Metal airplanes remained more costly than mixed wood-and-metal types throughout the 1930s, even though costs gradually decreased as airplane manufacturers gained experience with metal. A 1930 German study, which examined both European and American airplane construction, found "all-metal construction . . . much more expensive than mixed construction." In 1932 the plant manager of the Boeing Airplane Company reported that all-metal fuselages cost Boeing twice as much as fabric-covered types when produced in the same quantities. Preliminary studies at Boeing indicated that all-metal wings would also cost twice as much as those built of wood and fabric.[42] As late as 1939, large metal airplanes

[41]R. H. Fleet to Col. Bane, December 12, 1921, RD3103, file 452.1-Gallaudet Type I Airplane/1921, National Archives and Records Administration, Suitland Reference Branch, R.G. 342, Records of the U.S. Air Force Command, Activities, and Organizations, Sarah Clark Collection (hereafter Sarah Clark Collection); Dichman (n. 25 above), pp. 3, 8–9, 11–12, 27; "Technical Bulletin No. 29: Status of Aviation Material under Development for United States Air Service," *Air Service Information Circular*, no. 379 (October 1922), p. 8. Total contract cost for the DB-1 is from the *Congressional Record*, 68th Cong., 2d sess., 1925, 66, pt. 2:1399–1404. On Hall's relationship with the navy, see JCH [Cmdr. Jerome C. Hunsaker] to Scientific Section [Bureau of Aeronautics], May 27, 1922; H. C. Mustin to Charles Ward Hall, November 28, 1922, box 4374, QM (58) vol. 1, C. W. Hall, Inc., National Archives, R.G. 72, Records of the Navy Bureau of Aeronautics, General Correspondence, 1925–42. See also Trimble (n. 13 above), pp. 87–88.

[42]H. Herrmann, "Relative Economy of Different Methods of Airplane Construction," U.S. NACA *Technical Memorandums*, no. 618 (1931), p. 1 (translated from *Zeitschrift für Flugtechnik und Motorluftschiffahrt*, November 14 and 28, 1930); Gardner W. Carr, "Evolution of Metal Construction," preprint of paper for ASME Aeronautic Meeting, June 6-8, 1932, box 158, file "Metal Construction General; Beall: All Metal Airplane Const," Alexander Klemin Papers, Department of Special Collections, University Library, UCLA (hereafter Klemin Papers). Quote from Herrmann.

required twice as many hours of labor per airframe pound as typical wood-and-fabric biplanes of 1922, despite the widespread application of machine tools to metal airplane production during the 1930s.[43] Metal airplanes clearly failed to fulfill expectations for cheaper production, although this failure only became clear during the late 1930s.

Advocates of metal repeatedly invoked fire safety, weight efficiency, and production costs as arguments in favor of metal, even though practical experience proved equivocal. But another issue provided the most potent argument for metal—durability.

Proponents of metal had always claimed durability as its greatest advantage. Wood, said Junkers, "is subject to . . . fire and decay, and splinters when breaking; it bursts and warps from the effect of humidity . . . and the glued joints split; finally it is attacked by insects. . . . Metal is free from all such drawbacks."[44] Indeed, metals are in general more durable than wood, though both deteriorate when left unprotected. Metal's true advantage in durability could be answered only by practical experience, ideally through studies involving careful observation of comparable wood and metal airplanes under similar operating conditions. If such studies were ever done, they were never made public. On the other hand, duralumin, the most promising metal, had severe corrosion problems comparable to the durability problems of wood.

Proponents of duralumin were initially very sanguine about its corrosion resistance.[45] As the use of duralumin spread, however, reports of corrosion problems began to accumulate, and by 1925 evidence of an especially insidious type of corrosion began to appear—intercrystalline embrittlement. In common types of corrosion, chemical reactions eat away the surface of the metal while leaving the properties of the underlying material unchanged. Intercrystalline embrittlement, on the other hand, produces little change on the surface; rather, it proceeds into the metal along the grain boundaries of the alloy's crystalline structure. This process changes the physical structure of the metal, producing a marked reduction in ductility and a significant reduction in tensile strength. Embrittled duralumin gives

[43]Charles D. Bright, "Machine Tools and the Aircraft Industry: The Boeing Case" (paper presented at the annual meeting of the Society for the History of Technology, Sacramento, Calif., October 24, 1989).

[44]Junkers (n. 8 above), p. 417. See also DeWoitine (n. 26 above), pp. 9–10; McDarment (n. 26 above), p. 24; Miller and Seiler (n. 13 above), p. 210.

[45]Stout, "The Modern Airplane and All-Metal Construction" (n. 26 above), p. 503; F. O. Carroll, "Metals Used in Airplane Construction," *Iron Age* 113 (April 24, 1924): 1206.

little warning of impending failure, an especially dangerous situation for airplanes in flight.[46]

Intercrystalline embrittlement was first recognized by the Bureau of Standards in 1925 during tests of duralumin airplane parts for the navy. The bureau's findings caused considerable concern in the federal aeronautics establishment. When the navy airship *Shenandoah* (the ZR-1) crashed in September 1925, the Bureau of Standards found widespread intercrystalline corrosion in parts of the wreckage. Although the corrosion did not contribute to the accident, the bureau's study gave further publicity to the embrittlement problem.[47]

Duralumin's corrosion problems led a number of former supporters to form a distinctly negative assessment of its durability. Because the salt environment accelerated corrosion, duralumin corrosion was especially troublesome for the navy. In 1930 Lieutenant Lloyd Harrison of the Navy Bureau of Aeronautics summarized his years of experience with metal airplanes: "We have found that wood . . . was much more reliable than metal during the same period, with regard to the main structural elements." In another context, Massachusetts Institute of Technology professor Joseph Newell repeated with approval a comment made to him by the chief engineer of "one of our most progressive airplane companies. . . . 'For durability and dependability I'll have my all-metal airplanes made of wood.' " Newell admitted, however, that a recently developed duralumin product known as Alclad promised improved corrosion resistance, and Harrison thought that the navy's problems with duralumin corrosion were "in the way of being solved."[48]

The aviation community did indeed mobilize its resources to solve the duralumin corrosion problem, most successfully through the development of Alclad, an aluminum alloy bonded to a coating of pure aluminum. Alclad was the result of a concerted effort by the federal government and the Aluminum Company of America (Alcoa) to solve the problem of intercrystalline corrosion. As will be shown below, no similar effort was made to solve the durability problems of wood aircraft.

[46]Henry S. Rawdon, "Corrosion Embrittlement of Duralumin, I: Practical Aspects of the Problem," U.S. NACA *Technical Note*, no. 282 (April 1928), pp. 6–8; William Nelson, "Duralumin and Its Corrosion," *Aviation* 21 (November 1, 1926): 738.

[47]Rexmond C. Cochrane, *Measures for Progress: A History of the National Bureau of Standards* (Washington, D.C., 1966), p. 284; [C. P. Burgess], "Report of the Chairman, Committee on Materials for Aircraft," April 16, 1925, box 219, file 42-6E, NACA Numeric File; U.S. NACA, *Eleventh Annual Report* (Washington, D.C., 1925), p. 34.

[48]Harrison is quoted in "Discussion on Aircraft Materials," American Society for Testing Materials, *Proceedings* 30, pt. 2 (1930): 183, 188; Newell (n. 21 above), p. 171.

Thus neither theory nor experience seem to justify the enthusiastic support given to metal airplanes during the 1920s. In fire safety, weight, production costs, and durability, metal failed to demonstrate any marked advantage over wood. Nevertheless, support for metal construction enabled it to spread despite its problems, so that by the mid-1930s wood had been completely eliminated from major classes of American aircraft, including multimotored passenger airplanes and all U.S. combat aircraft.[49]

Given the indeterminacy of the technical case for metal, how can one explain its success? Determinist explanations that rely solely on technical factors are clearly inadequate. Historians of technology have gone far beyond technological determinism and now employ a variety of more sophisticated approaches, using concepts such as systems, presumptive anomaly, and entrepreneurial strategies. But these approaches fail to provide satisfactory accounts of the shift from wood to metal. Although civil aviation became a large-scale technological system during the interwar period, the problems with wood did not constitute a "reverse salient" preventing the continued expansion of the system. Commercial aviation grew rapidly in the late 1920s, its growth in no way inhibited by the use of wood structures.[50] Nor was support for metal the result of farsighted engineers who recognized presumptive anomalies that demonstrated the inability of wood structures to meet future requirements. There is no evidence that engineers based their support for metal on any clear conception of future requirements.[51] Neither do entrepreneurial strategies of supplier

[49]My argument for the technical indeterminacy of the choice between wood and metal has obvious parallels with Walter Vincenti's article ("The Retractable Airplane Landing Gear and the Northrop 'Anomaly': Variation-Selection and the Shaping of Technology," in this issue, pp. 1–33). Vincenti has demonstrated, in effect, that the choice between the fixed and retractable landing gear was indeterminate in the early 1930s. I would argue that technical indeterminacy is a general phenomenon: all choices among alternative technical paths are indeterminate at some point in their history. Vincenti's discussion of his variation and selection model supports the generality of indeterminacy because variations would have no room to compete if one choice were clearly superior on the basis of existing technical knowledge.

[50]On the systems approach and reverse salients, see Thomas P. Hughes, "The Evolution of Large Technological Systems," in *The Social Construction of Technological Systems*, ed. Wiebe E. Bijker, Thomas P. Hughes, and Trevor J. Pinch (Cambridge, Mass., 1987), pp. 73–74. On the growth of civil aviation, see *Historical Statistics of the United States: Colonial Times to 1970* (Washington, D.C., 1975), pt. 2, pp. 770, 772.

[51]For the "presumptive anomaly" model, see Edward W. Constant, *The Origins of the Turbojet Revolution* (Baltimore, 1980). Almost no one before the mid-1930s expected the increases in wing loading and engine power that made possible the large, high-speed metal airliners of the postwar era. Even if some aviation engineers did anticipate extremely large aircraft or supersonic flight, they did not connect these beliefs with

firms explain the success of metal. Admittedly, the capital-intensive, vertically integrated metals producers had considerably more market power and better research facilities than the suppliers of aircraft lumber. Yet all the evidence suggests that the metals firms were followers rather than leaders, responding to the aeronautical community's enthusiasm for metal aircraft.[52]

Metal did appear to offer clear advantages for certain specific applications, such as flying-boat hulls, due to the considerable weight in moisture absorbed by wood hulls.[53] But this local advantage cannot explain the nearly universal support for metal structures in all types of aircraft. To understand the aviation community's support for metal airplanes one must go beyond technical criteria and examine the culture of the aeronautical community and the specific ideology that helped justify its support for metal.

Progress Ideology and the Neglect of Wood

The clue to the aviation community's support for metal lies in the symbolic meanings that this community associated with various materials. Wood symbolized preindustrial technologies and craft traditions while metal represented the industrial age, technical progress, and the primacy of science. These symbolic meanings were not just vague, implicit assumptions. Leading figures in the aviation community made their beliefs quite explicit, articulating these symbolic associations into a specific ideology of technical progress that I term the progress ideology of metal. According to this ideology, the shift to metal was an inevitable consequence of technical progress, part of the shift of engineering from art to science. This ideology was a key factor in the demise of wooden aircraft.

---

support for metal. A good example is Igor Sikorsky, who advocated and built large passenger aircraft even before World War I, and was one of the earliest American manufacturers to switch to metal. I know of no instance in which Sikorsky linked his use of metal with his advocacy of large airplanes. See Igor I. Sikorsky, *The Story of the Winged-S* (New York, 1948); Frank Delear, *Igor Sikorsky: His Three Careers in Aviation* (New York, 1976).

[52]An example of a supplier as follower is Alcoa's development of Alclad, discussed above. See also Schatzberg (n. 8 above), pp. 218–23. On Alcoa research in general, see Graham and Pruitt (n. 13 above). I have found only one observer from within the aviation community who blamed lumber suppliers for the popularity of metal. See John F. Hardecker, "Specializing in the Production of Wooden Parts," *Aviation* 28 (1930): 20–21. John K. Smith first pointed out to me the argument about the relative ability of wood and metals suppliers to innovate.

[53]Trimble (n. 13 above), p. 85. In any case, improved waterproof coatings were a potential solution to this problem, one that certainly might have seemed attractive considering the corrosive effect of salt water on duralumin.

My claim that ideology shapes technical choice conflicts with the dominant conception of technology as an archetype of rational discourse and action. For many people, ideology implies bias, irrationality, and dogmatism. This view centers on a definition of ideology as false belief. Ideology becomes a more useful concept if the historian begins with a nonevaluative definition, one that does not attribute truth or falsehood to the beliefs articulated in ideologies. In this definition, ideology is an explicitly formulated system of beliefs that helps define a community and provide a common program of action. Ideology guides action by allowing a community to make sense of a situation when myth and tradition prove inadequate, as they so often do with rapidly changing modern technologies.[54] The progress ideology of metal represented an attempt by the aviation community to make sense of the symbolic contradiction inherent in the wooden airplane, the clash between the modernity of aviation and the traditionalism of wood. The belief in the inevitability of metal helped resolve this contradiction by defining the wooden airplane as a transitional technology on the path to metal construction.

Two themes dominated the progress ideology of metal, the first linking metal to progress and the second associating metal with science. Proponents of metal were especially vocal in their insistence that the shift to metal was the inevitable consequence of technical progress. "All the history of engineering relates the gradual displacement of timber by lighter and more durable structures of steel," argued a prominent British advocate of metal construction in 1923. Lieutenant Corley McDarment, an army spokesman, expressed the same view in 1925: "Just as the trend of engineering has always been toward replacing wood with metal, so has the new branch followed tradition. . . . The use of metal in aircraft construction is therefore only a natural consequence in the growth of engineering."[55]

[54]Clifford Geertz, "Ideology as a Cultural System," in his *Interpretation of Cultures* (New York, 1973), pp. 193–233; John Higham, "Hanging Together: Divergent Unities in American History," *Journal of American History* 61 (1974): 10. Note that I am not abandoning the idea of ideological critique: the nonevaluative definition is only the beginning, and I address the question of critique in the conclusion. The concept of ideology has itself become the object of ideological struggle, with many divergent meanings competing for dominance. In addition to Geertz, my conception of ideology has been influenced by Paul Ricoeur, *Lectures on Ideology and Utopia*, ed. George H. Taylor (New York, 1986). Hughie Mackay and Gareth Gillespie have recently called for more attention to the role of ideology in shaping technical choice. See their article, "Extending the Social Shaping of Technology Approach: Ideology and Appropriation," *Social Studies of Science* 22 (1992): 685–716.

[55]North (n. 24 above), p. 3; McDarment (n. 26 above), p. 19.

Proponents supported this claim of inevitability by frequent analogies to past triumphs of metal. M. E. DeWoitine argued that the replacement of wood by metal was a repeating pattern in new technologies, which the airplane was destined to recapitulate. Wood dominated early airplanes because "wood is by nature essentially a material for new industries, . . . ideal for the inventor, who . . . obtained results with but little design and calculation." As the technology advanced, metal would become dominant, just as it had in other industries. "I cannot conceive that the ultimate airplane can be in anything else but metal," concluded DeWoitine, "in the same way that metal ships today completely replace the wooden ships of days gone by." The analogy of the wooden ship also inspired William Stout: "In a comparatively few years from now [1922], wooden airplanes in the air will be scarcer than wooden ships on the sea."[56]

As an object lesson in the "struggle between metal and wood," two navy engineers invoked a very specific analogy, that of the steel railway coach: "At first it was a question of sacrificing low structural weight to satisfy the public demand for a safe vehicle; but later when the standard structural shapes gave way to special shapes developed for the purpose and when designers became more experienced and specialized[,] the steel railway coach became lighter than the wooden coach." Thus, even though the first metal planes might be heavier than comparable wood designs, this shortcoming would soon pass with the progress of metal construction.[57]

The second aspect of this progress ideology concerns the "scientific" character of metal in comparison to wood. This theme was rarely fully articulated, but it found expression in the argument that metal permitted greater accuracy in stress calculations.[58] Design in metal was considered more scientific because metal better met the assumptions of the theory of elasticity. More refined calculations were of little use in design, however, because safety standards were based on ultimate (breaking) load, which occurred when stresses had passed far beyond the elastic limit. Due to the limitations in structural theory, designers assumed elastic behavior when calculating ultimate loads, thus limiting any advantage to be gained from increased accuracy in the elastic range.[59]

[56]DeWoitine (n. 26 above), pp. 5–6, 26; Stout, "The Modern Airplane and All-Metal Construction" (n. 26 above), p. 500.

[57]Miller and Seiler (n. 13 above), p. 210. On wood vs. metal in railroad freight cars, see John H. White, "More than an Idea Whose Time Has Come: The Beginnings of Steel Freight Cars," *History of Technology* 11 (1986): 181–207.

[58]Miller and Seiler (n. 13 above), p. 210; DeWoitine (n. 26 above), pp. 11–12; William B. Stout, "Wood versus Metal for Airplanes," *U.S. Air Service* 8 (May 1923): 16.

[59]Younger, *Structural Design of Metal Airplanes* (n. 23 above), p. 7. See also "Exaggerated Refinement in Stress Analysis," *Aviation* 10 (February 21, 1921): 227.

The "scientific" argument for metal derived its force from the assumption that technological progress involves a trend from art to science. This assumption was made explicit in a semiofficial article written by Corley McDarment, a lieutenant in the War Department's information division. Although "flying started as an art," argued McDarment, "aviation is now crying out to science" to solve its problems. Aviation must wait a while "before the pure art in airplane construction gives way to pure science." Nevertheless, he continued, science had already assumed a major role in airplane design and had promoted the shift to metal. "It was the finger of science that pointed to metal in airplane construction." Wood and fabric construction do not "enable a manufacturer to say: 'This is true, and that is true.' " Metal, on the other hand, permits accurate predictions, claimed McDarment—presumably referring to stress calculations. "The scientific mind likes to build upon the most reliable figures obtainable. And these are certainly to be found among metal workers." According to this logic, the ineluctable movement of engineering from art to science dictated the use of metal.[60]

"Science" in this context did not refer to a logically coherent system of ideas or an epistemological method. Rather, science was an attribute of a technological style, one that valued the use of theoretical models, complex calculational procedures, and extensive, systematic empirical research.[61] The belief that metal was more scientific than wood was itself a cultural prejudice, as was the belief that "science" in technology was preferable to nonscience. The techniques subsumed under the concept of science did offer practical advantages to airplane design, but only in certain applications.[62] Most aviation engineers recognized the limited, instrumental role that the techniques of science played in engineering.[63] Nevertheless, science as a cultural concept did have symbolic power, which the advocates of metal appropriated to their cause.

[60] McDarment (n. 26 above), pp. 20–21, 23.

[61] For example, Temple N. Joyce, "Successful Commercial Aviation Analyzed," *Aviation* 14 (April 16, 1923): 420. On technological style, see Thomas P. Hughes, *Networks of Power: Electrification in Western Society, 1880–1930* (Baltimore, 1983), pp. 404–5.

[62] For example, "Airplane Research Work through Flight Tests," *Aviation* 15 (August 6, 1923): 154–55.

[63] Many aviation engineers in the 1920s, particularly in Britain, had a sophisticated understanding of the limited role of scientific theory in airplane design. See North (n. 24 above), pp. 11, 20; Stanley H. Evans, comment to A. J. Sutton Pippard, "The Training of an Aeronautical Engineer," *Journal of the Royal Aeronautical Society* 39 (1935): 85. The best analysis of science-technology historiography remains Otto Mayr, "The Science-Technology Relationship as a Historiographic Problem," *Technology and Culture* 17 (1976): 663–73.

The very terms of the debate about aircraft materials give further evidence of the influence of ideology. This debate posed a choice between two broad classes of materials, wood and metal, rather than between specific materials, such as Sitka spruce and 17ST aluminum alloy. When designing real airplanes, engineers had to choose specific materials, rather than simply deciding to use metal instead of wood. Within each of these categories were hundreds of materials with an extraordinary range of physical properties. But proponents of steel or aluminum framed the debate in terms of the dichotomy between wood and metal. Among advocates of metal there were some who favored steel and others committed to duralumin, but they gave this debate none of the passion reserved for the question of wood versus metal. The commitment to metal reflected, as Robert Friedel has observed, "a general attraction to the use of the inorganic over the organic." This general preference for metal reflected the dominant prejudices of the engineering community, not some objective technical logic.[64]

The progress ideology of metal was no mere epiphenomenon, but had demonstrable effects on the aviation community as a whole. Its first effect was to inhibit the public defense of wooden construction, producing a one-sided debate in the aviation press. Second, it undermined the arguments in support of wood that did appear, since even wood's defenders acknowledged the inevitable triumph of metal. Third, it provided a cognitive framework that encouraged basic research and practical efforts to improve metal construction while discouraging attempts to solve the problems of wood.

Given the continued widespread use of wood in aircraft until the late 1920s, one would expect to find a vigorous defense of it by some part of the aviation community. But such a defense never emerged. Progress ideology tends to stifle debate by making certain choices appear inevitable. The paucity of support for wood in the aeronautical literature reveals the strength of this ideology in aviation. Not a single article appeared in the American aviation press in the 1920s in defense of wooden construction.[65] Lieutenant McDarment noted this silence in 1925: "The wood and fabric people are not doing much

[64]Robert Friedel, "The Coming of the All-Metal Airplane: A Study in Ideas," NASA Historical Office Summer Seminar 1974 (September 13, 1974), pp. 7–8 (typescript). On the aluminum-steel debate, see, e.g., Adolf Rohrbach, "Materials and Methods of Construction in Light Structures," U.S. NACA *Technical Memorandum*, no. 515 (1929), pp. 6a–9 (translation of "Entwurf und Aufgaben des Leichtbaues," *Wissenschaftliche Gesellschaft für Luftfahrt, Jahrbuch* [1926]: 64–78).

[65]Based on articles listed in U.S. NACA, *Bibliography of Aeronautics*, annual volumes for 1922–29.

talking in defense of these materials." There were occasional spirited defenses of wood against the claims of metal's advocates, but these rarely appeared in the aviation press. One defense appeared in 1924 in the *Journal of Forestry,* hardly standard reading for aeronautical engineers. Other arguments in support of wood were scattered through the pages of a 1930 handbook on wooden aircraft construction, published by the National Lumber Manufacturers Association. Both these publications suggested possibly fruitful paths for further research in wood construction.[66] Yet few voices were heard from within the aviation industry for continued research and development work on wooden construction, despite the continued dependence of commercial aircraft on wood.

Even when manufacturers were willing to defend wood in public, progress ideology structured the debate in a way that handicapped the supporters of wood. A publicity newsletter for the Fokker Aircraft Corporation illustrates this point. The company's founder, Anthony Fokker, was one of the most successful and innovative designers of American transport aircraft in the interwar period.[67] But his company's progressive image seemed threatened by its continued use of wood wings. The 1926 newsletter, entitled "Why Are There No Fokker 'All Metal' Airplanes?" is written in a defensive tone. The newsletter acknowledged the historical trend from wood to metal, as evidenced by ships, railroad cars, and automobiles. It accepted the logic of those who "feel that the airplane is bound some day to go through this same process. . . . Against the eventual prospect of such development *nothing can be said.*" If Fokker airplanes continued to use wood, this indicated not conservatism but "that all is not well with all metal construction." The newsletter enumerated these disadvantages, arguing against metal wings on the grounds of safety and ease of repair. At the same time, the newsletter defended Fokker's reputation as a technically progressive company, despite its continued use of wood. "In the Fokker factories, both in the United States and abroad, the spirit of progress, of constant improvement, dominates."[68]

[66]McDarment (n. 26 above), p. 20; Walter M. Moore, "Some Recent Developments in the Use of Wood in Airplane Construction," *Journal of Forestry* 22 (1924): 366–71; Trayer (n. 30 above), pp. 142, 149, 153, 157–58.

[67]Hoff (n. 7 above), p. 221; Brooks (n. 4 above), p. 57.

[68]"Why Are There No Fokker 'All Metal' Airplanes?" *Fokker Bulletin* no. 13 (September 1926), p. 2, box 15, Clement M. Keys Papers, National Air and Space Museum, Archival Support Center, Suitland, Md. (emphasis added). See also Anthony H. G. Fokker, "Air Transportation," *Annals of the American Academy of Political and Social Science* 131 (May 1927): 185–86; and Anthony Fokker, "An Answer to Mr. Mayo," *Western Flying* 10 (October 1931): 31.

The Fokker company's defense of wood actually served to undermine its continued use. The argument for metal was based not only on technical comparisons but also on historical analogies shaped by the progress ideology of metal. The newsletter accepted the logic of this argument, which dictated the inevitable triumph of metal. The Fokker company merely quibbled over the timing. It argued, in essence, that metal suffered from a few teething problems, which made the continued use of wood necessary for the time being. This defense of wood actually helped justify the concentration of research and development work on metal construction, and the subsequent neglect of wood. And this newsletter appeared in 1926, when every American commercial airplane in production used wood except for the Fords.

The dominant belief in the inevitable triumph of metal did more than just vitiate arguments in favor of wood; it also directly undermined the planning and research necessary for the continued use of wood. This effect appears starkly in the army's response to fears of timber shortages, and in federal research on the durability of airplane materials.

The army's approach to potential timber shortages clearly demonstrates the asymmetrical response to the problems of wood and metal. Advocates of metal often claimed that insufficient timber supplies favored the adoption of metal. Indeed, timber shortages during World War I had sparked the U.S. Army's interest in metal aircraft, although these shortages were due to production bottlenecks rather than shortages of suitable trees.[69] The prospect of timber shortages continued to trouble some army officers throughout the 1920s. These officers were worried about commercial loggers, who were rapidly depleting the best stands of the virgin Sitka spruce favored for aircraft lumber, turning these huge trees into sawdust to supply paper mills.[70] The army responded to these concerns by burying its collective head in the sand, ritualistically invoking the inevitable triumph of metal construction to justify lack of planning for wartime timber production.

As early as 1921, the army began citing the prospective shift to metal as an excuse to avoid planning for spruce production. Fear of timber shortages resurfaced in the mid-1920s, when mobilization planners realized that the army was ill equipped to supply the large quantities of spruce that manufacturers would need in the event of war. An Air Corps study of spruce supplies recommended that the

[69]DeWoitine (n. 26 above), p. 18; North (n. 24 above), p. 3; McDarment (n. 26 above), p. 20. On the army's problems with spruce production, see the correspondence for 1917 and 1918 in box 863, file 411.1A-Spruce, National Archives, R.G. 18, Records of the Army Air Force, Entry 166, General Correspondence 1917–1938 (hereafter AAF/GenCor).

[70]Trayer (n. 30 above), pp. 34–36, 38–40.

army establish a reserve of cut aircraft lumber to meet initial mobilization requirements. In early 1928 the Air Corps' Materiel Division rejected this recommendation, arguing instead that the army "expedite the development of all-metal airplane structures." After some bickering within the Air Corps, Assistant Secretary of War C. B. Robbins endorsed the rejection of the spruce reserve, pledging instead "to support any reasonable program of the Air Corps in research and experimental development of the all metal plane."[71]

While the Air Corps was using the threat of timber shortages to justify support for metal construction, its officers expressed no similar concern for supplies of metal. Military planners simply assumed that adequate resources would be available to meet wartime requirements. As early as 1924, however, an Air Corps study warned that domestic reserves of high-grade bauxite would last no more than twenty years at 1922 production rates. By 1925, imported bauxite was supplying over half of U.S. consumption. Dependence on foreign supplies for crucial materials usually induces severe anxiety among military planners. However, no such concerns arose in the development of an air force built predominantly of aluminum. When the U.S. began mobilizing for World War II, the aircraft industry faced a severe aluminum shortage that lasted through 1942. Many merchant seamen lost their lives to German submarines while transporting South American bauxite to the United States.[72]

The imprint of the progress ideology of metal is clearly evident in this dramatic contrast between the concern over spruce supplies and the lack of attention paid to the corresponding problem with bauxite. A similar contrast exists between the aviation community's vigorous response to the durability problems of duralumin and lack of research on the durability of wood structures.

[71]John W. Weeks to Mr. C. E. Arney (Seattle Chamber of Commerce and Commercial Club), June 8, 1921; W. E. Gillmore (C/MatDiv) to C/AC, "Specific Procurement Plan—Emergency Production of Spruce Aircraft Lumber," January 27, 1928; J. E. Fechet to Assistant Secretary of War, February 4, 1928; Maj. Jacob E. Fickel (for C/MatDiv) to Materiel Liaison Secretary, OCAC, "Replacement of Spruce Requirements by Metal Construction," March 2, 1928; Col. W. P. Wooten (Director of Procurement, Corps of Engineers), by direction of Assistant Secretary of War to C/AC, "Plan for Procurement of Aircraft Spruce," March 19, 1928, box 863, file 411.1A, Spruce, AAF/GenCor (first quote from Gillmore, second from Wooten).

[72]Lt. A. J. Lyon to Capt. Robert L. Walsh (OCAS), February 12, 1924, RD3134, 452.1-All Metal Planes [1924], Sarah Clark Collection; *Historical Statistics of the United States* (n. 50 above), pt. 1, p. 605; Alfred Goldberg, "Equipment and Services," in *Men and Planes*, vol. 6 of *The Army Air Forces in World War II*, ed. Wesley F. Craven and James L. Cate (Chicago, 1955; reprint, Washington, D.C., 1983), pp. 442–44; George David Smith, *From Monopoly to Competition: The Transformations of Alcoa, 1888–1986* (Cambridge, 1988), p. 218.

The greatest problem with the durability of wooden airplanes lay not in the wood itself, but rather in the glued joints.[73] Researchers working on wood structures recognized the need to improve the durability of glues. Within the federal government, research on wooden airplanes, including glue research, was coordinated through the NACA's Subcommittee on Woods and Glues. The Forest Products Laboratory (FPL) at the Department of Agriculture conducted most of this research, mainly with funding from the army and navy.[74]

When enthusiasm for metal airplanes took hold of the Army Air Service in the summer of 1920, the army almost immediately began to limit funding for the FPL's wood research in order to devote more money to metal airplanes.[75] Support declined further in 1925, when the army and navy decided to eliminate all funding for the modest program of glue research at the FPL. This action alarmed George W. Trayer, a senior FPL researcher and the chairman of the NACA Subcommittee on Woods and Glues. According to Trayer, "the future status of wood in aircraft construction" depended on a better understanding of glue durability. Trayer requested NACA support for a comprehensive, long-term study of aircraft glues.[76]

The NACA responded favorably to Trayer's request, approving a modest, three-year study at the FPL on the durability of glued joints. The study was conducted with no sense of urgency. The FPL researchers had little contact with airplane manufacturers, who could have guided them toward problems of immediate practical import. Instead, the researchers proceeded methodically, carefully refining their experimental procedures before beginning the exposure tests that provided a practical measure of durability.[77]

[73]Advocates of metal correctly identified glued joints as the main source of deterioration in wood structures. See, e.g., William B. Stout, "Veneer or Metal Construction," *Aviation* 10 (February 21, 1921): 232.

[74]Alex Roland, *Model Research: The National Advisory Committee for Aeronautics, 1915–1958*, NASA SP-4103 (Washington, D.C., 1985), pt. 2, p. 438; "Progress Report of the Work of the Forest Products Laboratory for the Aeronautic Industry," February 1, 1921, p. 1, box 222, file 42-8A, NACA Numeric File; S. W. Allen and T. R. Truax, "Glues Used in Airplane Parts," NACA *Report* no. 66, in U.S. NACA, *Sixth Annual Report* (1920), pp. 387, 391, 396.

[75]Secretary of War to Secretary of Agriculture, June 20, 1919, July 11, 1919; Thurman H. Bane (C/EngrDiv) to C/AS, "Allotment of Funds to Forest Products Laboratory," August 16, 1920, box 100, file 112.4—Allotment for Forest Products Laboratory, AAF/GenCor.

[76]George W. Trayer to Members of Subcommittee on Woods and Glues, October 2, 1925, enclosing "Report of Subcommittee of Woods and Glues Relative to Two Fundamental Problems Which, to Be Carried on Properly, Need Financial Assistance," October 2, 1925, box 222, file 42-8A, NACA Numeric File.

[77]"Report of Committee on Materials for Aircraft for Annual Meeting," October 22, 1925, p. 24, box 219, file 42-6E, NACA Numeric File; "Minutes of Meeting of

Unfortunately for the glue study, the NACA rapidly lost interest in research related to wooden airplanes. George W. Lewis, the NACA's director of research, did his best to discourage continued research on wood structures. In May 1928, Lewis refused to publish an FPL manual on aircraft gluing practices, claiming that the rapid spread of metal construction made its publication unnecessary. In April 1931, he recommended replacing the Subcommittee on Woods and Glues with a Subcommittee on Miscellaneous Materials. Lewis justified his recommendation by claiming that "practically all" military airplanes were of all-metal construction while admitting that wood remained important for "many" commercial models. In fact, most commercial aircraft in 1931 continued to use wooden wings. Even the army was still buying wooden-winged fighters, and the majority of army planes in service used wood wing spars. Nevertheless, the NACA approved Lewis's recommendation and disbanded the Subcommittee on Woods and Glues.[78] The FPL never completed the glue durability study and published no results.

The NACA's lukewarm pursuit of the glue study contrasts sharply with its vigorous response to the intercrystalline embrittlement of duralumin. The two problems were in many ways comparable, since both raised doubts about the safety of aircraft structures. The NACA first became aware of intercrystalline embrittlement in February 1925 and immediately enlisted the Bureau of Standards to conduct a major research program on the problem. In 1927 the NACA published the bureau's preliminary recommendations on protective coatings. That same year Alcoa announced its solution to intercrystalline embrittlement, Alclad, which was developed using preliminary results from the bureau's corrosion study. By 1928 the NACA had published a series of reports detailing the results of the corrosion study. The federal research establishment thus required only three years to develop a

---

Committee on Materials for Aircraft," June 10, 1927, p. 6; "Progress Report on Permanence of Glues for the National Advisory Committee for Aeronautics," filed June 6, 1928, transmitting F. L. Browne, Don Brouse, and C. E. Hrubesky, "Summary of Progress Report on Chemical and Physical Studies of Water Resistant Glues," n.d.; G. W. Trayer, "Report of Subcommittee on Woods and Glues," February 15, 1928, box 222, file 42-8A, NACA Numeric File.

[78]G. W. Lewis to director of Forest Products Laboratory, May 18, 1928, and "Memorandum for the Chairman [J. S. Ames]," April 9, 1931; J. S. Ames to G. W. Trayer, April 29, 1931, box 222, file 42-8A, NACA Numeric File; "Airplanes of the World," *Automotive Industries* 64 (February 28, 1931): 383; long-distance telephone conversation between Col. Arnold and Lt. Haddon, May 5, 1931, with report, RD3202, file 400.112-Experimental Projects, Sarah Clark Collection.

theoretical understanding of duralumin embrittlement and the practical means to mitigate it.[79]

The contrast with the FPL glue research could not be more clear. The NACA responded to the embrittlement problem as if it posed an immediate threat to the aviation industry, while treating glue durability as though it were an issue of incidental academic interest. In fact, the situation was almost reversed. In the mid-1920s duralumin was of little importance to the American aircraft industry, and only one major manufacturer, Ford, had committed itself to duralumin airplanes. On the other hand, the overwhelming majority of aircraft depended for their safety on the reliability of glued joints. By discouraging research on the problems of wood, the belief in the inevitable triumph of metal became a self-fulfilling prophecy.

In 1931, wood wings were on the wane among new military models and the largest commercial airplanes. Yet wood still remained the most important material in the wings of commercial airplanes. Then on March 31, a three-engined Fokker F-10A crashed in Kansas, killing all seven occupants, including famed Notre Dame football coach Knute Rockne. This fatal accident would also prove fatal to the wooden airliner.

The crash made front-page news throughout the country. Department of Commerce investigators immediately began searching for the cause of the accident. When inspection of other F-10As revealed serious deterioration in the glued wing joints, apparently due to moisture inside the wing, the Department of Commerce ordered the planes removed from passenger service, even though investigators failed to establish a clear link between the glue deterioration and the Rockne crash. The Commerce Department permitted the planes to resume service only if the airlines undertook periodic inspection of the interiors of the wings, which required the removal of the plywood covering, a costly procedure. But the public and the industry had lost

[79]George K. Burgess to G. W. Lewis, February 5, 1925; G. W. Lewis to George K. Burgess, February 10, 1925, box 218, file 42-6C, NACA Numeric File; [Bureau of Standards], "Outline Study of Intercrystalline Embrittlement of Duralumin," June 5, 1925, box 1132, file 470.1A, Aluminum, Duraluminum, AAF/GenCor; U.S. NACA, *Annual Reports: Eleventh* (1925), p. 35; *Twelfth* (1926), p. 39; *Thirteenth* (1927), p. 45; and *Fourteenth* (1928), p. 50; E. H. Dix, " 'Alclad': A New Corrosion Resistant Aluminum Product," U.S. NACA *Technical Note* no. 259 (August 1927); Robert H. Brown, "Aluminum Alloy Laminates: Alclad and Clad Aluminum Alloy Products," in *Composite Engineering Laminates*, ed. Albert G. H. Dietz (Cambridge, Mass., 1969), p. 227. See also Bettye H. Pruitt and George David Smith, "The Corporate Management of Innovation: Alcoa Research, Aircraft Alloys, and the Problem of Stress-Corrosion Cracking," in *Research on Technological Innovation, Management and Policy*, ed. Richard S. Rosenbloom (Greenwich, Conn., 1986), 3:44.

confidence in Fokker's wooden wings, and few Fokkers remained in service. Anthony Fokker was forced to resign from the company that bore his name, and he soon returned to the Netherlands.[80]

Because the aeronautical community accepted the logic of the progress ideology of metal, the Rockne crash was interpreted as revealing a fundamental flaw in wood construction. The *New York Times* editorialized that the new inspection requirements signaled the end of wooden aircraft.[81] Yet these consequences did not follow ineluctably from the technical circumstances of the Rockne crash. Both metal and wood airplanes had suffered accidents due to the deterioration of materials, but in metal these failures served to locate problems to be solved. Engineers could have behaved similarly with regard to the glue deterioration in the Fokker wings, seeing it as a solvable problem in an otherwise sound design. Had the NACA pursued research on glue durability with as much vigor as research on duralumin corrosion, perhaps the gluing problems in the Fokker planes could have been avoided. Even if the FPL research had not been able to prevent the Rockne crash, a better understanding of airplane glues might have given manufacturers enough confidence to continue with wood construction.

In a 1938 retrospective lecture, E. P. Warner, MIT aeronautics professor and former editor of *Aviation*, questioned the consequences of the Rockne crash: "The condemnation [of wooden structures] may have been too hasty and too severe, for it is quite possible that protective treatments and water-proof adhesives can yet be found that will overcome the liability to deterioration that is wood's weakest point."[82] In 1938, such improvements were more than just a possibility. Several manufacturers were already developing new wooden airplanes based on synthetic resin adhesives, and one company even had a model in production. Nevertheless, the aviation community's persistent prejudice against wooden airplanes inhibited the widespread application of these innovations.

The Neglected Alternative: Synthetic Resin Adhesives

While the aviation community was hard at work improving metal construction, it was ignoring a potential solution to the problems of glue durability—synthetic adhesives. The first such adhesives used

[80] *New York Times*, April 1, 1931, p. 1; May 5, 1931, pp. 1, 25; June 25, 1931, p. 5; May 9, 1931, pp. 6, 16; July 11, 1931, p. 24; August 22, 1931, p. 18; Komons (n. 15 above), pp. 183–89; Edward P. Warner, *Technical Development and Its Effect on Air Transportation* (Northfield, Vt., 1938), p. 28; H. L. Smith (n. 14 above), p. 332.

[81] *New York Times*, May 9, 1931, p. 16.

[82] Warner (n. 80 above), p. 28.

phenol-formaldehyde resins, which formed durable, waterproof, and fire-resistant bonds when subjected to heat and pressure. Leo Baekeland, the inventor of Bakelite, patented a method for making plywood with these resins in 1912. A German firm began commercial production of resin-bonded plywoods in 1930, and American firms followed a year later. These new plywoods, often termed "plastic" plywoods, were dramatically superior to plywoods made with organic glues, especially with regard to durability in damp conditions. Resin adhesives pointed the way toward eliminating the problems that had contributed to the Rockne crash. Initially the resins were only suited to plywood, but in the late 1930s chemical companies developed cold-setting resins suitable for assembly work.[83] Despite these advances, the aviation press carried almost no stories on the new adhesives until the late 1930s, even though detailed discussions appeared in other technical publications.[84]

Plastic plywoods offered more than just improved durability; they also promised major advances in aircraft production through the use of molding techniques. All-metal aircraft remained costly to produce due to the complex network of stiffeners riveted to the external skin. Molded plywood airframes required no rivets and needed far fewer stiffeners due to wood's greater buckling strength. In the late 1930s, a few engineers began to examine the advantages of plywood molding techniques using the new adhesives. The prominent aviation engineer Virginius E. Clark, in collaboration with the Fairchild Engine and Airplane Corporation, developed the "Duramold" system, which produced large, precisely shaped plywood panels. Clark used a Duramold fuselage in his design of the Fairchild Model 46, a single-engine passenger plane that first flew in 1937.[85]

[83]Wiebe Bijker, "The Social Construction of Bakelite: Toward a Theory of Invention," in Bijker, Hughes, and Pinch, eds. (n. 50 above), pp. 170–71, 174–77; F. E. Brill, "Wood-Veneering—a New Use for Phenolic Plastics as an Adhesive," *Plastics and Molded Products* 7 (December 1931): 689–90, "Phenol Resin Gives New Tool to Marine Engineers," *Plastic Products* 9 (July 1933): 189–90; Ray Sorensen, "Dry Film Gluing in Plywood Manufacture," *ASME Transactions, Wood Industries* 56 (1934): 41, 44; L. A. Sontag and A. J. Norton, "Phenolic Resin Adhesives in the Plywood Industry," *Industrial and Engineering Chemistry* 27 (October 1935): 1115, 1118; Andrew Dick Wood and Thomas Grey Linn, *Plywoods: Their Development, Manufacture and Application* (Edinburgh, 1942), pp. 83–84.

[84]An article by the usually well-informed Alexander Klemin provides a clear example of the industry's ignorance of synthetic adhesives, "Metal Airplane Construction," *Aero Digest* 27 (July 1935): 42, 113.

[85]V. E. Clark, "Low-Density Structural Material," *Aero Digest* 35 (July 1939): 101–2; Marschner (n. 31 above), pp. 41–42; "Molded Airplanes for Defense," *Modern Plastics* 17 (July 1940): 27, 29; Charles Barton, *Howard Hughes and His Flying Boat* (Fallbrook, Calif., 1982), p. 83. Plywood molding techniques had been used successfully in the

Clark believed that Duramold promised a vast reduction in airplane production costs, and he turned to the military for funds to develop his method. The army rejected his request for a major development project, arguing that "the present highly satisfactory all-metal airplane is the result of a long period of development at considerable expense. We should concentrate on the perfection of metal airplanes."[86] This response was not unreasonable, though such reasoning could also have justified the rejection of research in metal construction in the early 1920s. On the other hand, by 1938 the rising threat of a European war should have made the army think carefully about the need for rapid production of inexpensive airplanes.

President Roosevelt was giving considerable thought to airplane production, and in November 1938 he launched a major expansion in American airpower. By early 1941, the aircraft industry faced a severe aluminum shortage, and the army belatedly asked the industry to design noncombat airplanes using other materials, especially plywood. The industry built roughly 200 all-wood airplanes for the army. When the aluminum shortage eased in 1943, the army canceled plans for large-scale production of wooden airplanes and returned to well-tested aluminum designs.[87]

The aluminum shortage also spawned the most notorious American wooden airplane, the Hughes HK-1 Hercules, derisively termed the *Spruce Goose,* although it consisted primarily of birch. The *Spruce Goose* was a daring exercise of technical hubris, an airplane far larger than any previously constructed, relying on novel materials and new production techniques. Not surprisingly, technical problems prevented Hughes from completing the airplane before the end of the war.[88]

While the Hughes HK-1 was mired in technical problems, a British wooden airplane, the de Havilland Mosquito, was proving tremendously successful. The Mosquito was conceived by de Havilland in 1938 as a fast unarmed bomber, protected only by its speed and

---

1920s, most notably on the Lockheed Vega, but in the early 1930s the industry abandoned these designs in the headlong rush to metal construction (Gerard F. Vultee, "Fabrication of the Lockeed 'Vega' Airplane-Fuselage," *SAE Journal* 23 [November 1928]: 451–52).

[86]V. E. Clark to C/AC, February 7, 1938; quotation in 1st Ind., E. R. Householder (Adjutant General) to C/AC, February 18, 1938, box 985, file 452.1, All Metal Planes; O. Westover (C/AC) to Adjutant General, "Plastic Materials," February 15, 1938; Brig. Gen. George R. Spalding (Asst. Chief of Staff) to Chief of Staff, "Plastic Materials," February 17, 1938, box 939, file 423A, Plastic Materials, AAF/GenCor.

[87]On the aluminum shortage, see Goldberg (n. 72 above), pp. 442–44. On the wartime wooden airplanes, see Schatzberg (n. 8 above), pp. 417–24.

[88]A good popular account of the *Spruce Goose* based on archival sources is Barton (n. 85 above).

maneuverability. The British Air Ministry initially objected to the Mosquito's wooden construction and unorthodox arrangement, but in 1940 it agreed to go ahead with the project in order to take advantage of unused capacity in the woodworking industry. The Mosquito fuselage consisted of two layers of birch plywood over a balsa-wood core, formed in two halves over concrete forms. The wings were a complex structure of spruce and plywood.[89]

The combat performance of the Mosquito exceeded all expectations. This airplane became "one of the most outstandingly successful products of the British aircraft industry during the Second World War." The Mosquito was far more efficient at placing bombs on target than the large all-metal bombers that formed the backbone of the Allied strategic bombing campaign. The British built over 7,700 Mosquitos, showing that quantity production was perfectly feasible for wood airplanes.[90]

The success of the Mosquito proves that, even in the 1940s, no fundamental technical obstacles prevented the widespread use of wood in airplanes. Nevertheless, two decades of prejudice against wood construction gave metal airplanes a massive advantage over wood in design, manufacturing, and operation. Even if the aeronautical community had abandoned its prejudice against wood during World War II, metal airplanes would have remained dominant. By World War II, the choice of materials was already firmly cast in metal.[91]

Conclusion

I have argued that the progress ideology of metal was a central factor in the shift from wood to metal airplanes. This ideology helps explain the words and deeds of aviation engineers, aircraft manufacturers, and government aviation officials from the early 1920s to the early 1940s. It makes sense of the army's enthusiastic embrace of the

[89]"New War Planes for the United Nations," *Modern Plastics* 20 (March 1943): 64–67; M. W. Bourdon, "The de Havilland Mosquito," *Automotive and Aviation Industries* 88 (June 15, 1943): 31, 89; Edward Bishop, *The Wooden Wonder: The Story of the de Havilland Mosquito* (London, 1959), pp. 21, 27. Initially the Mosquito used organic casein glue, but durability problems in tropical locations prompted a switch to synthetic resins.

[90]Quote from Owen Thetford, *Aircraft of the Royal Air Force since 1918,* 6th ed. (London, 1976), p. 192 (see also pp. 187–203, 225–26); C. Martin Sharp, *D.H., a History of de Havilland,* 2d ed. (St. John's Hill, Shrewsbury, 1982), p. 191; Robin Higham, *Air Power: A Concise History* (London, 1972), p. 130; Gordon (n. 27 above), p. 165.

[91]Designers of wooden airplanes during World War II recognized the advantages that accumulated practical experience gave to builders of metal airplanes. See F. R. Shanley, "Problems in Aircraft Structural Research," preprint for ASME annual meeting, November 30—December 4, 1942, p. 20, box 207, file "Stress Analysis: Research," min Papers.

Junkers JL-6 in 1920 despite the lack of reliable technical information, and also the NACA's endorsement of metal in its 1920 *Annual Report*, before it had conducted the research to justify its stand. This ideology also helps explain why so many aviation engineers failed to recognize the indeterminacy of the choice between wood and metal, and why they persisted in their confident claims for metal's superiority. The determinist rhetoric that linked metal with progress can only be understood as the articulation of this ideology. The progress ideology of metal also makes sense of the federal government's allocation of resources in favor of metal airplanes. It explains why the army used its concerns about spruce supplies to divert funds to metal construction while ignoring comparable problems with supplies of aluminum. It explains why the NACA treated intercrystalline corrosion as a crisis requiring immediate action while regarding the durability of wood glues as a minor issue. Finally, this ideology explains the aviation community's lack of interest in the development of synthetic wood adhesives in the 1930s, even though these adhesives promised to mitigate the most serious problems of wood structures.

But my explanation of the shift from wood to metal remains incomplete in an important respect: it fails to explain the historical origins of the progress ideology of metal itself, and why this ideology proved so attractive to the aeronautical community. The origins of this ideology lie in its links to broader cultural currents, specifically the popular faith in technical progress and the widespread conception of modern industrial society as the result of a shift from the organic to the inorganic.

The progress ideology of metal drew its main strength from the general ideology of technical progress that pervaded American culture in this era. By the late 19th century almost all Americans embraced technical change as the key to human progress. By emphasizing technical change, Americans displaced earlier moral and spiritual conceptions of progress.[92]

By itself, faith in technical progress did not forge the link between metal and progress. This link was based on the widespread belief that modern industrial society was, in its essence, a shift from the organic to the inorganic. Writing at the turn of the century, Werner Sombart gave this idea formal expression in his *Moderne Kapitalismus*. According to an American interpreter of this book, Sombart viewed the

[92]Merritt Roe Smith, "Technology, Industrialization, and the Idea of Progress in America," in *Responsible Science: The Impact of Technology on Society*, ed. Kevin B. Byrne (San Francisco, 1986), pp. 9–10; John Staudenmaier, "Perils of Progress Talk: Some Historical Considerations," in *Science, Technology, and Social Progress*, ed. Steven L. Goldman (Bethlehem, Pa., 1989), pp. 270–74.

displacement of wood by metal as part of the general trend "toward the economic emancipation of men from the limits of organic nature."[93] This idea also found expression in the literature of technological utopias, which prominently featured new materials, especially aluminum.[94]

Both the general ideology of technical progress and the specific link between metal and progress proved attractive to engineers. As engineers professionalized, they seized on the idea of technical progress to legitimate their role in an industrial system dominated by large corporations. At the same time, engineers wrapped themselves in the mantle of "science" to distinguish their expertise from the traditional knowledge of craft workers.[95]

Engineers proved particularly receptive to the rhetoric linking metal and progress. For mechanical engineers especially, "wood was anathema to the ideals of precision, power and production" that defined the profession and clearly distinguished engineers from millwrights and carpenters. Craft skills remained essential for working with metal as well as wood, but metal's uniformity made it attractive to engineers designing tasks for less skilled workers, especially in mass-production industries.[96] For civil engineers, the highly visible monuments of Victorian engineering, especially the great metal bridges, created prominent symbols linking metal with technical progress. When American engineers entered aviation in large numbers during World War I, displacing the self-taught designers who had previously domi-

[93]Frederick L. Nussbaum, *A History of the Economic Institutions of Modern Europe: An Introduction to Der Moderne Kapitalismus of Werner Sombart* (New York, 1933), p. 285. Lewis Mumford, although heavily influenced by Sombart, objected to Sombart's emphasis on the shift from the organic to the inorganic; the emerging neotechnic age, claimed Mumford, was witnessing a "returning to the organic" (*Technics and Civilization* [New York, 1963], pp. 371–72).

[94]Kenneth M. Roemer, *The Obsolete Necessity: American Utopian Writings, 1888–1900* ([Kent, Ohio], 1976), p. 111. See also Howard P. Segal, *Technological Utopianism in American Culture* (Chicago, 1985). According to Roemer, high-speed railroads and electricity also featured prominently in late-19th-century utopian writings. In the early 20th century, aviation should certainly be added to this list. See Joseph J. Corn, *The Winged Gospel: America's Romance with Aviation, 1900–1950* (New York, 1983), esp. chaps. 1 and 2.

[95]Edwin T. Layton, *The Revolt of the Engineers: Social Responsibility and the American Engineering Profession*, 2d ed. (Baltimore, 1986), esp. chap. 3, "The Ideology of Engineering"; Ruth Oldenziel, "Gender and the Meanings of Technology: Engineering in the U.S., 1880–1945" (Ph.D. diss., Yale University, 1992), pp. 18–22. See also David F. Noble, *America by Design: Science, Technology, and the Rise of Corporate Capitalism* (New York, 1977). For a good example of this rhetoric from a prominent engineer, see R. H. Thurston, "The Border-land of Science," *North American Review* 150 (1890): 67–79, reprinted in *Changing Attitudes toward American Technology*, ed. Thomas P. Hughes (New York, 1975), pp. 178–90.

[96]F. T. Evans, "Wood since the Industrial Revolution: A Strategic Retreat?" *History of Technology* 7 (1982): 51. See also Gordon (n. 27 above), pp. 256–58.

nated the fledgling industry, they brought with them their prejudices against wood in engineering structures.

It was these engineers, when faced with a clash between the modernity of aviation and the traditionalism of wood, who clearly articulated the progress ideology of metal. This ideology provided aviation engineers with an interpretive framework that made sense of the contradictory symbols of the wooden airplane, while also resolving the indeterminacy of the choice between wood and metal. These engineers decided, in effect, that the future of airplanes lay with metal, and they took the necessary steps to make this future a reality.

When I first introduced the concept of ideology, I advocated a nonevaluative approach that rejected the definition of ideology as false belief in favor of Clifford Geertz's definition of ideology as an interpretive system. The nonevaluative approach, however, blunts the critical edge of ideology, undermining the idea that ideologies distort reality. When using a nonevaluative approach, one cannot say that aviation engineers made the wrong choice in supporting metal; their choice simply made sense in terms of their own interpretive framework. Different ideologies might have led to different choices, but there is no nonideological position that would allow one to determine the objectively correct choice.[97]

But on another level, the definition of ideology as distortion does apply to the progress ideology of metal.[98] The progress ideology of metal helped engineers identify the technical advantages inherent in metal airplanes, but it blinded them to the advantages of wood structures, preventing them from appropriating the useful and valid aspects of the wood tradition. In recent years, nonmetallic materials have again found a place in airplane structures in the form of fiber-reinforced composites, of which wood is a natural example.[99] Few aviation engineers, however, recognize the kinship between composite structures and the wooden airplanes of the past.

[97]This is, I gather, the essence of Terry Eagleton's objection to the poststructuralist approach to ideology. See his *Ideology: An Introduction* (London, 1991).

[98]I am here following Ricoeur's concept of ideology as consisting of three levels of meaning: symbolization (or integration), legitimation, and distortion (Paul Ricoeur, "Ideology and Utopia," in *From Text to Action: Essays in Hermeneutics*, trans. Kathleen Blamey and John B. Thompson [Evanston, Ill., 1991], 2:318). See also Ricoeur, *Lectures on Ideology* (n. 54 above).

[99]On composite materials, see Gordon (n. 27 above), chap. 8. Some supporters of wooden construction in the late 1930s recognized the potential of fiber-reinforced resins. See G. de Havilland, " 'Filled' Resins and Aircraft Construction," *Journal of the Aeronautical Sciences* 3 (1936): 356–57. De Havilland concluded, based on preliminary research, that it was "likely that synthetic resins may one day play an important part in aircraft construction" (p. 357).

[14]

Social Choice in Machine Design: The Case of Automatically Controlled Machine Tools, and a Challenge for Labor

DAVID F. NOBLE

PREFACE: A NEW (OLD) CHALLENGE FOR LABOR

One morning in September of 1977, the Lykes Company, a diversified conglomerate, announced that it was closing down the Campbell Steel Works of the Youngstown Sheet and Tube Company. Taking the town and the union locals by surprise, the move threw almost five thousand workers out of their jobs and seriously threatened the economy of a community that had depended upon iron and steel manufacture since 1803. Throughout the following year there was a collective effort on the part of community officials, union locals, and outside agencies to reopen the steel works under "worker-community" control. As of September 1978, the effort was unsuccessful, for several reasons.

Of course, there has been opposition on the part of Lykes stockholders to worker-community control, and the city council, which hoped to purchase the property from Lykes, has found it difficult to enlist the support of investors to raise the necessary capital. More recently, the attorney general of the U.S. has sanctioned a merger between Lykes and the LTV Corporation (owner of Jones and Laughlin Steel) that would make the possibilities of local ownership and control, or even a reopening of the plant under present ownership, more remote than ever. But these are not the only reasons for the lack of success of the worker-community program.

Although the Youngstown situation has given rise to creative proposals by the worker-community coalition—proposals that embrace problems of control ranging from the routines of the shop floor to

This is an extended version of an article written for the anthology, *Case Studies in the Labor Process* (Monthly Review Press, forthcoming).

large-scale finance and strategy (including plant sitings and closings) —it has also dramatized some of the weaknesses that undermine such efforts, shortcomings that are by no means restricted to the people of Youngstown or to labor. A recent editorial in *In These Times* (August 23, 1978, p. 14) described these weaknesses.

> After long conditioning under capitalism to forego self-reliance and responsibility for the enterprise as a whole, most workers, as became evident in Youngstown, have difficulty with the idea of self-management; lacking self-confidence, many positively fear it. Workers' uncertainty and fear of exercising their own capacities in the workplace extend into the wider political arena, where they tend to defer to the authority and prestige of their "betters" rather than trusting themselves to "run the country."
>
> In essence, democracy is self-government. Without a self-confident working class, ready to govern itself at work and in society at large, there can be no democratic socialism in the U.S. The current mutilation of workers' self-esteem is a basic ideological problem confronting American socialists as well as others seriously concerned for the prospects of democracy in the U.S.

There are many things that account for this lack of self-confidence —and, again, it is by no means restricted to labor (academia is renowned for it, despite its cover of arrogance). One of them, the focus here, is an ideology of technological determinism. Technological determinism is an enervating view of history as an automatic, inevitable process, dictated by the ineluctable, self-determining march of technology. It is a view promoted by the powerful to obscure and legitimize their own handiwork and their relations with the less powerful, and it is a view that fosters passivity and cynicism, a general sense of helplessness in the face of "technology" or "change."

In the present climate, when labor is on the defensive against mounting, well-financed opposition and the rationalization and automation of production is being intensified in all industries (in a desperate attempt by U.S. firms to shore up their deteriorating international market position in manufacturing), it is becoming imperative for labor to shake the habit of technological determinism, if only better to defend itself against what is being done in the name of "technology." Labor, in essence, faces three important challenges. First, it must transcend the ideology of technological determinism by demystifying technology itself. It must convince itself that human choice, not technology, moves history—despite the seeming automaticity of the machinery in the shop—and thus become reawakened to the possibilities. Second, it must regain its confidence by preparing itself, through study of the possibilities, to articulate its own choices in its own class interest. Third, and finally, with this new-found self-reliance, it must struggle, on the shop floor and politically, to get into the position to make these choices and thereby steer the course of "progress" in a more humane direction.

In this paper, I will focus upon the first of these challenges, the demystification of technology and the transcendence of technological determinism. But all three tasks are fundamentally interrelated and guide the discussion throughout. The heart of the paper is a case history of automatically controlled machine tools, a recent and very significant production technology. It shows how the intentions and choices of particular people have informed the design and deployment of the new technology, given it shape, direction, and purpose. It shows also what happens when this technology, with its frozen dreams, confronts people with different purposes and different dreams. In short, the paper looks at the ways in which technological development reflects the present social relations of production, and it implies that such development might be different, were the social relations to be changed.

BEYOND TECHNOLOGICAL DETERMINISM

The hard-sounding authority of the phrase *technological development* belies its ambiguity. However popular as an explanatory or justificatory device, the notion is but a convenient catch-all, signifying nothing and everything. With reference to human labor, our focus here, the term is commonly used, in a descriptive mode, to suggest the grand evolution of the material artifacts of "work" that set human beings off from the animal world. Alternatively, in a more modern setting, it is used to refer to an integral aspect of the capitalist process of accumulation, wherein costs are minimized and productivity is enhanced through the profitable substitution of machinery for labor. In a normative and teleological mode, the phrase is used to connote "progress," the steady upswing of human society characterized by freedom from want and toil and by greater human dignity, individual creativity, and autonomy. Or, in a less sanguine view, the same process is seen as the dehumanization of people through the progressive rationalization of labor, either as a routine aspect of capitalism (and thus the managerial quest for control, class hegemony, and intensified exploitation) or as an independent process in itself, called industrialization or modernization. The variations are endless.

As can readily be seen, each use of the term is pregnant with intent as well as content. More important for present purposes, all uses tend to have something in common; they reflect the habit of thought that will be referred to here as "technological determinism."

However much technological development is understood to be mediated by (unwitting) human agency or tied, in its use, to an economic or social process like capital accumulation, it itself is nevertheless commonly comprehended as an essentially autonomous, unilinear, and causal process.

Technological determinism actually embraces two basic, interrelated notions: first, that technological development itself is a "given" that is self-generating and follows a single course; and second, that this process has effects "outside" it, usually referred to as "social impacts." In the first instance, technological development is seen to be essentially independent of social setting; however much it may be embedded in a particular time and place, its form reflects less these historical particulars than the extrapolation of its own immanent dynamic. In the second instance, technological cause is understood to be irreducible and to have singular effects and not others; moreover, the relation between cause and effect is seen to be automatic and inevitable (necessary)—and is more often assumed than demonstrated.[1] Of course, few would claim to be consistent technological determinists, in the precise philosophical meaning of the term; we are talking rather about a prevalent tendency, a subtle although commonplace habit of thought. Like all other determinisms, this habit denies a realm of human freedom; it views history not as a domain of possibilities but as a sequence of necessities.

Why such a tendency should exist is a subject for historical inquiry in itself. Perhaps foremost among the reasons is the common alienation from, and thus ignorance of, technology, on the part of those people who so often write about its supposed impacts. This is not anyone's fault, of course, but rather the result of the professional monopolization of technical knowledge, the industrial monopolization of professionals, and the general specialization and fragmentation of intellect. Thus, for many observers, technology remains mysterious, alien, and impenetrable, simply a given. And this appearance itself is reinforced by the market context in which technology often develops, where everything appears to happen by itself, automatically. But there are reasons why observers remain content with such a superficial view.

For apologists of the status quo, clearly, such a notion of automaticity absolves everyone of blame and responsibility and ratifies things as they are: they could not be any other way given the ineluc-

1. This assumption is not always correct. See, for example, Stephen Marglin, "What Do Bosses Do? The Origins and Functions of Hierarchy in Capitalist Production," and Katherine Stone, "The Origins of Job Structures in the Steel Industry," *The Review of Radical Political Economy*, vol. 6 (Summer 1974).

table march of technological progress. For lazy revolutionaries, who proclaim liberation through technology, and prophets of doom, who forecast ultimate disaster through the same medium, such notions offer justifications for inactivity. And, for the vast majority of us, numbed into a passive complicity in "progress" by the consumption of goods and by daily chants echoing the slogan of the 1933 World's Fair—SCIENCE FINDS, INDUSTRY APPLIES, MAN CONFORMS—such conceptions provide a convenient, albeit often unhappy, excuse for resignation, for avoiding the always difficult task of critically evaluating our circumstances, of exercising our imagination and freedom. Finally, the habit of exaggerating the causal role of technology in history is readily, if subtly, adopted by historians for the simple reason that it makes their work, the writing of history, easier.

But technological determinism is not simply an intellectual shortcoming; it is politically dangerous. Depicting an inexorable, disembodied, and omnipotent historical force that rules out human action and choice, it fosters passivity, quiescence, and resignation, and their correlates, cynicism and despair. Moreover, as a confusion in thought, it generates artificial contradictions that confound effective action. For example, the labor movement's relation to technological development is an unnecessarily contradictory one. Union officials tend readily to welcome rationalization so long as such technological progress guarantees a bigger slice of the pie without jeopardizing jobs and membership. Workers on the shop floor, however—even those whose jobs have been "red circled" (guaranteed)—quite often see such rationalization less as a lightening of their load or as a bigger paycheck than as a loss of control and a tightening of managerial authority over them. Thus, a conflict arises: in general, union spokesmen tend to be "for it" while people on the shop floor tend to be "against it." The "it," however, is taken as a given, in the habit of technological determinism. Rarely, if ever, is the "it" viewed as a range of possibilities, a domain of human choices—at present the reserve of management—that might potentially provide for *both* increased productivity *and* greater shop floor worker control, depending upon the design. The contradictions, arise, then, out of a false notion of a fixed technology, a notion symptomatic of technological determinism. (In this case, the notion is in part a reflection of the virtual exclusion of labor from the realm of technological decision-making, which thus leaves it in no position to recognize the extent to which technological development is the special product of managerial choices made to increase productivity through the enhancement of managerial, not worker, control.)

To the extent that social commentators and scholars simply accept this technology as inevitable and necessary, they ratify and legitimate the managerial choices that inform it. However unwitting they may be, their studies are thus partisan acts. The purpose of this brief paper is to try to overcome the tendency of technological determinism, and the confusion and impotence it generates, by pointing up the possibilities inherent in technological development. It is a deliberately partisan effort, launched from the other side. Before we can ever hope confidently to strive to intervene in this important process, in order to reshape it according to different choices, we must convince ourselves that such choice is possible. And to do this, we must look at the choices that have already been made. Such an investigation, moreover, should make it quite apparent that, first, technological development itself is neither autonomous nor unilinear, and that, second, technological development as a "cause" is neither irreducible nor automatic.

To begin with, we must recognize that technological development is a social process in itself, that it is not only man-made but made of men (and, increasingly, of women, too). Elsewhere, in trying to understand the relationship between modern technology and capitalism, I have examined that social process in some detail, to identify the people who tend most often to give direction and shape to technological development (engineers), to explain how they obtained this important prerogative (professionalization), to describe the social context in which they learn to use it (technical education), to map out the institutional framework within which they exercise it (corporate research organizations, university laboratories, the patent system, a competitive market environment, etc.), and, finally, to articulate a dominant ideology of technical advance (an engineering ideology of profit-making and management).[2] In that study, it became clear that, as Seymour Melman observed, "if one wants to alter our technologies, the place to look is not to molecular structure but to social structure, not to the chemistry of materials but to the rules of man, especially the economic rules of who decides on technology."[3] Technological development, in short, is not an independent force impinging upon society from the outside, according to its own internal logic; rather, is is a social activity in itself, which cannot but reflect the

2. David F. Noble, *America by Design: Science, Technology and the Rise of Corporate Capitalism* (New York: Alfred A. Knopf, 1977).

3. Seymour Melman, "The Impact of Economics on Technology," *Journal of Economic Issues* 9, no. 1 (March 1975): 71.

particulars of its setting: the time, the place, the dreams and purposes, the relations between people.[4]

But how does the technology reflect the social context, to what extent, and in what specific ways? Where is the imprint of the social activity that gave issue to it? Marx asserted that "instruments of labor not only supply a standard of the degree of development to which human labor has attained, but they are also indicators of the social conditions under which that labor is carried on."[5] How might we go about reading those indicators, and what do they indicate? One way, beyond merely examining the artifact and guessing at its intended use, is to try to recreate the social history of the technology, to reconstruct the choices that became frozen in the hardware. This would then make it possible to examine the ways in which these choices reflect the ideology and thus the social position of the designers. But to look for choices, of course, requires that the assumption be made that technological development is not unilinear, that it could, potentially, follow any number of chosen courses. And the validity of this assumption rests upon the identification, and thus awareness, of possible alternatives.

If it is assumed that technological development is neither autonomous nor unilinear, what are the implications of these assumptions for the causal view of technological development? We discover now that the "impacts" of technology must follow not simply from the technology but from the design choices it embodies, choices that (fully intended or not) define or at least constrain subsequent choices in deployment. In short, technological development, however much it is a cause of social change, is never an irreducible, first cause. It too has been caused, or, more correctly, chosen, as thus indirectly so have its "impacts."

4. It perhaps ought to be pointed out that it is a mistake to try to eclipse technological determinism by simply substituting for it another mechanical, say economic, determinism. Market factors are critical ingredients in the process of technological development, but they are not the only ones, nor, at times, even the most important ones. Sweeping economic explanations for technical advance, however logically consistent, provide only a framework for historical analysis, not a substitute for it. This history, like any other, is a complex of scientific, technical, sociological, economic, political, ideological, and cultural threads, all of which give direction to it. Thus we see the role of the government often rendering market considerations beside the point, fascination with remote control on the part of designers and their sponsors outweighing simple cost considerations, and a fetish for order, predictability, continuous-flow, power, or mathematical elegance having more to do with the outcome than other, more practical economic concerns. In short, here the principle of parsimony can be too avidly applied.

5. Karl Marx, *Capital* (Moscow: Foreign Languages Press, 1959), 1: 200.

However, and finally, to say that technologies embody human choices and that these choices reflect the intentions (desired impacts) of the designers—intentions that reflect the ideology (often inarticulate and prereflective) and thus social position of the designers—is not to say that these intentions (desired impacts) are automatically fully realized in the simple construction and use of the technology. Far from it. To end the investigation here, by simply extrapolating the impacts from the choices, would be to beg some of the most important questions and confuse mere intentions for the whole of reality (a common and convenient error). In actuality, the "impacts" are always determined subsequent to the introduction of the technology and in ways not altogether consistent with the intentions of designers. That is, if technology does not automatically "cause" a social effect, neither do the choices (intentions) that lie behind it.

The relation between cause and effect is always mediated by a complex and often conflictive social process, and it is this process, and not the technology or the choices that informed the technology, that determines the final outcome. This social process can be broken down into several aspects. Two of these have received considerable attention from industrial engineers and sociologists: the debugging (and possible minor redesign) of the physical technology following trial use, and the psychological, social, and organizational "adaptation" to the technology of those affected by it. A third aspect, often mistaken for (or dismissed as) "irrational fear of change" is the quite rational struggle on the part of those affected by the technology against the choices embodied in it and the impacts implied by them, a struggle that accurately reflects an objective conflict of interest between, say, machine operators and designers. It is here that we find the only point of entry for labor to register its choices in the process of technological development.

The following brief case history of the design, deployment, and subsequent use of a particular production technology—automatically controlled machine tools—is not meant to be a full account. Rather, it is presented in order to illustrate the points made above and, more importantly, to ground them in the historical and the concrete. Theorizing, however vital to our understanding, is never in itself sufficient to comprehend reality (much less to change it). In this social history, we will endeavor to: place the technology in its social setting, identify the designers, and indicate the loci of decision-making wherein its shape was determined; identify the choices made in the *design* of the hardware and software systems, by examining selected and abandoned alternatives; explore how these choices reflected the social relations of production (the "horizontal" relations between large and small firms and between firms and the state, and the "vertical" relations between management and labor within the firm); examine the choices made in *deployment* subsequent to design (since in the case of a relatively portable technology like machine tools, deployment—physical and institutional installations—is separable from design, and choices enter in at both stages of development to successively delimit possibilities and contribute to the final outcome); and examine closely the social process subsequent to installation in order to contrast the realities of that experience with the expectations implicit in the original design and deployment choices. Again, the whole purpose of this exercise is to move us beyond the facile and mystifying habit of technological determinism, and the sense of hopelessness that it mirrors and reinforces, by reawakening us to the possibilities.

THE TECHNOLOGY: AUTOMATICALLY CONTROLLED MACHINE TOOLS

The focus here is a particular production technology, numerically controlled contouring machine tools.[6] It appears that the numerical control (N.C.) revolution in manufacturing, as it has been called, is leading, on one hand, to increased concentration in the metalworking industry and, on the other, to a reorganization of shop-floor activities within the larger firms in the direction of greater managerial control over production and, thus, over the work force. These changes are most often attributed to the advent of the new technology and are seen to follow logically, inevitably, as a consequence of it. At a Small Business Administration hearing on the impact of N.C. on small business in 1971, for example, one expert, the president of Data Systems Corporation, intimated that it was inevitable that "we will see some companies die, but I think we will see other companies grow very rapidly."[7] James Childs, president of the Numerical Control Society, was far less sanguine in his assessment of the situation, pointing out that "the technological gap is definitely growing and unless proper

6. The discussion here is restricted to the first few generations of N.C. (tape control); it does not include later developments such as Computer Numerical Control (CNC), Direct Numerical Control (DNC) or Flexible Manufacturing Systems (FMS), or any other aspects of Computer Aided Design/Computer Aided Manufacture (CAD/CAM). These are subjects for another article.

7. U.S., Congress, Senate, Hearing before the Subcommittee on Science and Technology of the Select Committee on Small Business, *Introduction to Numerical Control and Its Impact on Small Business*, "Statement of Kenneth Stephanz," 92nd Cong., 1st sess., June 24, 1971, p. 76.

action is taken, will continue to grow until the small shop is extinct. Unless we are willing to allow a relatively few large shops to control the market . . . we must support the small shop."[8] Whatever the orientation, there was consensus among those in attendance that the metalworking industry was undergoing a change as a consequence of the new technology, a change that seemed to spell disaster for many of the smaller firms, which, in 1971, constituted 83 percent of the firms in the industry.[9] The most common reason cited for this trend was the inaccessibility of the new technology to the smaller shops, and this was explained in terms of the expense of N.C. machines (hardware) and the difficulties associated with programming them (software).

If the new technology was seen as having an important effect on the structure of a vital industry, that is, on the relations of production between firms, it was also seen as the cause of significant changes within shops. Indeed, it has been argued that smaller shops could not take advantage of N.C. because they were unwilling to undertake the social reorganization efforts that the new technology required.[10] Earl Lundgren, a sociologist who has studied the impact of N.C. on shop reorganization, insists, for example, that the successful employment of the new technology demands extensive training programs for parts programmers, the shift of shop-floor skills from the machinists to the parts programmers and methods personnel, the intensification of centralized control over all aspects of the production process, and a corollary, the detailed coordination of effort throughout the shop.[11]

8. Ibid., letter from James Childs, p. 87.

9. Ibid. See statements of John C. Williams, Edward Miller, and Senator David Gambrell. For more information about the diffusion of N.C. tools, see Clifford Fawcett, "Factors and Issues in the Survival and Growth of the U.S. Machine Tool Industry" (Ph.D. diss., George Washington University, 1976); Jacob Sonny, "Technological Change in the U.S. Machine Tool Industry, 1947-1966" (Ph.D. diss., New School for Social Research, 1971); A. Romeo, "Interindustry Differences in the Diffusion of Innovations" (Ph.D. diss., University of Pennsylvania, 1973); "1976 American Machinist Inventory of Metalworking Equipment," *American Machinist,* December 1976; S. Kurlat, "The Diffusion of N.C. Machine Tools," Technical Memorandum, EIKONIX Corporation, April 1977; and Jack Rosenberg, "A History of Numerical Control, 1949-1973: The Technical Development, Transfer to Industry, and Assimilation," unpublished manuscript, October 1973.

10. Small Business Administration, "The Impact of N.C. on Small Business," (Washington, D.C.: Government Printing Office, 1971), p. 129. On organizational impacts of N.C., see: Ervin M. Birt, "Organizational and Behavioral Aspects of N.C. Machine Tool Innovation" (M.S. thesis, Course XV, M.I.T., 1959); James J. Childs, "Organization of a Numerical Control Operation," *Mechanical Engineering* (May 1959), p. 61; and Earl F. Lundgren, "Effects of N.C. on Organizational Structure," *Automation* 76 (January 1969): 44-47.

11. Lundgren, "Effects of N.C. on Organizational Structure." In his study, Lundgren observed that, among other things, "there was little doubt in all cases that management fully

In the course of a site-visit survey of twenty-four N.C.-using large- and medium-sized firms in 1978, I found much evidence that such a reorganization effort was underway, that in nearly every case management had attempted to transfer skill from the shop floor to the programming office, to tighten up lines of authority, and to extend control over all aspects of production, that is, to meet the social requirements of N.C., requirements that signaled profound changes in the relations of production between management and labor.[12]

For the technological determinist, the story is pretty much told: N.C. leads to concentration in the metalworking industry and to a tightening of managerial control on the shop floor. All that remains to be done is to sit back and watch the inevitable process unfold. But for the critical historian of technology, the problem has merely been defined. The first question to be asked is, do these consequences really follow from this technological cause? Is the causal connection real? Do the cost and complexity of the new technology really have anything to do with the diffusion of it in the metalworking industry? Might a better explanation be that smaller firms have never even heard of N.C. because of their distance from the locus of technological change in the industry?[13] Along the same lines, is it really necessary to divide the programming and machine operating functions within the shop? Could programming, like other tooling, be done closer to the floor or by people on the floor? Does the new technology itself make such reorganization impossible?[14] For purposes of brevity, we will here assume that there *is* a sufficient degree of causal connection between the technology and its supposed effects to warrant further investigation.

This leads us to ask the next question: If *this* technology has these significant consequences, why *this* technology? Here I will limit myself

---

intended to transfer as much planning from the shop floor to the staff offices as possible," and that "under N.C. the operator is no longer required to take part in planning activities." He concluded that therefore "a change in organization structure to define and formalize the relationships required by the N.C. technology is strongly indicated." See also L. K. Williams and C. B. Williams, "Impact of N.C. Equipment on Factory Organization," *California Management Review* 7 (Winter 1964): 25-34; R. C. Brewer, "Organizational and Economic Aspects of N.C.," *Engineers Digest* 146 (August 1965): 103; William H. Bentley, "Management Aspects of N.C.," *Automation* 7 (October 1960): 64-70; Earl Lundgren and M. H. Sageger, "Impact of N.C. on First Line Supervision," *Personnel Journal* 46 (December 1967): 715.

12. This survey included primarily large- and medium-sized plants in the Midwest and New England of manufacturers of machine tools, farm implements, heavy construction equipment, jet engines, aircraft parts, and specialized industrial machinery.

13. On this point, see the letter from James Childs in "Introduction to Numerical Control and Its Impact on Small Business," p. 87.

14. See, for example, Harry Braverman's discussion of the social possibilities of N.C. in his *Labor and Monopoly Capital* (New York: Monthly Review Press, 1974), p. 199.

to a single aspect of this problem. All would agree that technology affects the social relations of production; here I am turning that around to ask, how have the social relations of production defined the technology? Could it be the case that the new technology, which took shape within the context of these social relations, not only affects them but also reflects them? If the new technology was developed under the auspices of large firms within the metalworking industry, and under the auspices of management within these firms, is it just a coincidence that the technology serves to extend the market control of large firms within the metalworking industry and enhance managerial authority within the shop? Why did N.C. take the particular *form* that it did, a form that seems to have rendered it inaccessible to small firms, and *why only N.C.?* Are there other ways of automating machine tools?

Let us begin with the technology. A machine tool (say, a lathe or milling machine) is a machine used to cut away surplus material from a piece of metal to produce a part with the desired shape, size, and finish. Machine tools are really the guts of machine-based industry because they are the means whereby all machinery, including machine tools themselves, are made. Traditionally, the machine tool is operated by a machinist who transmits his skill and his purpose to the machine by means of cranks, levers, and handles. Feedback is achieved through hands, ears, and eyes. Throughout the nineteenth century, technical advances in machining developed by innovative machinists built some intelligence into the machine tools themselves, making them partially "self-acting": automatic feeds, stops, throw-out dogs, mechanical cams. These mechanical devices relieved the machinists of certain manual tasks; the machinist, however, retained control over the operation of the machine. Together with elaborate tooling—fixtures for holding the workpiece in the proper cutting position and jigs for guiding the path of the cutting tool—these design innovations made it possible for less-skilled operators to use the machines to cut parts after they had been properly "set up" by more skilled men.[15] Still,

15. The use of jigs and fixtures in metalworking date back to the early nineteenth century and were the heart of interchangeable parts manufacture, as Merritt Roe Smith has shown in *Harpers Ferry Armory and the New Technology* (Ithaca: Cornell University Press, 1976). But it was not until much later, around the turn of the century, that the "toolmaker" as such became a specialist, distinguished from machinists. The new function was a product of scientific management, which aimed to shift the locus of skill, and control, from the production floor, and the operators, to the toolroom, as a matter of basic shop reorganization. But however much the new tools allowed management to employ less skilled, and thus cheaper, machine operators, they were nevertheless very expensive to manufacture and store, and they lent to manufacture a heavy burden of inflexibility—shortcomings that one Taylorite, Sterling

the source of the intelligence was the skilled machinist on the floor. Another step in machine design was the advent of tracer technology, perfected during the 1930s and 40s, which involved the use of patterns, or templates; these were traced by an hydraulic or electronic sensing device, which in turn conveyed the information to a cutting tool, which reproduced the pattern in the workpiece. The template, then, was, in essence, the cam that contained the intelligence for the machine, intelligence that came from a patternmaker. Tracer technology made possible elaborate contour cutting, but it was still only a partial form of automation. You needed different templates for different surfaces on the same workpiece. During the 1940s, with the war-spurred development of a whole host of new sensing and measuring devices as well as precision servomotors, which made possible the accurate control of mechanical motion, people began to think about the complete automation of contour machining.

Automating a machine tool is different from automating, say automotive manufacturing equipment, which is single-purpose, fixed automation and cost effective only given a high volume of product and thus high demand. Machine tools are general purpose, versatile machines, used primarily for small-batch, low-volume production of parts. The challenge of automating machine tools, then, is to retain the versatility of the machine. The answer that evolved was the use of some medium—film, lines on paper, magnetic tape, punched cards of tape—that would store the intelligence and control the machine. To change the part produced, you had only to change the tape. The machine remained the same.

The automating of machine tools, then, involves two separate processes. You need machine controls, a means of transmitting information to the machine to make the tables and cutting tool move as desired. And you need a means of getting the information on the medium, the tape, in the first place. The real challenge was the latter. Machine controls were just another step in a known direction, an

Bunnell, warned about as early as 1914 (cited in David Montgomery, "Fall of the House of Labor," unpublished manuscript). The cost savings that resulted from the use of cheaper labor were thus offset by the expense of tooling. Numerical control, as we will see, was developed, in part, to eliminate the costs and inflexibility of jigs and fixtures and, equally important, to take skill, and the control it implied, off the floor altogether. Here again, however, the expense of the solution was equal to or greater than the problem. It is interesting to note that in both cases, jigs and fixtures and numerical control, where very expensive new technologies were introduced to make it possible to hire cheaper labor, the tab for the conversion was picked up by the state, from the Ordnance Department in the early nineteenth century, to the Departments of the Army and Navy in World War I, to the Air Force in the second half of the twentieth century.

extension of gunfire control technology developed during the war. The tape preparation was something new. The first viable solution was record playback, a system developed in 1946-47 by General Electric and Gisholt and a few smaller firms.[16] It involved having a machinist make a part while the motions of the machine under his command were recorded on magnetic tape. After the first piece was made, subsequent identical parts could be made automatically by playing back the tape and reproducing the machine motions. John Diebold, the management consultant, among the first to write about such "flexible automation" in his book *Automation*, heralded record playback as "no small achievement . . . it means that automatic operation of machine tools is possible for the job shop—normally the last place in which anyone would expect even partial automation."[17] But record playback enjoyed only a brief existence, for reasons we shall explore. (It was nevertheless immortalized as the inspiration for Kurt Vonnegut's *Player Piano*.[18] Vonnegut was a publicist at G.E. at the time and saw the record-playback lathe he describes in the novel.)

The other solution to the medium-preparation problem was numerical control (N.C.). Although some trace its history back to the Jacquard loom of 1804, N.C. was actually of more recent vintage, the brainchild of a defense subcontractor for Bell Aircraft, John Parsons, and engineers at M.I.T. subcontracted by Parsons.[19] Numerical control,

16. The discussion of the record-playback technology is based upon extensive interviews and correspondence with the engineers who participated in the projects at General Electric (Schenectady) and Gisholt (Madison, Wisconsin). For a brief description of the G.E. technology, see Lawrence R. Peaslee, "Tape-Controlled Machines," *Electrical Manufacturing Magazine*, November 1953, pp. 102-8. Key patents for the G.E. system include Orrin W. Livingston, "Record-Reproduce Programming Control System for Electric Motors," patent no. 2,755,422 (July 17, 1956); Lawrence R. Peaslee, "Programming Control System," patent no. 2,937,365 (May 17, 1960); and Lowell L. Holmes, "Magnetic Tape Recording Device," patent no. 2,755,160 (July 17, 1956). Key patents for the Gisholt system: Leif Eric de Neergaard, Frederic W. Olmstead, and Hans Trechsel, "Control System for Machine Tools Utilizing Magnetic Recording," patent no. 3,296,606 (January 3, 1967), and Leif Eric de Neergaard "Method and Means for Recording and Reproducing Displacements," patent no. 2,628,539 (February 17, 1953). For a post-N.C. view of the advantage of record-playback control, see Veljko Milenkovic, "Single-Channel Programmed Tape Motor Control for Machine Tools," patent no. 3,241,020 (March 15, 1966). Milenkovic's patent was used by the American Machine and Foundry Company (AMF) in the development of their "Versatran" robot. Industrial robots, like "unimates," are record-playback controlled.

17. John Diebold, *Automation* (New York: Van Nostrand, 1952), p. 88.

18. Kurt Vonnegut, *Player Piano* (New York: Delacorte Press), 1952. Kurt Vonnegut, letter to author, February 19, 1977.

19. This is why the Numerical Control Society's award given for significant advancements in the technology is called the Jacquard Award. In 1970, this award went to the developers of the first N.C. machine at M.I.T. See "How It All Began," *Metalworking Economics*, June 1970, p. 43. For general historical discussion of such automation, see Diebold, *Automation*;

first demonstrated in 1952, represented a dramatic departure in machine tool technology.[20] Record playback was, in reality, a multiplier of skill, a means of obtaining repeatability. The intelligence still came from the machinist who made the tape. Numerical control was based upon an entirely different philosophy of manufacturing. Here the specifications for a part, the information contained in the engineering blueprint, is broken down, first, into a mathematical representation of the part, then into a mathematical description of the desired path of the cutting tool, and ultimately, into hundreds or thousands of discrete instructions, translated for economy into a numerical code, which is read by the machine controls. The N.C. tape, in short, is a means of formally synthesizing the skill of a machinist, circumventing his role as the source of the intelligence of production (in theory). This new approach to machining was heralded by the National Commission on Technology, Automation and Economic Progress, as "probably the most significant development in manufacturing since the introduction of the moving assembly line."[21]

CHOICE IN DESIGN: HORIZONTAL RELATIONS OF PRODUCTION

The short history above of the automation of machine tools describes the evolution of a new technology as if it were simply a technical, and thus a logical, development. Hence it tells us very little about why the technology took the form that it did, why numerical control was developed while record playback was not and why N.C. as it was designed proved difficult for the metalworking industry as a whole to absorb. Answers to questions such as these require a closer look

James R. Bright, "The Development of Automation," in *Technology in Western Civilization*, ed., Kranzberg and Pursell, vol. 2, chap. 41; Braverman's chapter on "Machinery," in *Labor and Monopoly Capital;* and Ben Seligman, *Most Notorious Victory: Man in an Age of Automation* (New York: Free Press, 1966), esp. pp. 122-27. For a more detailed history of N.C., see Donald P. Hunt, "The Evolution of a Machine Tool Control System and Succeeding Developments" (M.S. thesis, course XV, M.I.T., 1959) and "A Numerically Controlled Milling Machine: Final Report to the U.S. Air Force in Construction and Initial Operation" (Servomechanisms Laboratory, M.I.T., July 30, 1952), pt. 1.

20. No effort will be made here to trace the early history of N.C., which includes the work of Emanuel Scheyer, Max Schenker, Eric de Neergaard, and Fred Cunningham, among many others. Parsons is generally credited with having initiated the first successful hardware project, the one that, with the help of the Air Force, gave rise to the commercial development of N.C.

21. Frank Lynn, Thomas Roseberry, and Victor Babich, "A History of Recent Technological Innovations," in *Technology and the Emerican Economy*, ed. National Commission on Technology, Automation, and Economic Progress (Washington D.C.: Government Printing Office, 1966), vol. 2, *The Employment Impact of Technological Change*, app., p. 89.

at the social context in which the design of the N.C. technology was developed. Here we will look first at the ways in which the design of the N.C. technology reflected the horizontal relations of production, that is, the relations between large and small firms and the independent role of the state. In the following section, we will explore why N.C. was chosen and record playback was not, by looking at the vertical relations of production between labor and management.

To begin with, we must examine the nature of the machine-tool industry itself. This tiny industry, which produces capital goods for the nation's manufacturers, is a boom-and-bust industry. It is very sensitive to fluctuations in the business cycle, experiencing an exaggerated impact of good times—when everybody buys new equipment—and of bad times—when nobody buys. Moreover, it tends to emphasize the production of "special" machines, which are essentially custom-made for users. These two factors combine to explain much of the cost of machine tools: manufacturers devote their attention to the requirements of larger users so that they can cash in on high-performance, specialized machinery, which are very expensive due to high labor costs and relatively inefficient low-volume production methods.[22] The development of N.C. exaggerated these tendencies.

Parsons got the idea for the technology while trying to figure out a way of having a computer control the positioning of a drill; he extended the idea to two-axis milling when he examined the specifications for a wing panel for a new combat fighter. The new high-performance, high-speed aircraft required a great deal of difficult and expensive machining operations to produce airfoils (wing surfaces, engine blades), integrally stiffened wing sections for greater tensile strength and less weight, and variable thickness skins.[23] Parsons took his idea for N.C. to Wright-Patterson Air Force Base and there convinced people at the Air Material Command (AMC) that the Air Force should underwrite

22. In addition to studies cited in n. 10, see Nathan Rosenberg, "Technical Change in the Machine Tool Industry, 1940-1910," *Journal of Economic History* 23 (1963): 414-43; Harliss Wagoner, *The United States Machine Tool Industry from 1900 to 1950* (Cambridge: M.I.T. Press, 1968); Murray Brown and Nathan Rosenberg, "Patents, Research and Technology in the Machine Tool Industry," *The Patent, Trademark and Copyright Journal of Research and Education*, vol. 5, no. 1 (Spring 1961). See also Seymour Melman, "Report on the Productivity of Operations in the Machine Tool Industry in Western Europe," European Productivity Agency Project no. 420, October 1959.

23. For a discussion of the initial conceptualization of N.C. by Parsons, see "How It All Began," *Metalworking Economics*, June 1970, p. 43; Rosenberg, "A History of Numerical Control, 1947-1973," Hunt, "Evolution of a Machine Tool Control System"; and Seligman's discussion of N.C. in his *Most Notorious Victory*. This brief history is based upon interviews with participants in the M.I.T. N.C. project and examination of project files.

the development of this potent new technology.[24] When Parsons got the contract, he subcontracted with M.I.T.'s Servomechanism Laboratory, which had experience in gunfire control systems.[25] Between the signing of that initial contract in 1949 and 1959, when the Air Force ceased its formal support for the development of software, the military spent at least $62,000,000 on the research, development, and transfer of N.C.[26] The major support came in 1955. Up to that time, the Air Force and M.I.T. had tried in vain to interest machine tool builders and the aircraft industry in the new technology; except for one company, Giddings and Lewis, no one would have anything to do with it—that is, they wouldn't put their own money into it. In 1955, N.C. promoters succeeded in changing the specifications in the AMC budget allocation for the stockpiling of machine tools from tracer-controlled machines to N.C. machines. At the time, the only such machine in existence was in the Servo Lab at M.I.T., but the Air Force now undertook to pay for the purchase, installation, and maintenance of over one hundred such machines to be installed at government expense in the factories of prime- and subcontractors; the contractors, aircraft manufacturers, and their suppliers would also be paid to learn to use and maintain the new technology.[27] That is, the Air Force created a market for N.C. Not surprisingly, machine tool builders got into action; research and development expenditure in the industry multiplied eightfold between 1951 and 1957.[28]

The point of this story is that the same thing that made N.C. possible, massive Air Force support, also quite possibly determined the shape the technology would take. Criteria for design of machinery normally include cost to the user. Here cost was hardly a major consideration; machine tool builders were simply competing to meet performance specifications for government funded users in the aircraft industry. They had little concern with cost effectiveness and absolutely no incentive to produce less expensive machinery for the commercial market. (This lack of cost effectiveness is illustrated by a story told

24. William M. Webster, "Air Force Participation," *Mechanical Engineering*, May 1959, pp. 68-69.

25. The details of the origins of M.I.T. involvement in N.C. are provided in letters from Gordon Brown, William Pease, and James McDonough to Mr. R.A. Jerue of the N.C. Society in 1970, pertaining the the Jacquard Award presentation. These are in the possession of Professor Karl Wildes, who has written an unpublished history of the M.I.T. Servomechanisms Laboratory. These accounts have been supplemented by interviews.

26. Rosenberg, "History of Numerical Control."

27. Details of the contract scheme are provided by Rosenberg and Wildes.

28. Edwin Mansfield, *The Economics of Technological Change* (New York: Norton, 1968), p. 56.

by one of the design engineers at Kearney and Tracker, one of the major builders of N.C. machines. Kearney and Tracker had one of their machines installed on their own production floor in order to demonstrate to prospective customers that they themselves used the new technology. In fact, the one machine, known as the "turkey," was never used except for demonstrations.)[29]

But the development of the machinery itself was only part of the story; there was also the separate evolution of the software. Here, too, the requirements of the Air Force dictated the shape of the technology. At the outset, no one fully appreciated the difficulty of getting the intelligence of production on tape, least of all the M.I.T. engineers on the N.C. project (few of whom had ever had any machining experience prior to their involvement in the project).[30] Essentially control engineers and mathematicians, they nevertheless had sufficient hubris to imagine that they could readily synthesize the skill of a mere machinist. It did not take them too long to discover their error. Programming became the major focus of the project quite early once it was determined, in an economic study, that tape preparation was the big stumbling block to N.C.'s economic viability.[31] The first programs were prepared manually, a tedious, time-consuming operation performed by graduate students, but thereafter efforts were made to enlist the aid of Whirlwind, M.I.T.'s first digital computer.[32] The earliest programs were essentially subroutines for particular geometric surfaces that were compiled by an executive program developed by Arnold Siegel.

In 1956, after M.I.T. had received another Air Force contract for software development, a young engineer and mathematician, Douglas Ross, came up with an entirely new, and more fundamental approach to the programming problem. Rather than treating each separate problem with a separate subroutine, the new system called APT (Automatically Programmed Tools) was essentially a skeleton program—"a sys-

29. Interview with a former Kearney and Tracker design engineer, 1978. For an extended discussion of the "cost maximizing" effect of military subsidization of industry, see Seymour Melman, *Pentagon Capitalism: The Political Economy of War* (New York: McGraw Hill, 1970), and idem, *The Permanent War Economy: American Capitalism in Decline* (New York: Simon and Schuster, 1974).

30. Interview with project participants and machine shop director, Department of Electrical Engineering, M.I.T., May 1977.

31. Robert H. Gregory and Thomas V. Atwater, Jr., "Economic Studies of Work Performed on an N.C. Milling Machine," Engineering Report no. 18 (Servomechanism Lab), M.I.T., March 1, 1956.

32. A brief history of software development for N.C. is traced in Hunt, "Evolution of a Machine Tool Control System." A more detailed account is given in Douglas T. Ross, "Origins of the APT Language for Automatically Programmed Tools," SofTech, Inc., February 1978.

tematized solution" as it was called—for moving a cutting tool through space; this skeleton was to be "fleshed out" for every particular application.[33] The APT system was flexible, fundamental, a basic system for up to five-axis control. The Air Force loved it because of its flexibility; it seemed to allow for quick mobilization, rapid design change, and interchangeability between machines in a plant, between users and vendors, and between contractors and subcontractors. With these ends in mind, the Air Force pushed for standardization of the APT system; the Air Material Command cooperated with the Aircraft Industries Association Committee on N.C. to make APT the industry standard.[34]

Before too long, the APT computer language had indeed become the industry standard, despite initial resistance within the plants of aircraft companies. Many of these companies had developed their own languages to program their N.C. equipment, and these in-house languages, while less flexible than APT, were nevertheless proven, relatively simple to use, and perfectly suited to the particular needs of the company. APT was something else entirely. For all its advantages, indeed because of them, the APT system had decided disadvantages. The more fundamental a system is, the more cumbersome it is, and the more complex, the more skilled the programmer must be and the bigger the computer must be to handle the larger amount of information. In addition, the greater the amount of information, the greater the chance for error. In the aircraft plants, initial resistance was overcome by higher level management, who had come to believe it necessary to learn how to use the new system "for business reasons" (cost-plus contracts with the Air Force).[35] The exclusive use of APT was enforced. Thus began what Doug Ross himself described as "the tremendous turmoil of practicalities of the APT system development"; the system proved "erratic and unreliable" for a long time, a major headache for the aircraft industry.[36]

The standardization on APT, at the behest of the Air Force, also had two other interrelated consequences. First, it inhibited the development of alternative, simpler languages, such as "NUFORM," created by A. S. Thomas, Inc., which might have rendered contour programming

33. Ross, "Origins of the APT Language."

34. Ibid. See also Douglas T. Ross, "The APT Joint Effort," *Mechanical Engineering*, May 1959, p. 70; George E. Kinney, "Summary of AIA Activities in Numerical Control," (Paper delivered at the Electronics Industries Association Symposium on N.C. Systems for Machine Tools, September 17, 1957), pp. 91-92.

35. Rosenberg, "History of Numerical Control," p. 17.

36. Ross, "Origins of the APT Language," p. 35.

more accessible to smaller shops. Second, it forced those who ventured into N.C. into a dependence upon those who controlled the development of APT, upon large computers, and upon mathematically sophisticated programmers. The aircraft companies, for all their headaches, could afford to grapple with APT thanks to Air Force subsidy. Commercial users were not so lucky. Those who wanted military contracts were compelled to adopt the APT system; those who could not afford the system, with its training requirements, its computer demands, and its headaches, were deprived of government jobs.[37] The point here is that the software system that became the de facto standard in industry had been designed with a particular user, the Air Force, in mind. As Ross explained, "the universal factor throughout the design process is the *economics* involved. The advantage to be derived from a given aspect of the language must be balanced against the difficulties in incorporating that aspect into a complete and working system."[38] APT served the Air Force and the aircraft industry well, but at the possible expense of less-endowed competitors. (Incidentally, as is the case with many programming languages, the termination of a statement in the APT language is indicated by a double dollar sign: $$.)

CHOICE IN DESIGN: VERTICAL RELATIONS OF PRODUCTION

Thus far we have talked only about the form of N.C.—its hardware and software and how these reflected the horizontal relations of production. But what about the precursor to N.C., record playback? Here was a technology apparently perfectly suited to the small shop; tapes could be prepared by recording the motions of a machine tool, guided by a machinist or a tracer template, without programmers, mathematics, languages, or computers.[39] What happened to this technology? Essentially, it was abandoned in favor of N.C. by the aircraft industry and General Electric. Small firms never saw it. The Gisholt system, designed by Hans Trechsel to be fully accessible to machinists on the floor, was shelved once that company was bought out by Giddings and Lewis, one of the major N.C. manufacturers.[40] The

37. Rosenberg, "History of Numerically Control"; statement of A. S. Thomas in Small Business Administration, "Impact of N.C."; interview with A. S. Thomas, Westwood, Massachusetts, February 1978.

38. Ross, "Origins of APT Language," p. 13.

39. The advantages of record playback are discussed in Peaslee, "Tape-Controlled Machines"; and Milenkovic, patent no. 3,241,020.

40. Interview with Hans Trechsel, designer of Gisholt's "Factrol" system, October 1977.

G.E. system was never really marketed; demonstrations of the system elicited little enthusiasm from either machine tool builders or aircraft companies.[41] Giddings and Lewis bought the only full record-playback control, for a profile mill, but switched to a modified N.C. before the machinery was actually installed.[42] G.E.'s magnetic-tape N.C. machine control for milling—the most popular control in the 1950s and 1960s—was actually advertised initially as having a "record play-back" option, but the description of this feature disappeared from the manual in a few years, even though the magnetic-tape control system retained the record-playback capacity.[43] Why was there so little interest in this technology on the part of those fortunate larger users who got to see it?

The answer to this question is complicated; certainly it must include the fact that aircraft industry requirements for high-complexity, four- and five-axis machining jobs were simply beyond the capacity of record-playback (or manual) methods, and neither could eliminate expensive jigs, fixtures, and templates (and patternmakers) to the extent made possible with N.C. But it would be a mistake to assume that numerical control was selected, and record playback was abandoned, for such strictly technical reasons. Part of the explanation has to do with the nature of labor-management relations in metalworking. Here we will emphasize this aspect of the explanation simply because it is the one most often left out of the story, namely, how technological development reflects the vertical relations of production.

Capitalist production is the activity of a Divided Humanity, of two separate and antagonistic classes of human beings. The one possess nothing but their labor power (of which skill is a part), which they must barter for survival. The other owns the means of production and possesses the wherewithal to buy the labor-power, and thus skill, of the first. Only when these two classes combine their resources—labor power (skill) and the means of production—does productive

41. Interviews and correspondence with participating engineering and sales personnel, G.E., Schenectady.

42. This system was called "Numericord"; for a brief description, see Harry Ankene, "The Numericord System" (Paper delivered at the EIA Symposium on N.C. Systems for Machine Tools, September 17, 1957). The history of this system is traced in Harry Ankene and John L. Dutcher, "Record Playback Control of a HYPRO Skin Mill" (Paper presented at AIEE Conference on Machine Tools, October 26, 1954); and idem, "Computer-Prepared Numerical Data for Machine Control," *Electrical Manufacturing*, January 1956.

43. Original G.E. "Three Motion Magnetic Tape Contouring Control" (GE 7-2584) contained a "Record Operation" option that later disappeared. See, for example, the discussion of the G.E. 3-D milling system in *The Iron Age*, May 1, 1958, p. 84. No mention is made of the option by this time.

activity actually take place. From the capitalist's point of view, labor is merely a factor of production, which he purchases (or "eliminates," through "labor-saving" technologies). Unfortunately for the capitalist, however, labor is also people, people who do not view themselves simply as a factor of production.

Capital in reality does not buy labor but rather rents the services of those who possess it, and must thereafter encourage them to exercise it, to convert labor power into labor. For only then does the capitalist begin to get a return on his wage investment; only thereby does the worker contribute to the production and marketing of commodities and thus to the capitalist's accumulation of capital.

The overall task of capital vis-à-vis labor, therefore, is twofold: having separated labor from its tools and reunited the two through its own profit-rendering mediation, it must now try to motivate labor to get workers to work. This situation, then, is one marked by the de facto dependence of capital upon labor, a dependence that capital has, for the last eighty years or so, been trying to minimize as much as possible. Thus, we have two central characteristics of the relations of production under capitalism: management/capital owns the means of production and hence has the prerogative and wherewithal to direct the development of production technology; and capital/management is forever striving to minimize its dependence upon those who labor for it, in part by appropriating for itself the "intelligence of production" (the skills of the workshop) and by separating conception and control from execution, thereby enhancing managerial authority over the process of production. Capital is opposed in this effort by labor, which has nothing to gain and everything to lose if the effort succeeds.

Ever since the nineteenth century, the labor-intensive machine shops have been a bastion of skilled labor and the locus of considerable shop-floor struggle. Frederick Winslow Taylor introduced his system of scientific management to try to put a stop to what he called "systematic soldiering," the control of production by the workers who controlled their machines ("pacing," restriction of output).[44] Workers practiced pacing in order to protect themselves, to avoid killing "gravy" piece-rate jobs and avoid rate cuts, to retain control over their own work, to stretch out the work for fear of layoffs, to keep some time

44. Frederick W. Taylor, *Scientific Management* (New York: Harper & Brothers, 1947). See the excellent discussion of Taylor in Braverman's chapter on Scientific Management, *Labor and Monopoly Capital*, chap. 4.

for themselves, and to "make out" and at the same time exercise their creativity by working the "angles" (not to mention to express hostility to management). Their major means of achieving such shop-floor control was their actual control over the machines; they alone possessed the skill required to produce finished parts. Taylor tried to eliminate pacing by transferring these skills from the hands of machinists to the handbooks of management. He failed. There is still no science of cutting metals, and process engineers and time-study men have never succeeded in putting a stop to worker control of production. Thus, in the 1940s, when sociologist Donald Roy went to work in a machine shop, he found pacing alive and well; the men on the floor still controlled production. He recounted an incident that neatly summed up his experience. " 'I want 25 or 30 of those by 11 o'clock,' Steve the superintendent said sharply, a couple of minutes after the 7:15 whistle blew. I smiled at him agreeably. 'I mean it,' said Steve, half smiling himself, as McCann and Smith, who were standing near us, laughed aloud. Steve had to grin in spite of himself and walked away. 'What he wants and what he is going to get are two different things,' said McCann."[45] Thirty years later, sociologist Michael Burawoy returned to the same shop, which had yet to experience the full impact of N.C., and concluded, in his own study of shop-floor relations, that "in a machine shop, the nature of the relationship of workers to their machines rules out coercion as a means of extracting surplus.[46]

This was the context in which the automation of machine tools took place, and it should be seen, in part, as a further managerial attempt to wrest control over production from the work force. As Peter Drucker put it, "what is today called 'automation' is conceptually a logical extension of Taylor's scientific management."[47]

G.E. launched its automation project following the biggest strike in the company's history, one upheaval among many across the nation.

45. Donald F. Roy, "Work Satisfaction and Social Reward in Quota Achievement: An Analysis of Piecework Incentive," *American Sociological Review* 18 (1953): 513. See also idem, "Quota Restriction and Gold-Bricking in a Machine Shop," *American Journal of Sociology* 57 (1951-52): 427; and idem, "Efficiency and 'The Fix': Informal Intergroup Relations in a Piecework Machine Shop," *American Journal of Sociology* 60 (1954-55): 255. For more on worker restriction of output, see Stanley B. Mathewson, *Restriction of Output among Unorganized Workers* (Carbondale: Southern Illinois University Press, 1969); and idem, "The Toolmaker," *New Left Review*, no. 53 (January 1969), p. 63.

46. Michael Burawoy, "The Organization of Consent: Changing Patterns of Conflict on the Shop Floor, 1945-1975" (Ph.D. diss., University of Chicago, 1976).

47. Peter F. Drucker, "Technology and Society in the Twentieth Century," in *Technology in Western Civilization*, ed. Kranzberg and Pursell (New York: Oxford University Press, 1967), 2: 26.

The culmination of a decade of work stoppages over lay-offs, rate cuts, speed-ups (accompanying the introduction of tungsten carbide tools), and the replacement of skilled machinists with apprentices and women, the G.E. Schenectady strike was led by United Electrical Workers machinists (and was bitterly opposed by the G.E. Engineers Association).[48] Not surprisingly, the project was secret; publicist Kurt Vonnegut recalled, with characteristic understatement, that "they wanted no publicity this time."[49] Between that year and 1956, that is, during the first decade of machine-tool automation development, the aircraft industry—the major user—also experienced serious labor trouble as the IAM (International Association of Machinists) and UAW (United Automobile, Aerospace, and Agricultural Workers) competed to organize the plants. The postwar depression had created discontent and resentment on the part of workers faced with lay-offs, company claims of inability to pay, and mass reclassifications downward.[50] Major strikes took place at Boeing, Bell Aircraft (Parson's prime contractor), McDonnell Douglas, Wright Aeronautical, G.E., Evendale (jet engines), North American Aviation, and Republic Aircraft.[51] It is hardly difficult, then, to explain the popularity among management and technical men, of a 1946 *Fortune* article entitled "Machines without Men."[52] Surveying the technological fruits of the war (sensing and measuring devices, servomechanisms, computers, etc.), two Canadian physicists promised that "these devices are not subject to any human limitations. They do not mind working around the clock. They never feel hunger or fatigue. They are always satisfied with working conditions, and never demand higher wages based on the company's ability to pay." In short, "they cause much less trouble than humans doing comparable work."[53]

One of the people who was inspired by this article was Lowell Holmes, the young electrical engineer who directed the G.E. automation project. However, in record playback, he developed a system for

48. The story of postwar labor relations at G.E. Schenectady is gleaned from the *UE News*, *Schenectady Gazette*, the *Union Star*, and the *Electrical Union News*, 1940-55.

49. Kurt Vonnegut, letter to author, February 1977.

50. Arthur P. Allen and Betty V. H. Schneider, *Industrial Relations in the California Aircraft Industry* (Berkeley: Institute of Industrial Relations, University of California, 1956).

51. U.S., Department of Labor, Bureau of Labor Statistics; and "Work Stoppages in Industry, 1946-1956," *Monthly Labor Review*.

52. E. W. Leaver and J. J. Brown, "Machines without Men," *Fortune*, November 1946, p. 190. The popularity of this article was reflected in the fact that most interviewees remembered the piece. Lowell Holmes, project engineer on the record-playback project, recalled specifically that the article was the inspiration for his work and that it was widely read in technical circles. Interview, March 1977.

53. "Machines without Men," *Fortune* (1946), p. 203.

replacing machinists that ultimately *retained* machinist and shop-floor control over production because of the method of tape preparation. This defect was recognized immediately by those who attended the demonstration of the system; they showed little interest in the technology. "Give us something that will do what we say, not what we do," one of them said.[54] The defects of record playback were conceptual, not technical; the system simply did not meet the needs of the larger firms for managerial control over production. N.C. did. "Managers like N.C. because it means they can sit in their offices, write down what they want and give it to someone and say, 'do it,' " the chief G.E. consulting engineer on both the record playback and N.C. projects explained, "With N.C. there is no need to get your hands dirty, or argue."[55] Another consulting engineer, head of the Industrial Applications Group, which served as intermediary between the research department and the sales department at G.E. Schenectady, and a key figure in the development of both technologies, explained the shift from record playback to N.C. "Look, with record-playback the control of the machine remains with the machinist—control of feeds, speeds, number of cuts, output; with N.C. there is a shift of control to management. Management is no longer dependent upon the operator and can thus optimize the use of their machines. With N.C., control over the process is placed firmly in the hands of management—and why shouldn't we have it?"[56] It is no wonder, then, that at G.E., N.C. was often referred to as a management system, not as a technology for cutting metals.[57]

Numerical control appeared to potential users in the large aircraft contracting firms to be a valuable means of taking the intelligence of production, and thus control over production, off the shop floor. The new technology thus dovetailed nicely with their larger efforts to computerize company operations.[58] At the same time, in the intensely anti-Communist 1950s, N.C. looked like a solution to security problems, enabling management to remove the blueprints from the

54. This statement was quoted by a number of the participants and was attributed to the president of a "leading machine tool company."

55. Interview, G.E., Schenectady, May 1977.

56. Interview, G.E., Schenectady, May 1977.

57. G.E. Company, Management Consulting Services Division, "Next Step in Management: An Appraisal of Cybernetics" (1953). See also J. Forrester et al., "Strengthening Management for the New Technology," *American Management Association* (1955).

58. Like most large companies that entered the computer age early, the aircraft firms were faced with a number of problems: first, how fully to utilize the computer (after payroll, then what?), and second, how to extract the information for data banks from the personnel who possessed it (whose jobs depended upon their access to or control over informa-

floor so that Communists could not get their hands on them. N.C. did appear to eliminate the need for costly jigs and fixtures and the templates required by tracer machining, and it did make possible the cutting of complex shapes that defied manual methods. But, additionally, and of equal importance, N.C. seemed to eliminate once and for all the problems of "pacing," to afford management greater control over production by replacing problematic time-study methods with "tape-time" (using the time it takes to run a part tape as the base for calculating rates) and replacing skilled machinists with more tractable "button-pushers," who would simply load and unload the automatic machinery. The intention is the important thing here. If, with hindsight, N.C. seems to have required organizational changes in the factory, changes that enhanced managerial control, it is because the technology was chosen, in part for just that purpose. This becomes even more clear when one looks at how the new technology was deployed.

CHOICE IN DEPLOYMENT: MANAGERIAL INTENTIONS

There is no question but that management saw in N.C. the potential to enhance their authority over production and seized upon it, despite questionable cost effectiveness.[59] Machine tool builders and control

---

tion). N.C. contributed to the solution of both problems: the use of APT absorbed 30 percent of their computer capacity throughout the 1960s, and N.C. itself appeared to allow managers to circumvent shop floor personnel considerably.

59. The cost effectiveness of N.C. depends upon many factors, including training costs, programming costs, and computer costs, beyond mere time saved in actual chip-cutting or reduction in direct labor costs. The M.I.T. staff who conducted the early studies on the economics of N.C. focused on the savings in cutting time and waxed eloquent about the new revolution. At the same time, however, they warned that the key to the economic viability of N.C. was a reduction in programming costs. Machine tool company salesmen were not disposed to emphasize these potential drawbacks, though, and numerous users went bankrupt because they believed what they were told. In the early days, however, most users were buffered against such tragedy by state subsidy. Today potential users are somewhat more cautious, and machine tool builders are more restrained in their advertising, tempering their promise of economic success with qualifiers about proper use, the right lot and batch size, sufficient training, and so on.

For the independent investigator, it is extremely difficult to assess the economic viability of such a technology. There are many reasons for this. First, the data is rarely available or accessible. Whatever the motivation—technical fascination, keeping up with competitors, and so on—the purchase of new capital equipment must be justified in economic terms. But justifications are not too difficult to come by if the item is desired enough by the right people. They are self-interested anticipations and thus usually optimistic ones. More importantly, firms rarely conduct postaudits on their purchases, to see if their justifications were warranted. Nobody wants to document his errors, and if the machinery is fixed in its foundation, that is where it will stay, whatever a postaudit reveals; you learn to live with it. The point here is

manufacturers, of course, also promoted their wares along these lines; well attuned to the needs of their customers, they promised an end to traditional managerial problems. Thus the president of the Landis Machine Company, in a trade journal article entitled "How Can New Machines Cut Costs?" stressed that "with modern automatic controls, the production pace is set by the machine, not by the operator."[60] The advertising copy of the MOOG Machine Company of Buffalo, New York, similarly described how their new machining center "has allowed management to plan and schedule jobs more effectively," while pointing out, benevolently, that "operators are no longer faced with making critical production decisions."[61]

Machine tool manufacturers peddled their wares and the trade journals echoed their pitch.[62] Initially, potential customers believed

---

that the economics of capital equipment is not nearly so tidy as economists would sometimes have us believe. The invisible hand has to do quite a bit of sweeping up after the fact.

If the data does exist, it is very difficult to get a hold of. Companies have a proprietary interest in the information and are wary about disclosing it for fear of revealing (and thus jeopardizing) their position vis-à-vis labor unions (wages), competitors (prices), and government (regulations and taxes). Moreover, the data, if it were accessible, is not all tabulated and in a drawer somewhere. It is distributed among departments, with separate budgets, and the costs of one are the hidden costs to the others. Also there is every reason to believe that the data that does exist is self-serving information provided by each operating unit to enhance its position in the firm. And, finally, there is the tricky question of how "viability" is defined in the first place. Sometimes, machines make money for a company whether they are used productively or not.

The purpose of this aside is to emphasize that "bottom line" explanations for complex historical developments, like the introduction of new capital equipment, are never in themselves sufficient, nor necessarily to be trusted. If a company wants to introduce something new, it must justify it in terms of making a profit. This is not to say, however, that profit making was its real motive or that a profit was ever made. In the case of automation, steps are taken less out of careful calculation than on the faith that it is always good to replace capital with labor, a faith kindled deep in the soul of manufacturing engineers and managers—as economist Michael Piore, for example, has shown. See his "The Impact of the Labor Market upon the Design and Selection of Productive Techniques within the Manufacturing Plant," *Quarterly Journal of Economics*, vol. 82 (1968). Thus, automation is driven forward, not simply by the profit motive, but by the ideology of automation itself, which reflects the social relations of production.

60. Grayson M. Stickell, "How Can New Machines Cut Costs?" *Tooling and Production*, August 1960, p. 61.

61. *MOOG Hydra-Point News*, Buffalo, New York, 1975.

62. Charles Weiner, former associate editor of *Tooling and Production Magazine* (one of the leading trade journals for the metalworking industry), authored several articles on numerical control in the early 1960s, one of which was based upon a "comprehensive survey of machine tool builders and control manufacturers." According to Weiner, all of his information for these articles came from the manufacturers themselves, and no effort was ever made to make an independent evaluation of N.C. for the readers of the magazine. Although the editors strove for credibility with production men in the metalworking industry, they were guided primarily, as are the editors of all controlled circulation magazines in the trade press, by the need to secure advertisements. Thus they tended to exaggerate the promise of a new technology,

the hype; they very much wanted to. Earl Lundgren, a sociologist, conducted surveys of user plants in the 1960s and concluded that "the prime interest in each subject company was the transfer of as much planning and control from the shop to the office as possible; . . . there was little doubt in all cases that management fully intended to transfer as much planning from the shopfloor to the staff offices [methods engineers and programmers] as possible." Moreover, management believed that "under numerical control the operator is no longer required to take part in planning activities."[63]

In my own survey in 1978 of twenty-five plants in the Midwest and New England—including manufacturers of machine tools, farm implements, heavy construction equipment, jet engines and aircraft parts, and specialized industrial machinery—I observed the same phenomenon. Everywhere, management initially believed in the promises of N.C. promoters and attempted to remove all decision-making from the floor and assign unskilled people to N.C. machines, to substitute "tape-time" for problematic time studies to set base rates for piecework and measure output quotas, to tighten up authority by concentrating all mental activity in the office, and otherwise to extend detail control over all aspects of the production process.

This is not to say, however, that I drew the same conclusions that Lundgren did in his earlier survey. Characteristically for an industrial sociologist, he viewed such changes as *requirements* of the new technology whereas, in reality, they reflected simply the *possibilities* of the technology that were "seized upon," to use Harry Braverman's phrase, by management to realize particular objectives, social as well as technical. There is nothing inherent in N.C. technology, for example, that makes it *necessary* to assign programming and machine tending different people (that is, to management and workers, respectively): the technology merely makes it *possible*. Management philosophy and motives alone make it necessary that the technology be deployed in this way, and this is another form of necessity entirely, namely, the (to their minds, forced) social choices of the powerful.

One illustration of managerial choice in machine deployment is provided by the experience of a large manufacturing firm near Boston. In 1968, owing to low worker morale, turnover, absenteeism, and the general unreliability of programming and machinery, the company faced what it termed a "bottleneck" in its N.C. lathe section. Plant

---

seconding the claims of advertisers, in order to appeal to a particular readership and, by so doing, to elicit further support from the advertisers of the new technology, the manufacturers themselves.

63. Lundgren, "Effects of N.C. on Organizational Structure."

managers were frantic to figure out a way to achieve the expected output from this expensive equipment. In that prosperous and reform-minded period they decided upon a job enlargement/enrichment experiment wherein machine operators would be organized into groups and their individual tasks would be extended. Although it was the hope of the company that such a reorganization would boost the morale of the men on the floor and motivate them to "optimize the utilization" of the machinery, the union was at first reluctant to cooperate, fearing a speed-up. The company was thus hard-pressed to secure union support for their program and instituted a bonus for all participants. At one of the earliest meetings between management and union representatives on the new program, the company spokesman for the plan began his discussion of the job-enlargement issue with the question (and thinly veiled threat), "Should we make the hourly people button-pushers or responsible people?" Given the new technology, management believed it now had that choice.[64]

A second illustration of the managerial imperative behind technological determinism is provided in an interview I had with shop managers of a plant in Connecticut. Here, as elsewhere, much of the N.C. programming is relatively simple—the technology is used to machine simple parts formerly produced on conventional machinery (as opposed to complex three-to five-axis contour cutting such as of airfoils, etc.). Observing this, I asked the two men why the operators couldn't do their own programming. At first, they dismissed the suggestion as ridiculous, arguing that the operators would have to know how to set feeds and speeds, that is, be "industrial engineers." I countered this by pointing out that the same people probably set the feeds and speeds on conventional machinery, routinely making adjustments on the process sheet provided by the methods engineers in order to make out. They nodded. Then they said that the operators could not understand the programming language. This time I pointed out that the operators could often be seen reading the mylar tape—twice-removed information describing the machining being done—in order to know what was coming (especially to anticipate programming errors that could mess things up). Again, they nodded. Finally, they looked at each other and smiled, and one of them leaned over and

64. This is not to say that the new technology actually enabled management to reduce people to button-pushers, a centuries-old dream of managers everywhere, and still get quality production out the door; it didn't, as we will see in the following section on "shop floor reality." The point here is that, with the advent of N.C., management began in earnest to think about (and make decisions about) redesigning jobs and reorganizing the production process on the assumption that the new technology gave them this capability.

confided, "we don't want them to." Here is the reality behind technological determinism.[65]

REALITY ON THE SHOP FLOOR

In the discussions above on choice in design and deployment, we have endeavored to show why it is a mistake to view technological development as an autonomous or unilinear process. The evolution of any technology reflects the social relations that gave birth to it and follows from the human choices that inform it. But, as we argued at the outset, it would be an error to assume that we can now simply substitute one form of determinism for another, to suppose that, in having exposed the choices, we can now simply deduce the rest of reality (rather than just describe a part of it). For it would be as much a serious flaw in reasoning to extrapolate reality from the intentions that underlie the technology[66] as from the technology itself. Intentions are never so automatically realized, no more than desire is identical to satisfaction.

"In the conflict between employer and employed," John G. Brooks observed in 1903, "The 'storm centre' is largely at this point where science and invention are applied to industry."[67] It is here, in the the storm center, that the reality of N.C. is hammered out; it is here, in the storm center, where those who choose the technology finally come face to face with those who do not. It is their first confrontation with what philosophers call "other minds."

The introduction of N.C. control machinery has not been altogether uneventful, especially in plants where the machinists' union (such as the International Association of Machinists, the United Auto Workers,

65. This reality can be seen even more clearly in the case of the latest generation of N.C. machines, which come equipped with a small computer (hence the name: computer numerical control, CNC). With these machines, it has become possible to store programs right at the machine; more important, it is now possible to edit programs (or create them from scratch) on the shop floor by simply punching information into storage numerically (reminiscent of the old Gisholt Factrol system) or by moving the machine manually to the desired position and feeding the information into storage (a digitized version of record playback). In short, these new machines, created to overcome the programming problems of N.C., are ideal for shop floor and operator control over the whole process. However, they are almost never used in this way. It is characteristic of the larger metalworking shops that operators are not permitted to edit the programs, much less to create new ones themselves. Management usually defends this policy by arguing that too many cooks spoil the stew.

66. This is an error that Harry Braverman tended to make in his path-breaking *Labor and Monopoly Capital.* As a result, the potential for class struggle was minimized. One is left merely with a juggernaut on the one side and impotence and despair on the other.

67. Brooks quoted by David Montgomery,' "The Fall of the House of Labor," unpublished manuscript, chap. 4, p. 1.

the United Electrical Workers, or the International Union of Electrical Workers) have had a long history. Work stoppages and strikes over rates for the new machinery were common in the 1960s. At General Electric, for example, there were strikes at several plants and one during the winter of 1965 shut down the entire Lynn, Massachusetts, plant for a month. Here we will focus less upon these overt expressions of conflict than upon more subtle indications that management dreams have tended to remain dreams: the use of "tape time" to replace time-studies; the deskilling of machine operators; and the question of control over production.

Early dreams of the use of tape time to set base rates and determine output proved fanciful. The machines could not produce parts to specifications without the repeated manual intervention of the operator, in the early machines to make tool offset adjustments and to correct for program errors and even now to insure a good finish to tolerance (especially with rough castings). At one large factory in New England, for example, where N.C. has been in use for nearly a quarter of a century, the machines are often used almost like manual machines. There is, in reality, a spectrum of manual intervention requirements depending upon the machine, the product, and the like, but the mythology that the tape ran the machine without the need for people was dispelled pretty early on—and along with it went the hope of using the tape to measure performance.

The de-skilling of machine operators has, on the whole, not (yet) taken place, for two reasons. First, as was mentioned, the assigning of labor grades and thus rates for the new machinery was, and is, a hotly contested issue, yet to be finally resolved one way or the other. Second (and the reason why skill levels on numerical control equipment are as high in nonunion shops as in union shops), any determination of skill requirements must take into account the actual degree of automaticity of the machinery and, as we have seen, this was not up to expectations. That is, management has had to have people on their machines who know what they are doing simply because the machines do *not* run by themselves, and they are very expensive. While it is true that many manufacturers initially tried to use unskilled people on the new equipment, they rather quickly saw their error and upgraded the classification (although in some places they merely gave the skilled operators a premium while retaining the lower formal classification, presumably in the hope that some day the skill requirements would actually drop to match the classification). It has been argued, moreover, that, in some ways, N.C. requires greater skill—as was suggested by the United Electrical Workers in their *UE Guide to Automation* of 1960—because of the need to oversee the complex and rapid machining of parts, to constantly watch for "nonevents." Whether this is true or not is certainly debatable. One thing is not, though. The intelligence of production in machine shops has not yet been built into the machine or taken off the floor.

And this brings us, once again, to the question of control. In theory, the programmer prepares the tape (and thus sets feeds and speeds, thereby determining the rate of production), proofs it out on the machine, and then turns the show over to the operator who from then on simply presses start and stop buttons and loads and unloads the machine (using standard fixtures). This rarely happens in reality as was pointed out above. Machining to tolerances more often than not requires close attention to the details of the operation and frequent manual interventions through manual feed and speed overrides. This aspect of the technology, of course, reintroduces the control problem for management. Just as in the conventional shop, where operators are able to modify the settings specified on the work sheet (prepared by the methods engineer) in order to restrict output or otherwise "make out" (by running the machine harder), so in the N.C. shop, the operators are able to adjust feeds and speeds, for similar purposes.

Thus, if you walk into a shop you will often find feedrate override dials set uniformly at, say, 70 percent or 80 percent of tape-determined feedrate. In some places, this is called the "70 percent syndrome"; everywhere it is known as pacing, the collective restriction of output on the floor. To combat it, management sometimes programs the machines at 130 percent; sometimes they actually lock the overrides altogether to keep the operators out of the "planning process." But this, of course, gets management into serious trouble since interventions are required in many cases to get parts out the door.

To what extent the considerable amount of human intervention required for machining (and repairs) is attributable to the inherent unreliability of the technology itself is hard to assess. But it is certainly true that the technology develops shortcomings once it is placed on the shop floor, whether they were there in the original designs or not. Machines often do not do the job they are supposed to, and downtime is still excessive. Whether this is due to operator "adjustments" (or failures to adjust) or technical defects is hard to determine, but sabotage is an acknowledged problem. "I don't care how many computers you have they'll still have a thousand ways to beat you," lamented one manager in a Connecticut plant. "When you put a guy on an N.C. machine, he gets temperamental," another manager in

Rhode Island complained. "And then, through a process of osmosis, the machine gets temperamental."

On the shop floor, it is not only the choices of management that have effects. The same antagonistic social relations that, in their reflection in the minds of designers, gave issue to the new technology, now subvert it in reality. This contradiction of capitalist production, of a Divided Humanity, presents itself most clearly to management as a problem of "worker motivation," and management's acceptance of this challenge is its own tacit acknowledgment that it does not have control over production, that it is still dependent upon the work force to turn a profit.

Thus, in evaluating the work of those whose intentions to wrest control over production from the work force informed the design and deployment of N.C., we must take into account an article written by two industrial engineers in 1971. Appearing in *The Manufacturing Engineering and Management Journal,* it was entitled "A Case for Wage Incentives in the N.C. Age." It makes it quite clear that the contradiction of capitalist production has not been eclipsed—computers or no computers.

> Under automation, it is argued, the machine basically controls the manufacturing cycle, and therefore, the worker's role diminishes in importance. The fallacy in this reasoning is that if the operator malingers or fails to service the machine for a variety of reasons, both utilization and subsequent return on investment suffer drastically.
>
> Basic premises underlying the design and development of N.C. machines aim at providing the capability of machining configurations beyond the scope of conventional machines. Additionally, they "de-skill" the operator. Surprisingly, however, the human element continues to be a major factor in the realization of optimum utilization or yield of these machines. This poses a continuing problem for management, because a maximum level of utilization is necessary to assure a satisfactory return on investment.[68]

This same contradiction, faced by every capitalist since the emergence of the capitalist mode of production, was more succinctly, if inadvertently, expressed by another plant manager in Connecticut. With a colleague chiming in, he proudly described the elaborate procedure they had developed whereby every production change, even the most minor, had to be okayed by an industrial engineer. "We want absolutely no decisions made on the floor," he insisted; no operator was to make any change from the process sheets without the written authorization of a supervisor. A moment later, however,

68. Martin R. Doring and Raymond C. Salling, "A Case for Wage Incentives in the N.C. Age," *Manufacturing Engineering and Management* 66, no. 6 (1971): 31.

looking out onto the floor from his glass-enclosed office, he reflected upon the reliability of the machinery and the expense of parts and equipment and emphasized, with equal conviction, that "we need guys out there who can think."

CONCLUSION: LABOR'S CHALLENGE REVISITED

We have here tried to get beyond the ideology of technological determinism by demystifying the development of a particular technology. The purpose of this exercise, as was indicated at the outset, was to regain confidence in the fact that technology consists of a range of possibilities; if certain human choices guided the course of its development in the past, other choices could steer it in the future. This done, we must turn again to the remaining two tasks: to study the possibilities and clarify new choices and to struggle to get in the position to make them.

It is time, it is imperative, that the labor movement as a whole begin to assume the responsibility for the design and deployment of technology—the organization of our shops, the structure of our jobs, the shape of our lives. For only by becoming self-reliant and assuming this responsibility can we regain the confidence to secure the power to exercise it. This challenge, now faced by all unions throughout industry (due, in part, to the homogenizing effect of management-designed automation, which diminishes the differences between industries and between shops and offices) was clearly, if rather modestly, summarized by an N.C. machine tool operator in a large shop in Massachusetts:

> The introduction of automation means that our skills are being downgraded and instead of having the prospect of moving up to a more interesting job we now have the prospect of either unemployment or a dead-end job. [But] there are alternatives that unions can explore. We have to establish the position that the fruits of technological change can be divided up—some to the workers, not all to the management as is the case today. We must demand that the machinist rise with the complexity of the machine. Thus, rather than dividing his job up, the machinist should be trained to program and repair his new equipment—a task well within the grasp of most people in the industry.
>
> Demands such as this strike at the heart of most management prerogative clauses which are in many collective bargaining contracts. Thus, to deal with automation effectively, one has to strike at another prime ingredient of business unionism: the idea of "let the management run the business." The introduction of N.C. equipment makes it imperative that we fight such ideas.[69]

This, in essence, is the challenge. Overt political and covert shop-

69. Frank Emspak, "Crisis and Austerity in the Seventies," unpublished manuscript.

floor resistance to the choices of management, which exploit the contradictions inherent in the capitalist mode of production, must be encouraged and enlarged; but these are not in themselves sufficient actions, given the current desperate corporate drive to rationalize industry and fight organized labor. It is no longer enough merely to respond to new methods nor even to anticipate the changes that are "coming down." The goal must be to gain control of, and thereafter redirect, the social process of technological development itself. And this end, of course, requires—as it contributes to—a larger transformation of the social relations of production and, thus, the eclipse of the capitalist system as a whole.

Politics & Society 8, nos. 3-4 (1978): 313-47

Part IV
International and Comparative Studies

[15]

THE PLASTICS INDUSTRY: A COMPARATIVE STUDY OF RESEARCH AND INNOVATION

Square-bracketed references, thus [1], *are to the list of sources in Appendix* III. *Other references, thus* (1), *are to the footnotes at the bottom of the page.*

Summary

Plastics are already one of the world's main groups of industrial materials; in volume terms, world plastics consumption is now greater than that of all non-ferrous metals combined. The main point that emerges from international comparisons is West Germany's predominance in production and exports (and also in exports of plastics machinery). German production of plastic materials in 1961 was 19 kilograms per head, compared to 17½ in the United States and 12 in Britain. This is not due to lower costs. In material costs, Germany has been, if anything, at a disadvantage: she had only a small petro-chemicals industry in the early post-war years, and she has very little natural gas. In labour costs (which in any case are not very important in this industry) and in plant costs, again there is no evidence of any marked German advantage over her competitors. In general her quoted prices were not below those of other countries.

Technical progress seems to be the main explanation of Germany's lead: this needs a long time-span for analysis, since scientific advances may need a decade or more before they show themselves in commercial production. Three measures of technical progress are attempted in this article: research expenditure; the analysis of patents; and the analysis of innovations. On all three counts, the most noticeable fact is the dominating position of the German firm of I.G. Farben in the inter-war years and up to 1945. There is little doubt that this firm spent more on plastics research than any other chemicals firm; in the period 1931-45 it took out more than twice as many patents as its nearest competitor—and this is still true if the analysis is restricted to key patents only; finally, I.G. Farben was the innovator of a large number of the materials which now make up the bulk of plastics output. The success of this firm was no doubt helped by the fact that in the inter-war years as well as in wartime the development and production of substitute raw materials was an important strand in German economic policy.

From 1945 to 1952 I.G. Farben was being reorganised. In this period, American firms took the lead in technical progress in plastics, and British firms—notably ICI—also appeared among the leaders. By the second half of the nineteen-fifties, the successor firms to I.G. Farben were again coming back into prominent positions in research, development and innovation.

Technical progress results in leadership in production in this industry, because patents and commercial secrecy together can give the innovator a head start of as much as 10-15 years. Other countries, it is true, may shorten the catching-up process if they are in a position to purchase the technical know-how, or if they are countries in which the innovating firms set up subsidiaries (as American firms have done in Japan). Even after patents expire, accumulated experience will help to keep the innovator in the lead, and he will be in a better position to produce new and improved grades of material. But, for the standard grades, new producers with cost advantages may after 15-30 years eventually challenge the innovating firms. Thus, some thirty years after the United States and West Germany first produced PVC, Italy and Japan are now overtaking these countries in per capita production; for they have advantages in costs of production and are able to quote lower prices.

In explaining the country pattern of consumption, other factors come in; it is not wholly a matter of technical progress in production—for instance, Sweden, with no production of polyethylene, has a higher level of consumption per head than either Britain or Germany. The progressiveness of the user industries is important here. On this point, the comparison of the British and German patterns of use is instructive—in so far as our inadequate statistics of plastics consumption by industry permit a comparison to be made. The differences in the use of plastics between the two countries are much more marked in the old industries, such as construction, than in the new industries, such as vehicles and electrical engineering. German plastics materials manufacturers have stimulated applications in these old industries: for instance, they encouraged the development of chipboard, whose output in Germany in 1961 was over 1 million cubic metres, compared to 85 thousand cubic metres in Britain. This industry consumes enormous quantities of glue made from plastic

---

This article was prepared by C. Freeman, assisted by Miss A. Young and Mrs. J. Fuller, of the National Institute. The study was made possible by financial contributions from the Organisation for Economic Co-operation and Development, who hold the copyright.

materials which were developed by I.G. Farben. Apart from this, plastics are used much more widely in the German construction industry for roofing, panelling, thermal insulation and interior fittings of all kinds.

Two other general factors are briefly considered. First, there is the possible connection between faster economic growth and high plastics production—there is some evidence that in fast-growing economies the switch to new materials is made more easily. Secondly, there is the question of the importance given to the plastics industry in those countries with economic planning. In the USSR it seems that the planning agencies, by excessive devotion to steel, have hitherto held back the growth of the plastics industry, but in Japan they have apparently provided a stimulus.

Introduction

Although its ' centenary ' is being celebrated this year, the plastics(1) industry is really not much more than 50 years old. The production of cellulosic plastics (celluloid) began in the latter part of the 19th century, and the commercial production of protein plastics (Galalith) began at the turn of the century. But these early materials were based on naturally occurring polymers, which today account for less than 10 per cent of world production. The truly synthetic phenol-formaldehyde plastics (better known as Bakelite) were introduced just before the First World War. The three best-known plastic materials today are probably PVC (polyvinyl chloride), polyethylene, and polystyrene, and all of them are less than 35 years old. Together they now account for over half of total world plastics output.

Plastics, although comparatively new, are already one of the world's main groups of industrial materials. World plastics consumption, by weight, is now larger than that of either copper or aluminium. However, the low specific gravity of most plastic materials gives them a weight advantage over most metals, so that in volume terms world consumption is greater than that of all non-ferrous metals combined, although still less than a quarter of world steel consumption. Plastics, judging from past growth rates (table 1), will continue to gain rapidly on conventional materials. Their production, although not necessarily their fabrication, is now undertaken mainly by chemical firms, and they are the fastest growing main sector of the chemicals industry, which is itself a ' growth industry '.

Their growth rate is high mainly because plastic materials have outstanding technical and cost advantages in a wide range of applications.[1] Plastics are light, easy to fabricate and install ; they have good electrical insulation, excellent resistance to corrosion and pests and low maintenance costs ; many can be made in a wide range of colours, or transparent. Plastics have some disadvantages—for example, the rather limited temperature range within which many of them can be used. Nevertheless their remarkable success in a very short space of time suggests that they will be increasingly important in the later part of this century.

(1)Plastics are man-made materials which can be made to flow on the application of adequate heat and pressure, and take up a desired shape. This shape is retained when the applied pressure and heat are withdrawn. They differ from similar man-made materials, such as glass and ceramics, in their organic origin. They are composed of giant molecules of organic substances based on chains of carbon atoms. For casein and cellulosics, these chains (polymers) are of natural origin, but the great majority of polymers are now synthesised from simple chemical units, or monomers.

The classification of plastics is necessarily somewhat arbitrary and there are synthetic materials which correspond to this definition but are still not considered as plastics here. An OECD Working Party has been attempting for some years to establish an internationally acceptable classification and for the purpose of this article plastics are defined on the same basis as in the Chemical Reports of the OECD.[2] Synthetic fibres and synthetic rubber, although belonging essentially to the same group of materials, are excluded from this definition. Where materials such as nylon can be used both as a fibre and as a plastic, that part of the production which is for fibres is excluded. The OECD Classification (and the Brussels nomenclature) divides plastics into four groups :

(i) Condensation products (39.01 *in the Brussels nomenclature*), including poly-condensation and poly-addition products ; these are mainly but not exclusively thermo-setting products—that is to say, they become soft and plastic on the first application of heat, but then undergo a chemical change and set hard. The most important of these are phenolics made from phenol and formaldehyde (best known as Bakelite) and aminoplastics made from urea and formaldehyde and from melamine and formaldehyde.

(ii) Polymerisation products (39.02 *in the Brussels nomenclature*), including co-polymerisation products. These are mainly thermo-plastic : that is to say that although they will harden on cooling, they will re-soften on re-heating. The best known are PVC (polyvinyl chloride), polyethylene (of which ' Polythene ' is a brand name) and polystyrene. Still small in volume of production is polypropylene, which is often classed together with polyethylene under the single heading of poly-olefins. Other important polymerisation products are acrylics (for example Perspex), polyvinyl acetate and poly-tetra-fluorethylene (PTFE).

(iii) Cellulosics (39.03 *in the Brussels nomenclature*), of which the best known are celluloid and regenerated cellulose film (Cellophane).

(iv) Hardened proteins (39.04 *in the Brussels nomenclature*) such as casein.

This classification is used in this article. Fuller details of the materials included in each sub-division are shown in the Appendix on Sources and Methods (page 50).

The producers of plastic materials use basic raw materials—such as petroleum products, natural gas, coal, cellulose, etc.—or intermediates derived from these materials, such as ethylene, propylene and acetylene—to produce plastic materials. These materials may be in the form of moulding and extrusion compounds, solid or liquid resins, emulsions, dispersions, and so forth. The materials producers may themselves turn out film, sheet, rods, tubes, mouldings, extrusions, and other fabricated products ; or this may be done by specialised fabricators, who purchase the materials from the basic producers. The materials producers and the fabricators are together described as the plastics industry.

Table 1. World[a] production and growth rates of various materials

	Production				Compound growth rates				
	Thousand metric tons[b]				*Per cent per year*				
	1913	1938	1950	1960	1913 to 1938	1938 to 1950	1950 to 1960	1938 to 1960	1913 to 1960
Plastics	35	300	1,500	5,700	9.0	14.3	14.3	14.3	11.5
Rubber, synthetic	—	6	543	1,914	—	45.0	13.4	30.0	..
Rubber, natural	122	925	1,890	2,010	8.4	6.1	0.6	3.6	6.1
Copper[c]	1,000	1,840	2,280	3,660	2.5	1.8	4.8	3.2	2.8
Aluminium[c]	70	530	1,280	3,610	8.4	7.6	10.9	9.1	8.8
Zinc[c]	800[d]	1,400	1,810	2,420	2.0	2.1	3.0	2.5	2.3[d]
Steel[b]	53[e]	88	153	241	1.8	4.7	4.7	4.7	3.1[e]

Source: UN Yearbook of Statistics, H. Saechtling, *Werkstoffe aus Menschenhand.*

(a) Excluding USSR, China and Eastern Europe.
(b) Million metric tons for steel.
(c) Primary refined production.
(d) Zinc 1909.
(e) Steel 1910.

The willingness to change from traditional to new methods and materials is one of the factors in any country which make for faster economic growth and greater competitiveness. A country which is ahead in the use of plastics, for example, may have significant technical and cost advantages over countries which rely on older materials and techniques. So it is illuminating to establish which countries are in fact ahead in the production and consumption of plastic materials, and why; that is the purpose of this article. The comparison concentrates on West Germany and Britain.

The figures

Partly because it is a young industry, it is not easy to obtain figures for plastics production, consumption and trade which are consistent from year to year and which are internationally comparable. (The statistical problems, and the methods we have used, are discussed in more detail in the Appendix, page 50.) The figures for the principal countries have been much improved in the last few years, and those for 1960-62 are probably reasonably reliable and comparable. Those for earlier years have a wider margin of error. However, this margin is not wide enough to vitiate the conclusions drawn in the article. It is significant that if we compare our estimates with those made independently by others,[1] the differences are not very big.

One of the problems is that the basic materials are often mixed with reinforcing agents—'fillers' or 'plasticisers'. For production or sales, figures are in general[2] available which exclude these fillers: that is, they are on a 'net resin basis'. But in the statistics of international trade the fillers are normally included, and we have had to make estimates of the net resin content of imports and exports of plastics. All the figures in the article are in terms of net resin content, unless otherwise stated.

Another problem is the general inadequacy of the figures, particularly for the end-uses of plastics. At the moment nobody, inside or outside the industry, can make useful estimates of the future size and pattern of the demand for plastic materials in Britain; for there is inadequate detailed knowledge about which industries use plastics now, how much they use and for what purpose. Such figures as there are (some are given in table 14) are based on trade estimates and market research surveys.

Throughout, we have relied to a considerable extent on the advice and assistance of the firms involved in the industry; and we are grateful to all the firms who helped us—and in particular to the principal producers of plastic materials in the United Kingdom, ICI, Shell, Distillers, Monsanto.[3] The analysis and conclusions of the article are the responsibility of the National Institute alone.

International comparisons

A few advanced countries predominate in world[4] production and exports. The United States, West Germany, Japan and Britain—in that order—account for 80 per cent of world output (table 2). Japan, in spite of her large volume of production, is still a net importer; and the other three countries between them

(1) Appendix, page 52, table 17.
(2) With certain exceptions noted in Appendix I, page 50.

(3) A list of all firms and organisations who have co-operated in providing information, views and advice, is given in Appendix IV, page 62.
(4) The term 'world' throughout the article excludes the USSR, Eastern Europe and China.

Table 2. Production, consumption and trade in plastics materials

		United States	United Kingdom	West Germany	France	Italy	Japan	Sweden	Netherlands	Belg./Lux.
Total figures (*thousand metric tons*)										
Production	1955	1,744.2	323.4	355.2	110.1	89.7	113.7	29.5	24.1	16.4
	1961	3,239.2	615.0	1,028.8	367.0	384.4	781.0	65.7	76.1	46.6
Per cent increase ..	*1955-61*	*86*	*90*	*190*	*233*	*328*	*587*	*123*	*216*	*184*
Consumption	1955	1,644.8	263.9	294.4	119.0	81.8	120.3	38.2	32.6	21.9
	1961	2,866.6	507.7	829.0	367.8	333.8	793.7	95.0	92.7	97.7
Per cent increase ..	*1955-61*	*74*	*92*	*182*	*209*	*308*	*560*	*149*	*183*	*346*
Exports	1955	131.6	85.6	78.8	14.6	19.7	4.9	7.5	13.1	6.6
	1961	383.2	192.3	298.1	69.5	111.9	44.4	31.1	61.6	24.5
Exports as per cent of production	*1961*	*12*	*31*	*29*	*19*	*29*	*6*	*47*	*81*	*53*
Imports	1955	2.2	26.1	18.1	23.5	11.8	11.5	16.2	20.6	12.2
	1961	10.6	77.3	98.4	70.3	61.3	57.1	60.5	78.2	75.6
Imports as per cent of consumption	*1961*	*1*	*15*	*12*	*19*	*18*	*7*	*64*	*84*	*77*
Net exports	1955	129.4	59.5	60.7	− 8.9	7.9	− 6.6	− 8.7	− 7.5	− 5.5
	1961	372.6	115.0	199.8	− 0.9	51.1	− 12.7	− 29.3	− 16.6	− 51.2
Per head figures (*Kg*)										
Production	1955	10.70	6.34	6.95	2.54	1.87	1.28	4.06	2.24	1.78
	1961	17.64	11.80	19.00	7.98	7.76	8.31	8.74	6.54	4.90
Consumption	1955	9.91	5.19	5.87	2.75	1.70	1.35	5.26	3.03	2.37
	1961	15.61	9.61	15.29	8.00	6.74	8.44	12.64	7.93	10.28

Source : See Appendix II, page 52.

provide 70 per cent of world exports. There are still a large number of countries where important materials are not produced at all.

In absolute terms, the United States is of course the largest producer, consumer and exporter ; but on a per head basis, West Germany comes first in production and exports, and is a close second to the United States in consumption (table 3). If we relate consumption to national product, again the German figure (together with that of Japan) is exceptionally high (chart 1). West German consumption in 1961 was almost as big as that of France and Britain together, and her production was larger.

This German pre-eminence might be explained by lower costs or prices, or by a different industry structure—if plastics-consuming industries were much more important than elsewhere in Europe. Or it might be due to technological factors : earlier discovery and development of new plastic materials, and earlier innovation in finding applications for existing materials. This, in turn, might be related to the degree of conservatism among users and potential users of plastics, and to the rate of change in the economy as a whole. We discuss these possible explanations in turn.

Costs of production

It was the almost unanimous opinion of the firms and organisations interviewed that the West German plastics industry did not have any major input-cost advantages over her main competitors—that there was nothing equivalent, for instance, to the advantage Canada derives for her aluminium production from cheap electric power. This consensus of opinion is not proof, of course ; and proof is hardly possible since, for the plastics industry as for most new industries, detailed production cost figures are hard to come by.

Table 3. Estimates of per capita production and consumption of plastics materials, by type, 1961

Kg. per head

		USA	UK	West Germany	France	Italy	Japan	Sweden	Netherlands	Belg./Lux.
Condensation (*39.01*)										
Phenolic	Production ..	1.31	0.97	1.01	0.40	0.46	0.59	1.23	0.56	..
	Consumption ..	1.25	0.73	0.85	0.43	0.44	0.60	1.31	0.46	..
Aminoplasts	Production ..	0.98	1.39	2.76	0.61	0.83	1.83	2.14	0.24	..
	Consumption ..	0.93	1.12	2.56	0.79	0.82	1.83	2.03	0.94	..
Other	Production ..	2.18	1.56	3.51	1.21	0.74	0.21	1.54	1.95	..
	Consumption ..	2.10	1.61	2.86	1.20	0.89	0.18	1.31	1.30	..
Total	Production ..	4.47	3.92	7.28	2.22	2.03	2.63	4.91	2.75	1.27
	Consumption ..	4.28	3.46	6.27	2.42	2.15	2.61	4.65	2.70	3.84
Polymerisation (*39.02*)										
PVC	Production ..	2.41	2.06	3.61	2.60	3.03	3.29	1.53	0.77	..
	Consumption ..	2.26	1.90	3.08	2.28	1.97	2.97	2.13	2.02	..
Polystyrene	Production ..	2.83	1.00	2.03	0.92	0.79	0.31	0.54	—	..
	Consumption ..	2.48	0.77	1.29	0.87	0.43	0.41	0.99	0.43	..
Polyolefins	Production ..	4.21	2.59	1.94	0.65	1.13	0.62	—	1.55	..
	Consumption ..	3.31	1.53	1.55	1.08	1.15	1.07	2.23	1.29	..
Other	Production ..	1.99	0.93	2.43	0.57	0.48	0.97	0.59	0.84	..
	Consumption ..	1.68	0.89	1.43	0.52	0.66	0.93	1.77	0.97	..
Total	Production ..	11.44	6.58	10.00	4.74	5.43	5.19	2.66	3.16	3.64[a]
	Consumption ..	9.73	5.09	7.36	4.75	4.21	5.38	7.11	4.72	..
Cellulosic (*39.03*)										
Total	Production ..	1.73	1.27	1.69	1.02	0.28	0.49	1.17	0.94	..
	Consumption ..	1.60	1.03	1.65	0.82	0.37	0.44	0.88	0.50	..
Protein (*39.04*)										
Total	Production ..	..	0.03	0.02	0.01	0.02	—	—	..	..
	Consumption ..	..	0.03	0.01	0.01	0.01	..	..	..	..
Total, all materials										
	Production ..	17.64	11.80	19.00	7.98	7.76	8.31	8.74	6.54	4.84
	Consumption ..	15.61	9.61	15.29	8.00	6.74	8.44	12.64	7.93	10.28

Source : See Appendix II, page 52.
(a) Including 39.03 and 39.04.

But such evidence as there is on costs, and such price figures as are published, support this conclusion.

The pattern of costs varies considerably between the different plastic materials, and for the same material between different firms and processes, depending largely on the degree of integration. But typically costs of operating labour are low, and raw material and capital costs are high. For most plastic materials the range of costs (excluding return on capital) would be roughly as follows :

	Per cent of cost
Materials and process chemicals	40–70
Fuel and power	3–7
Operating labour	3–7
Maintenance	4–8
Depreciation	5–15
Research and royalties	3–7
Overheads, including administration, supervision, testing, sales	15–25

For PVC, polyvinylacetate and melamine-formaldehyde, electric power costs are much higher than the range shown above. Capital costs are particularly high for polyethylene, polypropylene and acrylics, but relatively low for some condensation products, where plant economies of scale are not very important. Average capital employed per man in the United States plastics industry has been estimated at $22 thousand,[3] but it is higher than $40 thousand in some poly-olefin plants, where an investment in one plant of over $20 million is not uncommon. Overhead costs are generally high, because of the need for expensive facilities for testing and control, and for a technical sales service. They are particularly high for new plastic products which have not yet reached a high volume of production.

Chart 1. Plastics consumption and national product, 1961[(a)]

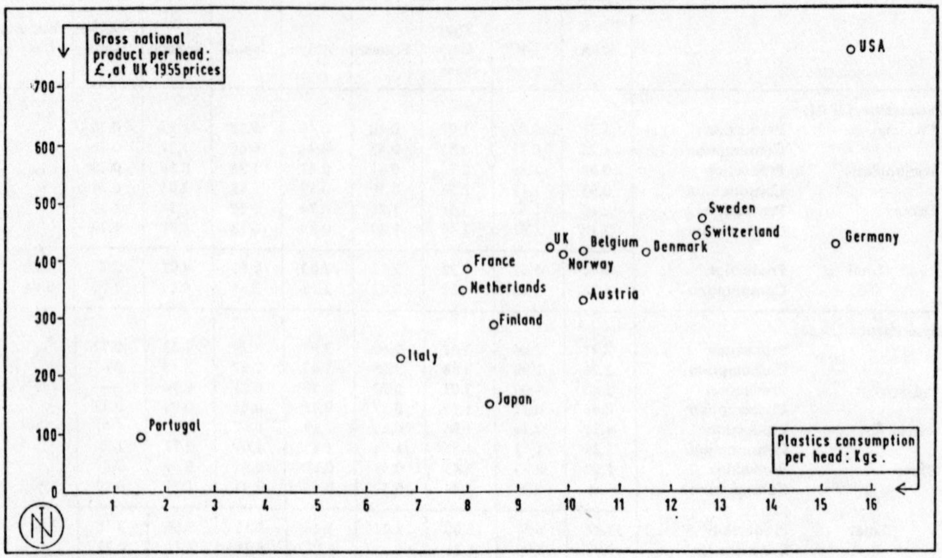

Source: Tables 2 and 3, OECD *Chemical Industry in Europe* and NIESR estimates.
(a) 1960 for Finland, Denmark and Portugal.

Material costs

The most important single element in costs for almost all plastics is material costs, including chemicals, such as catalysts, which are consumed in the production process, although not embodied in the end-product. It appears that West Germany has had no general advantage over her competitors in material costs, and indeed was at a disadvantage compared with some of them.

Originally coal and cellulose were the most important sources of the basic materials and other chemicals needed for plastics manufacture ; but since the Second World War they have been increasingly replaced by oil and natural gas. As long as coal was the source, Germany was not at a disadvantage—although she had no special advantage either over other coal-producing countries, except in know-how. But as oil became more important, she was at a disadvantage compared with the United States, and to a lesser extent with Britain, where the post-war refinery-building programme started earlier than in Germany.[(1)] The United States had a big lead in petro-chemicals, and Britain's output was ahead of West Germany's up to 1960 (table 4). As late as 1957, only 12 per cent of West German production of ethylene and acetylene, as against 74 per cent of American production, came from oil or natural gas.

Such gases as ethane, propane, ethylene, propylene and acetylene can sometimes be made available from oil refineries very cheaply. In the United States they were in fact often ' waste ' products which were flared before petro-chemical complexes were built to use them.[4] Manufacturers of plastic materials who could site their plants within pipe-line distance of refineries

Table 4. Output of petro-chemicals

Thousand metric tons of carbon content

	1953	1954	1955	1958	1959	1961
West Germany	55	70	110	246	335	858
France	16	28	37	118	260	510
Italy	12	16	29	82	180	246
Netherlands ..	12	27	28	50	54	85
United Kingdom	124	164	197	302	378	618
Others	..	..	..	15	18	156

Source: *Chemische Industrie XIII* 1961, December, p. 767.

(1) However, Britain was also handicapped in the early post-war years by a lack of refinery feedstock : thus she used an expensive process for producing ethylene from alcohol made from molasses.

had an advantage compared with those still based on coal-derived chemicals, imported petro-chemicals or other materials. More recently, there has been a tendency for chemical firms to set up their own specialised cracking plants to provide their feedstock, instead of relying on oil refinery by-products.

In the last decade, natural gas has become an increasingly important source, as it is often an even cheaper feedstock than oil. Here again, West Germany (together with Britain) has been at a disadvantage, compared with Italy, France and the United States.[1] Italy in particular has been exploiting this natural advantage, with big new plastic plants in Ravenna and Ferrara ; the recent rapid growth of the Italian plastics industry and the rise in exports may well be due in part to this factor, as well as to generous Government support of new investment projects in the South.

Consequently, the success of the German plastics industry in the 1950s cannot be explained by lower material costs.[2] Indeed, the industry was at some disadvantage in this respect—particularly for polyethylene, polystyrene, acrylics and epoxy resins. The disadvantages were less serious for PVC, which can be manufactured by various alternative processes ; however, the manufacture of carbide both for PVC and polyvinylacetate needs considerable amounts of electric power, and West Germany is not a cheap power producer. Japan and Italy have advantages over her in power costs, and this is one reason for their relatively high output of PVC (table 3).

(1)West Germany has in fact a very small amount of natural gas, some of which is used in the polystyrene plant at Hüls.

(2)This does not mean, of course, that none of the materials used was cheaper in Germany. Her coke and acetylene technology were particularly advanced, and she undoubtedly derived some benefits from this. There is also evidence of differences in pricing policy between chemical firms in Britain and West Germany which may have resulted in higher prices in Britain for such chemicals as formaldehyde.

Other costs

Although she had no advantage in raw material costs, West Germany did have a slight advantage in capital costs over Britain, but not over most other European countries or Japan. It is not at all easy to compare the costs of erecting chemical plants in different countries : even for similar products, plants vary in the proportion of process equipment delivered to the site, in the plant's specifications, in the volume of on-site building work, in design costs and in many other respects. General building cost indices or labour cost indices are not much use in an industry such as chemical engineering. However, a useful comparison has been made by Mr. E. A. Stallworthy of Bataafsche Internationale Chemie Maatschappij N.V. (Shell)[5]; this has the advantage that it is based on actual experience of erecting a similar standard styrene plant in eight different countries (table 5). He suggests that the capital cost of erecting plant in West Germany was about 10 per cent lower than in Britain or Sweden, but higher than in most other European countries. Some other estimates put British costs in a slightly more favourable light, but it is probable that West Germany did enjoy some advantage in the 1950s. However, in the last year or two, there is evidence that costs have tended to level in Western Europe (including Britain), although in very advanced types of equipment, the United States and West Germany may still have advantages, based on chemical engineering know-how.

In labour costs too, although West Germany had an advantage some years ago, this has now disappeared, and her labour costs are higher than Britain's. It was in any case a relatively unimportant advantage, since labour costs are so small a part of total costs. It is true that there is also a labour cost element in overhead costs ; but German sources believe that for this type of labour her advantage was

Table 5. Comparative costs of building a styrene production unit in various European countries, 1959-1961

Percentage of total cost

Country	United Kingdom	Belgium	France	Holland	Italy	Spain	Sweden	West Germany
Total materials	50.6	55.4	55.8	59.3	59.3	62.5	50.6	55.8
Freight, purchasing, import duties, etc.	5.5	6.6	6.5	7.0	7.0	9.4	9.3	6.5
Erection	34.8	27.7	29.1	22.2	24.1	18.4	32.4	29.1
Piling	0.3	1.7	—	3.0	0.4	—	—	—
Design	8.8	8.7	8.6	8.5	9.3	9.8	7.7	8.6
Total estimated cost	100	100	100	100	100	100	100	100
Ratio, with UK = 100	**100**	**87**	**88**	**82**	**82**	**77**	**98**	**88**

Source : E. A. Stallworthy, *Comparative Investment Costs in Western Europe* (Paper presented to the Association of Cost Engineers, and A.A.C.E., May 1963).

less, and not significant. In fact, overhead costs may have been higher in Germany than in other countries : American estimates place German sales and administrative costs a little higher than Britain's but lower than those of the United States.[6] German chemical firms spent more on research and development (in relation to turnover) than the largest British firms did, and probably more than most other European producers as well.

The analysis of employment at BASF, Ludwigshafen, gives some indication of the importance of overhead costs in general. In 1956, only 28 per cent of their employees were engaged on production, and in 1961 only 23 per cent ; the rest—72 and 77 per cent respectively—were in technical departments, in research and development, in administration, or were trainees or apprentices.(1) These figures are for total employment in all types of chemicals ; the overheads in plastics are probably well above the average.

In sum, it does not seem at all likely that lower costs explain the success of the German plastics industry. In the early fifties, when the German industry did have lower capital and labour costs than Britain, its raw material costs were higher. By the time these had fallen—as the German petro-chemicals industry caught up—capital and labour costs had risen to the British level. On balance, there was probably no significant net advantage.

Prices of plastic materials

Price figures can also provide some suggestive evidence on costs ; for if West Germany had enjoyed a general advantage over other producers in lower production costs, it might well have shown itself in a lower price level for German plastic materials.

In fact, from 1950 to 1955 prices of most plastic materials appear to have been a little higher in Germany than in Britain or the United States. Nonferrous and steel prices were also slightly higher, [7] so that the ratio of plastics prices to those of the principal competing materials was probably similar in all three countries. German polystyrene prices were about 10 per cent lower than in Britain ; but about 10 per cent higher than in the United States (table 6). Conversely, prices of PVC were higher than in Britain, lower than in the United States. The prices of most other materials were a little higher than either in Britain or the United States ; polyethylene prices were far higher. These comparisons are summed up in an index of the prices of six principal standard materials (table 6) ; this index, it is true, does not take account of all specialised grades, and—particularly later on—quoted prices may not be an entirely reliable guide. Even so, it is fairly clear that the recovery of German plastics consumption in the first half of the 1950s cannot be attributed primarily to lower relative prices than those prevailing in Britain and the United States.

From about 1955 onwards, prices of many plastic materials began to fall more sharply ; trade barriers were diminished, tariffs and quotas were reduced, and international competition became more intense. Surplus capacity began to be a serious problem and there was widespread ' dumping ' of temporary surpluses on the international market. For some plastics this caused prices to fall very sharply, particularly in unprotected markets ; so the quoted prices of domestic manufacturers may not reveal the full extent of these price falls. The markets most severely affected were Belgium, the Netherlands, Scandinavia and Britain, all of which have low tariffs. For example, the real price of PVC in Britain in 1961 and 1962 was probably 10-15 per cent below the quoted prices. In the United States, where tariffs are very high, domestic competition became much more severe, and prices fell even more sharply, partly as a result of the entry of a number of new producers.[9] The West German market was affected by the more acute international competition, but not in such a way as to provide a price stimulus to consumption greater than that in other European countries. German prices fell with the rest, but were not below the general international level, except for polystyrene.(2)

The price index for six materials (table 6) suggests that by 1960 German prices were only fractionally higher than British prices, but were still significantly higher than American prices. By this time also French prices had fallen close to the German level, and Italian prices were below them. In 1961 and 1962 prices fell still lower, and both British and German producers complained of dumping in their home markets ; there is little doubt that, because of higher tariffs,(3) West Germany suffered less. However, by 1963, German quoted prices had fallen slightly below Britain's.

(1)The breakdown of the numbers employed at BASF Ludwigshafen,[8] was as follows (in thousands) :—

	1956	*1961*
Production..	10.0	10.4
Technical department, including maintenance	12.0	16.0
Research and development ..	3.6	6.5
Administration, managerial, trainees and apprentices ..	10.4	13.1
	36.0	46.0

(2)However, by this time the greater part of German polystyrene production was no longer of standard grades but of more specialised high impact materials, for which her prices were not particularly low.

(3)A comparison of tariffs on plastic materials is given in Appendix II, table 26.

Table 6. Prices of plastic materials.

Quantity lots, annual averages, DM per kilo

		USA	UK	West Germany	France
PVC, granular	1953	3.55	2.30	2.50	..
	1955	3.20	2.25	2.25	2.80
	1957	2.70	2.20	1.90	2.50
	1960	1.85	1.65	1.75	1.90
Polyethylene, high pressure	1953	4.30	4.80	7.00	..
	1955	3.75	4.10	5.40	5.20
	1957	3.20	3.65	4.35	4.75
	1960	2.40	2.70	*3.00*	3.55
Polystyrene, standard crystal	1953	3.00	3.65	3.20	..
	1955	2.70	3.35	3.00	3.15
	1957	2.35	3.35	2.90	2.85
	1960	1.60	2.60	2.10	2.30
Phenol-formaldehyde	1953	1.80	2.05	2.30	..
	1955	1.75	1.90	1.90	2.50
	1957	1.95	1.85	1.90	2.25
	1960	1.95	2.00	1.90	1.90
Urea-formaldehyde	1953	3.00	*3.00*	3.00	..
	1955	3.00	*3.00*	3.00	3.20
	1957	3.10	*3.00*	3.00	2.80
	1960	2.90	*2.80*	*2.80*	2.35
Melamine formaldehyde	1953	4.10	*5.15*	5.15	..
	1955	4.10	*5.15*	5.15	4.60
	1957	4.30	*5.15*	5.15	4.15
	1960	4.30	*4.90*	4.90	3.45
Materials index[a] .. .. West Germany = 100	1955	96.3	96.1	100.0	109.2
	1960	90.5	99.7	100.0	103.2

Source : Information from firms, and published prices ; H. Saechtling, *Chemie und Technologie der Kunststoffe*. Italicised figures are estimates.
(a) Weighted by 1955 and 1960 combined consumption of each material in USA, UK, West Germany and France.

So far, the evidence rests on international comparisons of quoted prices. There is no way of knowing what special prices may have been made in the various countries. But there is no reason to think that there was any particular pressure forcing prices down more in Germany than elsewhere : domestic competition in Germany does not seem to have been exceptionally severe. Typically, the manufacture of plastic materials is undertaken by large or very large firms. This is especially true of thermoplastic products, which in most manufacturing countries are made by fewer than half a dozen producers, with often only two or three of any importance (table 7). In small countries there is not usually room for more than one or two producers, because of the economies of scale and the great importance of manufacturing know-how, research facilities, and technical sales service. It is true that there are a few more producers of the main thermoplastics in West Germany than in Britain or in France (except for polystyrene) ; but the firms are not by any means independent of each other. Taking account of interlocking shareholdings,[1] the German market structure was no less oligopolistic than that of the other main producing countries. There are slightly more producers of most thermo-setting plastics in all manufacturing countries ; these materials need less capital, and—since they are older—technical know-how is more widely diffused.

There is, in sum, no evidence that Germany's leading position in plastics was explained either by lower costs of materials and other inputs, or by lower prices. It is true that there are individual plastic materials, for Germany and other countries, where

[1] For instance, Rheinische Olefinwerke is jointly owned by BASF and Shell ; Hoechst owns 50 per cent of the capital of Wacker and 33 per cent of the capital of Ruhrchemie ; and BASF, Hoechst, Dynamit Nobel, Bayer and Hüls were all part of I.G. Farben before the post-war reorganisation of the German chemical industry.[10]

Table 7. Production capacity for principal thermoplastics in various West European countries, end 1961

	PVC		Polystyrene		High pressure polyethylene		Low pressure polyethylene		Polypropylene	
West Germany ..	Wacker-Chemie	A	BASF	A	Rheinische Olefin	A	Hoechst	B	Hoechst	D
	Hoechst	B	Hüls	B	BASF	D	Rheinische Olefin	C	BASF	D
	BASF	B	Dynamit Nobel	D			Scholven-Chemie	C	Hüls	D
	Hüls	B					Hüls	D	Hibernia	D
	Dynamit Nobel	C					Hibernia	D	(Rheinische Olefin)	..
	Deutsche Solvay	D					Ruhrchemie	D		
	Anorgana	D					(BASF)	(D)		
United Kingdom ..	ICI	A	BX (DCL)	D	ICI	A	British Hydrocarbon	C	ICI	C
	British Geon (DCL)	B	Distrene (DCL)	C	Union Carbide	B	Shell	D	Shell	D
	Bakelite	D	Monsanto	C	Monsanto	C				
			Sterling	D	(Shell)	(C)				
			Styrene Products	C						
			Kaylis	D						
France	Pechiney	B	Plastichimie	B	Ethylene Plastique	B	Manufacture Normande	C	Pechiney	D
	Solvic	B	Dispersions plastiques	D	(Petroplastique)	(B)	Soc. Normande	D	Soc. Normande	D
	St. Gobain	C	Huiles Goudrons	D			Naphthachimie	D		
	Kuhlmann	D	Lorraine-Kuhlmann	C						
	Rhone-Poulenc	D	Monsanto-Boussois	C						
			Resines et Vernis	D						
Italy	Montecatini	A	Montecatini	C	Montecatini	B	Montecatini	D	Montecatini	C
	Sicedison	B	Italiana Celluloide	C	Celene	C	Solvay	D		
	Chimica Ravenna (ENI)	C	Sicedison	C	Asfalti Bitumi	C	(SIR)	(B)		
	Solvic	C	(Dow)	(C)	(Anic-Gela)	(B)				
	(Chimiche Meridionali)	(B)	SIR	D	(SIR)	(C)				
	(SAICI)	..								
	(Rumianca)	..								
Sweden	Stockholme Superfosfat	C	Svenska Polystyren	D	(Uni-Kemi)	(C)			(Esso)	(D)
					(Esso)	..				
Netherlands	Shell	C	(Dow)	..	Staatsmijnen	B	(Staatsmijnen)	(D)	(Rotterdamse Po)	(D)
	(Caltex)	..			(Du Pont)	(B)				
Belgium	Solvic	B			(Cobenam)	(C)			(Argus)	..
					(Argus)	..				
Austria	Halvic	D			(OMV/Hoechst)	(D)			(Danubia)	(D)
	(Salzburger Aluminium)	..								

Source : Information from firms. A = Over 50,000 tons a year ; B = 21,000 to 50,000 tons a year ; C = 11,000 to 20,000 tons a year ; D = Up to 10,000 tons a year. See Appendix 1, page 51, for the spelling out of the abbreviations. Firms shown in brackets are those which have set up or announced plants in 1962 or 1963.

prices were relatively low and consumption particularly high, and conversely. This was true, for example, of polystyrene in Germany. But here the relatively low price seems to have been the consequence, not of lower costs of materials, but of technical advance ; quantity production of this material began earlier in Germany than elsewhere. Indeed, it may be more generally true that, although German costs of inputs and prices were no lower than in Britain, costs per unit of output may still have been below those of other countries because of more advanced production methods—and that consequently German profits were higher. There is no direct evidence on profits from production of plastics—nearly all the big firms are large chemical combines which do not show profits of their plastics divisions separately. But it may be perhaps significant that it is British firms, rather than German, which have recently emphasised very strongly the danger of 'profitless prosperity'—a high level of output at unremunerative prices. In their view this would jeopardise the whole future growth of the industry by compelling firms to cut back on research, development and technical services. It is, incidentally, interesting that this view implies that research and technical factors are likely to be more important for long run growth than temporary price reductions.[11]

Technical factors

Cost and price factors do not explain the German lead in plastics production and consumption. Technical factors appear to have been more significant. Since the industry is based entirely on the discovery and application of completely new man-made materials, scientific research and development are its foundation, and their importance does not stop with the discovery and first commercial production of a new material. This is only the beginning, and must be followed by years of intensive applied research and development work to explore potential applications, to modify the material and create a variety of grades suited to each application, to blend it with other materials, to improve and cheapen the production process, and to find the best ways of processing the material, which may involve the design of new machinery. The benefits of this research and development are to some extent cumulative. A country which has a large number of scientists and engineers experienced in applied research and development with some plastic materials, will probably find it easier to develop others. Furthermore, the chemistry and technology of other materials such as synthetic rubbers and synthetic fibres are so closely related to plastics, that advances in any one of these fields will help progress elsewhere.[12]

Partly because of this, and partly because of the high costs of research and development, there is a strong tendency for research to be concentrated in a few very large concerns in each country. Since the early part of the 20th century, the importance of individual inventors in the research and development process has been declining. The patent statistics for the industry suggest this (table 8). The proportion of patents issued to individuals declined from nearly half in the period 1791-1930, to fewer than 10 per cent in the postwar period, although strategic patenting (see page 35) by large firms may to some extent inflate their figures. Of course, gifted individuals still play an extremely important part in the inventive process. But they do so more and more as employees of large corporations (Carothers, for instance, discovered nylon while working for Du Pont) or in association with large corporations—thus Professor Natta co-operated with Montecatini in the development of polypropylene. Frequently, the costs of building and operating the pilot plant required for the development of synthetic materials are far beyond the capacity of individual inventors or small firms, and research into applications also needs substantial resources. Furthermore, those small firms which have patented new inventions have often subsequently been absorbed ; for example, the Naugatuck Chemical Company, which did important research work in the field of polystyrene, sold its patents and know-how to the

Table 8. Patents issued for the principal groups of plastics, 1791-1955

	1791-1930		1931-1945		1946-1955	
	Number	Per cent of total	Number	Per cent of total	Number	Per cent of total
Patents taken out by :						
Individuals	1,803	43	791	15	489	8
Firms	2,436	57	4,341	85	5,749	92
Total	4,239	100	5,132	100	6,238	100

Source : J. Delorme, *Anthologie des Brevets sur les matières plastiques*, (3 vols.), Amphora, Paris, 1962.

Dow Chemical Company in 1937, which then successfully launched commercial production. The British firm of Petrochemicals, which had done some work in the field of styrene polymerisation and had acquired the patents for the new Ziegler low pressure polyethylene process, was taken over by Shell in 1955. So, because large firms dominate the industry, the examination here of the process of technical advance concentrates on them.

The analysis of technical factors needs a much longer time span than the analysis of prices or costs. There is usually a period of five to twelve years between research and normal commercial production ; and it is sometimes even longer before production is large enough to show up significantly in the statistics. High pressure[1] polyethylene was discovered by ICI and first successfully produced in the 1930s ; it had extremely important radar and other applications in the war ; but it was not until the 1950s that it came into large volume consumption in Britain. The vital research and development on low pressure polyethylene and on polypropylene was done in the late '40s and early '50s, and the patents taken out between 1951 and 1957, but it is only in the '60s that they are beginning to be produced in large quantities. Thus the pattern of plastics production and consumption in the 1950s depended largely on the research and development of previous decades.

The problem of measuring the resources which go into research and development, and of assessing their output, has only recently begun to attract economists' attention. There are considerable conceptual difficulties, and not many figures to work on. Three different measures are used here : first, statistics of expenditure on research and development ; secondly, patent statistics ; and thirdly, the dates and origins of major innovations. These measure three distinct but overlapping aspects of a complex process. Expenditure on research is usually necessary before an innovation can be made—but it also continues after the innovation has become successful commercially. Patents may be taken out at any stage of a research project, and will usually be followed by ' patents of addition ' after commercial production has begun. So events do not often follow the simple time sequence—research : development : patents : innovation : production.

The three methods of measurement supplement each other. The figure for the sums spent on research and development is inadequate by itself, since research can be quite fruitless. In the same way a large number of patented processes fail to reach the stage of commercial production. If we take the three methods together, then we have first an assessment of the resources committed to the search for new inventions and the improvement of old ones ; secondly—in patent statistics—a measure of the actual output of inventions ; and thirdly, in the figures for innovations, a measure of success in passing from the development stage to full-scale commercial production. This triple assessment should provide a reasonable indicator of the capacity of a firm, or a country, for undertaking successful research.

[1]The high pressure method is used mainly for manufacturing low density polyethylene, and the low pressure method is used for high density (or linear) polyethylene.

Research expenditure

The figures available on research and development in the chemical industries are of limited usefulness, because they do not separate expenditure on research into plastic materials from other chemicals expenditure. Nevertheless they can be used to provide clues to the scale of plastics research and development activity in various firms and countries.

The most striking feature of the figures is the outstanding position of the German chemical combine I.G. Farben from the time of its formation in 1925 until it was dissolved after the Second World War. This firm's research expenditure averaged just over 7 per cent of its turnover from 1925 until 1939. From 1926 to 1931 it was between 7 and 10 per cent. It cut back research spending fairly drastically in the world recession—to 4.9 per cent of turnover in 1933. The main economies were in development costs, and the research staff was maintained throughout at over 1,000 qualified scientists and engineers.[13] From 1934 to 1939 research expenditure rose again to between 5 and 6 per cent of turnover. During the war, the level of research was maintained, but it probably fell just below 4 per cent of turnover, since turnover was rising rapidly.

Throughout this period the firm spent more on research than it distributed in dividends. These figures exclude extra-mural expenditure, technical services and donations to universities, etc. I.G. Farben must have been one of the most research-intensive large firms in the world before 1939, with a ratio of expenditure to turnover more than twice as high as that of most comparable firms. Moreover this firm seems to have devoted a far higher proportion of its resources to research in synthetic materials than multi-product chemical firms in other countries. There is the evidence of the patent statistics (page 36) ; there is also the evidence of the wide range of new materials, such as PVC and polystyrene, which this firm initiated. According to one account,[14] it was the need to concentrate research resources and make large investments in high polymer chemistry and synthetic materials that finally persuaded the individual firms to form I.G. Farben itself. Later, the whole trend of the economic policy of autarky in Germany under the Hitler regime favoured the production of ' ersatz ' or synthetic materials. Altogether, there are strong grounds for believing that I.G. Farben spent more on plastics research and development than any other firm in the world.

This research programme was disrupted from 1945 to 1952, when the whole future structure of I.G. Farben was in doubt, but it was not completely broken off, and the research teams remained in existence.(1) The successor firms, especially BASF, Bayer and Hoechst continued to spend heavily on research throughout the 1950s, averaging between 4 and 5 per cent turnover(2) (or between 5 and 6 per cent if capital expenditure is included).

The leading American firms, Du Pont, Union Carbide, Dow and Eastman Kodak also now spend over 4 per cent of turnover on their research and development,[15] but among British firms only ICI reaches this level. (I.C.I's. ratio in 1961-2 was about 4 per cent(2) calculated on the sales and research of U K companies only—that is, excluding the sales and research of overseas subsidiaries). There are no data for French firms but there is little doubt that they spend significantly less[16] and the same is probably true for most other firms in Europe with the exception of CIBA.

Turning from the figures for individual firms to those for the chemical industry as a whole, it seems that now the United States is well ahead, in absolute terms; the amount of research done in the United States chemicals industry is probably about $4\frac{1}{2}$ times the British figure(3) and 3 times the German. But the United States' lead in plastics is probably less than this, for two main reasons. First, a higher proportion of American research goes into rather specialised military/space applications. Secondly, plastics research is almost certainly a larger part of total chemicals research in Germany (and Japan) than in the United States—simply because the plastics industry is a larger part of the total chemical industry in those two countries. Research expenditure is usually a higher proportion of the value of plastics output than of the output of the general run of chemicals.(4)

Then we must take into the reckoning as well the 'infra-structure' of research: the work of universities and other research institutions. Teaching and basic research in chemistry were particularly advanced in Germany in the first part of this century, and the theoretical work of Staudinger and his school in high polymer chemistry were an essential foundation for the achievements of I.G. Farben.(5)[17] However, the fundamental research undertaken in universities and such government-aided laboratories as the Kaiser Wilhelm Research Institutes or DSIR stations is generally available throughout the world; industrial firms anywhere may benefit from it, so long as they are alert to scientific developments outside their own country. Consequently the advanced state of basic chemical research in Germany may not be as important, in explaining German pre-eminence, as the development work in the chemical firms themselves. Industrial or co-operative institutes outside the chemical firms do not appear to have made any notable contribution in Germany until after the war. Since then two organisations have been set up specifically to serve the plastics industry. An Institute for Plastics Fabrication, established in Aachen in 1951, concerns itself mainly with the processing and testing of materials including welding, extruding, moulding and glueing. In 1953 a larger Plastics Institute was set up in Darmstadt, which is concerned primarily with the fundamental properties and structure of plastic materials and with methods of testing and documentation. Together these Institutes employ a staff of over a hundred, but their total research expenditure is of course only a small fraction of the annual expenditure of the chemical firms themselves. In Britain there has for some time been anxiety over the lack of adequate fundamental research into plastics materials and technology.[19] It was not until 1962 that the Rubber Research Association also took in plastics and became the Rubber and Plastics Research Association, so that its work in this field has only just begun.(6) It seems that co-operative or state research institutes are not important in other countries, except for the TNO Plastics Research Institute at Delft in the Netherlands.

Patent statistics

There are one or two big differences between the industrial patterns of patent figures and of research expenditure figures. In defence-oriented industries,

(1) 'The German research teams, as exemplified in the I.G. laboratories, were outstanding instruments of accomplishment. What could be done with these groups? Should they be given picks and shovels? Should they be broken up and transferred as individuals to laboratories in other countries? Should they be allowed to continue in their present location? Many hours were spent arguing the pros and cons of this situation and we inevitably arrived at the conclusion that humanity's interest would best be served by putting these groups to work in the surroundings and with the associations to which they were accustomed, but under a competent allied commissioner who would merely insist that no direct war projects be worked upon. Certainly some of the results would have war applications; but the tremendous good which could come from these able groups should outweigh that risk.' *German Plastics Practice*—J. DeBell, W. C. Goggin, W. E. Gloor, pages 10-12.

(2) These figures exclude technical services and capital expenditure.

(3) The absolute figures are: United States £390 million, West Germany £55 million, Britain £37 million, Japan £18 million in 1961. But because of differences in research costs a 'research exchange rate' must be used.[18]

(4) Thus for example the plastics section of one of the larger British chemical firms spends about 5 per cent of turnover on research and development, which is above the general average for the firm.

(5) Among the distinguished scientists who worked for this firm were two Nobel prize-winners.

(6) A small amount of work was however already being done at the National Chemical Laboratory.

Table 9. Patents delivered in various branches of British and French industry, compared with research expenditure, 1961

	France	United Kingdom	United Kingdom	
	Percentage of total number of patents delivered		*Percentage of total research expenditure, manufacturing industry*	
			Excluding aircraft	*Including aircraft*
Aircraft	1.8	1.7	—	38.4
Electrical engineering	17.3	22.2	38.3	23.5
Instruments	10.6	6.3	4.1	2.5
Chemicals	20.6	24.0	20.3	12.5
Vehicles	6.6	5.0	4.3	2.7
Engineering	16.3	18.0	13.1	8.0
Metals and metal products	7.5	9.5	6.1	3.7
Building materials, wood and furniture, building	11.0	6.4	6.0	3.7
Textiles and clothing	6.3	5.4	3.8	2.4
Food, drink and tobacco	1.9	1.5	3.2	2.0
	100.0	100.0	100.0	100.0

Source : Y. Fabian, *Measures of output of R and D*, OECD 1963 ; *Report of the Comptroller General of Patents*, 1961 ; *Bulletin de la Propriete Industrielle—Statistiques*, 1961 ; *Report of Advisory Council on Scientific Policy*, Cmnd. 1920, 1963.

such as aircraft and electronics, the exigencies of security and the effect of government contracts is to produce a very low ratio of patents to expenditure. But if some allowance is made for these industries, then there is a broad correspondence between research expenditure and patenting in the principal branches of industry (table 9).[1] The similarity of pattern suggests that patent statistics, if used with care, can help to build up a rather more detailed picture of the structure and direction of research activity than the expenditure figures provide. Further, the industrial pattern of British and French patents is similar : this suggests that although international differences in patent procedure may affect the total number of patents granted in each country, they probably do not affect the distribution between industries.[20]

The main difficulty of patent statistics is one of weighting : some patents are extremely important whilst others are negligible. Some are taken out partly for strategic commercial reasons, rather than to register any original invention. For example after taking out patents on moisture-proof cellophane in the 1920s, Du Pont then proceeded to take out a large number of patents on alternative methods to block the possibility of competitors circumventing their key patents.[21] This procedure is not, of course, confined to any one firm or country.

But for a large group of patents over a fairly long period, the distortions arising from this factor are less important. An experimental attempt has therefore been made to compare the research effort of the principal firms by using the numbers of patents they have taken out for plastic materials at various stages in the growth of the industry[2] (tables 10 and 11).

The outstanding feature of this analysis, as of the analysis of research expenditure, is the dominant position of the German chemical combine I.G. Farben in the period before and during the Second World War. From 1930 to 1945 this one firm accounted for nearly a third of all the patents taken out by the 30 leading firms, and for over 17 per cent of the patents

[1]An exact correspondence should not in any case be expected because of some variations in the propensity to patent between firms and industries and because patent statistics are analysed on a product field basis while research expenditure statistics are on a company classification.

[2]For the analysis made here the main source has been a French publication[22] which systematically lists all the patents delivered for plastic materials from 1791-1955, dating them from the year of their acceptance. Since there is often a delay of one to four years between application, acceptance and publication, this covers most patents published up to 1959. This French source covers American, German, French and British patents (without double counting) and classifies them into various groups (Appendix table 27). It can be assumed that almost all patents of any significance would be published in one (and in many cases all) of these four countries. Unfortunately it has not been possible to analyse the patents according to the number of years which they have remained in force.

In order to make some comparison with the most recent period an analysis has also been made of patents taken out in Britain in the field of plastic materials up to 1962. This could be expected to have some bias towards British firms, since there will be a number of patents taken out by them in London but not elsewhere.

Table 10. The share of some leading firms in patents delivered for plastics materials

Percentage of total for 30 leading firms[a]

Country	Firm	1791–1930	1931–1945	1946–1955
United States	Du Pont	8.4	13.0	17.6
	Monsanto	0.1	1.5	7.8
	American Cyanamid	0.7	2.4	7.4
	Dow	0.4	4.7	5.2
	B.F. Goodrich	0.3	1.3	4.4
	U.S. Rubber	—	—	4.3
	Rohm & Haas	0.3	0.8	4.3
	Eastman Kodak	18.3	4.9	3.9
	Standard Oil	0.4	1.7	3.4
	Celanese Corp.	0.8	2.7	2.0
	Union Carbide (Carbide & Carbon)	1.6	3.6	1.9
	Hercules Powder	1.4	5.3	1.8
	Phillips Petroleum	—	0.1	1.5
	GEC	4.1	5.2	1.2
	Total	36.8	47.2	66.7
West Germany	I.G. Farben	37.4	36.0	0.7
	BASF	4.1		3.2
	Bayer	6.0		3.1
	Hoechst (Meister, Lucius & Brüning)	6.0		1.6
	Ruhrchemie	0.1		
	Hüls	—	—	1.1
	I.G. Farben and successors	53.6	36.0	9.7
	Wacker	—	2.6	0.7
	Röhm & Haas	0.9	4.9	—
	Total	54.5	43.5	10.4
United Kingdom	ICI	2.7	3.6	7.0
	Distillers	—	0.8	2.0
	Total	2.7	4.4	9.0
France	St. Gobain	—	0.4	2.1
	Rhone Poulenc	1.4	1.1	1.2
	Total	1.4	1.5	3.3
Italy	Montecatini	—	0.1	0.1
Switzerland	CIBA	4.5	2.3	2.8
International	Shell/N.V. de Bataafsche	0.1	1.0	7.8
	Total of above	100.0	100.0	100.0

Source : J. Delorme : *Anthologie des Brevets sur les Matières Plastiques*, Paris 1962. For the spelling-out of the abbreviations, see Appendix I, page 51.
(a) These are thirty of the leading firms in 1946-55 in patents taken out in UK, USA, France and Germany : see table 11.

from all sources. At this time the most important new developments were taking place in the field of thermoplastics, especially the vinyl resins. In this group I.G. Farben were alone responsible for a quarter of all the patents taken out in the world. But they were also the leading firm in every other group of plastic materials.

American and German firms dominated the picture in this period, being responsible between them for 90 per cent of the patents taken out. Amongst other firms, only two British firms (ICI and British Celanese) were in the first 30, only two French (Rhone Poulenc and Cie Fr. Thomson-Houston) and one Swiss (CIBA). The remaining 25 were all American and

Table 11. Patents in plastics materials taken out by leading firms

30 leading firms' patents taken out in UK, USA, France, Germany						30 leading firms' patents taken out in UK only(c)			
1791-1930	No.	1931-1945	No.	1946-1955	No.	1954-1958	No.	1959-1962	No.
1. I.G. Farben	346	1. I.G. Farben	889	1. Du Pont	637	1. ICI	299	1. ICI	485
2. Eastman Kodak	169	2. Du Pont	321	2. Monsanto	283	2. Du Pont	288	2. Du Pont	428
3. Du Pont	78	3. Röhm & Haas(b)	145	3. American Cyanamid	266	3. Standard Oil/Esso	243	3. F. Bayer*	346
4. Ceulluloid Mfg. Co.	66	4. Hercules Powder	132	4. Shell/N.V. de Bataaf.	263	4. F. Bayer*	199	4. Union Carbide	327
5. Bakelite Corp.	59	5. {GEC Eastman Kodak	120 120	5. ICI	253	5. US Rubber	170	5. Standard Oil/Esso	246
6. {Bayer & Co.* Meister, Lucius & Brüning*	55 55			6. Röhm & Haas(b)	210	6. Midland Silicones	168	6. Hoechst*	207
		7. {Dow Kodak-Pathe & Kodak Co.	115 115	7. Dow	187	7. Monsanto Co.	143	7. Montecatini	205
8. CIBA	42			8. B. F. Goodrich	160	8. GEC	128	8. CIBA	168
9. Bakelite GmbH	40	9. ICI	90	9. US Rubber	156	9. Celanese Corp.	113	9. Dow Chemical Co.	162
10. {BASF* GEC	38 38	10. Carbide & Carbon	88	10. Eastman Kodak	140	10. Courtaulds	109	10. Phillips Petroleum	153
		11. Phrix Arbeitsgemeinschaft	73	11. Standard Oil/Esso	131	11. Shell/N.V. de Bataaf.	108	11. BASF*	145
12. Br. Thomson-Houston	35	12. Celanese Corp.	67	12. BASF*	115	12. Union Carbide	107	12. Röhm & Haas Co.	132
13. Consortium für Elek.(a)	33	13. A. Wacker Ges.	65	13. F. Bayer*	111	13. BASF*	106	13. Shell/N.V. de Bataaf.	118
14. Br. Celanese	28	14. American Cyanamid	60	14. CIBA	101	14. Dow Chemical Co.	100	14. American Cyanamid	112
15. Chem. Fab. Albert	26	15. CIBA	56	15. St. Gobain	77	15. Röhm & Haas Co.	87	15. US Rubber	110
16. {Barrett Co. Ellis-Foster ICI	25 25 25	16. Ellis-Foster	51	16. Distillers	74	16. {American Cyanamid Dow Corning	76 76	16. Midland Silicones	109
		17. Bakelite Corp.	48	17. {Gen. Aniline & Film Co. Celanese Corp.	73 73			17. Monsanto Co.	100
		18. Deutsche Hydrierwerke	46			18. B. F. Goodrich	75	18. GEC	92
19. {Cie. Fr. Thomson-Houston Naugatuck	24 24	19. Pittsburg Plate Gl.	44	19. Wingfoot	72	19. Hoechst*	68	19. B. F. Goodrich	83
		20. Br. Celanese	42	20. {Cie. Fr. Thomson-Houston Carbide & Carbon	69 69	20. {Hercules Powder Distillers	64 64	20. Dunlop Rubber	79
21. {Kroll & Co. Canadian Electric Pathe et Cie	23 23 23	21. Standard Oil	41					21. Minnesota Mining	77
		22. {Bakelite Ges. Cie Fr. Thomson-Houston	40 40	22. Hercules Powder	65	22. Dunlop Rubber	60	22. Courtaulds	73
				23. Hoechst*	57	23. Phillips Petroleum	57	23. Celanese Corp.	63
24. {A.G. für A.F. Kunstharz Pollak	21 21	24. Monsanto	37	24. Phillips Petroleum	55	24. CIBA	56	24. Hercules Powder	60
		25. Deutsche Kelluloid	35	25. Kodak-Pathe	47	25. Wingfoot	38	25. Distillers	53
26. Chem. Fab. Griesheim	16	26. B. F. Goodrich	32	26. American Viscose	45	26. Minnesota Mining	28	26. Chemstrand	51
27. {Carbide & Carbon PF Instruments	15 15	27. Thuringische Zellwolle	30	27. Chemstrand	44	27. {Chemstrand Rhone-Poulenc Wacker-Chemie	23 23 23	27. Rhone-Poulenc	47
		28. Rhone-Poulenc	26	28. {Rhone-Poulenc GEC	43 43			28. Chem. Werke Hüls*	42
29. E. Schering	14	29. {Harvel Research Consortium für Elek.(a)	24 24					29. Wacker-Chemie	41
30. {Hercules Powder Soc. Chem. des usines du Rhone	13 13			30. {Chem. Werke Hüls* Koppers Co.	42 42	30. Gen. Aniline & Film Corp.	20	30. Dow Corning	34

Sources: Delorme, *Anthologie des Brevets sur les Matières Plastiques ; Abridgements of Specifications*, Patent Office, London. For the spelling-out of the abbreviations, see Appendix I, page 51.

* — Part of I.G. Farben.

(a) Undertaking research in association with A. Wacker.

(b) The German and American parts of Röhm and Haas are listed together here. For breakdown see Appendix table 27.

(c) See footnote (2), page 35.

German. In the period before 1930 American and German firms were also responsible for over 80 per cent of the patents taken out by firms. The I.G. Farben combine was only formed in 1925 but by 1930 it had registered more than twice as many patents in this field as any other firm in the world in the whole period from 1791-1930.[1] Moreover its research was already at this time spread over the whole field of materials from cellulosics to the new vinyl and acrylic plastics.

With the exception of Du Pont, the leading American chemical firms came relatively late into this field, although Eastman Kodak (from the film side) and G.E.C. (through their interest in insulating materials) both played an important part in the earlier period and are still amongst the leaders, for example in poly-acetals and poly-carbonates. Röhm and Haas was at this time based primarily in Germany and was principally concerned with the development of acrylic materials. By the 1930s ICI was among the leaders, although still far behind I.G. Farben in the range of its plastics research and production. Over a long period, the Swiss firm CIBA, has been consistently among the leading firms in numbers of patents taken out. But with these exceptions it was not really until the post-war period that British, French and other European firms began to compare with the American and German leaders.

From 1946 to 1952 I.G. Farben was being reorganised by the Allied Military Governments, and was not in a position to take out any patents. Moreover many of its secrets were compulsorily made available to British, French and American firms in 1945 to 1946. It was not until 1952 that the successor firms to the dissolved combine—principally BASF, Bayer, Hoechst, and Hüls—were able to resume normal production and research activity. Consequently the figures for 1946 to 1955 show American firms in a dominant position with 8 of the 10 leading firms, and Du Pont well ahead of the field. However, the combined total of patents taken out by the I.G. Farben successor firms even in this period was greater than that of any firm except Du Pont. By this time British firms had begun to catch up with the Americans and Germans, and the three big British producers—ICI, Shell and Distillers—were all among the first 20 firms in terms of numbers of patents.

The most recent patent statistics suggest a continued recovery by German firms, with all three of the principal I.G. Farben successors in the top twelve. From 1959 to 1962 ICI's position was extremely strong, but the other British producers were still somewhat behind the leading American and German firms and behind the Italian chemical combine, Montecatini, which has risen very rapidly in the 1950s. Again, with the exception of CIBA, other European chemical firms were still well behind and Japanese patents were only just starting to appear.

The patent figures were also analysed in another way, to meet the objection that the simple total of patent numbers might be so distorted by strategic patenting that they did not genuinely represent successful research. With the assistance of a scientific consultant, Dr. C. A. Redfarn, all the patents taken out from 1790-1955 were scrutinised, and the most important ones—those which marked the major technical advances necessary for the successful large-scale production of 30 plastic materials—were selected. These key patents were then entered to the firm and country responsible for them. Out of 117 major technical advances,[2] 51 originated with German firms or individuals (30 from I.G. Farben), 43 with American (12 from Du Pont) and 15 with British (7 from ICI). Only 8 came from all other countries (France, Italy, Switzerland and Netherlands). This suggests that if the total number of patents could be weighted by quality of achievement, the result would not differ very much from the analysis of unweighted patent numbers, provided the sample is sufficiently large and the period sufficiently long. The much larger numbers of patents taken out by American firms in the post-war period would not show up fully in the analysis of key patents, as it is still too early to assess the importance of some of them.

An analysis was also made of the first mention of these materials in scientific literature, as they were often known on a laboratory scale before they could be successfully produced commercially. For example, polystyrene is first mentioned in 1831 and vinyl chloride in 1835 ; Professor Kipping devoted a life's work to basic research on organic silicon compounds before the silicone resins were produced commercially. In these early mentions, which reflect basic research findings rather than applied research and development work, the names of British and French scientists occur more frequently. This gives some substance to the view that in these countries a disproportionately large part of the scientific effort was devoted to fundamental research—or perhaps more correctly that too small an effort was directed to the commercial development of promising fundamental work. The social barriers between the academic world and industry may also have been lower in Germany and the United States than in France and Britain.[23]

[1]Already during the First World War, before the formation of I.G. Farben, there had been a considerable amount of research in Germany on synthetic materials.

[2]An 'advance' may be marked by one or several related patents.

Principal innovations

A third way of measuring technical capability is to compare the dates at which the principal new inventions in the industry were introduced in the various countries. The main problem here, as in listing key patents, is to distinguish the innovations and assess their relative importance. The most obvious milestones in the history of the plastics industry are the dates at which commercial production of the various new materials began. A country with a strong independent research tradition would frequently be the first to produce new materials, whereas one lacking in research capability would often be ten or more years behind the leaders, since it depended upon licensing and know-how agreements. Hufbauer has prepared a list of over thirty plastic materials and dated the first commercial production of these materials in a number of countries[24] (table 12). His study shows clearly the German and American lead. Each of these countries has been responsible for the first commercial production of fourteen materials ; Britain has been responsible for two (high pressure polyethylene and urea-formaldehyde) and France (cellophane), Italy (polypropylene) and Switzerland (epoxy resins) for one each. A technical ' imitation lag ' can be calculated for each country, based on the number of years between first commercial production anywhere in the world, and first commercial production in that country.(1) On this basis, Hufbauer estimates that the German and American weighted average imitation lag was two to three years in 1939, 1950 and 1960.[25] This includes the products which they were the first to manufacture, for which the lag was of course nil. The British lag was 5 to 7 years and the French lag a little longer. In 1950 and in 1960 the Japanese lag is 12 to 13 years, and the Italian 14 to 15 years. Most other countries, except Canada, Sweden and Switzerland, show a lag of more than 20 years both in 1950 and 1960.(2)

This brings out once again the key role of a few large firms in this industry. Firms such as Du Pont, I.G. Farben and ICI, when they were not the world's first producers of a new material, were frequently the first imitators or producers via a new process. This is possible because research in all technically advanced countries is proceeding to some extent on similar lines, so that when a major new discovery is made in one country, research teams in other large firms are much more quickly able to assimilate and imitate it. Frequently the obstacle to imitation may be a difficult patent situation rather than lack of technical know-how. Thus during the war I.G. Farben were able independently to launch production of polyethylene (ICI discovery) and of nylon (Du Pont), while ICI was able to launch production of PVC in 1940. More recently several American firms were able to start production of polypropylene, challenging the Montecatini patent position within a few years of the first discovery. The larger firms sometimes makes agreements with one another for the exchange of know-how, and a firm which has something to offer in the way of its own research and development may be able to obtain more favourable terms. Thus Distillers were able to enter into mutually advantageous arrangements with American firms such as B. F. Goodrich (for PVC), Dow (for polystyrene) and later with Union Carbide (for the whole range of plastic materials). GEC (US) and Bayer cross-licensed each other in carbonates.

German firms were not in a position to launch innovations in the early years after the war. But since the early nineteen-fifties, the successor firms to I.G. Farben have again become prominent. For example, German firms have led in the introduction of low pressure (linear) polyethylene so that, although ICI was the world leader in high pressure polyethylene, West German consumption of polyethylene (taking both types together) is now as high as that of the UK, and her per capita production of low pressure polyethylene is the highest in the world.

But although an independent research and development effort is a good basis for speedy imitation, it is not the only way. Although British firms have been among the leaders in research and development since the war, they appear to have been rather rather slower to imitate some important new advances—for example, the production of high impact polystyrene, of acrylonitrile/butadiene/styrene (ABS) and of polycarbonates. More than half of total polystyrene consumption now consists of hardened grades of material. These new types of polystyrene, based on the incorporation of synthetic rubbers, were developed first of all by I.G. Farben during the war in Germany, and after the war by various American companies. The very strong material ABS was first produced in the United States soon after the war and by Bayer in Germany in 1955, but not until 1962 in Britain when Distillers started production, followed a year later by two other firms. Similarly polycarbonates and acetal resins have not yet come into production in Britain although they may do so in 1964 (ICI and Celanese). These materials have great potentialities in engineering applications

(1)The materials are weighted by their relative importance in world trade.

(2)It may be argued that the date of first polystyrene production in Britain and France should be set somewhat earlier as there was a small war-time production by Distillers in England which was discontinued in 1945 and not re-started until 1950 (the date which Hufbauer gives). Rhone-Poulenc had a similar small-scale production in France. There are other minor alterations which might be made to the dates, as it is difficult to draw the borderline between experimental and commercial production. But changing a date here or there by a year or two would not significantly modify the main outlines of the picture which emerges from his analysis.

Table 12. First commercial production of various plastic materials

I. Family group 1. Product (a) Form	United States	Germany (West Germany after 1945)	United Kingdom	France	Italy	Japan
I. Cellulosic plastics						
1. Cellulose Nitrate :						
(a) Celluloid	1870 Albany Dental Plate	1878 Rheinische Gummi & Zelluloid	1877 British Xylonite	1875 Cie Franco-Americaine	1924 SIC	1908 Nippon Celluloid & Artificial Silk
(b) Photographic film base	1884 Eastman Kodak		1932 ? BX Plastics	?	1920 Soc. Italiana Polvere & Esplosivi	1927 Dai Nippon Celluloid
(c) Low viscosity lacquer	1923 Du Pont	1923? many firms	1923? Nobel Inds.	?	1924 SIC	1928 Dai Nippon Celluloid
2. Cellulose Acetate :						
(a) Photographic film base	1908 Celluloid Corp.	1905 Bayer	1938 ? BX Plastics	1928 Soc. Chimique Usines du Rhone	1954 SIC	1953 Dai Nippon Celluloid
(b) Lacquer, glue, or air-plane dope	1917 Eastman Kodak	1905 Bayer	1916 Br. Celanese	1912 Blanchisserie & Teinture		1927 Dai Nippon Celluloid
(c) Thermoplastic ..	1927 Celanese of America	1925 Kalle	1923 Br. Xylonite	1922 Soc. Chimique Usines du Rhone	1936 SIC	192 7Dai Nippon Celluloid
3. Cellulose Acetate Buty-rate	1931 Eastman Kodak	1935 ? I. G. Farben (Bayer)				
4. Ethyl Cellulose ..	1935 Hercules Powder	1932 I. G. Farben			1939 Fab. Italiana Prodotti Speciali	
5. Methyl Cellulose ..	1939 Dow	1932 Kalle	1961 ICI			1957 Matsumoto Oil & Fat
6. Cellophane	1924 Du Pont	1925 Kalle	1930 Br. Cellophane	1917 La Cellophane	1946 SIC	1929 Tokyo Cellophane
II. Thermoset plastics						
7. Galalith	1919 Aladdinite	1899 Vereinigten Gummiwaren	1912 Erinoid	1900 Fr. Pellerin & Orosdi	1921 La Societa Polenghi	1927 Dai Nippon Celluloid
8. Phenol-formaldehyde :						
(a) Moulding powder ..	1909 General Bakelite	1910 F. Raschig ; Rutgerswerke	1916 Damard Lacquer	1919 Fr. du Ferodo	1922 Soc. Italiana Bakelite	1923 Nippon Bakel
(b) Adhesive or coating..	1909 Gen. Bakelite	1910 F. Raschig ; Rutgerswerke	1910 Damard Lacquer	1916 Fr. du Ferodo	1922 Soc. Italiana Bakelite	1923 Nippon Bakel
(c) Cast phenolic ..	1920 Gen. Bakelite	?	1919 Damard Lacquer	1916 Fr. du Ferodo		
9. Urea-formaldehyde :						
(a) Moulding powder ..	1929 American Cyanamid	1929 I. G. Farben (Dynamit AG)	1928 Br. Cyanides	1930 Kuhlmann	1936 Montecatini	1935 Keikoku Chemical
(b) Adhesive or coating..	1936 American Cyanamid	1932 I. G. Farben (BASF)	1934 Br. Cyanides	1936 ? Kuhlmann	1942 Montecatini	?
10. Melamine-formaldehyde:						
(a) Moulding powder ..	1939 American Cyanamid	1938 I. G. Farben (Dynamit AG)	1946 Br. Industrial Plastics	1955 Resines & Vernis Artificiels	1951 Montecatini	1951 DenkiKagaku ; Nippon Carbide
(b) Adhesive or coating..	1939 American Cyanamid	1935 Henkel	1938 Br. Industrial Plastics		1951 Montecatini	?
11. Alkyd	1926 General Electric	1927 ? I. G. Farben (Bayer)	1929 Nobel Chemical Finishes	1928 Kuhlmann	1927 Montecatini	1931 Kansai Paint
12. Polyester	1942 Pittsburgh Plate Glass	1953 Reichhold Chemie	1950 Catalin ; Bakelite	1950 Nobel-Bozel	1949 Montecatini	1953 Japan Cataly Riken Synthetic
13. Silicone	1941 General Electric	1950 Wacker	1952 Midland Silicones	1954 Rhone-Poulenc ; St. Gobain		1951 Shin Etsu
14. Epoxy	1947 Devoe & Reynolds	1955 Reichhold Chemie	1955 Bakelite		1958 ? Soc. Italiana Resina	

I. Family group 1. Product (a) Form	United States	Germany (West Germany after 1945)	United Kingdom	France	Italy	Japan
III. Thermoplastic plastics						
15. Polyvinyl acetate ..	1928 Union Carbide	1928 Wacker	1949 Dunlop Rubber	1937 Rhone-Poulenc	1954 Edison Settore Chimico	1936 Shin Nippon Chisso Hiryo
16. PVC	1933 Union Carbide	1931 I. G. Farben (BASF)	1940 ICI	1940 St. Gobain	1951 Montecatini	1939 Shin Nippon Chisso Hiryo
17. Polyvinyl alcohol ..	1938 Du Pont	1928 Wacker		1938 Nobel Fr.	1958 Edison Settore Chimico	1949 Nippon Synthetic Chemical
18. Polyvinyl butyral ..	1937 Shawinigan Resins	1948 Hoechst		1942 Rhone-Poulenc		1944 Shin Nippon Chisso Hiryo
19. Saran	1940 Dow			1961 Plastichimie		1950 Kureha Chemical
20. Polystyrene	1933 Naugatuck Chemical	1930 I. G. Farben (BASF)	1950 Distillers(a) Monsanto	1951 Pechiney(a)	1942 Montecatini	1957 Asahi-Dow ; Mitsubishi-Monsanto
21. Polystyrene/styrene-butadiene	1947 Dow	1942 I. G. Farben (Bayer)	1954 British Xylonite	1954 Monsanto	1954 Montecatini	
22. Polystyrene/styrene/acrylonitrile	1948 Rohm & Haas	1942 I. G. Farben (Bayer)		1962 Monsanto	1962 Sicedison	
23. Acrylonitrile-butadiene-styrene	1946 Naugatuck Chemical	1955 Bayer	1962 Distillers	1960 Kuhlmann		1963 Mitsubishi-Monsanto
24. High pressure polyethylene	1941 Du Pont	1944 I. G. Farben (BASF)	1937 ICI	1954 Ethylene Plastique	1952 Montecatini	1954 Sumitomo Chemical
25. Linear polyethylene ..	1956 Phillips Petroleum	1955 Hoechst	1959 Hydrocarbon Chemicals	1956 Rhone-Poulenc	1954 Montecatini	1958 Mitsui Chemical
26. Polypropylene	1957 Hercules Powder	1957 Hoechst	1959 ICI	1960 Pechiney ; Societe-Normande	1957 Montecatini	1961 Tokuyama Soda
27. Polymethyl methacrylate:						
(a) Adhesive or emulsion	1931 Rohm & Haas	1927 Rohm & Haas	1932 ICI	1937 Nobel Francaise	1957 Montecatini	1962 Mitsubishi Rayon
(b) Thermoplastic ..	1936 Rohm & Haas	1930 I. G. Farben (Bayer)	1933 ICI	1938 Nobel Fr.	1937 Italiana Plexiglas	1938 Asahi Glass ; Mitsubishi Rayon
28. Nylon	1941 Du Pont	1943 I. G. Farben (BASF)	1950 ICI	1943 Rhodiaceta	1946 ? Rhodiaceta	1953 Toyo Rayon
29. PTFE	1943 Du Pont ; M. W. Kellogg	1958 Hoechst	1945 ICI	1958 Soc. Resines Fluorees	1955 Montecatini	1963 Nitto Fluoro Chemicals
30. Acetal	1953 Du Pont					1963 Nippon Shokubai
31. Polycarbonate	1957 General Electric	1957 Bayer				1959 Kunoshima Chemical

Source : G. Hufbauer, *Synthetic Materials ; a study in international trade* (with minor amendments in agreement with the author). For the spelling-out of the abbreviations, see Appendix I, page 50.

(a) See footnote (2), page 39.

because of their hardness, good temperature range and resistance to creep. However ICI was well ahead of other European firms in the production of poly-tetrafluorethylene (PTFE). Japan, which was a long way behind Britain, France and Germany in the introduction of the older plastic materials, was ahead of both Britain and France in acetals and polycarbonates (table 12). Until recently Japan had a much smaller research effort than some European countries and did not innovate any of the principal materials, but she has been very successful recently in two of the various methods of imitation : the purchase of technical know-how from foreign firms and production by foreign-owned subsidiaries. Almost all the plastic materials were first produced in Japan under licence or by American subsidiaries. Thus although the evidence of key innovations confirms the German and American technical leadership suggested by the research and patent data, it also shows the importance of successful imitation by all available methods. In Japanese manufacturing industry as a whole, expenditure on royalties was rather less than half as big as research expenditure ; but in petrochemicals more was spent on royalties than on research.[26]

Plastics machinery

German pre-eminence in materials seems to have been accompanied by a similar lead in machinery and processing techniques. The methods of fabrication were at first based mainly on imitating the techniques used for metal-working and for processing rubber and ceramics, but it became increasingly necessary to develop new specialised machines. The most important of these are injection moulding machines. The first injection moulding machines suitable for large-scale production of thermoplastic articles were made by Eckert und Ziegler in Germany in 1926, and this firm together with F. Braun (Isoma) remained the principal European producers of plastics machinery until the war.[27] Eckert und Ziegler were owned by I.G. Farben who encouraged their development work in plastics machinery, and also undertook some research work in their own laboratories. Perhaps the most important single advance in design of machinery was made by a British firm, R. H. Windsor who in 1953 produced and marketed a twin-screw injection moulding machine. (Before the war, Windsor had acted as importers of Eckert und Ziegler machines). The use of the screw principle for plasticising in injection moulding machines was patented by Beck of I.G. Farben in 1943 and about the same time in France. But in Germany this patent was not actually followed up by the development of a prototype or a machine until 1956, when Ankerwerk, Gebr. Goller made the first single screw in-line injection moulding machine in collaboration with Beck, who was still working in the research laboratories of BASF. As a result of Windsor's work, that of BIP Engineering in compression moulding machines, and others, British firms were able to overtake Germany's technical lead to some extent after the war. In the immediate postwar period American firms also made important advances, but failed to recognise the importance of screw plasticising for injection moulding. This provides an interesting example of 'technological gap' trade[28] between advanced countries, as the technical lead of European producers enabled them to increase their exports to the United States and also to license American producers of screw plasticising machines.[29] United States imports of moulding machinery from West Germany increased from $1.2 million in 1960 to $4.9 million in 1962 ; from Britain they rose from $0.2 million to $0.5 million. In spite of the technical successes of a few outstanding firms, the British industry as a whole still lags a little behind the German. There appear to be only three British producers spending as much as 3 per cent of their turnover on research and development, and many of the others are producing under licence. In Germany there are half a dozen or more producers with a strong independent research and technical effort, and spending 3-5 per cent of turnover on research and development. But the largest producer, Battenfeld, has concentrated on simple, low-priced machines.

The evidence suggests that the technical leadership of Germany, United States and Britain was even more pronounced in plastics machinery than in plastic materials. These three countries account for 90 per cent of world exports (table 13) of rubber and plastics machinery, and Germany is particularly strong in plastics machinery. Japanese production of plastic machinery, as of plastic materials, is still largely under licence from American, German and British firms[30] and she is still a net importer.

Advances in machine design make possible new applications of plastic materials. For example, the present large-scale use of rigid PVC for rainwater systems in several European countries has been made possible because screw injection moulding machines are now available which can fabricate satisfactory joints on a large scale. Another example is the collaboration of ICI, Windsor and Rollex which enabled acrylic polymers to be used on a large scale for television screens and for natural skylighting.[31] These advances are, of course, not necessarily limited to the country producing the machines. For example, PVC injection moulded telephones are now being produced in Japan on the basis of development work by Ankerwerk (Germany) in co-operation with Japanese firms.[32] But generally speaking a technically advanced

Table 13. Exports of rubber and plastics machinery

$ million

	1954	1955	1956	1957	1958	1959	1960	1961	1962
United States	21.3	27.1	31.0	37.2	30.8	34.2	47.7	47.2	55.0
West Germany	7.9	11.2	15.7	20.5	23.9	30.5	39.3	67.1	66.7
United Kingdom	7.5	7.1	7.1	9.1	16.4	31.2	31.3	38.4	34.5
France[(a)]	1.4	2.0	0.8	1.5	2.6	5.6	6.6	7.9	8.5
Italy[(b)]	..	..	..	..	..	..	..	..	4.5
Total of above	38.1	47.4	54.6	68.3	73.7	101.5	124.9	160.6	169.2

Source : National Trade Statistics.

(a) Includes machinery for soap, stearin, perfume and pharmaceutical products.
(b) Plastics machinery only.

and progressive machine industry will principally benefit the country in which it is located, because material-suppliers, machine-makers and fabricators can more easily co-operate there in experiment, development and design. This tripartite co-operation appears to be closer and more satisfactory in Germany than in Britain.[33]

Technical progress : the consequences for production

Summing up the evidence from all three measures of technical progress—research expenditure, patents and innovations—it seems that in the years up to 1945 Germany, or one might say I.G. Farben, was well in the lead. Before the Second World War, only the United States was a serious rival ; after the war she was able to draw ahead, when German industry was dislocated and disorganised. But in the last five or six years the successor firms to I.G. Farben in Germany have again been prominent in patents and innovations. Britain's research effort in plastics has been very much greater since the war than before it.[(1)] The French and Italian research effort in plastics seems to come some way behind that of Germany, the United States and Britain.

In the analysis of the effect this research and innovation had on production and trade, it is useful to distinguish three phases : the research-development phase, the technical-commercial phase, and the open competition phase. During the first phase, which lasts typically between five and ten years, a new product is brought to the stage of successful pilot plant production. Most of the development work is done by specialised research and development teams, and typically towards the end of this phase, patents are taken out.[(2)] During the second phase, which lasts from 15 to 25 years, large-scale commercial production is launched ; research and development work continues, but its trend shifts to applications and process research. In this phase the innovating country has a decisive advantage, and is likely to be the leading country in per capita production and exports and often also in consumption. It has a head-start over its rivals and its lead in technical know-how is buttressed for at least 15 years by patents and commercial secrecy. Production may be licensed to other countries, but this is usually done on a limited scale only after a number of years.[34] Licensing is restricted in the early period so that prices may be held high enough to recoup development costs. Furthermore, the arrangements usually carry restrictive export provisions.

Acetal resins are an example of materials in this early phase. Although first produced by Du Pont in 1953, they were not licensed for production in Japan until ten years later, and are still not produced in Britain. More serious competition may develop at an early stage if several countries have simultaneously been doing the research and development necessary for a new product.

Soon after patents expire, the third phase begins. Imitation is easier and technical know-how begins to spread more quickly. The number of producers increases, attracted by the high profits, and prices begin to fall more sharply. But even now, 20 or 25 years after successful pilot plant production, the innovating country or countries still normally enjoy some major advantages in accumulated production experience and technical know-how. Thus, for example, 25 years after first producing high pressure polyethylene, ICI was still able to make arrangements for the sale of technical know-how to an industrially and scientifically developed country such as the USSR, as well as to many other countries.[35] With the exception of the United States, Britain still has the highest per capita production and exports of high

(1)The Annual Report of ICI for 1962, commenting on the firm's research expenditure, says ' The largest single component of this total was work associated with the invention, development, manufacture and use of organic polymers.'

(2)Some patents may of course be taken out at any stage of the project.

pressure polyethylene in the world.[36](1)

However, in this third phase of open competition, directly technical factors will become less important, and other factors—such as material costs—will matter much more. Thus, for example, some 30 years after the United States and West Germany first produced PVC, Italy and Japan are now overtaking these countries in per capita production and consumption : for they have cost of production advantages and are able to quote lower prices.[37](2) So they are able to challenge the innovators in export markets—or indeed in the innovators' home market : for example Italian sales of plastics in West Germany.

But even in this final phase the innovators will still benefit to some extent from their long experience, from their accumulated knowledge and from their still-continuing research and development. Other producers may overtake them in production costs and prices for standard grades, but they are able to introduce new and improved qualities in the old materials and in this way open up new markets and retain old ones. This is happening with all the three major thermoplastic materials—PVC, polystyrene, and polyethylene. The old producers are also able to some extent to offset other cost disadvantages by continuing technical progress in processing and sometimes also by economies of scale. Germany and the United States are still leaders in per capita production and consumption of condensation products and cellulosics (table 3), groups which contain a high proportion of materials now 40 years old or more.

The pattern of consumption

Technical progress, then, seems to throw a good deal of light on the country pattern of production. It cannot, of course, in the same way explain the country pattern of consumption. To take a straightforward example, Sweden, in 1961, had no production of polyolefins, and consequently no technical progress in it ; but her consumption per head was higher than in either Britain or West Germany (table 3). The level of consumption in a country will depend, amongst other things, on the level of national income, and possibly also on its rate of growth ; on differences in industry structure, since some industries use more plastics than others ; on whether user industries are themselves progressive and research minded ; on the level of tariffs ; and also on the extent to which producers of plastics are energetic in promoting its use.

Plastics consumption is in fact quite high in a number of countries which have little production of their own, but which have high incomes and liberal import policies. This is the position in Switzerland, Belgium and the Netherlands, as well as in Sweden ; all these countries have benefited from the export efforts of innovating countries, and indeed from the low prices of dumped materials. Importing countries like Australia, on the other hand, which protect their infant industries with fairly high tariffs, may take longer to reach high levels of consumption in spite of their high income levels.

Before assessing the effect of the structure of industry, it is useful to divide plastics consumption into two groups : consumption by fabricators of plastic products,(3) and consumption by other users. In the first group, which accounts for 60-80 per cent of consumption in the leading countries, packaging, construction and electrical engineering are the big users. In the second group, plastics are mainly used in the manufactures of glues and adhesives, paints, coatings, and auxiliaries used in the textile industries (table 14).

The electrical engineering industry and the paint industry are much bigger in Germany than in France or Italy : so this would explain part of the difference in per capita consumption. But in Germany and Britain the two industries were about the same size in 1961. The only significant structural differences between these two countries is that Germany has a bigger construction industry (both absolutely and relatively) and, partly as an indirect consequence of this, a bigger glue and adhesives industry.

Roughly speaking, German glue and adhesives industry consumes nearly 3½ times as much plastics materials as the British : the difference is about 120 thousand tons. Its output is about 2½ times as big. So, of this higher consumption of 120 thousand tons, about 75 thousand may be explained by structure, and 45 thousand by the fact that glue in Germany is made from plastics rather than from other materials. This higher utilisation is largely due to the technical progress of the plastics industry. I.G. Farben pioneered the urea-formaldehyde syrups and polyvinyl acetate adhesives which today contribute a large part of its total output.

But even the apparent ' structural ' difference is in a way misleading. A good part of it can be explained by the German chipboard industry, which consumes

(1)There are some special reasons for the high United States production of polyethylene. Under a special war-time agreement, ICI transferred all its technical know-how to two American companies, Du Pont and Union Carbide, so that they could launch large-scale production for allied military requirements. Furthermore, under the Ryan judgment arising from the United States Government anti-trust action in 1952, high pressure polyethylene was compulsorily licensed to half a dozen other American companies, and these were given the right to export to those countries in which ICI's basic patent was still in force.

(2)See Appendix table 25, page 59, for lower Japanese and Italian prices of PVC polymer.

(3)This includes cables.

Table 14. The industrial pattern of plastics consumption, 1961

	Thousand metric tons			As per cent of total consumption[(a)]		
	UK	Germany	Italy	UK	Germany	Italy
Plastics fabricating						
Packaging	86	75	48	17	9	14
Electrical engineering[(b)]	87	107	37	17	13	11
Domestic ware and other consumer goods	74	108	67	15	13	20
Construction, including flooring and interior fittings	60	154	64	12	19	19
Engineering, vehicles, instruments, belting and miscellaneous industrial	48	70	33	10	8	10
Total, plastics fabricating	**355**	**522**	**250**	**70**	**62**	**75**
Other plastics consumption (principally chemical industry)						
Paint	82	106	..	16	13	..
Adhesives, glue	50	168	..	10	20	..
Other	20	41	..	4	5	..
Total, other uses	**152**	**315**	**84**	**30**	**38**	**25**
Total, all uses	**507**	**829**	**334**	**100**	**100**	**100**

Source : NIESR, based on trade estimates ; *Chemische Industrie*, October 1961 ; Saechtling, *Werkstoffe aus Menschenhand* ; and Gesamtverband Kunststoffverarbeitende Industrie Reports and Statistics ; Instituto Italiano per gli studi sui consumi, *Consumi e impieghi di materie plastiche.*

(a) In terms of weight. In terms of value the share of construction would be much lower and that of engineering higher.

(b) Including cables, etc.

an enormous volume of glue.[(1)] Its output in 1961 was just over 1 million cubic metres, compared to 85 thousand cubic metres in Britain. The chipboard industry was the result of an intensive effort during and after the war, encouraged by the manufacturers of plastics materials, to utilise waste wood commercially.

So perhaps a better comparison is to consider the glue and adhesives industry as an input of the construction industry (which it is, to a large extent), and to compare Britain and Germany in this way. The German construction industry then—including the plastics content of glue—consumed over 300 thousand metric tons of plastics in 1961, compared to just over 100 thousand metric tons in Britain : and, at most, the construction industry in Germany is only one third greater than the British industry. On this way of assessing the effect of structure, it only explains some 30 thousand metric tons, of the difference of 200 thousand tons between the two countries.

Most of the difference, then, between the two countries can be put down to a higher utilisation of plastics within German industries. This was already well advanced before and during the War. Among the numerous applications which were already developed in Germany before 1945 were vinyl flooring, oriented polystyrene film, PVC film, PVC covered wires and cables, rainwear, a variety of PVC pipes and fittings for the chemical industry, vinyl coated bags, PVC shoe soles, hospital sheeting, curtains, food containers, insulating foams of PVC, polystyrene, and urea formaldehyde and magnetophone tapes.[38] In construction, also, plastics were and are much more widely used in roofing, panelling, thermal insulation, and interior fittings of all kinds.

(1)This explains the very high German consumption of urea-formaldehyde resins (table 3).

Producers, fabricators and users

The extent to which plastics are used in any industry—the utilisation factor—will depend on the total research and development effort put in by materials producers, fabricators, and users to extend its use. In most countries, the bulk of such research expenditure, perhaps four-fifths of it, is done by the materials producers themselves ; and, of the rest, the user industries seem to do more than the fabricators, whose share is very small.

Most of the fabricators are small and medium-sized firms who have no facilities for research and development. There were 789 fabricating firms in Britain in 1958, employing on average about 50 people and 1,100 in West Germany in 1960, employing an average of about 75. In both countries there are a few exceptional firms, such as Commercial Plastics, which themselves do research and development into new applications. But the great majority do little or no research. Some of the fabricators are 'trade-moulders', taking on work for a variety of customers as well as selling their own final products and varying their output continuously. Others are producing typically a small range of products for only one or a few customers, and are virtually the plastics departments of large consumers of plastic components. In both Britain and West Germany some of the largest users, especially in the electrical field, mould and extrude their own plastics, for example cable-makers and radio firms. But an electrical engineering firm like Lucas in Britain, which is one of the largest fabricators and users in the country, may also put out work to many other smaller trade-moulders ; and the plastics departments of large user firms may also take on trade moulding for outside customers when they have spare capacity. The fabricating side of the industry resembles the engineering industry, where sub-contracting also provides flexibility.

The large consumers, and those firms which are both fabricators and consumers, contribute more to the technical progress of the industry than the independent fabricators, because they have more facilities. But there is no evidence that such consumers are much more active in this way in West Germany than in Britain or the United States. As between Britain and West Germany, the difference in the use of plastics appears to be least in technically advanced industries such as electrical engineering, cables, paint, automobiles, and packaging, and greatest in relatively backward industries such as construction, domestic ware and furniture (table 14).

Thus, for example, in the car industry, it is true that some West German firms were earlier than the British in introducing new applications for plastics in vehicle manufactures. Volkswagen were already using over 22 lbs. per vehicle in 1959,[39] and now use up to 29 lbs. in over a hundred different parts. But there is probably now not much difference between British and German use in cars. The new BMC 1100 (August 1963) uses 35.6 lbs. of plastics per car, and the much lighter Mini uses 23.4 lbs. (table 15). British manufacturers have been ahead in the use of reinforced plastics for the cabs of commercial vehicles and other applications.[(1)]

Table 15. Weight of plastics in various British Motor Corporation vehicles

lbs.

Material	A.60 December 1961	Mini August 1963	1100 August 1963
PVC in leathercloth ..	13.3	10.5	11.9
PVC unsupported ..	3.5	2.75	5.0
Acrylics	2.0	0.5	1.1
Nylon	} 0.5	0.16	0.23
Polyacetal		0.016	0.016
Polypropylene		0.38	0.5
Polyethylene HD ..		0.33	—
Polyethylene MD ..		0.27	0.35
Styrene	} 0.7	0.27	0.34
Phenolics		1.0	0.95
ABS		0.35	0.38
Cellulose acetate ..		1.1	0.9
Polyurethane foam ..	—	5.75	14.0
Total	26.3	23.35	35.6
Weight of vehicle ..	21.5 cwt.	12.5 cwt.	16 cwt.

Source : K. C. Waldron (British Motor Corporation) 'Plastics in the car industry', paper presented to the Plastics Institute, London Section, October 8, 1963.

The cable industry is another industry where there is probably no great difference between Britain and West Germany in plastics utilisation factors, although here again German manufacturers switched over to plastics earlier. In packaging, Britain is ahead of West Germany in total plastics consumption (table 14). Chain-stores, multiples and other large firms have a larger share of the distribution network in Britain, and some of them have, on their own account, promoted new types of plastics packaging.[40] Also, the largest manufacturers of metal containers in Britain (Metal Box) recognised the potentialities of plastics very early, and undertook large-scale development work. Further, the materials producers themselves have ventured into this field.

The clock and watch industries provide an example of development work done by users. The British firm of Smith's has been able to make important savings by replacing expensive machined metal parts with plastics. Other firms had previously failed to do this, because their development work was inadequate, and they did not redesign components specifically for plastics. Smith's plastics development section succeeded in overcoming these problems, and the even more difficult problem of improving injection moulding techniques to the precise limits required for this industry.[42] When user-industries are technically advanced and alive to the possibilities of new materials and techniques, they may often initiate new applications either independently or in co-operation with material producers, fabricators, and machine-makers.[43]

Plastics in the construction industry

It is in the less technically advanced industries and especially in construction that the German lead is greatest. This is not primarily due to the efforts of the building industry itself, which is backward in

(1)Car manufacturers in all countries need to be alive to the possibilities of economy in materials as these account for 70 per cent of manufacturing costs. Savings of up to 80 per cent have been possible for some components (Appendix table 16) but plastics still account for only about 1 per cent of the weight of a Volkswagen to over 80 per cent iron and steel and 6 per cent non-ferrous metals. Development work which is now in progress in several countries is likely soon to lead to a radical change in these percentages.[41]

research in all countries. It accounted for less than half of one per cent of the total research expenditure by British private industry in 1961,[44] and the proportion is probably no higher in the United States and West Germany.

The German lead appears to be due to a greater experience of plastics over a longer period, and the applications research of plastics producers. No doubt also shortages of traditional materials before, during and after the war made users more willing to accept alterations. The pressure of demand on the German building industry was also very strong during the period of postwar reconstruction.

The potential applications of plastics in building and civil engineering are enormous, and even in West Germany and Japan, which have gone further than most countries, consumption is still relatively small. There are a number of reasons for this. The construction industry is mainly made up of small firms, which cannot afford the type of large-scale development work which is required. The responsibility for design and fabrication are separated—one is with the architect, the other with the contractor ; this tends to inhibit innovation and weaken the incentives to cost reduction through technical change.[45] National standards, and the specifications and regulations laid down by public authorities, all serve to slow down, sometimes to a standstill, the adoption of new materials. For instance, it is taking a very long time to get plastic pipes accepted for water supply in Britain. These institutional barriers to new materials are also important in the United States. There too—although firms like Monsanto produced experimental all-plastic houses several years ago—the building industry does not use plastics on any scale. This may be the biggest single reason for the low United States ratio of plastics consumption to national product, compared with West Germany and Japan.

This backwardness in the use of plastics in building has given rise to concern in several countries, and a number of proposals has been made for reform. These include co-operative research and development facilities for the building industry, and the use of Government contracts to promote innovation.[46] In Japan, building firms contribute to the cost of research carried out by chemical firms, and the Ministry of Construction also subsidises research programmes in building applications of plastic materials.[47] In Sweden building firms also contribute an annual levy to the National Council for Building Research which is co-operating with the Swedish Plastics Federation in the organisation of a joint research programme. In Britain the Royal Institute of British Architects convened a conference in October 1963 to work out a policy for the development of plastics as major building materials : the conference brought together builders, architects, plastics manufacturers and government representatives. The Minister of Works has already proposed a new building research and information association to be financed partly by the Government and partly by the industry.

Other factors

In an analysis of this kind, it is not practicable to discuss all possible factors, or to pursue every matter up the chain of causation to the end. For instance, it could be argued that the business decisions discussed here—to expand research, or to increase investment, or to experiment with new materials—are all management decisions, and that the comparative study of technical progress should be taken further into a comparative study of management. This is a much wider question, and beyond the brief of this article. Equally, we have not gone into questions of differences in taste, and psychological barriers to the adoption of plastics ; it is difficult to find any conclusive evidence on these matters.

Two other questions are perhaps worth raising. First, to what extent has the adoption of these new materials been linked to the rate of economic growth ? There does seem to be some link : it is perhaps significant that the first four countries in rate of economic growth in the last decade—Japan, Italy, West Germany and Austria—also rank in the same order in the intensity of their use of plastics (chart 2) : that is, in the ratio of plastics consumption to gross national product. It might indeed be expected that a faster rate of economic growth would both stimulate using industries to accept new materials, and also influence the investment decision of producers. Conversely, producers who have become accustomed to a rather sluggish rate of growth will naturally be less ambitious with their plans. The President of the European Council of ICI pointed out in 1962 that investment plans in the Common Market countries called for a 60 per cent increase in plastics capacity, whilst those in Britain provided for a 30 per cent increase, and commented : ' In Britain we are naturally conservative and cautious. It is rare to find a situation where substantial over-capacity has been deliberately provided . . . The British petro-chemical industry while planning ahead with reasonable boldness has avoided the problem of unsold surpluses and hence has not had to face the temptation of dumping because of too large plants'.[48]

Another question of some interest is the influence of the central plan in those countries which have one.

[1]One of the most important materials used in construction is rigid PVC. Appendix table 24, page 59, shows that Japan, West Germany and Italy all make much more extensive use of this material than the United States does.

Chart 2. Growth-rates and plastics 'intensity'[(a)]

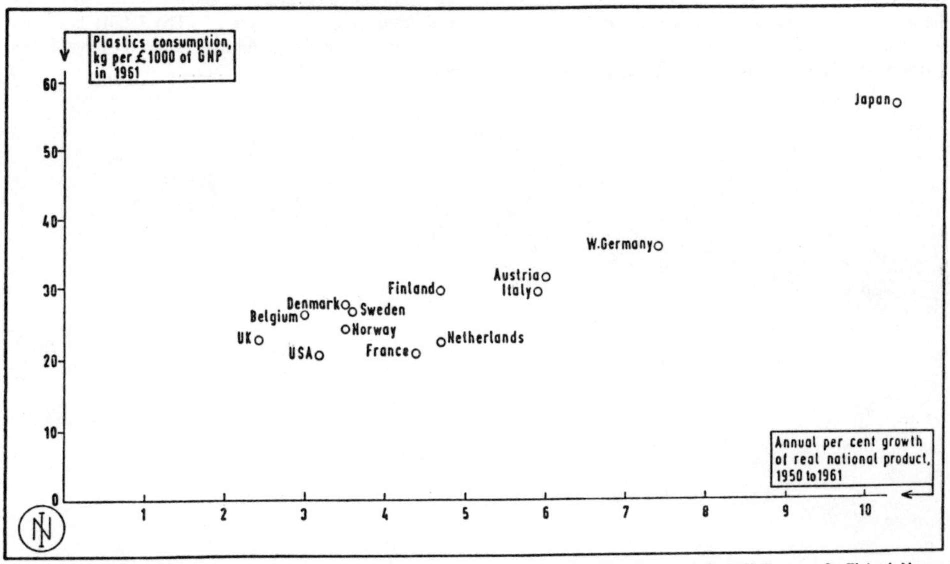

(a) Plastics consumption is for 1961, except for Finland and Denmark, when it is for 1960. Growth-rates are for 1950-61, except for Finland, Norway and Belgium, where they are for 1950-60.

Here, Mr. Khrushchev's criticism of the planning organisations for their backwardness in recognising the importance of plastics is interesting. He said :[(1)][49] 'Had planning and economic organs studied economic problems more profoundly they should have determined when and in what quantities ferrous and non-ferrous metals and other materials should be replaced by synthetic materials, and developed their production in every way. Unfortunately, economic and planning organs do not take sufficient account of the achievements of science and technology, do not use these achievements for the accelerated development of those categories of production and branches which are economically the most advantageous and promising...

'Production of steel is, so to speak, a well-worn road with deep tracks, and here even a blind horse will not swerve because the wheels will break. A material has appeared which is superior to steel and cheaper, but they still shout : Steel ! Steel ! If we had only fulfilled the plan for the smelting of steel but overfulfilled it for polyethylene we would have done better and would be much richer. But this is hard

(1)'At my request the State Planning Committee and the Committee on Chemistry have submitted a memorandum with calculations of the economic efficiency of using plastics in the economy as substitutes for lead, copper, zinc, ferrous metals, fabrics and timber materials. Here is what this memorandum shows :

'In the cable industry 67,000 tons of lead could have been replaced by polyethylene in 1962. Capital investments needed to organise the production of one ton of lead amounting to 1,630 rubles and for one ton of polyethylene, to 1,000 rubles. Each ton of polyethylene replaces three tons of lead. While capital investments totalling 108 million rubles are needed for building up capacities of 67,000 tons of lead, the building up of polyethylene capacities to replace this quantity of lead would require only 23 million rubles, or nearly 80 per cent less.

'In addition to a big saving in capital investments the national economy would also have a big saving by reducing the cost of cables, since the cost of insulating a cable when polyethylene is used is cut by half.

'Pipes also made of polyethylene could be used instead of steel gas pipes and waterpipes, in housing and industrial construction. Capital investments for building up production capacity of 1,000 metres of steel pipes amount to about 1,300 rubles and polyethylene pipes, to about 600 rubles, or nearly 54.6 per cent less.

'Calculations show that in organising the production of 100 million metres of pipes from polyethylene, instead of steel pipes, the saving on capital investments would total 72 million rubles ; moreover the cost of producing pipes from polyethylene would be 30 per cent less than of metal pipes. Flooring from polymeric materials is 30-40 per cent cheaper than wooden flooring.

'In the production of high-voltage transformers the use of one ton of epoxide resin makes it possible to release up to two tons of copper and nine tons of hot-rolled stock ; one ton of polyamide resin replaces about five tons of bronze in the manufacture of sanitary equipment.

'The use of plastics in engineering, and various other industries and construction reduces the weight of articles, and their size, cuts operating outlays and raises labour productivity.'

to do because there are people in the State Planning Committee who stop those who sensibly want to change the steel production targets in favour of synthetic materials.'

Japanese planning on the other hand probably helped the industry's rapid growth in that country as the industry was given a high priority and very ambitious targets were set.[50] The USSR has now also set extremely ambitious revised targets for plastics.

APPENDIX I. SOURCES AND METHODS

A. Definition of plastics materials

Item 599.01 of the Standard International Trade Classification (synthetic plastics materials in blocks, sheet, rods, tubes, powder and other primary forms) is subdivided into the following headings of the Brussels Tariff Nomenclature (the revision agreed on 30th April, 1954):

39.01 :—Condensation, polycondensation and polyaddition products, whether or not linear (for example, phenoplasts, aminoplasts, alkyds, polyallyl esters and other unsaturated polyesters, silicones).

39.02 :—Polymerisation and copolymerisation products (for example, polyethylene, polytetrahaleoethylene, polyisoacetylene, polystyrene, polyvinyl chloride, polyvinyl acetate, polyvinylchloracetate and other polyvinyl derivatives, polyacrylic and polymethacrylic derivatives, coumarone indene resins).

39.03 :—Regenerated cellulose, cellulose nitrate, acetate and other cellulose esters, cellulose ethers and other chemical derivatives of cellulose plasticised or not but excluding collodians and vulcanised fibre.

39.04 :—Hardened proteins (for example, hardened casein and gelatine.

Headings 39.01-39.04 are taken to apply to materials answering to the description in the respective headings if in the following forms only:

(a) Liquid or pasty (including emulsions, dispersions and solutions;

(b) Blocks, lumps, powders (including moulding powder), granules, flake and similar bulk form;

(c) Monofil not meant for textiles, tubes, rods, sticks and profile shapes;

(d) Plates, sheets, strip, film and foil, whether or not printed or otherwise surface-worked, and rectangular articles cut therefrom not further worked;

(e) Waste and scrap.

Headings 39.01 and 39.02 apply only to goods of a kind produced by chemical synthesis answering to one of the following descriptions:

(a) Artificial plastics including artificial resins;

(b) Silicones;

(c) Resols, liquid polyisobutylene and similar artificial high polymers of very high molecular weight. Artificial waxes are excluded.

B. Production, trade and consumption figures

Production figures are quoted wherever possible. Where satisfactory sources could not be found, sales figures from the OECD annual publication, *The Chemical Industry in Europe*, 1955 and 1956, and *The Chemical Industry*, 1961/1962 are used.

Trade figures (quantities) were collected from national sources in conjunction with OECD, *Trade Statistics Bulletins*: Series IV Supplements, 1957, and Series B Supplement, 1960.

Production figures are used in preference to sales figures together with net trade to calculate consumption, as in several countries there is a considerable margin between sales and production representing production by fabricators rather than stock change.

Ideally it might have been desirable in estimating plastics consumption to make some adjustment for plastic materials embodied in finished products which enter international trade. This was not practicable; but it is unlikely that such an adjustment would make any big difference to the ranking of countries. Whereas West German exports of plastics materials amounted to about 30 per cent of total output over the last few years, exports of fabricated plastic products amounted to only 12 per cent of total output.

An attempt was made to reduce production and trade figures to their net main content. (Figures in *The Chemical Industry in Europe* and certain other sources were already on a net resin basis, except for phenolic and aminoplastic moulding materials). The following factors were used:

	Net resin: per cent of gross weight
Phenolics and cresylics	
Modified solid and liquid	85
Moulding and extrusion compounds	50
Laminates	40
All others (except unmodified solid and liquid)	90
Aminoplasts	
Moulding and extrusion	60
Laminates	50
All others	90
Unspecified	
Laminates	50
Polyvinyl chloride	
Plasticised	65
Average all forms	75
All vinyls	
Average	75
All others	100

Polyvinyl chloride and polystyrene occur both in pure forms and compounded with other resins. Generally speaking, their copolymers are included with them. The main exception is for the copolymers of polyvinyl chloride in West Germany and the United Kingdom. They are reported under 'other polymerisation products' for both production and trade in West Germany and for production only in the United Kingdom.

West German production and exports of polystyrene are estimated. United Kingdom sales figures for polyolefins are used in preference to production for 1961 as there were considerable sales from stock during that year. United Kingdom trade in polyolefins is estimated.

C. Source of individual country figures[1]

Production

United States: *Synthetic Organic Chemicals, United States Sales & Production*, 1955 and 1961.

---

(1)Some detailed notes on the statistical treatment of the various country figures are available on request.

United Kingdom : *Synthetic Resins & Plastics Materials*, Board of Trade Business Monitor Series.

West Germany : *Industrie und Handwerke Reihe 3, Industrielle Produktion*, 1957 and 1961.

France : *Chemical Industry in Europe*, 1956 and 1961. *Industrie des Plastiques Modernes*, April 1961.

Italy : *Chemical Industry in Europe*, 1956 and 1961. *Annuario Statista Italiano*, 1957 and 1961.

Japan : *Statistics of Japanese Industrial Production*, 1957-1961. *Annual Statistical Yearbook*, 1955 and 1961.

Sweden : *Chemical Industry in Europe*, 1955 and 1961. *Industrie*, 1955 and 1961.

Netherlands : *Chemical Industry in Europe*, 1956 and 1961. *Plastics*, January 1961 and 1962.

Belgium : *Chemical Industry in Europe*, 1955 and 1961. *Plastics*, January 1961 and 1962.

Trade

United States : *Exports of Foreign & Domestic Merchandise*, 1955 and 1961. *Imports of Merchandise for Domestic Consumption*, 1955 and 1961.

United Kingdom : *Annual Statement of the Trade of the United Kingdom*, 1955 and 1961.

West Germany : *Der Aussenhandel der Bundesrepublik Deutschland*, 1955 and 1961, and further details from Statistisches Bundesamt, Wiesbaden.

France : *Tableau Générale du Commerce Extérieur*, 1955. *Statistiques du Commerce Extérieur*, 1961.

Italy : *Statistica Annuale del Commercio con l'Estero*, 1955. *Statistica Mensile del Commercio con l'Estero*, December 1961.

Japan : *Annual Return of the Foriegn Trade of Japan, Commodity by Country*, 1955 and 1961.

Sweden : *Handel I and II*, 1955 and 1961.

Netherlands : *Maandstatistiek van die in-uit en doervoer per goedensoort*, January-December 1955 and 1961.

Belgium : *Bulletin du Commerce Extérieur*, 1955 and 1961.

D. Sources of the tables on patents : tables 8, 10, 11 and 27

The numbers of patents in plastics materials for the years up to 1955 are taken from J. Delorme, *Anthologie des Brevets sur les Matieres Plastiques*, Editions Amphora, Paris. There are three volumes covering the periods 1791-1930, 1931-1945 and 1946-1955, and these periods were those used for the analysis. The publication lists all important patents taken out in the field of plastics and divides these into twelve main sections. Each section is subdivided into chapters dealing with particular plastics materials, such as polyethylene, cellulose acetate, etc. For tables 8, 10, 11, and 27 seven main sections were studied :

- I Non-ethylene hydrocarbon resins
- II Vinyl resins
- V Resins derived from alcohols, hydrocarbon resins, their ethers and esters
- VI Aldehyde and ketone resins
- VII Resins derived from acids and their esters
- VIII Phenolic resins
- IX Amino resins

The following sections were excluded :

- III Diolofin resins
- IV Acetylene resins
- X Plastic materials derived from proteins
- XI Resins derived from heterocyclic compounds
- XII Inorganic plastic materials

Most of the patents quoted in the volumes have been compiled from American and French records, as most important patents are taken out in one of these two countries ; but some patents from other countries—Germany, Britain, Belgium, Holland, Switzerland—are also included. Full references are given for each patent, including the date of publication (of American patents) or that of application (of French patents), the name of the firm and/or individual who took out the patent, and a short description of the subject matter. Generally speaking, patents taken out by foreign subsidiaries have been classified to the parent firm but it was not always possible to identify these.[1]

As there is no volume in this series covering the post-1955 period, supplementary information about plastics patents from 1954-1962 was obtained from the Patent Office, London. For the period 1954-1956 the publication used was *Patents for inventions—abridgements of specifications—allotted to Group IV(a) : synthetic resins, cellulose.* In the patent classification system this comprises classes :

- 2(ii) Cellulose, non-fibrous, and cellulose derivatives (including artificial filaments, sheets, and the like, containing same) ;
- 2(v) Synthetic resins.

For the period 1956-1962 the coverage of Group IV(a) altered slightly : *Patents for inventions—abridgments of specifications—allotted to Group IV(a) : Synthetic resins, cellulose, organo-silicon compounds and indiarubber.* In the patent classification system this comprises classes :

- 2(2) Non-fibrous cellulose ; cellulose derivatives ; artificial filaments, sheets, and the like containing cellulose derivatives ;
- 2(5) Condensation polymers ;
- 2(6) Resinous polymerisation products ;
- 2(7) Organo-silicon compounds ;
- 70 Indiarubber, guttapercha ; miscellaneous organic substances.

The publication lists all the patents taken out in Britain for plastics materials (Group IV(a)) and, although most of the important foreign patents are represented, there is some bias towards British firms in the resulting figures. This publication includes some materials outside the strict definition of plastics.

E. The full names of firms abbreviated in the text and tables

A.G. für A.F.	Aktien Gesellschaft für Anilin-Fabrikation
B.A.S.F.	Badische Anilin & Soda Fabrik, A.G.
BX	British Xylonite
DCL	Distillers' Co., Ltd.
E.N.I.	Ente Nazionale Idrocarburi
G.E.C.	General Electric Company
I.C.I.	Imperial Chemical Industries, Ltd.
S.I.C.	Societa Italiana della Celluloide
S.I.R.	Societa Italiana Resine

[1]Companies like Shell or CIBA, which operate on a large scale in many countries, are treated as one large concern.

APPENDIX II SUPPLEMENTARY TABLES

Table 16. Plastics in motor vehicles : selected plastics components now in use in British Motor Corporation vehicles

Component	Original material	Plastic	Weight	Percentage saving	Weekly production	Annual saving
			Grams	*Per cent*	*Thousands*	*£ thousand*
Bush, selector lever	Phosphor-bronze	Nylon 66	3	80	2	8.5
Speedometer wheel	Steel	Nylon 66	46	60	4	15.5
Bush, speedo pinion	Manganese bronze	Nylon 66	10	45	2	60.0
Bush, speedo pinion	Cast iron	Nylon 66	18	80	4	7.7
Oil filler cap	Steel assembly	Polypropylene	20	40	16	7.3
Strap, oil filler cap	Steel wire	Nylon II	5	40	16	
Battery tray	Steel	Polyethylene MD	171	18	1	1.6
Armrest	Wood/cotton/leather	Polystyrene/ABS/PU foam	135	50	4	100.0
Dipstick tube	Die-cast aluminium	Nylon 6/10	14	40	4	1.0
Instrument cowl	Wood fibre[a]	Polyethylene HD	150	50	8	32.0
Cover, steering column	New component	Polystyrene, H.I.	115	—	8	—

Source: K. C. Waldron (British Motor Corporation), 'Plastics in the car industry', Paper presented to the Plastics Institute, London section, October 8, 1963.
(a) Painted.

Table 17. Various estimates of per capita consumption of plastics

Kilograms per head

		USA	UK	West Germany	France	Italy	Japan	Sweden	Netherlands	Belgium
NIESR	1955	9.9	5.1	5.9	2.8	1.7	1.4	5.3	3.0	2.4
	1961	15.6	9.6	15.3	8.0	6.7	8.4	12.6	7.9	10.3
OECD	1961	..	9.1	15.7	8.1	6.6	..	12.1	9.8	10.6
Saechtling	1959	14.0	7.6	12.5	5.9	4.0	4.5	10.0	7.1	..
Hufbauer	1960	12.8	8.0	14.0	7.6	6.0	5.0	11.7	9.3	8.1
TNO	59/60	10.0	9.0	15.0	7.5	5.0	..	11.5	9.0	8.5
Chemische Industrie	1950	6.4	2.5	1.8	0.9	0.6	0.2	2.3	1.2	1.0
	1953	7.6	3.2	4.1	1.5	1.1	0.8	3.5	1.8	1.7
	1959	11.1	7.5	12.2	5.9	4.1	3.3	9.8	7.1	7.0

Source: NIESR: table 3; OECD: *Chemical Industry in Europe*, 1962, p. 95; Saechtling: *Werkstoffe aus Menschenhand*, Munich, 1961, Table 27. Hufbauer: *Synthetic materials, a study in international trade* (Cambridge University doctoral thesis), p. 73. TNO: *Plastics in figures*, Plastics Research Institute, Delft, Netherlands. *Chemische Industrie*, Vol. XIII, Dusseldorf, 1961.

Table 18. Production of plastics materials, 1955

Thousand metric tons

	USA	UK	West Germany	France	Italy	Japan	Sweden	Nether-lands	Bel/Lux.
Condensation (*39.01*)									
Phenolic	192.7	45.3	44.6	9.6	8.5	11.0	5.0	..	..
Aminoplast	132.6	57.1	57.0	9.5	14.5	43.3	5.0	..	..
Other	289.1	40.7	53.2	15.4	5.9	0.4	8.0	..	..
Total	**614.3**	**143.0**	**154.7**	**34.5**	**28.8**	**54.7**	**18.0**	**16.5**	**3.6**
Polymerisation (*39.02*)									
PVC	239.1	48.2	64.3	32.2	34.0	32.4	4.1	..	..
Polystyrene	235.5	22.8	21.7	11.6	7.3	—	1.4	..	..
Polyolefins	182.5	29.2	6.4	4.0	4.7	—	—	..	..
Other	272.1	31.2	52.5	8.0	3.9	0.6	0.2	4.9	..
Total	**929.2**	**131.3**	**144.9**	**55.8**	**50.0**	**33.0**	**5.7**	**4.9**	..
Cellulosic (*39.03*)									
Total	230.7	46.2	54.5	19.3	10.1	26.0	5.8	2.7	..
Protein (*39.04*)									
Total	..	2.8	1.1	0.5	0.7	..	(a)	(a)	..
Total of above	**1,774.2**	**323.4**	**355.2**	**110.1**	**89.7**	**113.7**	**29.5**	**24.1**	**16.4**

Percentages of total

	USA	UK	West Germany	France	Italy	Japan	Sweden	Nether-lands	Bel/Lux.
Condensation (*39.01*)									
Phenolic	10.9	14.0	12.5	8.7	9.4	9.7	16.9	..	..
Aminoplast	7.5	17.6	16.0	8.6	16.1	38.1	16.9	..	..
Other	16.3	12.6	15.0	14.0	6.6	0.3	27.1	..	..
Total	**34.6**	**44.2**	**43.6**	**31.3**	**32.2**	**48.1**	**61.0**	**68.5**	**21.9**
Polymerisation (*39.02*)									
PVC	13.4	14.9	18.1	29.2	37.9	28.5	13.9	..	..
Polystyrene	13.3	7.0	6.1	10.5	8.2	—	4.7	..	..
Polyolefins	10.3	9.0	1.8	3.6	5.2	—	..	..	..
Other	15.3	9.7	14.8	7.3	4.4	0.5	0.7	..	..
Total	**52.4**	**40.6**	**40.7**	**50.7**	**55.8**	**29.0**	**19.3**	**20.2**	..
Cellulosic (*39.03*)									
Total	13.0	14.3	15.3	17.5	11.3	22.9	19.7	11.3	..
Protein (*39.04*)									
Total	..	0.9	0.4	0.5	0.8	..	..	..	..
Total of above	**100.0**	**100.0**	**100.0**	**100.0**	**100.0**	**100.0**	**100.0**	**100.0**	**100.0**

Source : See Appendix I.
(a) Incorporated in 39.02.

Table 19. Production of plastics materials, 1961

Thousand metric tons

	USA	UK	West Germany	France	Italy	Japan	Sweden	Nether-lands	Bel/Lux.
Condensation									
(39.01)									
Phenolic	241.3	51.4	54.5	18.2	22.8	55.6	9.3	6.5	..
Aminoplast ..	179.7	73.2	149.4	28.1	41.4	172.3	16.0	2.8	..
Other	400.3	82.5	190.1	55.4	36.5	19.6	11.6	22.7	..
Total	**821.3**	**207.1**	**394.0**	**101.8**	**100.6**	**247.6**	**36.9**	**32.0**	**12.0**
Polymerisation									
(39.02)									
PVC	443.3	108.6	195.5	119.4	150.0	309.0	11.5	9.0	..
Polystyrene ..	519.6	52.5	110.0	42.4	39.0	29.0	4.1	—	..
Polyolefins ..	772.5	136.6	105.0	29.0	56.0	58.1	—	18.0	..
Other	365.6	48.9	131.6	26.0	24.0	91.4	4.4	9.8	..
Total	**2,101.0**	**346.6**	**542.1**	**217.6**	**269.0**	**487.4**	**20.0**	**36.8**	**34.0**
Cellulosic									
(39.03)									
Total	317.0	67.2	91.5	47.1	14.0	46.0	8.8	7.3	(a)
Protein									
(39.04)									
Total	..	1.7	1.2	0.5	0 7	..	..	..	..
Total of above ..	**3,239.2**	**622.6**	**1028.8**	**367.0**	**384.3**	**781.0**	**65.7**	**76.1**	**46.6**

Percentages of total

	USA	UK	West Germany	France	Italy	Japan	Sweden	Nether-lands	Bel/Lux.
Condensation									
(39.01)									
Phenolic	7 4	8 3	5.3	5.0	5.9	7.1	14.0	8.5	..
Aminoplast ..	5.5	11.7	14.5	7.7	10.7	22.1	24.4	3.7	..
Other	12.4	13.3	18.5	15.1	9.5	2.5	17.7	29.8	..
Total	**25.3**	**33.3**	**38.3**	**27.8**	**26.2**	**31.7**	**56.2**	**42.0**	**26.1**
Polymerisation									
(39.02)									
PVC	13.7	17.4	19.0	32.5	39.0	39.6	17.5	11.8	..
Polystyrene ..	15.8	8.4	10.7	11.6	10.1	3.7	6.2	..	..
Polyolefins ..	23.8	21.9	10.2	8.1	14.6	7.4	..	23.6	..
Other	11.5	7.8	12.8	7.1	6.2	11.7	6.7	12.9	..
Total	**64.8**	**55.7**	**52.7**	**59.3**	**70.0**	**62.4**	**30.4**	**48.4**	**73.0**
Cellulosic									
(39.03)									
Total	9.8	10.8	8.9	12.8	3.6	5.9	13.3	9.6	(a)
Protein									
(39.04)									
Total	..	0.3	0.1	0.1	0.2	..	..	..	..
Total of above ..	**100.0**	**100.0**	**100.0**	**100.0**	**100.0**	**100.0**	**100.0**	**100.0**	**100.0**

Source: See Appendix I.
(a) Incorporated in 39.02.

Table 20. Consumption of plastics materials, 1955

Thousand metric tons

	USA	UK	West Germany	France	Italy	Japan	Sweden	Nether-lands	Bel/Lux.
Condensation (*39.01*)									
Phenolic	184.8	41.2	39.2	10.8	8.8	..	6.1	..	..
Aminoplast ..	125.9	42.6	55.4	13.5	16.4	..	5.0	..	..
Other	278.0	39.1	47.8	16.6	6.4	..	5.8	..	..
Total	**588.7**	**122.9**	**142.4**	**40.8**	**31.7**	**55.0**	**16.9**	**14.4**	**7.9**
Polymerisation (*39.02*)									
PVC	219.8	44.1	51.3	31.1	..	30.1	6.8	..	..
Polystyrene ..	215.6	13.9	16.8	12.3	..	..	2.0	..	..
Polyolefins ..	152.3	17.7	7.8	6.6	..	..	2.3	..	..
Other	255.9	25.3	29.5	11.1	..	..	4.3	..	..
Total	**843.5**	**100.9**	**105.4**	**61.1**	**37.0**	**41.4**	**15.3**	**12.8**	..
Cellulosic (*39.03*)									
Total	212.6	37.0	45.1	16.5	12.5	23.8	6.0	5.4	..
Protein (*39.04*)									
Total	..	3.0	1.5	0.5	0.7	(a)	(a)	(a)	..
Total of above ..	**1,644.8**	**263.9**	**294.4**	**119.0**	**81.8**	**120.3**	**38.2**	**32.6**	**21.9**

Percentages of total

	USA	UK	West Germany	France	Italy	Japan	Sweden	Nether-lands	Bel/Lux.
Condensation (*39.01*)									
Phenolic	11.2	15.6	13.3	9.1	10.7	..	15.9	..	..
Aminoplast ..	7.7	16.1	18.8	11.3	20.1	..	13.0	..	..
Other	16.9	14.8	16.2	13.9	7.9	..	15.3	..	..
Total	**35.8**	**46.6**	**48.4**	**34.3**	**38.7**	**45.7**	**44.2**	**44.2**	**36.1**
Polymerisation (*39.02*)									
PVC	13.4	16.7	17.4	26.2	..	..	17.7	..	..
Polystyrene ..	13.1	5.3	5.7	10.4	..	..	5.4	..	..
Polyolefins ..	9.3	6.7	2.7	5.5	..	..	5.9	..	..
Other	15.6	9.6	10.0	9.3	..	..	11.2	..	..
Total	**51.3**	**38.3**	**35.8**	**51.4**	**45.2**	**34.5**	**40.1**	**39.3**	..
Cellulosic (*39.03*)									
Total	12.9	14.0	15.3	13.9	15.2	19.8	15.7	16.5	..
Protein (*39.04*)									
Total	..	1.1	0.5	0.4	0.9	..	..	..	..
Total of above ..	**100.0**	**100.0**	**100.0**	**100.0**	**100.0**	**100.0**	**100.0**	**100.0**	**100.0**

Source : See Appendix I.
(a) Incorporated in 39.03.

Table 21. Consumption of plastics materials, 1961

Thousand metric tons

	USA	UK	West Germany	France	Italy	Japan	Sweden	Nether-lands	Bel/Lux.
Condensation (*39.01*)									
Phenolic ..	229.6	38.6	45.9	19.9	21.6	56.3	9.9	5.4	..
Aminoplast ..	170.1	59.3	139.0	36.3	40.6	172.0	15.2	10.9	..
Other	386.8	85.0	155.2	55.3	44.3	16.9	9.8	15.2	..
Total	**786.5**	**182.9**	**340.2**	**111.4**	**106.5**	**245.2**	**35.0**	**31.5**	**36.5**
Polymerisation (*39.02*)									
PVC	414.3	100.5	167.0	104.7	97.4	279.2	16.0	23.5	..
Polystyrene ..	455.8	40.5	70.0	39.8	21.5	38.9	7.4	5.1	..
Polyolefins ..	607.7	80.8	84.2	49.6	57.2	100.2	16.8	15.0	..
Other	309.2	47.0	77.4	23.8	32.6	87.5	13.3	11.3	..
Total	**1,787.0**	**268.8**	**398.7**	**217.9**	**208.6**	**505.8**	**53.5**	**55.0**	..
Cellulosic (*39.03*)									
Total	293.1	54.3	89.4	37.8	18.2		6.6	5.8	..
Protein (*39.04*)									
Total	..	1.8	0.8	0.7	0.5	..	..	..	..
Total of above ..	**2,866.6**	**507.7**	**829.0**	**367.8**	**333.8**	**793.7**	**95.0**	**92.2**	**97.7**

Percentages of total

	USA	UK	West Germany	France	Italy	Japan	Sweden	Nether-lands	Bel/Lux.
Condensation (*39.01*)									
Phenolic	8.0	7.6	5.5	5.4	6.5	7.1	10.4	5.8	..
Aminoplast ..	5.9	11.7	16.8	9.9	12.2	21.7	16.0	11.8	..
Other	13.5	16.7	18.7	15.0	13.3	2.1	10.4	16.4	..
Total	**27.4**	**36.0**	**41.0**	**30.3**	**31.0**	**30.9**	**36.8**	**34.1**	**37.4**
Polymerisation (*39.02*)									
PVC	14.5	19.8	20.1	28.5	32.4	35.2	16.9	25.5	..
Polystyrene ..	15.9	8.0	8.4	10.8	6.4	4.9	7.8	5.5	..
Polyolefins ..	21.2	15.9	10.2	13.5	17.1	12.6	17.6	16.3	..
Other	10.8	9.3	9.3	6.5	9.8	11.0	14.0	12.3	..
Total	**62.3**	**53.0**	**48.1**	**59.2**	**62.5**	**63.7**	**56.3**	**59.6**	..
Cellulosic (*39.03*)									
Total	10.2	10.7	10.8	10.3	5.5	5.4	7.0	6.3	..
Protein (*39.04*)									
Total	..	0.3	0.1	0.2	0.1	..	..	..	..
Total of above ..	**100.0**	**100.0**	**100.0**	**100.0**	**100.0**	**100.0**	**100.0**	**100.0**	**100.0**

Source: See Appendix I.

Table 22. International trade in plastics materials, 1955

	USA	UK	West Germany	France	Italy	Japan	Sweden	Netherlands	Bel/Lux.
	Exports (estimated net resin content), *thousand metric tons*								
Condensation (39.01)	26.0	23.6	15.0	2.5	0.5	0.3	3.6	7.9	1.0
Polymerisation (39.02)	86.5	49.7	52.4	6.2	17.3	2.2	2.5	2.9	1.6
Cellulosic (39.03)	19.1	12.3	10.6	5.8	1.9	2.5	1.4	2.0	4.0
Protein (39.04)	—	..	0.8	—	—	—	—	0.2	..
Total	131.6	85.6	78.8	14.6	19.7	4.9	7.5	13.1	6.6
	Imports (estimated net resin content), *thousand metric tons*								
Condensation (39.01)	0.4	3.4	2.7	8.9	3.4	0.6	2.5	5.8	4.7
Polymerisation (39.02)	0.8	19.4	12.9	11.5	4.2	10.6	12.2	9.8	5.9
Cellulosic (39.03)	1.1	3.1	1.2	3.0	4.2	0.3	1.6	4.6	1.5
Protein (39.04)	..	0.2	1.3	—	—	—	—	0.3	..
Total	2.2	26.1	18.1	23.5	11.8	11.5	16.2	20.6	12.2

	USA	UK	West Germany	France	Italy	Japan	Sweden	Netherlands	Bel/Lux.
	Net foreign trade (estimated net resin content), *thousand metric tons*								
Condensation (39.01)	25.6	20.1	12.3	−6.4	−2.8	−0.3	−1.1	2.1	−3.7
Polymerisation (39.02)	85.7	30.4	39.5	−5.3	13.0	−8.5	−9.6	−6.9	−4.3
Cellulosic (39.03)	18.1	9.1	9.4	2.8	−2.3	2.2	−0.2	−2.6	2.4
Protein (39.04)	—	−0.2	−0.5	—	—	—	—	—	..
Total	129.4	59.5	60.7	−8.9	7.9	−6.6	−8.7	−7.5	−5.5
	Value of foreign trade, SITC 599.01, $ *million*								
Total exports ..	133.9	74.4	70.2	17.6	18.3	4.9	6.7	12.3	8.5
Total imports ..	3.9	36.2	15.4	21.4	10.7	10.0	16.2	20.6	13.6
Net foreign trade ..	130.0	38.2	54.8	−3.8	7.6	−5.1	−9.5	−8.3	−5.1

Source : See Appendix I.

Table 23. International trade in plastics materials, 1961

	USA	UK	West Germany	France	Italy	Japan	Sweden	Nether-lands	Bel/Lux.
	Exports (estimated net resin content), ***thousand metric tons***								
Condensation (39.01)	35.6	37.3	73.9	12.3	18.4	4.5	13.6	18.9	4.9
Polymerisation (39.02)	317.2	134.7	214.4	42.4	90.0	35.8	12.2	33.7	12.1
Cellulosic (39.03)	30.4	20.1	9.4	14.8	3.3	4.1	5.2	8.7	7.1
Protein (39.04)	—	0.2	0.6	—	0.2	—	0.1	0.3	..
Total	**383.2**	**192.3**	**298.1**	**69.5**	**111.9**	**44.4**	**31.1**	**61.6**	**24.1**
	Imports (estimated net resin content), ***thousand metric tons***								
Condensation (39.01)	0.8	13.1	20.0	22.0	24.3	2.1	11.6	18.4	29.4
Polymerisation (39.02)	3.2	56.8	70.9	42.7	29.6	54.1	45.7	52.3	40.8
Cellulosic (39.03)	6.5	7.2	7.4	5.4	7.5	0.9	3.1	7.1	5.5
Protein (39.04)	—	0.2	0.2	0.2	—	—	—	0.4	..
Total	10.6	77.3	98.4	70.3	61.3	57.1	60.5	78.2	75.8

	USA	UK	West Germany	France	Italy	Japan	Sweden	Nether-lands	Bel/Lux.
	Net foreign trade (estimated net resin content), ***thousand metric tons***								
Condensation (39.01)	34.8	24.2	53.9	−9.7	−5.9	2.4	2.0	0.5	24.5
Polymerisation (39.02)	314.0	77.9	143.4	−0.1	60.4	−18.3	−33.5	−18.6	..
Cellulosic (39.03)	23.9	12.9	2.1	9.0	−4.2	3.2	2.1	1.5	..
Protein (39.04)	—	−0.1	0.4	−0.2	0.2	—	—	—	..
Total	**372.6**	**115.0**	**199.8**	**−0.9**	**51.1**	**−12.7**	**−29.3**	**−16.6**	**−51.7**
	Value of foreign trade, SITC 581, ***$million***								
Total exports ..	285.4	132.4	237.9	63.3	58.1	31.0	22.5	38.4	19.0[(a)]
Total imports ..	6.2	79.6	57.9	54.3	40.7	43.5	41.7	55.4	47.5
Net foreign trade ..	**279.2**	**52.8**	**180.0**	**9.0**	**17.4**	**−12.5**	**−19.2**	**−17.0**	**−28.5**

Source : See Appendix I.
(a) Excluding some polyvinyls.

Table 24. Comparison of PVC usage by application in 1960

Lb per head

Country	Total	Calendering	Cables	Coated fabrics	Flexible extrusion	Records	Flooring	Belting	Rigid applications[(a)]
West Germany ..	6.3	2.3	1.7	0.4	0.7	—	0.4	—	0.9
Japan	5.5	1.3	0.6	0.4	0.4	—	0.1	—	2.5
UK	4.7	1.2	0.8	0.5	0.5	0.2	0.3	0.3	0.2
USA	4.6	1.2	0.5	0.4	0.4	0.3	0.9	—	0.1
France	4.0	1.2	0.6	0.3	0.3	0.1	0.4	—	0.3
Italy..	2.8	0.6	0.5	0.1	0.7	0.1	—	—	0.9

Source : D. G. Owen, 'The commercial development of PVC', Plastics Progress, London 1961, page 80.
(a) Mainly in construction.

Table 25. PVC pipes—comparative costs with cast iron and cement asbestos

	PVC pipe production	PVC polymer price in 1960	Extruder's selling price of pipe	Pipe prices: PVC	Pipe prices: Cast iron	Pipe prices: Cement asbestos
	(tons a year)	*(pence/lb)*	*(pence/ft.)*	*(pence/ft.)*	*(pence/ft.)*	*(pence/ft.)*
UK						
2½ in. rain water ..	..	..	42	24[(a)]	34[(a)]	18[(a)]
3 in. water mains class "B" (wall thickness 0.136 in.)	1,000	15	41	36	42	33
Italy						
2 in. rain water	15,000	10.5–11	32	12[(a)]	..	12.5[(a)]
4 in. stack	..	..	32	31[(a)]	..	24[(a)]
Japan						
3 in. water mains (wall thickness 0.20 in.) ..	65,000	13–14	20–22	36[(a)]	65[(a)]	30[(a)]
Holland	6,000	15	28	..	..	..

Source: D. G. Owen : 'The commercial development of PVC', Plastics Progress, London 1961, page 85.
(a) Builders' merchants prices to builders.

Table 26. UK and Common Market Tariffs on certain plastics materials, 1962

Percentages

	UK full	Benelux EEC	Benelux Other	France EEC	France Other	Germany EEC	Germany Other	Italy EEC	Italy Other	Common Market Final
PVC	10	1.5	6.9	15	21	9.5	16	11	(a)	20
Polythene	10	1.5	6.9	15	22	9.5	16	13.5	(b)	20
Acrylics	10	2.5	8.6	15	23	7.5	16.8	13.5	(c)	21
Urea Powders ..	20	2.5	7.1	7.5	15	7.5	12	9.0	16.2	15
PF Resin	10	1.5	5.7	7.5	15	7.5	12	9	16.2	15
PF Powders ..	20	1.5	5.7	7.5	15	7.5	12	9	16.2	15
Nylon Moulding Powders	10	2.5	8.8	17.5	26.2	7.5	17.6	11	22	22
PTFE Polymer ..	10	2.5	9.1	17.5	26.5	9.5	18.4	13.5	24.4	18
Polyester Film ..	10	10	20	15	22.8	7.5	16	10	20	20
Polystyrene.. ..	10	2.5	8.3	15	22.8	7.5	16	12.1	21.3	20

Source : British Plastics Federation.
(a) PVC Resins 20 ; PVC Moulding Compounds 21.3.
(b) Polythene Moulding Compounds 21.3 ; Other 20.
(c) Aqueous emulsions, solutions, dispersions 14.2 ; Solutions containing volatile organic solvents 21 ; Moulding materials and sheet 21.5.

Table 27. The share of some leading firms in patents delivered for various classes of plastic materials

	Total number of patents delivered			Number of patents in vinyl plastics		Number of patents in phenol and amino plastics		Number of patents in cellulosics		Number of patents in other plastics	
	1791–1930	1931–1945	1946–1955	1931–1945	1946–1955	1931–1945	1946–1955	1931–1945	1946–1955	1931–1945	1946–1955
United States											
Du Pont	78	321	637	84	328	75	113	67	35	95	161
Monsanto	1	37	283	17	174	6	53	8	8	6	48
American Cyanamid	6	60	266	3	23	38	142	6	8	13	93
Dow	4	115	187	85	143	5	8	19	14	6	22
B. F. Goodrich	3	32	160	31	120	1	9	—	—	—	31
US Rubber	—	—	156	—	64	—	19	—	10	—	63
Rohm & Haas Co.	3	20	154	2	41	2	31	7	4	9	78
Eastman Kodak	169	120	140	59	32	—	3	61	32	—	73
Standard Oil	4	41	123	29	94	1	6	3	1	8	22
Celanese Corp.	7	67	73	—	9	19	22	45	28	3	14
Union Carbide (Carbide & Carbon)	15	88	69	77	47	1	9	5	3	5	10
Hercules Powder	13	132	65	1	10	2	3	27	28	102	24
Phillips Petroleum	—	4	55	4	44	—	2	—	3	—	6
GEC	38	128	43	23	17	83	9	2	—	20	17
West Germany											
I.G. Farben	346	889	26	365	7	268	12	99	—	157	7
BASF	38		115		51		25		3		36
Bayer	55		111		20		53		4		34
Hoechst (Meister, Lucius & Brüning)	55	Part of I.G. Farben	57	I.G.	27	I.G.	4	I.G.	—	I.G.	26
Ruhrchemie	1		9		8		—		—		1
Hüls	—		39		23		16		—		—
Wacker	—	65	25	43	19	2	5	11	1	9	—
Röhm & Haas A.G.	8	122	—	17	—	20	—	2	—	83	—
United Kingdom											
ICI	25	90	253	40	92	23	51	15	6	12	104
Distillers	—	20	74	14	47	—	10	5	1	1	16
France											
St. Gobain	—	10	77	10	51	—	22	—	3	—	1
Rhone Poulenc	13	26	43	8	25	1	3	14	—	3	15
Italy											
Montecatini	—	3	24	1	9	2	3	—	—	—	12
Switzerland											
CIBA	42	56	101	—	5	48	66	6	1	2	29
International											
Shell/N.V. de Bataafsche	1	25	263	13	94	1	73	1	10	10	86

Source : Jean Delorme, *Anthologie des Brevets sur les Matières Plastiques*, Vols. I-III, Amphora, Paris.

APPENDIX III. SOURCES FOR REFERENCES

[1] See for example :
D. M. Wells, 'Uses of unplasticised PVC pipe for water supply', *British Plastics*, March 1960 ;
B. S. Dyer and W. Dawson, 'Plastics in building', *Royal Society of Health Journal*, July 1962 ;
and for automobile components Table 16 and :
K. C. Waldron, 'Plastics in motor vehicles', *Transactions of the Plastics Institute*, June 1963.

[2] OECD, *Chemical Industry in Europe*, Annuals from 1954 to 1962.

[3] G. Hufbauer, *Synthetic materials : a study in international trade*, Cambridge University (Ph.D. thesis), page 123.

[4] V. E. Yarsley and E. G. Couzens, *Plastics in the service of man*, London, 1956, page 71 ;
Modern Plastics, January 1963 ;
First Century of Plastics (Ed. Kaufman), London, 1963, page 90.

[5] E. A. Stallworthy (Bataafsche Internationale Chemie Maatschapij N.V.), *Comparative investment costs in Western Europe*, paper presented to a joint meeting of the British Section A.A.C.E. and the Association of Cost Engineers, May 1963.

[6] National Industrial Conference Board, Study of comparative costs in American and European manufacturing industry (quoted in *World Petroleum*, June 1963).

[7] F. J. Glover (NIESR), paper to be published on trends in supply and demand for non-ferrous metals.

[8] R. Malik, 'European Chemicals, an appraisal', *Times Review of Industry*, 1962.

[9] *Modern Plastics*, January 1962 and January 1963.

[10] Annual Reports of Hoechst and BASF.

[11] See for example :
A. Renfrew (ICI), 'The British Plastics Industry and the Common Market', *British Plastics*, August 1963, page 432.

[12] H. Saechtling, *Werkstoffe aus Menschenhand*, Munich, 1961, Chapter 3.

[13] F. ter Meer, *Die I.G. Farben*, Düsseldorf, 1953.

[14] *Ibid*, page 24.

[15] Annual Reports of Companies and *European Chemical News*, April 27th, 1962.

[16] See for example :
'The weakness of industrial research', *Some aspects of scientific research in France*, page 11, Service de presse et d'information, French Embassy in London, 1963.

[17] H. Saechtling, *op. cit.*, pages 12-14.

[18] C. Freeman, 'Research and Development : a comparison between British and American Industry', *National Institute Economic Review*, No. 20, page 21.

[19] This applies particularly to polymer *physics* (as opposed to polymer *chemistry*). See :
W. J. Carswell and D. Broyd, 'Non-metallic materials in engineering', *NEL Report*, No. 101 ;
Department of Scientific and Industrial Research, 1963.

[20] For a general discussion on the value of patent statistics as an indication of research output see the report of the Conference organised by the National Bureau of Economic Research on *The Rate and Direction of Inventive Activity*, Princeton, 1962, particularly the papers of S. Kuznets and J. Schmookler ;
and also Y. Fabian, *Measures of output of R and D*, OECD, Paris, 1963.

[21] S. F. Mueller, 'The origins of the basic inventions underlying Du Pont's major product and process innovations, 1920 to 1950' in *The Rate and Direction of Inventive Activity*, Princeton, 1962, page 330.

[22] J. Delorme, *Anthologie des Brevets sur les Matieres Plastiques*, Vols. I-III, Paris, 1962.

[23] For example, B. V. Bowden, 'The higher education of scientists and engineers' paper delivered to a Conference of the Institute of Personnel Management, London, 1961.

[24] G. Hufbauer, *op. cit.*, Appendix A.

[25] G. Hufbauer, *op. cit.*, Chapter 6.

[26] *European Chemical News*, 2nd March, 1962, page 27.

[27] H. Saechtling, *op. cit.*, page 30 ; information from firms, and *First Century of Plastics*, (Ed. Kaufman), page 103.

[28] For full elaboration of the 'technological gap' theory of foreign trade as applied to plastics, see G. Hufbauer, *op. cit.* This thesis is an application and elaboration of concepts put forward by M. V. Posner in *Oxford Economic Papers*, October 1961.

[29] to [34] Information from interviews.

[35], [36] J. C. Swallow (ICI), 'Polyolefins today and tomorrow', *Plastics*, June, 1959 ; and 'Ethylene polymers, past and future', *Transactions of the Plastics Institute*, February 1963.

[37] See for example :
D. G. Owen, 'The commercial development of PVC', *Plastics Progress*, London, 1961.

[38] J. M. De Bell *et al*, *German Plastics Practice*, Springfield, 1946.

[39] H. P. DeHaan and G. N. Freund, (Plastics Division, Allied Chemical Corporation) 'New plastics applications overseas', *Automotive Industries*, 1st July, 1963 ;
H. Saechtling, 'Kunststoffe werden technische Werkstoffe', *Technik und Forschung*, April 1963 ;
K. C. Waldron, *op. cit.*

[40] A recent example is the development of PVC fruit squash bottles by Mr. F. Marzillier in association with Marks and Spencers and Staveley Industries. (*Evening Standard*, 2nd May, 1963.)

[41] Note[39] and Stuart Marshall, 'Plastics', *The Guardian*, 16th October, 1963.

[42] *Times Review of Industry and Technology*, October 1963, page 43.

[43] The cable industry is another good example. See A. Renfrew and P. Morgan, *Polythene*, Iliffe, 1960.

[44] Cmnd 1920, *Annual Report of the Advisory Council on Scientific Policy* 1961-1962, HMSO, 1963.

[45] B. N. P. Hutcheson (Distillers Co. Ltd.), contribution to symposium held by the International Union of Architects, July 1961.

[46] National Academy of Sciences, *A program for building research in the United States*, Washington, 1962.

[47] *Plastics Industry News*, Tokyo, 1963.

[48] D. M. Bell, 'Britain's petrochemical industry looks forward to becoming part of the Common Market', *Chemische Industrie*, June 1963, page 301.

[49] N. S. Khrushchev, *The economic development of the USSR and party guidance on the national economy*, report to Central Committee of Soviet Communist Party, Moscow, November 1962.

[50] 'Japan auf dem Wege zum drittgroessten Chemieland', *Chemische Industrie*, December 1961, page 796.

APPENDIX IV. FIRMS AND ORGANISATIONS WHO ASSISTED IN THE INQUIRY

Allied Chemical Corp., Plastics Division
Ankerwerk Gebr. Goller
Arbeitsgemeinschaft Deutsche Kunststoff-Industrie
Badische Anilin & Soda Fabrik, A.G.
Bakelite, Ltd.
Baker Perkins Granbull, Ltd.
Bettenfeld Maschinenfabriken, GmbH.
British Insulated Callender's Cables, Ltd.
British Motor Corporation, Ltd.
'British Plastics'
British Plastics Federation
J. Brockhouse & Co., Ltd.
Commercial Plastics, Ltd.
Department of Scientific & Industrial Research
Distillers, Ltd.
East Anglia Plastics, Ltd.
'European Chemical News'
Ford Motor Co., Ltd.
Forest Products Research Laboratory
Formica, Ltd.
Gesamtverband Kunststoff Verarbeitende Industrie
Imperial Chemical Industries, Ltd., Plastics Division
'Kunststoff Rundschau'
Joseph Lucas Ltd.
The Marley Tile Co., Ltd.
Metal Box, Ltd., Plastics Group
Monsanto Chemicals, Ltd.
National Science Foundation
Organisation for Economic Cooperation and Development (OECD) :—
  Directorate for Scientific Affairs
  Special Committee for Chemical Products
Osma Plastics, Ltd.
'Plastics'
The Projectile & Engineering Co., Ltd.
A. E. Reed, Ltd.
'Rubber & Plastics Age'
Shell Chemical Co., Ltd.
Shorko Packaging (U.K.), Ltd.
Tecalemit, Ltd.
Thermo Plastics, Ltd.
T.N.O., Economic-Technical Department, The Hague
R. H. Windsor, Ltd.
Wissenschaftsrat, Cologne
Yarsley Research Laboratories, Ltd.

Acknowledgements are also due to Dr. C. A. Redfarn and Mr. A. Wilson of Technical Commercial Surveys for their assistance as consultants.

[16]

INDUSTRIAL R & D IN PHARMACEUTICAL FIRMS IN THE EARLY TWENTIETH CENTURY

By JONATHAN LIEBENAU

What is outstanding about the development of pharmaceutical laboratories in German, British and American drug companies is their differences. Here is a case where broadly the 'same' technical goals were being pursued: to produce new products which would make profits, and to provide therapeutics. In these three countries the state of the chemical, biological, pharmacological and medical knowledge was roughly similar. Medically the problems which the industry confronted were much the same in all three. All suffered increasingly from urban epidemics and in most cities tuberculosis and diphtheria were endemic. The outstanding differences stem from the organisation of medical science and practice, and differences in the structure and evolution of the industry.

Before the turn of this century in Germany the industry was growing rapidly and changing quickly. It was part of a troubled economy searching for novel products. New drugs were tested and introduced through academic institutions and hospitals, and then sold by independent pharmacies.

In the United States the outstanding characteristic of manufacturing pharmacy was its recent development out of the practice of pharmacy. Ethical pharmaceuticals production was boosted because of the particular interest in the new medicines of the 1890s. Production was stimulated by a public health system which was on the lookout to incorporate esoteric medical science. Transportation and advertising were transforming the production and sale of drugs nationwide.

The major British manufacturers of medicines were usually first and foremost importers and wholesalers of materia medica. With a well-integrated and hierarchically organised medical profession there was not a great pull for novelty in therapeutics, and new drugs were rarely introduced through hospitals. The leading firms of the 1890s were long-established, most dated from the early part of the century.

Attitudes and practices for innovation reflected these differences. The use to which a company put its laboratory depended upon the advantage the firm saw in investing in scientists, and the extent to which the company was linked to the medical profession. The industry's relation to government was also crucial. These factors were quite different in Germany from Britain, and different again from the United States.

I

Overall the growth rate of the German industry in this sector accelerated rapidly, overtaking the British and spectacularly reducing their foreign trade. The impact became dramatic when during the First World War the British medical community realised that the nation had become dependent upon German products in a number of important areas. This brought an immediate response, but one inadequate to compensate fully for the German lead.

The German pharmaceutical industry was made up of a few old, well-established firms which had grown out of pharmacies and expanded with new production techniques in the mid-nineteenth century. Companies such as Merck and Schering conducted a traditional business in galenicals, basic chemicals and other common medicines. Their scope was limited to the drug business and they maintained traditional ties with pharmacists. This changed in the 1880s when the new dyestuffs manufacturers began to produce synthetic therapeutics. Although England and France held the dominant positions in coal-tar dye production in the 1860s, by the mid-1870s German production accounted for six times the value of English production, and overshadowed French output by over seven times.[1] The German industry was also becoming composed mainly of big firms.[2]

In the mid-1880s two problems coincided to throw the dye industry into crisis. The price and marketing convention governing the crucial red dyes broke up in 1885, forcing prices down to half their previous level within a year.[3] At the same time the price of coal tar rose as new methods of producing coal gas reduced the quantity of byproduct made.[4] Laboratories helped to resolve the crisis with new dyestuffs and later by opening up the new field of synthetic medicines. Rapid development continued throughout the 1880s and 1890s and yielded high profits. The annual average dividend rates of the German coal-tar dyestuffs firms were consistently higher than 20 per cent in the 1890s, although the dividend rates of the German iron and coal firms often fell below 10 per cent.[5] Not that industrial research can be used as a full explanation for these profits; it was rather a contributing factor, and of course there was a need for sufficient funds to finance R & D staff and facilities.

At the Bayer company in Leverkusen there was no commitment to organised research until the late 1880s. The crisis in the dyestuffs trade of the early 1880s made them receptive to change. Research offered Bayer a strategy that might reduce vulnerability to the fluctuations of the business cycle. For chemists too the depressed economy stimulated change. Their numbers had been growing steadily due to expanded opportunities for training in the 1870s.[6] When that expansion slowed in the 1880s academic careers were blocked and as mobility reduced, chemists more readily accepted industrial employment. It was at that time that synthetic organic colours were first produced in academic laboratories, and that the connections were made between the chemical intermediates used for the production of dyestuffs and the manufacture of pharmaceutical products.

Aside from scientific opportunities and economic stimuli, the first German patent law encouraged the transplantation of research from academe to industry.[7] Enacted in 1877,[8] this patent law had three remarkable effects. The publication of patents provided information to all companies on advances in the state of the art. Since for chemicals the patents were limited to processes, chemists were stimulated to find and patent all possible processes, thereby blocking their competitors. Further, the resulting endless patent litigation constituted a sort of legal warfare that intensified the connections between manufacturers and scientists called upon to provide expert testimony.[9]

The need to invest in research staff and facilities was only slowly accepted in the 1880s. However, the number of chemists employed grew and the sources of chemists shifted from employees of rival firms where they had been trained to perform industrial functions to recruits from universities and medical institutions.[10] Academic chemistry was better able to supply appropriately trained scientists, and the companies were themselves better able to make use of graduates. Bayer established a pattern of hiring Ph.D. chemists for one year, and then sending them to a university or Technische Hochschule laboratory, where they worked on research problems related to the company production programme.[11] Not until 1886 did they build their own laboratory for academic-style research; even then they kept it at a safe distance from the manufacturing site.

The critical change at Bayer came as a result of the influence of Carl Duisberg. Duisberg had joined the company in 1884 after having done research for Bayer at the University of Strassburg. As his work on artificial organic colours expanded he gathered a large team of chemists which performed many functions. Its members had to make rapid shifts among research, development, production, and sales, and began to have increasing influence in the company. The company responded in 1890 by recognising the research chemist as a distinctive full-time role, and in 1891 opened a new, centralised, integrated laboratory. As the laboratory grew, its functions became more clearly defined, and the general purpose of transforming findings into profitable products was clarified. Testing and process improvements became a separate area, as did the patent bureau. Chemical engineering was the concern of the technical room, where laboratory procedures were scaled up into manufacturing processes. Duisberg joined the Bayer board of directors in 1900, championing his laboratories and ensuring that they received due attention from the management. By that time the chemical staff had grown to over 130 and R & D was being conducted on many fronts.[12]

The Hoechst company near Frankfurt had started off small and late. In 1863 it consisted of three founders, five workers, one chemist, and an office girl.[13] They established their centralised company laboratory in 1883 in a building peripheral to the main plant.[14] Its director, Dr von Gerichten, was charged with a variety of tasks growing out of the dyestuffs manufacturing process, including producing new colours and testing for constant quality. He had a small staff of chemists but maintained very frequent contact with academic scientists. Paul Ehrlich, later to found modern chemotherapy, was one such young academic who came to the attention of von Gerichten on account of his doctoral thesis on the use of methylenblue.[15] But as attention became directed to synthetic medicinals, the main interest at Hoechst turned to the antipyretic effects of alkaloids.[16] Friedrich Stolz, working in von Gerichten's laboratory, developed a number of synthetic alkaloids, finally settling on one preparation for production which he called Pyramidon.[17] In the late 1880s Pyramidon became a major money-maker for Hoechst.[18]

The second major influence on the decision to commit resources to research on pharmaceuticals was the stress on finding a treatment for tuberculosis. Robert Koch, the leading medical scientist of the day, had contacted Hoechst about producing tuberculin and this was seen as a marvellous opportunity to break into a huge new area of the drug business.[19] He suggested that Emil Behring be contacted about a possible role in the development of tuberculin and other bacteriologically based medicines.[20] Behring continued to work in Berlin until 1892 when he was able to leave the armed forces, where he had been obliged to serve as a medical officer.[21] When he moved to Frankfurt to work on Hoechst-sponsored research into the immune response in animals, he found a staff of around a dozen chemists and pharmacists testing drugs and monitoring the production of tuberculin and Pyramidon.[22] The level of sophistication varied greatly within the laboratory. Stolz himself was an imaginative scientist, and a very well-respected research director. His staff included a number of young graduates who conducted routine tasks using only the most standard equipment. The level of sophistication of the equipment was not particularly high, but the application of chemical theory was considerable.

At neither Bayer nor Hoechst is there any evidence that much long-term planning went into the early decisions to fund the laboratories, nor were the cost/benefit calculations done in anything more than a rudimentary way. No systematic analysis was done which would clearly lead to the conclusion that it might be worth some investment. The companies were in any case making good profits on their dyestuffs and could easily afford a slightly speculative investment.

In Germany drugs were tested in hospitals by academic physicians and marketed through pharmacies. The medical profession was less conservative than in Britain and there was a general willingness to experiment with new products. Physicians were also excited generally about the promise of medical science and were on the lookout for new, science-based products. In German pharmaceutical firms the production of new medicines was essential for competitive advantage. The industry was growing quickly, especially with the entry of the dye manufacturers into pharmaceuticals, and established markets were constantly being challenged. The best way for these firms to distinguish themselves and open new markets was for them to introduce new medicines. This they did with great competitive zeal. Contacts with universities were maintained at a very high level and every potential new product was immediately patented. Although few of these patents were used in the manufacture of new drugs, they did delineate areas of enquiry in which the company was specialising.[23] The number of products which each firm then offered for sale became tremendous. While there was considerable overlap among standard medicines on offer, there were also many branded or speciality products which were more or less unique to the firm.[24] The stimulating industrial environment where new firms such as Hoechst and Bayer competed with older manufacturers with traditional markets such as Merck,[25] encouraged a very different style of laboratory and radically different links with universities.

Just before the outbreak of the First World War, Carl Duisberg summed up his feelings about trends in the industry:

> What an organization, what boundless intelligence is necessary, and what immense energy has to be expended in order to discover a new synthetic remedy and to smooth its path through the obstacles of commerce! First, we need a fully equipped chemical laboratory, then a pharmacological institute with a staff of men trained in medicine and chemistry, an abundance of animals to experiment upon, and finally – the latest development in this field – a chemotherapeutic and bacteriological department, equipped according to the ideas of Paul Ehrlich; all these must be in close connection with one another. Whatever has been evolved, and after much painstaking effort selected as useful, finds it way into the manufacturing department, there to be elaborated in the most minute details and brought to the highest possible pitch of perfection. Now begins the arduous work of the scientific department! Here the right sponsors must be found, here all prejudices must be brushed aside and an extensive propaganda initiated. Next, a host of clinicians and practitioners must be called into requisition so that what has been evolved in the silent workshop will be conducted on a staunch ship into the wild sea of publicity. And, finally, it is the calculating salesman's turn; he must bring in enough to cover all the expenses of the innumerable experiments that have been made, if the new drug, which has swallowed so much money, is to survive and prosper. . . . If, however, a great hit is made . . . then the envious, the patent- and trademark-violator, and even the smuggler, cling to our heels and seek to rob us of our profits, which, taking everything into consideration, are really not large.[26]

The German industry then began with company-sponsored R & D in a haphazard, reactive way. No strategy for corporate laboratory development was discussed, and even the notion that there should be an adequate return on investment was missing. The people brought in were for the most part either chemists or general medical scientists, and rarely specialists in pharmaceutical chemistry. By the period immediately before the First World War this had changed, as it became apparent that laboratories capable of conducting academic style research were necessary. For competitive advantage and to be able to keep up with the raised technical standards of both industry and the medical profession, laboratories had to be integrated into the newly expanded corporate structures of large fine chemicals and drug companies.

II

This commitment to academic chemistry and a strategy based on patenting stands in contrast to trends in the United States. There, only a few major pharmaceutical firms had any kind of laboratory by the early 1890s. They varied in size and use depending on the products being manufactured and the relative importance of introducing new products as compared with the importance of importing, re-packaging and marketing products which were produced elsewhere.[27] Most companies, even those doing a minimal amount of manufacturing, employed at least one chemist, whereas others, such as Smith Kline and French of Philadelphia, had a staff of around a dozen chemists and laboratory assistants[28] in addition to occasional contracts or coordinated research projects with members of the faculty at the local pharmacy college.[29]

Smith Kline established its 'analytical laboratory' in 1893 when Lyman F. Kebler was instructed to work on the chemistry of drugs.[30] Kebler had been trained in pharmacy at the University of Michigan, in a programme which was institutionally linked with Parke Davis and Co.[31] His familiarity with the latter's practices must have been highly appreciated when he joined its rival, Smith Kline, on earning his degree in 1892. He organised the laboratory the following year, adopting the title of 'chief chemist'. That laboratory became specialised almost immediately, concentrating on the analysis of possibly adulterated goods which the company bought from suppliers. The company advertised widely that any drugs which did not meet its standards would be sent back. With Smith Kline's expanding trade, investment in the laboratory probably repaid itself quickly in the value of diluted and adulterated drugs exposed as well as the advertising advantages it brought.[32]

Parke Davis & Co. of Detroit, taking advantage of their close ties with the University of Michigan School of Medicine, began advertising already in the early 1890s that they had a large qualified scientific staff whose work included standardising their products and assuring purity.[33] At that time Parke Davis was almost alone in using this as an outstanding selling point, but by 1910 this advertising practice became one of the hallmarks of those manufacturers who were trying to distinguish themselves from the poor reputation earned by hucksters in the trade. Distinct change was taking place in the mid-1890s stimulated by the simultaneous availability of a new stock of trained bacteriologists and medical scientists, and an optimism that new science-based therapeutics might be available to cure a wide range of diseases.

In addition to varying in size, company laboratories varied in level of sophistication. This was to some extent the function of the products which the company was dealing with, but it was also related to the availability of scientific manpower. A more isolated company, for example, Eli Lilly in Indianapolis, Indiana, had very few contacts with the scientific and medical community other than a couple of local doctors who were then friends of the Lilly family.[34] From the early 1890s, Lilly employed one pharmacy graduate at a time, first a general pharmacist trained at Purdue University School of Pharmacy, and then later a graduate of the higher quality Philadelphia College of Pharmacy. Only in the mid-1890s was a second trained member of the scientific staff added, a pharmaceutical botanist.[35] The American pattern then was for companies which had long specialised in manufacturing pharmacy to develop their ties with local pharmacy colleges, medical scientists who were specifically interested in therapeutics, and with the region's medical community. These people were brought in more for their role as adjuncts to the marketing function than as a means of building a product-development staff.

III

In Britain the use of trained chemists in the many smaller pharmaceutical firms was not on a very large scale. A few of the companies had been formed by scientifically-minded pharmacists or who had at least a tradition of encouraging scientific work. Companies whose proprietors were members of the gentlemanly businessmen's scientific circle often encouraged chemical research to a limited extent.[36] What is striking about this early British company-supported scientific work was the manner in which it resembled amateur chemistry as one might find described in the diaries of a naturalist. Allen & Hanburys, which since the late eighteenth century had a reputation for encouraging chemical experimentation, maintained a well-stocked laboratory.[37] However, neither Allen & Hanburys nor the similarly inclined T & H Morsons & Sons[38] ever organised a concerted development programme for new products. Correspondingly the expectations of both the company and of their customers did not include the introduction of a large number of novel products or the application of systematic chemical testing techniques.

At Allen & Hanburys, William Allen indulged his broad-ranging chemical interests in his pharmaceutical laboratory from the 1790s.[39] His manufacturing pharmacy in the City of London formed one of the most influential loci for metropolitan science of the age. In addition to providing basic facilities and good company for the nucleus of groups later to be influential in the formation of a number of cultural and intellectual societies, the company set a style of enquiry for serious pharmacists.[40] Daniel Hanbury, who later became a partner, was strongly influenced, as were other members of the Quaker pharmacy community such as Daniel

Bell and Luke Howard. Bell's pharmacy and Howard's business both grew to significant proportions as manufacturing businesses, and both continued their tradition of gentlemanly dignified puttering in chemistry.

Thomas Morson similarly engaged his scientific pretensions in his London pharmacy established in 1821. But Morson looked beyond the putterings of the older generation and applied some of the more systematic chemistry he had learned in Paris, then a centre of pharmaceutical chemistry. In his work on alkaloids in particular, Morson was able to gain a quick lead with quinine imports.[41] In the year he started business, Morson produced six pharmaceutical products but within four years he was using his background as apprentice surgeon and French-trained chemist to expand markedly.[42] From new facilities he prepared over 50 products and sold them either as fine chemicals to be used as intermediaries, or as medicines.[43] Morson's preparations were included in a guide 'for administration of common medicines, by indicating their respective doses, and thereby enable those persons resident at a distance from their Medical Attendent to employ them without danger'. The guide[44] represented one of two ways Morson tried to reach his new clients. The other way was to foster contacts with other fine chemical manufacturers, dyers, perfumers and suchlike, in order to sell chemical intermediaries to them. Between his expertise as a trained pharmacist, and the lucrative medical business, pharmaceuticals soon came to dominate his business, as importer and wholesaler.

The style of business development exemplified by Allen & Hanburys and by Morson came to characterise the British pharmaceutical industry during the nineteenth century. Most of the firms were centred around their reprocessing laboratories, and in so far as they introduced new products, it was by combining standard medications. By the 1890s in any case, the market for innovative chemical intermediaries had been usurped by Germany, leaving British firms with the business they did best, international trade and marketing.

For a general description of pharmaceutical product development laboratories in Britain in the mid-1890s we have to consider the combination of very small companies and their limited research efforts. The outstanding characteristic of laboratories in British pharmaceutical companies was the variety of forms they took and the unsure position that they held within the company structure. May & Baker, which was concentrating on antipyretics among the few new products which they were marketing, employed at best one chemist at a time up to the First World War.[45] They did maintain contacts with some university scientists, and occasionally asked for and paid laboratory staff within the universities to conduct tests. But they never expanded their research effort or systematised a programme linked to academics. The Evans Medical Co. in contrast maintained formal and direct contact with medical scientists at the University of Liverpool.[46] In their case the contact developed after the introduction of biological products which grew out of the new sciences of bacteriology and immunology and led to the production of serums and antitoxins against diphtheria, tetanus, meningitis, and other contagious diseases.[47] Evans, shortly after the turn of the century, began producing biologicals and were, along with the Wellcome Co. and the Lister Institute, the only British producers of diphtheria antitoxin, tetanus antitoxin, and the related serums and vaccines. The Lister Institute was able to do that because they had a large staff of bacteriologists;[48] Evans conversely chose to maintain contacts with bacteriologists and in particular physiologists who could perform the routine toxicity tests and analyse the level of concentration in the serums.[49]

The Wellcome Co. was unique among British firms in maintaining a large well-funded company laboratory.[50] From 1894 the Wellcome Physiological Research Laboratories, sponsored by and directly integrated into the company,[51] began to build up the facilities and to hire the talented staff necessary to produce antitoxins, serums and vaccines.[52] Wellcome's direct investment in a laboratory and its recognition of the need for scientific staff, followed the pattern of leading German and American producers.

This was not a chance occurrence. Both Sir Henry Wellcome and his partner, Silas Burroughs, were originally from the United States and had trained at the Philadelphia College of Pharmacy, the leading American school of pharmacy at the time.[53] Wellcome's move to Britain as an agent for the Wyeth Brothers Co. started him on a venture first to sell American products in England, and later with his own company, to apply many US trends to the British scene. His introduction of a research and development laboratory was a direct effort to adopt methods which his American colleagues had originally brought across the Atlantic from Germany and France.[54] Wellcome's plan was simple in conception and clearly planned. While the Lister Institute was having problems raising money and securing the most appropriate staff,[55] Wellcome saw that he could finance a similar operation but concentrate on development problems directed at potentially marketable products. He took a personal interest in appointing the staff and initiated programmes in the physiological standardisation of drugs. His first efforts were to reproduce the results of the French and German developers of diphtheria antitoxin.

Diphtheria antitoxin was one of the first medical products which demonstrated the potential profitability of a company research establishment. Developed in Berlin and produced first by laboratories there, in Frankfurt and by the Pasteur Institute, diphtheria antitoxin was quickly recognised to be of value for treating the disease which had long been endemic in urban areas.[56] When the Lister Institute was unable to meet the challenge, opportunities opened for Wellcome.

The Evans Medical Co. (Evans, Sons, Lescher & Webb of Speke) followed Wellcome in some important respects.[57] Members of the scientific staff there had maintained close contact with the expanding Liverpool University Medical School and its new scientific endeavours. Edward Evans (knighted in 1906 for 'political services') was in particular closely linked to the leading social, political, and industrial interests in the region. Other members of the family frequently served on committees in public service, for commercial organisations, and associated with the University. One such committee, which in fact served all these roles, was that of the Incorporated Liverpool Institute of Comparative Pathology.[58] Jointly sponsored by the university, the medical faculty, the city medical institutions, the veterinary department of the city, the Chamber of Commerce and a committee of shipowners, the Institute was conceived of as a combination of research institute for the university, a service to the animal products importers, and as a potential source of new human and veterinary medical products.[59] The Evans company had a number of projects in mind for the new Institute when they provided the main grant to support the construction of laboratories. Within the first few months plans were made to manufacture a variety of vaccines, sera and other products used in the treatment, prevention and diagnosis of disease. Calf Vaccine lymph (for smallpox vaccine) and antidiphtheritic serum were to be two of the major products, and would be marketed by Evans.[60] In additions, plague prophylactics and plague serum, anthrax vaccine and anthrax serum, blackleg vaccine, tuberculin, and mallein (for glanders in horses) were to be standard products from the Institute. In addition, Professor Annett of the University Medical School, the director, had developed a kind of 'virus' to kill rats, mice and vermin which quickly became a big seller.[61]

In the new areas of drug production, the biologically based serums and vaccines, the Liverpool Institute of Comparative Pathology served the Evans company admirably. There the company had the resources of a fully staffed serum department headed by a university professor and a lecturer in bacteriology in the School of Hygiene at the University, Dr Stenhouse Williams,[62] along with a chief laboratory assistant, three laboratory and office assistants, 'a boy, 2 lady packers, a sterilising room and medium room assistant', and a farm staff employing a bailiff and two farm hands.[63] The positioning of the Institute, further, promised excellent links to a wide range of contacts who would obviously be beneficial to Evans' business. Throughout the public health and medical facilities of the city in particular, but also 'The County Councils of the District, private companies with interests in common with the Institute, as well as some Colonial Governments, have engaged the services of the staff, and used the products of the Institute from time to time'.[64]

In fact, sales for all products were extremely low. Only the 'Liverpool [rat] Virus' ever sold over £1,100 worth in a year, and antidiphtheria serum, which it was thought would be the biggest product, rose only to £420 (in 1907) and then fell to 30 per cent of that level after the cancellation of an order by Liverpool City Hospitals.[65] The failure in this area did not deter the Evans company significantly, despite the fact that it led to closure of the Institute. In fact the final demise of the Institute seems not to have been much of a disappointment to the principals. Drs Annett and Stenhouse Williams had secured their positions within the Medical School, and the directors, Sir Robert Boyce and Professor C. S. Sherrington were deeply involved in many other activities and had not taken much of a part in operating the Institute.[66]

The Evans company positively benefited from the dissolution. They purchased the plant and facilities in 1911 after it had gone into voluntary liquidation, a move presumably supported by the Evans representatives in the administration of the Institute.[67] In fact Professor Annett turned over the details of his production procedures to the company, and continued to receive a retainer from them, first at the level he had received from the Institute (£300 p.a.) soon raised by an additional £100 per year.[68]

The years of association with the Institute marred the beginnings of in-house science at the Evans company proper. From around 1904 the board began implementing a scheme to allocate over £12,000 for a 'laboratory extension' with the purpose to 'commence fresh manufactures'.[69] The extension was to include a new building, complete with electricity for light, power and a lift, sprinklers and, in 1905, two steam dynamos.[70] The work there was to be chemical and not to overlap with the biological researches and production of the Institute of Comparative Pathology. Staff was assembled when the construction was completed, the first priority being to find someone to direct the new product development work. The board agreed that what was needed was an 'Expert and research chemist for [the] laboratory ... advertisements should be inserted at once with a view of securing [an] expert chemist specially to watch and take advantage of Duty-free Alcohol developments for manufacturing purposes'.[71] The purpose, then, was clear. Product development and innovation might be considered some part of the job, but the main concern was to modify existing preparations and production processes taking advantage of new, cheaper sources of alcohol.[72] New products for Evans, as with many other British companies, were introduced mainly from Germany, although the patent infringement suit which the Pasteur Institute threatened over the sales of blackleg vaccine is indicative of another source of new products.[73]

So we can see from the examples of two of the leading innovative firms that while some efforts were being made to develop new products, to establish research laboratories, and to foster ties with academic workers in key areas, British companies were still far smaller, and notably less successful in this endeavour, than their German and American counterparts.

It is only fair, then, to ask what potential the British companies had to recruit good people to develop laboratories. In the case of Evans' advertised position, the company had to recruit a Dr Moore from Canada to fill the job.[74] The statistics show that the number of trained people in relevant fields was lower in Britain than in Germany and even the American universities were already training chemists and medical scientists.[75] Furthermore, Americans had already long had a tradition of studying in Germany and a large proportion of leading scientists had been to German or Viennese institutions for training.[76] This appears less the case for British scientists.

This impression seems to be confirmed by one notable British exception, Edmund Muspratt, the Mancunian chemical manufacturer who studied with Liebig in the 1870s. He remarks on his American colleagues in his memoirs,[77] but makes little mention of the few British chemists studying in Germany. Of course British students had other opportunities for chemical and advanced medical training. The number of chemistry graduates from Edinburgh and London universities in particular grew quickly during the period.[78] There were opportunities to study chemistry at other institutions also. At the Royal Institution and at the Royal School of Mines, in addition to other places such as the Surrey and London Institutions, there were permanent positions for lecturers or professors of chemistry.[79] Furthermore, the quality of medical education offered was quite high, although it seldom included much medical chemistry, and had no pharmacology.[80] What stands out most prominently, however, is the lack of apparent connection between university science training and the evident needs of industry.

British companies such as Whiffen & Morson, and May & Baker had by the turn of the century settled into a set pattern of business. Their main concern was to import raw materials under the best financial terms, process them as quickly and efficiently as possible and sell or export them, maintaining a flow as much as possible. They were small firms whose range of business was narrow and whose complex contractual and marketing arrangements with other companies were finely balanced. Although they ostensibly intended to introduce new products, they actually had an aversion to innovate, for fear that it would interfere with their business practices.[81] With the introduction of clearly efficacious analytical tools, they were quite willing to use them as they saw best. This was in testing the quality of shipments, compensating for inadequately potent mixtures by adding more of the active ingredient, and in maintaining a reasonable, but undemanding level of consistency in the final products.[82] Chemists were rarely if ever used in the pre-war period to test for new potential medicines, and only occasionally were their analytical skills used even to discover what ingredients were in their competitors' products.

IV

In Germany and the United States competition based on the availability of new products, and a general urge for novelty in the medical profession stimulated the drug industry to investigate unusual preparations. It is not evident that there was a similar market-pull in Britain which could be used as a stimulus for extended laboratory work in the British drug companies. The proliferation of new products in Germany and the United States created a number of problems. Most significant was the perceived need to distinguish between useful products and products which were either useless or dangerous.

In the period between the turn of the century and the First World War, both countries began to look very seriously at the need for some form of drug regulation. In Germany a partnership emerged between the government hygiene laboratories, the companies, universities, and philanthropic research laboratories. These various forces were all brought together in Frankfurt where new products from Hoechst were often stimulated by research at the University of Frankfurt or other universities in the area, and medicines were tested, analysed, regulated and licensed in the co-operative involvement of the Hoechst Co., the University of Frankfurt medical school, the Hygiene Institute in Frankfurt (government-funded) and the private research laboratory, the Georg-Speyer-Haus.[83] Begun in 1902, by 1906 this combination of institutions was working in informal co-operation and had been involved with the production procedures for a number of new products. The structure as it existed in the first decade of the century was tested and proved satisfactory in the long procedures leading up to the famous development of the anti-syphilitic drug Salvarsan by Paul Ehrlich and Hoechst.[84]

Those two dates were also the key years for drug regulation in the United States. In 1902 the first American Act regulating the manufacture of new medicines and licensing their producers was passed after two incidents involving in one case contaminated diphtheria antitoxin which led to the deaths of a number of children in St Louis,[85] and in the other case the suspicion that large quantities of smallpox vaccine were contaminated with tetanus leading to a number of deaths in and around Camden, New Jersey.[86] The 1902 Act for the first time called on the government to monitor the production of new medicines by a permanent laboratory of trained medical scientists who would have inspection and licensing powers over all producers who sold biological products across state lines.[87] This Act was possibly the single most influential stimulus for the establishment of pharmaceutical laboratories in American drug companies. Its first and most obvious significance is that laboratories were required to qualify for licenses.

It quickly became clear to American drug producers that the Act made possible a new means of competition: competition based upon the quality of laboratory facilities that the company could provide and the stringency of the standards which they could meet. It also implied that the leading producers could benefit by extremely tough standards which would be difficult for smaller competitors to equal. Laboratories of the American producers of these innovative medicines quickly grew and their functions began to expand with their size. No longer were they used simply to standardise basic products or to fiddle with new combinations of drugs, but rather they were now in the business of setting standards, meeting the requirements of the regulators, and producing as many novel products as they were capable of doing which would distinguish their product lines from those of their competitors.

The regulators in turn quickly became dependent upon company laboratories for the expertise they needed to assess new products and new production procedures. A partnership between regulators and the leading producers such as Smith Kline and the H.K. Mulford Co. emerged whereby frequent contact was maintained with the government laboratory and standards were set based upon the trends within those companies.[88] In this way the government, through the public health bodies, became directly involved in the affairs of commercial laboratories. Laboratories quickly became central figures of the firms, even though in many cases their roles had not been determined and their level of efficiency was low.

A similar but not identical development was taking place in Germany, where fluid co-operation between the universities, the government and the companies had been important for a longer period of time. The German medical profession was less faddishly excited about novel preparations for the sake of their novelty than appears to have been the case in the United States. Whereas on the one hand German producers were taking out new patents on potential medicines they in fact marketed a more limited variety of preparations than their smaller American counterparts.[89] Novelty alone was not an incentive for the German producers. They concentrated on a small number of well-defined, specifically targeted and medical conditions for which they produced drugs.

In Britain the public health establishment was not particularly interested in distributing or regulating the new medicines through their offices. There arose no great interest among British physicians in novel products. British drug companies maintained their market position by securing their traditional control over British retailers, as well as lucrative sources of raw materia medica from tropical colonies.[90] They also maintained their market control of exports to those colonies. Firms such as Howards, Morsons, Whiffen and others held this position through a myriad of legal and sub-legal contracts and agreements with, in particular, German drug companies.[91] They also obtained licenses to produce a small number of the most attractive German medicines such as the antipyretics. Regulation was not stringent in Britain and there was not the expectation among the British medical or pharmaceutical companies that drugs would need tough standards.

The one exception, Burroughs Wellcome, seemed not to prosper particularly from their having an investment in a laboratory. On the contrary, it seemed that Wellcome's two marketing innovations: the tabloid formulation and the extensive use of detail men, would account for their success.[92] It appears to have been a reasonable choice on the part of Wellcome's competitors not to follow their example and divert funds which might otherwise be used to extend marketing or to make the conduct of business more congenial.

Conclusion

It was easy for German companies to have laboratories. They had a tradition of scientific research; they had an amenable academic community; they had the co-operation of medical men and medical institutions; and they had very active and fruitful government support. For American companies the role of the laboratory was not clear to the manufacturers, but from a relatively early stage they maintained them to use primarily for standardising. Later they were important in producing a variety of novel and apparently sophisticated products for a medical profession newly self-consciously 'scientific'. The influence of the public health bodies was also significant in the US both as a market for medicines which were intended for epidemic diseases and urban endemic diseases such as diphtheria, tetanus and tuberculosis, and also because they were a visible endorsement for new scientific methods. With regulation, a strategy for competition between American drug producers began to focus more and more upon the level of technical sophistication and the establishment of standards which could be used as weapons against smaller and less highly capitalised companies.

British manufacturers did not establish laboratories for R & D and retained their small staffs of chemists almost exclusively for the limited tasks of checking the quality of raw materials the company purchased and of maintaining a rudimentary level of standardised production. The Wellcome Laboratory stands as an exception. Established in 1894, it maintained its stature as Britain's foremost commercial pharmaceutical laboratory and it was not copied. However, in the inter-war period it became evident that British companies were seriously under-invested in R & D laboratories, and there were many attempts to compensate for this.

Business History Unit
London School of Economics

NOTES

This work comprises some preliminary results from research under way into the history of the pharmaceutical industry in Britain, Germany and the United States. I would like to acknowledge help from Dr Gary Werskey, Judy Slinn and librarians and archivists in the following companies: Smith Kline Corp., Merck & Co., Hoechst AG, Merck AG, Schering AG, Behringwerke AG, the Wellcome Foundation, Glaxo plc, Merck, Sharp & Dohme (UK). The papers of Parke-Davis laboratories are at the Smithsonian Institution. Other major collections are at the University of Liverpool Archives, the Kremers Reference Files of the University of Wisconsin School of Pharmacy, the College of Physicians of Philadelphia, the National Library of Medicine, the Library of Congress, and the National Archives. Support for this research came from an SSRC/DRG travel grant, the Pharmaceutical Manufacturers Association, the Smithsonian Institution, and the Leverhulme Trust through the Business History Unit, London University.

1. S. Kaku, 'The Development and Structure of the German Coal-Tar Dyestuffs Firms', in Akio Okochi and Hoshimi Uchida, *Development and Diffusion of Technology, Electrical and Chemical Industries* (International Conference on Business History 6, 1980). See also L. F. Haber, *The Chemical Industry during the Nineteenth Century. A Study of the Economic Aspect of Applied Chemistry in Europe and America* (Oxford, 1958).
2. R. Grabower, *Die Finanzielle Entwicklung der Aktiengesellschaft der deutschen chemi-*

schen Industrie und ihre Beziehungen zur Bankwelt (Leipzig, 1910), p. 135, cited also in S. Kaku, op. cit. See J. Kocka, 'Strategy and Structure in German Industrial Development', in A. Chandler and H. Daems (eds.), *Managerial Hierarchies: Comparative Perspectives on the Rise of the Modern Industrial Enterprise* (Cambridge, MT, 1980).

3. G. Meyer-Thurow, 'The Industrialisation of Invention: A Case Study from the German Chemical Industry', *Isis* 73 (1982), 363–81. See also J.J. Beer, 'Coal-Tar Dye Manufacture and the Origins of the Modern Industrial Research Laboratory', *Isis* 49 (1958), 123–31.
4. T.I. Williams, *A History of the British Gas Industry* (Oxford, 1982).
5. Meyer-Thurow, op. cit.
6. J. Ben-David, *The Scientist's Role in Society: A Comparative Study* (Englewood Cliffs, NJ, 1971).
7. Meyer-Thurow, op. cit.
8. German Patent Act, 1877, in reprint of 'Foreign Patent Law, German Patent Law, 1877', *Commissioners of Patents Journal*, 25 May 1877.
9. A similar situation is described in L.S. Reich, 'Radio electronics and the development of industrial research in the Bell system', Ph.D. dissertation, Johns Hopkins University, 1977, pp. 171–4.
10. Hoechst Archive, laboratory records.
11. Meyer-Thurow, op. cit.
12. Ibid.
13. Hoechst Archive, records of central laboratory.
14. Ibid.
15. Dokumente aus Hoechster Archive 'Die zusammenarbeit Paul Ehrlich und Hoechst' (Hoechst, 1972); M. Marquardt, *Paul Ehrlich* (London, 1949); C.E. Dolman, 'Paul Ehrlich', *Dictionary of Scientific Biography* Vol. 4 (New York, 1974), pp. 295–305.
16. Dokumente aus Hoechster Archive, 'Antipyrine' (Hoechst, 1974).
17. Dokumente aus Hoechster Archive, 'Stolz und Pyramidon' (Hoechst, 1976).
18. Hoechst Archive, Pyramidon records.
19. Hoechst Archive, Tuberculin records.
20. Dokumente aus Hoeschster Archive.
21. Ibid. See also H. Zeiss and R. Bieling, *Behring, Gestalt und Werk* (Berlin, 1941).
22. Hoechst Archive.
23. J. Liebenau, 'Patents and the Chemical Industry: Germany, Britain and the United States, 1890–1930', paper presented to the Royal Society of Chemistry Annual Convention, Exeter 1984.
24. Trade catalogues, the best source for information about marketed products are little used, but survive in most company archives.
25. Merck AG, *E. Merck* (Darmstadt, 1957), Merck Archives.
26. *Chemical News*, 23 May 1913, pp. 246–7.
27. J. Liebenau, 'Medical Science and Medical Industry, 1890–1929', Ph.D. dissertation, University of Pennsylvania, 1981. See Chapter 1.
28. Smith Kline Archive, 'Analytical Laboratory Notes', *c.* 1908.
29. Ibid. See also 'Smith Kline and French', *Pharmaceutical Era*, 31 (December 1896), p. 957.
30. Ibid. See also T. Wagner, 'Smith Kline and French', unpublished manuscript, *c.* 1961 in Smith Kline Archive.
31. M.K. Weikel, 'Research as a function of the pharmateutical industry, the American formative period', M.A. thesis, University of Wisconsin, 1962.
32. Smith Kline Archive, 'Analytical Report, 1908', p. 3.
33. Parke-Davis Records, Smithsonian Institution; see also M.K. Weikel, op. cit.
34. E.J. Kahn, Jr., *All in a Century* (Indianapolis, IN, 1976), pp. 25–6.
35. Ibid., p. 27.
36. This is best exemplified by the circle of Quaker pharmacists, most notably William Allen, Thomas Bell, Daniel Hanbury. See A. Raistrick, *Quakers in Science and Industry* (London, 1950); J.M. Liebenau, 'Businessmen and Scientists in Nineteenth Century London', paper presented at the Royal Institution, 1982.

37. Ibid. See also Desmond Chapman-Huston and Ernest C. Cripps, *Through a City Archway* (London, 1956); Ernest C. Cripps, *Plough Court* (London, 1936).
38. T. & H. Morson Papers, in private collection.
39. See note 37, above.
40. Allen & Hanburys Papers, Glaxo Archive.
41. *Chemist & Druggist*, 29 June 1916, pp. 135–8; *Pharmacy Journal*, 7 March 1874, pp. 726–7; *Pharmacy Journal*, 18 April 1981, p. 416; *Chemist & Druggist*, 14 May 1870, pp. 142–3.
42. Ibid.; T. Morson Diary, Morson Papers.
43. T. & H. Morson Catalogues, Morson Papers.
44. First Morson Catalogue, *c.* 1825, Morson Papers.
45. See J. Slinn, *History of May & Baker* (forthcoming).
46. Evans Sons, Lescher & Webb papers are at their former plant, now Glaxo, Speke. See also Liverpool Institute of Comparative Pathology papers, University of Liverpool Archives.
47. See Liebenau, 'Medical Science', op. cit., pp. 136–40.
48. Annual Reports, Lister Institute (London, 1908); H. Chick, *et al.*, *War on Disease* (London, 1971).
49. Liverpool Institute of Comparative Pathology papers, 5/3/8.
50. 'Wellcome Physiological Research Laboratories' (London, *c.* 1899), public relations brochure.
51. H.J. Parish, 'History of the Wellcome Physiological Research Laboratories', unpublished manuscript (*c.* 1980), property of author.
52. Ibid.; see also H.J. Parish, *Victory with Vaccines: The Story of Immunization* (Edinburgh, 1968); and interview with Dr Parish.
53. J.W. England, *The First Century of the Philadelphia College of Physicians, 1821–1921* (Philadelphia, PA, 1922).
54. H. Turner, *Henry Wellcome: The Man, his Collection and his Legacy* (London, 1980).
55. H. Chick, *et al.*, op. cit.
56. See for example, F.W. Andrews, *et al.*, *Diphtheria, its Bacteriology, Pathology and Immunology* (London, 1923); pp. 13ff; and H.J. Parish, *History of Immunization*, op. cit., pp. 118–40.
57. *Evans Medical* (Liverpool, *c.* 1955), a public relations history of Evans Sons, Lescher & Webb; and Evans papers, Glaxo Archive.
58. Letter Book, Evans papers.
59. Ibid. Liverpool Institute of Comparative Pathology papers.
60. Evans papers.
61. Ibid.
62. T. Kelly, *For the Advancement of Learning, University of Liverpool 1881–1981* (Liverpool, 1981).
63. Evans papers.
64. Liverpool Institute of Comparative Pathology papers.
65. Ibid.
66. Ibid. See also J.C. Eccles and W.C. Gibson, *Charles Sherrington: His Life and Thought* (Berlin, 1979).
67. Liverpool Institute of Comparative Pathology papers.
68. Ibid. See also Letter Book, Evans papers.
69. Letter Book, Evans papers.
70. Ibid.
71. Ibid.
72. Alcohol Duty Controls, 1902.
73. Letter Book, Evans papers.
74. Ibid.
75. D. Cardwell, *The Organisation of Science in Britain* (London, 1968).
76. T.N. Bonner, *American Doctors and German Universities, A Chapter in International Intellectual Relations, 1870–1914* (Lincoln, NB, 1963).
77. E. Muspratt, *My Life and Work* (London, 1917), pp. 18ff.

78. D. Cardwell, op. cit.
79. C. Russell, *The Social History of Science in Britain* (London, 1983).
80. J. Parascandola, *History of Pharmacology* (in preparation).
81. See, for example, D. Schon, 'The Fear of Innovation', in R.M. Hoiner, S. Kingsbury, D.B. Gleicher (eds.), *Uncertainty in Research Management and New Product Development* (London, 1966).
82. See Whiffen papers, Greater London Record Office.
83. 'Geschichte des Georg-Speyer Haus', in *Berichte a. d. Georg-Speyer Haus* (Frankfurt a-M, 1952).
84. Ibid.; Hoechst papers; see also P. Ehrlich and S. Hata, *Die Experimentelle chemotherapie der Spirilossen* (Berlin, 1910).
85. US Congress, Senate, 58th Congress Report No. 1209, *Hearings Before the Committee on Manufacturers ... for preventing the adulteration of foods ... drugs* (Washington, DC, 1904), pp. 78–97.
86. J. McFarland, 'Tetanus and Vaccination: An Analytical Study of 95 Cases of the Complication', *Lancet* (1902ii), 730–35; see also H. K. Mulford Co., 'The Exposure of a Disreputable Proceeding', pamphlet, *c.* 1902 in J. McFarland papers, College of Physicians of Philadelphia.
87. Federal Act of 1 July 1902; 32 Stat L.; 728; see Liebenau, 'Medical Science', op. cit., pp. 237–58 for analysis of the Camden, NJ incident and the 1902 Act.
88. US Hygienic Laboratory Records, National Archives; see also logbook for Hygienic Laboratory, 1902–4, Historical Collection, National Library of Medicine.
89. Compare trade catalogues of Hoechst, Merck AG, and Schering AG with Smith Kline & French, Merck & Co., Sharp & Dohme, H.K. Mulford for 1895–1914.
90. Howards and Whiffen papers in Greater London Records Office, Morson papers in private collection.
91. The Whiffen papers contain extensive evidence of agreements of various kinds from *c.* 1880–1916.
92. H.J. Parish, 'History of the Wellcome Physiological Research Laboratories', op. cit. 'Detail men' were salesmen who visited physicians to describe new products.

[17]

The 1993 Newcomen Prize Essay

IBM and its Imitators: Organizational Capabilities and the Emergence of the International Computer Industry

Steven W. Usselman
Department of History
University of North Carolina, Charlotte

In this paper I will examine the utility of the theory of organizational capabilities and evolutionary economic change, as conceptualized by Richard Nelson and Sidney Winter, for explaining the emergence of the computer industry. In particular, I will discuss whether this theory fits the historical experience of IBM during its first two decades of involvement with computers from 1945 to 1965. To provide additional perspective, I will draw comparisons between the IBM case and several efforts to imitate its success. These include the failed attempts of GE and RCA to enter the computer business, the mixed record of the British in fostering an institution comparable to IBM, and the spectacular success of the Japanese in mimicking the American colossus. My treatment of these imitators will not be exhaustive. I wish simply to draw attention to common elements that help explain success and failure in the computer industry and to examine whether those elements fit the theory of Nelson and Winter.

The point of this exercise is to provide an empirical test of a theory that holds great promise for those of us who believe bureaucratic organizations and technical innovation must be brought to the center of discussions of economic change. Of the many attempts to construct such a theory, I find that of Nelson and Winter by far the most attractive. Thus far, however, Nelson and Winter have addressed their work primarily to economists, for whom they have a rather devastating message. Expecting resistance, they quite sensibly have steered clear of anecdotal evidence and have sought to keep the discussion as much as possible on the theoretical grounds economists prefer. Models, not examples, predominate. I would find the theory more persuasive, and

BUSINESS AND ECONOMIC HISTORY, Volume 22, no. 2, Winter 1993.
 ISSN 0849-6825.

S. Usselman / 2

perhaps more useful, if it were supported by some empirical studies. Since any theory that purports to explain modern economic growth must be able to explain plausibly the course of development in the computer industry, it is a logical candidate for study.

The central argument of this paper is that, on the whole, the theory of Nelson and Winter holds up remarkably well to this test, and that the computer industry provides an outstanding example of the process of economic change they describe. During its formative stages in each of the three countries studied, the industry followed a course that can be explained in terms of the organizational capabilities of existing business institutions and the market pressures that acted as a selection mechanism upon them. IBM succeeded because the hothouse of the Cold War computer industry rewarded it for what it already was; GE and RCA stumbled because their established strengths mattered less in that environment. The British industry languished under the twin burdens of a sluggish market and a deceptively weak leading firm; the Japanese flourished when state-stimulated markets drew out capabilities of firms whose histories resembled those of IBM.

The bulk of this paper seeks to provide a basis for these generalizations, but in the process it also points up certain aspects of the theory that might require further elaboration and perhaps even some revision. At the end, I will discuss some of these and ponder their implications for the role of theory in business history.

A Brief Synopsis of the Theory

Before turning to the computer industry, let me summarize briefly the salient features of the theory. I draw here primarily on ideas presented by Nelson and Winter in their 1982 book *An Evolutionary Theory of Economic Change*. There Nelson and Winter replace the classicist's image of a sea of free producers with a portrait of the economy as consisting of a fairly limited number of established firms or organizations. Firms, like organisms, possess various attributes and abilities. Nelson and Winter refer to these as "organizational capabilities," which ordinarily are expressed in a set of "decision rules." These capabilities involve a great deal of routine, but they also undergo constant modification, in part as a result of deliberate efforts to solve problems and in part because of random events (exogenous shifts in demand caused by events such as a war or a radical technical breakthrough). The market constantly winnows out those modifications or adaptations that do not result in profits.

At its core, this theory rests upon an analogy between firms and skilled individuals. It asks us to think of firms as possessing certain deeply ingrained habits or routines, much like skilled practitioners. Such routines exhibit three key features. First, they are "programmatic," meaning that they generally involve a sequence of actions. In a firm, the sequence might link together many people or subunits. Second, they are undergirded by "tacit knowledge"--knowledge that one has but does not think much about and often cannot articulate fully. Third, they involve many choices, but usually these are made automatically, without awareness that a choice is being made [Nelson and Winter, 1982, p. 7]. Just as craftsmen develop their skills through practice, firms acquire these routines by doing. Over time, they come to possess a body of routine behaviors that is so deeply embedded in the firm that it is taken for granted. No one may even be able to articulate it or explain how it is passed on and maintained.

A firm *is* its routines, for better or worse. Market conditions may reward a firm for its routines or penalize it. But we should not expect a firm to break dramatically from its routines, any more than a skilled craftsman can suddenly change course in midcareer. Indeed, change for an organization might be even more difficult than for an individual because the comparative weakness of central control makes conscious, deliberative choice difficult. Nelson and Winter are emphatic on this point:

> One cannot infer from the fact that an organization functions smoothly that it is a rational and "intelligent" organism that will cope successfully with novel challenges. If anything, one should expect environmental change to make manifest the sacrifice of flexibility that is the price paid for highly effective capabilities of limited scope [Nelson and Winter, 1982, p. 126].

Even when a firm does change in response to the environment, those changes will strongly resemble what came before. Chiding those who operate under the assumption that a firm can pursue any response and become anything it wants to be, Nelson and Winter conclude that

> it is quite inappropriate to conceive of firm behavior in terms of deliberate choice from a broad menu of alternatives that some external observer considers to be 'available' opportunities for the organization. The menu is not broad, but narrow and idiosyncratic; it is built from the firm's routines, and most of the 'choosing' is also accomplished automatically by those routines. This does not mean that individual firms cannot be brilliant successes for a short or long period: success or failure depend on the state of the environment . . . . Efforts to understand the functioning of industries and larger systems should come to grips with the fact that highly flexible adaptation to change is not likely to characterize the behavior of individual firms. Evolutionary theory does this [Nelson and Winter, 1982, pp. 134-35].

These passages capture the essence of the argument as I understand it and will apply it to the computer industry. When I conclude that the industry fits the theory, this is what I have in mind. Even in an industry often characterized as experiencing revolutionary change, we can detect substantial elements of continuity. The shape of the new can be seen in what came before if one looks closely at the embedded capabilities of the firms involved and pays particular attention to what Nelson and Winter refer to as the "programmatic" nature of their routines. Much of the apparent change resulted not so much from conscious choice as from the "automatic" functioning of routines that had a built-in capacity for innovation along certain lines.

The United States

The story of computing in the United States through the mid 1960s can largely be boiled down to one issue: how and why did IBM come to occupy its dominant position?[1] By the mid fifties, IBM had secured approximately 85 percent of the domestic market, a share it

1 This section of the paper is based primarily on my own monograph, which is currently in progress [Usselman, 1993]. Full, detailed references are not yet available.

would retain (if we treat personal computers as a separate entity) for the next three decades. The company enjoyed a similar position in much of Europe and Asia as well [Brock, 1975; Flamm, 1988]. How did this happen? My answer is simple. The war brought forth a "new" technology--the computer--and dramatically increased the demand for it. Much of this demand came initially from the American government for reasons having to do with the Cold War, but private-sector demand proved vigorous as well. Like most "new" technology, the computer had important antecedents, and existing firms had developed capabilities related to those antecedents. None had done so more thoroughly than IBM. Despite some superficial differences between computers and the earlier tabulating equipment that had formed the core of IBM's business, computers involved a mix of knowledge and capabilities that matched those existing at IBM extraordinarily well.

To appreciate this match, we need to understand some things about the nature of computers and the nature of IBM's business prior to the war. Let's begin with computers.

What we think of as the modern, programmable computer had its proximate roots in the 1930s and World War II. A number of scientists, mathematicians, and engineers, working as individuals or in small groups, built one-of-a-kind machines that could, with some adjustment, perform various complex calculations. These efforts took place in many countries, almost always in university laboratories. With the onset of war, most of the calculations arose in the course of military projects, such as the compilation of ballistics firing tables, the processing of information obtained from radar, and the breaking of codes. In the United States, the ballistics problem gave rise to the most conspicuous efforts, most notably the ENIAC machine developed by Eckert and Mauchly at the University of Pennsylvania and the Mark I designed by Howard Aiken at Harvard. For reasons of security, the work associated with code-breaking and radar remained out of public view, but key figures within the emerging military/university/industry research complex were well aware of it. The American radar effort, located in the giant Radiation Laboratory that occupied much of the MIT campus during the war, had drawn together engineers and scientists from throughout industry, academia, and the military. After the war the participants scattered, bringing knowledge of electronics to many corners [Flamm, 1988; Katz and Phillips,

S. Usselman / 6

1982].

Apart from their staggering speed, perhaps the single most important feature of the new computers was that they had no single, specific use. They could be programmed to perform different tasks. Indeed, their expense made it essential that they be programmed to perform different tasks. Admittedly, those tasks initially did not seem all that diverse. Most involved complex calculations based on differential equations. Code-breaking represented a different application with important implications, but it was done in secret by the same sorts of people--mathematicians--using similar thought processes. It took a truly brilliant and prescient individual, such as Alan Turing, to recognize that those methods of reasoning could be used to resolve all sorts of problems [Hodges, 1983]. Even when applied to calculations, however, the computer had to be tailored or programmed to receive certain information, manipulate it in particular ways, and print out or store the results in a specific format. The tailoring process involved many things--the logical arrangement of circuits and switches, instructions encoded in language read by the machine's memory, input devices, storage, and printers (the latter three known collectively as peripherals). Over time, as innovations in magnetics reduced the cost of memory and users continually discovered new applications for computers, more and more of this programming came to involve language [Pugh, 1984]. Early on, switches and circuitry were very important, and they remained an influential factor even with the rise of languages. Peripherals, though often neglected by historians, figured prominently throughout.

The task of arranging circuits logically, programming machines, and combining peripherals in various configurations to perform different tasks would itself have sustained a dynamic technological frontier in the postwar era. Before scientists and engineers went far down that path, however, a second frontier opened. This was solid-state technology. The most famous example, the transistor, was announced in 1947 by Bell Labs. But the transistor was really just the tip of a gigantic iceberg. Drawing on an extraordinary base of knowledge about the electrical, magnetic, and photochemical properties of materials, electronic componentry underwent continual innovation that together produced a "revolution in miniature," culminating in the integrated circuit of the early 1960s (an innovation whose performance would subsequently be improved by orders of magnitude through a sustained effort at manufacturing engineering) [Braun, 1978; Flamm, 1988; Lewin, 1982; Mowery, 1983]. This ongoing revolution had obvious implications for computing. Now the balancing act inherent to computer installations would not only involve a mix of circuitry, programming, and peripherals; it would also have to incorporate continually changing componentry.

In terms of the evolutionary model, the period from the late thirties through the late forties had produced a series of random events. Two technical mutations had appeared--the electronic computer and the beginnings of solid state--and military considerations had prompted the American government to provide a vibrant market for both. With the exception of AT&T, these random events had occurred almost independently of established business organizations. Because of its unique relationship with the United States government, moreover, AT&T would not itself attempt to exploit the mutations commercially, but it would facilitate their diffusion to others through licensing agreements and a program of technical symposia.

Now, what of IBM, the firm that would prove the most important of the "others"? At the onset of war, IBM was a solid, moderate-sized corporation that leased electromechanical accounting equipment manufactured at its plant in Endicott, New York. Led by Thomas Watson, a graduate of John Patterson's NCR school of management, the company had a reputation as an outstanding sales organization. IBM salesmen worked continually to build their "installed base" of leased machines, which each month earned them and IBM rental income. Salesmen could, of course, increase their base by attracting new customers, but they could also do so by persuading existing customers to use novel arrangements of IBM equipment to perform new tasks. The production facility in Endicott operated as a mechanical job shop, responding to requests from the field for solutions to particular problems. It constantly took gears, ratchets, and relays obtained from outside suppliers from which it produced novel machines, and it devised numerous ways of joining counters, printers, and other machines in complex installations. Naturally, the mechanics at Endicott routinely looked for opportunities to reduce the variations and build in volume. Sales statistics and education programs helped the company strike a balance between novelty, which generated revenue, and standardization, which produced economy. The production facility also worked in close collaboration with engineers who installed and maintained the equipment in the field [Cortada, 1993; Sobel, 1981; Usselman, 1993].

In sum, IBM was an organization whose business had naturally fostered these qualities: salesmanship that required close attention both to technology and to the particular requirements of each customer, regular exchange of information between the field and the plant, flexibility in production, and a willingness to compromise. In my view, these qualities put IBM in an excellent position to adapt to the electronic computer and the solid-state revolution. Some observers, reflecting on this situation, have argued that IBM behaved in a highly unusual (and remarkably enlightened) fashion in making this transition because the electronic computer ultimately made the company's installed base of electromechanical accounting equipment obsolete. I believe this argument represents a profound misreading of the situation--one that fails to comprehend the nature of technical knowledge in the data processing business, overlooks the importance of existing organizational capabilities, and neglects market conditions. A brief synopsis of IBM's history from 1949 to 1965 will bear this out.

IBM made the transition from electromechanical accounting equipment to electronic computers by pursuing two paths. One was to go after the emerging market for large, scientific machines, which was funded largely through defense contracts. The other was to begin to convert some of its established electromechanical equipment to electronics and to build some degree of electronic programmability into it. Through much of the 1950s, these efforts remained conspicuously separate. The former was centered in Poughkeepsie in a new facility built during the war; the latter remained anchored in the original plant at Endicott. But despite this separation, which many at Poughkeepsie actively tried to foster at the time and which many observers have subsequently exaggerated, the two facilities shared important traits. Each took basic components and arranged them in complex machines that were leased to customers and maintained by IBM in the field. Working in collaboration with their customers and their assemblers, the field force tailored the machines to perform a variety of specialized tasks. Computers, in other words, called forth many of the same qualities as the older technology.

The more significant differences between the activities at Poughkeepsie and Endicott during the early fifties had to do with the market. Most of Poughkeepsie's customers were sophisticated scientific and engineering organizations who leased their machines with defense funds. (The first major product initiative, later named the 701, was originally called the Defense Calculator.) These consumers differed considerably from those in IBM's traditional business accounting market, and one might reasonably ask why IBM pursued them. Several factors help explain this move. First, many of the early scientific computers had been built from modified IBM punched-card equipment. Second, the company had a long tradition of doing business with government, and its chief executive officer was a close friend of the Roosevelts. During the war it had sponsored the Mark I project. Third, IBM was a market-oriented company, and scientific computing represented an obvious opportunity, one that did not threaten its established base at all [Bashe, 1986; Usselman, 1993].

That market was especially attractive because of the approach the government took to computing. From the beginning, government did not attempt to target firms with the most impressive research organizations. In other words, it did not pursue a "supply-side" approach, in which it assumed that money spent on research would ultimately yield computers. Instead, the vast majority of its support came in the form of purchase orders for computing power. The government, acting as an "informed first user," set goals and put out bids to have them met. Moreover, it did so not through a single, coordinated plan, but by placing money in the hands of many different organizations that each put out their own contracts for bid. Each of the armed services acted as a consumer, as did the laboratories of the Atomic Energy Commission, as did the large manufacturers of aircraft operating under government contract. In effect, the government set up a market for powerful computers--a market of informed users who expected to exert sufficient input into the design of their machine to insure that it performed the particular operations they desired [Bashe, 1986; Broack, 1975; Flamm, 1988; Katz and Phillips, 1982; Lewin, 1982; Nelson, 1982a; Mowery, 1983].

This was a market that suited IBM perfectly. The company's entire culture was dedicated to the task of meeting specific data-processing problems in the field. The only significant difference between large electromechanical data-processing installations and these machines was that the computers would use vacuum tubes instead of electromechanical relays and would involve a staggering amount of wiring. But these differences appear trivial when placed in the total context of the task. Computers called for extensive sales, maintenance, programming, and field engineering. Within the plant, the company would look as always for ways to build standardization into the machines while retaining sufficient flexibility to meet the demands of each user. At this point, before the advent of lower-cost memory enabled users to reprogram the machines easily, such custom tailoring often involved the wiring itself, just as customizing the older equipment involved unique arrangements of gears and ratchets. One of the greatest challenges in such work, IBM had long since learned, was to keep track of design changes as the machine moved into production and out into the field. This task required massive record keeping and close cooperation among engineers responsible for design, assembly, and service. In tackling those jobs, IBM drew freely on personnel who had performed similar tasks at Endicott.

Purchasers recognized these qualities in IBM and favored the company because of them. This was certainly true in the case of the SAGE contract, a massive anti-aircraft project funded by the Air Force that called for 23 pairs of computers operating in real time. The Air Force relied on Jay Forrester of MIT to design these novel machines, but when it came time to build them it chose IBM over several other firms favored by the academic designer. The Air Force cited IBM's experience with assembly and service as the critical factors influencing its choice.

The SAGE contract proved extraordinarily important to IBM, because it introduced the company to a variety of military-sponsored technical efforts aimed at reducing the costs of assembling or packaging electronic circuits [Lewin, 1982; Pugh, 1984; Usselman, 1993]. Especially prior to the advent of the integrated circuit, packaging was perhaps the most important element of computer production. It was where logical design, componentry, and custom-tailoring intersected. SAGE and subsequent government contracts for state-of-the-art machines, such as STRETCH, helped IBM build on its established expertise as an assembler and stay abreast of the latest developments in solid state techniques [Bashe, 1986].

Throughout this critical early period of government support, IBM benefited from a quality that might at first seem a detriment in an environment of rapidly changing technology. That quality was humility. In short order, computing had opened two technological frontiers--logical design and solid-state componentry. It was very easy for people working at those frontiers to feel a certain hubris. Many of the practitioners were physicists and mathematicians. The work they did was new and scientific, and yielded fame and Nobel prizes. IBM filled a far less glamorous middle ground. It purchased components, as it always had, and it let its customers have input into the logical designs. This attitude permitted IBM to move to the center of knowledge in the industry.

Humility also aided IBM during this period by helping prevent it from ignoring potential customers. Many organizations working on computers focused almost exclusively on the high end. Perhaps the most apt comparison is Engineering Research Associates (ERA), a company that concentrated on building computers for the most sophisticated users. "Because of the nature of its market," Kenneth Flamm has noted in his excellent history of the computer industry,

> engineering considerations dominated ERA's business orientation . . . . In sharp contrast with firms seeking a commercial market, ERA experienced little feedback from users and little direct contact with what remained a relatively unknown market. The emphasis on technical sophistication over marketing, it may be argued, persisted in the computer companies that the engineers brought up in ERA went on to found [Flamm, 1988, p. 46].

As IBM entered the scientific market, it never lost sight of the commercial market and the potential connections between the two. Most significantly, the company did not isolate work on the large computers in a separate scientific or defense wing. As it worked on the 701, IBM simultaneously developed the 702 for business purposes. The two programs shared many of the same personnel and the same technology. Here again, one can see that an established characteristic of the firm--its tradition of entering many markets and seeking to transfer lessons learned in one market to the others--ultimately contributed to success in the computing business.

In actuality, however, IBM's efforts to transfer the fruits of its work on scientific machines directly to the commercial market seldom worked as well as planned. The real growth in business computing came instead from the second path, the operations at Endicott, where engineers developed programmable electronic calculators. Their first big success was the 650, which ultimately sold in the thousands; later they generated the 1400 series, a spectacular success of the early sixties that made computing far more common in business than it had ever been before. The Endicott facility also produced a series of input-output devices that helped develop the market for both large and small computers. Though these products made use of electronics, they also drew extensively on the mechanical skills available at Endicott. Printers and disk storage devices in particular were as distinguished as much for their rapid, precise mechanical motions as for their logical design [Bashe, 1986; Usselman, 1993].

The introduction of these products perhaps lends some support to the contention that IBM showed extraordinary daring in making its established line obsolete. But again, I would urge that the developments at Endicott be seen in the context of IBM's previous history and the emergent market for data processing. By the time of its move into electronics, IBM had a long history of making its own machines obsolete. Its sales force had long since learned that continual change, if it produced some new capabilities, was the surest path to larger contracts. Greater calculating power would almost certainly lead customers to spend more on novel methods of printing or to do additional tasks. Someone might have drawn a sharp division between electromechanical devices and electronic ones, but to do so would have been out of the ordinary, at least at IBM. In assessing this situation, moreover, we should remember that the market for data processing was growing rapidly. Machines taken out of service at one installation could readily be placed in another. Many were shipped overseas, where depression and war had created an enormous pent-up demand that domestic manufacturers could not meet. No firm was better positioned to perceive these marketing opportunities than IBM, the largest accounting machine company, and one with established outlets around the world [Sobel, 1981; Usselman, 1993].

By the mid fifties, IBM had achieved a commanding position in computing. Drawing on its established capabilities, it had responded to all segments of a vibrant market. Flamm has noted that by 1950 one could clearly identify four separate approaches to computing: 1) large commercial machines such as Eckert and Mauchly's Univac, the successor to ENIAC; 2) large scientific machines such as that designed by the mathematician John von Neumann for Princeton's Institute for Advanced Studies; 3) computers for use in real-time control applications, such as that under development at MIT in Project Whirlwind; and 4) small machines that might appeal to cost-conscious consumers [Flamm, 1988, p. 105]. I would add that IBM had at that time already become involved in all four, and that by 1955 it had assumed the lead in all four, with its 702/705, 701/704, SAGE, and the 650, respectively. Each of these machines, moreover, had given birth to sustained efforts to generate further development.

Because IBM ended up with development efforts aimed at all segments of the market, it was then in a position to see and feel pressures from what would soon emerge as the central recurrent dynamic force in the computer industry: the convergence of machines designed for one market with those designed for another as the availability of new memory increased programming capacity and as changes in componentry improved processing power. From the mid fifties on, this issue continually created problems within IBM as its machines competed with one another in the marketplace and its development efforts overlapped. The problems within development were compounded by the continually advancing technology of solid state, which constantly blurred divisions between component manufacture and logical design. Mervin Kelly, the former research director of Bell Labs whom Tom Watson, Jr., had hired as a consultant, warned Watson that IBM would lose the capacity to design computers if it failed to integrate backward into the components business. Kelly predicted that established component producers that also had experience designing, building, and marketing electronic products, such as RCA and GE, would eventually dominate the industry. Though few did so as early as Kelly, in time many observers in the business press expressed similar judgments. Managers at GE and RCA laid plans for the move [Usselman, 1993].

During the late fifties IBM addressed these pressing matters. It struggled to sort out its development efforts and to master solid-state manufacturing. Steps taken during this period would eventually culminate in System/360, a single line of computers that would replace all other IBM machines, run the same programs, and contain solid-state circuits of the same standard design manufactured from scratch entirely within IBM. No other product announcement would have a more profound effect on the computer industry at least until the coming of the personal computer [Pugh, 1991; Usselman, 1993].

Space does not permit a full accounting of the tortuous course that culminated in the production of System/360, but one feature of the process deserves emphasis. Even as IBM integrated backward into component production, its traditions of assembly, packaging, flexible production, and feedback from the sales force and field engineers remained essential factors in its success. The key remained not simply to master components but to strike balances among componentry, logical design, and markets. Within componentry, moreover, one needed to strike balances between performance and manufacturability. Once again, IBM's tradition of product engineering and its lack of technical hubris proved extraordinarily useful.

Drawing on its established capabilities in circuit assembly and packaging, IBM moved in two steps. First it developed a new solid-state package using transistors obtained under license from Texas Instruments. (IBM's established position in the marketplace no doubt made it an attractive plum for Texas Instruments.) The new package, known as Standard Modular System or SMS, introduced IBM to the world of chemical or "wet process" manufacturing. This technology formed the basis of its 1400 series computers and other large-scale machines. The line operated at Endicott, where its designers again took advantage of the available mechanical skills to build the necessary conveyers and other materials-handling equipment.

Then, for its System/360 series announced in 1964, IBM took the expertise acquired from TI and developed its own internal component production facility. Though constructed near the Poughkeepsie plant, a management team from the SMS production area ultimately took charge of running this operation, with ample assistance from personnel borrowed from TI. Significantly, IBM in building this facility struck a fundamental compromise, choosing not to develop the new integrated circuits and instead concentrating on building a production line of great flexibility that could readily respond to shifts in demand and keep track of design changes. True to its heritage, even as it moved into extraordinarily capital-intensive process manufacturing, IBM did not want to sacrifice the flexibility it had come to rely on as an assembler of customized machines. (It gained additional flexibility by continuing to rely on the SMS format for many of the peripherals.) Though System/360 is often lauded as bringing a high degree of order to the market by consolidating IBM's offerings in a few standard models, in reality the system included machines of countless variations.

Contrary to predictions, RCA and GE never competed successfully with System/360, despite conspicuous efforts to market similar lines. Though my conclusions are speculative, since I know of no detailed studies of computer production at either firm, I would suggest that the poor match between these firms' organizational capabilities and the tasks inherent to computer production hold the key. Neither RCA nor GE had much experience marketing complex products to business people who were not scientists and engineers [Graham, 1986]. Since feedback from the business machines market remained an important ingredient in IBM's success, the absence of an established marketing and support team almost certainly handicapped its two competitors. As Kelly's remarks to Watson suggest, RCA and GE no doubt hoped that their traditions of manufacturing their own components would compensate for deficiencies in marketing. But IBM's experience with backward integration into component production suggests to me that those firms' experience with vacuum tube production might well have worked to their detriment. Manufacturers of vacuum tubes did not generally achieve success with semiconductors, an industry that came to be dominated by new firms [Dosi, 1982; Lewin, 1982; Mowery, 1983]. Companies that manufactured components, moreover, often tended to exaggerate their importance and neglect packaging. Again, the IBM case suggests that packaging remained essential even with the rise of solid state. Though the materials changed, the outlook and approach remained much the same. And the most critical new technical processes, such as photoetching, would have been as unfamiliar to RCA and GE as they were to IBM. In sum, like many other observers of the industry, Kelly exaggerated the importance of one technical feature and underestimated the importance of organizational capabilities that facilitated coordinated action on many features at once.

S. Usselman / 16

Britain

In 1965, at virtually the same moment IBM delivered its first System/360 computers, the economist Christopher Freeman published a report on the international electronics capital goods sector in Britain's *National Institute Economic Review* [Freeman, 1965]. In his extensive discussion of computing, Freeman in effect addressed the same question I have just considered--why had IBM succeeded?--but with an added concern for how the British might best respond. To a remarkable degree, Freeman's analysis lends support to the thesis that the combination of a robust market and existing organizational capabilities explained IBM's remarkable emergence.

In preparing his report, Freeman interviewed managers of many of IBM's competitors around the world. "Almost all firms interviewed," he reported, "attributed the United States' success to two principal factors: a technical lead in many products, and the vast American military and space budgets" [Freeman, 1965, p. 51]. As I and others have emphasized, the key to American government support appeared to rest more in its procurement policy than in its direct support for R&D. "Some [European] firms believe," reported Freeman, "that the much larger American home market and the stimulus to demand from the government programs are more important in explaining the American lead than the government support for R&D as such" [Freeman, 1965, p. 72]. Freeman concurred with this opinion. "Almost all the early demand in the United States was from the military market. Few people then envisaged the large-scale use of computers for data processing, and both government and industry thought mainly in terms of military and scientific applications" [Freeman, 1965, p. 59].

While freely embracing the view that military demand had played a major role, Freeman spent the bulk of his article explaining why military demand alone did not account for IBM's success. The "military-space market is very different from the normal commercial market," Freeman cautioned.

> The products are more specialized and expensive, being made to more exacting specifications. Both in America and Britain some firms which have concen-

> trated on the military market and have the largest military R&D contracts have a less satisfactory record in the civil market and as exporters. Thus the supposed benefits of American government programs are not so straightforward as appears at first sight" [Freeman, 1965, pp. 51-52].

These cautionary remarks led naturally to a discussion of IBM, which clearly could feel plenty of satisfaction with its record in the civil market and as an exporter. Freeman described IBM as "a very successful and fast-growing office equipment company with a strong tradition of product innovation before it began to manufacture Electronic Data Processing (EDP) equipment. It already had a strong world-wide sales organization and field force of engineers" [Freeman, 1965, p. 59]. Freeman also drew attention to IBM's personnel policies and to its customer relations, which he saw as all of a piece. "Few companies can have given so much attention to the selection and training of their own employees," he observed, "and to the training of their customers" [Freeman, 1965, p. 59].

The significance of these attributes became clear when Freeman assessed the nature of the computer business. Significantly, Freeman downplayed the importance of the inventor and emphasized the firm. "American firms' technical and commercial lead since the Second World War has depended not so much on their capacity for original invention or completely new products as on their success in developing a series of greatly improved models embodying new features in design and far higher standards of performance," he wrote. "The nature of the industry is such that it is not necessary to invent everything in order to secure a strong technical or commercial lead . . . . What is necessary is to have a strong development and engineering capacity, so that inventions made elsewhere may be rapidly assimilated, imitated, utilized, and improved upon" [Freeman, 1965, p. 63]. "The transition from laboratory prototype to successful commercial production and sales calls for rather different resources and skills [than those of the inventor]," Freeman noted in another passage. "This transition is difficult and expensive for such complex products as electronic capital goods. The firms which succeeded combined a well-organized research, development, and test programme with good production planning and technical service" [Freeman, 1965, p. 62].

S. Usselman / 18

Again and again, Freeman stressed the importance of information flows in the computer industry. "The development process in the electronic capital goods industry consists largely of devising methods of assembling components in new ways, or incorporating new components to make a new design, or developing new components to meet design requirements," he explained. "Consequently, there must be close collaboration between end-product makers and component makers." Significantly, Freeman noted that "this does not necessarily mean that there must be vertical integration," since "specialist component makers may offer some economic advantages." "In part," he wrote, "this may be done by licensing and know-how agreements with other firms in the industry, or with governments, in part by recruitment or outside consultancy" [Freeman, 1965, p. 63]. "Successful development," he concluded, "depends on a good deal of give and take between firms" [Freeman, 1965, p. 65].

Writing at the time of IBM's integration backward into components, Freeman devoted more attention to cooperation and information flows in the design and production stages than in the area of customer relations. But he clearly appreciated the importance of field engineering and sales as well. Freeman viewed customer relations as especially important in the business market, and in discussing them he reiterated his belief that success in the military or scientific market did not necessarily lead to commercial success. "The distinction between the 'scientific' and the commercial 'EDP' markets is important," he noted. "Sometimes they may use the same machines, although with different configurations and peripherals. But, whereas the 'scientific' customer usually knows a good deal about the machine and can do a lot of his own 'software' and maintenance, the 'commercial' customer usually needs a great deal of training, advice, and assistance from the manufacturer. The field force in the EDP market must be much larger, and firms which are successful in one market will not necessarily be successful in the other" [Freeman, 1965, p. 61].

I have dwelled at some length on Freeman's analysis because I believe it represents a contemporary account that lends considerable support to my thesis, without itself being directly influenced by the theory of organizational capabilities. Interestingly, however, Freeman had much more to say about American developments than he did about the performance of British computer firms. In effect, he argued by implication, accounting for British failure not so much by considering the subject directly but by analyzing the success of its rivals. Presumably British firms and the British government did not do what IBM and the American government did.

Two recent historical studies of British institutions lend considerable support to that presumption. The first is Martin Campbell-Kelly's fine institutional history of ICL, the "British IBM" [Campbell-Kelly, 1989]. Because ICL has absorbed so many of the data-processing firms in Britain, including BTM (British Tabulating Machines) and Powers-Samas, the two British manufacturers of electromechanical accounting equipment before World War II, Campbell-Kelly's book provides an overview of the data-processing industry in Britain during the twentieth century. The second study is John Hendry's *Innovating for Failure*, which traces the actions taken by the British government's National Research and Development Corporation (NRDC) to foster the computer industry [Hendry, 1990]. These two books converge on the critical period from 1949 to 1959, when the British computing industry fell dramatically behind the American despite a comparatively similar starting position. In other words, they focus on British institutions during the period for which Freeman has so perceptively analyzed American performance. The books also converge on a common theme: failure.

Space does not permit a full discussion of these accounts. Here I wish only to sketch the authors' answers to two questions: why was BTM unable to grow at anything like the pace of its American counterpart, IBM, and why did NRDC have so little success in fostering the computer industry?

Campbell-Kelly's history suggests to me that much of the explanation for BTM's sluggish performance and its inability to follow IBM with vigor into electronics and computers rests in the company's prewar roots. BTM had operated under license from IBM. It had manufactured machines like those built in the States by IBM and leased them in the British market. On the surface, it possessed the same organizational capabilities as IBM, and for that reason many in Britain always expected it to follow a similar course in computing. In reality, however, BTM had never developed the vitality that characterized IBM before the war. In particular, it lacked the vibrant production engineering mechanism that proved so instrumental to

IBM's success with computing.

Another problem BTM faced was that it stood outside the university research establishment. This was true initially of IBM in the United States as well, but after the war that changed under the influence of the liberally funded, procurement-driven American military. In the United States, moreover, universities and government had never established very firm connections, so IBM and other corporations faced little resistance from entrenched interests. In Britain, however, universities and government had established close links by the time of the war, with business drawn into the alliance only through the mediating agency of the consulting engineer. After the war, Britain's much smaller defense budget did not have the strength to dissolve those traditional bonds. University research remained closely tied to government and the military. This skewed efforts away from development and manufacturing [Mowery and Rosenberg, 1989]. In the specific case of computing, this had the effect of keeping Britain's remarkable code-breaking machines and the pioneering computers developed at Manchester University separate from BTM and other manufacturers. In 1949, moreover, BTM severed its ties to IBM, thus depriving itself of ready access to an alternative source of computer designs.

The British government created NRDC in large part to redress this situation. This publicly owned corporation was supposed to insure that patents generated through publicly supported research were made available to private corporations for rapid commercial exploitation. NRDC would offer financial support for firms willing to use the patents. Lord Halsbury, the director of NRDC during its first decade, identified the computer industry as an obvious candidate for support. Halsbury, who made frequent trips to the United States and closely monitored IBM's early successes in marketing electronic calculators and computers to its traditional business customers, became convinced that the British computer industry would succeed only if one of its electronics firms merged its technical expertise with the marketing experience of either BTM or Powers-Samas. But Halsbury lacked the resources to force the issue. Out of concerns for fairness, the government had restricted the funds available from NRDC and had insisted that they be repaid. Firms accepting these funds, moreover, had to agree to turn all patents generated in the course of the subsidized project back to NRDC. With such scant bait, Halsbury had little chance of luring firms into the organizational alliances he wisely desired.

The combination of its research traditions, NRDC's limitations, and BTM's incompetence left Britain without an organization with the capabilities of IBM. Behind these organizational matters, I believe, lurked another more basic difference with the American situation. The British market for data processing lagged behind the American market, even when adjusted on a per capita basis, in large part because the British lacked enthusiasm for data processing. I cannot provide extensive evidence directly supporting this hypothesis or even offer many suggestions about what caused the lack of enthusiasm.[2] The subject of public attitudes toward computing remains largely unexplored. (Here is an important opportunity to explore cultural inputs into the process of technical change.) I would, however, point out that IBM has long perceived the British market as sluggish. During IBM's association with BTM, the British firm never met its parent's sales objectives for it. Though correspondence among IBM and BTM managers cited by Campbell-Kelly suggests that IBM found ample cause for blame within BTM, my guess is that consumer taste accounted for much of the failure to meet quota. It is much easier to exert pressure on one's subsidiary than to alter conditions in its market. The comparative lack of vitality within BTM before the war stemed in large measure, I suspect, from the fact that it did not enjoy nearly the degree of feedback from customers as that which propelled so much change within IBM. The sluggish market deprived BTM of the engine that drove IBM to develop its organizational capabilities, setting in motion a vicious cycle because BTM then lacked the ability to generate products that might have spurred demand.

2 At this point my discussion bears considerable similarity to the wider issue of the role of organization in Britain's relative economic decline, growing out of the work of Chandler and Lazonick [Chandler, 1990 and 1992; Elbaum and Lazonick, 1986; Lazonick, 1991]. Most critics agree that organization figured prominently, but some have argued that cultural factors, including a general resistance to innovation, must also be taken into account [Broadberry and Crafts, 1990 and 1992; Coleman, 1987; Davenport-Hines and Jones, 1987; Kirby, 1992; Walker, 1980; Weiner, 1981]. Curiously, the literature on British decline contains virtually no references to the computer industry [above references, plus Pollard, 1982].

S. Usselman / 22

In Hendry's account of NRDC, one can catch glimpses of Lord Halsbury's frustration with British consumers. Returning from his trips to the United States, Halsbury repeatedly tried to convince people in Britain that business could provide a market for computers. When these efforts failed, he even suggested that British computer companies pursue the American market. Halsbury's one significant accomplishment during the decade was in arranging for government to purchase ten computers.

Freeman, too, lends some support to this contention. Freeman believed British computer firms had stayed too closely tied to the scientific/government market. This orientation had skewed their development efforts and retarded growth. His comments about the difficulty of moving between the scientific and commercial markets reflect this concern. Apparently Freeman did not attribute this behavior primarily to decisions made by managers and government officials, but to the environment in which those managers and officials operated. For when Freeman turned from description to proscription, he did not call for direct support of institutions that would follow a different course. Instead, he called for government to fund projects that would create new markets for computers. Freeman suggested that government establish a national network and support the development of computer applications in education. He believed this approach would help nurture institutions with the capabilities he admired.

British public policy did change in response to System/360, but not in the direction Freeman advocated. Instead, the British government directed most of its energies toward encouraging a series of mergers that culminated in the formation of ICL in 1968. The new firm combined BTM, which by then had merged with its old competitor Powers-Samas and taken the name ICT, with most of the electronic computing and components manufacturers in Britain. More than a decade late, it fulfilled Lord Halsbury's goal. ICL was clearly modeled after IBM. It was a fully integrated manufacturer with a desire to pursue all markets. One observer has aptly characterized it as a "national champion," a company anointed by government to provide leadership. Several other European nations pursued a similar strategy in response to System/360 [Flamm, 1988].

The strategy has not worked, for reasons Freeman appreciated at the time and Campbell-Kelly makes clear in his history. IBM had grown organically in soil enriched by government. It had established capabilities before the computer came along, and it had moved gradually to acquire additional ones as the technology changed. Though BTM on its surface appeared to possess the same original capabilities as IBM, it had never developed them as fully. Operating in a much less vibrant market, it had done little to develop the additional abilities necessary to develop and manufacture computers in a timely fashion. Now government had attached to it firms that appeared to possess the necessary skills in componentry, design, and manufacture. Unlike IBM, however, the recently merged British firms were expected to master the critical balancing act among those functions virtually overnight. And government had still taken no measures to sweeten the market. Not surprisingly, then, Campbell-Kelly's book ends with a series of chapters detailing one disappointment after another as each new ICL product fails to meet expectations.

Japan

As is so often the case in the economic history of the past half-century, in studying British failure we find the keys to Japanese success. The Japanese pursued a course remarkably similar to that suggested by Freeman, with some additional public policies that took advantage of the established capabilities of their government organizations. In part by design and in part by accident, the Japanese replicated the conditions that had given rise to IBM. As a result, Japan ended up with an internationally competitive industry led by a company, Fujitsu, that more genuinely possessed the capabilities of IBM than did ICL.

In pursuing this course, Japan may very well have benefited from its comparative backwardness. Prior to World War II, the Japanese had no domestic accounting machine industry. There was no BTM to muddy its waters and confuse the situation. Aside from some experiments at the Nippon Electric Company (NEC), the country had little experience in electronics, either [Flamm, 1988; Fransman, 1990; Sobel, 1986]. The absence of a continuing tradition of relationships between the research establishment and the military may have provided further benefits. Though such relationships had proved quite beneficial in the United States, where the military had received a dramatic increase in funding, the British case had shown that close links between researchers and the military could distort development when defense was funded at a lower level.

Starting with a comparatively clean slate, the Japanese spent the decade after the occupation monitoring developments and gathering information. Lacking an established producer in the data-processing industry, they felt little urgency to respond immediately to the extraordinary changes taking place in the United States and Europe. Japan's established electrical manufacturers concentrated on the more urgent task of rebuilding and expanding its electric power grid and on the less capital-intensive consumer products sector. Meanwhile, two institutions within Japan provided some tentative support for research. The first was Nippon Telegraph and Telephone (NTT), the state-owned telephone monopoly, which operated a research facility known as the Electrical Communications Laboratory (ECL). ECL monitored developments in componentry and circuit design that might have implications for the telephone network. These activities fit comfortably within the lab's traditional role, which was to keep the giant monopoly abreast of developments so that it could make intelligent purchasing decisions. (Unlike AT&T, which built much of its own equipment, NTT traditionally acquired equipment from electrical suppliers such as Fujitsu and NEC, though it often had considerable input into design specifications.) The second institution supporting research was the government's Ministry of International Trade and Industry (MITI), whose Electrotechnical Laboratory (ETL) pursued a program much like that at ECL for the economy as a whole. Together, ETL and ECL accomplished three tasks that would later prove critical. They fostered reservoirs of knowledge about componentry; they put NTT and the Japanese government in a position to be informed, intelligent consumers of electronic machines; and they helped further traditions of cooperation among potential consumers, electrical equipment manufacturers, and those with a knowledge of components [Flamm, 1988; Fransman, 1990; Johnson, 1982].

As the fifties progressed, pressure built within the Japanese business community to acquire data-processing equipment. With no established domestic producer, the government agreed in 1954 to let

IBM sell equipment through a subsidiary known as IBM Japan. For six years IBM Japan sold equipment manufactured outside Japan to Japanese firms, for the first time establishing an installed base of data-processing equipment in Japanese business. Much of this base consisted of refurbished electromechanical equipment, but toward the end of the 1950s IBM introduced some electronic machines into Japan as well. Meanwhile, established Japanese equipment manufacturers moved tentatively into the market with machines that, like those sold by IBM Japan, would have appeared old-fashioned by American standards. As I have emphasized in discussing the American case, however, old machines would help foster many of the basic skills and organizational abilities as well as new electronic computers would have. By 1961, Japanese firms held 18 percent of the market [Fransman, 1990, p. 27].

Beginning in 1960, the Japanese government began a series of measures designed to bolster the domestic market for computers and insure that Japanese firms enjoyed advantages in pursuing that market. One measure, which might at first seem to contradict my general characterization of the new policies, was to permit IBM Japan to manufacture computers within Japan. In exchange for this concession, however, the Japanese government obtained rights to license all IBM technology. Though many students of the computer industry have argued that such licensing arrangements have little economic significance because the rapid pace of technical change makes licensed techniques obsolete before they can be exploited [Flamm, 1988; Mowery and Rosenberg, 1989], I would suggest that this particular licensing arrangement at the very least sent an important signal to Japanese firms to go ahead and compete without fear. At the same time, moreover, the government hiked the duty on imported computers (including those manufactured by IBM Japan) to 25 percent and established rules requiring the Japanese government (including giant NTT) to buy domestic. These rules applied even to universities, which as in other nations had consistently provided a demand for the most sophisticated machines and an important source of user feedback. In addition, MITI established a new leasing company, the Japanese Electronic Computer Company (JECC), which agreed to purchase machines from the manufacturers and place them in industry. Though the Japanese manufacturers had to buy back at book value any machines returned to JECC, the leasing firm almost certainly helped reduce risks. JECC had access to low-interest loans from the government, which it in effect passed on to computer makers and users; by agreeing to depreciate machines rapidly and lowering their book value, it assumed some of the losses from returned machines [Anchordougy, 1989].

This combination of policies clearly helped build Japan's domestic manufacturers. By 1966, their share of the Japanese computing market had risen to 54 percent [Fransman, 1990, p. 27]. Perhaps more importantly, the government had accomplished this not by targeting particular firms and creating a "national champion" but by seeding the market for computers. Interestingly, when MITI had first proposed the creation of JECC, it had hoped to use the company to shape research and development activities directly. NTT and other consumers of computers had objected, however, so that MITI had to settle instead for a mechanism that worked through the market. As ultimately structured, JECC subsidized the market but did not foreclose competition among Fujitsu, NEC, Hitachi, and others [Anchordougy, 1989]. As Martin Fransman has demonstrated convincingly in his recent study, cooperative initiatives such as JECC have often retained more substantial elements of competition than outside observers of the Japanese computer industry imagine [Fransman, 1990]. By stabilizing prices, moreover, JECC skewed that competition toward technology and performance--much as the military had done earlier in the United States.

After 1966, most government policy affecting computers in Japan came in the form of large procurement contracts issued by MITI and NTT. From 1966 through 1972, MITI conducted the Very High Speed Computer Project, and in 1968 NTT launched the Dendenkosha Information Processing System (DIPS) [Fransman, 1990]. In placing these orders, Japan pursued a strategy that David Mowery and Nathan Rosenberg have called that of a "fast second" [Mowery and Rosenberg, 1989]. It did not attempt to leapfrog the competition; it simply tried to match IBM as quickly as possible. MITI aimed to bring Japan's established electronics manufacturers up to speed with IBM, and the DIPS program set a goal of surpassing the performance of IBM's System/360 in telecommunications networks applications. Another MITI program launched in 1971, known as the Mainframe Computer Project or 3.5 Generation, expressly set a goal of matching the performance of IBM's recently announced System/370 line, the successors to its notorious 360s [Fransman, 1990]. System/370 had led RCA and GE to give up the computer business and had prompted the British to redouble their commitment to the national champion approach [Campbell-Kelly, 1989; Flamm, 1988].

Though these government procurement programs encouraged co-operation, they still retained significant elements of competition [Fransman, 1990]. The government set the technical goals and promised a market, but firms remained free to work out the details. As a result, one could observe in the Japanese industry of the 1970s much the same jockeying among firms that had occurred in the American industry during the 1950s as companies with various organizational capabilities struggled to win contracts and gain control of projects. Fujitsu, Hitachi, and NEC all participated in DIPS, but they pursued separate approaches to the problems posed by telecommunications. Fujitsu, a vertically integrated manufacturing and design organization, sometimes obtained components from Hitachi, but in 1970 it broke off that relationship in favor of one with the American firm headed by Gene Amdahl. Later Fujitsu and Hitachi cooperated in the 3.5 Generation program. Unlike the British and some other European nations, the Japanese government had not anointed a national champion in hopes it would possess the abilities to compete with IBM. It had set a goal of competing with IBM and created a competition that produced organizations with those abilities.

Conclusions and Cautions

In conclusion, let me briefly summarize the course of development in the computer industry as I understand it, then assess how well it conforms to the evolutionary theory of Nelson and Winter.

My story takes as its starting point the simultaneous invention of the electronic programmable computer on the eve of and during World War II. Like most inventions, it had a long way to go before it would become the innovation most of us think of when we hear the word computer. Because of its very nature, the computer would ultimately be defined through a process of compromise. Feedback from users was especially important. The complexity of this process was compounded by a second revolution (or mutation) in componentry, which

S. Usselman / 28

itself involved compromise between considerations such as manufacturability and performance. More than the computer itself, the solid-state revolution occurred outside of established traditions. The gap between an integrated circuit and a vacuum tube was greater than that between a programmable electronic computer and an electromechanical accounting machine.

What factors most influenced the process of compromise and shaped the course of events that defined the computer? Two stand out: first, the nature of the domestic market and second, existing organizations with established capabilities that made them likely to respond to the market and facilitate compromise. At the end of the war, the United States had by far the most vibrant market, in large part because of military demand but also because of enthusiasm for data processing in the business sector. In that hothouse environment, IBM flourished because its established traditions--especially those in marketing and product engineering--meshed extraordinarily well with the nature of the two new technologies. The British, though holding a similar position at the time of invention, lacked both the market and the organizational capabilities. Britain's policies and organizational traditions, moreover, tended to impede the development of both a vibrant market for computers and the organization best suited to reach such a market. The Japanese, though starting from a position behind both the Americans and the British, ultimately recreated the conditions that existed in the United States after the war. Their government fostered a domestic market and let firms respond to it. Because Japan lacked an established producer of electromechanical accounting equipment, the subsequent struggle unfolded somewhat differently than the American case, but again those firms with established organizational capabilities that best matched the tasks posed by computing technology emerged as the dominant players.

How well does this story mesh with the theory of Nelson and Winter? In its emphasis on established organizational capabilities, especially those that incorporate some degree of change into the routine operations of the firm, my history of computing resonates with the theory. Superior performance occurred where firms that possessed such capabilities encountered a market context that provided a vibrant selection mechanism. My findings also lend considerable support to the efforts of scholars like William Lazonick and Alfred Chandler, who recently have attempted to build a theory of organizational capabilities into historical explanations of comparative economic performance among nations [Chandler, 1990 and 1992; Elbaum and Lazonick, 1986; Lazonick, 1991].

While in these fundamental respects my history of computing supports the evolutionary theory, it also highlights some features of that theory that perhaps require further illumination. In their discussion of routines, for instance, Nelson and Winter acknowledge that their thinking is informed primarily by situations in which a few firms manufacture a single product. Their discussion of innovation and learning-by-doing is heavily influenced by the literature on manufacturing, especially the perceptive studies of Nathan Rosenberg. Similarly, though Nelson and Winter note that their theory can consider organizational innovation as well as technological change and in several passages point out the difficulty of separating the two, they pay attention almost exclusively to technological innovation.

The history of computing strains against each of these simplifying assumptions. Though IBM's established capabilities in manufacturing were important to its success in computing, the real value of those capabilities came from their fluid connections with other parts of the organization. The flow of information between manufacturing and marketing was critically important. This was especially true because IBM and other computer makers were *not* producing a single, uniform product. Product differentiation remained a key to strategy and performance even with System/360, an initiative ostensibly intended to bring uniformity to the market.

The prominence of marketing and its linkages to other parts of the organization relate to a feature of Nelson and Winter's theory that thus far I have not addressed. As they well recognize and acknowledge, the biological process of natural selection cannot strictly be applied to economic affairs because economic species (firms) can actively seek to influence the environment that selects upon them (the market), whereas biological species are selected passively and unknowingly.[3]

3 For purposes of simplification, I will ignore here the growing body of thinking among evolutionary biologists that suggests species might also possess such abilities.

S. Usselman / 30

Firms, moreover, can pass on new behaviors they acquire in response to selection pressures, while biological species must rely on unconscious processes of mutation and reproduction to generate change.

Nelson and Winter downplay these inherent limitations in the evolutionary analogy by stressing that much of the response of firms to selective pressures is in fact unconscious, too, but they acknowledge that two types of behaviors pose special challenges to their interpretation. The first is marketing, especially advertising that is intended to support strategies of product differentiation. The other is reseach and development. Nelson and Winter have little to say about the former, though they identify marketing strategy as an important area for further study, but they do devote quite a bit of attention to R&D. Here again, their primary objective is to demonstrate that even research is usually constrained in identifiable ways by established routines. Firms develop certain ways of searching for solutions, and those habits tend to bias searches in certain directions.

Does the case of computing, in which research and marketing seem so prominent, strain the inherent limits of the theory to the point of breaking? I think not. The experience of computing firms lends considerable support to the argument that much of the most critical research and development occurs in the ordinary course of routine operations. Indeed, my conception of the industry indicates that research and development might be even more deeply ingrained in a firm's routines than Nelson and Winter suggest. At IBM and elsewhere, the research that mattered most was linked inextricably to manufacturing and marketing. The marketing force did much of the "searching" that Nelson and Winter see as central to R&D. Just as I would caution against placing too much emphasis on learning and innovation in manufacturing at the expense of that which takes place in other areas of the organization, so I would caution against isolating research and development from the rest of the firm. Searching, like learning-by-doing, seems to have taken place continually throughout the firm.

My other words of warning pertain not to the internal characteristics of firms but to the environment in which they function. For all the attention given in this paper to organizational capabilities, the summary at the beginning of this conclusion began with references to two things that existed outside the firm: the nature of the technology and the nature of the market. Two giant externalities remain. This is, I believe, inherent in the evolutionary theory.

As Nelson and Winter note at the outset, their theory assumes that the market performs as the neoclassical orthodoxy suggests--large numbers of consumers with access to all information make optimizing decisions [Nelson and Winter, 1982, p. 39]. I am not at all certain this assumption applies to the computer industry. Obviously, government policy affected the market for computing. Nelson and Winter are of little direct help in this regard; they explicitly dismiss from consideration situations in which government acted as a procurer [Nelson and Winter, 1982, pp. 269 and 392]. To be sure, my comparison lends some justification to their assumption; it suggests that government policies in the United States and Japan fostered the development of computing by encouraging several consumers, thus creating demand more like that envisioned by neoclassicists than one might first assume. But still, the market for computing can hardly be said to have fit the neoclassical image, especially during its early stages when the "first movers" took hold. Suppliers dealt with a few exceptionally well-informed consumers with whom they developed close relationships [Flamm, 1988]. The policies of various governments, moreover, created several distinct markets within the international economy. Though the firms I discussed were international, they did not compete in a single global marketplace for computing. The character of each firm's domestic market mattered a great deal to its performance.

My concerns about the role of the market in the evolutionary theory go beyond the matter of public policy and beyond the single case of computers. As I suggested with regard to the British case, the demand for computers was shaped not just by policy but by cultural values and consumer preferences. If all we desire as business historians is to explain why some firms failed and others prospered from within a group facing the same cultural inputs, then we might justifiably ignore those inputs. But how often can we safely assume that all firms faced the same inputs? Clearly not when we make international comparisons. But I also question whether the assumption holds even within a domestic market. In cases involving the emergence of a new technology, success often comes to those who find themselves cultivating a particular niche, either as result of previous activities or because of superior entrepreneurship. And do we want to confine our-

S. Usselman / 32

selves to analyzing the performance of firms within industries? As historians, don't we also want to address the question of why some industries emerge and others do not?

This question brings me to the second externality--the nature of the technology. For all of the subtlety Nelson and Winter bring to the subject of technical change and for all of the attention they devote to the role of economic institutions in innovation, technology remains an independent, exogenous variable in the theory. This becomes clear in their discussion of "natural trajectories," in which they note that new technologies often give rise to regimes that follow logical patterns of development. Two common patterns or trajectories are toward economies of scale and mechanization. Generally, a body of broadly shared knowledge underlies such trajectories [Nelson and Winter, 1982, pp. 258-59].

Nelson and Winter's analysis of this phenomenon closely parallels that of Giovanni Dosi, who in a perceptive article couples the concept of natural trajectories with that of the technical paradigm. Borrowing from theories of scientific change, Dosi suggests that technology moves forward in waves, with a major breakthrough followed by a succession of modifications that move naturally toward a readily perceptible end. Like Nelson and Winter, he cites the semiconductor industry as an example [Dosi, 1982]. My study of IBM and the leading Japanese computer makers suggests that one could extend such a line of analysis to the computer industry as a whole. The recent experiences of IBM would certainly seem to lend credence to such thinking. The firm has ridden a trajectory to its logical end.

The notion of technological paradigms and trajectories raises two serious implications. First, it directs attention away from economic institutions and toward bodies of knowledge that transcend the firm. At its root, technology remains something of a black box--a matter of chance when viewed from an economic standpoint. We can try to make sense of it, but as Joel Mokyr's recent effort shows, any attempt to do so must go beyond economics and consider culture; and even then we may end up despairing of true understanding [Mokyr, 1990]. Second, it smacks of determinism. Given an adequate market, do technologies inherently drive institutions to perform certain tasks and assume certain forms [Klepper, 1992]? Nelson and Winter, like Chandler, have not escaped this question.

Ultimately, these comments about markets and technology serve to remind us that business is a vehicle, not an end in itself. IBM tells us a great deal *about* the computer revolution; it was not *itself* the revolution. Like other business institutions, IBM functioned as an intermediary between the market and technical innovation. We can learn a great deal by studying such an intermediary. At times, the vehicle itself becomes the most important element in historical change. Chandler has documented an entire epoc in American history when this was true [Chandler, 1977]. Nelson and Winter have equipped us with a powerful tool for analyzing similar situations of narrower scope. They have provided useful, if agonizing, discussions of how the intermediaries connect with the market and with the technical community and come to absorb some of those realms into themselves. But in the end, technical innovation and the market remain external and unexplained, with complex histories of their own. Perhaps inevitably, even the best theories in business history will tell us more about business than about history.

References

Anchordoguy, Marie, *Computers Inc.: Japan's Challenge to IBM* (Cambridge, MA, 1989).

Bashe, Charles J. et.al., *IBM's Early Computers* (Cambridge, MA, 1986).

Braun, Ernest, *Revolution in Miniature: The History and Impact of Semiconductor Electronics* (New York, 1978).

Broadberry, S.N. and N.F.R. Crafts, "Britain's Productivity Gap in the 1930s: Some Neglected Factors," *Journal of Economic History*, 52 (1992), 531-558.

Broadberry, S.N., "The Impact of the Depression of the 1930s on Productive Potential in the United Kingdom," *European Economic Review*, 34 (1990), 599-607.

Brock, Gerald W., *The U.S. Computer Industry: A Study of Market Power* (Cambridge, MA, 1975).

Campbell-Kelly, Martin, *ICL: A Business and Technical History* (New York, 1989).

S. Usselman / 34

Chandler, Alfred D., "Organizational Capabilities and the Economic History of the Industrial Enterprise," *Journal of Economic Perspectives*, 6 (Summer 1992), 79-100.

Chandler, Alfred D., *Scale and Scope* (Cambridge, MA, 1990).

Chandler, Alfred D., *The Visible Hand* (Cambridge, MA, 1977).

Coleman, D.C., "Failings and Achievements in Some British Businesses, 1910-80," in Davenport-Hines and Jones, pp. 1-17.

Cortada, James W., *Before the Computer: IBM, NCR, Burroughs, & Remington Rand & the Industry They Created* (Princeton, 1993).

Davenport-Hines, R.P.T. and Geoffrey Jones, eds., *Enterprise, Management, and Innovation in British Business, 1914-1980*, a special issue of *Business History*, 29 (October 1987).

Dosi, Giovanni, "Technological Paradigms and Technological Trajectories," *Research Policy*, 11 (1982), 147-162.

Elbaum, B. and Lazonick, W., eds., *The Decline of the British Economy* (Oxford, 1986).

Flamm, Kenneth, *Creating the Computer: Government, Industry, and High Technology* (Washington, 1988).

Fransman, Martin, *The Market and Beyond: Cooperation and Competition in Information Technology Development in the Japanese System* (Cambridge, ENG, 1990).

Freeman, C., "Research and Development in Electronic Capital Goods," *National Institute Economic Review*, 34 (Nov. 1965), 40-97.

Graham, Margaret B.W., *RCA and the Videodisc* (Cambridge, ENG, 1986).

Hendry, John, *Innovating for Failure: Government Policy and the Early British Computer Industry* (Cambridge, MA, 1990).

Hodges, Andrew, *Alan Turing: The Enigma* (New York, 1983).

Johnson, Chalmers, *MITI and the Japanese Miracle* (Palo Alto, 1982).

Katz, Barbara Goody and Almarin Phillips, "The Computer Industry," in Nelson (1982), ed., pp. 162-232.

Kirby, M.W., "Institutional Rigidities and Economic Decline: Reflections on the British Experience," *Economic History Review*, 45 (1992), 637-660.

Klepper, Steven, "Entry, Exit, and Innovation over the Product Life Cycle," mimeo, 1992.

Lazonick, William, *Business Organization and the Myth of the Market Economy* (Cambridge, ENG, 1991).

Lewin, Richard C., "The Semiconductor Industry," in Nelson (1982), ed., pp. 9-100.

Mokyr, Joel, *The Lever of Riches: Technological Creativity and Economic Progress* (New York, 1990).

Mowery, David C., "Innovation, Market Structure, and Government Policy in the American Semiconductor Electronics Industry: A Survey," *Research Policy,* 12 (1983), 183-197.

Mowery, David C. and Nathan Rosenberg, *Technology and the Pursuit of Economic Growth* (Cambridge, ENG, 1989).

Nelson, Richard R., ed., *Government and Technical Progress: A Cross-Industry Analysis* (New York, 1982).

Nelson, Richard R.,"Government Stimulus of Technological Progress: Lessons from American History," in Nelson (1982), ed., 451-482.

Nelson, Richard R., "Introduction," in Nelson (1982), ed., 1-9.

Nelson, Richard R. and Sidney G. Winter, *An Evolutionary Theory of Economic Change* (Cambridge, MA., 1982).

Pollard, Sidney, *The Wasting of the British Economy: British Economic Policy 1945 to the Present* (London, 1982).

Pugh, Emerson, *Memories that Shaped an Industry* (Cambridge, MA, 1984).

Pugh, Emerson, et. al., *IBM's 360 and Early 370 Systems* (Cambridge, MA, 1991).

Sobel, Robert, *IBM: Colossus in Transition* (New York, 1981).

Sobel, Robert, *IBM vs. Japan: The Struggle for the Future* (New York, 1986).

Usselman, Steven W., "Creating System/360: The Origins of Modern Computing at IBM," monograph in progress, 1993.

Walker, W.B., "Britain's Industrial Performance 1850-1950: a Failure to Adjust," in Keith Pavitt, ed., *Technical Innovation and British Economic Performance* (London, 1980), 19-37.

Weiner, Michael J., *English Culture and the Decline of the Industrial Spirit, 1850-1980* (Cambridge, ENG, 1981).

[18]

Cambridge Journal of Economics 1995, 19, 155–174

The globalisation of technology: what remains of the product cycle model?

John Cantwell*

This paper re-examines two hypotheses associated with earlier versions of the product cycle model. The first hypothesis, that innovations are almost always located in the home country of the parent company, is rejected on the basis of evidence drawn from 100 years of US Patent Office data. The second hypothesis, that the international dispersion of activity is led by technology leaders, is historically valid. However, over the last 20 years, technology leaders have been ahead instead in the globalisation of technology—that is, in developing internal international networks to exploit the locationally differentiated potential of foreign centres of excellence.

1. Introduction

This paper re-examines two hypotheses associated with earlier versions of the product cycle model (Vernon, 1966), using new historical evidence on the international dispersion of corporate technological activity by large American and European industrial firms, based on patents granted to these companies at the US Patent Office since the turn of the century. The first hypothesis states that innovations are almost always located in the home country of the parent company, and usually close to the site of the corporate technological headquarters. The second hypothesis is that international investment is led by technology leaders, as a means by which they increase their share of world markets and world production. If the first hypothesis does not necessarily hold, then the second can be interpreted more specifically to state that the internationalisation of technological development is led by firms with the strongest records in innovation.

Contrary to the first hypothesis, a new literature on the recent internationalisation of industrial research (for example, Ronstadt, 1977; Lall, 1979; Pearce and Singh, 1991; Granstrand, Håkanson and Sjölander, 1992) has suggested that innovations may be geographically dispersed within multinational corporations (MNCs). US-owned MNCs have recently witnessed an increase in the degree of internationalisation of their research, from low levels in the 1960s. It is perhaps no accident that empirical support for the product cycle hypothesis and subsequent criticism of it was largely derived from data on US MNCs. The new trend towards the internationalisation of research and development

Manuscript received 31 May 1994; final version received 9 November 1994.

*University of Reading. I am grateful to the referees for helpful comments on an earlier draft, and to the Economic and Social Research Council (ESRC) for their support under research grant R000232250. I wish to thank also Pilar Barrera who worked on related aspects of the project, Jane Myers and Jim Hirabayashi of the US Patent and Trademark Office for their invaluable assistance, and Cathy Jones and her many fellow students for their tremendous efforts during the data collection.

0309-166X/95/010155+20 $08.00/0

(R&D) in US MNCs became evident by the 1970s (see, for example, Mansfield, Teece and Romeo, 1979).

Three kinds of theoretical justification were provided for the first product cycle hypothesis. First, there are economies of scale in the R&D function, and if they are strong enough R&D will be concentrated in a single centre. Second, there are locational economies of integration and agglomeration in innovation. In new product development a close interaction is normally required between research and production facilities and users, while if several firms are engaged in the development of related new products (or processes) geographical proximity will encourage a greater volume of beneficial spillovers between them. Third, the original product cycle model viewed innovation as a demand-led process (see Schmookler, 1966, for another contemporary account), and thus for US MNCs innovation would be located in the more conducive environment of their home country, stimulated by the particular characteristics of the demand of high income consumers and skill-intensive downstream production facilities.

However, more recent evidence has demonstrated that the internationalisation of technological activity in large MNCs is not simply the outcome of a new trend (Cantwell and Hodson, 1991; Patel and Pavitt, 1991). As early as the 1960s many European MNCs (especially those originating from the UK, The Netherlands, Belgium and Switzerland) had already been quite highly internationalised in their technological activity. In Europe, the largest French and German industrial MNCs come closest to the US pattern, recently increasing the extent of internationalisation of their R&D from comparatively low levels in the 1960s. By contrast, large Japanese firms owned few research facilities outside Japan in the 1960s. Although their international R&D has been expanding, research in Japan itself has been growing at least as quickly, so the degree of internationalisation of research in Japanese-owned MNCs has remained low.

Here, it is argued that the first product cycle hypothesis should be rejected, not so much as a result of recent trends, as on the basis of historical evidence. In particular, the leading US electrical equipment companies and European chemical firms enjoyed a significant international dispersal of their technological activity in the interwar period. The source of evidence for this view is the US patenting of the largest US and European industrial companies over time, which identifies the location of technological activity at the corporate level, and thus provides a measure of the geographical dispersion of corporate invention. It is this same source of evidence that has been used often in other similar international comparisons that relate to the recent period, and which has helped to create the re-evaluation of trends in the international R&D of MNCs just described. What is new about the data in this study is the extension of records of corporate patenting back in time: here they are considered back to 1920.

The second early product cycle hypothesis states that international investment (and here, the international dispersion of technological activity) is led by technology leaders. The theoretical justification for this hypothesis is that the most technologically competent companies enjoy lower operating costs than their competitors and provide higher product quality, which generates higher profits and rising international market shares. In their turn, different degrees of technological competence across companies are a consequence of the firm-specific and path-dependent characteristics of technological change (Cantwell, 1991B, 1994; Nelson, 1991, 1992; Dosi, Teece and Winter, 1992; Pavitt, 1992; Teece, Pisano and Shuen, 1990). The greater capability of the most competent or technologically leading firms (or their greater capability to release able management resources and team expertise to plan for and organise growth, as argued by

Penrose, 1959) enables them better to expand their activity in new fields or environments, and higher profits provide them with the financial wherewithal to offset the costs of doing so. An ancillary argument is especially important in the case of early internationalisation and is featured in the contemporary explanation of the product cycle hypothesis; that is, a technologically leading group of firms from a particular location may—past some point—find that the continued increase in its penetration of foreign markets is challenged by protectionist barriers, and by the gradual learning of local competitors in those markets. This raises the incentive to produce locally, and insofar as indigenous companies develop their own lines of related technological development there is also a greater incentive to internationalise innovation (in addition to the need to adapt products to the distinctive features of host country demand).

This hypothesis appears to be consistent with the historical evidence, except perhaps during phases of technological hegemony when innovative development is concentrated in the home country. However, the explanatory power of this hypothesis has been eroded by recent trends; the composition of firms involved in the internationalisation of technological activity has now been broadened, extending to cover a much wider range of companies. Therefore, an alternative version of the second hypothesis is proposed to fit the current situation. It is suggested that technology leaders are now ahead instead in the globalisation of technology—that is, in the development of international intra-firm networks to exploit the locationally differentiated potential of foreign centres of excellence. These networks are internal to the firm in order to build upon or extend its core technological competence through an internally coordinated learning process, but they are complementary to external inter-firm networks whose role is the exchange of knowledge and occasionally cooperation in learning through technology-based joint ventures (Cantwell, 1991B, 1994; Cantwell and Barrera, 1995).

The data employed are described further in Section 2, while Sections 3 and 4 relate to the re-assessment of the first and second product cycle hypotheses respectively. Section 3 examines the historical importance of the international dispersion of technological development in leading American and European firms. Particular attention is paid to firms in the chemical and electrical equipment industries broadly defined, as these are the industries which are most reliant on science-based technologies. Historically, US firms were strongest in the electrical equipment industry, while European (and especially German) firms were stronger in chemicals (Chandler, 1990; Cantwell and Barrera, 1993). Among leading companies such as these, technological activity is not always much more widely geographically dispersed today than it was in the interwar or early post-war periods. However, Section 4 considers what has changed recently for these firms, and terms this a new trend towards the globalisation of technology—that is, the emergence of internationally integrated structures for technological development. It is suggested that globalisation in this sense is a modern characteristic of corporate technology leaders. Some conclusions are set out in Section 5.

2. The data

Patenting is a measure of invention, and so corporate patenting is more a measure of wider technological activity (changes in production methods) in firms, and not just R&D as such. For larger firms like those covered here, R&D is the most important source of new knowledge and skills, and so for them the internationalisation of technological activity revolves around the internationalisation of research. However, production engineering is often an important complementary source of new inventions that are incorporated into technology. It should also be noted that the location of basic R&D that feeds into the development of productive applications at some other site may not itself be picked up separately in the patent statistics. For these reasons the title of the paper refers to the internationalisation of technological activity (the development of new methods of production), and not to the internationalisation of R&D as such.

Two types of information have been collected manually from the *US Index of Patents* and the *US Patent Office Gazette*. First, all patents were recorded that were assigned to a selection of large US-owned and European-owned firms between 1890 and 1968. From 1969 onwards equivalent information has been computerised by the US Patent Office. The firms selected for the historical patent search were identified in one of three ways. The first group consisted of those firms which have accounted for the highest levels of US patenting after 1969; the second group comprised other US, German or British firms which were historically among the largest 200 industrial corporations in each of these countries (derived from lists in Chandler, 1990); and the third group was made up of other companies which featured prominently in the US patent records of earlier years (a method that proved most significant for a number of French firms that had not been identified from other sources).

In each case, patents were counted as belonging to a common corporate group where they were assigned to affiliates of a parent company. Affiliate names were normally taken from individual company histories. In all, the US patenting of 857 companies or affiliates was traced historically; together these comprise 284 corporate groups. Owing to historical changes in ownership, 17 of the affiliates were allocated to more than one corporate group over the period as a whole. No significance has been attached to the location of the particular affiliate to which each patent is assigned, since this may be different from the location of the inventive activity that gave rise to the patent. However, the location of the parent company is an important dimension in the analysis, as this is treated as the home country or the country of origin of the corporate group. By consolidating patents attributable to international corporate groups, it is then feasible to examine the geographical distribution of technological activity within groups, and the possible formation of internal intra-group international networks (as opposed to external inter-firm networks, which are not the subject of this paper). Each corporate group is also allocated to an industry on the basis of its primary field of production; occasionally, firms have moved between industries historically, sometimes associated with changes in ownership, and this has been allowed for.

The company to which a patent has been assigned (if any), and the name and location of residence of the inventor responsible for the underlying invention, are both recorded separately in the US Patent Office data, including the earliest data. Where patents have been assigned to firms, the inventor is normally an employee of the company or is directly associated with it in some other way, but occasionally independent individual inventors do choose to assign their patents to firms (Schmookler, 1966). Assignments by independent individuals were more common in the nineteenth century but, at least from the inter-war years onwards, the typical assignor was a prominent member of a corporate research laboratory, or some other similar in-house company facility. Although it is normally difficult to trace these named individuals in secondary sources on the firms concerned (as they are not usually also senior managers), the location of assignors can be checked against business history sources on the international location of activity in particular firms. Such checks on a selection of large firms have confirmed that whenever a location has been responsible for significant numbers of patents being assigned to a company, that firm did indeed have some in-house facility in the location in question at the relevant time. Companies checked in this fashion include various US firms active abroad and European companies in the US (Stocking and Watkins, 1946; Beaton, 1957; Wilkins, 1974, 1989; Chandler, 1990), Courtaulds and British Celanese (Coleman, 1969), Du Pont and ICI (Hounshell and Smith, 1988), and General Electric and GEC (Reich, 1986; Jones and Marriot, 1971).

Second, using once again the *US Index of Patents* and the *US Patent Office Gazette*, for every patent granted in years between 1890 and 1962 the country of residence of the inventor has been recorded. From 1963 this information has already been computerised by the US Patent Office. Where patents are assigned to companies, these data on the country of origin of invention indicate the location of the R&D facilities (or other sources of technological improvement) that gave rise to each patent. As this information on the location of invention relates to individually numbered patents, it can be combined with a sectoral classification of the technological activity with which the patent is associated. This employs the system of patent classes used by the Patent Office; fortunately, as these classes change the Office reclassifies all earlier patents accordingly, so the classification is historically consistent. Although patents may be assigned to several fields, the primary classification was used in all cases. Various broad categories of technological activity were derived by allocating classes or sub-classes to common groups of activity.

Two distinctions between different aspects of these classifications of the data are worth re-emphasising. First, the sectoral classification of patents, in terms of the type of technological activity with which each patent is associated, is distinguished from the main industrial output or markets of the companies to which patents may be assigned, both of which have been recorded separately. Most large companies have engaged in at least some development in most of the general spheres of technological activity (for instance, chemical firms develop many mechanical technologies, including chemical machinery and equipment), irrespective of the industry in which they operate.

Second, the country of location of the invention, which for large firms typically represents or is allied to the location of corporate R&D, is distinguished from the location of the firm to which a patent is assigned, and from the location of the parent company which owns this firm. While no significance is attached to the location of the assignee, the location of research and the location of ultimate ownership (the parent company) are critical. The extent to which these locations differ over the total patenting of each corporate group is the measure of the degree of internationalisation of that group's technological activity. These distinctions are crucial to understanding that the measurement of the degree of internationalisation of technological activity in what follows is not to be mistaken for a measure of the international spread of patents taken out by firms (the patents counted were all those—and only those—granted by the US Patent Office), nor should it be mistaken for a measure of the international dispersion of the legal departments or agents responsible for making patent applications on behalf of the group (the actual assignee is ignored once it has been linked to a parent company of a large group in the dataset). Instead, what is measured is the internationalisation of the underlying technological activity that gave rise to the knowledge which subsequently led to a patent being granted to the group.

Table 1. *Shares of US patenting of the largest nationally owned industrial firms due to research located abroad (%)*

	1920–1939	1940–1968	1969–1990
US	6·81	3·57	6·82
Europe	12·03	26·65	27·13
UK	27·71	41·95	43·17
Germany	4·03	8·68	13·72
Italy	29·03	24·76	14·24
France	3·35	8·19	9·55
Netherlands	15·57	29·51	52·97
Belgium	95·00	53·90	60·60
Switzerland	5·67	28·33	43·76
Sweden	31·04	13·18	25·51
Total	7·91	8·08	14·52

Source: US patent data compiled at the University of Reading, with assistance from the US Patent and Trademark Office, US Department of Commerce.

3. The historical role of the internationalisation of technological activity

As just described, the degree of internationalisation of corporate technological activity is measured by the share of patenting that is attributable to research (or other sources of invention) located outside the home country of the corporate group or groups in question. The broad changes in the internationalisation of technological activity between 1920 and 1990 measured in this way are set out in Table 1, organised by national groups of firms according to the location of ultimate ownership. The share of foreign research in the total corporate technological activity of the largest US and European industrial firms considered together averaged about 8% in 1920–1939 and 1940–1968, before rising significantly to roughly 14·5% in 1969–1990. However, this recent average trend increase in the internationalisation of activity as measured by corporate patenting reflects primarily the rising share of US patenting accounted for by European firms (which are more internationalised) *vis-à-vis* their US counterparts.

There appear to be three categories of national groups of firms. In the first, German- and French-owned firms come closest to the standard view, in the sense that the degree of internationalisation of their technological activity was very low historically, but has been on a slowly rising trend, and has increased significantly recently. The second category comprises the historically more multinational British, Swiss and Dutch companies, whose technological activity became substantially internationalised after the Second World War, achieving an early increase in international scope that is reflected in the picture for large European firms as a whole. The third category is very different from the others. The technological activity of the largest US and Swedish firms was as highly internationalised historically as it is today; and, of course, for some individual companies (such as the American General Electric) the extent of internationalisation was much higher in the interwar period than it has been recently. The largest American and Swedish firms retreated from their international research operations after the Second World War, and they have only recently regained the position they held before that time.

Table 2. *Shares of US patenting of the largest nationally owned industrial firms due to research located abroad (%)*

	US	Europe	UK	Germany	France	Switzerland	Sweden	Total
1920–1924	3·32	8·45	7·61	2·89	3·02	4·00	39·77	4·03
1925–1929	6·61	8·19	14·26	5·42	2·67	3·73	29·88	6·87
1930–1934	8·17	10·83	25·88	4·79	4·12	6·38	27·08	8·80
1935–1939	6·94	15·11	34·63	2·83	3·33	6·04	30·23	8·86
1940–1959	3·79	27·10	40·84	9·46	8·90	22·24	11·98	7·67
1960–1964	2·85	24·22	39·52	8·26	7·46	27·28	13·24	7·62
1965–1968	3·69	28·02	46·44	8·21	7·76	39·56	15·22	9·42
1969–1972	5·22	28·21	42·66	12·65	7·71	45·86	18·14	12·21
1973–1977	5·98	25·42	40·09	11·03	6·44	44·67	21·15	13·18
1978–1982	6·65	24·60	39·17	12·14	6·66	43·68	27·25	13·82
1983–1986	8·51	27·13	45·32	14·83	9·20	41·02	29·84	15·92
1987–1990	8·95	30·86	50·55	17·76	18·26	42·74	31·49	18·56

Source: As for Table 1.

A more detailed periodisation of the records of each of these three groups of large firms can be gleaned from Table 2. For French and German companies this shows that the upward trend in the internationalisation of technological activity is not as gradual as might have been supposed from Table 1, but in fact is based on some discrete jumps. As for other European firms, there was an increase in internationalisation immediately after the Second World War, following the nationalist retreat of the late 1930s, and the expropriations of German-owned firms abroad in the 1940s. There was then a further rise in the early 1970s in German firms, and in the late 1980s for German companies again, but more especially for French firms (Cantwell and Kotecha, 1994). British and Swiss firms increased to much higher rates of internationalisation of technological activity after the war than did French or German companies. In the British case the origins of the process can be traced to a trend increase throughout the interwar period, which was consolidated after the war, rising close to a peak by the late 1960s, before recovering again in the 1980s. The degree of internationalisation of Swiss firms reached a peak in the early 1970s, which has not quite been matched since.

For the largest US firms the internationalisation of technological activity peaked in the early 1930s (at a level they have only very recently equalled), while Swedish firms began in the early 1920s from a high point that they have not yet since recaptured. The internationalisation of research in both the US and Swedish groups fell sharply after the war, but picked up again from the late 1960s onwards. It is this latter trend that has been quite widely commented upon, perhaps not surprisingly, especially by American and Swedish scholars (such as Ronstadt, 1977; Mansfield, Teece and Romeo, 1979; or Granstrand, Håkanson and Sjölander, 1992). However, this contrasts with the more general European experience. The internationalisation of technological activity in large European firms considered as a whole increased after the Second World War (as did all the constituent European national groups except the Swedish), and again from the mid-1980s (again this applies to all groups except the Swiss).

Besides their national groups, firms may also be allocated to industrial groups. The broad industrial classification adopted here groups industries in accordance with the prevailing type of technological activity in the sector in question. Thus, firms in the

Table 3. ***Shares of US patenting of the largest nationally owned industrial firms due to research located abroad, grouped by industry (%)***

	US	Europe	UK	Germany	France	Switzerland	Sweden	Total
				1920–1939				
Chemical	2·75	12·42	41·48	4·88	4·64	5·02	4·44	6·88
Electrical	10·13	3·21	1·98	2·58	0·63	8·20	8·89	9·42
Mechanical	5·15	20·18	29·27	3·56	2·93	1·94	36·45	9·17
Transport	1·61	4·95	8·02	2·76	4·89	n.a.	0·00	2·01
Total	6·81	12·03	27·71	4·03	3·35	5·67	31·04	7·91
				1940–1968				
Chemical	2·24	39·91	66·42	12·51	15·65	31·65	12·40	11·74
Electrical	6·03	14·15	8·81	6·01	5·16	14·31	8·40	7·26
Mechanical	3·27	23·54	37·94	4·15	12·72	17·12	15·70	7·85
Transport	1·41	4·26	5·52	2·36	2·22	n.a.	5·50	1·77
Total	3·57	26·65	41·95	8·68	8·19	28·33	13·18	8·08
				1969–1990				
Chemical	5·65	32·65	55·33	18·08	9·03	46·84	14·52	17·91
Electrical	9·09	27·32	27·11	11·01	8·43	34·22	29·32	14·65
Mechanical	6·51	23·94	50·31	7·67	16·83	33·23	27·75	13·58
Transport	4·95	10·16	10·55	8·47	4·98	n.a.	12·36	6·53
Total	6·82	27·13	43·17	13·72	9·55	43·76	25·51	14·52

Note: n.a.=not applicable.
Source: As for Table 1.

chemical, pharmaceutical and coal and petroleum products industries all rely mainly on chemical and related technologies; firms in the electrical equipment and office equipment or computer industries base themselves principally on electrical technologies; and firms in the motor vehicle, aircraft and rubber and plastic products industries are concerned with the major transport technologies (engines and tyres). For ease of exposition these are each referred to collectively as the chemical, electrical equipment and transport industries. Firms in all other industries rely mainly on more traditional mechanical technologies, so the mechanical group is a much more heterogeneous mixture.

As shown in Table 3, the US electrical equipment firms were much more internationalised in their research than were European firms in the equivalent industry in the interwar period, despite the much higher overall rate of internationalisation of technological activity in the European group. In Britain, Germany and France, large chemical firms were more internationalised historically than those in the electrical equipment sector, although in Switzerland the electrical companies featured more strongly, and in Sweden it was mechanically-based firms that were responsible for the very high overall internationalisation of research in the largest nationally owned firms at that time. The mechanical group were also highly internationalised in the UK (notably the textile companies British Celanese and Courtaulds), and British and French transport firms (particularly the tyre companies Dunlop and Michelin) appear more prominently than the other members of that group.

A similar discrepancy between the relative extent of internationalisation of research in chemical and electrical equipment firms in the US and Europe still existed in recent years, but the difference is less marked than it was. Among European companies the stronger internationalisation of chemical than electrical firms also holds now for Switzerland, but not for Sweden. For Swedish firms the mechanical group does not dominate international research as it once did, but in the UK and France the mechanical group is relatively more prominent than it used to be. What this reflects in each case is a broadening of the range of firms engaged in the internationalisation of technological activity across a wider spectrum; for example, while the mechanical engineering firm Alfa-Laval accounted for the bulk of the foreign research of Swedish companies historically, its share has steadily declined as other firms have begun to engage in technological development abroad (Zander, 1994).

For US and European firms as a whole a more detailed periodisation for the major industrial groups can be found in Tables 4 and 5. Table 4 shows that US-owned electrical equipment firms in the 1930s had a higher degree of internationalisation of technological activity than they have had at any time since, including the 1980s. About 12% of their research was located abroad in the 1930s, compared to roughly 11% in the 1980s. Two American-owned companies that contributed especially heavily to European-located research and production in the interwar years were General Electric and RCA. The degree of internationalisation of technological activity in General Electric was 18·3% in 1920–1939 but only 2·4% in 1968–1990, while in RCA the equivalent proportions were 20·4% and 5·4%.

In contrast, the largest European firms sustained a sizeable increase in the extent of the internationalisation of their research between the early 1930s and the 1950s. This increase was led by companies in the science-based sectors, and especially those in chemicals. The share of foreign research in the leading European chemical firms rose from 8·2% in 1930–1934 to 42·6% in 1940–1959, while for electrical equipment companies it increased from 5·1% to 11·0% (see Table 5). For the largest European chemical firms the immediate post-war rate of internationalisation proved to be a peak that has not since been surpassed (for them the foreign share stood at 34·8% in 1987–1990), but in the post-war period the major European electrical equipment companies steadily increased their international research, to a point where the foreign share is now similar to that in chemicals.

The European story partly reflects the British experience, which is described in Table 6. For UK-owned companies the increase in the internationalisation of technological activity between the early 1930s and the 1950s (from 25·9% to 40·8%) was also largely associated with a very strong internationalisation of research in the chemical industry (from 23·3% to 66·7%). The prominent contribution of the chemical companies is consistent with the view that, historically, internationalisation was linked to technological competence. European firms were technologically strongest in the chemical fields, in which areas they internationalised their research early; while American firms were relatively stronger in the development of electrical technologies, and in this field they led the early internationalisation of technological activity.

Thus, at least for technology leaders, the internationalisation of technological activity is not a new phenomenon. Of course, the first product cycle hypothesis might be rescued by a restatement to the effect that as a general rule the home country centre has been and remains the single most important site for the technological development of MNCs. Given that, as reported in Tables 1 to 6, the foreign share of technological activity has rarely been greater than two-thirds and is usually much less, MNCs are not 'stateless corporations' (Patel and Pavitt, 1991). The national origins of MNCs have been and

Table 4. *Shares of US patenting of the largest US-owned industrial firms due to research located abroad, grouped by industry (%)*

	1920–24	1925–29	1930–34	1935–39	1940–59	1960–64	1965–68	1969–72	1973–77	1978–82	1983–86	1987–90
Chemical	0·96	1·90	4·03	2·36	1·87	2·12	3·08	4·37	5·21	5·33	6·36	7·71
Electrical	4·14	8·50	12·27	11·99	7·01	3·81	5·23	7·29	7·94	9·00	11·60	10·90
Mechanical	3·31	5·72	5·83	5·06	2·87	4·00	3·58	4·61	5·39	6·71	9·36	9·34
Transport	0·12	1·37	1·94	1·73	0·90	1·69	2·36	3·45	4·21	4·51	5·81	7·59
Total	3·32	6·61	8·17	6·94	3·79	2·85	3·69	5·22	5·98	6·65	8·51	8·95

Source: As for Table 1.

Table 5. *Shares of US patenting of the largest European-owned industrial firms due to research located abroad, grouped by industry (%)*

	1920–24	1925–29	1930–34	1935–39	1940–59	1960–64	1965–68	1969–72	1973–77	1978–82	1983–86	1987–90
Chemical	7·07	7·52	8·15	18·22	42·55	37·17	38·46	35·40	31·25	30·51	32·30	34·77
Electrical	4·12	2·81	5·14	1·73	11·01	13·87	19·96	26·90	24·51	21·44	28·17	34·10
Mechanical	10·84	12·91	23·00	24·74	25·73	18·73	23·31	25·45	22·25	23·98	23·87	24·59
Transport	3·75	9·15	4·69	3·04	4·77	3·34	4·45	7·12	7·40	7·69	12·76	16·19
Total	8·45	8·19	10·83	15·11	27·10	24·22	28·02	28·21	25·42	24·60	27·13	30·86

Source: As for Table 1.

Table 6. *Shares of US patenting of the largest UK-owned industrial firms due to research located abroad, grouped by industry (%)*

	1920–24	1925–29	1930–34	1935–39	1940–59	1960–64	1965–68	1969–72	1973–77	1978–82	1983–86	1987–90
Chemical	22·22	19·44	23·27	52·12	66·66	64·19	67·90	57·79	49·61	50·03	58·82	62·50
Electrical	7·14	4·67	0·55	0·67	5·61	6·94	17·37	27·09	33·56	24·85	19·53	26·53
Mechanical	6·55	16·34	36·51	32·13	35·57	38·53	44·70	50·06	48·10	45·47	51·14	59·88
Transport	5·45	14·46	8·13	5·68	5·75	4·37	6·03	7·75	7·65	7·77	13·29	20·31
Total	7·61	14·26	25·88	34·63	40·84	39·52	46·44	42·66	40·09	39·17	45·32	50·55

Source: As for Table 1.

continue to be critical in determining the geographical and sectoral composition of their technological activity, based on their path-dependent evolution from a nationally differentiated expertise (Kogut, 1987, 1990). While globalisation has been defined here as the international integration of MNC networks, some authors have used the term to mean the loss of national identity by companies, and when defined in their unhelpful way, globalisation has not taken place.

It is also clear that from the perspective of the MNC as a whole, or from that of its home country, the internationalisation of research is generally less than the internationalisation of production (Patel and Pavitt, 1991). One possible way of thinking about this is that most MNCs constitute an internal locational hierarchy of activity (Hymer, 1975). To simplify matters for the sake of exposition, suppose that the production of each firm is divided into the technologically sophisticated and the simple or assembly-type. Then, technologically sophisticated production tends to become geographically concentrated in certain locations, and assembly-type production tends to agglomerate in others (Cantwell, 1987). The home country operations of the MNC stand at the pinnacle of its hierarchy, as a base for technologically sophisticated production, as hinted at by both Vernon and Hymer. Since the siting of research facilities is normally linked to the local support of technologically sophisticated production while little or no research accompanies assembly-type production, it follows that production in total is more widely geographically dispersed than research. So, while the home country is the single most important site for innovation, it does not necessarily follow that production in the home centre is more research-intensive than in the other advanced centres in which the MNC establishes technologically sophisticated production.

This argument suggests that, of the original theoretical underpinnings of the first product cycle hypothesis, locational economies of integration and agglomeration play a greater role than do economies of scale in the R&D function. Previous criticisms of the first product cycle hypothesis had shown that, while there is a minimum efficient scale for R&D facilities, the effect of firm size on the degree of internationalisation of R&D is ambiguous (Mansfield, Teece and Romeo, 1979). By contrast, a good deal of evidence has been gathered in support of the importance of economies of agglomeration or local clustering in the location of production (Krugman, 1991; Porter, 1990; Dahmén, 1970; Sölvell, Zander and Porter, 1991), and especially in the geographical location of innovation (Jaffe, 1989; Jaffe, Trajtenberg and Henderson, 1993; Feldman, 1993; Cantwell, 1991A) and in the location of the technologically sophisticated production of MNCs (Cantwell, 1987). The lesson seems to be that economies of locational agglomeration are important, but that for MNCs they may occur in various centres and not exclusively in the home country, although the home base is the most significant single such centre.

4. The new globalisation of technology by corporate leaders

It has been shown that technological activity in some US MNCs was more widely dispersed internationally in the interwar period than it is today, while many European MNCs were already geared up to technology creation abroad by the 1950s. The companies that achieved the greatest internationalisation of activity historically were generally technology leaders. This is what is suggested by the second product cycle hypothesis as formulated above, although there is a qualification to an acceptance of the hypothesis. This is, that where technology leaders are in a very strong or hegemonic

Table 7. *The cross-firm coefficient of variation (expressed as a percentage) of the share of patenting due to research located abroad*

	US	UK	Germany	France	Switzerland	Sweden
Chemical industry group						
1920–1939	217·5	135·1	177·3	89·0	88·6	190·9
1969–1990	121·1	84·6	47·4	90·4	25·1	21·1
Electrical equipment industry group						
1920–1939	144·7	232·5	165·6	223·6	38·9	141·4
1969–1990	148·5	31·7	95·9	74·7	86·1	73·5
Mechanical industry group						
1920–1939	289·2	181·3	193·6	189·3	141·4	219·7
1969–1990	106·0	63·3	119·8	118·3	36·1	115·3
Transport industry group						
1920–1939	202·4	219·4	117·4	264·6	n.a.	n.a.
1969–1990	101·0	157·1	126·2	207·1	n.a.	99·1

Note: n.a. = not applicable.
Source: As for Table 1.

position through rapid innovation at home, the home centre may exercise such a strong attraction for further research that the internationalisation of technological activity is weak—a situation that seems to have applied to US companies in the early post-war years, and to Japanese firms today. Yet despite the historical significance of the geographical dispersion of innovation by MNCs, it is only relatively recently that the literature on MNCs has devoted much attention to the international creation of technology, as opposed to international technology transfer (Cantwell, 1994).

The more recent broadening of the internationalisation of technological development to a wider range of firms is partly responsible for finally drawing attention to this issue. The reduction in the industrial focus of internationalisation mentioned previously offers some indirect evidence consistent with the broadening of internationalisation across firms, but Table 7 provides direct evidence of this process within industrial groups. This shows that the extent of variation across firms in the degree of internationalisation of their technological activity has been generally on a downward trend. A wider range of companies has now engaged in foreign technological development, in what was once mainly the province of a smaller number of leading firms. However, there are occasional exceptions to this trend. In the US electrical equipment group, ITT has remained highly international, and Sperry (prior to the formation of Unisys) also became strongly committed to foreign research, leaving others in their wake. In the Swiss electrical equipment group it was Brown Boveri (prior to the formation of ABB) that pulled ahead of the pack in its foreign operations, and in the German transport group the motor vehicle component producer Robert Bosch played this role, in each case sustaining the extent of cross-firm variation in the degree of internationalisation. Elsewhere, lower transport and communication costs contributed to a general expansion across large firms in the internationalisation of technological activity.

The other recent change which has often been discussed in other contexts is that formerly local market oriented affiliates have been increasingly integrated into international networks within their respective MNCs, such networks coming to resemble 'heterarchies' more than hierarchies (Hedlund, 1986; Doz, 1986; Porter, 1986; Bartlett and Ghoshal, 1989; Dunning, 1992). In technological activity, too, the location-specific capabilities of internationally dispersed MNC affiliates may have become more closely integrated than in the past, linked to a strategy for technology creation in the MNC as a whole, and not only with separate reference to each of the geographical parts of the company's business. This can be termed a new globalisation of technological innovation.

The theoretical rationale for the recent international integration of productive activity is that the economic benefits attributable to a more refined locational division of labour within the MNC have often come to outweigh the costs of being less nationally responsive in each market, costs associated with adverse political repercussions and the continued national differentiation of demand (Doz, 1986). In an integrated MNC network each affiliate specialises in accordance with the specific characteristics of local production conditions, technological capabilities and user requirements. The network benefits from economies of scale through the local concentration of particular lines of activity (increasing returns from local research in a specialised field as opposed to research in general), economies of locational agglomeration through an interchange with others operating in the same vicinity in technologically allied fields as suggested earlier, and economies of scope through the international intra-firm coordination of related but geographically separated activities. The experience acquired in a specialised activity in one location creates technological spillovers that can be passed on to other parts of the MNC network elsewhere. It has been shown that since the 1970s, in industries in which such net advantages to multinational integration were available, multinationality has been a source of competitive success and faster growth (Cantwell and Sanna-Randaccio, 1993).

The transformation of the MNC, from a mainly multi-domestic structure of separate affiliates each serving their local markets, to an integrated internal network structure, has relevance to other aspects of the product cycle model apart from the hypotheses mentioned earlier. The wider product cycle model extended beyond the two hypotheses discussed here, to various other hypotheses and assumptions, such as that foreign direct investment can be treated as essentially local market-oriented, and that firms can be thought of as akin to single product producers. It can be argued that this wider model broke down largely because of globalisation in the sense just described, entailing the international integration of MNC networks. A global scanning for (new sources of) innovation and greater international linkages between production facilities imply an interactive flow of products and technological knowledge between countries (Vernon, 1979). This contrasts strongly with the original product cycle perspective of innovation and technological knowledge essentially flowing outwards from a single major centre, namely the home country.

The new globalisation of the responsibilities of affiliates to the MNC as a whole can be illustrated with reference to the shift that has occurred in the pattern of technological specialisation of foreign-owned research facilities. To examine this issue the patent data were used to construct a measure of the distribution of technological specialisation across various types of activity, for different groups of firms within their industry. It is also possible to distinguish between the pattern of specialisation in domestic technological activity and in research abroad, with reference to a corporate group or groups. The index of technological specialisation across different fields of activity that has been calculated is often termed an index of revealed technological advantage (RTA). The RTA value of a selected group of firms in a particular sector of technological activity is given by its share of US patents in that sector granted to companies in the same industry, relative to that

Table 8. *The RTA values in selected sectors of technological activity, of US-owned firms in the electrical equipment industry, and German-owned firms in the chemicals industry*

	1920–1968		1969–1990	
	At home	Abroad	At home	Abroad
(a) US-owned firms in the electrical equipment industry				
Lighting and wiring	0·96	1·30	0·94	0·77
General industrial equipment	1·07	0·67	0·97	1·96
(b) German-owned firms in the chemicals industry				
Bleaching and dyeing processes	2·43	1·43	1·91	0·71
Organic chemicals (dyestuffs)	1·46	1·47	1·33	0·97
Pharmaceuticals	1·22	3·16	1·22	2·04

Source: As for Table 1.

group's overall share of all US patents assigned to firms in the industry in question. Denoting as P_{ij} the number of US patents granted in the field of technological activity i (defined with reference to the patent class system, as described earlier) to the selected group of firms j in a particular industry, then the RTA index is defined as follows:

$$RTA_{ij} = (P_{ij} / \Sigma_j P_{ij}) / (\Sigma_i P_{ij} / \Sigma_{ij} P_{ij})$$

The index varies around unity, such that values greater than one suggest that a group of firms is comparatively specialised in the activity in question relative to other firms in the same industry, while values less than one are indicative of a lack of specialisation by the standards of the industry (see Cantwell, 1993, for a further discussion). For the purposes of historical comparison, attention is focused on the two groups of corporate technology leaders most prominent historically, each originating from major centres of innovation in their respective industries—that is, US-owned firms in the electrical equipment industry, and German-owned companies in the chemical industry. For each group the RTA index is separately calculated for activity located in its home research and its foreign research facilities, and for each of the two broad periods 1920–1968 and 1969–1990. In order to avoid the problems associated with low numbers of patents, the analysis is restricted to sectors of technological activity in which all large firms in the industry in question were granted 900 US patents or more in 1920–1968. This involved 19 sectors in the electrical equipment industry, and 20 in the case of chemicals. The values of the RTA index calculated for a few selected sectors are shown in Table 8.

The European-located research of US-owned electrical equipment firms was not historically (1920–1968) geared to local European strengths, but instead represented the local development of fields related to the core technologies that had been pioneered at home (in telecommunications and general electrical systems, including lighting). For these purposes, the pattern of technological specialisation of countries and broader regions such as Europe relative to one another was observed using an analogous RTA measure for countries rather than national groups of firms, grouping all patents (and not just those assigned to the largest firms) by the location of invention—indeed, it was in this country-specific (as opposed to firm-specific) form that the RTA index was originally proposed (Soete, 1987; Cantwell, 1989). In more recent years (1969–1990), while the domestic activity of US-owned firms in this sector has continued to concentrate on general electrical equipment, the focus of their foreign research has shifted to specialised machinery and general industrial equipment, both of which are areas of European advantage. Thus, the technological activity of US-owned foreign affiliates (which is still mainly conducted in Europe) has shifted towards an attempt to exploit the technological potential of the location in which it is carried out.

Similarly, the largest German chemical firms were strongly specialised historically (between 1920 and 1968) in the same fields in developing technology both at home and abroad—namely, in bleaching and dyeing processes, in organic chemicals including dyestuffs, and to a lesser extent in pharmaceuticals. Thus, these large German companies quite directly exploited their major strengths abroad, further developing these new products and techniques for local industries and markets in other countries. In foreign research today (since 1969) these firms are no longer specialised in their corporate strengths of dyes and dyeing, but they retain a stronger focus on pharmaceuticals in which they are not quite so strong at home. Rather like the American electrical companies, the German chemical firms have shifted their international research strategies away from the pure exploitation of their own strengths adapted to the needs of each particular local market, and towards an attempt to tap into foreign centres of expertise, in their case mainly in the development of pharmaceuticals in the US, Britain and Switzerland.

Other recent evidence also suggests that this type of internationally integrated or globalised strategy for innovation characterises corporate technology leaders today. The extent to which the affiliates of MNCs specialise within their industry across national boundaries in accordance with the comparative advantage of local expertise seems to depend upon the pattern of locational hierarchy that exists between alternative centres. In the European chemical industry Germany is the dominant centre, the UK is a second-order centre, and Italy is of the third order. In this sector, German MNCs (the leaders) are technologically specialised in the other European centres in line with host country strengths, and the same is true of British chemical MNCs in Italy (Cantwell and Sanna-Randaccio, 1992).

However, when operating in Germany, British chemical companies follow a pattern of technological specialisation that accords with their own comparative advantages in the industry and those of their home centre, the UK. They do not appear to be especially prone to try and tap into the areas in which Germany expertise is relatively greatest, but rather treat Germany as a general reservoir of skills that can be used principally to extend those lines of operation on which they are already focused in their home base. Technology leaders originating from higher-order centres tend to establish a more extensive locationally specialised network of technological activity in support of an international innovation strategy than has been developed as yet by firms that originate from lower-order centres.

It is true that this result depends upon an analysis of cross-border technological specialisation conducted at a fairly broad level of aggregation—across 20 or 30 sectors of technological activity, as described earlier. It may well be that when companies from a lower-order centre locate research in a higher-order one to engage in the same broad lines of activity in which they are specialised at home, there is still some geographical specialisation at the more detailed level of particular products or processes, and at the more finely disaggregated patent class level (Zander, 1994). In this event, the distinction between corporate technology leaders and other firms would be more a matter of the degree of locational specialisation that they have managed to achieve in technological development, rather than the existence of such a strategy of specialisation across centres.

The globalisation of technological innovation in MNCs, in the sense here of an international integration of geographically dispersed and locally specialised activities, tends to reinforce and not to dismantle nationally distinctive patterns of development or national systems of innovation (Nelson, 1993). Contrary to what is sometimes alleged, globalisation and national specialisation are complementary parts of a common process, and not conflicting trends (see Archibugi and Michie, 1993, for a further discussion). The incentive to organise affiliate specialisation is the desire to tap into the locally specific and differentiated stream of innovation in each centre, but by specialising in accordance with these local strengths the latter are reinforced. The creation of tacit capability is localised and embedded in social organisations (Nelson and Winter, 1982), and this organisational distinctiveness has a location-specific as well as a firm-specific dimension. The particular path of innovation followed in each country or region has historical origins (Rosenberg, 1976, 1982). In the period of globalisation since the late 1960s the general tendency has been for MNCs to become more technologically diversified as they establish newly-integrated technological systems, while countries or locations have become more specialised in their technological activity (Cantwell, 1993).

5. Conclusions

One aspect of the product cycle model is now rather discredited, namely the idea that demand-led innovation (together with economies of scale in R&D) in the home country dictates the geographical restriction of corporate research and the most technologically sophisticated production to the site of the parent company. For one thing, the demand-led view of innovation that was prevalent in the 1960s is now more widely acknowledged to be one-sided and potentially misleading (Mowery and Rosenberg, 1979). Innovations generally rely on a firm-specific learning process that interacts with both the growth of demand and the creation of new scientific and technological knowledge. In a region or country that enjoys technological leadership, high incomes and demand are as much a consequence of that leadership (high technological capability and thus high productivity) as they are a cause of it. For another thing, the peculiarities of foreign production conditions and demand have required leading MNCs historically to develop innovations abroad, related to those that had been pioneered at home.

For this latter reason, another aspect of the model can be extended to provide a further historical application that remains relevant. That is, the product cycle view that outward industrial investment is most widely associated with technology leaders, generally in conjunction with their holding a strong export position; so, too, the earliest internationalisation of technological development was largely due to such leaders. Historically, some highly competent US electrical equipment firms were considerably more international in their technology creation then than they are now. In Europe large chemical firms were relatively technologically stronger, and it was they that led the historical internationalisation of corporate research facilities.

In more recent times technology leaders have altered the nature of international technology creation by pioneering the international integration of MNC facilities into regional or global networks. Globalisation in this sense involves the establishment of new international structures for technology creation. In the past, foreign technological activity exploited domestic strengths abroad, it was located in response to local demand conditions, it assisted in the growth of other high income areas, and its role ranged from the adaptation of products to suit local tastes through to the establishment of new local industries. At that time the capacity to develop internationally dispersed innovations derived from a position of technological strength in the firm's home country base, and led to similar lines of technological development being established abroad. By contrast, today, for companies of the leading centres, foreign technological activity now increasingly aims to tap into local fields of expertise, and to provide a further source of new technology that can be utilised internationally in the other operations of the MNC. In this respect, innovation in the leading MNCs is now more genuinely international or, in the terminology used here, it has become 'globalised'.

There are two similarities between the theoretical rationale provided for the product cycle model (which applied best to the US and to US-owned MNCs in the early post-war period) and that suggested for the current globalisation of technological activity in MNCs. Both explanations rely on the role of the economies of locational agglomeration, and on the leadership exercised by the most technologically competent firms. The essential difference is that in the product cycle model just one pre-eminent centre for innovation was recognised, whereas in the globalisation story there are multiple locations for innovation, and even lower-order or less developed centres can still be sources of innovation. Hence, the theoretical concepts also used in the product cycle approach apply differently in the account of globalisation; locational agglomeration occurs in the clusters of distinctive innovations that occur in many centres and not only in one unique centre, while the greater capability of the most competent firms manifests itself not just in the wider geographical dispersion of their investments (a more important consideration historically), but in the broader degree of cross-border specialisation that they are able to manage.

The product cycle view of the MNC as a locational hierarchy also remains relevant, although this too needs extending. It is still true that the home country is generally the single most important site for corporate technological development. More interestingly, the form of locational hierarchy in the leading firms of the most advanced centres is now much more complex than it used to be, and more complex than is the equivalent hierarchy of other MNCs. The affiliates of the leading companies in other major centres may be thought of as constituting an interactive network. Cross-investments between the major centres in the most technologically dynamic industries (Dunning, 1988; Cantwell, 1989; Cantwell and Sanna-Randaccio, 1992) have probably helped to reinforce the existing pattern of geographical specialisation, and the importance of these centres as locations for innovation. Having been the first to establish an international spread of technological activity, MNCs from the leading centres in a given industry now exploit locational diversity in paths of innovation to a greater extent than do other firms.

Bibliography

Archibugi, D. and Michie, J. 1993. 'The Globalisation of Technology: Myths and Realities', University of Cambridge Research Papers in Management Studies, No. 18, May

Bartlett, C. A. and Ghoshal, S. 1989. *Managing Across Borders: The Transnational Solution*, Boston, MA, Harvard Business School Press

Beaton, K. 1957. *Enterprise in Oil: A History of Shell in the United States*, New York, Appleton-Century-Crofts

Cantwell, J. A. 1987. The reorganisation of European industries after integration, *Journal of Common Market Studies*, vol. 26, 127–151

Cantwell, J. A. 1989. *Technological Innovation and Multinational Corporations*, Oxford, Basil Blackwell

Cantwell, J. A. 1991A. The international agglomeration of R&D, in Casson, M. C. (ed.), *Global Research Strategy and International Competitiveness*, Oxford, Basil Blackwell

Cantwell, J. A. 1991B. The theory of technological competence and its application to international production, in McFetridge, D. G. (ed.), *Foreign Investment, Technology and Economic Growth*, Calgary, University of Calgary Press

Cantwell, J. A. 1993. Corporate technological specialisation in international industries, in Casson, M. C. and Creedy, J. (eds), *Industrial Concentration and Economic Inequality*, Aldershot, Edward Elgar

Cantwell, J. A. 1994. Introduction, in Cantwell, J. A. (ed.), *Transnational Corporations and Innovatory Activities*, London, Routledge

Cantwell, J. A. and Barrera, M. P. 1993. 'The Rise of Corporate R&D and the Technological Performance of the Largest European Firms from the Interwar Years Onwards', University of Reading Discussion Papers in Economics, no. 271, September

Cantwell, J. A. and Barrera, M. P. 1995. Inter-company agreements for technological development: lessons from international cartels, *International Studies in Management and Organisation*, forthcoming

Cantwell, J. A. and Hodson, C. 1991. Global R&D and UK competitiveness, in Casson, M. C. (ed.), *Global Research Strategy and International Competitiveness*, Oxford, Basil Blackwell

Cantwell, J. A. and Kotecha, U. 1994. L'internationalisation des activités technologiques: le cas français en pespective, in Sachwald, F. (ed.), *Les Défis de la Mondialisation: Innovation et Concurrence*, Paris, Masson

Cantwell, J. A. and Sanna-Randaccio, F. 1992. Intra-industry direct investment in the European Community: oligopolistic rivalry and technological competition, in Cantwell, J. A. (ed.), *Multinational Investment in Modern Europe*, Aldershot, Edward Elgar

Cantwell, J. A. and Sanna-Randaccio, F. 1993. Multinationality and firm growth, *Weltwirtschaftliches Archiv*, vol. 129, 275–299

Chandler, A. D. 1990. *Scale and Scope: The Dynamics of Industrial Capitalism*, Cambridge, MA, Harvard University Press

Coleman, D. C. 1969. *Courtaulds: An Economic and Social History (Volume II: Rayon)*, Oxford, Oxford University Press

Dahmén, E. 1970. *Entrepreneurial Activity and the Development of Swedish Industry, 1919–1939*, Homewood, American Economic Association Translation Series

Dosi, G., Teece, D. J. and Winter, S. G. 1992. Towards a theory of corporate coherence: preliminary remarks, in Dosi, G., Giannetti, R. and Toninelli, P. A. (eds), *Technology and Enterprise in a Historical Perspective*, Oxford, Oxford University Press

Doz, Y. 1986. *Strategic Management in Multinational Companies*, Oxford, Pergamon

Dunning, J. H. 1988. *Multinationals, Technology and Competitiveness*, London, Unwin Hyman

Dunning, J. H. 1992. *Multinational Enterprises and the Global Economy*, Wokingham, Addison-Wesley

Feldman, M. P. 1993. An examination of the geography of innovation, *Industrial and Corporate Change*, vol. 2, 451–470

Granstrand, O., Håkanson, L. and Sjölander, S. (eds) 1992. *Technology Management and International Business: Internationalisation of R&D and Technology*, Chichester, John Wiley

Hedlund, G. 1986. The hypermodern MNC: a heterarchy?, *Human Resource Management*, vol. 25, 9–25

Hounshell, D. A. and Smith, J. K. 1988. *Science and Corporate Strategy: Du Pont R&D, 1902–1980*, Cambridge, Cambridge University Press

Hymer, S. 1975. The multinational corporation and the law of uneven development, in Radice, H. (ed.), *International Firms and Modern Imperialism*, Harmondsworth, Penguin

Jaffe, A. B. 1989. Real effects of academic research, *American Economic Review*, vol. 79, 957–970

Jaffe, A. B., Trajtenberg, M. and Henderson, R. 1993. Geographical localization of knowledge spillovers as evidenced by patent citations, *Quarterly Journal of Economics*, vol. 108, 577–598

Jones, R. and Marriot, O. 1971. *Anatomy of a Merger: A History of GEC, AEI and English Electric*, London, Cape Ltd

Kogut, B. 1987. Country patterns in international competition: appropriability and oligopolistic agreement, in Hood, N. and Vahlne, J. E. (eds), *Strategies in Global Competition*, London, Croom Helm

Kogut, B. 1990. The permeability of borders and the speed of learning among countries, in Dunning, J. H., Kogut, B. and Blomström, M. (eds), *Globalisation of Firms and the Competitiveness of Nations*, Lund, Lund University Press

Krugman, P. R. 1991. *Geography and Trade*, Cambridge, MA, MIT Press

Lall, S. 1979. The international allocation of research activity by US multinationals, *Oxford Bulletin of Economics and Statistics*, vol. 41, 313–331

Mansfield, E., Teece, D. J. and Romeo, A. 1979. Overseas research and development by US based firms, *Economica*, vol. 46, 187–196

Mowery, D. C. and Rosenberg, N. 1979. The influence of market demand upon innovation: a critical review of some recent empirical studies, *Research Policy*, vol. 8, 103–153

Nelson, R. R. 1991. Why do firms differ and how does it matter?, *Strategic Management Journal*, vol. 12, 61–74

Nelson, R. R. 1992. The role of firms in technical advance, in Dosi, G., Giannetti, R. and Toninelli, P. A. (eds), *Technology and Enterprise in a Historical Perspective*, Oxford, Oxford University Press

Nelson, R. R. (ed.) 1993. *National Innovation Systems*, Oxford, Oxford University Press

Nelson, R. R. and Winter, S. G. 1982. *An Evolutionary Theory of Economic Change*, Cambridge, MA, Harvard University Press

Patel, P. and Pavitt, K. L. R. 1991. Large firms in the production of the world's technology: an important case of 'non-globalisation', *Journal of International Business Studies*, vol. 22, 1–21

Pavitt, K. L. R. 1992. Some foundations for a theory of the large innovating firm, in Dosi, G., Giannetti, R. and Toninelli, P. A. (eds), *Technology and Enterprise in a Historical Perspective*, Oxford, Oxford University Press

Pearce, R. D. and Singh, S. 1991 *Globalising Research and Development*, London, Macmillan

Penrose, E. T. 1959. *The Theory of the Growth of the Firm*, Oxford, Basil Blackwell

Porter, M. E. 1986. Competition in global industries: a conceptual framework, in Porter, M. E. (ed.), *Competition in Global Industries*, Boston, Harvard Business School Press

Porter, M. E. 1990. *The Competitive Advantage of Nations*, New York, The Free Press

Reich, L. S. 1986. *The Making of American Industrial Research: Science and Business at GE and Bell, 1876–1926*, Cambridge, Cambridge University Press

Ronstadt, R. C. 1977. *Research and Development Abroad by US Multinationals*, New York, Praeger

Rosenberg, N. 1976. *Perspectives on Technology*, Cambridge, Cambridge University Press

Rosenberg, N. 1982. *Inside the Black Box: Technology and Economics*, Cambridge, Cambridge University Press

Schmookler, J. 1966. *Invention and Economic Growth*, Cambridge, MA, Harvard University Press

Soete, L. L. G. 1987. The impact of technological innovation on international trade patterns: the evidence reconsidered, *Research Policy*, vol. 16, 101–130

Stocking, G. W. and Watkins, M. W. 1946. *Cartels in Action*, New York, Twentieth Century Fund

Sölvell, Ö., Zander, I. and Porter, M. E. 1991. *Advantage Sweden*, Stockholm, Norstedts

Teece, D. J., Pisano, G. and Shuen, A. 1990. 'Firm Capabilities, Resources and the Concept of Strategy', University of California at Berkeley Consortium on Competitiveness and Cooperation Working Papers, No. 90-8

Vernon, R. 1966. International investment and international trade in the product cycle, *Quarterly Journal of Economics*, vol. 80, 190–207

Vernon, R. 1979. The product cycle hypothesis in a new international environment, *Oxford Bulletin of Economics and Statistics*, vol. 41, 255–267

Wilkins, M. 1974. *The Maturing of Multinational Enterprise*, Cambridge, MA, Harvard University Press

Wilkins, M. 1989. *The History of Foreign Investment in the United States to 1914*, Cambridge, MA, Harvard University Press

Zander, I. 1994. *The Tortoise Evolution of the Multinational Corporation*, Stockholm, Institute of International Business, Stockholm School of Economics

Part V
The Not-So-Peculiar British Case

[19]

Firm Structure, Government Policy, and the Organization of Industrial Research: Great Britain and the United States, 1900–1950

DAVID C. MOWERY

¶ *An important factor in the poor performance of the British economy in the twentieth century is the low level of investment by British firms in research and development. In this article, Professor Mowery compares the development of industrial research in Great Britain and the United States between 1900 and 1950, focusing on the reasons for the much lower levels of industrial research investment in Britain. He concludes that the substantial British lag can be attributed to differences in corporate structure, educational systems, and government policy.*

A central factor in Britain's uneven economic fortunes since 1900 is the poor innovative record of its industry. Partly because innovation lagged, British manufacturing productivity and exports grew more slowly than those of its industrial competitors.[1] The innovative performance of British firms was affected by low levels of industrial research, in which British firms invested fewer resources than did American enterprises. This article analyzes the development of British industrial research from 1900 to 1950, focusing particularly on the reasons for the lower level of research and development (R&D) investment in British industry. The 1900–50 period is of interest for several reasons. During this time the gap between British economic performance and that of other nations (notably the United States) became apparent; concurrently, investment in industrial research grew rapidly in the United States and Germany. In addition, since both the poor British economic performance and the growth in foreign R&D con-

David C. Mowery is assistant professor of economics and social science at Carnegie-Mellon University. His article has benefited from the comments of Louis Cain, Alfred D. Chandler, Jr., Bernard Elbaum, William Lazonick, Derek Morris, Richard Nelson, Nathan Rosenberg, Richard Tedlow, and two anonymous *Review* referees, as well as participants in the Anglo-American Conference and the Berkeley-Stanford Economic History Colloquium. He is grateful to Professor Chandler for the use of unpublished data. An earlier version of this article was presented at the Anglo-American Conference on the Decline of the British Economy, Boston University, 30 September to 1 October 1983. Research for the paper was supported by the National Science Foundation (PRA77-21852), the Division of Research of the Harvard Business School, the Andrew W. Mellon Foundation, and the Program in Technology and Society at Carnegie-Mellon University.

[1] A recent exploration of this argument may be found in K. L. R. Pavitt, "Introduction and Summary," in *Technical Innovation and British Economic Performance*, ed. K. L. R. Pavitt (London, 1980).

Business History Review 58 (winter 1984).

cerned many official British observers and policy makers, it is possible to trace the British government's policy responses, as well as their impact.

Recent empirical research provides a strong basis for the conclusion that R&D investment enhances national economic performance, especially export growth and productivity.[2] In both these areas, Britain fell behind other industrial nations during the 1900–50 period. During this period, the share of British exports in total world trade declined substantially more than did the British share of world manufacturing output; even more dramatic was the absolute decline in British exports during the 1920s. Declining exports reflected the continued dependence of British foreign trade on staple commodities such as coal, iron, and textiles.[3] The development of alternative export industries was impeded by low levels of R&D. Productivity growth in the British economy during this period also lagged behind that of other European countries and the United States.[4] Clearly, influences other than R&D investment and innovative performance affected these trends. Nonetheless, the weight of empirical evidence suggests that a more vigorous R&D effort might have offset this spiral of economic decline. The persistence of low levels of R&D investment in British industry is especially striking in view of the fact that many official and unofficial reports acknowledged the technological failings of British industry.

Compounding the problem, the organization of the British industrial research system had the effect of reducing the returns on this investment. Reliance on government-funded cooperative research organizations, the scarcity of engineers (as opposed to scientists) in British

[2] Innovative activity at the firm and industry level—measured in terms of either inputs (e.g., expenditures or employment in research and development) or outputs (e.g., patents)—has been shown in several studies to influence productivity growth and export performance. E. Mansfield, *The Economics of Technological Change* (New York, 1968) and N. Terleckyj, *Effects of R&D on Productivity: An Exploratory Inquiry* (Washington, D.C., 1974) present empirical evidence on the links between industry and firm productivity growth and R&D activity. D. B. Keesing, "The Impact of Research and Development on United States Trade," *Journal of Political Economy* 75 (Jan./Feb. 1967): 38–48, and W. Gruber, D. Mehta, and R. Vernon, "The R&D Factor in International Trade and International Investment of United States Industries," *Journal of Political Economy* 75 (Jan./Feb. 1967): 20–37, both found a significant statistical linkage between export performance and R&D investment in a number of industrialized nations during the postwar period. While their findings cover only the period after 1950 (because of data limitations), similar links between research activity and trade performance probably also were present earlier. Statements such as that of Arthur Lewis in note 3 below are consistent with this view.

[3] See R. C. O. Matthews, C. H. Feinstein, and J. C. Odling-Smee, *British Economic Growth 1856–1973* (Stanford, Calif., 1982), esp. 436–37. W. A. Lewis, *Economic Survey, 1919–1939* (London, 1949), 79, noted that "the largest category of British exports was in those commodities expanding least in world trade. The leader par excellence in world trade was the United States; only 17.1 per cent of her manufactures exports were in the lowest category (in terms of growth in world trade during 1913–29) compared with Britain's 42.1 per cent, and 28.6 per cent were in the highest category, compared with Britain's 4.3 per cent. Germany also was well ahead."

[4] See L. Rostas, *Comparative Productivity in British and American Industry* (Cambridge, 1948) and E. H. Phelps Brown, "Levels and Movements of Industrial Productivity and Real Wages Internationally Compared, 1860–1970," *Economic Journal* 83 (March 1973): 58–71.

industrial research, the weak links between academic and industrial research, and individual firms' failure to organize industrial research effectively and coordinate it with other functions, all undermined the effectiveness of the modest British investment in R&D.

Three sets of related factors are central in explaining the low level and inefficient organization of British industrial research during this period. The first is the structure of the British manufacturing firm. In the United States, the development of industrial research was closely associated with the reorganization of the American corporation described by Alfred D. Chandler, Jr.[5] Chandler's "managerial revolution" in American business included the absorption into the firm of a range of activities formerly carried out through the market. Among the functions absorbed and expanded by the firm was industrial research. In Britain, such a transformation and reorganization of major firms occurred later than in the United States and was less complete in certain respects. British firms thus tended to be more dependent on market institutions for the organization and conduct of industrial research. Where research was carried out within the British firm, its effectiveness was often limited by the incomplete rationalization of internal firm structure. In short, the structural development of American industrial enterprises ultimately allowed for a more effective exploitation of the complementarities between research activity and production activity. These complementarities are exploited most effectively in a nonmarket (i.e., intrafirm) setting.

Another important influence, affecting both the development of the British firm and the organization of British industrial research, was government policy. A weak British antitrust policy influenced the development of R&D in two ways. Anticompetitive price and market-sharing agreements among British firms, particularly after the end of free trade, undercut the incentives for the pursuit of competitive advantage through innovative activity. Moreover, the British government's rationalization policies of the late 1920s and 1930s worsened the situation, by supporting cartel agreements among small and inefficient firms.[6] British antitrust policy also may have reduced the incentives of British firms to pursue large, multifirm mergers, which in the United States were closely associated with the development of modern corporate management structures. It is striking that the two countries differed both in their antitrust policies and in the evolution of firm structure; however, evidence of a causal connection between antitrust

[5] A. D. Chandler, Jr., *The Visible Hand* (Cambridge, Mass., 1977).

[6] G. C. Allen, *British Industry and Economic Policy* (London, 1979), 74, argued that "the measures taken to deal with this condition of excessive capacity must be classed among the noteworthy failures of economic policy of the interwar period."

policies and levels of R&D investment is not conclusive.[7] The British government did, it should be added, undertake a novel program of direct support for cooperative research within industry just after World War I. Given current policy debates over proposals to improve the innovative performance of American industry, a historical evaluation of British and American government industrial research policies is of considerable topical interest.[8]

A third major influence on the development of British R&D was the structure of technical and managerial education at the secondary and university levels—and here again government policy played a central role. A tradition of scant public financial support for such education in Britain did nothing to improve the number, or the professional training, of engineers, who are still much scarcer there than in Germany or the United States.[9] In addition, Britain did not develop the less formal links between higher and technical education and industry that proved to be of great importance in the industrial research systems of the United States and Germany. Both as a provider of technical expertise and a producer of technical manpower, the British educational network was less well suited to the support and promotion of industrial research than the American system, which received much more public financial support.

These three explanatory factors interacted in a complex fashion. The structure of British firms and industries enforced dependence on the market (rather than the firm) to coordinate transactions. This industrial structure in turn reduced the demand of British industry for technical and managerial personnel. Government policy evolved in an accommodating fashion, emphasizing above all else (especially before World War I) noninterference with the operations of the market. It is possible that a different set of government policies could have substantially enhanced the level and effectiveness of British industrial research, especially given the modest (and largely ineffective) policies employed. Yet even the more interventionist policies of British governments since 1945 have not raised industrial research activity and technological per-

[7] In Germany, antitrust policy was also weak, while industrial research activity in many industries was substantial. But Germany differed from Britain in other respects, notably the educational system, the involvement of banks in the organization and financing of mergers, and the resulting pattern of firm structure. The interactions among these three factors are of great importance in explaining international differences in the level of R&D activity. Moreover, the impact of industrial research on performance may depend as much on its organization as on the level of R&D investment. German firms appear generally to have been more successful than British enterprises in organizing and managing industrial research.

[8] For example, a number of proposals have been made for the establishment within the United States of publicly funded cooperative industrial research organizations. For a discussion of the theoretical basis for and results of some of these programs, see D. C. Mowery, "Economic Theory and Government Technology Policy," *Policy Sciences* 16 (winter 1983): 27–43.

[9] See M. J. Peck, "Science and Technology," in *Britain's Economic Prospects*, ed. R. E. Caves (Washington, D.C. 1968).

formance to American levels. Indeed, a striking aspect of this comparative analysis is the stability, over a long time period, of both the key explanatory factors and the gap in R&D investment. Throughout the 1900–50 period, the nature and degree of the differences between British and American firm structure, educational systems, and government policies remained unchanged.

COMPARATIVE LEVELS OF R&D IN GREAT BRITAIN AND THE UNITED STATES

Neither the U.S. nor the British data on research activity permits a quantitative comparison covering the entire 1900–50 period. However, the existing data is consistent in suggesting that during and after this period, British manufacturing firms lagged substantially behind their American counterparts in average research intensity (measured as either the share of research employees in total firm employment, or the share of R&D expenditures in total firm sales). C.H. Freeman, analyzing data on expenditures from the mid 1950s, and M. Sanderson, using employment data from the 1930s, concluded that the level of research intensity in British firms averaged about one-third that of American manufacturing firms. Lower levels of research activity in British manufacturing are not well explained by transatlantic differences in the mix of industries in the manufacturing sectors of these two nations.[10] Other data on American and British industrial research from the 1930s and 1940s also suggests that R&D investment in British manufacturing was well below that of American manufacturing.

Estimates of research employment in U.S. manufacturing during the 1930s and 1940s are taken from the surveys of the National Research Council.[11] Employment of research professionals in the laboratories of American manufacturing firms stood at 10,900 in 1933 (0.18 percent of total manufacturing employment), rising to nearly 28,000 in 1940 (0.35

[10] C. H. Freeman, "Research and Development: A Comparison between British and American Industry," *National Institute Economic Review* 20 (Feb. 1962): 21–38, stated that U.S. industry's R&D expenditures per employee were three times as great as British industry's. Industrial research employed 0.13 percent of the U.S. work force, compared with 0.05 percent in Britain in 1961. See also M. Sanderson, "Research and the Firm in British Industry," *Science Studies* 2 (April 1972): 107–51. Both Freeman and Sanderson found electrical machinery, chemicals, and petroleum products to be among the most research-intensive industries in the United States and Great Britain during the 1930s and 1950s. Freeman concluded in addition that the most research-intensive industries in his analysis (including aircraft, chemicals, electrical and nonelectrical machinery, and instruments) accounted for a larger percentage of net manufacturing output in Great Britain (32 percent in 1935 and 45 percent in 1958) than in the United States (28 percent in 1935 and 40 percent in 1958).

[11] See the data in National Research Council, *Industrial Research Laboratories of the United States*, bulletins 91, 104, and 113 (Washington, D. C., 1933, 1940, 1946), reported and analyzed in D. C. Mowery, "The Emergence and Growth of Industrial Research in American Manufacturing, 1899–1946" (Ph.D. diss., Stanford University, 1981).

percent), and 45,900 in 1946 (0.39 percent). On the basis of surveys conducted by the Federation of British Industries, research employment in manufacturing can be estimated at 1,724 in 1933 (0.030 percent of the total manufacturing work force), 2,575 in 1935 (0.042 percent), 4,505 in 1938 (0.066 percent), and 5,200 in 1945–46 (0.080 percent). Research intensity in American manufacturing overall thus was roughly four to five times that of British manufacturing during the 1930s and 1940s.[12]

These differences in manufacturing research intensity reflect different levels of public funding of research in the two countries, as well as differences in the institutional location of publicly funded research. In Britain, government research installations, such as the National Physical Laboratory and the quasi-public research associations, accounted for a substantially higher percentage of total research expenditures than they did in the United States. J.D. Bernal estimated in 1939 that R&D expenditures by British industry amounted to £2.2 million, while total research expenditures by the British government (including financial support of the research associations) totalled £2.95 million. In contrast, George Perazich and Philip M. Field estimated total expenditures on research within U.S. industry at $175 million in 1937; total federal research expenditures in this year were $72 million, of which $38 million was spent within industry.[13] It is possible that publicly funded research in Britain partially compensated for the lower levels of in-house research employment in British manufacturing. However, very little British government-supported R&D was conducted within industrial firms during this period. The government's focus on research associations (discussed further below) and national laboratories may, in fact, have weakened the impact of public R&D funding on industrial innovation.

While the available data suggests that the research intensity of British manufacturing was well below that of the United States, this comparison does not control for transnational differences in the size distribution of firms. If anything, however, the largest British firms appear to have been even further behind their American counterparts in research intensity than were the smaller British manufacturers. A comparison of the 200 largest manufacturing enterprises in each country is

[12] The American data for 1946 reflects the impact of high levels of wartime federal support of industrial research. Nevertheless, the prewar data suggests that the gap between British and American manufacturing in in-house research activity was somewhat greater then than in the postwar period.

[13] See J. D. Bernal, *The Social Function of Science* (London, 1939); G. Perazich and P. M. Field, *Industrial Research and Changing Technology*, Works Progress Administration National Research Project Report M-4 (Philadelphia, 1940); and Senate Military Affairs Committee, Subcommittee on War Mobilization, *The Government's Wartime Research and Development* (Washington, D.C., 1945), 8.

instructive.[14] In 1933, 116 of the firms in a sample of 160 American firms from the top 200 had research laboratories. In 1936, only 20 of the largest 200 British firms had such in-house research facilities. The postwar comparison is similarly dramatic. Of the 200 largest firms in American manufacturing as of 1948, 164 maintained industrial laboratories, according to the National Research Council. In Britain, only 40 of the top 200 had such facilities. A comparison of American and British firms above the estimated threshold size necessary for the support of in-house research also reveals an American lead. In a sample of 33 American firms lying above this estimated threshold in 1933, all but two maintained research facilities. In a sample of eight British firms lying above this threshold in the 1930s, only one appears to have had an in-house research laboratory.[15]

If, as suggested earlier, the level of industrial research can influence the performance of the national economy, how did it affect the performance of individual firms? The available evidence is most abundant and convincing for U.S. manufacturing firms, and does not cover the entire 1900–50 period, but it permits some comparisons of the effects of research activity on the performance of British and American firms.

Research employment was a significant factor in the ability of an American manufacturer to remain within the ranks of the 200 largest firms during 1921–46 (the only period for which consistent data on research activity at the firm level is available for U.S. firms).[16] A cross-sectional analysis has also shown that growth (in the book value of assets) was higher among more research-intensive American firms during 1933–46 (although not during 1921–33). Between 1921 and 1946, there occurred a significant decline in turnover and instability within the ranks of the largest firms in the American economy—due in part to the very rapid growth of industrial research during that period.[17] Many large American firms relied on industrial research to preserve positions

[14] Data on the identity of the 200 largest British manufacturing firms was compiled from various sources by Alfred Chandler and associates, while data on the existence of in-house research facilities was taken from a 1936 survey conducted by the British Association of Scientific Workers, *Industrial Research Laboratories* (London, 1936), and E. N. da C. Andrade, *Industrial Research: 1946* (London, 1946). Similar data for the 200 largest American manufacturing firms in 1930 and 1948 was assembled by Chandler, and data on research employment in these firms was extracted from the National Research Council surveys of industrial research for 1933 and 1946 (see note 11).

[15] For a more detailed discussion of the data, assumptions, and statistical techniques involved in this analysis, see D. C. Mowery, "Industrial Research and Firm Size, Survival, and Growth in American Manufacturing, 1921–46: An Assessment," *Journal of Economic History* 43 (Dec. 1983): 953–80, and idem., "British and American Industrial Research: A Comparison, 1900–1950," presented at the Anglo-American Conference on the Decline of the British Economy, Boston, 30 September to 1 October 1983.

[16] See Mowery, "Industrial Research and Firm Size, Survival, and Growth."

[17] See R. C. Edwards, "Stages in Corporate Stability and the Risks of Corporate Failure," *Journal of Economic History* 35 (June 1975): 428–57; N. R. Collins and L. E. Preston, "The Size Structure of the Largest Industrial Firms, 1909–1938," *American Economic Review* 51 (Dec. 1961): 986–1011, or A. D. H. Kaplan, *Big Business in a Competitive System* (Washington, D.C., 1964).

of market power through aggressive and extensive patenting policies, and employed research to support product diversification and the acquisition of new technologies.[18] In British industry, there was no comparable transition from instability to stability in market structure just after World War I. In the early twentieth century the survival rate of large British firms appears to have been substantially lower than that of the largest American firms; only in the 1940s does a comparable increase in survival rates appear.[19] Lower levels of industrial research in British manufacturing thus are associated with lower levels of firm survival and a more unstable market structure during a portion of the 1900–50 period.

Research activity also influenced the growth of industrial productivity, although the data again does not cover the entire 1900–50 period. Within U.S. manufacturing, interindustry differences in total factor productivity growth between 1921 and 1946 were significantly and positively related to interindustry differences in research intensity.[20] These results are of interest in view of the contrasts between British and American labor productivity growth during the early twentieth century. American manufacturing substantially outperformed British manufacturing in this respect for much of the 1900–50 period. The gap in labor productivity increased during the 1920s, the period of most rapid growth in the number and size of American industrial research laboratories.[21]

THE STRUCTURE OF THE BRITISH FIRM: THE DELAYED MANAGERIAL REVOLUTION AND ITS CONSEQUENCES

The structure of British firms had much to do with the lower level of industrial research in British manufacturing, as well as with its fre-

[18] For further discussion, see L. S. Reich, "Research, Patents, and the Struggle to Control Radio: A Study of the Big Business and the Uses of Industrial Research," *Business History Review* 51 (summer 1977): 208–35; idem., "Industrial Research and the Pursuit of Corporate Security: The Early Years of Bell Labs," *Business History Review* 54 (winter 1980): 504–29; W. F. Mueller, "Du Pont: A Study in Firm Growth" (Ph.D. diss., Vanderbilt University, 1955); and G. W. Brock, *The Telecommunications Industry* (Cambridge, Mass., 1981).

[19] See L. Hannah, *The Rise of the Corporate Economy* (London, 1976), 117, 167.

[20] See N. Terleckyj, "Sources of Productivity Advance: A Pilot Study of Manufacturing Industries, 1899–1953" (Ph.D. diss., Columbia University, 1961), as well as the results presented in Mowery, "British and American Industrial Research."

[21] See Rostas, *Comparative Productivity in British and American Industry*. J. W. Kendrick, *Productivity Trends in the United States* (Princeton, 1961), 136, concludes that total factor productivity in U.S. manufacturing grew at an average annual rate of 5.3 percent from 1919 to 1929, and 1.9 percent from 1929 to 1937. Matthews et al., *British Economic Growth*, 229, estimate that total factor productivity in British manufacturing grew at an average annual rate of 1.9 percent from 1924 to 1937. While some tendency for British productivity growth to catch up with that of American manufacturing thus is indicated for the 1930s (possibly because of the severe impact of the Great Depression on output growth in the United States), these estimates are broadly consistent with those of Rostas in pointing to a large gap in the 1920s.

quently dysfunctional organization. Whereas many American firms underwent a transformation during the late nineteenth and early twentieth centuries that affected their internal structure, product lines, and range of activities, in British industry this reorganization was delayed and, when it did occur, less dramatic.

The link between this transformation and the development of industrial research stems from the characteristics of technical and scientific knowledge as commodities and industrial innovation as a process. For all but the most prosaic and standardized activities, industrial research is most efficient and effective when conducted on an in-house basis. Commercially successful innovation requires the combination of skills and information from a wide range of functions within the firm, and often exploits firm-specific knowledge emerging from complex production processes.[22] To obtain information about such processes at arms-length is exceedingly difficult, fraught as it is with uncertainties and moral hazard, and frequently does not permit the effective exploitation of firm-specific knowledge. Furthermore, the efficient organization of industrial research has historically been associated with the growth of a strong central staff in charge of functions such as marketing or production engineering. To be effective, industrial research thus requires complementary changes in the structure and organization of firms and markets; where these changes have not occurred, industrial research has been more modest and less effective.

The development of industrial research within the American firm was closely linked, as both cause and effect, with the reorganization of the American corporation during the late nineteenth and early twentieth centuries. The growth of industrial research was encouraged by the presence of technically trained managers, a strong central office staff able to focus on strategic, rather than operating, decisions, and the integration within the firm of functions such as marketing. In-house research was better able to combine the heterogeneous inputs necessary for commercially successful innovation, to use and increase the stock of highly firm-specific knowledge gleaned from marketing and production personnel, and to exploit the joint-product nature of manufacturing activity and certain forms of technical knowledge.

Conversely, industrial research spurred the development of the modern firm. As research conducted in-house proved superior to that

[22] For further discussion, see D. C. Mowery, "The Relationship between the Contractual and Intrafirm Forms of Industrial Research in American Manufacturing, 1900–1940," *Explorations in Economic History* 20 (Oct. 1983): 351–74; or W. M. Cohen and D. C. Mowery, "Firm Heterogeneity and R&D: An Agenda for Research," in *Strategic Management of Industrial R&D*, ed. M. Crow, B. Bozeman, and A. N. Link (Lexington, Mass., 1984).

conducted by contract, independent research organizations' share of total U.S. research employment declined. At the same time relatively research-intensive firms performed particularly well. Their increasingly dominant role in the American and the world economy, as well as their key role in spurring product diversification confirmed the significance of industrial research for firm and industry structure.[23]

The merger wave of the late nineteenth and early twentieth centuries hastened these trends. The firms emerging from the numerous horizontal mergers of the period were among the largest in the American economy, and they gradually undertook a systematic internal reorganization and centralization, developing a strong central office staff to coordinate the activities of their many component parts. This internal rationalization was of critical importance to the subsequent incorporation within the firm of major nonmanufacturing activities, such as marketing and industrial research. In firms like American Telephone and Telegraph, U.S. Steel, or Du Pont, the development of a strong central office was closely associated with the establishment or significant expansion of a central industrial research facility.[24]

These giant firms were among the leaders in the establishment of in-house research facilities within American manufacturing. Between 1921 and 1946, in-house research employment, which was originally concentrated among the very largest U.S. manufacturing firms (especially in chemicals), gradually diffused among smaller firms and across industries.[25] Among the 200 largest manufacturing firms, the relative research intensity of the smaller firms grew rapidly, while the firm size threshold associated with the existence of an in-house laboratory declined.

In Britain, the merger wave of the late nineteenth and early twentieth centuries was less significant. Fewer firms, and a smaller total amount of assets, were absorbed than in the United States.[26] The Brit-

[23] In-house research grew rapidly within American manufacturing between 1900 and 1940, and independent contract research organizations' share of total industrial research employment declined from 15 percent in 1921 to 7 percent in 1946; see Mowery, "The Relationship between the Contractual and Intrafirm Forms of Industrial Research." By 1940, over 60 percent of the clients of such independent research organizations as the Mellon Institute had in-house research facilities, and contract research was operating primarily as a complement to, rather than as a substitute for, in-house research. An example of the importance for technical change of the incorporation of several functions within the firm is Courtaulds' exploitation of the technology for rayon production in the early 1900s. As D. C. Coleman, *Courtaulds: An Economic and Social History* (Oxford, 1969), 2:66, points out, the fact that the firm was a textile, as well as an aspiring rayon yarn, producer, was of considerable help in refining and exploiting the rayon production technology.

[24] For additional discussion see Mowery, "Industrial Research and Firm Size, Survival and Growth."

[25] Ibid. Note, however, that while the largest firms were the leaders in the early years, they were no more research-intensive, relative to their size, during the 1921–46 period than during the postwar period.

[26] See L. Hannah, "Mergers in British Manufacturing Industry, 1880–1919." *Oxford Economic Papers*, new ser., 26 (March 1974): 1–20, esp. 10.

ish merger wave also involved a much narrower range of industries, primarily brewing (reflecting the development of tied houses in response to licensing restrictions) and textiles. Both British and American manufacturing experienced another wave of mergers in the aftermath of World War I. In the United States this merger wave had a more modest impact on firm and market structure than its predecessor.[27] The British merger wave of the 1920s was substantially more significant, and gave rise to such major modern firms as ICI and Unilever. The truly giant centralized firm thus did not begin to develop in Great Britain until well after the first American merger wave. Since the development of such corporate forms is linked to the rise of industrial research, the slower evolution of British firm structure almost certainly contributed significantly to the observed differences between American and British industrial research activity during this period.

Mergers between British firms frequently resulted in an awkward collection of poorly coordinated, fiercely independent subsidiaries, rather than a streamlined or efficient structure.[28] Inefficient and excess capacity often remained, as some directors opposed the extinction of their firms or the closing of their factories. One of the most famous examples of incomplete rationalization of internal firm structure is the Calico Printers' Association, a "lumbering leviathan" formed in 1902 with a board of directors numbering no fewer than eighty-four.[29] Interestingly, a committee appointed by the firm in 1907 to recommend ways of improving its poor performance strongly advocated the establishment of a central research department

> equipped with necessary appliances for research and experimental work, conducted by the ablest and best-trained chemists, specially qualified to pursue investigations, in which chemical processes and mechanical and electrical appliances are involved. This department would form a "Clearing House" for difficulties experienced in any branch works in carrying out complicated processes, and in the introduction of new colours and methods of production. It would inquire into, and test if necessary, all new inventions, and obtain information from other countries of new developments in calico printing. It would be the brains of the business.[30]

[27] See M. S. Salter and W. A. Weinhold, *Merger Trends and Prospects*, report for the Office of Policy, U.S. Department of Commerce (Washington, D.C. 1980), esp. 4.

[28] See A. D. Chandler, Jr., "The Growth of the Transnational Industrial Firm in the United States and the United Kingdom: A Comparative Analysis," *Economic History Review*, 2d ser., 33 (Aug. 1980): 396–410; P. L. Payne, "The Emergence of the Large-Scale Company in Great Britain, 1870–1914," *Economic History Review*, 2d ser., 20 (winter 1967): 519–42; and L. Hannah, "Managerial Innovation and the Rise of the Large-Scale Company in Great Britain," *Economic History Review*, 2d ser., 28 (May 1974): 252–71.

[29] H. W. Macrosty, *The Trust Movement in British Industry* (London, 1907), 147.

[30] Quoted in ibid., 152.

Despite this eloquent recommendation, no central research facility was established.

This pattern of incomplete internal rationalization was by no means confined to firms in the older staple industries but was common among large British firms as well. William Lever was known for supporting competition among the subsidiaries of his firm—a leader in the rapidly growing consumer goods market of twentieth-century Britain.[31] The development of United Steel, one of the more progressive firms in the British steel industry, suggests that the formation of that firm was in no way motivated by a desire for greater efficiency or a rationalization of industry structure.[32] Nor was the firm destined to acquire a strong central office. Only in the early 1930s, some seven years after the United Steel merger, did the firm establish a central office and a central research department, and adopt a more systematic policy of internal rationalization. Despite these reforms, the entire central sales and administrative staff of the company numbered only 150 persons in 1950, when the work force totaled more than 28,000.

Major new firms in two of the "new industries"—chemicals and electrical machinery—experienced severe difficulties in the central coordination of their newly merged subsidiaries, and frequently failed in the strategic management and coordination of research activities. In a technically progressive industry, Imperial Chemical Industries—a firm that consciously emulated many of the management techniques of Du Pont—achieved a unified organizational structure only in the 1950s. As a result, according to ICI's historian, even after "ICI had been in existence for a dozen years . . . the central research organization was still far from fully developed. Indeed, its fundamental principles had hardly been laid down, apart from the somewhat negative one of leaving the Groups alone as much as possible."[33] Similarly, Associated Electrical Industries (AEI)—the product of a 1928 merger of British Thomson Houston and Metropolitan Vickers—was wracked by fierce internal rivalries between its formerly independent constituent units. Indeed, the firm's historians attribute British industry's slow pace in

[31] "Within his own business he [Lever] salved his Victorian economic conscience by the application of competitive principles; all associated companies should 'compete strongly with Lever Brothers in every line. . . .' " C.M. Wilson, *The History of Lever* (London, 1954), 1:273.

[32] "United Steel was not intended to be a very centralized business. . . . There is no record of the founders of the business having as their object the achievement of increased productive efficiency horizontally as between the businesses which were to become branches of the combine through what was later to be called 'rationalization,' and an enforced specialization upon different parts of their joint product." P. W. S. Andrews and E. Brunner, *Capital Development in Iron and Steel* (Oxford, 1951), 121.

[33] W. J. Reader, *Imperial Chemical Industries: A History* (Oxford, 1975), 2:93. Reader argues further that ICI, in sharp contrast to Du Pont, failed to master the integration of product diversification with its internal structure until the late 1950s, because of an inability to reorganize existing divisions within the firm according to new product opportunities and strategies (ibid., 461).

developing a commercially practical jet engine in part to the lack of effective cooperation between its two components, each of which had made significant progress in jet engine development in the late 1930s.[34] The Marconi firm, which was absorbed in 1946 by English Electric after being controlled for many years by Cable and Wireless, operated with a similarly chaotic internal structure through the late 1940s.[35]

Even within large British firms in technically progressive industries, then, centrifugal forces were extremely powerful and worked against the development of the strong central staff associated with industrial research in American manufacturing firms. Where such firms were, by American standards, research-intensive (in the late 1930s, ICI was spending 2.8 percent of its total sales on R&D, compared with 2.4 percent at Du Pont), the lack of central managerial control often allowed inefficient duplication and internal rivalries in the research effort.[36] Of possibly greater significance were the weak links between research and the other staff functions, and between the research laboratory and the production units. These links are critical to innovative performance. The fact that such structural problems existed within firms in both technically progressive and stagnant industries suggests that these difficulties were not industry-specific, but were instead related to the overall structure and institutional environment of the British economy.

An important feature of the economic environment was the absence of a strong antitrust policy, which slowed the emergence of multifunction firms in Great Britain. The British firm or firms interested in reaping the benefits of industrywide coordination of price or output could choose from a wide range of alternatives from trade associations to a thoroughgoing merger. In Great Britain throughout this period, legal sanctions against such interfirm cooperation were minimal.[37] Trade associations were a primary vehicle for such arrangements, and during

[34] R. Jones and O. Marriott, *Anatomy of a Merger: A History of GEC, AEI, and English Electric* (London, 1970), 155.

[35] See W. J. Baker, *A History of the Marconi Company* (London, 1970), 332. Yet another large firm formed in the 1920s, Metal Box, also was characterized by a lack of internal cohesion (see W. J. Reader, *Metal Box: A History* [London, 1976], 40–41).

[36] See Reader, *Imperial Chemical Industries: A History*, 2:414.

[37] P. Mathias, "Visible and Invisible Hands in Great Britain," in *Managerial Hierarchies*, ed. A. D. Chandler, Jr., and H. Daems (Cambridge, Mass., 1980), 67, states that "in important respects—particularly in coordinating sales policies and securing monopolistic control over prices—cartels were an alternative to merger that was open to European entrepreneurs but closed to their counterparts in the United States. Firms that preferred to maintain a single-unit structure in Britain were therefore free to do so while, at the same time, reducing competition by joining a cartel; in the 1930s and during World War II, in fact, these arrangements were actively encouraged by the government. United States industrialists like Gerald [sic] Swope, the dynamic president of the American General Electric Company, accustomed to the antitrust tradition, were advised by bankers in Britain that they need not create large, centralized corporations through mergers for their European operations; market competition could be regulated

the 1920s and 1930s became commonplace in much of the manufacturing sector, frequently with official encouragement as part of programs of industrial rationalization.

Louis Galambos and other scholars have noted that such trade associations were also important in the United States in the late nineteenth century.[38] After the passage and increasingly stringent judicial interpretation of the Sherman Antitrust Act, however, the number and the effectiveness of such interfirm agreements declined. Some scholars have suggested that the merger wave of the 1892–1904 period, particularly the surge in mergers after 1897, was in part a response to this new juridical environment. Finding that informal or formal price-fixing and market-sharing agreements were of dubious legality, firms resorted to mergers, creating (with the aid of the rapidly expanding New York financial community) entirely new firms that substituted intrafirm for interfirm means of controlling prices and markets.[39]

It may be claiming too much to argue that the mild antitrust climate prevented mergers or the rationalization of such mergers in British industry as did occur. However, it seems likely that the pursuit of technical efficiency was blunted by the cartelized environment, especially after the demise of free trade. The effect of such informal or formal interfirm cooperation, of course, was the preservation of firms of smaller size and lower efficiency than otherwise would have been the case. Schumpeterian competition (i.e., the aggressive introduction of new products or improved versions of older ones—a process heavily dependent on research) was discouraged by such agreements among firms.[40]

---

through agreement with other firms, and there was thus no need to acquire them." Macrosty, *The Trust Movement in British Industry*, and A. F. Lucas, *Industrial Reconstruction and the Control of Competition* (London, 1937), among others, provide detailed accounts of price and output agreements among firms in a wide range of industries.

[38] See L. Galambos, *Competition and Cooperation* (Baltimore, 1966), 35.

[39] This argument is stressed in G. J. Stigler, "Monopoly and Oligopoly by Merger," in *The Organization of Industry*, ed. G. J. Stigler (Homewood, Ill., 1968). The Supreme Court ruled in the *Trans-Missouri Association* case in 1898 and the *Addyston Pipe* case in 1899 that the Sherman Act outlawed all agreements among firms on prices or market sharing. According to H. B. Thorelli, *Federal Antitrust Policy* (Baltimore, 1954), a total of 84 mergers occurred from 1890 to 1897, followed by 24 in 1898, 105 in 1899, and a total of 83 from 1900 to 1902. The evidence is circumstantial, but still strong, in support of the argument that the Supreme Court decisions had an impact on merger activity. See also Chandler, *The Visible Hand*, 331–36.

[40] W. J. Reader, "The Chemicals Industry," in *British Industry between the Wars*, ed. N. K. Buxton and D. H. Aldcroft (London, 1979), 174, notes that "it was settled policy in ICI (Imperial Chemical Industries) to avoid competition with customers or suppliers which meant, in effect, avoiding competition with virtually every manufacturing company of any importance in the United Kingdom. For many years, accordingly, ICI kept clear of the range of chemical products, associated with industrial alcohol, which were supplied by the Distillers' Company Ltd and which were of considerable importance in the fields of organic activity opening up in the late 1920s and early 1930s. Similarly, in order to avoid giving offence to Courtaulds, important customers for caustic soda, ICI, unlike Du Pont, kept well clear of rayon and cellophane."

Another factor contributing to the relatively slow pace of the British managerial revolution before 1950 was the limited involvement of financial institutions in corporate management and mergers. In the United States, by contrast, financiers and banks played an important role in the finance and organization of forerunners of the modern corporation like the railroads. Financial institutions were also of great importance, especially after 1900, in the development of industrial corporations. Outside financiers played a central role in facilitating many large American mergers, and often oversaw the design of new organizational and management structures. The size of the firms resulting from these mergers, the mergers' financial requirements, and their organization by major financial houses all hastened the replacement of owners by managers. In Great Britain, on the other hand, family control of major firms remained a significant factor far longer. Through much of the early twentieth century, British industrial firms seeking long-term financing had little alternative to floating new issues on the stock market, or relying on nonbank sources for borrowing. The supervision of corporate growth and organization by financial institutions, which was even more pronounced in Germany than in the United States, was nowhere to be found in Britain.[41]

British firms of all sizes thus remained far more dependent on the market for the organization and control of their activities than was true of the largest American firms. In industries such as textiles, the production process was carried on through a highly developed interfirm division of labor. The nonelectrical machinery and steel industries also were populated by relatively small firms that avoided any role in product marketing or distribution.[42]

This British reliance on market mechanisms was also reflected in the importance of the consulting engineer, whose baleful influence on innovation and export performance has been cited by a number of observers and official committees. Whereas the importance of the American independent research organization declined during this period,

[41] The Committee on Finance and Industry's *Report* (London, 1931), 165, remarked on the absence of such close links.

[42] See W. A. Lazonick, "Factor Costs and the Diffusion of Ring Spinning in Britain prior to World War I," *Quarterly Journal of Economics* 96 (Feb. 1981): 89–109; and idem., "Industrial Organization and Technological Change: The Decline of the British Cotton Industry," *Business History Review* 57 (summer 1983): 195–236. The share of industry capacity accounted for by "combined" firms (i.e., firms that carried out both weaving and spinning operations) declined substantially between 1884 and 1914, according to data in R. E. Tyson, "The Cotton Industry," in *The Development of British Industry and Foreign Competition*, ed. D. H. Aldcroft, (Glasgow, 1968). See the Committee on Commercial and Industrial Policy after the War, *Final Report* (London, 1918); S. B. Saul, "The Engineering Industry," in *The Development of British Industry and Foreign Competition;* K. Warren, "Iron and Steel," in *British Industry between the Wars;* and S. Tolliday, "Industry, Finance and the State: An Analysis of the British Steel Industry during the Inter-War Years" (D.Phil. thesis, University of Cambridge, 1979).

the independent consulting engineer continued to play a significant part in such British industries as heavy electrical and nonelectrical machinery. The consulting engineer, who designed and supervised the production of complex capital goods for a particular customer, worked against the development of more efficient corporate organization and in-house research in at least two ways. First, the continued reliance on custom design and production of machinery prevented sufficient standardization of product lines to support the development of larger firms capable of exploiting scale economies in design and production. In addition, the consulting engineer assumed an important mediating role in the research and development process, which reduced the incentive for engineering firms to invest in research within their boundaries, and made it more difficult for them to exploit information received directly from customers or suppliers in the course of such intrafirm innovation as did occur. S.B. Saul's comment illustrates these difficulties:

> The consulting engineer all too often was deprived of the opportunity of combining his theoretical knowledge with a deep practical understanding. The industrialist, for his part, called in the consulting engineer whenever calculations had to be made—fixing the proper thickness of wall for a hydraulic cylinder, or determining the strains in a simple lattice girder or workshop crane. Things had so fossilized that if the poor consultant tried to give the manufacturer a free hand, there was a chorus of protest. . . . There is little point in arguing that German locomotive makers were more advanced in ideas of design than were the British, for by and large the use of consulting engineers meant that the British maker was concerned with organizing his works to the best of his ability, not with design. The consultant's aim tended to be to design a product of high technical quality which would do him credit in his profession. Executing the work in the most economical manner was not the prime object by any means. Not being in close contact with any one firm, he failed to design in a manner which fitted in with any particular manufacturing facilities. Standardization was extremely difficult to achieve and the method inhibited continued development work on plant when it was in operation.[43]

This analysis suggests the dangers of exclusive reliance on contractual arrangements in the innovation and design processes. Both directly, by offsetting incentives for firms to invest in in-house research, and indirectly, by supporting an industry structure characterized by small firms and low levels of product standardization, the consulting engineer worked against the emergence of large corporations that in-

[43] Saul, "The Engineering Industry." I. C. R. Byatt, *The British Electrical Industry 1875–1914* (Oxford, 1979), 177–78, argues that the reliance on consulting engineers in the British electrical machinery industry reduced the flow of incremental innovations and adaptations resulting from the actual operation of the machinery. On the northeast coast, where consulting engineers were not widely used, this feedback channel was not obstructed, and important innovations resulted.

corporated research and other functions within their boundaries. Furthermore, the efficiency and effectiveness of the innovation process was impaired.

In the United States, the interaction between evolving firm structure and in-house research was of great importance to the development of the modern industrial firm. In Britain, this interaction was largely absent. The modest importance of intrafirm industrial research reflected the lack of dramatic change in the structure of the British firm. As a result, the independent consulting engineer continued to play a major role in such industries as electrical and nonelectrical machinery. Firms in other industries were also unlikely to establish large-scale, centralized research facilities covering the full range of current and future activities, in part because of the weakness of the central office structure and high-level managers' lack of technical expertise.

Neither British nor American firms developed in a vacuum, of course. The structure of the British firm, and its modest commitment to industrial research, were both responses to, and influences on, the broader institutional environment. Important elements of this environment were the higher educational system and the government, both of which are considered below.

SCIENTIFIC AND MANAGERIAL EDUCATION IN GREAT BRITAIN AND THE UNITED STATES

A key factor in the growth of industrial research in U.S. manufacturing was the parallel growth of a large higher education network. The development of research as a recognized professional activity within academia in the late nineteenth century was closely followed by the emergence of industrial research. A complex web of connections linked the two kinds of research, including personnel, ideas, and financial, as well as political, support.[44]

In Britain, on the other hand, few strong links developed between academia and industry. In comparison with the United States, Britain suffered a shortage of professional engineers for industrial research and a shortage of professionally trained managers; moreover, there was no opportunity to benefit from the kind of informal involvement of aca-

[44] For a discussion, see A. Thackray, "University-Industry Connections and Chemical Research: An Historical Perspective," in U.S. National Science Board, *The University-Industry Connection: A Collection of Papers* (Washington, D.C., 1982). M. B. W. Graham, "Industrial Research in the Age of Big Science," presented at the meetings of the Society for the History of Technology, Washington, D.C., 22 October 1983, provides a description of the relationship between academic and industrial research during and after the interwar period in the United States.

demic personnel in industrial research that developed during the 1920s and 1930s in the United States. The educational system clearly was not the sole factor behind these developments; the lack of professional training for managers reflected the family control of many British firms. Indeed, the educational system may be viewed more generally as the supply response to the demand for research investment by firms. Because British industry's demand for technical personnel was weak, changes in the structure of the institutions supplying research and managerial personnel were unlikely.[45]

In considering the influence of higher education on the development of industrial research, the most important difference between the British and American systems was their size. In the early 1920s, there were roughly 24,000 British university students; the figure rose to between 50,000 and 60,000 by the late 1930s. By contrast, American institutions of higher learning awarded over 48,000 degrees in 1914 and more than 216,000 in 1940. With a total population 35 percent that of the United States, Britain had only about 6 percent as many students in its higher educational system in the late 1930s.[46]

The small size of the higher education sector in Great Britain reflected the modest level of government support for higher education during the first half of the twentieth century. In the United States, on the other hand, state government provided substantial support for higher education. Using data from the University Grants Committee, J.D. Bernal estimated total government support of British university education at £2.7 million (slightly more than $9 million) in 1934–35. In contrast, total government appropriations for American higher education in 1929–30 are estimated to have been $340 million.[47] One result was the development of a much larger higher education system in the United States which, whatever its shortcomings, was in many cases better suited to provide technical education for the needs of industry. In particular, the American system of higher education placed much

[45] M. W. Sanderson, *The Universities and British Industry, 1850–1970* (London, 1972), 117, argues that the demand on the part of industry for technically trained personnel was weak, especially before 1914.

[46] See A. Briggs, "Social History 1900–1945," in *The Economic History of Britain Since 1700*, ed. R. Floud and D. W. McCloskey, vol. 2 (Cambridge, 1981); U.S. Bureau of the Census, *Historical Statistics of the United States: Colonial Times to 1970* (Washington, D.C., 1975), 1:386. The size of the higher educational system was an important "supply-side" influence on the growth of industrial research; J. H. Beer, *The Emergence of the German Dye Industry* (Urbana, Ill., 1959) cites the high rate of production of chemistry Ph.D.s by German higher education in the late nineteenth century as an important influence on the growth of industrial research in the German chemicals industry. As the supply of professional chemists exceeded available academic employment opportunities, emigration or industrial research were the only alternatives open to the German graduate chemist.

[47] Bernal, *The Social Function of Science*; Carnegie Foundation for the Advancement of Teaching, *The States and Higher Education* (San Francisco, 1976), 35.

more emphasis on the training of engineers, a group of great importance for the industrial research and innovation process.

Britain lagged far behind other industrialized nations in its output of engineers. One scholar noted in 1968 that "no other country trains as engineers such a low proportion of its total professional manpower."[48] The chronic shortage of university-trained engineers is well illustrated by the experience of Nobel Industries, a key component of ICI. As late as World War II, there were no engineers in the firm's chief research laboratory or (of equal or greater importance) in its management. According to its historian:

> This was to some extent the result of the decision made long before and hardened into a tradition at the end of the century, that the factory should be in the hands of chemists; but even if the attempt to employ qualified engineers had been made it must have failed, for engineers trained in the same way and to the same degree as chemists could not have been found in any number. Despite the fact that British industries were in the main engineering industries or conducted by.engineers, the engineering profession had been very little concerned with full-time academic training, and indeed had a prejudice against any form of training other than that of the workshop and the evening school.[49]

It was only after World War II that Nobels established a research facility devoted primarily to engineering research on manufacturing processes.

To compound the problem, the training of engineers in Britain tended to reinforce, rather than bridge, the barriers between science and industrial practice. The dominant form of British engineering education well into the twentieth century was a combination of night school and apprenticeship, both of which were conducted outside the major research universities. This system facilitated an important element of vertical mobility between the skilled trades and professional engineering, but at the cost of perpetuating the gap between the universities and the engineering profession.[50] In the United States, by contrast, engineers functioned as critical links between advanced sci-

[48] Peck, "Science and Technology," 450.

[49] F. D. Miles, *A History of Research in the Nobel Division of I.C.I.* (London, 1955), 58.

[50] See W. E. Wickenden, *A Comparative Study of Engineering Education in the United States and in Europe*, Bulletin 16 of the Investigation of Engineering Education (New York, 1929), esp. 140–42. R. Floud, "Technical Education and Economic Performance: Engineering in the Late Nineteenth Century," unpublished ms, 1978, focuses largely on the ability of this sytem of engineering education to respond to increasing industrial demand during the late nineteenth century, rather than considering the content and the implications for innovation of this form of engineering education, in arguing that British technical education during this period was not deficient.

entific research and industrial practice.[51] The content of British engineering education, as well as its numerically small output, made such links less effective.

The American engineer also performed the important function of fusing managerial and technical skills. Engineer-managers such as Frederick W. Taylor, Charles Kettering and Alfred Sloan of General Motors, and Pierre Du Pont played a major role in the integration of industrial research with the business practices of the American firm. Engineers trained in the United States received far more managerial instruction than their British or European counterparts.[52]

A major factor in the close links between American higher education and industrial research was the state university, founded with federal support (through the 1862 Morrill Act) and supported by local legislatures. These universities sought, for obvious political reasons, to be helpful to local industry. Many state universities established organizations—analogous to their agricultural extension and research services—to provide industry with applied engineering skills and expertise on a consulting basis. Examples of such organizations include the Bureau of Industrial Research at the University of Washington, founded in 1916, and the Industrial Research Department at the University of Oklahoma, founded in 1917.[53] Such organizations were of direct service to industry and also trained engineers and, increasingly, Ph.D.s in chemistry or physics for industrial research positions. U.S. public funds for the support of higher education were supplemented by the grants of the large private foundations that grew rapidly in the early twentieth century. Much of the funding from such institutions as the Rockefeller Foundation was intended to bolster the research links

[51] The importance of this linkage is underlined in L. S. Reich, "Irving Langmuir and the Pursuit of Science and Technology in the Corporate Environment," *Technology and Culture* 24 (April 1983): 200. He observes that at General Electric's central research facility in the early years, "Many industrial researchers were both scientists and engineers by training. Five of the best people at Bell in the early years had taken undergraduate degrees in engineering, then proceeded to Ph.D.'s in science. Irving Langmuir received a B.S. in metallurgical engineering before acquiring his Ph.D. in physical chemistry, and William Collidge, his well-known contemporary at GE, began with a B.S. in electrical engineering, then went on to a physics Ph.D."

[52] See A. D. Chandler, Jr., *Strategy and Structure* (Cambridge, 1962), 317; Wickenden, "A Comparative Study of Engineering Education," 258.

[53] See for additional discussion, Thackray, "University-Industry Connections and Chemical Industry Research." J. W. Servos, "The Industrial Relations of Science: Chemistry at M.I.T.," *Isis* 71 (Dec. 1980): 531–49, discusses the rise and decline of a similar organization, the Research Laboratory for Applied Chemistry, at the Massachusetts Institute of Technology during the period 1910–35, arguing that advocates of pure academic research eventually eclipsed such industrially oriented research activity at MIT. While such conflicts undoubtedly took place elsewhere, several points should be noted. Even after the reduction in influence of the applied chemistry laboratory, links between MIT and industry remained strong, and certainly were far stronger than those between British industry and elite institutions of higher education. The laboratory's decline was also a reflection of the fact that such applied research facilities were much more widespread in American industry in the 1930s than before World War I, as noted above in the discussion of private independent research organizations in the United States.

between university and industry. In Britain, cooperation between academia and industry was more informal and less widespread, and received much less public or private financial support.

The higher education complex in the United States also played a much more important role in training managers than the British system did. As Charlotte Erickson and Alfred Chandler have noted, American managers were more likely to have received a university degree than their British counterparts.[54] Moreover, American managers often had greater technical training and expertise; those British managers who received any postsecondary education tended to be arts majors. Indeed, it has been argued that the technical background of top-level British managers actually declined in the 1900–50 period—a period in which the proportion of this group trained in either the public schools or the ancient universities increased sharply.[55] In neither of these training grounds were future managers given a strong background in either management techniques or the technological basis of modern manufacturing industry.

The lack of professional and technical training among British managers reflects the persistence of family or owner control of British manufacturing well into the 1940s, as well as the absence within British manufacturing of an extensive middle-management hierarchy. As Chandler has observed, it was the growth of middle management, in response to the increasing size and range of functions performed by the American manufacturing firm, that created a demand for professional management education, as well as shifting control of the large firm from owners to managers.[56] The lack of industry demand for professional engineers was a significant factor in Britain's failure to achieve more far-reaching reforms and greater state support for higher education. The continued weakness of Britain's technical and professional management training capacity, then, cannot be separated from the delayed and incomplete nature of the corporate reorganization of British manufacturing.

THE ROLE OF GOVERNMENT

As the many studies of British industry, education, and industrial research cited above make clear, it was generally recognized during the early twentieth century that British industry was lagging behind

[54] See C. Erickson, *British Industrialists: Steel and Hosiery 1850–1950* (Cambridge, 1959), and Chandler, *Visible Hand*.

[55] See D. C. Coleman "Gentlemen and Players," *Economic History Review*, 2d ser., 26 (May 1973): 92–116.

[56] Chandler, *Visible Hand*.

German and American practice in the exploitation of new technologies and scientific knowledge, and in the development of new institutional mechanisms to encourage such exploitation. Popular and official discussion of foreign (especially German) competition reached a local peak of sorts immediately before the outbreak of World War I. The extent of British dependence on foreign sources of supply for various critical chemical and optical products became excruciatingly clear in wartime:

> The outbreak of war found us unable to produce at home many essential materials and articles. We were making less than a dozen kinds of optical glass out of over a hundred made by our enemies. We could hardly make a tithe of the various dyestuffs needed for our textile industries with an annual output of over £250,000,000. We were dependent on Germany for magnetos, countless drugs and pharmaceutical products, even for the tungsten used in our great steel works and for the zinc smelted from the ores which our own empire produced.[57]

The British government's response to the supply crisis in technologically sophisticated materiel took several forms. A state-owned corporation, British Dyestuffs, was established to expand the production of synthetic dyestuffs and intermediates. This new corporation proved to be a mixed success. Events (notably the 1915 shell crisis) forced British Dyestuffs to devote much of its effort to the production of explosives, rather than dyestuffs. It remained for two private firms, British Alizarine and Levinsteins Ltd., to produce most of the synthetic dyes for the war effort. While British Dyestuffs did establish a research department, the firm's in-house technical staff devoted most of its efforts to process engineering and improvement (concerned primarily with the manufacture of explosives). The firm attempted to use research contracts with senior chemistry faculty at Oxford, Cambridge, Leeds, and Liverpool universities for more fundamental research, but "as late as October 1917 nothing much had come of it and apparently nothing ever did."[58] Rather than encouraging research within the firm, government policy promoted reliance by British Dyestuffs on external research organizations. British Dyestuffs did not prosper after the war, despite the imposition of a prohibitive tariff on imports of dyestuffs, and was merged with Brunner Mond, Nobels, and United Alkali to form Imperial Chemical Industries in 1926.

The second major wartime policy initiative in industrial research was the establishment in 1916 of the Department of Scientific and Industrial Research (DSIR), which assumed responsibility for the control of

[57] Advisory Council to the Committee for Scientific and Industrial Research, *Report* (London, 1916), 8.

[58] W. J. Reader, *Imperial Chemical Industries: A History* (Oxford, 1970), 1:275.

such major civilian governmental research establishments as the National Physical Laboratory. An important new policy associated with the DSIR was financial support for industrial research associations (RAs)—cooperative research organizations that firms within a given industry were free to join. According to P.S. Johnson, a central concern in the discussions leading to the RA scheme was the ability of cooperative research to substitute for the in-house research facilities absent in British industry at the time:

> One of the principal impediments to the efficient organisation of industrial research was the small scale on which much of British industry operated. The investigation, for instance, of the fundamental properties of cotton fibres was outside the reach of all except the very large firms, and even in the latter management might only be prepared to spend a very small amount on it. . . . The implications were that, even apart from the traditional attitudes of British management to research, a certain minimum turnover was required before a firm could support a research department or even, at the lowest level of indivisibility, a research worker.[59]

First established in 1918, the RAs were to be funded jointly by government and industry for a five-year period, after which sufficient industry support was expected to allow the RAs to continue with private funding alone.

The research association scheme initially appeared to be a considerable success. Over the first five years of the program (1918–23), twenty-four RAs were established in industries ranging from woolen textiles to laundering. However, the expected groundswell of industry financial support failed to materialize. By 1923, when state aid was expected to cease, only one RA, devoted to research on portland cement, was able to continue on the basis of industry support alone. Public support of the remaining associations was renewed for an additional five years, but the level of funding was intended to shrink over the period.

The 1920s were a difficult period for the RAs. A number of the associations failed as their income declined and business support dwindled. By the end of the second five-year trial in 1928, it was clear that the RAs would collapse without increased public support. Financial aid was increased for a period ending in the early 1930s, when budgetary crises forced major reductions. In 1933, D.W. Hill observed, "The average expenditure of the associations was only £14,500 a year and this average included a few very large associations so that the average for thirteen associations was less than £7,000 a year."[60] The DSIR sig-

[59] P. S. Johnson, *Co-operative Research in Industry* (New York, 1973), 19.
[60] D. W. Hill, *Co-operative Research in Industry* (London, 1947), 54.

nificantly increased its expenditures in support of the RAs through the rearmament years of the late 1930s. During and after World War II, state support of the cooperative research scheme expanded dramatically, and the RAs' future was assured.

Despite their long history, the RAs had a modest impact on the innovative performance of British manufacturing. As early as 1929, the Balfour Committee remarked on "the admittedly inadequate extent to which the results [of the RAs' research work] have been utilised up to the present by British industries." The committee argued that "at present, it is too often the case that the information, compiled with great skill and labour, simply runs to waste for want of a properly equipped 'receiving mechanism' within the works."[61] With this statement, the committee accurately diagnosed one of the central difficulties of the RA scheme. Cooperative or extramural research institutions could not entirely substitute for in-house research facilities because client firms needed a significant level of in-house technical expertise in order to use the results of such extramural research effectively. Without such a "receiving mechanism," the firm was far less likely to benefit from the activities of the research associations.[62]

Thus the RAs could do little to solve the central problem to which they were addressed—the lack of in-house research activity within British industry. As Johnson argued,

> The government was mistaken in its view that the RAs' work would be primarily concerned with research for the small firm which was unable to do any for itself. . . . What has happened is that the RAs now derive their main support from the larger firms. Thus, one of the original aims of the scheme—to overcome the difficulties associated with the small firm—has gone largely unmet.[63]

The RA scheme was intended to deal with the supply of research and development, rather than the use of the results of research. Inasmuch as the utilization, rather than the supply, of R&D was critical to improving the technical performance of British industry during this period, the program was not a great success.

The research associations, the British government's primary policy

[61] Committee on Industry and Trade, *Final Report* (London, 1929), 215, 217.

[62] A. Albu's description of the research association in marine engineering is illustrative: "By the time of the last war Parsons [marine engine and turbine] designs had become conservative. . . . The anxiety of the Admiralty led to the setting up of the Parsons and Marine Turbine Research and Development Association (PAMATRADA) in 1944; but although it had a well-staffed research department the marine engine builders, most of whom were adjacent to the shipyards, were too small to be able to support it, and lacked the level of technical management to be interested in the results of research. The Association declined into a licensing organization for Parsons designs." A. Albu, "Merchant Shipbuilding and Marine Engineering," in *Technical Innovation and British Economic Performance*, 175.

[63] Johnson, *Co-operative Research in Industry*, 51–52.

response to the perceived technical shortcomings of manufacturing, essentially became a more formal version of the interfirm cooperation that characterized the behavior of British industrial firms in other areas. The contrast with the developing American research system is particularly sharp. In a period during which industrial research within American manufacturing increasingly was organized on an in-house, rather than extramural, basis, British government policy encouraged the opposite trend. The research associations were an extension of the organizational structure typified by the consulting engineer, rather than a significant change in the technological organization or foundations of British industry.

British government policies, both those directly concerned with industrial research and those focusing on factors indirectly affecting research, did little to advance, and may have retarded, the development of modern industrial research within British manufacturing. Policies influencing the structure of British industry, which was admittedly antiquated, had little impact. Weak antitrust policies and the government rationalization policies in Britain removed an important motivation for mergers and the internal governance of transactions that played a major role in the development of large American manufacturing firms with in-house research facilities. The competitive environment resulting from the mild antitrust posture of the British government also reduced the incentives of British firms to pursue product and process innovation. The small size of the higher education complex in England and the modest level of state financial support for this system impeded the development of links between academia and industry, and produced a system that trained too few professional engineers for the needs of British industry.

The DSIR and its associated research associations also do not appear to have had a major impact on either the research activity or the innovative performance of British industry. It is difficult to know what would have occurred had the RAs not been established, or had they been much more substantially funded. There is widespread agreement, however, that the RAs were not effective as substitutes for in-house research facilities.

The weak policy response of the British government is remarkable in view of the many studies and reports that had identified, with great unanimity, the technical and structural shortcomings of British industry that underlay the research deficiencies. Why did government not move more forcefully? S.W. Tolliday has suggested that one reason for the character of British industrial rationalization policies in the 1930s was the desire (especially on the part of the Bank of England) to forestall any move toward nationalization.[64] The lack of a strong antitrust policy may also have been motivated by government officials' desire to preserve the existing structure of industry ownership, in order to avoid alienating political supporters of the Liberal and Conservative governments that held power for most of the 1900–50 period. Education policy reveals a similar tendency even more clearly. The small size of the English university system in the first half of the century reflected the stratified character of the secondary educational system. The state had assumed responsibility for public secondary education only in the late nineteenth century. Although the system was much criticized, financial support and reform were rather modest throughout the interwar period (indeed, the financial crisis of the early 1930s resulted in severe reductions in central government support of secondary education).[65]

Rather than addressing the underlying causes of lagging technical performance and underinvestment in R&D by British industry, the British government used public funds to support industrial research directly, in the research associations. The RAs may be seen as an industrial policy that would disturb none of the institutions that hampered industrial research and innovation within British industry.[66] In view of the great obstacles to the development of a more progressive industrial and educational system, a more inventive set of government policies would have been necessary to improve industrial research and innovative performance in British industry. Such policy initiatives did not emerge.

The contrast with the role of government in the United States is dramatic. The federal policies that were crucial for the development of a vigorous American industrial research infrastructure evolved in an atmosphere in which the federal government paid relatively little attention to the technical shortcomings of industry. Federal antitrust policy, while not consistently pursued with great energy, arose as a major policy initiative in the late nineteenth century with little explicit consideration of its implications for dynamic efficiency. However, the complex structure of higher education in the United States, with its

[64] See Tolliday, "Industry, Finance, and the State,"276.

[65] W. J. Reader, *Professional Men* (London, 1966), 115, argues, in terms similar to those of Coleman, "Gentlemen and Players," that the slow pace of reform reflected the English middle and upper middle classes' lack of interest in such issues during the late nineteenth and early twentieth centuries.

[66] British public policy toward R&D since World War II bears some striking similarities to that of the interwar period. A high level of government spending on R&D has been overwhelmingly concentrated (or overcommitted—see Peck, "Science and Technology") in two sectors that have important military applications but are isolated from the rest of the British economy: nuclear energy and aircraft. This policy has reduced the resources (both financial and human) available for R&D elsewhere in the economy.

vast size and numerous local sources of financial support, was originally motivated in part by a desire to meet the technical needs of industry and agriculture. At the state level especially, the support of state universities and other institutions of higher education was increasingly seen as a means of providing technical services and support to local industry. At the federal level, government policies explicitly directed at the support of industrial research were conspicuously absent before 1941, with the possible exception of the National Advisory Committee on Aeronautics. The fortunes of British and American industrial research suggest that American government policy, which affected the proximate influences on industrial research investment, ultimately proved more effective than that of the British government, which attempted to deal directly with a perceived shortfall of industrial research activity without disturbing the foundations of the existing industrial and educational systems.

CONCLUSION

During the first half of the twentieth century, the British manufacturing sector invested significantly less in industrial research than did the American manufacturing sector. The low level of industrial research, and the fact that much of it was performed outside of manufacturing firms, reflected British firms' continued reliance on markets as a primary means for organizing economic activity. In the United States, on the other hand, more and more products and functions were included within the boundaries of the developing firm. The establishment of in-house research was a central component of this reorganization of the American firm. The fact that British firms did not undergo a similar evolution before 1950 is critical in explaining the substantially higher levels of research activity in American manufacturing. Several factors contributed to the delay of the managerial revolution in British industry: a weak antitrust policy, government rationalization policies, a lack of involvement of financial institutions and operatives in the internal management and reorganization of British firms, and persistent family control of British firms. Between 1940 and 1960, as the structure of the British firm and manufacturing sector changed, the level of industrial research activity rose, but remained well below the American level.

Firm structure was not the sole factor influencing the low level of industrial research in Britain. The educational system, particularly higher education, remained small, in part because of low levels of government financial support. The training of engineers in particular was deficient. The lack of engineers in research and management impaired the effectiveness of existing industrial research activity, and may have constrained its expansion, particularly that of applied research and development. The small supply of engineers, and the nonacademic character of their training, perpetuated the separation of British laboratory science and British technological practice. The former was of extremely high quality, while the latter often was severely deficient.

The typical practice of organizing British industrial research on a contractual, interfirm basis meant that even the low level of activity that existed was less effective than it might have been. The advantages resulting from the combination of research and other activities within a single firm could not be realized by either the consulting engineer or the research associations.

The structural development of the British firm, the lack of educational support for industrial research, and the growth of the research associations all were influenced by government policy, which was highly conservative in design and modest in impact. Although many perceptive critical analyses of British industry's technical performance were available, major policy shifts aimed at improving this performance were not attempted. In the United States, the fundamental factors influencing the structure of industry and the supply of technical personnel were considered to be either beyond the direct control of the federal government or already so well established that no activist policy was needed. State government support of education remained important, however.

Britain's failure to match the level of industrial research achieved in U.S. manufacturing thus reflected the institutional legacy of an earlier era of industrial activity and organization—as well as the failure of government policy to change this institutional environment. In the United States the private firm had essentially preempted any government role in direct control or funding of industrial research before World War II. British firms did not assume this role to the same extent. Both during and after this period examined, the British government's efforts to fund and direct industrial research and innovation have been largely unsuccessful.

[20]

Science Studies, 2 (1972), 107–151.

Research and the Firm in British Industry, 1919-39

MICHAEL SANDERSON
School of Social Studies, University of East Anglia

Today the far-seeing industrialist no longer relies solely on the accumulated knowledge of the workers in his own factory, or even those in his own industry and on the chance possibilities of their producing new knowledge at a rate that will meet the needs of industrial expansion, but he sets up organizations for scientific research ... (Sir Arthur Fleming, 1928).[1]

I

In the years between the wars, British industry, in spite of its generally depressed condition, experienced rapid scientific development. The universities and industry drew closer together, the graduate was drawn increasingly into the firm and there was a very considerable rise of scientific research both in industry as a whole, but more importantly within the firm itself. This phenomenon of the rise of research within the firm has been widely misunderstood in three ways. Firstly, it has largely been ignored because contemporaries, whenever they discussed industrial research, did so almost exclusively in terms of the DSIR and its attendant Research Associations and Government research organizations. They were forced to do so because most firms were then understandably loth to disclose private details of their own research activities and expenditure, in sharp contrast to the plethora of information supplied about the public sector. The Balfour Committee on Industry and Trade in the 1920s, for example, discussed industrial research at fair length but concentrated almost entirely on the Research Associations, deliberately ignoring research within the firm on the grounds that 'there are no means by which accurate and complete information can be obtained'.[2] The very full discussion by Heath and

[1] Typescript, 'The Application of Science to Industry: Research Control of Processes by Technologists' (1928), in the papers of Sir Arthur Fleming of Metropolitan Vickers (GEC), Trafford Park.

[2] *Report of the Committee on Industry and Trade: Factors in Industrial and Commercial Efficiency*. Part I of the *Survey of Industries*, chapter IV, 320.

Hetherington likewise saw research as virtually confined to the Associations, with scarcely a passing mention of the work of individual companies.[3] Secondly, some literature has taken a totally opposite view, arguing that the most important invention came neither from group research nor the firm but from the individual inventor detached from the research team of the firm.[4] Thirdly, others have taken an avowedly left-wing standpoint, arguing that research within the firm by its very capitalist nature was stifled by secrecy and lack of freedom, sterile in its results, and even part of a racket to defraud the public.[5] There has accordingly arisen a distorted picture among historians of the interwar years that exaggerates the role of bodies whose contribution to industrial research, though honourable, was quite marginal to the total activity and which virtually ignores, belittles or even slanders the vast bulk of industrial research which was carried out privately within the firm.

It is now timely to try to redress this balance. In the first place, the interest in scientific research on the part of economists is enjoying something of a vogue, and this prompts the historian, with his inevitably cruder evidence, to try to dispel some of the darkness surrounding the early stages of this development in industry. Secondly, with the passing of time, many of the first generation of scientific workers in the early laboratories have just retired or are on the point of retiring and are consequently free to discuss issues and to disclose evidence about their early work.[6] Likewise, private minutes and notebooks of research laboratories, once, in an earlier technology, confidential, may without commercial harm be made available for study in a way that they could not in the days of the Balfour Committee. Accordingly, it is the purpose of this essay to try to use some of this new evidence to throw light on this most important area of industrial research, to consider some of its problems and to reveal it as one of the great hidden engines of technological change in the interwar years.

Firstly, however, it must be said that research within the firm was not entirely a new phenomenon in the interwar years. In one sense all innovating entrepreneurs had always made their own experiments. But this should be distinguished from the use of teams of scientists, other than pro-

[3] Sir Frank Heath and A. L. Hetherington, *Industrial Research and Development in the United Kingdom: a Survey* (London, 1946).

[4] J. Jewkes, D. Sawers and R. Stillerman, *The Source of Invention* (London, 1962).

[5] J. D. Bernal, *The Social Function of Science* (London, 1939), 153–5.

[6] A list of industrial scientists who have made possible the writing of this article by granting me interviews and making private documentation available will be found in the Appendix. This essay will tend not to deal with chemical firms (with the exception of those in the food industry).

prietors or entrepreneurs, engaging in experiment as opposed to normal production. Some of the earliest developments must have been in calico printing, one of the most scientific industries of the early nineteenth century. It was noted in the 1840s, for example, that 'in every large print works there is either a partner or a manager thoroughly versed in practical chemistry . . . there is generally a laboratory fitted up with all the requisite apparatus . . . for making chemical experiments'.[7] In engineering, Crossley had a research department in the late 1870s to match the efforts of continental firms at improving the gas engine,[8] and in chemistry, the Nobels works at Ardeer began their research department in 1871.[9]

It was, however, the 1890s and 1900s that saw the first flush in the creation of this new department of the firm. After the German Patent Act of 1876, the 1880s saw the rise of the large research laboratories of the German chemical firms BASF and Bayer, which expanded at a phenomenal rate from the mid-1890s to the outbreak of war.[10] Moreover, it was provable and well publicized that the heavy research spenders among German firms were also those with vast profits and 20–30 per cent dividends.[11] As well as the chemical firms the research laboratory activities of Krupps, the Pennsylvania Railroad Co. and of GEC Schenectady were all well known and admired in England. In the atmosphere of intense trade rivalry of the generation before the outbreak of war, the establishment of research departments by British firms was but a necessary imitative measure to attempt to keep up with their foreign competitors and with each other. Moreover, the time was ripe because the civic universities were expanding in the 1880s and had greatly increased their degree work in the 1890s as a prelude to seeking charters in the 1900s. Consequently, a new breed of higher educated graduate scientist was appearing on the scene suitable for staffing the new laboratories.

Firms dealing with metals and chemicals frequently had laboratories for testing materials supplied to them and for routine examination of defects, and this testing, analysis and sampling often imperceptibly developed into experiment. This, with the advent of the graduate, developed into more

7 George Dodd, *The Textile Manufactures of Great Britain* (London, 1844), 55.

8 S. B. Saul, 'The Engineering Industries', in D. H. Aldcroft (ed.), *The Development of British Industry and Foreign Competition 1870–1914* (London, 1968), 216.

9 F. D. Miles, *A History of Research in the Nobel Division of ICI* (ICI, 1955).

10 J. J. Beer, *The Emergence of the German Dye Industry* (Urbana, 1959), chapter VIII, 'The Rise of the Industrial Research Laboratory'.

11 W. S. Turner, 'Scientific Research, does it Pay in Business?', *Sheffield Telegraph* (2 April 1909). Turner wrote a series of articles (23 March–9 April) on this topic; he was the late Professor of Glass Technology at Sheffield.

genuine research. United Alkali on its formation in 1890 started its laboratory in 1891 'for chemical research . . . with a view to improving the processes of manufacture and where thought advisable of investigating and adopting new inventions'.[12] Thereafter the formation of laboratories and beginning of research in firms followed fairly rapidly. Burroughs Wellcome in 1894,[13] the Cambridge Scientific Instrument Company in 1895,[14] Ilford in 1898,[15] Crosfield[16] and the Viscose Spinning Syndicate in the 1890s[17] were early starters. In the next decade or so they were followed by British Westinghouse[18] and Rudge Whitworth in 1901,[19] the Calico Printers Association in 1902,[20] Firth Brown in 1908,[21] Cadbury formalized their Experimental Committee in 1909,[22] and both Daimler[23] and Laurence Scott were active in the 1900s,[24] and W. H. Allen in 1911.[25] The rapidity of this development is well illustrated by the contrasting opinions of two observers divided by only a few years. For example in 1902 Sir James Dewar considered that 'I do not know of any firms in England where chemists are employed in research work',[26] though this was simply untrue as Nobel, ICI, Crosfield, Ilford for example showed. But in 1907 S. H. Higgins of Manchester University considered that 'in English works a good deal of research is carried out . . .'.[27] Likewise it was noted in the leather trade that 'several enlightened tanners had provided themselves with small experimental manufacturing plants to carry forward the

12 E. K. Muspratt, *My Life and Work* (London, 1917), 185.

13 Information from the firm.

14 MS Plans of the factory 1895 at the offices of the firm.

15 Information from the firm, this is the date of their earliest research notebook.

16 A. E. Musson, *Enterprise in Soap and Chemicals, Joseph Crosfield and Sons Ltd., 1815–1965* (Manchester, 1965).

17 Edwin J. Beer, *The Beginning of Rayon* (Paignton, 1962; written 1955).

18 *Bridging the Gap . . . . The Research Organisation of Metropolitan Vickers Electrical Co. Ltd.* (n.d., early 1930s).

19 A. E. Harrison, 'The Competitiveness of the British Cycle Industry, 1890–1914', *Economic History Review*, Second Series, **22** (August 1969), 302.

20 P. L. Payne, 'The Emergence of the Large Scale Company in Great Britain', 1870–1914, *Econ. Hist. Rev.*, Second series, **20** (December 1967), 529.

21 *The Firth Brown Research Laboratories* (1937).

22 Iolo A. Williams, *The Firm of Cadbury, 1831–1931* (London, 1931), 87.

23 *Bristol Times and Mirror* (1 January 1929), referring to W. Morgan's research at Daimlers 1906.

24 T. J. Barfield, *Scott built a Dynamo* (Norwich, n.d.), 18, 28.

25 'The New Laboratory', *The Queens Engineering Works Magazine*, January 1912 (W. H. Allen, Bedford).

26 *L.C.C. Technical Education Board Sub-Committee on the Application of Science to Industry* (1902): evidence of Sir James Dewar, 19.

27 S. H. Higgins, *The Dyeing Industry* (Manchester, 1919; 3rd edition of his book which was first published 1907), 79. His comment was made in the 1900s.

experiments of their laboratories',[28] and in Sheffield on the eve of the war it was said that 'today most Sheffield firms have a fine chemical laboratory with a competent staff',[29] in contrast to the 1880s when 'there were very few works possessing works laboratories'.[30]

It is worth stressing this because it is often assumed that before the First World War British industry did virtually no research and that this was an element in her retardation and backwardness, whereas in fact there was a good deal more than was realized. A speaker in 1919 looking back observed, 'I agree that a great deal more research work has been done in this country than is commonly supposed, but that it has been kept secret or hidden'.[31] A rare revealing of this secrecy is provided in a letter from the professor of metallurgy at Sheffield, indicating that while there was no objection to the publication of some work by the NPL, the Sheffield firms would merely smile at it, because what they did in their works far exceeded what the NPL regarded as a piece of new research. The firms, however, would make no comment and enter into no discussion, since they did not care to have their own scientific work realized by others.[32] It will be appreciated in these circumstances how easy it was for contemporaries and, indeed, for historians to underestimate the amount of research within the firm before 1914; this underestimate was just what many of the firms desired.

And yet, with this proviso, complaints that more ought to be done in industrial research increased as the research itself increased. Influential professors like Ambrose Fleming,[33] Silvanus Thompson[34] and Alfred Marshall[35] variously deplored that more was not done. The excuse usually

28 *Yorkshire College Application to the Company of Skinners for the establishment of a Leather Industries Department* (*circa* 1896) (in the files of the Proctor Department of Leather Technology, University of Leeds).

29 K. V. Jones, *Life of John Viriamu Jones* (London, 1915), 84.

30 W. Ripper, 'Appreciation of Professor J. O. Arnold,' *Floreamus* (Sheffield Student Union Magazine) (June 1930).

31 'Planning a Works Research Organisation', *Institution of Electrical Engineers* (February, May 1919): W. A. Chamen in the Bristol branch discussion on a paper by Arthur Fleming. This will henceforth be referred to as 'Research Symposium'.

32 MS Letter, Professor O. Arnold to Bennet H. Brough, 23 May 1906, in Letterbooks of Arnold in Sheffield University Library.

33 Sir Ambrose Fleming's evidence to the Gresham Committee on the University of London, QQ 15,195–15,203, 643.

34 *The Times* (3 March 1906). His letter on 'Science and the Industries' deplored that experimental departments were even being abandoned and that no single British firm was spending one-tenth of what a single American firm was spending on lamp research.

35 Letter, Alfred Marshall to Sir William Ramsay (1914), in A. C. Pigou (ed.), *Memorials of Alfred Marshall* (London, 1925), 488–9, complaining that the Dundesland Iron Co. had no 'experimental station'.

made before the war for not engaging in research was that competition in those years was so fierce and profits accordingly so pared back that nothing could be spared for so speculative an activity.[36] The technically stagnant and hard pressed tinplate industry argued thus,[37] but so also did the more thrusting GEC,[38] which lends some plausibility to a suspiciously self indulgent argument that was to be rendered quite unfounded by the war. On the eve of the war Sir Gerard Muntz considered that there were not more than half a dozen firms in the non-ferrous trades with laboratories.[39] Sir Arthur Fleming, with rather rigorous American-influenced standards of what constituted a research laboratory, considered that 'not more than half a dozen manufacturing concerns had established research organizations' by 1913,[40] and Sir Harold Hartley considered that in the early years of the century the total expenditure on industrial research was 'well below £1,000,000'.[41] Thus the war broke on a position in which there had been rapid imitative development in the formation of early laboratories from the 1890s with gathering complaints that more ought to be done, mingled with pleas that the severity of competition rendered this financially impossible. As Raphael Meldola put it mildly in his presidential address to the Chemical Society, 'Research as a "cult" is not understood'.[42]

II

If such was the modest development before the war, to what extent was there a marked development in the formation of research laboratories within firms in the 1920s and why? The later years of the war and the early 1920s saw another flush of formation of research teams and laboratories in firms of first rate importance. Dunlop began in 1915,[43] Metro-

36 *The Times* (16 March 1906): Sir Joseph Lawrence's reply to Silvanus Thompson.

37 *R. C. University Education in Wales* (Haldane), Cd., 8699. 1917 Evidence of H. S. Thomas, Q.10,955.

38 A. G. Whyte, *Forty Years of Electrical Progress* (London, 1903), 95 *et seq.*

39 *Natural Science in Education: the Report of the Committee on the Position of Natural Science in the Educational System of Great Britain* (Sir Joseph Thomson) (1918), 225–6.

40 Sir Arthur Fleming's notes on a speech to the Chemical Plant Manufacturers Association (1938), in Fleming's papers. *op. cit.*, note 1.

41 Sir Harold Hartley, 'Science and Industry—the Pattern for the Future' (Fawley Lecture, 1954), 4. Sir Harold was already well embarked on his own scientific career in the 1900s and his observation is consequently a first-hand one.

42 *The Pharmaceutical Journal*, Fourth series, **24** (30 March 1907), 390.

43 Information from the firm.

politan Vickers[44] and the CWS[45] in 1917, Tootal[46] and Courtaulds[47] in 1918, Glaxo[48] and Hadfield[49] in 1919, while by 1922, in addition to firms already mentioned, Kynoch, Lever, Boots, Chance and Adam Hilger had all established research departments doing work of major significance.[50] The following two years saw the beginnings of the great GEC laboratory at Wembley[51] in 1923 and that of Siemens at Preston[52] in the same year, while Colman[53] and British Thomson Houston followed suit in 1924[54] Following the giants, others caught the craze; Fenner, the belting firm, established an experimental weaving room in 1925,[55] Clark shoes in 1926.[56] Hollins did research in their Seedhill works from 1928,[57] and by the end of the 1920s some four hundred firms were involved in research and development[58] and scientific research within the firm was being canvassed as a new and promising career for the university graduate.[59] Other prominent developments followed in the 1930s seemingly undeterred by the depression; Wellworthy piston rings between 1930 and 1932,[60] United Dairies in 1931,[61] Dickinson's paper in 1932,[62] the United Steel Companies

44 John Dummelow, *Metropolitan Vickers Electrical Co. Ltd., 1899–1949* (Manchester, 1949), 58.

45 Information from the firm.

46 Jewkes, Sawers, Stillerman *op. cit.*, note 4, 155.

47 C. H. Ward Jackson, *A History of Courtaulds* (London, 1941), 121.

48 Sir Harry Jephcott, *The First Fifty Years* (Glaxo, 1969), 90.

49 Sir Robert Hadfield in 'Research Symposium' (1919), *op. cit.*, note 31, 172.

50 A. P. M. Fleming and J. G. Pearce, *Research in Industry* (London, 1922), 212.

51 'The Research Laboratories of GEC Wembley'. *The Woman Engineer* (March 1926). The GEC research organization began in 1919.

52 J. D. Scott, *Siemens Brothers, 1858–1958* (London, 1958).

53 Information from the firm.

54 H. A. Price Hughes, *BTH Reminiscences* (BTH, 1946).

55 Ralph Davis, *Twenty One and a Half Bishop Lane, a History of J. H. Fenner and Co. Ltd., 1861–1961* (London, 1961), 56.

56 Kenneth Hudson, *Towards Precision Shoemaking: C. and J. Clark Limited and the Development of the British Shoe Industry* (Newton Abbot, 1968), 32.

57 F. A. Wells, *Hollins and Viyella, a study in Business History* (Newton Abbot, 1968), 144.

58 *Industry and Research: Report of the FBI Industrial Research Committee* (October, 1943), 7.

59 For example, *Careers to which Subjects may Lead: Handbook of the Faculty of Science* (Birmingham University, 1924): 'The larger industrial firms also maintain staffs of research chemists ...'; also *Careers for University trained Men and Women* (University of Liverpool, *circa* 1925): 'Many of the larger chemical firms already possess well equipped research laboratories'.

60 D. Pearce and D. I. Hodges, *Wellworthy: the first Fifty Years* (n.p., n.d.; 1969?), 82.

61 MS Minutes of the Advisory Scientific Committee of United Dairies (1931).

62 Joan Evans, *The Endless Webb: John Dickinson and Co. Ltd., 1804–1954* (London, 1955), 209.

in 1934[63] and LMS[64] and David Brown in 1935.[65] At the same time it will be appreciated that in many cases the establishment of a research laboratory or committee merely formalized developments already in progress. For example, Bibby[66] and Pilkington[67] were both highly experimental and science-minded firms, but it is difficult with both cases to establish a clear date of starting research, although both formalized this activity in 1938–9.

Once the leaders had set an example, the movement gathered momentum by imitation. Early starters in research received a powerful stimulus from the later years of the war in two ways. The recognition of Britain's backwardness in certain strategic materials prompted a massive programme of research undertaken by the universities in conjunction with industry, bringing home to the latter the value of science in meeting their practical problems. Moreover, the war induced the 'cult of science' whose absence Meldola had earlier deplored, finding its expression in such bodies as the 'Neglect of Science Committee', and in attempts to urge State intervention in the formation of the DSIR in 1915. With the DSIR, the Research Associations and the government research stations all coming into being in the later years of the war and the early 1920s,[68] 'research' was in the air, and indeed the word itself more consistently replaced the older 'experiment' in this connection. At the same time there appeared a heavy barrage of propaganda on behalf of the virtues of research, presenting now not so much the old paragon of Germany, but the United States as the exemplar. There was for instance the example of Dr C. E. K. Mees, the distinguished British pioneer of photographic chemistry in the firm of Wratten and Wainwright before the war, who in 1913 went as director of research to Kodak in the United States. Mees was very well known and respected in British industrial chemistry circles, and his advocacy of research and description of his methods were well publicized in Britain.[69] The Board of

63 P. W. S. Andrews and Elizabeth Brunner, *Capital Development and Steel* (Oxford, 1952), 200.

64 'The LMS Research Laboratory', *Journal of the Society of the Chemical Industry*, (20 December 1935). *Chemistry and Industry Reviews*, **13**, 1097.

65 Desmond Donnelly, *David Browns: the story of a Family Business, 1860–1960* (London, 1960), 61.

66 Information from the firm.

67 T. C. Barker, *Pilkington Brothers and the Glass Industry* (London, 1938), 214.

68 For example, the Fuel Research Station 1918, Food Investigation Board 1918, Radio Research Board 1919, appointment of a Director of Scientific Research to the Admiralty, Building Research Station 1921, Forest Products Research Board 1921, Low Temperature Research Station, Cambridge 1922, and so forth.

69 C. E. K. Mees, *The Organisation of an Industrial Research Laboratory* (New York, 1920) The importance of Dr Mees was more than once impressed upon me. It is also not without interest to note here that Dr Mees' granddaughter married the son of

Education in 1915 also was keen to make universities and industrialists aware of American industrial research,[70] and Sir Joseph Thomson's Committee on Natural Science in 1918 inevitably urged more research in industry.[71] Perhaps most important, however, was the persistent influence of Sir Arthur Fleming, the head of research and education at Metropolitan Vickers, who more than anyone in England in the interwar years preached the gospel of research and exemplified it in his own works at Trafford Park.[72] This wealth of example, both foreign and domestic, created an atmosphere highly conducive to encouraging firms to undertake their own scientific investigation, if for no other reason than imitation, irrespective of any rational motive or any deeply considered view of the importance of science.

The war years, as well as creating a suitable ambience, also had more direct effects in pushing firms into research. For example, Dunlop began their research as a direct result of the enormously increased activity and multiplication of problems created by the rapid advance of vehicles and tyres during the war. Tootal also began theirs, as Kenneth Lee said, 'as a result of experience gained during the war',[73] and paradoxically Siemens had to begin theirs as a result of their loss of the benefits of research carried out by the German-based part of the company. But probably most important, the war provided not only the stimulus and the example of science, but also the profits with which to pursue this new development. Indeed it was frankly said that research laboratories were a good way of spending excess war profits,[74] and it was surely no coincidence that Metro-Vickers, CWS, Tootal, Courtauld, Glaxo, Hadfield should all have been either

---

one prominent British director of chemical research whose other son was also a research manager with another leading chemical company, though this is not to argue the existence of any 'dynastic' element in British industrial research.

70 T. L. Humberstone, *An Experiment in Industrial Research* (Board of Education Pamphlet No. 30, 1915).

71 *Op. cit.*, note 39, 161, 235.

72 Notably in his important publications, 'Planning a Works Research Organization', *Institution of Electrical Engineers* (February, May 1919), and *Research in Industry* (London, 1922), and other materials to be found in the footnotes of this essay. Fleming had also worked on research in the United States for Westinghouse. Metro-Vickers and BTH were by their research reflecting the American practice of their parent companies. Thus, the American influence was twofold, both directly through propaganda and then indirectly through the example of the great British electrical companies with American origins.

73 Kenneth Lee, 'Industrial Research: a Businessman's View', *Proceedings of the Royal Institution*, **28** (15 December 1933), 115.

74 R. M. Longman in the Newcastle discussion, 20 January 1919, in the 'Research Symposium', *op. cit.*, note 31.

starting research departments or building laboratories often on a lavish scale precisely in this 1917–19 late war and postwar boom time.

Behind this movement, apart from imitation, there lay a number of sound reasons. Firstly, the separation of research as a distinct entity from the modification of production and process control arose partly from the conditions of large scale production. This was seen with some insight by Hugh Quigley:

> We have reached the stage where standardization in manufacture and the application of mass production to almost every element in industrial activity have made it impossible to experiment with new ideas in the course of manufacture itself and some substitute has been found to be necessary for this loss of experimental facilities. This substitute can only be found in research. . . .[75]

Different variations were, however, possible on this theme. In some industries, research and routine process control testing could be carried on side by side with one little different from the other, whereas in other industries they had, by the nature of the processes, to be distinct. In the steel industry, for example, the variety of routine tests was small, and it was disturbing to the flow of work to have research requesting non-standard tests. In electrical works on the other hand, the variety of routine tests was large and extra irregular tests could more easily be accommodated.[76] It was thus easier for electrical firms to carry out some research without having a separate research organization than for steel firms, the danger for the latter being that the alternative to having separate research facilities was to do no research at all.

Secondly, the creation of formal research organizations in the 1920s was often necessary to replace some heroic inventor–founder–entrepreneur of the previous century. Many firms grew to prominence in the nineteenth century through the technical and even scientific abilities of their founders; since they fed the firm with their own inventions, there was no need for a separate, independent research organization. This was obviously the case with E. J. Houston and George Westinghouse, who both died in 1914. J. B. Dunlop died in 1921 and W. H. Allen in 1926. Sir Horace Darwin, the founder of the Cambridge Scientific Instrument Company, died in 1928, Joseph Bibby entered his seventies in the 1920s, and Hans Renold who had formerly done all his own design retired in 1928, so that the

[75] Hugh Quigley, 'Engineering in the Economic Development of Manchester', *The Woman Engineer* (March 1924).

[76] J. G. Pearce in the 'Research Symposium', *op. cit.*, note 31.

design for the retooling of the firm in the 1930s was done by a Development Department.[77] Clarks started research in the awareness that the family could no longer supply all the high ability necessary for growth. When all these individuals died or reached advanced age, leaving behind largish firms which were often leaders of their own particular technology, this lead—sometimes even survival—could be maintained only by replacing the founder with a research team as the new fount of scientific expertise.

Thirdly, the necessity of research was closely related to the choice of manufacturable products presented to a firm, and to the increasing ramification of chemical knowledge. The more chemistry-based firms became, the more varieties of chemical combination appeared to open an increasingly widening range of product differentiation. For example, the concern to spin filament for electric light bulbs was connected with the creation of artificial fibres and, in turn, with the development of film and with colour photography. In this situation, research became necessary to explore the possibilities of new products in adjacent areas of chemistry and to ensure the quality of new lines. It was put to me, for example, that firms like Colman, Crosse and Blackwell, and Heinz could easily have continued with their main products without research, but when they began increasingly to develop fresh lines, then research became essential. Colman[78] began their research department in an attempt to diversify into infant foods to use the by-products of their starch factory. Glaxo began research initially because of problems of process control with a vastly increasing output, but its marked expansion came within a few years as the firm began to diversify from dairy products into pharmaceuticals. Courtauld began their new laboratory to explore the new cellulose acetate dope produced during the war, and United Dairies partly to discover how to use their whey. In all these cases research was made necessary by the concern to diversify products by exploiting adjacent areas of the chemistry of vitamins, fats, starches and hydrocarbons in fields where custom and tradition were not sufficient to ensure technical mastery. The ever shrewd Arthur Fleming saw clearly that 'those (firms) having a wide range of products will always find it convenient to establish their own research organizations . . .'.[79]

[77] Basil Tripp, *Renold Chains: a History of the Company and the Rise of the Precision Chain Industry, 1879–1955*, **125** (London, 1956), 141.

[78] J. W. Corran and S. L. Lombe, 'The Research Department Carrow Works, 1924–1949' (private typescript for circulation to the Board of Directors of Colmans).

[79] Fleming's typescript notes, 'Research in Industry', an address to Manchester University Physics Society, 8 February 1924, in Fleming's papers.

Fourthly, if the good times of the war were a stimulus to research, so also for some were the adversities caused by the loss of markets. For example, Clarks began research as part of a reappraisal of their production in the 1920s as a way out of the collapse of their traditional export trade and also under the influence of the example of Bally laboratories in Switzerland. Likewise Fenner, the belting manufacturers, lost their overseas markets during the war and realized that to survive they would have to break into the home market with a product markedly in advance of those already entrenched there. Hence their experimental department and the change from leather to woven belting. In both cases, the reorientation of markets forced on them by war losses prompted two courageous firms to turn to research as a means of leapfrogging rivals.

Fifthly, new techniques of scientific investigation also forced firms in traditional industries like mechanical engineering to undertake research. W. H. Allen, for example, found in 1912 that such was the multiplication of metals with which they were working that they had to adopt a scientific approach to their various qualities, and Zeiss photomicrographic apparatus now became necessary for the metallurgical side of advanced mechanical engineering. Likewise David Brown set up their laboratories in the 1930s to take advantage of the application of X-ray techniques to metallurgy. With the application of physics to engineering, the rise of instrument makers like the Cambridge Scientific Instrument Company, and the devising of apparatus in the universities,[80] the mechanical sciences followed the chemical in their recognition of research.

III

What, then, was the extent in quantitative terms of industrial research being undertaken by firms in Britain in the interwar years? In 1936, the Association of Scientific Workers circularized 450 firms known to be undertaking research, and in 1939 Professor Bernal considered there were 300–600 such laboratories in Britain, though this estimate seems to suggest a confusion between firms and factories. The first clear evidence, however, was provided by an FBI survey[81] in 1943 which found:

[80] For example, Professor Arnold's alternating stress apparatus, and Bragg's X-ray techniques, which became applicable to textile fibres.

[81] *Industry and Research: Report of the FBI Industrial Research Committee* (October 1943).

1930: 422 firms were spending £1,736,000 on research and development
1935: 484 ,, ,, ,, £2,696,000 ,, ,, ,, ,,
1938: 566 ,, ,, ,, £5,442,000 ,, ,, ,, ,,

However, even this report considered that 'these results do not by any means cover all firms engaged in research and development', and that 'the figures considerably understate the true position', though how considerably it seemed impossible to say. Data was first available to the FBI in 1930, but we have indicated that the early 1920s saw a considerable rise in formation of research laboratories, and it is possible to infer the scale of this growth by indirect means. For example, the number of patents sealed doubled in the 1920s, from 12,301 in 1919 to 20,765 in 1930.[82] Likewise, laboratory technicians increased in the work force from 5,000 to 11,000 between 1921 and 1931.[83] Obviously these were spread between private firms, universities, and research associations and some government departments, but there were far more firm laboratories than in all the other sectors put together, and this increase must to a great extent reflect a rise in research activity within the firm. To put this in perspective it may be noted that 20 out of 225 firms of 50 or more employees in the Manchester area had research laboratories by the end of the 1930s.[84]

The relative spread of this activity across the range of industries can be calculated from the first industrial research directory of 1946 which was compiled just before the end of the war.[85] Since it is doubtful that many new firms could have started private research during the war itself, it is fair to assume that this survey represents the situation substantially as it was in 1939. Of the most prominent industrial research laboratories listed. 231 were distributed between industries in the following manner:

Chemicals	56	Textiles	12
Engineering	37	Glass	6
Food	34	Oil	5
Electrical engineering	34	Paper	4
Metallurgy	26	Civil engineering	3
Steel	14		
		Total	231

82 B. Mitchell, *Abstract of British Historical Statistics* (Cambridge, 1962), 269.

83 Guy Routh, *Occupation and Pay in Great Britain, 1906–60* (Cambridge, 1965). 17–18.

84 *Industry and Science: a Study of their Relationship ... in the Great Manchester Area, 1950–3* (Manchester, 1954), 49, gives evidence from which this is calculable.

85 E. N. da C. Andrade (ed.), *Industrial Research* (London, 1946), 'List of Industrial Research Laboratories', compiled 'before the end of hostilities'.

Needless to say, these 231 firms were not the sum total of firms doing research, as the FBI figures show. We can explore differences between industries further by estimating the relative spending on research by firms in different industries both in absolute terms and as a proportion of turnover. From a survey of 1936, it is possible to calculate average annual expenditures on research in 32 firms for that year, and these may be grouped in the following way:[86]

	£		£
Drugs	8,250	Metallurgy	3,060
Electrical engineering	7,500 (1/11)	Chemicals	2,700 (2/11)
Oil	7,500 (4/11)	Textiles	2,375
Food	5,500 (11/11)	Engineering	2,293
Iron and Steel	5,200 (8/11)		

As it is of interest to compare this order with the situation in the United States, the above table has inserted in brackets the ranking order of American firms measured by their relative spending on research in proportion to sales for 1937.[87] In the British case, it is evident that while engineering and chemical firms made up the large bulk of those doing research, their expenditure per firm was relatively low and the individual expenditure of chemical firms seemed relatively lower in Britain than in the United States. Conversely, the oil firms, though obviously few in number, had a much higher relative expenditure and, indeed, their ranking was much the same in Britain and the United States. In Britain the food and electrical engineering industries had many firms of fairly high expenditure. The position of electrical engineering was much the same in Britain as in the United States, but food chemistry research was considerably more important here than there. Finally, the textile industry had fairly few firms engaging in research, and a low average expenditure.

To establish some estimates of relative spending on research in relation to turnover, a number of informants made estimates for their particular firms. These give some idea of differences between industries, but the survey was confidential and the firms are not identified:

[86] Calculated from *Industrial Research Laboratories: a list compiled by the Association of Scientific Workers*, Sir Halley Stewart Trust Publications III (London, 1936).
[87] Edwin Mansfield, *Industrial Research and Technological Innovation, an Econometric Analysis* (London, 1969), 11.

Research Expenditure as a Percentage of Turnover in the 1930s

Metropolitan Vickers (as representing the ideal of a highly research-conscious electrical engineering firm)[a]	2%
A firm in chemical and engineering products	less than 1%
A firm with food and other mixed interests	about 0.5%
A food firm	less than 0.1%
A food firm	less than 0.05%

[a] Sir Arthur Fleming's typescript notes of a speech to the British Chemical Plant Manufacturers' Association Annual Dinner, 25 February 1938, in Fleming's Papers. He refers to 2 per cent as the amount that ought to be spent by a large progressive engineering firm. Since Metro-Vickers' turnover was then £10 million, this would make their research expenditure about £20,000, which was the same as United Steel's. The orders of magnitude are thus realistic.

It is worth putting this research expenditure by firms into perspective, both with other types of research and with that of Britain's overseas competitors. Firstly, expenditure by firms on research was vastly more important than similar expenditure by other bodies and this makes its neglect by writers on the subject particularly misleading. The FBI underestimate of private firm research was £5.4 million for 1938, and yet the subscription income of the twenty-two research associations was only £326,000 and the DSIR grants thereto only £192,000, giving a total income of £560,000 for 1938 or, say, a tenth of the hidden private expenditure at the most.[88] Secondly, the Royal Society calculated that total expenditure on fundamental science for an average pre-1939 year was of the order of £366,000, again but a slight sum compared with private firm spending.[89] How then did this compare with other countries?

It was estimated that in the United States £50 million a year was spent on research, constituting 0.5 per cent of their national income, and in the USSR 1,000 mililon roubles, or 1 per cent of national income,[90] these presumably representing all kinds of research since research within the firm in the Soviet context would be meaningless. The same writer estimated that British spending on research was one-tenth of a per cent of the national income. However, the FBI survey considered that about £5.5 million was being spent by firms on research but admitted that these figures 'considerably understated' the true position. Another half million

[88] FBI (1943), *Report, op. cit.*, note 81.

[89] *Report on the Needs of Research in Fundamental Science after the War*, Royal Society (January 1945), 9.

[90] 'The Finance of Research', *The Scientific Worker* **10** (November–December 1938), 124.

was being spent by Research Associations and just under half a million on pure research. Accordingly, the British figure should be at least £6.5 million a year for the later 1930s, and that an underestimate. With the net national income at current prices being of the order of £4,000 million for the later 1930s, this gives a percentage spending on industrial research of the order of 0.15 per cent of the national income, or 0.16 per cent if pure research is included. The FBI figures are admitted to be an underestimate. If, for the sake of argument, we consider them to represent only a half of the true position, then this would raise the proportion of spending on industrial research to 0.29 per cent. The first figure of 0.15 is certainly too low, the second of 0.29 is possibly too high, and the true proportion lies somewhere in between but nearer the second than the first. In either case the British proportions were higher than the 0.1 per cent assumed by the ASW but lower than the 0.5 per cent of the United States.

Britain's relative backwardness in industrial research may be expressed in other ways. In the United States, by 1938, 42,000 persons were employed in industrial research establishments,[91] compared with the FBI figure of 4,382 for Britain. This latter figure is, of course, an underestimate, but since the American population was then about three times larger than that of Britain, the British figure for industrial research personnel would have had to be 14,000 to achieve any proportionate parity with the United States. This could only have been so if the FBI figures had been such an underestimate as to be only a third of the true figure. This seems rather unlikely, and we may still maintain that research personnel were fewer in Britain than in the United States both absolutely and proportionately. In other terms, patents given by the three advanced countries to their own nationals, 1930–2, showed the United Kingdom in a somewhat lagging position. The UK gave 27,204, Germany 57,284 and the United States 130,134, again a discrepancy out of line with relative differences in population.[92] Thus British industrial research activity may be regarded as somewhat behind that of the other two major industrial nations. On the other hand, we should not allow this to disguise the very real contribution that research within the firm did make to British industry. To this we will now turn.

[91] D. Landes, *The Unbound Prometheus* (Cambridge, 1969), 482, citing *Fortune*, **20** (October 1939).

[92] Sir Kenneth Lee, *op. cit.*, note 73.

IV

What did research within the firm contribute to the advancing technology of British industry in these years and what were the main research and development achievements of the firms we are concerned with in this essay?

Chief Discoveries, Developments and Special Pioneering Fortes of Certain Company Research Departments, 1919–39

ENGINEERING	
W. H. Allen[a]	Development of Perlit cast iron for high temperature and pressures. N33 sea water corrosion-resistant alloy. Leading manufacturer of ship ventilation equipment.
David Brown	Centrifugal process for increasing density of metal in castings.
Wellworthy	Heat forming piston rings replacing peening. Thermocron iron DTD 233 for aircraft engines, later used by Rolls Royce for the Spitfire Merlin engine.
Boulton and Paul[b]	First twin-engine all-steel aeroplane for RAF. Pneumatic gun turret.
Brown Firth	Staybrite and its variants. Nitriding process for surface hardening of steels.
ELECTRICAL ENGINEERING	
BTH[c]	Sound film apparatus for talking pictures. High-voltage white fluorescent light.
GEC[d]	Osram light bulbs, radio valves, photoelectric cell, road lighting.

[a] *Queens Engineering Works Magazine* (1918–1938) and MS Research Record Books of R. T. Rolfe (1918–1939).

[b] *The Leaf and the Tree* (Boulton and Paul, Norwich, 1947).

[c] Information from the firm; also Price Hughes, *op. cit.*, and H. Warren, 'The British Thomson Houston Research Laboratory', *Journal of Careers* (May 1939).

[d] 'The Research Laboratories of GEC Wembley', *The Woman Engineer* (March 1926), 'Account of a Visit to GEC Laboratory, Wembley', *Queens Engineering Works Magazine* (January 1928); also A. G. Whyte, *op. cit.*

Metropolitan Vickers[e]	Outside broadcasting, radio valves, Apiezon low vapour pressure oils for sealing joints making possible high vacua, high voltage work, electron microscope, deep therapy X-ray, engineering acoustics.
Siemens	Fluorescent lighting, sun ray lamps.
TEXTILES	
Courtauld	Cellulose acetate rayon. Fibro artificial cotton.
Tootal	Crease-resistant cotton.
Fenner	'Pyramid' woven belting for power transmission.
RUBBER	
Dunlop[f]	Centrifugal process for separating latex cream, Dunlopillo.
C. and J. Clark	'Pussyfoot' springy soling.
FOOD	
Glaxo	Ostelin Vitamin D extract.
Colman	'Almata' baby food, Robinson's Barley Water, 'Krusto' pastry mix, strained vegetables.
CWS[g]	Preservation of sheep skins, physical-chemical control of flour milling, comminuted fruit drinks.
Bibby[h]	'Trex' vegetable cooking fat. Continuous extrusion of oil from seed, scientific cattle food.
United Dairies[i]	'Lacta' vitaminized milk powder, prevention of mould on cheese.
SCIENTIFIC APPARATUS	
Ilford[j]	Photographic plate for scientific work. X-ray film.
Cambridge Scientific[k] Instrument Company	Gas analysis, carbon monoxide and dioxide measurement, *p*H measurement, Mallock electric calculating machine.

[e] *Bridging the Gap: a brief outline of some of the activities of the Research Organization of Metropolitan Vickers Electrical Co. Ltd.* (n.d.; early 1930s); B. G. Churcher, 'The Acoustics Laboratory of the Metropolitan Vickers Electrical Company's Research Department', *Engineering* (26 May 1933); various pamphlets of the firm, *The High Voltage Research Laboratory of Metropolitan Vickers, Vacuum Distillation*. J. Dummelow, *op. cit.*

[f] Information from the firm.

[g] Information from the firm.

[h] Information from the firm.

[i] Information from the firm, MS Minutes of the Advisory Scientific Committee 1931–9, Reports of the Chief Chemist 1936–9.

[j] Information from the firm.

[k] MS Reports of the Research Department CSIC 1923–39. *Seventy-Five Years of Successful Endeavour 1881–1956. Cambridge Gas Analysers* (CSIC, 1929).

If we consider the developments and achievements listed in the table, we may ask what problems they were trying to overcome and what their fundamental purposes and effects were. Firstly, there were innovations that may be regarded as labour saving, either in response to a shortage of labour or, more usually, as an attempt to reduce costs through savings on existing labour. Labour cost saving as a motive for research was particularly talked about in the postwar boom of 1919 but not generally thereafter.[93] In this category may be classified seed oil winning by continuous extrusion, the work of the works and engineering departments of Colman and Dunlop in devising labour saving machinery for the manufacture of fruit juices and tyres, and possibly the replacing of peening by heat forming by Wellworthy. More important, however, seem to have been those innovations that were capital saving in their effects, the pickling of sheep skins, the mould prevention on cheese, the comminuted drink (using the goodness of the peel as well as the pulp), Clark's 'Springbok' insole board using waste leather, and the concern of various firms about dried milk and its potential uses, all using commodities that would otherwise have been wasted, or else cutting spoilage and loss of capital in that form. We are more impressed with capital saving than labour saving as a recurrent theme of many of these research activities. Thirdly, there were those developments which were responses to the non-economic demands of the law and public interest embodied in the person of the Public Analyst, especially in matters such as effluent treatment, and all aspects of food production. Many of the food chemists regarded this aspect of public safety as probably the most important motive of their work.

Finally, and probably most important of all, much innovation seems to have arisen not as a response to labour or capital availability or even to the demands of public safety, but from a certain logic of technical advance in which the balance of the factors of production had little relevance. The metallurgical advances Thermocron, Perlit, Staybrite and so forth were responses to the demands of increasing heat, pressure and corrosion with which engineering was dealing. The X-ray demanded an (Ilford) plate on which it could be recorded, the presence of gas in new types of reactions demanded the (Cambridge) instruments to trace and measure them. As technical advance presented its own problems whose solution dictated the

93 For example, A. P. M. Fleming, 'Industrial Research', in *Lectures on Industrial Administration*, ed. B. Muscio (London, 1920), 237; the lectures were given in Cambridge in 1919. See also the comments of H. I. Rogers in the Bristol discussion in the 'Research Symposium'. *op. cit.*, note 31.

next advance so it presented fresh opportunities. The centrifugal separation of latex cream led to Dunlopillo, while the work at Metropolitan Vickers on low vapour pressure oils led to the possibility of sealing high vacuum joints and consequently to continuously evacuated high vacuum apparatus for radio valves, high voltage X-ray tubes and the electron microscope.

We must examine further how responsive research was to considerations of factor prices and how far this was an issue in the minds of the scientists concerned. But even when we consider those developments that seem to be labour saving, there are ambiguities and cross currents between firms. At Colman there was certainly a concern for labour saving machinery on the fruit juice side, but this was devised by their own works department rather than the research laboratory. In the chemical aspects of their work, there is little evidence that labour saving or even considerable cost reduction was a powerful motivation; indeed, two of their main new products, 'Almata' and 'Krusto', eventually faded out not because of their quality but because of their cost. At Dunlop, similarly, the actual labour saving aspects of innovation arose, as at Colman, not so much from the work of the research laboratory as from that of their central engineering equipment department, a large and active unit. A feature of this industry was the very heavy cost of fixed capital, and indeed the effect of research was to increase the capital requirements in relation to those of labour. This was not, however, a reaction to any shortage of labour at home or even to the high cost of labour, though much was highly skilled, but was rather a response to American conditions. Since their chief competitors were in the United States and since the technology of the British firm had to keep pace, so its balance of factors of production came to assume American capital-intensive characteristics. The change in the balance of labour was incidental and technical changes were not prompted by considerations of the price of labour at home. Bibby's continuous extrusion again appears to be a straight labour-saving device, but was not prompted by any shortages of labour in that depressed port. Much of the effect of the new process was to make the work less cruelly arduous than it had been in traditional pressing, and the low cost and ampleness of labour for this grim occupation did not impede the introduction of the more advanced method. Thus, with these three firms, motives were somewhat varied and mixed, and no commentator saw labour saving as an important motive behind their research activities.

The most direct examples of research as a response to changing factor prices of raw materials were in the field of metallurgy which lends itself to this kind of precise calculation and substitution. The rising tin prices of summer 1926, for example, prompted Allen to experiment with different compositions of gunmetal, devising a new No. 2 gunmetal at £67.9.0 a ton compared with their original No. 1 at £83.1.0. It helps to explain why firm research was normally more interested in saving on raw materials than labour costs when it is appreciated that, in this particular situation, the skilled labour costs per ton of metal were calculated at £15, whereas the cost of a ton of tin was £293.25. They made some similar experiments on the relative costs of PMG alloy and gunmetal in 1930.[94] This then was clearly an industry where research was often highly responsive to price changes. On the other hand, in photography and food it did not seem to be so. In photography, research in the interwar years was not responsive to the price changes of chemical constituents in the way that with the increased price of silver it has since become. Technical quality was a far more important consideration, especially in Ilford's fortes of X-ray and scientific plates. The same attitude seemed very much to pervade the food industry, where there was far more concern among chemists for quality and rising standards than in seeking by chemical means to devise substitutes for expensive raw materials. Indeed, the very depression of world trade and especially of primary products in food seemed to remove this pressure from research in a situation where raw materials for jams, mincemeats, fruit juices, flour and so forth were in any case low in price. Consequently, changes in quality were more evident than changes in price. It became evident that response to factor prices was not a potent motivation for much research activity except metallurgy, but that responses to prices of raw materials seemed to be stronger and more fruitful than concern about the availability or cost of labour.

A second point to emerge concerning these research achievements was that they were often necessary to the survival of certain firms or of various of their branches, and that this research formed the basis of their success and prestige. Research lay behind the very considerable lead, if not outright monopoly, enjoyed, for example, by W. H. Allen ventilation equipment for shipping (a fact that the Luftwaffe fortunately failed to appreciate in the last war), Wellworthy piston rings, Bibby's cooking fat, and Ilford's photographic materials especially for the Far East. The very names Dunlop, Courtauld, Tootal, Clarks' sandals, Colman's mustard, Robinson's

94 MS Research Notebooks of R. T. Rolfe, p. 3 of book beginning 1926, and notes on experiments, 3 October 1930.

Barley Water are virtually synonymous with their products, due to their acknowledged technical excellence or even supremacy which was due in turn to their research expertise. Moreover, although these were the successful firms, there is ample evidence that a lack of research often led to failure. For example, Colman took over responsibility for Farrow's Tomato Ketchup in the 1920s when both chemically and commercially it was in poor condition. It was only the intervention of a newly arrived PhD biochemist that saved the product by reconstituting it as he likewise later reconstituted Robinson's Barley Water. Here, rather dramatically, research proved the difference between commercial life and death for two products of a food group formerly unaccustomed to this approach. In the highly vulnerable area of textiles, there were around thirty firms in the 1930s trying with various degrees of success to spin rayon, all of whom except Courtauld were doomed to extinction. Likewise, Fenner's devising of new belting saved them from the collapse that attended their former rival Cockill, which had failed to experiment and to create or even to follow new technologies. In all the cases listed above, there is only one where research excellence did not materially contribute to the firm's strength and survival. Ironically, it was not Boulton and Paul's valuable aeronautical work that proved to be the backbone of their commercial strength, for this had to be abandoned, but rather the metal construction for two other symbols of the thirties—the pylons for the National Grid and the dome of Blackpool Pleasure Beach Ice Drome.

The research departments were valuable to firms not only in making their own innovations but also in reaching out beyond the firm to draw in ideas from outside. This point is of some importance because Jewkes, Sawers and Stillerman put it forward, as one of their arguments against firm research, that such teams are unresponsive to outside ideas. In fact, there is ample evidence to the contrary. For example both Colman and Glaxo were very interested in the Steenbock Vitamin D patent of the University of Wisconsin which the latter applied with notable success. Without a research department such firms could have neither known, understood nor appreciated the importance of such academic work for their own products. Similarly, it was a visit of the chief research chemist of the CWS to the Flour Physics Symposium at Duisburg in the early 1930s that led to the adoption of advanced physical chemical control equipment in flour milling and gave the CWS a special excellence in the field at that time. Also, David Brown were able to pick up the French Randupson process of moulding and were able to attempt to imitate a

German system of casting aircraft steel, which they could not have done without a research team to devote to it. By this receptivity to outside ideas research laboratories in British industry were going some way to rectifying the pre-war situation when, for example, the importance of Haber's publicized work on nitrogen fixing was not taken up and appreciated for lack of research on gas reactions.[95]

Not only were teams receptive to ideas from outside, but by keeping a close watch on the products of other firms in the field they ensured that innovations could be diffused rapidly through industry as a whole. For example, when Bibby developed 'Trex', this forced Lever to retaliate with 'Spry'. The early pioneering of the comminuted drink by the CWS and others prompted other large and well established firms who had formerly been critical and antagonistic to the inclusion of peel and the whole fruit to follow suit. The vitaminizing of Glaxo milk powder in the 1920s meant that United Dairies 'Lacta' likewise was vitaminized, and the firm kept a very close analytical watch on the relative chemical compositions of 'Lacta' and Cow and Gate;[96] similarly the flour physics apparatus, whose reception had been readily adopted by the CWS, spread throughout the flour milling industry in the 1930s. The effect of all this was that, whereas before and just after the First World War there was often a considerable disparity in quality in, say, food products with one acknowledged front runner and a field of 'also rans', by the end of the 1930s there was a very much greater homogeneity and consistency of quality between firms in the same product. This was due partly to the demands of the law, but also to a very great extent to the activities of the research departments of the firms themselves, operating in a mixture of rivalry and co-operation to create a technology which was becoming increasingly advanced and, among their small circle, more widely understood.

V

How, then, did research within the firm fit into the broader context of industrial research in the universities and the Research Associations? What were the relationships of firms with these in their research and indeed between each other, and how far was there a shift in the location of main research activity?

95 Julian Huxley, *Scientific Research and Social Needs* (London, 1934), 135. This point was put to him by the ICI research laboratory at Northwich.

96 Minutes of the Nutrition Committee of the Research Advisory Committee of United Dairies (2 January 1935).

Some firms had either at crucial times or over a long period quite intimate research relationships with universities.[97] When Colmans first began their department to examine baby foods in 1924 they approached Professor Gowland Hopkins, the great nutritionist of Cambridge, for his advice on the product and one of his staff came to Norwich to help set up the department. But thereafter in the interwar years close contact with the universities was not as important to them. United Dairies had very close and continuing association with the University of London; indeed, professors of the University, including Sir Jack Drummond, staffed their Advisory Scientific Committee. This was far from a sinecure, and a great deal of work for United Dairies was done in the laboratories of the university. Their staff also tended to be London graduates, often recommended directly by Drummond. None of the other firms with which we are concerned had such close, formal and continuing links. The CWS was also especially enthusiastic about university links, using Leeds for advice on leather, the Cambridge Low Temperature Research Station (as did most of the food firms), the University of Manchester (for the geology of Cheshire salt), and Professor T. P. Hilditch of Liverpool, the great authority on the chemistry of fats, for advice on margarine. Although Bibbys had no regular relations with universities before 1939, this was much mitigated in that their chief chemist was a pupil of Hilditch and did advanced work under him on fats after further research under another of Hilditch's pupils, a former Lever chemist from Port Sunlight. It is worth stressing this point, for Liverpool University biochemistry both before and during Hilditch's time had an important influence on research personnel in Bibby, Colman, Lever, Glaxo and the CWS in food and fat matters.[98]

Ilford kept in touch with Sir William Pope, the Professor of Chemistry at Cambridge, and the Cambridge Scientific Instrument Company had close dealings with Professor Shakespear of Birmingham over the gas analyser. The food chemical and scientific chemical firms thus frequently tended to have some contact with university departments; where they did not consistently, it was often because their fields of interests were such that

[97] On the matter of university industrial research and the use of graduates by industry at this time, see Michael Sanderson, 'The Universities and Industry in England, 1919–39', *Yorkshire Bulletin of Economic and Social Research* **21** (May 1969). In that essay I gave the impression that United Dairies were not a significant user of graduates; after examining the papers of their research committee and talking with chemists who were there at the time I am happy to correct that misconception in this paragraph.

[98] On Hilditch see 'Thomas Percy Hilditch, 1886–1965', *Biographical Memoirs of Fellows of the Royal Society* (November 1966). He had formerly been chief research chemist at Crosfields.

the chemists within the firm were leading national or world authorities on their particular subjects and there was little they could learn from outside. Moreover, many took the view that this was a correct balance, with the universities concentrating on fundamental research and the teaching of future employees but not trying to do pieces of industrial work in which they had no experience and no application of plant conditions, and in which they were mere amateurs compared with specialist experts within the firm.

If this objection applied in some chemical based firms it was even more relevant in engineering. There were no particular university contacts with Allen, Wellworthy, Dunlop or BTH in these years, while Siemens' academic contacts were rather with the local Harris Technical College in Preston. The chief exception was Metropolitan Vickers, an exception indeed in most matters dealing with research. They had close contact with Manchester through Professor Miles Walker, T. E. Allibone did work for them at Sheffield, and they had close links with Cambridge through the provision of equipment and through some of the personnel at the Cavendish, notably Allibone and Cockcroft. The firm's Intelligence Section kept in systematic touch with leading university departments, and from 1928 ran a summer school for university staff. In short while there were fair contacts with universities, they were more evident in chemicals and food than in engineering, with the occasional brilliant exception in electrical engineering. On the other hand, advanced scientific firms with no apparent direct connection with universities would still remain closely in touch with their work through the survey of published research, which again was a special function and service of the firm research department.

This raised the question of how far industrial firms undertook fundamental scientific research and indeed how far they thought they ought to do so. None of my informants regarded themselves as engaging in fundamental research in these years, and in fact regarded this as the proper province of the universities. However, GEC and Metropolitan Vickers had already moved very considerably in that direction. Even in the early 1920s, GEC were praised by the Professor of Chemical Engineering at University College, London for creating 'what is in effect a university laboratory and a university atmosphere within their own organization',[99] and he thought:

99 *Science and Labour*, ed. T. L. Humberstone. Report of a conference, 30, 31 May 1924: 'Scientific Research in Relation to Industry', E. C. Williams, 53–4. Some managers at GEC thought that this was excessive. G. Cheliotti noted of GEC 1920–46 that 'the men who go into research seem to develop a mentality which is almost as foreign to industry as is that of those who stay in the universities . . .': report of a Joint Conference of the Management Associations and the AUT, 23 February 1946, 10 (in AUT file, 'Meetings with Other Bodies').

the tendency is indeed more and more for industrial research workers to explore a far wider domain than is apparently involved in the problem immediately ahead; that is a reversion to the view that the most valuable industrial developments often result from the pursuit of knowledge alone.

In fact this was over-sanguine, and very few firms indeed held this view of research in the interwar years. Metropolitan Vickers was one of the best exemplars of this approach in doing a great deal of avowedly fundamental research (which they identified as such) in acoustics, high voltage work and atomic physics.[100] Indeed, their research department was one of the best two or three 'university' departments of physics in the country with no less than ten of their research workers, including Sir John Cockcroft, eventually moving out to become university professors. Fleming considered that more fundamental research would be necessary by firms because more of the new discoveries in the engineering field were arising from the work of the physicist, who in turn required apparatus of great engineering magnitude; 'on this account . . . it is probable that the great manufacturing organizations will undertake more and more fundamental research'.[101] However, atypical as it was of British industry, Metropolitan Vickers research under Sir Arthur Fleming reached one of the high peaks of research within any firm in the interwar years.

Probably more important than these links with universities, however, were those with Research Associations and Government research establishments under the DSIR. The attitude to this linkage varied widely across the spectrum of industry. At one end were those industries which tended to rely heavily for their research on associations and for whom research within the firm was not important, notably coal, brewing, laundries, cotton and cast iron, generally industries not highly concentrated and within which the existence of a large number of small units did not, for most of them, make private research worth while. Other industries, such as glass and leather, drew heavily from outside the firm, but from university research rather than from associations, while iron and steel and coal appeared to rely on research associations which in turn relied on university departments to carry out the research for them. With these externalized forms we are not of course here concerned. In the middle, however, were

100 A. P. M. Fleming, 'Industrial Research', *The Times Trade and Engineering Supplement* (November 1933), and *Bridging the Gap, op. cit.*

101 Fleming's Presidential Address to the Institution of Electrical Engineers, *Journal of the Institution of Electrical Engineers,* **84** (January 1939), 6. It is interesting to note, however, that Fleming seems to have changed his mind in favour of this view; in 1919 he was not in favour of fundamental research in the firm: 'Research Symposium' (1919), *op. cit.,* note 31.

industries whose firms ran very active research departments within the firm but which also were enthusiastic supporters of association research, notably the electrical, food and scientific instrument industries.

British Thomson Houston, for example, was especially keen on the ERA and prided itself on being particularly astute in adopting the benefits of association research notably in switch gear. Colman, United Dairies and CWS greatly valued the Cambridge Low Temperature Station, and the DSIR played an important part in helping with the problems of United Dairies and the CWS in effluent disposal. The food industry was one in which a particularly good balance could be struck between firm and association research, as the effluent problem indicated. The Food Manufacturers' Association did not devise new food processes (this was the province of the firm) but did devise, for example, the potash-phosphate ratio test for the fruit content of jam, and tests for detecting insect fragments in flour acceptable and useful to the trade as a whole. The success of this balance derived partly from the fact that the food industries consisted of a large number of firms of high scientific standing willing to accept all-industry self-imposed standards. They have a high sense of public obligation, and also a rather close knit personal relationship between the personalities in the trade, focused to a great extent through membership and meetings of the Society of the Chemical Industry and the committees of the various Research Associations concerned.

At the other end of the spectrum were those industries where virtually all the research was carried on privately within the firm, notably in rubber tyres and photography. These were both industries where there was one particularly dominant British firm—Dunlop and Ilford—whose competition came not so much from within Britain as from the United States. These leading firms carried the bulk of the burden of research within their industries which, unlike much engineering and food, for example, had a certain element of secrecy attached to them. The similarity between the positions of these two firms is further extended by a consideration of the role of research associations within their industries. In photography, although a British Photographic Research Association had been one of the first in existence, yet Kodak, whose research was American-backed, did not join. The result of this was that Ilford increasingly came to bear the vast bulk of the expenditure of what was intended to be research for the whole industry. Accordingly, the association closed down and its staff dispersed to Ilford and Kodak as research came to be re-located within the firm. A similar situation arose with rubber where, from 1919, there had been a

Rubber and Tyre Manufacturers Research Association, which never had enough funds or facilities for research. Attempts to obtain a bill to levy the industry in 1932 for the Association failed, and research within this industry in Britain rested almost entirely upon Dunlop's research within the firm, though Dunlops had also been a main prop of the association. Portland Cement was a third example where the firms arranged their own research, in this case by detaching their association from the DSIR. In these concentrated industries, research within the firm was the dominant form. In those of relatively little concentration, there was strong if not total reliance upon associations.

It may be questioned why the electrical industry did not display more of the characteristics of rubber and photography. In electricals, however, although there were few very large scientific researching firms, no one attained a clear mastery over the others. This was partly due to the policy of the Central Electricity Board, which spread its orders between the firms to avoid undue dependence on any one; this had the effect of preventing great discrepancies in technical capability. This made them in some ways more like food firms rather than like those industries dominated by one large excellent producer. In these circumstances, the mixture of firm and association activity was relevant and could flourish.

We have indicated that industries varied in the sources from which they drew their research, but there are two interesting examples in the interwar years where the firm research department came to replace the university as the fount of innovation, notably dyestuffs and electrical engineering. Before 1914 firms relied a good deal for new products and processes in dyes on chemists in the universities, such as the work of Weizmann for Clayton Aniline Works at Manchester in the 1900s,[102] the sulphide dye research of Imperial College, and that of the Tinctorial Chemistry department at Leeds. The work of the internal firm research departments was largely confined to existing products and processes. With the formation of British Dyes and then the Dyestuffs Division of ICI, research moved out of the universities and into the firm. It was noted in 1930 that 'the tendency during the recent years has been towards concentration of all research work within the firms' own works, though some firms still make use of colonies of research workers at the universities.'[103] This is in no way to belittle the university contribution—ICI Dyestuffs, for example, ran

[102] Chaim Weizmann, *Trial and Error* (London, 1949), chapter VII, 146–8.

[103] *Report of the Dyestuffs Industry Development Committee* (1930), Cmd. 3658, 30–1.

monthly research committees of professors and their own research heads.[104] But the great developments of the industry in the interwar years, the Caledon Jade Green and the Monastral Blue, came from within the firm and not from an outside academic body. Furthermore, when ICI opened their new dyestuffs laboratories at Blackley in 1938 they claimed them to be 'probably the main focus of organic research in the United Kingdom'.[105] What is interesting about this observation is that if any institution in the later nineteenth century merited this description it would probably have been Schorlemmer's laboratory at Manchester University. Such was the importance of the rise of research within the firm and the change in location of the source of science in this field that a private company research laboratory and not a university department could seriously be put forward as the 'main focus'.

The same case could be made out for the electrical industry. Before 1914 the great bulk of the post-Faraday advances in electrical science in Britain came from the universities, from Wheatstone, Clerk Maxwell, Rayleigh, Kelvin, Ambrose Fleming, Perry and Ayrton, Lodge, John Hopkinson and so forth. With the founding of the great research laboratories of Westinghouse (later Metropolitan Vickers), GEC, BTH and Siemens, these came to replace the universities as the source of new technology, and we cannot point to any set of achievements in the universities in this field in the interwar years remotely comparable either to their own record before 1914 or to the contemporary interwar record of the research laboratories of electrical firms. This was partly associated with the disengagement of British university physics from industrial application in the 1920s and 1930s as the Cavendish physics of Thomson and Rutherford successively conquered the provincial universities, replacing in fashion the electrotechnic interests of the pre-1914 generation of physics professors. Indeed, this may have been all to the good, since it led the universities back to fundamental research now that the firms were mature enough to carry out their technological research for themselves. There are signs that the increasing maturity of the firm in carrying out its own research would pose problems of adjustment for the future. As industries became more mature, so they preferred to engage in their own research rather than rely on university departments,[106]

104 C. J. T. Cronshaw, 'Industry and the Universities', *Manchester University Chemical Society Magazine* (1932). Cronshaw was joint managing director of the Dyestuffs Group, ICI.

105 'New ICI Laboratories at Manchester', *Journal of the Society of the Chemical Industry* (15 January 1938). *Chemistry and Industry Review.* **16**, 56.

and as firms increased their internal research so they became less willing to continue supporting research associations.[107] It may be there is a logical succession, university–research associations–firm, in the changing locus of research activity in certain industries, and these shifts in dyestuffs and electricals in the interwar years were but early movements themes in an ongoing movement.

If there were variations in the extent to which firms' research departments cooperated with outside research bodies, how far did firms cooperate between themselves in research? This may be considered at two levels—firstly, did firms in the same industry cooperate on similar lines? and secondly, how far was there cooperation across industries between product developers and users? Most importantly, how far in certain industries could research functions and obligations be removed from the firm immediately concerned and pushed back vertically to suppliers?

Research cooperation between firms closely involved in the same industry was not common but was found in food and mechanical engineering. For example, Colman advised both Crosse and Blackwell and Heinz about their rival salad creams. Also, United Dairies even more directly did work on whipped cream, and on the enzymes of milk-catalese (catalese being an indicator of udder disease) with Dr Lampitt of J. Lyon (a doyen of the science and chairman of the Food Section of the Society of the Chemical Industry). In engineering, the Fox Club was formed in 1927, by which W. H. Allen, Mavor and Coulson, the Stanton Iron Works, Mirlees, Bickerton and Day met to discuss technical problems and to visit one another's works. On the other hand direct cooperation between product competitors was not very common and we need not stress it. The food and especially light mechanical engineering industries were particularly open in their work, and in the latter there could be no secrecy about designs peculiar to certain firms. In other industries, like tyres and photography, firms generally operated in ignorance of the research of their rivals. An element of secrecy sometimes led to inevitable duplication of research, as in the parallel work on gold as a sensitizer in photography in the late 1930s. We found two examples, even in the food trade, of areas of work where chemists were not aware of very similar developments in other prominent

[106] This point was put to me by the professor of a scientific department, whose chair had long been founded to provide much of the scientific research for a certain industry, and who was now finding that the industry preferred to do its own research and was less interested in some kind of special relationship with a university.

[107] This point was put to me by the head of a certain large firm, whose own internal research activities had expanded such that they were seriously thinking of considerably cutting down their commitments to Research Associations.

firms. For example, Bibby, who were researching on poultry feed, were not aware in the 1930s that Colman had a large colony of canaries in Norwich on which they were experimenting with a similar product. Similarly, the CWS, which did much original work on effluent, were not aware of Dr Muer's work on the same problem for United Dairies, though both benefited through their common connection with the DSIR unit working on effluent. The implication was sometimes drawn, with half an eye on the Trust Research Institutes of the USSR of the thirties, that research within the private firm, in so far as it entailed a measure of secrecy and in so far as that in turn entailed a measure of duplication of research, was inefficient.[108] Even with the one or two instances cited, one scarcely finds this to be so. In any case, although it was admitted that there was duplication of research among the electrical firms, yet this was justified on the grounds that it was necessary for each to build up its own background experience that could not totally be learned from reading at second hand of the activities of others. Moreover, having a large number of different institutions all working on the same problem may prove a perfectly valid strategy of research, as the broad front approach to local anaesthetic research in the First World War indicated.

Across industries, however, frank cooperation between research departments of firms proved to be an enormously fruitful way of advancing technical knowledge. This is best illustrated by a closer examination of W. H. Allen's relations with other research departments over two areas of problems in metallury and oil. Allen were, and are, prominent manufacturers of pumps, diesel engines and turbines and their own research department had thus to work closely with those of their metal suppliers and those of the oil companies whose products were in turn used in Allen's machines. For example, John Brown were anxious to expand the uses of stainless steel and their research department advised Allen's on its machining. Allen replied that they found it unsawable and unturnable and it was found in turn that the material needed further heat treatment which resulted in the modification of earlier researches by Dr Hatfield of the Brown Firth laboratory.[109] Close and frequent contact with Brown persisted throughout the 1920s, and Brown sent them details of their nickel chromium steel which Allen were using for the Admiralty.[110] In the 1930s

108 Bernal, *op. cit.*, note 5, 57, 107. He certainly exaggerates the secrecy and underestimates the amount of cooperation in research by firms on these years.

109 MS Research Notebooks and Papers of R. T. Rolfe of W. H. Allen Ltd., Bedford; letters, 18 and 19 June 1920.

110 Letter, 22 February 1927.

there was continuous collaboration with Firth Vickers and Firth Brown over the R323 manganese molybdenum steel and the Duratlas anti-creep steel for high temperature steam installations. Most importantly, Allen played an important part in helping to develop Staybrite, making welding tests, and ascertaining its resistance to corrosion in sea-going conditions between 1930-6, which resulted in the modification Staybrite FMB.[111] They also inevitably had dealings with the United Steel Companies, advising Steel, Peech and Tozer that their steel was too sulphurous for welding,[112] and, finding that one of their steels was not suitable for the exhaust valves of diesel engines, they were recommended to use a modified silicon chromium steel.[113]

This same kind of ceaseless cooperation and modification and cross-referring of problems between research departments was also evident in oil. Allen worked in close cooperation with Shell concerning the development of oils for diesels and turbines. Indeed, the Nestle's turbine at Ashbourne became virtually a working extension of both their research laboratories (as BTH tried out equipment at the Odeon Cinema, Leicester Square) as they sought to combat the problem of the emulsifying of oil. It was also collaboration between the two that developed Shell's C3 cylinder and bearing oil for diesels.[114] We have paused to consider this example that Allen provides, because it well illustrates the realities of technical advance in the new era of the research laboratory as apparently small pieces of problems were passed about between two or three different firms, modifying here and adjusting there, accommodating the needs of final users, continually watching over new materials and products actually in use to improve them for the future, all totally divorced from any consideration of advertisement or of passing the blame for failure.

Allen were not unusual in this. The same was true of the Cambridge Scientific Instrument Company.[115] We find among its clients major research firms like Dunlop, ICI, BTH, GEC, Lever and a score of others. But these firms had not merely to be customers but often research collaborators as well—for example, in work on automatically controlled

[111] Letter, Firth Brown Ltd., 22 December 1936.

[112] Letter, 1 May 1936.

[113] Letter, United Steel Companies' Central Research Laboratory, Stocksbridge, 22 November 1934.

[114] For oil references, see 8 January 1930, 20 November 1933, 30 April 1934, 25 November 1935, 17 February 1936, but also very commonly elsewhere throughout Rolfe's papers.

[115] MS Research Reports of the Cambridge Scientific Instrument Company (1923–30) *passim.*

exposure when making enlargements from negatives (which was done in conjunction with Adam Hilger and Kodak), work with ICI on the glass electrode of *p*H equipment, with Boots on tablets to absorb sulphur dioxide, with the Norman Cement Co. on the testing of the oxygen recorder, and so forth. Again, as with Allen, technical advance was coming through the cross-fertilization of the different expertise of a number of research departments working together.

It is important to stress that this was a totally new approach to innovation, and not one available before the interwar years. With such limited development of the research laboratory before 1914, the burden of discovery rested so much more on the individual inventor and such contacts as he could establish in universities or in his own firm. The rise of the research department within the firm in the 1920s and 1930s not only changed this by bringing invention into the firm, divorced from the person of the entrepreneur; it also created a new situation whereby new developments came not so much from one firm or one research team as from the cumulative advance of knowledge through the interaction of many of them together. The possibilities of progress rose as a function not of the arithmetical increase of laboratories but of the vastly greater multiple of the possible combinations of pairs within their sum total.[116] It is this kind of new situation that contributes to an explanation of the marked technical advance of the interwar years, much of which was not in the form of easily dateable or clearly definable 'invention' attributable to one inventor.

We suggested earlier that it was possible for some firms and, indeed, whole industries not apparently engaging in research to benefit from their results by having their problems solved by the research of an adjacent industry and often a supplying firm. With some industries this was normal; for example, the cotton textile industry did not usually engage in research on spinning and weaving at the level of the firm because experimentation on this was the province of the machine builders who supplied them, notably Mather and Platt who were every active in research. The same was the case in boots and shoes where the BUSMC provided the machinery for the industry and took care of its research and development, thus maintaining a certain technical level amongst firms who did not themselves engage in

[116] For example, the possibilities of advance by having five laboratories (not of course in the same firm) instead of one are not increased by five but tenfold since there are ten possible combinations of pairs within that group. With eight laboratories, the increase is twenty-eight-fold.

research.[117] Or again, steel companies traditionally relied on refractory makers to do their research in this field until the Central Laboratories of the United Steel Corporation decided to include refractories within the scope of their steel research.[118] The Cambridge Scientific Instrument Company also seemed to be used sometimes to handle the research problems of firms either not engaging in research themselves or facing problems with which they could not cope. For example, they devised the train speed recorder for the LNER which, unlike the LMS, had no research laboratory of its own. Thus the results of firm research were often more widely diffused than would be evident by a consideration of certain apparently non-research-minded areas of industry, because sometimes problems central to one firm were being tackled, not within that firm, but by suppliers in an adjacent area of the industry.

Finally, research departments, apart from contributing to the technologies of firms, also performed a more pervasive and indirect but none the less important role in changing the social structure of the firm. The research department acted as an entrée for the graduate and was a means of increasing the graduate proportion of the labour force at a time when employers were somewhat unsure what his role ought to be. Secondly, the research department often acted as a pool whence personnel could be decanted out into executive positions, thus pervading management with a greater appreciation of science. Thirdly, research was a means by which the firm could usefully employ better educated women at a time of surplus and cheap female labour.

The proportion of graduates to total labour force in firms in the interwar years was low, for example: British Dyestuffs Corporation 4.15%,[119] a small leather company 0.5%,[120] Cadbury 2%;[121] while Maxwell Garnett's estimate for industry as a whole was 2%[122] and James Bowie's estimate for a desirable proportion was 7.5%.[123] Within research departments of firms, on the other hand, graduate proportions were normally high:

117 Roy A. Church, 'The Effect of the American Export Invasion on the British Boot and Shoe Industry, 1885–1914', *Journal of Economic History* **28** (June 1968), especially 249.

118 P. W. S. Andrews and Elizabeth Brunner, *Capital Development in Steel: a Study of the United Steel Companies Ltd.* (Oxford, 1952), 200.

119 *Journal of Careers* (henceforth *J.C.*) **5** (June 1926), 37.

120 Information from Stanley Briggs Esq. concerning his own firm in the 1920s.

121 *J.C.* **18** (February 1939) 114–15: Cadbury's intake ratios.

122 Maxwell Garnett, speaking at the *Second Congress of the Universities of the Empire*. A. Hull (ed.), (London: G. Bell and Sons, 1921), 206.

123 Bowie, in *J.C.* **14** (October 1935), 564.

GEC 1931[a]	33.3%	Colas Products[e]	50%
GEC 1938[b]	29.2%	Metropolitan Vickers[f]	53%
Gas Light and Coke Co.[c]	20%	Ilford[g]	40%
Associated Portland Cement Manufacturers[d]	33%	United Dairies[h]	about 30%
		BTH[i]	30–40%

[a] *J.C.* **10** (February 1931), 25–6.
[b] *J.C.* **17** (October 1938), 566–7.
[c] *J.C.* **18** (April 1939), 252–3.
[d] *Industrial Research Laboratories,* the ASW list (1936), *op. cit.,* note 86.
[e] *Ibid.*
[f] *Ibid.*
[g] Rough estimate made by the firm for 1930s.
[h] *Ibid.*
[i] *Ibid.*

It is interesting to note the consistency of these scattered figures. Graduate percentages for total labour force were almost always under 5%, but those for research departments were between a third to a half. The research department thus brought the graduate into the firm on a scale that would not otherwise have been reached. This had the effect of making firms not used to employing graduates familiar with their capabilities and led to a greater pervasion of the appreciation of science throughout the firm, especially important in companies with accountant- and solicitor-dominated boards. This was especially the case when graduates from the research section moved out into general management. For example, at BTH, personnel from the research department went out into production with processes that they had been responsible for developing in the laboratory; likewise, at Metropolitan Vickers, the Process Section of the Research Department would go out in the same way. One director of research in another company in a different industry said that, in the 1920s, he liked when possible to move some of his staff into general management, to make his firm in turn more receptive to the scientific work of the young research department. Fleming himself was keen on the idea of the research department as the training ground for people who were later to rise to high positions elsewhere in the firm :

> Often the research organisation of a manufacturing company particularly where the products are essentially the outcome of scientific research, forms the training ground for the leading personnel, one which is as yet insufficiently recognized by industrial concerns.[124]

The research department also enabled firms to benefit by absorbing the

[124] A. P. M. Fleming, in the *Daily Dispatch Industrial Survey* (1936).

excess of talented women whom they would not have used in executive positions. The woman graduate found her entry into industry through industrial research, not usually in engineering or heavy industry, but particularly in the finer chemical work. There was Dr Frances Heywood of the Monotype Corporation;[125] Dr Hamer of Ilford, a former assistant of Sir Eiger mountaineer and leather industry biochemist. Lyon, United Dairies, William Pope at Cambridge; and Dr Dorothy Jordon Lloyd, the noted Eiger mountaineer and leather industry biochemist. Lyon, United Dairies, CWS, Boots, Lever, May and Baker, Chivers, Glaxo, Peek Frean, Robertson, Schweppes, Kodak and Burroughs Wellcome were all firms with research establishments employing women graduates;[126] indeed, Lyon had four such by the end of the 1920s.[127] Here again, the research establishment in the firm was playing an ancillary role of both diversifying and improving the quality of labour in the firm.

VI

If then research within the firm yielded such benefits, why did not yet more firms espouse it, and what were the inhibitions and impediments to this activity? This reverse side of the question is worth examining, because some factors that were commonly argued to be impediments do not in fact turn out to be so, and this is important when the supposed existence of these impediments is put forward as an argument for the lack of effectiveness of firm research.

The first barrier to research within the firm was the size of the firm itself. There was clearly a level in the size of the firm at which research became feasible and below which it was not commonly worthwhile. In order to get some idea of the threshold of size at which the researching firm was commonly found, we may list fifteen firms known to be researching in order of their total employment—300, 446, 1,000, 1,000, 1,350, 1,500, 2,000, 2,000, 2,700, 4,000, 6,000, 10,000, 12,000, 15,000, 25,000. We do not know of one lower than 300 but, on the other hand, 1,000+ employees seems to be the level at which the researching firm was most commonly found. J. D. Bernal, writing in 1939, also seemed to think that only firms employing more than one thousand workers could afford a research team

125 D. W. Hughes, *Careers for our Daughters* (London, 1936), 223; also 'A Woman in Metallurgy', *The Woman Engineer* (December 1927).

126 'Prospects for Employment for Women Science Graduates—Industrial Research Laboratories', *J.C.* **17** (May 1938), 293.

127 Elsie Lang, *British Women in the Twentieth Century* (London, 1929), 261.

large enough for serious work.[128] Fleming considered that 6,000 square feet was the smallest viable size for a firm research laboratory, and this would entail a capital outlay of £8,950 and total annual maintenance of £8,700. We can calculate from his proposals of square feet per worker that such a laboratory would have about sixteen workers.[129] If we consider hypothetically that the proportion of research staff to total employees in the firm cannot be extravagantly different from the proportion of turnover spent on research, then this sixteen seems to assume a labour force of about 1,600, which has some relation to the actual figures cited above. Firms smaller than 1,000, and certainly smaller than a few hundreds, were far better advised to support research associations.

Above the threshold of size, two financial factors were regarded as impediments to company research—the attitude of banks and the problem of taxation. One of the arguments for the necessity of research associations in the early days had been that the replacing of private banks by joint stock banks had hampered the possibility of firms launching out into research, because:

> The modern bank authorities are not acquainted with the personnel and policy of local firms and are not prepared to finance new undertakings and developments in the way the private banks did.[130]

This may have been an issue, because we find in 1933 an organization being formed under the auspices of Lloyds Bank to enable industries to finance research.[131] However, we do not find that this was a serious drawback. Indeed, it was evident that the large-scale developments over the end of the war and the early 1920s must have been financed internally with profits from wartime activity. It was the fact that research was financed from this source that gave much more prominence to the issue of taxation which bulked much larger as a grievance in the finance of research than any concern about bank finance.

It was often argued at the time that the system of taxation impeded research within the firm. The anomolous position existed in the interwar years whereby expenditure on advertisement and even contributions to

128 Bernal, *op. cit.*, note 5, 56.

129 A. P. M. Fleming, 'Industrial Research', in Muscio, *op. cit.*, note 93.

130 *Report of the Committee of Privy Council for Scientific and Industrial Research 1915–16*, Cd. 8336 (1916), VIII, 26.

131 *Journal of the Society of the Chemical Industry*, **11** (1933). Chemistry and Industry Review, **2** (1933), 131.

research associations could be made out of untaxed income, but research within the firm had to be financed out of taxed income.

> With the exception of the contributions of industry to the Research Associations most if not all expenditure on research is treated by the Inland Revenue as Capital Expenditure, so that practically no items of expense debited to investigation can be included in profit and loss account prior to assessment for Income Tax. The effect of this is a loss of 22½p in every £ which private enterprise spends in making itself scientifically efficient.[132]

The fact that research expenditure did not entail income tax relief had, it was argued, various adverse side effects. It drove company expenditure into advertising rather than research. Also, it led to a kind of bogus research.

> Stereotyped operations which may be performed by workers ignorant of science; whilst research of the highest scientific character that which brings about the innovations that most help to expand the national revenue does not escape taxation.[133]

And yet this factor, whatever was said at the time, was not so important in impeding research as may be thought, for two reasons. Firstly, since research expenditures were generally so low, 1% of turnover or less, this extra tax consideration was quite marginal. Secondly, any firm with an accountant worth his salt could arrange matters quite legitimately, so that a good deal of research expenditure was spread among other tax-allowable activities of the firm with the result that, in fact, it was quite impossible to distinguish sharply between what was spent on research and what on production.[134] Accordingly, none of the firms interviewed considered that taxation really inhibited their research in the 1930s. The anomaly was, however, removed, and it was indicated in the budget speech of 1944 that, after the war, research expenditure on buildings, plant and machinery would be allowable over a period of five years or over the life of the assets if shorter as a deduction from taxable profits.[135]

If the accountant appeared in the taxation problem as the remover of a potential inhibitor to research, he also appeared in another role as an inhibitor himself. It was complained that the accountant had impeded the

[132] R. W. Western, 'Memorandum on Private Enterprise as a Factor in Industrial Research', *The Scientific Worker* **10** (November–December, 1938), 143.

[133] 'Income Tax and Research', *The Scientific Worker* **11** (Winter 1939), 106. The ASW pressed this issue from 1933.

[134] The three firms that distinctly remembered this as an issue explained that this was the way in which it was overcome.

[135] Heath and Hetherington, *op. cit.*, note 3, 314.

development of research in firms before 1914 'chiefly on account of the fact that accountancy classified such capital outlay and expenses under the heading of unproductive labour'.[136] This was still a very real factor in the interwar years, especially for research departments that had to prove themselves in the face of boards whose members came from backgrounds in accountancy and law rather than technology. One vivid example was given to me of the overriding of technical by accountancy considerations. In a certain firm, fruit awaiting jamming throughout the year was put down in sulphites to preserve it. But this had the effect of bleaching out the natural colour, which meant in turn that it had to be redyed and in turn could not be designated as 'fresh fruit standard'. To overcome this situation, it was suggested by the research chemist that it be partially preserved with sugar, in which it was in any case eventually to be boiled. But this technical proposal was opposed by the accountants on the grounds that this for part of the year would entail showing an outlay on sugar for which no commensurate production was forthcoming and the sulphiting of fruit continued. In the same firm, accountancy proposals for charging individual factories for technical and research services received by them had likewise to be countered by the head of research on the grounds that the weakest parts of the firm required the greatest technical help, and if they were to be charged for it they would be all the less likely to seek it. It was clear from these two illustrative issues that situations could arise in firms where the views of the research department and those of the accountants as to how the best interests of the firm may be served could be at variance.

There was also a third accountancy issue, posed by Sir Arthur Fleming. Fleming came to the view that a firm would be totally justified in retarding the activities of its research department if its work was accelerating the speed of technological change to the extent that it was making expensive plant obsolete before its period of depreciation had been reached. Too rapid technical change, he thought, may

> mean that all the money sunk in productive plant and facilities fail to pay their way and go through a reasonable stage of depreciation before they have to be replaced due to this quick obsolescence resulting from a new process or a new product . . . then there is a grave risk that progress may be too fast.

He frankly thought certain developments were better left on one side either until the firm was faced with competition or until the existing plant had

136 A. H. Railing, in the 'Research Symposium' (1919) *op. cit.*, note 31, 179.

'served a useful life and has carried its proper period of depreciation'.[137] Here, again, an accountancy concept of depreciation was at variance with a research department's normal desire for the most rapid technological advance.

It might have been thought that recurrent depression, and especially the great crisis of the early 1930s, would have had a limiting effect on research. Indeed, Jewkes, Sawers and Stillerman put it forward as one of their arguments against corporation research that it is too vulnerable, since in a crisis firms will always cut down internal research. In fact, from the examples above, there is a good deal of evidence that this was not the case. The immediate postwar depression did reap a toll of research departments begun by firms in the wake of fashion without much sense of deep commitment. In 1921, 'firms which had set up research departments for the first time closed them as unnecessary luxuries and dismissed their staffs'.[138] The severe depression of 1929–32, however, had much less adverse effects on research than one would have thought. Dunlop felt hit by the depression but did not cut back research; nor did Bibby. Ilford did feel the effects of depression, but courageously expanded research, and were rewarded with a very rapid recovery from 1931, as the devaluation had an immediate effect on their export side. Colman scarcely felt the depression at all, and it had no adverse effects whatever on their research, which indeed enjoyed some of its most flourishing years over the turn of the decade. The same was the case with British Thomson Houston, whose most interesting technical achievements—the cinematographic apparatus and the beginning of lamp expansion—straddled the depression as their research expanded. At David Brown's, research was started as part of a general rationalizing policy, reducing staff on the one side, but beginning research and advertising on the other.

Thus it was perfectly possible for firms to respond to depression by retrenchment, but to see no paradox in not cutting back on research but actually starting or expanding it. The proportion of turnover spent on research must in the depression have risen quite markedly as a result. Some firms actually seemed to benefit from a slackening of production as an opportunity to concentrate on technical advance. For example, Scottish Dyes noted of the early 1920s' depression :

137 A. P. M. Fleming, 'The Economic Consequences of Industrial Research', his typescript of a speech at the LSE Department of Business Administration, 20 May 1935, in Fleming's papers. He repeated this view at the Oxford Management Conference, 1936.

138 E. D. Miles, *A History of Research in the Nobel Division of ICI* (ICI, 1955), 94

this hiatus, like adversity, was not without its uses. If our chemists were not needed for production they could research . . . they could venture out on new fields. So in these adverse years though we drew in our horns in many ways, we did not dispense with the service of a single chemist. Indeed, it was a time of precious research. . . .[139]

As a result they produced Caledon Jade Green, patented in 1922, one of the great technical achievements of the 1920s. At W. H. Allen's, likewise, research seemed unaffected and, if anything, stimulated by the depression of the early 1930s. They noted in 1932: 'this stagnation of actual marine business has been one of the means of stimulating developments in the design of new improvements on existing types of machinery'.[140] In this time the air injection diesel engine was almost entirely superseded by airless injection, diesel speeds increased within a year or two from 300 rpm to 1,200 rpm, and 1930 was also a year of considerable activity in replacing old steam-engines with steam turbo generators in a whole range of industries. These matters, real solid achievements of lasting value, were easily overlooked in the attention focused on the contemporary slumps, but the evils of the latter had but little effect on the virtues of the former.

The only one of the firms examined where research was definitely adversely affected by the slump was the CWS, which felt the depression in its variety of manufacturing industries and reacted by cutting down research which had previously been planned on too wide a scale to maintain. Moreover, the society, unlike many other firms, was totally committed to consumer goods demand, and at that time catered more for the working-class market whose consumption was, in turn, more likely to be affected by unemployment. At that period much of its strength was in any case in the vulnerable North. One would not expect to have a record of research departments that failed to come into existence because of the depression, but one such example may be Hollins who decided against starting a Central Research Department in 1933 and ceased to research at all when their chemist left, largely because they were at that time in a very depressed and financially shaky state. We are not inclined to regard the great depression as a severe inhibitor of research within the firm. Probably this was due to the fact that the really depressed basic industries of coal, textiles, shipbuilding, were not ones that undertook research in this form. Firm research was more commonly associated with the newer scientific and consumer industries whose firms, and indeed their research establishments, tended to

139 *Fifty Years of Sundour, 1906–56* (reprint of a speech by Sir James Morton, 1929, kindly lent by Standfast Dyes).

140 *Queens Engineering Works Magazine*, 'Review of Events' (January 1931), and 'Marine Work during 1932' (January 1933).

be located in the more prosperous Midlands and the South. The considerable rise of firm laboratories between 1930–38, as indicated by the FBI figures, also lends weight to this view.

VII

It has been the purpose of this essay to try to redress some of the neglect accorded by historians to research within the firm as opposed to that in universities, in research associations and by government departments. We have suggested that there was an initial spurt of development of company research in the 1890s and 1900s, but that this growth was especially marked around the end of the war and early 1920s, rising to some four hundred by the end of the 1920s and at least 566 by the end of the 1930s. This was made necessary and possible by the well-publicized example of Germany and the United States, by the demands of a rapidly increased war production, by the profits accruing from the war, by the loss of innovating founders, by the tendency towards mass production having impeded possibilities for modification on the job, and by the greatly extended choice of product differentiation afforded by the development of certain branches of chemistry. We have suggested that chemical, engineering and food firms made up the bulk of those engaging in research, but that the heaviest spenders per firm were in drugs, electrical engineering, oil and food. Spending as a proportion of turnover varied downwards from 2% in advanced electrical engineering to less than 0.5% in food. But expenditure on industrial research within the firm was at least ten times that of the DSIR supported bodies—hence its importance—while total scientific industrial research expenditure was of the order of 0.15–0.29% of the national income.

In considering the research and development achievements of a number of firms, we found a greater concern for capital saving than labour saving in their motivation, but probably greater than either, concern to diversify into new and adjacent products and to ensure the public safety and quality of the new products. In almost all cases, the new developments resulting from research had a strong bearing on the survival and commercial health of the firms concerned. We examined the differences between industries in their tendency to draw their research from outside the firm or from within it, suggesting that an almost exclusive reliance on research within the firm tended to be found in concentrated industries with one main British company bearing much of the responsibility for the scientific development of the industry, and pressed by American rather than native competitors. There was also discernible some shift from reliance on research outside the firm to research within the firm notably in dyes and electrical engineering, with the company replacing the university as the new focus of development. More important than collaboration with outside bodies was collaboration between the research departments of firms themselves across adjacent industries, and it was suggested that this in itself was an important innovation of the interwar years. Also it was suggested that the research department played a more pervasive social role in acting as an entrée for men and women graduates, thus raising the proportion of graduates and appreciation of science within firms and also acting as a pool of talent for general management. Finally, certain inhibitors were examined, with taxation and the depression being discounted relative to the effects of certain accountancy attitudes and smallness of scale as obstacles to research.

Throughout this essay, while urging the importance of research within the firm, we have been seeking to cast doubt on certain arguments presented by those who have seen this as of limited value. Jewkes, Sawers and Stillerman regarded as one of the drawbacks of firm research the supposition that such departments were not receptive to outside ideas. We have presented evidence that this was not the case in receiving ideas from outside this country, and certainly not the case as regards receiving ideas from universities, research associations and other firms. There is scarcely a firm that we have considered whose research department worked in any kind of scientific isolation; all benefited in some degree from outside contacts. Secondly, they assert that such departments are vulnerable because they tend to be closed down or restricted in depression. Again, we find, contrary to common sense, that this was not so; only the CWS of the firms we have considered experienced this, and a contributory cause of their problems was an excessively rapid development both inside and outside the CWS research department before the depression. Very many more firms, on the other hand, maintained and even expanded their research. Admittedly, we cannot consider here every firm in British industry, and doubtless there were a number of small, poorish firms that responded to depression in this way, but the significant firms we have considered across a range of industries did not react in this manner.

Thirdly, there was Professor Bernal's contention that research within the firm was stifled by secrecy. There were, no doubt, many instances in which constraints on the diffusion of the results of the firm's research impeded the most rapid development of science. Indeed, the example of the steelmasters of Sheffield before 1914, cited earlier, indicates this, and such attitudes may have continued in some quarters. However, we did not find for the firms which we examined closely that secrecy was important to their work. Moreover, in those industries where firms usually researched without knowledge of what their competitors were doing, including tyres and photography, one could not by any criterion call them stifled by secrecy, since they were successful innovators with research which enabled them to maintain a leading position despite fierce American competition. Fourthly, Bernal's assertion that research within the firm was 'turned to meretricious if not downright fraudulent ends'[141] is simply distorted and is based on no presented evidence. One cannot consider the enormous strivings of the food industry, in particular, after quality and safety in these years, and their creation of rigorous self-imposed standards without feeling that Bernal's accusation is wide of the mark. The great care taken by Colman and Glaxo over the nutritional content of baby food, the United Dairies' milk tests,[142] or the very frank exchanges of Allen, Firth, Shell and others between themselves, simply are not consistent with a conspiratorial view of company-sponsored science. Moreover, it is an important psychological and historical error to suggest that industrial scientists of the interwar years were acting under a totally different set of motivations or a different code of honour from that adhered to by pure scientists.

This essay has sought to draw attention to research within the firm as a neglected and underrated area of industrial and scientific advance in the interwar years; the full importance of such research was finally recognized in the following war. For if anything confirmed the value of these developments in the intervening years, it was the new modes of innovation in the Second World War compared with those of the First. In 1914–18, innovation relied almost entirely on university scientists working either in the universities or seconded temporarily to government establishments like Farnborough or occasionally to private firms. In 1939–45, the developments of radar and the atomic bomb were the product of a genuine partnership of the universities and the great research laboratories of Metro-Vickers, GEC, BTH and ICI, which had risen to prominence in the preceding decades. The creation of a stock of high level scientists outside the universities, especially in the 1920s and 1930s when the universities themselves were scarcely expanding, was of incalculable benefit to the war effort, as it had been to the diversification of industry in the preceding peace. In these

141 Bernal, *op. cit.*, note 5, 154.

142 It is worth putting on record in this regard that the United Dairies laboratories, for example, made no less than 1,706,308 sample tests on milk in 1938 (Report of the Chief Chemist, 1938).

real achievements, private industry showed an initiative and resiliance not altogether appreciated by contemporaries or historians whose vision of the period has been understandably overshadowed by the darker chimerae of depression, unemployment and the ineptitudes of public policy.

APPENDIX

This essay has been made possible by the kind cooperation of a number of heads of research who were active in various firms during the interwar years, and they have helped me by very full discussions on the issues raised here. I would like to express my thanks to Dr H. Jasperson of J. Bibby and Sons Ltd.; Mr A. G. Glover of the CWS; Mr E. A. Murphy of The Dunlop Company Ltd.; Dr R. C. Wright and Dr Muers of United Dairies (now Unigate); Mr W. H. Dimsdale and Dr Waller of Ilford Ltd.; Dr J. W. Corran of Colman; Dr K. J. R. Wilkinson of BTH (now Electric–AEI Machines Ltd.); and Mr J. R. Bryant of W. H. Allen (now APE).

I would also like to thank Mr H. Norman G. Allen, CBE, DSc, of W. H. Allen, for most generously allowing me the use of the unpublished papers of Allen's research department, and for lending a set of the invaluable Queens Engineering Works Magazines, as well as for illuminating discussion. The Cambridge Scientific Instrument Company and Unigate also made available the complete unpublished documents of their research departments for the period, for which I am most grateful. Mr J. S. Paton of GEC Power Engineering (formerly Metropolitan Vickers) lent me some unpublished papers of Sir Arthur Fleming. Mr R. L. Champion of the Wellcome Foundation, Mr K. Taylor of Standfast and Mr A. Swan of Glaxo all sent materials relating to their firms which might otherwise have proved difficult to trace.

Economic History Review, XLVII, 2(1994), pp. 213–238

T.S. ASHTON PRIZE ESSAY FOR 1991–2

British industrial research and development before 1945[1]

By D.E.H. EDGERTON and S.M. HORROCKS

It has long been almost an article of faith that the long-term 'decline' of British industry has been due, at least in part, to failings in research and development (R&D). Other possible causes of economic decline have been the subject of wide-ranging debate among historians, but R&D has remained outside the fray, and often serves as a refuge when other explanations fail. There is no coherent body of literature on industrial R&D, much less the sort of debate that characterizes mature historiographical traditions. Nevertheless, the 1980s have seen the forceful restatement by individual, influential, historians of the conventional view. In the introduction to their collection on the decline of the British economy Elbaum and Lazonick headed a section 'The underdevelopment of industrial research'. In it they stated that 'Government-sponsored co-operative research proved an inadequate replacement for the in-house research capabilities of modern corporations.'[2] Mowery, on whose paper they were drawing, argued that 'low levels of R&D activity . . . appear to be important factors in understanding British export and productivity performance during 1900-50'. He went on to explain the weakness of British R&D: 'the structure of the British industrial research system . . . undermined the effectiveness of the modest investment in R&D. British industry relied heavily on government-funded co-operative research organisations, engineers (as opposed to scientists) were scarce, and links between academic and industrial research were weak. Within firms, industrial research was often fragmented and poorly coordinated with other functions.'[3] Barnett wrote of the weakness of R&D in the British steel industry compared with the German industry in the late nineteenth century; he found 'grievous British weakness in research and development' in the interwar years, and cited an incomplete estimate for the late 1930s

[1] We are grateful to the following for permission to quote from their archives: the Confederation of British Industries, Cadbury Ltd., and the Société des Produits, Nestlé SA. We are very grateful to the staff of the Public Record Office, the Modern Records Centre, University of Warwick, Reading University Archive, and UMIST Archive. Elements of this article have been presented at a European Science Foundation workshop in Berlin and at a seminar at the Science Policy Research Unit, University of Sussex. Our former colleagues at the Centre for History of Science, Technology and Medicine, University of Manchester, John Pickstone and Geoffrey Bowker, made helpful comments on an earlier draft. Part of the research was supported by an ESRC research studentship.

[2] Elbaum and Lazonick, *Decline of the British economy*, pp. 11-2.

[3] Mowery, 'Industrial research', pp. 189-90. These arguments are repeated in Mowery and Rosenberg, *Technology*.

of £2 million spent on industrial R&D compared with $300 million in the United States.[4] More recently Chandler strongly suggested that British industry in the interwar years was weaker than US or German industry as a result of a failure to invest in each of his three prongs—production, distribution, and management. He discussed R&D only briefly, but its weakness is implied by much of his treatment of management failure. One of his few specific comments is 'the few firms that did make important investments in research, such as ICI, Burroughs Wellcome, and Dunlop, did begin to develop a broader line of products. They remained the exceptions, however.'[5] In all these works an unmistakable impression is given of failure, inadequacy, backwardness, and inefficiency.

In this article we will show that before 1945 British industry funded and performed R&D to a much greater extent than recent historical work has suggested.[6] We give detailed data for individual firms and revised estimates for total industrial R&D. Our second aim is to present an accurate picture of that effort: we emphasize the overwhelming importance of in-house R&D; its concentration within a few firms and sectors; the extent of warlike R&D; international and inter-firm links; the development of networks of technological experts; and the dominance of chemistry. Thirdly, we show that in the interwar years British firms spent at least as much on R&D as firms elsewhere, excepting IG Farben and some American firms. We draw our evidence both from existing studies and from new research, including a unique archival source here exploited for the first time, as well as unduly neglected technical journals.

Not all our conclusions are original. In 1972 Sanderson showed that the emphasis on research associations was misplaced, that research was done in a wide variety of industries, and that existing statistics were underestimates.[7] Saul's 1979 survey added useful information and argued for a very mixed record of interwar R&D.[8] Hannah gave an upbeat account of industrial R&D in his survey of British business history.[9] In 1987 Edgerton drew attention to the concentration of R&D in a few firms and R&D programmes, and highlighted the co-operative and competitive relations between firms, as well as indicating the limitations of business histories as sources for the study of industrial research.[10] Although recent years have seen the appearance of a number of studies of R&D in particular firms and industries,[11] the

[4] Barnett, *Audit of war*, pp. 93, 266, 270.

[5] Chandler, *Scale and scope*, p. 390.

[6] We shall use the term R&D anachronistically, and do not intend to discuss the many complex issues of measurement and comparability which inevitably arise. We are concerned with a particular type of input into invention and innovation, that of organized scientific and technological activity. We are not dealing with the output of inventions or innovations, and do not therefore deal with patenting activity, or with studies of British scientific and technological activity which are based on this somewhat unreliable measure, or with studies which discuss selected innovations. Neither are we concerned here with the efficiency with which innovation was produced from R&D.

[7] Sanderson, 'Research and the firm'.

[8] Saul, 'Research and development', which also deals with *innovation*, as does Sayers, 'Springs of technical progress'.

[9] Hannah, *Corporate economy*, ch. 8.

[10] Edgerton, 'Science and technology'.

[11] Liebenau, ed., *Challenge*, and papers in *Bus. Hist.*, 26 (1984) contain collections of case studies, although not all of them focus exclusively on R&D. See also Stranges, 'Birmingham to Billingham'.

published record still includes information on the R&D efforts of relatively few British firms. This, as we will show, is not due to a scarcity of R&D performing firms, but rather to a lack of collated information which is publicly available. We begin with an examination of contemporary debates and information on the R&D effort of British firms.

I

Before 1914, details of R&D efforts were typically kept secret by firms, and there was much criticism, especially by academic scientists, of an alleged lack of industrial interest in science.[12] The Great War led to state concern with industrial research and the creation of a Department of Scientific and Industrial Research (DSIR), along with industrial research associations, which it partially funded.[13] By the interwar years, however, firms were keen to advertise the fact that they did R&D: it became a well-known feature of modern large-scale industry.[14] The interwar technical press carried reports of the opening of new laboratories and details of the work carried out in them, as well as publishing papers by industrial researchers.[15] The research activities of major companies such as ICI, GEC, and Metropolitan-Vickers were well known and any avid reader of the technical press would be aware of the major achievements, both in introducing new products and in imposing new forms of scientific control over production processes. The academic scientific community also showed interest: the British Science Guild appointed a number of special committees to examine different aspects of the field and, after criticism, these included industrial scientists.[16] Industrial research became a popular lecture subject for scientific societies, and attracted the attention of major academic scientists.[17] There was also a debate between liberals and marxists who believed that the large corporation retarded the development and use of technology,[18] while advocates of the large corporation argued that it had beneficial effects because it financed research.[19]

During the Second World War the attention paid to the subject reached new heights. The Federation of British Industries (FBI), responding to its own report on reconstruction, formed an Industrial Research Committee

---

Other studies including much useful information about the R&D effort of particular firms are Allen, *Studies in innovation*; Briggs, *Broadcasting*, II; Burns, *Television*; Clayton and Algar, *GEC*; Fox, *Dyemakers*; Hardie, *Widnes*; Miles, *Nobel*.

[12] Sanderson, 'Research and the firm', p. 112; Turner, 'Public science'.

[13] Fleming, 'Science in industry'; MacLeod and MacLeod, 'Origins'; Varcoe, 'Scientists'. A useful summary of DSIR activities is Heath and Hetherington, *Industrial research*.

[14] Fleming and Pearce, *Research*. See also Mees, *Organisation of scientific industrial research*.

[15] *Chemistry and Industry*, one of the publications of the Society of the Chemical Industry, and *The Industrial Chemist* were major publishers of such articles.

[16] *Chemistry and Industry*, 44 (1924), p. 765 criticized a report of the British Science Guild Research and Inventions Committee. The subsequent Committee on the Position of the Technical Expert in the Public Services and Industry included A. P. M. Fleming among its members: British Science Guild, *Report*, p. vi, list of members.

[17] Smith, *Nation's balance sheet*; Bernal, *Social function*; Huxley, *Scientific research*.

[18] Bernal, *Social function*, esp. ch. 6. See Lilley, *Men, machines*.

[19] Niblett, 'Images of progress'.

(IRC) in late 1942.[20] In January and April of 1944 Nuffield College, Oxford, hosted private conferences attended by academics, industrialists, industrial researchers, and politicians, which emphasized the economic importance of extending and intensifying research and scientific techniques in a wider range of industries.[21] The Manchester Joint Research Council, established in 1944, brought together the city's Chamber of Commerce and its university.[22] Many other organizations and publications added their voices to the clamour for industrial research. Among these were *Nature*, the British Association for the Advancement of Science, the Institution of Electrical Engineers, the Association of Scientific Workers, and the London Chamber of Commerce. By the end of the war propaganda for industrial research was intense and in 1945 a conference organized by the FBI attracted 1,200 paying delegates whose enthusiasm was 'almost embarrassing';[23] regional conferences followed.[24] Articles concerning industrial research became a regular feature of the FBI's monthly circular.[25] The industrial and government interest in R&D was reciprocated by the scientific community. The British Association for the Advancement of Science held a major conference on 'The place of science in industry' in January 1945 and a second, on 'Scientific research and industrial planning', in December 1945.[26]

There was, however, very little information on the extent of British industrial research until 1943. The most important source of systematic information on industrial R&D in the interwar years was the DSIR. Despite having little knowledge of its extent,[27] the DSIR implied that industry was not doing enough,[28] but highlighted its own contributions, notably the funding of the research associations.[29] Although, as table 1 makes clear, research association spending was much lower than industry's internal spending, the impression could be gained, then and since, that research associations were the main funders of industrial research and that industrially funded R&D was deficient.[30] The Balfour Committee, after discussing

[20] Modern Records Centre, University of Warwick (hereafter MRC), FBI Archives, MSS 200/F/3/T2.12, press release of 11 Nov. 1942.

[21] Nuffield College, *Problems*.

[22] Dupree, ed., *Lancashire*, pp. 200, 240.

[23] Ibid., p. 332.

[24] MRC, FBI Archives, IRC minutes, MSS 200/F/1/1/166.

[25] *British Industries*, 1942 onwards.

[26] The transactions may be found in *The Advancement of Science*, III, no. 10 (May, 1945), pp. 106-55 and III, no. 12 (April 1946), pp. 286-333.

[27] MRC, FBI Archives, IRC minutes, MSS 200/F/1/1/166. Hetherington commented of the IRC 1947 report that, 'in his 20 years at the DSIR he had constantly sought statistics such as these and thought that they should be made widely available'.

[28] DSIR, *Annual report 1925/26*, pp. 15-45, 'Report of the advisory council'.

[29] A number of the RAs have been studied in detail: see Edgerton, 'Photographic industry'; Williams, 'Technical innovation'. For an overview see Varcoe, 'Co-operative research associations'. Other DSIR funded projects are covered in Hutchinson, 'Fruitful co-operation'; MacLeod and MacLeod, 'War and economic development'; Vernon, 'Microbiology', pp. 309-10.

[30] Sanderson, 'Research and the firm', pp. 107-8. Even in industries with a research association individual firms spent more than the research association did. Dunlop, for example, spent more than 10 times the budget of the RA of British Rubber Manufacturers on their own research, as well as being the largest single contributor to its finances (MRC, FBI Archives, MSS 200/F/T2/7/1). This was also a problem for both Cadbury and Rowntree in their dealings with their RA (P.R.O., DSIR 16/250, Report of inspection of the British Association of research for the cocoa, chocolate, sugar confectionery, and jam trades, 1925).

Table 1. *Comparison of expenditure by DSIR and industry for selected years*

DSIR	*1929/30 (£'000)*	*1934/5 (£'000)*	*1937/8 (£'000)*
Net expenditure excl. research associations	465.8	464.4	488.9
Grants to research associations	70.9	85.4	148.3
Total	536.7	549.8	637.2
Industry	*1930*	*1935*	*1938*
Own research	1,736	2,696	5,442
Research association contributions	150.75	230.25	260.5
Total	1,886.75	2,926.25	5,702.5

Sources: DSIR, *Report for 1929/30, 1934/5*, and *1937/8*, and FBI, *Research*, p. 7. The two main sources of research association income are shown. Estimates for industry are certainly underestimates, especially for 1930 and 1935. For an upward revision see table 4.

briefly some of the outstanding contributions of research in the US and Germany, noted:

> There is no doubt that, in Great Britain, research is extensively carried on in certain industries by private firms and corporations. No attempt is here made, however, to describe the work of private research laboratories in this country, since there are no means by which accurate and complete information can be obtained.[31]

The report, however, contained a detailed description of the work of the various branches of the DSIR and the research associations. In its final report, the Committee commented in damning fashion on the scientific outlook of British industry.[32]

Another widely cited source of information on interwar British R&D is provided by a list of public and private research laboratories compiled by the Association of Scientific Workers (AScW) in 1936. Mowery used it to conclude that only 20 of Britain's largest 200 companies undertook research.[33] The limitations of this survey were best summed up by Bernal: '450 firms undertaking research were circularized. Of these only 80 answered, of which only 35 gave particulars of expenditure, while 12 declined to give even the number of qualified workers employed.'[34] In fact, many of the major firms involved in research are missing from this list, including most of ICI's divisions, Dunlop, Courtaulds, Lucas, EMI, Kodak, Mond Nickel, Cadbury, and BP (see table 3). The low response rate was perhaps unsurprising given the leftish politics of the AScW at this time,[35] but surveys by more respectable agencies were almost as suspect. A 1946 list also omitted many private concerns but gave detailed information about government funded

[31] Committee on Trade and Industry, *Commercial efficiency*, p. 320. An editorial in *The Industrial Chemist* criticized the committee for failing to make any effort to obtain 'accurate details of the amount of research carried on by an industry to which it is the lifeblood'; it has 'wasted both its time and opportunities, and we must suppose, a certain amount of public money': (1927), p. 98.

[32] Committee on Trade and Industry, *Final report*, p. 218.

[33] Mowery, 'Industrial research', p. 192.

[34] Bernal, *Social function*, p. 108.

[35] MacLeod and MacLeod, 'Contradictions'.

industrial research, as well as organizations such as the BBC and the GPO which are surprisingly absent from other sources.[36]

Despite the paucity of information, reasonably accurate informed guesses about the extent and concentration of R&D could be made. In 1939 Bernal suggested that industrial expenditure had reached 'as much as two million pounds', that between 300 and 600 companies carried out research, and that 'serious contributions to industrial research probably come from less than a dozen big firms, which have large laboratories with a hundred to three hundred workers'.[37] But it was not until 1942 that the newly established IRC, which included a number of influential research directors, decided to survey research activity, both during the war and retrospectively.[38] The aggregated results for the 1930s were published in 1943 in a pamphlet entitled *Industry and research*.[39] The IRC asked for details of expenditure on 'industrial research and technical development (other than subscriptions to research associations etc)' and employment of technically qualified staff 'excluding if possible those engaged on purely routine testing', in the years 1930, 1935, 1938, and 1941.[40] Despite the war, the response rate appears to have been very high. However, many companies had difficulty because they kept no separate records of research expenditure; many firms were unable, in 1942, to provide details of their earlier effort.[41] Most companies involved in wartime weapons research on a major scale refused to divulge information, even for the 1930s. Our analysis suggests that the FBI were right to note that the figures 'considerably understate the true position'.[42] It is clear that, for the 1930s especially, the IRC figures left out some very important research spenders, and that the figures for each year are not comparable. On the other hand the IRC, to maximize the industrial effort, used a deliberately wide definition of R&D. A second survey, undertaken in 1945, was far more detailed, and included only firms spending in excess of £1,000 per annum on research. It also sought to establish to what extent firms were planning to expand their research programmes. The aggregated results of this survey were published in 1947 (see table 2).[43] The survey returns, most of which can be found in the FBI archives, give a unique picture of the R&D effort of individual British firms over a long period, and in the following sections we make much use of them.[44]

[36] Andrade, *Industrial research*.
[37] Bernal, *Social function*, p. 56.
[38] MRC, FBI Archives, minutes of the IRC for 11 Dec. 1942, MSS F/1/1/166.
[39] The results are widely cited. For example, Sanderson, 'Research and the firm', pp. 118-9; Mowery, 'Industrial research', p. 192.
[40] MRC, MSS 200/F/T2/7/1. Letter from J. Gough to firms, 1 June 1943.
[41] For example, Dunlop, Lucas, and Morris Motors. Morris gave only a figure for 1938 which was described as typical (MRC, FBI Archives, replies to 1942 survey, MSS 200/F/3/T2/7/1).
[42] FBI, *Research and industry*, p. 7.
[43] FBI, *Scientific and technical research*.
[44] Replies from individual firms are to be found in MRC, FBI Archives, MSS 200/F/3/T2/7/1, and the tabulated results in MSS 200/F/3/T1/127c.

Table 2. *Expenditure and employment of qualified technical staff for R&D in the UK, as revealed by FBI surveys*

	1930	1935	1938	1945	1950
Firms	422	484	566	420	301
Expenditure (£'000)	1,736	2,696	5,442	21,815	23,779
Firms	384	432	520	—	301
Staff	1,381	2,566	4,382	7,894	8,560

Sources: FBI, *Research*, p. 7; FBI, *Scientific and technical research*, p. 9; FBI, *Research and development*, pp. 6-7. The 1941 results were not aggregated. The results are not comparable across years, the samples and response rates were different, and the figures for employment do not correspond exactly with those for expenditure. For an upward revision of expenditure, see table 4.

II

Many British firms, in many sectors, were actively engaged in research well before the outbreak of the First World War. Sanderson notes that Sir Arthur Fleming, who had American standards in mind, said there were no more than six manufacturing firms with research organizations before 1913; and that Sir Harold Hartley estimated that total expenditure on industrial research was 'well below £1m' in the early years of the century.[45] Even £500,000 on *research* in industry was a lot of money: if we assume that £200,000 of this went on salaries, that would mean 200 scientists working on research paid £1,000 each, the salary of a university professor.[46] That such an estimate is reasonable is suggested by the fact that Sanderson listed 15 firms doing research before 1914 and that the list can now be augmented.[47] The scale of R&D in particular firms also shows that our total is plausible. From 1892, the United Alkali Company had a separate Central Research Laboratory employing six chemists under a scientist of distinction.[48] Cadbury reported a prewar budget of £5,138.[49] In the British dyestuffs industry all the significant firms did research and brought out new products, though on a small scale: two research chemists were employed in British Alizarin, perhaps more in Levinsteins and Read Holliday.[50] Dreaper, writing in 1914, and drawing on his own experiences, discussed the benefits a research chemist could bring to textile firms.[51] Donnelly's work indicates that chemists had long been employed by calico printers.[52] The submarine telegraph industry, in which Britain led the world, was based on extensive

[45] Sanderson, 'Research and the firm', p. 112.

[46] By contrast, a widely cited estimate gives between 180 and 225 graduate chemists working in British industry in all capacities in 1902 (Cardwell, *Organisation of science*, pp. 204-8). These figures are cited in such works as Russell, *Chemists by profession*, p. 191; Bud and Roberts, *Science versus practice*, p. 161; Alter, *Reluctant patron*, p. 225. For an analysis of the figures see Edgerton, 'Science and technology', app. 3. Recent work by Donnelly suggests a much larger number of trained chemists in industry in this period (Donnelly, 'Industrial recruitment').

[47] Sanderson, 'Research and the firm', pp. 109-12.

[48] Reader, *ICI*, I, pp. 116-9; Hardie, *Widnes*, p. 176.

[49] Cadbury Archives, Bournville, Research Committee Minutes of 11 April 1919.

[50] Fox, *Dye-makers*, p. 117. For the German industry see Meyer-Thurow, 'Industrialisation of invention'; Beer, *Dye industry*.

[51] Dreaper, *Research chemist*.

[52] Donnelly, 'Chemical education', ch. 2.

scientific research, although here the division between academic and industrial research is impossible to identify. The Sheffield steel industry developed analytical laboratories to service open hearth steel making; Robert A. Hadfield established a research laboratory in the 1880s which 'may have been the most advanced in the world';[53] the Brown-Firth Research Laboratory, founded in 1908, invented stainless steel.[54] Hadfield's research was related to armaments, which saw major innovations before 1914.[55] The major gun-makers invested some 6 to 12 per cent of profits in R&D at the beginning of the century; their interest in research 'set the armourers apart from the conservative mass of British industrialists'.[56] Vickers had around 300 to 400 design staff; the large armourers spent as much as £80,000 to £100,000 each on R&D before 1914.[57] Nobel, one of the major explosive and propellant makers, had one of the largest laboratories in the country, headed by a former War Office civil servant;[58] it employed nine research chemists in 1914.[59] Marconi, whose principal customer was the Admiralty, also did R&D.[60]

In the years before 1914, as later, research staff were only a small fraction of staff in industry who were technically qualified. A major function of scientists was analytical control of production, but such staff often did some research.[61] Brunner, Mond's 'bright young scientists' did research, though this was 'combined with their duties in the factory and was directed mainly, though not solely, at improving the ammonia-soda process'.[62] Cadbury's first chemist did some R&D from the start, but soon devoted himself full-time to such projects as the development of Cadbury's Dairy Milk chocolate. In 1911 the firm started a separate research section.[63] Many other firms were spending large sums of money on research long before it was considered necessary to establish a research department, and others never gave the activity any formal recognition. Moreover, many firms employed consultants for both analytical testing and R&D of new products before establishing their own in-house facilities.[64]

During the Great War the amount of research done in British industry increased dramatically, in response to the demand for products either new to Britain or entirely new. Many companies began to do their own research on a significant scale, existing research organizations were expanded and reorganized, and new research laboratories built. In 1914 Hadfield had 38

53 *Tweedale*, ' "Rule of thumb" ', p. 68; p. 82, n. 67.
54 *Idem*, *Sheffield steel*, p. 65.
55 *Idem*, ' "Rule of thumb" ', p. 73.
56 Trebilcock, ' "Spin-off" ', p. 48.
57 *Idem*, *Vickers*, p. 4.
58 Miles, *Nobel*, p. 43.
59 Haber, *Chemical industry*, p. 356.
60 Baker, *Marconi*; Pocock, *Radio industry*.
61 Watson, *Sugar refining*, pp. 100-8; 116-9; Liebenau, 'Corporate structure', p. 35; Edgerton, 'Photographic industry', pp. 108-9.
62 Reader, *ICI*, I, p. 231.
63 'The history of Cadbury's Dairy Milk', *Bournville Works Magazine* (1937), pp. 255-9, and Cadbury Archives, annual reports of the Chemists' Department, 1902-11.
64 University of Reading Library, Huntley & Palmer Archives, HP 143, correspondence with Alfred J. Shilton and Alfred Ashby, May-Sept. 1891.

research staff, in 1920 the figure had grown to 88.[65] Levinstein employed 75 research chemists by 1919. British Dyes, formed in 1915 on the basis of Read Holliday, expanded research very rapidly. Nobels ended the war with 21 research chemists and 13 analytical chemists.[66] Poison gas R&D involved 150 chemists.[67] Brunner, Mond put a great deal of effort into ammonium nitrate production, and new routes to phenol. Even at Cadbury and Rowntree new research projects aimed to save scarce raw materials and find substitutes.[68] R&D had proved its worth to many firms during the war, and this led to a strengthened position within the firm, and increased resources.

After the war the expansion continued in a context of cash-rich firms, some of which needed to replace founding inventor-entrepreneurs,[69] mergers between firms already important in R&D, and the safeguarding of infant industries.[70] The chemical industry is the best example. After the war the British Dyestuffs Corporation, a merger of Levinstein with British Dyes, spent £250,000 in 1919-20 on new laboratories and employed 80 research chemists. Cutbacks were soon made, to 30 in 1923, and then to only 15. Even so, in the period 1921-4, the company appears to have spent £50,000 per annum on research.[71] Nobel Industries, a merger of the major explosives makers, had a growing and increasingly diversified research programme by the late 1920s.[72] Brunner, Mond had by the 1920s its laboratory at Winnington, another acquired from Castner-Kellner in Runcorn, and new facilities at Billingham. The Billingham laboratory, completed in January 1922, could, according to Professor F. G. Donnan, 'bear comparison with anything to be found in Germany or the States'.[73] Imperial Chemical Industries, formed as a result of the 1926 merger of BDC, Brunner, Mond, United Alkali, and Nobels, became easily the largest spender on R&D in Britain. Alfred Mond sought to create 'a business at least equal in technology to IG', and research was expanded rapidly.[74] From an estimated £221,000 in 1926/7, expenditure rose to £350,000 in 1928, to around £1 million in 1930, concentrated in Billingham, Runcorn, Winnington, Widnes, and Ardeer.[75] By 1929, even the neglected dyestuffs effort used 70 research chemists at BDC and 12 more in Scottish Dyes.[76] The major new project was oil-from-coal on which more than £300,000 was spent in 1930 alone. Oil-from-coal alone represented about three times as much as Metropolitan-Vickers's entire research effort. In depression ICI laid off research staff,

[65] Tweedale, ' "Rule of thumb" ', p. 68, p. 82, n. 67.
[66] Haber, *Chemical industry*, p. 356.
[67] MacLeod, 'Arsenal', p. 55.
[68] Rowntrees Archives, York, Laboratories/Experimental Work, Box 1, Arch 1581, typescript by S. Allen, 'Some notes on R & Co's Chemical Dept', p. 7.
[69] Sanderson, 'Research and the firm', pp. 112-8.
[70] Haber, 'Government intervention'.
[71] Fox, *Dye-makers*, p. 171.
[72] By the late 1920s, 66 research chemists, 8 analytical chemists, and 11 physicists (Haber, *Chemical industry*, p. 356).
[73] Quoted in Reader, *ICI*, I, p. 367.
[74] *Idem*, *ICI*, II, p. 81.
[75] Ibid., p. 83. This figure differs substantially from that obtained by the FBI and which is reproduced in tabs. 3 and 4.
[76] Ibid., pp. 85-8, 103, 186.

but the cuts do not appear to have been as severe as in the US, or with the exception of IG, as in Germany.[77] It took until the late 1930s to reach spending levels of 1930. Now the concentration of effort was in organic chemistry. Whereas ICI dyestuffs were allocated only £26,960 for research in 1931, by 1937 £262,000 had been made available to develop organic chemistry.[78] ICI build new laboratories in Blackley, in the modernist style, a move emulated by a number of other firms.

The second major concentration of R&D was to be found in the electrical industry. International Telephone and Telegraph, which in 1925 purchased the foreign interests of AT&T, including its British operations—renamed Standard Telephones and Cables—built laboratories in Hendon with 420 employees by 1930, when they were closed.[79] The chairman of ITT, which had another laboratory in Paris, 'could boast that outside the USA he had the largest organisation in the world devoted exclusively to research and development in electrical communication'.[80] At both GEC and Metropolitan-Vickers, where research was formally established in the last years of the war, expenditure rose rapidly: GEC's expenditure grew from around £10,000 in 1920 to about £100,000 ten years later.[81] Two new laboratories were completed at Metropolitan-Vickers in 1921, with a high-voltage laboratory being added in 1924.[82] Expenditure increased from £71,000 in 1928 to £95,000 in 1930.[83] R&D was by no means confined to the 'new' industries. The total spent in the steel industry was estimated at £300,000 for the 1930s; between them United Steel, Colvilles, and Firth Brown accounted for approximately £120,000. The gas industry spent £295,000 in 1935.[84] In cotton textiles, the expenditure of the research association reached £84,000 in 1935, making it larger than any industrial laboratory, although the three largest firms combined outspent it.[85]

Concentration by firm was high. In 1938 ICI, Metropolitan-Vickers, GEC, British Thomson-Houston, Lucas, Dunlop, and United Steel employed more than 100 graduates or qualified staff on R&D. The FBI estimated that there were 4,382 qualified staff engaged in R&D in 1938, giving these firms about 30 per cent of total R&D employment. ICI alone spent about 19 per cent of the total in 1935 and 1938 (see table 4). Three food companies (excluding Unilever) spent more than £20,000 each in 1930, accounting for around 70 per cent of food R&D.[86] For industrial production as a whole, and within individual sectors, concentration of R&D by firm was considerably

[77] In the US between 1930 and 1933, there was a cut of one-third in scientific employment in industrial research: GE laid off 50 per cent, AT&T nearly 40 per cent, and Westinghouse stopped all pure research (Kuznick, *Beyond the laboratory*, p. 28). Du Pont cut back 20 per cent in 1932 but by 1934 expenditure was back at 1931 levels (Hounshell and Smith, *Du Pont R&D*, tab. 15.1b, p. 289). IG cut back expenditure, including salaries, but not staff (Reader, *ICI*, II, p. 33).

[78] Fox, *Dye-makers*, p. 188.

[79] Young, *Power of speech*, p. 67.

[80] Ibid., p. 60.

[81] Clayton and Algar, *GEC*, p. 413.

[82] Metropolitan-Vickers Electrical Co., *Fifty years*, p. 18.

[83] Niblett, 'Images of progress', p. 43.

[84] MRC, FBI Archives, MSS 200/F/3/T1/127c.

[85] DSIR, *Report 1934/5*.

[86] Horrocks, 'Consuming science'.

Table 3. *R&D employment and expenditure, ranked by expenditure in 1938 of firms spending over £25,000*

Rank R&D 1938		*1930*		*1935*		*1938*		*1941*		*1945*	
		Qualified staff	*£'000*	*Qualified staff*	*£'000*	*Qualified staff*	*£'000*	*Qualified staff*	*£'000*	*Qualified staff*	*£'000*
1	ICI*		420	464	625	510	1,075	680	1,190	1,396	3,500
2	GEC*	125	100	175					400	250	
3	BTH*	73		104						132	300
4	Metro-vick*	73	95	87	200	117				137	160
5	EMI				167		145		84		
6	Pilkingtons	13	109	20	114	37	120	34	130	63	
7	Lucas			84	70	103	102	155	197	171	450
8	Magnesium Metals						90		90		
9	Jas Williamson	1	1	9	23	14	85	9	13		
9	Unilever	30	35	57	65	79	85	85	100	111	200
11	Mond Nickel	29	29	39	41	59	72	73	82	64	90
12	Dunlop			85	50	100	65	105	70	77	158
13	Courtaulds	41	40	51	40	79	60	92	60	132	30
14	Brit Aluminium		32		40	48	56	50	73	54	80
14	United Steel*	34	19	86	44	113	56	135	70	32	50
16	Morris Motors					40	53				
17	Anglo-Iranian	50	35	50	35	85	50	100	120		
18	Boots*	29	34	29	34	38	45	68	80	50	90
19	Eastern Oil	22	30	27	35	35	43	47	57		
20	Brown Firth		35		35		35		35	30	50
21	BUSMC	9	18	9	27	16	32	9	13	22	35
21	Kodak	7	7	22	16	31	32	53	61	54	120
21	Rowntrees	28	31	34	27	40	32	34	26	20	35
24	Colvilles*	16	16	31	17	42	31	91	48		30
24	Calico Printers	36	23	48	29	51	31	51	39		
26	AEC		30		30		30		30	37	90
26	Bradford Dyers			25	25	25	30	29	40		
28	Morgan Crucible	27	19	46	22	58	29	53	28	38	50
29	Metal Box			4	4	28	28	19	24	19	32
29	Cadbury/Fry	21	26	21	27	23	28	15	14	20	30
29	LMS Railway	35	13	41	23	61	28	58	30	60	200
32	Bleachers Assoc	16	16	14	17	15	26	11	26		20
33	Lyons	71	28	70	22	84	25	80	24		

Notes: Ranking is by expenditure for 1938, for which the most complete figures are available. This list is not definitive, and there are clearly important companies missing; see table 4. The figures quoted for several companies differ from those given in other published sources, and they are not necessarily comparable between companies or even for different years for the same company.
Companies marked * appear in the AScW list.

Sources: All figures for 1930-41 are from the FBI archives (MSS 200/F/3/T1/127 c), except GEC, BTH, and Metro-vick. For BTH and Metro-vick see Edgerton, 'Science and technology', p. 101, nn. 1, 2. For GEC see Clayton and Algar, *GEC*, p. 413. All figures for 1945 are from the FBI archives (MSS 200/F/T2/7/4), except GEC as above.

higher than that for production, just as it is today. Another way of expressing this would be to say that there is a wide variation of R&D intensities within sectors as well as in industry as a whole.[87]

The character of the work being done is difficult to establish in the absence of detailed studies of individual laboratories. Nevertheless, in the case of Pilkington we know that the bulk of their large expenditure was for development rather than research, and at ICI more than half the effort of the Blackley laboratories went on the discovery of new products and the devising of processes for them.[88] We are not in a position to say anything about the extent of 'fundamental' research in industry.[89] Given this lack of detailed evidence it is important not to base any assessment of the character and effectiveness of this effort on models drawn from other places and other times. For example, we should not assume, without argument and evidence, that the scientific, technological, or economic importance of, say, product innovation is greater than that of process innovation, or that some industrial sectors or areas of research, say, 'chemicals and electricals' were more important than others. The effectiveness and efficiency of the formal R&D made by British firms is another issue which requires detailed study in itself. We have no comprehensive listings of the inventions and innovations that derived from formal R&D, or indeed from any source. Patent production is not a reliable measure of the output of formal R&D, because even within industrial companies carrying out R&D this was not the only source of patents, and we know that there were variations in the propensity to patent across technologies, industries, firms, and nations.

III

During the Second World War expenditure on R&D in industry increased much faster than the employment of qualified staff: this was probably due to the expansion of increasingly expensive aeronautical research, which in 1945 cost around £10,000 per member of qualified staff compared with about £2,000 in other sectors. A very large expansion in government funding for R&D in industry took place. While £5.4 million was an underestimate of spending in 1938, the FBI produced a figure of £20 million for 1945, which they suggested represented some three-quarters of the effort.[90] At least £10 million of the 1945 figure is likely to have come in the form of development contracts from the state; indeed it is possible that privately funded R&D decreased during the war. GEC, for example, was largely turned over to government-funded warlike work, which included pre-production making of essential new valves. By May 1940 GEC received some £30,000 per annum from the service departments; in 1942/3 the firm

[87] One issue we cannot deal with here concerns the different patterns of diversification in R&D and production for a given firm. See Edgerton, 'Science and technology'; Horrocks, 'Consuming science'.

[88] Fox, *Dye-makers*, p. 190.

[89] Bernal reported a study which showed that 6% of articles from the *Journal of the Chemical Society* and 42% in the proceedings of major engineering institutions came from industrial sources in 1929 (Bernal, *Social function*, p. 425).

[90] FBI, *Scientific and technical research*.

received about £180,000, roughly equivalent to its prewar private R&D.[91] ICI did £1.5 million of R&D for the government, £870,000 of which was for the 'Tube Alloys' atomic bomb project.[92] The executive director of 'Tube Alloys' was ICI's research director; ICI and Metropolitan-Vickers were the main industrial contractors. In 1942 atomic R&D cost £250,000, and increased very substantially in early 1943: 'The range of scientific and technical work on atomic energy in Britain during 1942 and 1943 was wide and deep. The base was certainly there for a take-off into full technological development.'[93] Some work with direct civil implications was expanded during the war, for example when the British pharmaceutical industry underwent a radical shift in its attitude towards research, and ICI became a major force in this area.[94] Development of new products was carried out in the framework dictated by wartime necessity, but this achieved results which proved to be of lasting importance. Lyons, drawing on prewar expertise in frozen cooked food, used idle ice cream plant to create 'Frood', which was used in airline catering and became a substantial dollar earner after the war.[95] Overall, however, most industrial research during the war was almost certainly devoted to warlike purposes.

It is significant that the FBI's 1942 survey did not elicit figures from the major performers of warlike R&D, even for the 1930s. Neither the airframe nor the aeroengine firms, nor the big electrical companies, supplied figures, but, as the 1945 survey shows, they spent on the largest scale during the war. Other evidence suggests that this continued the prewar situation; indeed, state support for defence R&D in the 1930s was almost certainly larger than before the Great War. The new Air Ministry was the largest single funder of R&D in interwar British industry, probably spending more in industry than ICI did. In the mid 1920s it was spending a total of £1.34 million on R&D, compared with £0.98 million by the Admiralty and £0.49 million by the War Office.[96] The engine department of Bristol Aeroplane (with Armstrong-Siddeley, Rolls-Royce, and Napier accounting for the bulk of engine production) received an average of £75,000 per annum for R&D and experimental engines between 1925 and 1930, though its expenditure was higher still.[97] The airship programme of 1924-30, which the Air Ministry largely financed, totalled £2.4 million, of which £0.41 million was for R&D. The work was divided between the Air Ministry and a subsidiary of Vickers.[98] We can be confident that in the 1930s both Bristol Aeroplane and Rolls-Royce were among the top 10 R&D performers in British industry. Furthermore, many of the electrical firms were closely involved in the development of equipment for the radar chain from the mid

[91] Paterson, *War diary*, pp. 37, 371-2.
[92] Reader, *ICI*, II, p. 293.
[93] Gowing, *Atomic energy*, pp. 108, 217, 229.
[94] Liebenau, 'British success'.
[95] Richardson, 'History of the catering industry', pp. 376-82.
[96] MacLeod and MacLeod, 'Committee of civil research', p. 699. In 1926 it was estimated that the Air Ministry itself employed 470 qualified scientists, followed by the Admiralty with 164 and the War Office with 100. MacLeod, 'Association of Scientific Workers', p. 131.
[97] Schlaifer and Heron, *Aircraft engines*, p. 153.
[98] Masefield, *Ride the storm*, p. 486.

Table 4. *R&D expenditure, 1945*

Rank R&D 1945		*R&D expenditure (£'000)*					*R&D employment 1945 qualified staff*
		1930	*1935*	*1938*	*1941*	*1945*	
1	ICI	420	625	1,075	1,190	3,500	1,396
2	Rolls-Royce					3,300	250
3	De Havilland					1,500	115
3	Napier				375	1,500	77
5	AV Roe					1,000	70
6	GEC	100			400		250
7	Lucas		70	102	197	450	171
7	STC					450	427
9	Philips					320	63
10	BTH					300	132
10	Courtaulds	40	40	60	60	300	132
10	Shell					300	151
13	Distillers					250	100
13	Marconi					250	144
13	Vickers Armstrong					250	58
16	Fairey					200	22
16	LMS Railway	13	23	28	30	200	60
16	Unilever	35	65	85	100	200	111
19	Metro-vick	95	200			160	137
20	Dunlop		50	65	70	158	77
21	Kodak	7	16	32	61	120	54
22	Gas Light & Coke					115	37
23	BDH			6	8	106	51
24	Turner & Newall			15	18	104	30
25	Anglo American Oil					100	39
25	RTT					100	21
25	Tele Cons & Main					100	22
25	Wellcome					100	141
FBI published total		1,736	2,696	5,442	4,896	21,815	
FBI total excl. electricals		1,736	2,696	4,095	4,896		
Estimated total incl. electricals		2,170	3,370	5,442	6,120		

Notes: Ranking is by expenditure for 1945. The extent to which the FBI totals for 1930-8 were underestimates is suggested by the number of the largest spenders, especially defence suppliers which were not included in these earlier figures. The totals for 1941 were not published, but can easily be calculated. Note, however, that these totals do include a number of collective totals for industries which cannot be broken down to the firm level: brewing, 10 companies, £50,000 for 1930-8 and £18,000 for 1941; gas, 12 companies, £210,000 for 1930, £295,000 for 1935, £390,000 for 1938, and £290,000 for 1941; plastics, £13,000 for 1930, £24,000 for 1935, £70,000 for 1938, and £73,000 for 1941; and electricals, 42 companies, £1,347,000 for 1938 only. The absence of electricals, fully 25% of the 1938 total, from the totals for other years enables us to produce a revised estimate for these earlier years, which, although closer to the actual level of expenditure, remains a significant underestimate because of the many companies still missing. Like table 1, this list should not be considered definitive; the figures may differ from those quoted elsewhere, and they are not necessarily comparable between companies, or between different years for any individual company.

Sources: For 1930-41 as table 1. All figures for 1945 from the FBI archives (MSS 200/F/3/T2/7/4) except GEC: Clayton and Algar, *GEC*, p. 413.

1930s. Marconi opened a new laboratory to house their expanding R&D efforts in 1939, a facility which was subsequently taken over by the Air Ministry and the Admiralty.[99] Lucas was also involved in the supply of technical equipment for aircraft. During the war warlike R&D was performed not just in private firms but in very substantially enlarged, or new, government research establishments. Indeed, while before the Great War the armourers tended to be ahead of government in R&D, and were still competitive in the interwar years, the Second World War saw many government establishments outpacing them in scientific research.[100] This led to conflict between government and industry, notably in the case of jet engines.[101]

IV

Having outlined the general development of British industrial R&D, let us turn to some specific aspects in more detail. British developments cannot be treated in isolation from world R&D. A very important feature of interwar R&D was that many of the largest British R&D performers were foreign-owned and/or were members of international and national cartels, patent pools, and scientific and technical agreements. Foreign-owned firms established research laboratories in Britain: as well as ITT, we may note Kodak in 1929,[102] Mond Nickel (a subsidiary of the American firm International Nickel), Boots, and BUSM. By the 1930s both Metropolitan-Vickers and BTH were under the control of the American GE, and English Electric was under that of Westinghouse.[103] English Electric created a research laboratory in the 1930s.[104] Philips, and Ericsson, Dutch and Swedish respectively, were significant spenders in 1945.[105] British firms also had laboratories abroad: in continental Europe in the case of Lever Bros/Unilever, and even as far away as Argentina in Glaxo's case (the Argentinian laboratory is credited with the development of Colloidal Calcium with Ostelin in the late 1920s).[106]

Just as important were international technical links between firms. The highly cartelized electric lamp industry, and the electric industry more generally, worked with patent pools and information-sharing agreements. English Electric, for example, had a patent, research, and market agreement with Westinghouse.[107] There is clear evidence that the electrical companies did a great deal of R&D for the purpose of patenting, since patents could be used to maintain monopoly. Patents, and patent pools, were used to

[99] Baker, *Marconi*, pp. 268-73, 308.
[100] Scott, *Vickers*, p. 302.
[101] Edgerton, 'Technical innovation'.
[102] *Idem*, 'Photographic industry', p. 121.
[103] See Jones, 'Foreign multinationals'.
[104] Jones and Marriott, *Anatomy of a merger*, p. 182.
[105] Figures for Ericsson for the 1930s (£12,000 in 1930 rising to £16,000 in 1941) suggest a significant increase in spending after 1941, reaching £33,000 in 1945 (MRC, FBI Archives, MSS 200/F/3/T1/127c and MSS 200/F/T2/7/4).
[106] Davenport-Hines, 'Glaxo', p. 147.
[107] Jones and Marriott, *Anatomy of a merger*, pp. 139-40.

sustain cartels by discriminatory granting of licences, and by threatening rogue firms with collective legal actions for infringement, sometimes of patents of dubious validity.[108] The international agreements between companies later drew criticism from those who felt that cartels crushed new enterprise, and on nationalistic grounds.[109] ICI joined the International Hydrogenation Patents cartel in 1931 (with Standard-IG and Royal Dutch Shell).[110] Much more important was the Du Pont/ICI Patents and Processes Agreement. This lasted from 1929 to 1948, and led to very extensive sharing of scientific and technical information between the two companies, with a few, very important exceptions. It prevented the duplication of research, and steered it away from dead ends encountered by one partner.[111] Another very important aspect of ICI's research was that it had demarcation agreements with major British companies such as Unilever and Distillers. ICI's research controller believed that these held back the company's research.[112] In the early 1930s Marconi formed a pool of scientific data, ideas, and inventions with BTH, EMI, and Philips; this stimulated Marconi's own R&D effort.[113] Collaboration between suppliers and customers was a very common occurrence, and an informal exchange of information between friendly firms in particular sectors often took place. Some companies stressed the availability of their research departments to their customers. Metal Box saw this as an important way of persuading food manufacturers to use cans; their laboratory was 'at the service' of all clients.[114]

No research department or laboratory within any firm worked in isolation from the wider scientific and technological world. Firms both contributed to and drew from this network, but most kept certain types of knowledge to themselves. Within each firm a high level of expertise relating to specific areas of operations existed, often very much specific to the individual company.[115] Nevertheless, the idea of a 'technological community', analogous to the communities of independent academic scientists, is useful.[116] Communities of practitioners developed in many areas of industrial science. As firms extended their research activities this informal exchange of information began to take place on a more systematic level, with the growth of specialized journals in some fields of industrial science and the formation of groups which met for the formal presentation of papers. Research workers were permitted, often actively encouraged, to publish papers concerning those aspects of their work not confidential to the firm, and they were given time off to attend meetings of relevant professional bodies. Most industrial scientists were members of one or more of the many general chemical and

108 See Monopolies and Restrictive Practices Commission, *Report on the supply of electric lamps*, esp. paras. 69 and 75.

109 Borkin and Welsh, *Germany's master plan*; Reimann, *Patents*.

110 Reader, *ICI*, II, chs. 9, 10, 11.

111 Hounshell and Smith, *Du Pont R&D*, ch. 10.

112 Reader, *ICI*, II, pp. 93-4, 303, 324-5, 369.

113 Baker, *Marconi company*, p. 272.

114 *Food Manufacture*, 14 (1939), p. xviii: Metal Box advertisement.

115 Edgerton, 'Photographic industry', pp. 124-5; *idem*, 'Science and technology'; Reich, *American industrial research*; Horrocks, 'Consuming science'.

116 Constant, *Turbojet revolution*, pp. 8-10.

engineering societies, and maintained a strong presence on their governing bodies, as well as belonging to more specialist organizations.[117] These institutions provided an effective means of communication not only between industrial scientists, but also with academics, as well as scientists and engineers employed by laboratories such as the DSIR research stations, the research associations, the National Physical Laboratory, and the Royal Aircraft Establishment.[118] Abstracting services, sometimes run by research associations, also had an important function. Staff in industrial laboratories also served as external examiners for university higher degree theses in subjects where much of the relevant expertise was to be found in industry, such as textile and electrical engineering.[119]

Into the interwar years independent 'inventors' and consultants enjoyed some importance. Inventors continued to bring their ideas to large firms, whose research departments then tested them to determine their usefulness for the company.[120] Consultants played an important role, especially for the smaller firms and in the provision of specialist services. Many consultants undertook mainly analytical work and process control, while others were more research oriented. In different areas of the country consulting chemists specialized in work which related to the local industry, with a heavy concentration of expertise in London.[121] The case of consultants Cross and Bevan in the field of cellulose chemistry indicates that they should not be ignored.[122] Neither can the role of universities and technical colleges. Sheffield University's Department of Metallurgy was, before the Great War 'one of the most—perhaps the most—advanced academic steel research facilities in the world', and by the decade after 1900 many of its graduates had helped to establish works laboratories in a number of Sheffield's firms.[123] University chemistry and engineering departments had a great many contacts with industrial R&D.[124]

Universities and technical colleges were linked to industrial research laboratories in an important way by providing trained personnel. In the late nineteenth century foreign, largely German and Swiss, chemists not uncommonly found work in Britain, notably in Lancashire. Thus the United Alkali Company, Levinstein, Brunner, Mond and Crosfields all employed foreign scientists, but by 1914 British staff were of overwhelming importance in these firms.[125] This is a very direct reflection of the increasing output of science graduates from British universities and technical colleges.[126] Most science graduates before the Second World War studied chemistry: it is

[117] Horrocks, 'Consuming science'.

[118] Heath and Hetherington, *Industrial research*; Sanderson, *Universities and British industry*.

[119] UMIST Archives, file F10, Faculty of Technology, reports on theses, 1925-36.

[120] The evaluation of such ideas was a stated function of the Cadbury Research Committee (Cadbury Archives, Research Committee minutes of 16 July 1920).

[121] A second FBI survey in 1945/6 included consultants, and its results reveal the extent to which such services remained essential: MRC, FBI Archives, MSS 200/F/T2/7/3 c contains their replies. See also the table in FBI, *Scientific and technical research*, p. 15.

[122] Allen, *Studies in innovation*, pp. 56-7.

[123] Tweedale, ' "Rule of thumb" ', p. 70.

[124] *Idem*, 'Geology'.

[125] Reader, *ICI*, I, pp. 93, 219; Fox, *Dye-makers*, pp. 14-5; Musson, *Enterprise*, pp. 79-83.

[126] Pollard, *Prime and decline*, ch. 3, section V, esp. pp. 193-4.

little wonder that in the interwar years the 'industrial chemist' was a well-known job description.[127] R&D, like scientific employment in industry, was dominated by chemists, even in industries outside the chemical industry proper, for example photography, food, and textiles. They were also found in significant numbers in iron and steel and even in electrical engineering. Engineers formed easily the second largest category of R&D employees; physicists, biologists, and mathematicians were a rarity.[128] By contrast with chemists, physicists, biologists, and mathematicians only about one-third of new entrants to the engineering profession were university graduates in the late 1930s, while the bulk of the remainder received their training in the rapidly expanding technical colleges.[129] By the late 1930s the total ouput of qualified scientists and technologists was some 5,000 per annum, about three-quarters of whom came from the universities.[130] We do not know what proportion were destined for industrial research. ICI employed around 1,000 chemists in 1938 and 615 qualified research staff.[131]A relevant comparison is that there were about 2,000 teaching staff in science and technology in British universities in the 1930s, fewer than the number of qualified staff engaged in industrial R&D in 1935.[132] Those firms which recruited chemists from universities often relied upon close contacts with individual members of staff to recommend suitable candidates to fill vacancies.[133] ICI's recruitment pattern varied across divisions. Alkali Group was dominated by Oxford-trained chemists, extending Brunner, Mond's connections from before the Great War.[134] Nobels drew on the Scottish universities, Dyestuffs and General Chemicals on grammar school boys who had attended university in London, Manchester, Leeds, or Liverpool.[135]

In the interwar years especially, there was clearly a demand from industry not just for science and engineering graduates, but for scientists with postgraduate research training. In 1927 ICI was looking to interest public school boys in scientific careers in the firm; they were expected not only to take a degree in science, but also to do one year of research.[136] The financial support of postgraduate science by the state was a wartime departure, and was one of the major activities of the DSIR. It is worth noting that there was no state support for postgraduate training in the arts, social sciences, medicine, or law. Many of the students who had received DSIR studentships went into industry, as DSIR surveys show (see table 5). Although we do

[127] An example is Harold Wilson's father, James Herbert Wilson, a dye chemist who started his career with Levinstein in 1907: Fox, *Dye-makers*, pp. 45-6. Carr-Saunders included chemists in his analysis of *Professions*, pp. 165-74, suggesting that there were 6,050 qualified members of this professional group in 1931, only 1,600 fewer than the figure given for civil servants, and about one-third that for accountants, p. 507.

[128] FBI, *Scientific and technical research*, tab. 2, p. 9. A similar distribution was found in the US, although there were probably more engineers to be found in US industry: Thackray et al., *Chemistry in America*, p. 347.

[129] Divall, 'Employers and engineering'.

[130] Poole and Andrews, *Government of science*, p. 258.

[131] Reader, *ICI*, II, pp. 77, 93.

[132] Bernal, *Social function*, app. I.

[133] Sanderson, 'Research and the firm', p. 130.

[134] Reader, *ICI*, I, p. 219.

[135] *Idem*, *ICI*, II, pp. 71-4.

[136] Ibid., p. 75.

not know what proportion of science and engineering postgraduate students were funded in this way, it was almost certainly a minority. University research in engineering was much less developed than in chemistry, but by the late 1930s pressure from both industry and academics helped put the issue on the agenda. Some sections of industry were demanding engineers with greater theoretical training.[137] One important characteristic of the personnel of industrial research laboratories was the fact that many obtained degrees and other qualifications through evening classes, some going on to take PhDs on the basis of work done in industry.[138] A number of companies employed women, especially during the Great War,[139] but never in large numbers and rarely on an equal basis with male researchers.[140]

Table 5. *Occupations after one year, and two years, of students whose DSIR postgraduate training allowance terminated in 1934/5*

Occupation	*1935/6* (%)	*1936/7* (%)
Scientific service under government	9	11
Scientific service in industry	24	41
Scholarships	34	6
University posts	10	21
Other	16	14

Source: DSIR, *Report for the year 1937/38*, app. IV, tab. 5, p. 177.

In the majority of cases even the senior R&D staff of firms were employees. In family firms active in research there are many instances where a member of the family was particularly responsible for encouraging its development. This was often as a result of personal contact with scientists, or because they were scientists or engineers themselves.[141] In most large firms senior research staff were paid far more than university teaching staff. In 1937 ICI had 37 technical staff paid more than £2,000, and 14 earning more than £3,000.[142] In the interwar years there were clearly recognized, powerful, and respected leaders of industrial research, most notably Clifford Paterson (Finsbury Technical College) of GEC; Arthur Fleming (Finsbury Technical College) of Metropolitan-Vickers; R. A. Freeth (Liverpool), R. E. Slade (Manchester), and W. Akers (Oxford) of ICI; Leslie Lampitt (Birmingham) of Lyons; Percy Dunsheath (Cambridge) of W. T. Henley, Isaac Schoenberg

[137] Divall, 'Employers and engineering', pp. 93-5.

[138] Edgerton, 'Photographic industry', p. 119, and Cadbury Archives, annual reports of the Chemists' Department, 1925/6 and 1926/7 record the award of M.Sc. degrees to A. W. Knapp, the chief chemist, and A. Churchman, senior biochemist.

[139] Cadbury Archives, Research Committee minutes, 1914-9 and annual reports of the Chemists' Department, 1914-20. Women scientists lost their jobs after the war. In some other companies women fared better: see Benson, 'Electrical engineering'.

[140] Sanderson, 'Research and the firm', p. 142. An account of the difficulties encountered by one woman brave enough to venture into this field is Bramley, 'Lathbury'.

[141] At Lyons personal contact was important (Richardson, 'History of the catering industry', ch. 12); at Chivers research was encouraged by William Chivers who had studied chemistry at Cambridge: *Chemistry and Industry*, 56 (1937), p. 759. His two brothers studied agriculture and engineering, again at Cambridge. Many similar examples could be cited. See also Edgerton, 'Science and technology', p. 93.

[142] Reader, *ICI*, II, p. 79.

(Kiev and Imperial College) of EMI, and Harold Hartley (Oxford) of the LMS Railway and the Gas, Coke and Light Co. (four of these men were engineers). Many of them would in due course receive knighthoods, and become board members of their companies,[143] and some were elected to the Royal Society. They were well respected by academics, and were consulted by the DSIR.[144] Fleming, one of the engineers, was probably the most prominent of them all, writing and lecturing extensively on the subject.[145] He served on many committees, and was consulted by the DSIR about plans for research after the Second World War.[146] There were undoubtedly major differences between academic research and industrial research careers: few scientists or engineers, once committed to industry, returned permanently to universities. There were exceptions: by 1949 Metropolitan-Vickers could boast that 10 university professors had passed through their research department.[147] T. P. Hilditch left Crosfields to take up the chair of industrial chemistry at Liverpool University, specializing in oils and fats. He maintained close links with his former industrial colleagues, and many of his pupils went on to work in industry.[148] Many other senior academics had formal connections with industrial research laboratories.

V

How did the British industrial R&D effort described above compare with that of other nations? For the years before 1914 the contrast has been drawn with the large and highly research-intensive German dyestuffs industry; but Pollard's recent comparative study of education, science, and technology in Britain and Germany before 1914 draws few firm conclusions about science and technology in industry.[149] In one sector—dyestuffs—there was a very clear German lead over the rest of the world, not just Britain.[150]

Historians have implied, or stated categorically, that British firms' support of research in the interwar years was, for example, 'well below those of comparable firms in the USA or Germany'.[151] More recently, Chandler has given the impression that German firms were ahead of British during that period.[152] None of these arguments result from systematic comparisons. Our firm-level evidence, when compared with that now becoming available for the US and the continent of Europe, forces us to reconsider (see table

[143] e.g. Lampitt at Lyons, Fairbrother at McDougalls, Fleming at Metro-vick, Renwick at Ilford, Akers at ICI, and Paterson at GEC.
[144] Lampitt was the only industrial scientist to serve on the Food Investigation Board, which he joined in 1934: *Chemistry and Industry*, 55 (1936), p. 353.
[145] Fleming, *Science and industry*. His views on industrial research are summarized in 'Impact of science'.
[146] P.R.O., DSIR 17/210, Committee on postwar developments, extracts from a letter from Fleming commenting on the draft report of the committee.
[147] Metropolitan-Vickers Electrical Co., *Fifty years*, p. 37.
[148] Morton, 'Hilditch', pp. 259-90.
[149] Pollard, *Prime and decline*, ch. 3.
[150] Homburg, 'Emergence of research laboratories'.
[151] Pollard, *British economy*, p. 57.
[152] Chandler, *Scale and scope*, chs. 13 and 14.

Table 6. *Comparison of some major R&D performers in the US and Europe, 1925-1940*

	R&D employment							
Firms	*1925*	*1926*	*1927*	*1928*	*1929*	*1930*	*1931*	*1932*
IG		958	1,000	1,050	1,100	1,100	1,050	1,000
Du Pont[a]						687	755	672
ICI								
GE	174	183	196	223	265	202	172	137
GEC						125		
Metro-vick						73		
BTH						73		
Philips	15	28	37	48	58	72	58	57
Siemens	78	78	82	92	88	68	51	47
AEG				30				
UK total						1,381		
US total							11,500	

	R&D employment							
Firms	*1933*	*1934*	*1935*	*1936*	*1937*	*1938*	*1939*	*1940*
IG	1,000	1,000	1,020	1,060	1,100	1,150		
Du Pont	725	785	847	912	1,051	979	1,034	1,261
ICI			464			510		
GE								
GEC			175					
Metro-vick			87			117		
BTH			104					
Philips								
Siemens								
AEG[a]	60							
UK total			2,566			4,382		
US total	27,567					44,292		70,033

Notes: [a] The figures refer to qualified staff except in the case of AEG ('staff') and Du Pont ('professional staff').

The companies are ranked by employment in 1930, for which the most complete figures are available. The table is by no means exhaustive, being based mainly on easily available published sources. Definitions of qualified staff differ between countries.

Sources: Du Pont: Hounshell and Smith, *Corporate strategy*, tabs. 15.1b and 15.2; ICI: FBI archives, MSS 200/F/3/T1/127 c; IG: Reader, *ICI*, II, p. 34; GE, Philips, and Siemens: Erker, 'Verwissenschaftlichung', p. 86; GEC: Clayton and Algar, *GEC*, p. 413; Metro-vick and BTH: Edgerton, 'Science and technology', p. 101, nn. 1, 2; AEG: Erker, 'Competition and cooperation', unpub. paper presented to the European Science Foundation workshop, Berlin, July 1991; UK totals from FBI, see tab. 4; US totals from Thackray et al., *Chemistry in America*, tab. 5.5, p. 347.

6). The comparison must be tentative: we do not know whether staff qualifications are entirely comparable across countries, or whether the implied definition of R&D is the same—for example, whether all research is included or only central laboratories—but it is worth making given the paucity of other evidence. As table 6 indicates, the American industrial R&D effort was, in absolute terms, some 10 times that of British industry in the late 1930s.[153] Individual American firms spent more, and employed

[153] Some of the difficulties of interpreting this difference are discussed in Edgerton, 'Science and technology'.

more researchers than did British firms. For example, Du Pont spent more than ICI, and GE more than GEC. There were exceptions to this general rule: despite starting three years after RCA, EMI concentrated more resources on television, enabling the BBC to start the world's first service three years ahead of the US.[154] Given that the American economy was much larger, it is not surprising that American industrial research was considerably less concentrated by firm than its British counterpart, or that there were more firms with the resources to invest heavily in R&D. Thus Du Pont, probably the largest spender in America, accounted for only some 5 per cent of expenditure; while Reich estimates that 13 American research laboratories accounted for over one-third of the total.[155] The basis of this calculation is not clear.

Comparison with Germany is more difficult. There are no aggregate figures for German industrial research in the interwar years. Chandler's work on German firms gives us some useful qualitative information, but no figures.[156] In table 6 we have brought together such information about German firms as is available to us. It appears that IG Farben spent more than ICI, though it should be noted that its R&D portfolio and production were probably more diversified; for example, it did notable work in pharmaceuticals, synthetic rubber, fibres, and photography which in Britain took place in companies other than ICI.[157] In the case of the electrical industry, however, the figures suggest that British electrical firms employed more researchers than their German equivalents. To our knowledge it has never been suggested that France, Italy or Japan did more R&D than Britain in this period, and such evidence as exists confirms this impression.[158] A safe conclusion, therefore, is that the overall levels of industrial R&D in Germany and Britain were comparable, at least before 1933, but much lower than in the US. This should come as little surprise, as it would correlate with levels of industrial productivity.[159]

VI

This article confirms the picture of significant and expanding R&D in many sectors of British industry put forward by Sanderson in 1972. We reject the picture implied in the more recent work of Mowery and Chandler, and restate the fact that the vast bulk of R&D in British industry was done by large companies in-house. The most significant element of government funding of R&D in industry came, in any case, not from the DSIR through the research associations, but from the service ministries. Historians have

[154] Saul, 'Research', p. 122.
[155] Reich, *American industrial research*, p. 256.
[156] Chandler, *Scale and scope*, chs. 13 and 14.
[157] Morris, 'Acetylene chemistry'; Plumpe, *I.G. Farbenindustrie*.
[158] *Culture technique*, 18 (1988); Paul, *Knowledge to power*; Russo, 'Italian science', pp. 285-6; Fukasaku, 'Japanese industrial research'.
[159] Broadberry and Crafts, 'Productivity differences'; Broadberry and Fremdling, 'Comparative productivity'. Note, however, that in discussing the role of R&D in productivity differences between Britain and the US, British spending is significantly underestimated, and the argument about concentration may not be convincing: Broadberry and Crafts, 'Productivity differences', p. 386.

too readily assumed that because they found no published evidence of major British companies doing research, this was not being undertaken. Our data suggest, and we recognize that they are incomplete, that it is safer to assume that large British companies were doing R&D by the 1930s, rather than not doing it. Historians should also accept that complaints about levels of research in British industry are not primarily evidence of its absence but of a belief that industry should do even more R&D. Although this has been a study of British industrial R&D, our evidence of interconnections between industrial R&D in different countries indicates that we should avoid purely nationalistic analyses. Nevertheless it is worth stressing that we find no evidence that German firms, other than IG, either individually or collectively spent more on R&D than British firms in the interwar years. Of course we await further detailed studies of R&D in German firms.

There is a tendency to regard R&D, patents, and innovation as interchangeable concepts and measures, and to assume a strong, positive correlation of each one with economic growth in all countries and at all times. Our study has been limited to R&D, which must be seen as just one input into invention and innovation, themselves only elements in economic growth. The relationships between these factors need to be explored historically, not merely assumed. It is regrettable that some historians have taken Britain's poor industrial record as itself indicative of low and/or inadequate R&D. This sort of argument is not confined to R&D: the alleged absence of other 'modern'—typically American—features is used in the same way. The result is a mutually reinforcing picture of British 'failure' in corporate structure, production technology, management, and R&D, and indeed a tendency to explain the absence of one feature by the absence of another. In this article we have not discussed these interconnected issues. But it is clear that the growth of British industrial R&D, not its absence, needs to be explained and the explanation may be that British firms were more like American firms, or German firms, than historians have allowed.[160] Moreover, if R&D can be considered an index of modernity, British industry in the interwar period showed at least some features of 'managerial' and indeed 'collective' capitalism, usually ascribed to the firms of more successful nations.[161] More generally our conclusions point once more to the fact that what might be called the technocratic critique of British institutions does not give us an adequate picture of the scientific and technological efforts they supported.[162]

Imperial College, London
University of Lancaster

[160] Hannah, 'European visible hand'.
[161] Lazonick, *Business organization*. It is also important to note that the whole picture of modernity as Chandlerian corporation is open to debate. See Sabel and Zeitlin, 'Alternatives'; Edgerton, 'Liberal militarism' for a brief discussion of shipbuilding and aircraft production.
[162] Edgerton, 'Peculiarities'.

Footnote references

Allen, J. A., *Studies in innovation in the steel and chemical industries* (Manchester, 1967).
Alter, P., *The reluctant patron: science and the state in Britain* (Oxford, 1987).
Andrade, E. N. da C., *Industrial research 1946* (1946).
Association of Scientific Workers, *Industrial research laboratories: a list* (1936).
Baker, W. J., *A history of the Marconi company* (1970).
Barnett, C., *The audit of war: the illusion and reality of Britain as a great nation* (1986).
Beer, J. J., *The emergence of the German dye industry* (Urbana, 1959).
Benson, M., 'The chemist in electrical engineering', *Woman Engineer*, 2 (1925), pp. 21-3.
Bernal, J. D., *The social function of science* (1939).
Borkin, J. and Welsh, C. A., *Germany's master plan: the story of an industrial offensive* (n.d., but 1939-45).
Bramley, R., 'Kathleen Culhane Lathbury', *Chemistry in Britain*, 27 (May 1991), pp. 428-31.
Briggs, A., *The history of broadcasting in the United Kingdom*, 4 vols. (1961-79).
British Science Guild, *Report of the committee on the position of the expert in the public services and industry* (1931).
Broadberry, S. N. and Crafts, N. F. R., 'Explaining Anglo-American productivity differences in the mid-twentieth century', *Oxf. Bull. Econ. & Stat.*, 52 (1990), pp. 375-402.
Broadberry, S. N. and Fremdling, R., 'Comparative productivity in British and German industry, 1907-37', *Oxf. Bull. Econ. & Stat.*, 52 (1990), pp. 403-21.
Bud, R. and Roberts, G. K., *Science versus practice* (Manchester, 1984).
Burns, R. W., *British television: the formative years* (1986).
Cardwell, D. S. L., *The organisation of science in England* (2nd edn. 1972).
Carr-Saunders, A. M. and Wilson, P. A., *The professions* (Oxford, 1933).
Chandler, A. D., Jr., *Scale and scope: the dynamics of industrial capitalism* (1990).
Clayton, R. and Algar, J., *The GEC research laboratories, 1919-84* (1989).
Constant, E. W., *The origins of the turbojet revolution* (Baltimore, 1980).
Culture Technique, 18 (1988). Special issue on 'Recherche, innovation, industrie'.
Davenport-Hines, R. P. T., 'Glaxo as a multinational before 1963', in G. Jones, ed., *British multinationals* (Aldershot, 1986), pp. 137-63.
Divall, C., 'A measure of agreement: employers and engineering studies in the universities of England and Wales, 1897-1939', *Soc. Stud. Sci.*, 20 (1990), pp. 65-112.
Donnelly, J., 'Chemical education and the chemical industry in late nineteenth- and early twentieth-century England' (unpub. Ph.D. thesis, Univ. of Leeds, 1988).
Donnelly, J., 'Industrial recruitment of chemists from English universities: a revaluation of its early importance', *Brit. J. Hist. Sci.*, 24 (1991), pp. 3-20.
Dreaper, W. P., *The research chemist in the works, with special reference to the textile industry* (1914).
Dupree, M., ed., *Lancashire and Whitehall: the diaries of Sir Raymond Streat*, 2 vols. (Manchester, 1987).
Edgerton, D. E. H., 'Technical innovation, industrial capacity and efficiency: public ownership and the British military aircraft industry, 1935-1948', *Bus. Hist.* 26 (1984), pp. 247-79.
Edgerton, D. E. H., 'Science and technology in British business history', *Bus. Hist.*, 29 (1987), pp. 84-103.
Edgerton, D. E. H., 'Industrial research in the British photographic industry, 1879-1939', in J. Liebenau, ed., *The challenge of new technology* (Aldershot, 1988), pp. 106-34.
Edgerton, D. E. H., 'Liberal militarism and the British state', *New Left Rev.*, 185 (1991), pp. 138-69.
Edgerton, D. E. H., 'The prophet militant and industrial: the peculiarities of Correlli Barnett', *Twentieth Cent. Brit. Hist.*, 2 (1991), pp. 360-79.
Elbaum, B. and Lazonick, W., eds., *The decline of the British economy* (Oxford, 1986).
Erker, P., 'Die Verwissenschaftlichung der Industrie', *Zeitschrift für Unternehmensgeschichte*, 35 (1990), pp. 73-94.
Federation of British Industries, Industrial Research Committee, *Research and industry: report of the FBI Industrial Research Committee* (1943).
Federation of British Industries, Industrial Research Committee, *Scientific and technical research in British industry: a statistical survey by the FBI Industrial Research Secretariat* (1947).
Federation of British Industries, Industrial Research Committee, *Research and development in British industry: a survey of expenditure on industrial research and development in the year 1950-51* (1952).
Fleming, A. P. M., 'The impact of science in industry', *Manchester Municipal College of Technology department of industrial administration monographs on higher management*, 2 (1946).
Fleming, A. P. M. and Pearce, F. G., *Research in industry* (1922).
Fox, M. R., *Dye-makers of Great Britain: a history of chemists, companies, products, and changes* (Manchester, 1987).
Fukasaku, Y., 'Origins of Japanese industrial research: prewar government policy and in-house research at Mitsubishi Nagasaki Shipyard', *Res. Pol.*, 21 (1992), pp. 197-213.
Gowing, M., *Britain and atomic energy, 1939-1945* (1964).

Haber, L. F., *The chemical industry, 1900-1930: international growth and technological change* (Oxford, 1971).

Haber, L. F., 'Government intervention at the frontiers of science: British dyestuffs and synthetic organic chemistry, 1914-1939', *Minerva*, XI (1973), pp. 79-94.

Hannah, L., *The rise of the corporate economy* (2nd edn., 1983).

Hannah, L., 'Scale and scope: towards a European visible hand?', *Bus. Hist.*, 32 (1991), pp. 297-309.

Hardie, D. W. F., *A history of the chemical industry in Widnes* (1950).

Heath, F. H. and Hetherington, A. L., *Industrial research and development in the United Kingdom: a survey* (1946).

Homburg, E., 'The emergence of research laboratories in the dyestuffs industry, 1870-1900', *Brit. J. Hist. Sci.*, 25 (1992), pp. 91-112.

Horrocks, S. M., 'Consuming science: science, technology and food in Britain, 1870-1939' (unpub. Ph.D. thesis, Univ. of Manchester, 1993).

Hounshell, D. and Smith, J. K., *Science and corporate strategy: Du Pont R&D, 1902-1980* (Cambridge, 1988).

Hutchinson, E., 'A fruitful co-operation between government and academic science: food research in the United Kingdom', *Minerva*, X (1972), pp. 19-50.

Huxley, J., *Scientific research and social needs* (1934).

Jones, G., 'Foreign multinationals and British industry before 1945', *Econ. Hist. Rev.*, 2nd ser., XLI (1988), pp. 429-53.

Jones, R. and Marriott, O., *Anatomy of a merger: a history of GEC, AEI, and English Electric* (1970).

Kuznick, P. J., *Beyond the laboratory: scientists as political activists in 1930s America* (Chicago, 1987).

Lazonick, W., *Business organization and the myth of the market economy* (Cambridge, 1991).

Liebenau, J., 'The British success with penicillin', *Soc. Stud. Sci.*, 17 (1987), pp. 69-86.

Liebenau, J., 'Corporate structure and research and development', in *idem*, ed., *The challenge of new technology: innovation in British business since 1850* (Aldershot, 1988), pp. 30-42.

Lilley, S., *Men, machines and history* (1948).

London Chamber of Commerce, *Report on scientific industrial research* (1944).

MacLeod, K., 'Politics, professionalisation and the organisation of scientists: the Association of Scientific Workers, 1917-42' (unpub. Ph. D. thesis, Univ. of Sussex, 1975).

MacLeod, R., 'The "Arsenal" in the Strand: Australian chemists and the British munitions effort, 1916-1919', *Annals Sci.*, 46 (1989), pp. 45-67.

MacLeod, R. and MacLeod, K., 'The committee of civil research: advice for economic development, 1925-1930', *Minerva*, VII (1969), pp. 680-705.

MacLeod, R. and MacLeod, K., 'The origins of the DSIR: reflections on ideas and men, 1915-16', *Public Administration*, 48 (1970), pp. 23-48.

MacLeod, R. and MacLeod, K., 'War and economic development: government and the optical glass industry in Britain, 1914-18', in J. M. Winter, ed., *War and economic development: essays in memory of David Joslin* (Cambridge, 1975), pp. 165-203.

MacLeod, R. and MacLeod, K., 'The contradictions of professionalism: scientists, trade unionism and the First World War', *Soc. Stud. Sci.*, 9 (1979), pp. 1-32.

Manchester Joint Research Council, *Industry and science* (Manchester, 1954).

Masefield, P., *To ride the storm: the story of airship R101* (1982).

Mees, C. E. K., *Organisation of scientific industrial research* (1920).

Metropolitan-Vickers Electrical Co. Ltd., *1899-1949: fifty years in brief* (Manchester, 1949).

Meyer-Thurow, G., 'The industrialisation of invention: a case study from the German chemical industry', *Isis*, 73 (1982), pp. 363-81.

Miles, F. D., *The history of research in the Nobel division of ICI* (1955).

Morris, P. J. T., 'The development of acetylene chemistry and synthetic rubber by IG Farbenindustrie AG, 1926-45' (unpub. D. Phil. thesis, Univ. of Oxford, 1982).

Morton, R. A., 'Thomas Percy Hilditch, 1886-1965', *Biographical Memoirs of Fellows of the Royal Society*, 12 (1966), pp. 259-89.

Mowery, D., 'Industrial research in Britain, 1900-1950', in B. Elbaum and W. Lazonick, eds., *The decline of the British economy* (Oxford, 1986), pp. 189-222.

Mowery, D. C. and Rosenberg, N., *Technology and the pursuit of economic growth* (Cambridge, 1989).

Musson, A. E., *Enterprise in soap and chemicals: Joseph Crosfield and Sons Ltd., 1815-1965* (Manchester, 1965).

Niblett, C. A., 'Images of progress' (unpub. Ph. D. thesis, Univ. of Manchester, 1980).

Nuffield College, *Problems of scientific and industrial research* (Oxford, 1944).

Paterson, C., *A scientist's war: the war diary of Sir Clifford Paterson, 1939-45*, ed. R. Clayton and J. Algar (1991).

Paul, H. W., *From knowledge to power: the rise of the science empire in France, 1860-1939* (Cambridge, 1985).

Plumpe, G., *Die I.G. Farbenindustrie AG: Wirtschaft, Technik und Politik, 1904-1945* (Berlin, 1990).

Pocock, R. F., *The early British radio industry* (Manchester, 1988).
Pollard, S., *Britain's prime and Britain's decline: the British economy, 1870-1914* (1989).
Poole, J. B. and Andrews, K., eds., *The government of science in Britain* (1972).
Reader, W. J., *Imperial Chemical Industries: a history*, 2 vols. (1970-5).
Reich, L., *The making of American industrial research: science and industry at GE and Bell, 1876-1926* (Cambridge, 1985).
Reimann, G., *Patents for Hitler* (1945).
Richardson, D. J., 'The history of the catering industry with special reference to J. Lyons & Co. to 1939' (unpub. Ph. D. thesis, Univ. of Kent, 1970).
Russell, C. A., *Chemists by profession* (Milton Keynes, 1977).
Russo, A., 'Science and industry in Italy between the wars', *Hist. Stud. Physical & Biological Sci.*, 16 (1986), pp. 281-320.
Sabel, C. and Zeitlin, J., 'Historical alternatives to mass production: politics, markets and technology in nineteenth-century industrialisation', *P. & P.*, 108 (1985), pp. 133-76.
Sanderson, M., *The universities and British industry, 1850-1970* (1972).
Sanderson, M., 'Research and the firm in British industry, 1919-39', *Sci. Stud.*, 2 (1972), pp. 107-51.
Saul, S. B., 'Research and development in British industry from the end of the nineteenth century to the 1960s', in T. C. Smout, ed., *The search for wealth and stability* (1979), pp. 114-38.
Sayers, R. S., 'The springs of technical progress in Britain, 1919-39', *Econ. J.*, 60 (1950), pp. 275-91.
Schlaifer, R. and Heron, R. D., *The development of aircraft engines and aviation fuels* (Boston, Mass., 1950).
Scott, J. D., *Vickers: a history* (1962).
Smith, F. E., *Industrial research and the nation's balance sheet* (1934).
Stranges, A. N., 'From Birmingham to Billingham: high-pressure coal hydrogenation in Great Britain', *Tech. & Cult.*, 26 (1985), pp. 726-57.
Thackray, A., Sturchio, J. L., Carroll, P. T., and Bud, R., *Chemistry in America, 1876-1976* (Dordrecht, 1985).
Trebilcock, R. C., *The Vickers brothers: armaments and enterprise, 1854-1914* (1977).
Trebilcock, R. C., ' "Spin-off" in British economic history: armaments and industry, 1760-1914', *Econ. Hist. Rev.*, 2nd ser., XXII (1969), pp. 474-90.
Turner, F., 'Public science in Britain, 1880-1919', *Isis*, 71 (1980), pp. 589-608.
Tweedale, G., *Sheffield steel and America* (Cambridge, 1987).
Tweedale, G., 'Science, innovation and the "rule of thumb": the development of British metallurgy to 1945', in J. Liebenau, ed., *The challenge of new technology: innovation in British business since 1850* (Aldershot, 1988), pp. 58-82.
Tweedale, G., 'Geology and industrial consultancy: Sir William Boyd Dawkins (1837-1929) and the Kent coalfield', *Brit. J. Hist. Sci.*, 24 (1991), pp. 435-52.
Varcoe, I., 'Co-operative research associations in British industry, 1918-34', *Minerva*, XIX (1981), pp. 433-63.
Varcoe, I., 'Scientists and organised research in Great Britain, 1914-1916: the early history of the DSIR', *Minerva*, IX (1970), pp. 192-217.
Vernon, K., 'Pus, sewage, beer and milk: microbiology in Britain, 1870-1940', *Hist. Sci.*, 27 (1990), pp. 289-325.
Watson, J. A., *A hundred years of sugar refining: the story of Love Lane Refinery, 1872-1972* (Liverpool, 1972).
Williams, M., 'Technical innovation: examples from the scientific instrument industry', in J. Liebenau, ed., *The challenge of new technology: innovation in British business since 1850* (Aldershot, 1988), pp. 8-29.
Young, P., *Power of speech: a history of Standard Telephone and Cables* (1983).

Official publications
Committee on Trade and Industry, *Factors affecting commercial efficiency* (1927).
Committee on Trade and Industry, *Final report* (1929).
DSIR, *Annual reports* (1925/6, 1929/30, 1934/5, 1937/8).
Fleming, A. P. M., *Science and industry: industrial research in the United States of America* (1917).
Monopolies and Restrictive Practices Commission, *Report on the supply of electric lamps* (1951).

Part VI
The Case of Japanese Industrial Research

[22]

Origins of Japanese industrial research: Prewar government policy and in-house research at Mitsubishi Nagasaki Shipyard

Yukiko Fukasaku

2 Avenue Romand, F-92210 Saint-Cloud, France

Final version received May 1991

The importance of accumulating technological capabilities has been stressed in the literature on technological change and international transfer of technology over the last decade. Although it has been pointed out that in Japan since World War II, technological capabilities were accumulated through in-house R&D which in turn played a significant role in facilitating assimilation of imported technology, little has been said about the origins of industrial research in Japan. The examination of prewar government policy toward research and in-house research at Mitsubishi Nagasaki Shipyard shows that in Japan, there existed a well-defined policy of placing priority on industry-related research since the beginning of the century. By this time large engineering firms like the Nagasaki Shipyard had set up research sections, and as in the postwar period, research activities were geared toward adapting and improving imported technology. This suggests that (1) research is important in accumulating suitable technological capabilities to facilitate the assimilation and improvement of foreign technology, (2) technological learning is a process that can be sustained and elaborated for a long time, the cumulation of which can lead to a degree of technological sophistication that is highly competitive in the international economy.

1. Introduction

Much of the literature on international transfer of technology in recent years has stressed the importance of accumulating capabilities in successfully assimilating imported technology [3,27]. This is because the effective utilization of foreign technology often requires its modification through minor technological changes; hence, suitable technological capabilities are needed to generate them. This in turn points to the importance of research and training within industrial enterprises. There has been a number of case studies on transfer of technology in which in-house research and training played key roles in bringing about technological changes to adapt, modify or improve imported technology [7]. This is in line with the recently emerging view that minor incremental technological changes are often as economically significant as major innovations, and rather than being public information, technology is cumulative and often specific to the firm.

The studies on technological innovation and technology transfer in Japan in the postwar phase [22–26] have revealed that there is a high correlation between the growth of R&D expenditures, technology imports and economic growth, indicating that technology was indeed an important component in economic growth, and that in this process, R&D and technology imports were complementary activities. More importantly, a major part of the R&D effort, especially in the 1950s and 1960s, was directed at "adaptive" research [26], or more appropriately, "improvement engineering" [28] which aimed at upgrading product quality and cost reduction, in short, making industrial products produced by originally imported technology more competitive. Blumenthal [1] has used the term "absorptive R&D" to refer to the dominant type of research activities in postwar Japan which facilitated the absorption of technological know-how from other countries, its adaptation to the country's specific needs, and the introduction of minor innovations.

Because these works focus on postwar developments in Japan, they give the impression that the process of building technological capabilities in Japan has taken place over a relatively short period of time. The examination of prewar government research policy and in-house research at a major engineering enterprise, Mitsubishi Nagasaki Shipyard, indicates rather that technological learning in Japan was a long-range and arduous process which started well before the war. The postwar technological development and the remarkable economic growth, in no small degree, owed to this long-range technological learning process. Despite the material devastation of World War II, knowledge and experiences accumulated before the war survived, and there were important continuities, especially with respect to human resources and institutions. This implies the value of investing in technological learning. Inquiry into prewar technological development in Japan is significant also for the following reasons. First, because of similarities in the level of economic growth, prewar Japanese experiences are more relevant than those of the postwar period for the contemporary developing countries. Secondly, as Ohkawa and Rosovsky [20] have pointed out, Japanese economy shows an accelerated growth or what they term "trend acceleration", characterized by the rapid increase in productivity and investment in the modern sector, starting in the first decade of this century. Significant growth, thus, had already started in the prewar period, and this warrants investigation with respect to the role of technology.

2. Government policy toward industrial research in prewar Japan

The Meiji government at first was an ardent advocate of rapid industrialization and directly undertook modern industrial enterprises in the 1870s. It constructed lighthouses, telegraph lines and railroads; modernized mines, shipyards, iron and steel mills; and set up model factories for textiles, machines, glass, cement and bricks. Some of these were strategic enterprises of the Tokugawa era which has been set up and operated since the 1850s with knowledge gathered through *rangaku* (Dutch learning), the earliest agent of technology transfer from Europe. However, because of the lack of technical and managerial competence, most of the enterprises proved to be failures by the early 1880s, and the government turned to a more realistic policy of promoting traditional industries, public works, agriculture, forestry and fishery. Its policy toward industrial research was along these lines, and the government founded experimental stations and research

Table 1
Government research institutes founded before 1900

	Name	Notes
1871	Hokkaido Development Research Institute	
1872	*Naimusho*	
	Naito Shinjuku Laboratory	Experimental station in agriculture and sericulture
1874	Tokyo Hygienic Laboratory	
1875	Tokyo Meteorological Observatory	
1875	*Kobusho*	
	Telegraph Department Electric & Glass Laboratory	Tests on telegraph and telephone equipment and glass. Renamed Electric Research Institute in 1891 and started research on industrial use of electricity in 1897.
1878	Tokyo Astronomical Observatory	
1882	Geological Survey Institute	
1888	Geographical Survey Institute	
1892	Research Unit on Prevention of Earthquake Damage	Became Tokyo University Seismological Institute in 1925
1893	Fishery Research Institute	
1895	Yokohama Raw Silk Laboratory	
1899	Epidemics Research Institute	Became attached to Tokyo University in 1916
1899	Mizusawa Latitude Observatory	

Source: [6,33].

institutes in these areas as seen in table 1. This list is indicative of the government's lack of interest in promoting research on modern industries at this stage.

The policy line shifted in the first decade of this century, as the government started to take active interest in promoting modern industries. This is exemplified in the founding in 1900 of *Kogyo Shikensho* (Industrial Research Institute). The founding of this institute resulted from the recommendation to found a research institute to promote modern industries made by the *Noshomusho* (ministry of agriculture and commerce) High Conference in 1898. Its mission was industrial policy making for post Sino–Japanese War Japan, when the country faced the first real urge for modern industrialization. *Kogyo Shikensho* provided testing and research services for small chemical manufacturing enterprises and undertook research on lacquer and paper manufacturing; in short, its main interest in the beginning was the improvement of traditional chemical industries by adopting modern methods. Soon, however, research on modern chemical industry was added. In 1903, research on ceramics and glass was started; in 1906, synthetic dyes; and in 1909, electro-chemicals [32].

The real spur for the government to promote modern industrial research came with World War I as reflected in the number and kind of government institutes founded after 1914 (table 2). Although research institutes for traditional industries were still founded, the number of those related to modern industries increased signifi-

Table 2
Government research institutes founded during 1900–1935

	Name	Notes
1900	Industrial Research Institute	
1903	National Bureau of Weights and Measures	
1910	Railroad Research Institute	Attached to Ministry of Railroads
1914	Sericulture Research Institute	
1916	Livestock Research Institute	
1916	Yawata Steel Mill Research Department	Expanded in 1919 to become a research institute
1916	Inspection Office for Ship Machinery Instruments	
1916	Aeronautics Research Institute	Attached to Tokyo Imperial University
1917	Research Institute for Prevention of Coal Mine Explosions	
1917	Institute of Physical and Chemical Research (*Riken*)	
1918	Silk Industry Research Institute	
1918	Temporary Nitrogen Research Institute	Attached to *Noshomusho*
1918	Osaka Industrial Research Institute	
1919	Tea Industry Research Institute	
1919	China and Porcelain Research Institute	
1920	Fuel Research Institute	Attached to *Noshomusho*
1920	Ocean Meteorological Observatory	
1920	High Altitude Meteorological Observatory	
1920	Nutrition Research Institute	
1920	Monopoly Bureau Research Institute	
1920	Animal Disease Research Institute	
1922	Forestry Research Institute	
1922	Metal Research Institute	Attached to Tohoku Imperial University, originally founded in 1919 as Steel Research Institute
1924	Printing Office Research Institute	
1925	Seismological Research Institute	Attached to Tokyo Imperial University
1926	Chemical Research Institute	Attached to Kyoto Imperial University
1928	Industrial Arts Education Institute	
1931	Thermal Spring Therapy Research Institute	Attached to Kyushu Imperial University
1932	Research Institute for Japanese Spiritual Culture	
1934	Building Materials Research Institute	Attached to Tokyo Polytechnique
1934	Microbial Disease Research Institute	Attached to Osaka Imperial University
1935	Telecommunications Research Institute	Attached to Tohoku Imperial University

Does not include research institutes attached to colonial governments. *Source:* [6,33].

cantly. Due to the stoppage of imports from Germany and other European countries, especially in chemicals and iron and steel, measures had to be taken to stimulate and facilitate the domestic production of these materials. By promoting industrial research, domestic production capability was expected to improve. In accordance to this policy, the Temporary Nitrogen Research Institute was founded to do research on the industrial production of ammonia using Haber's nitrogen fixation method. [1] The government added a research department to the Yawata Steel Mill, and founded a Steel Research Institute at Tohoku Imperial University in 1919. The latter expanded its scope to include research on metals other than steel to become the Metal Research Institute in 1922. This was made possible by donations from Sumitomo and Mitsubishi firms and the Kobe Steel Company. This illustrates one significant outcome of the expansion of research activities at this time, the creation of a strong government–university–industry link in research. National universities were increasingly given the rôle of doing research of interest to the government as reflected in the number of research institutes founded as part of national universities after 1916. Private industries generously gave financial support to these institutes when their interests were also served.

Another body which exemplified the close government–university–industry link was the *Rikagaku Kenkyusho* (Institute of Physical and Chemical Research, hereafter referred to by its short name, *Riken*). The founding of a "national scientific research institute to enhance national wealth by industrial development" [6, p. 92] was suggested to the government in 1913. The idea was taken up eagerly after the start of World War I. In 1915, a planning committee for the founding of *Riken* consisting of government officials, scholars and industrialists was set up. Private industries and the government as well as the imperial household contributed to the founding fund. Founded in 1917, *Riken* was to "...integrate pure and applied research for the development of industries; also, by training researchers and encouraging inventions, enhance intellectual productivity to consolidate the basis of wealth and strength of our country..." [33, p. 236]. National university professors were appointed as researchers at *Riken*. The institute was indeed a "national enterprise" which responded to the policy of strengthening government–university–industry links for industrial development. Fundamental research was done at *Riken*, but greater emphasis was placed on industrial research which could be commercialized. By 1937, more than 60 percent of *Riken*'s incomes came from subcontracting firms which commercialized its innovations.

Links with the army and navy were also strengthened. The Temporary Balloon Research Committee, set up jointly by the army and IJN (Imperial Japanese Navy) in 1909, became the Aeronautic Research Institute of the Tokyo Imperial University in 1916. Officers and engineers of the army and navy could be appointed as researchers at this institute, and military research could be done. The industry–military link also became close. The Nippon Kogaku Kogyo was set up by Mitsubishi upon the request of the navy to develop and manufacture optical weapons and instruments. The Nagasaki Arms Manufacturing Company branched out of the Nagasaki Shipyard in 1917 in response to the naval need for torpedoes. The Mitsubishi Shipbuilding Company's involvement in starting the manufacturing of submarines, aeroplanes and internal combustion engines during World War I was in accordance to the request of the navy. Other large manufacturing firms also developed links with the army and navy.

The expansion of research activities led the government to take measures to facilitate innovative activities. These included the revisions of standardization and patent laws. In order to simplify the situation in which three different systems of standards (traditional Japanese, yard-pound and metric) were in use, *Noshomusho* in 1919 recommended the adoption of the metric system only. This was enacted in 1921. The ministry at this time also launched a long-term project to establish industrial standards (Japan Engineering Standard) which was continued until 1941. Patent law which had been adopted in 1885 and once revised in 1909 was again revised in 1921, giving equal rights to the inventor and the employer of the inventor; in effect, making the law more "democratic" than before [31].

[1] The use of this patent became available as a result of wartime industrial property law.

Government financial support of scientific and engineering research also expanded during this period. The only grant given to scientific and engineering research before World War I was that of the Imperial Academy which was based on donations by large industrial concerns, Mitsui, Mitsubishi, Sumitomo and Furukawa. In 1917, *Noshomusho* set up a fund to encourage inventions. Exhibitions, competitions and lectures, as well as individual inventive and innovative activities, could receive grants under this programme. The Ministry of Education also started a grant scheme in 1918 for research in science, engineering, agriculture and medicine. A grant programme for encouraging industrial research was organized also by the cabinet in 1921. The original objective of the scheme was to encourage military industries in the civilian sector; however, from 1926 other industrial research could also receive this grant.

These government grants accounted for an average of 0.02 percent of government expenditure between 1916 and 1932 [6,20,32]. [2] The absolute amount fluctuated considerably during the period because of the recession. However, despite the economic situation, the government maintained interest in promoting research. This position is reflected in the recommendations of government committees and research councils which dealt with industrial policy making in the latter half of the 1920s. [3] These committees and research councils were concerned with stimulating and rationalizing industries in recession, and the fact that they were anxious to promote research as part of it indicates that, as stated by the Resource Research Council in 1929, there was by this time a clear realization that "...encouraging scientific research and promoting its application is a basic prerequisite for the development of industries..." [33, p. 352].

Given the economic situation of the time, the policy of research promotion was not easy to implement; however, in late 1932, *Nippon Gakujutsu Shinkokai* (Japan Association for Promotion of Science, hereafter referred by its short name, *Gakushin*) was founded. The direct move for setting up this body came from academic scientists and engineers, some of whom were members of the House of Peers, who argued that scientific research was indispensable for "improving the economic situation and to keep up in the world wide industrial and economic competition", and for "resource-poor Japan, the only way to survive the competition is to excel in 'original' (i.e., research-based) industries." They also argued that the promotion of research was a basis of national prosperity and a constituent element of national prestige [6, pp. 119–120]. Whether they were really nationalistic or not, many of them felt the need to make the research system more efficient, and here their interests coincided with that of the government.

Gakushin's research grant scheme was notable in several aspects. First, the total amount of grants made available was considerably larger than previous government schemes. In fact, it was double the total of other government grants in 1933. By 1940 *Gakushin* grant alone had quadrupled while the amount of other grants remained at the same level as 1933. After the introduction of this scheme the government grants for research continued to account for 0.04 percent of total government expenditure. Second, although research in natural sciences, engineering, agricultural and medical sciences as well as human and social sciences were subject to this grant scheme, a large percentage went to engineering (tables 3b and 3c). Third, an increasingly larger amount was given to cooperative research rather than individual research (table 3a). Committees were organized to coordinate cooperative research which brought together researchers from government,

[2] In 1979 there were 37 government grants in aid programmes for scientific and engineering research which accounted for 0.4 percent of total government expenditure [30].

[3] These were the following:

- *Shigen Shingikai* (Resources Research Council) organized in 1927 which made the following recommendations on research policy: (1) Improving communication and cooperation between research institutes; (2) Encouraging industrial application of research achievements; (3) Listing of nationally important research topics; (4) Protection of researchers, increasing research grants; (5) Creating a government body to coordinate research bodies.
- *Kokusan Shinko Iinkai* (Committee for Promotion of Domestic Production) in 1926 recommended the improvement of government and private research institutes as part of its general policy recommendation of import substitution and promotion of heavy and chemical industries.
- *Shoko Shingikai* (Research Council on Commerce and Industries) also made a recommendation to promote communication and cooperation among government and private research institutes as part of measures for industrial rationalization in 1929 [33].

Table 3a
Gakushin research grants (yen (%))

Year	(1) Total	(2) Cooperative research	(3) Individual research
1933	512,574	63,683 (12.4)	448,891 (89.6)
1934	676,170	218,592 (32.3)	457,578 (67.7)
1935	678,699	248,619 (36.6)	430,080 (63.4)
1936	727,112	327,200 (45.0)	399,912 (55.0)
1937	890,579	587,661 (66.0)	311,918 (35.0)
1938	1,325,428	1,015,128 (76.6)	310,300 (23.4)

private and military research institutes. Topics pursued were usually those linked to industrial and military application or those of interest to the government. [4] Thus, through *Gakushin*, the government succeeded in directing resources for industrial (and military) development. [5]

Research in the military sector, especially in the navy, became active and varied during this period. Up to the Russo–Japanese War (1904–1905) naval research was concerned chiefly with straightforward absorption of imported technology rather than going further to develop its own technology. After this war, however, IJN turned its interest increasingly toward the latter and set up research bodies to investigate various technical matters, such as wireless communication, electrical machinery, optical instruments, firearms, aeroplanes, fuels and engines. An experimental tank was built in 1908. The expansion of research activities continued through and after World War I. In the 1920s, despite disarmament agreements or, on the contrary, because of them, IJN further consolidated its research system. The experimental tank and the aeroplanes research institute, were merged to form the Naval Technical Research Institute which aimed to do basic technical research. At the same time, a research section was set up in all naval yards and factories to do research directly related to naval vessel construction and ordnance manufacturing and to complement research work of the Naval Technical Research Institute. Military research consumed a substantial part of government expenditure on research. A survey for the year 1930 estimated that research expenditure of the army and navy was roughly equal to the expenditure for all other government research institutes put together [6]. The navy, being more technology

[4] Some topics taken up were: Economic problems of Manchuria, Mongolia and China; movements of Pacific Islands; science of natural catastrophe; geographical methods of locating mines; plant fibre materials; aeronautic fuel; wireless communication equipment; cosmic rays, atomic science; steel for specialized use; solution to scarce resources; anti-corrosive materials; organic synthesis; electric welding [6, pp. 125–126].

[5] Hiroshige [6] points out that some notable post-World War II scientific and engineering achievements were originally started as *Gakushin* sponsored projects; e.g., atomic physics which produced two Nobel prize winners, and electric welding.

Table 3b
Break-up of individual research grants according to field (%)

Field	1933	1934	1935	1936	1937	1938
Human and social sciences	16.0	16.0	15.8	16.0	17.1	16.0
Natural sciences	29.7	23.6	29.2	18.4	20.3	19.2
Engineering sciences	40.7	44.2	37.1	37.3	36.6	37.8
Agricultural sciences	5.6	5.4	7.2	9.6	8.8	7.6
Medical sciences	7.1	9.3	10.3	13.6	12.2	9.7
Interdisciplinary fields	0.9	1.4	0.3	5.0	5.0	9.6

Table 3c
Break-up of cooperative research grants according to field (%)

Field	1933	1934	1935	1936	1937	1938
Human and social sciences	28.4	11.0	16.3	12.9	10.7	6.4
Natural sciences	0	20.2	10.9	12.9	28.3	10.3
Engineering sciences	39.5	43.0	40.8	47.5	42.0	62.2
Agricultural sciences	0	1.3	2.1	2.1	1.6	0.9
Medical sciences	32.1	24.7	29.9	24.9	17.4	20.0

Source: [6].

intensive than the army, accounted for a greater part of this sum.

Research also became an integrated activity in some private industrial enterprises. In-house research was started by large electricity, mining and chemical firms in the 1890s. Large shipyards and steel mills also set up research sections in the following decade. Research in the government sector could not yet respond to the needs of these enterprises. During World War I the number of in-house research departments increased steadily, as seen in table 4. However, the scale of most of them was very small and the facilities were generally poor. Since private industries were more sensitive to recession, some were closed down or reduced their activities during the 1920s; however, those industries which despite recession tended to expand, such as electricity and chemicals, strengthened their research capabilities. Although private industrial research in general was considerably limited compared to government and military R&D, a few like the Mitsubishi Shipbuilding Company succeeded in setting up research activities on a fairly large scale.

Finally, let us see in quantitative terms what all the national efforts in industrial research amounted to. Unfortunately, there are no statistics on R&D for this period; however, the extent of engineering research can be assessed by a government survey conducted in 1930. The figures in table 5 show how small in scale the research institutes and R&D expenditures in engineering were in 1930 compared to 1970. However, it is notable that in contrast to the great difference in absolute amounts, the difference in percent share of R&D expenditure to GNP is much smaller. [6]

Table 4
Number of research bodies founded by industrial enterprises

	No. founded	Cumulative
1881–1902	6	
1904–1913	35	
1912	34	34
1913	7	41
1914	6	47
1915	10	57
1916	15	72
1917	19	91
1918	18	109
1919	13	122
1920	17	139
1921	10	149
1922	8	157
1923	2	159

Source: [33].

Table 5
R&D efforts in engineering

	1930	1970
Number of research institutes		
total	349	11,595
national	73	
local public	83	
private	193	11,132
Number of researchers		
total	2,729	129,681
national	829	
local public	551	
private	1,349	93,679
R&D expenditure	15,720,000 yen (7.5 billion yen in 1970 value)	748.7 billion yen (627.1 billion yen in private)
As % of GNP	0.11%	1.05%

Source: [6,21,33]. Originally a survey conducted in 1930 by *Shigenkyoku* (Resource Bureau).

Although limited in scale, both the government and private enterprises made serious-attempts to promote research on modern industries before World War II. This was because there was a clear realization that research was conducive to industrial development and ultimately to national wealth and strength. Two features of the prewar government research policy are especially noteworthy. First is the overwhelming preoccupation with engineering and industrial research compared to arts and more basic scientific research. Second is that even during the recession in the 1920s and 1930s, the government tried to enhance industrial research as much as possible and to make it increasingly efficient by creating a strong government–university–industry–military link. However, government sponsored research could not meet all the research needs of private industrial enterprises; hence, large firms in ship-

[6] It is interesting to note that according to an estimate of research expenditures in Japan and USA for 1930, the percent share to national income was fairly comparable in these two countries already at this time at 0.15 percent for Japan and 0.22 percent for USA. This comparison is significant in view of the fact by 1930, USA had just finished the most active phase of the expansion of industrial research of the period of this century before the end of World War II [19,33].

building, mining, electricity and chemicals set up their own laboratories.

3. In-house research at Mitsubishi Nagasaki Shipyard

Mitsubishi Nagasaki Shipyard was the largest private shipyard in prewar Japan. Being the leading enterprise in the industry meant that its rôle was in competing with foreign shipbuilders in the market for large and sophisticated ships which the growing Japanese shipping industry as well as the navy required. This implied the imperative to close the technological gap by importing the most advanced technologies. Indeed the shipyard imported foreign technology by various means to achieve this end. In-house research played an important role in assimilating and improving foreign technology.

3.1. Jikkenba, the laboratory at Nagasaki and the Experimental Tank

Research as a distinct activity at the Nagasaki Shipyard started as a small chemical analysis laboratory in the foundry shop in 1904. This eventually developed into an independent research laboratory for the entire shipyard. The need for the chemical analysis laboratory arose out of the sudden expansion of heavy industries following the start of the Russo–Japanese war, which brought about certain technical changes in the industrial sector, one of which was that machinery parts which thus far had been made of cast iron started to be replaced by cast steel. Chemical analysis of steel was a necessary part of the production of cast steel [34].

For the new chemical analysis laboratory, the shipyard hired a university trained applied chemist, especially for the job. He was immediately sent to the Yawata Steel Mill to be trained in chemical analysis of steel for three months, and the laboratory started operation upon his return. Soon, to the chemical analysis work, material testing (tensile strength) was added. These two were the main activities of the laboratory until 1916. In the meantime, operations diversified to include chemical analysis of metals other than steel. For example, in 1907, the laboratory did analysis for Stone's manganese bronze, which the shipyard manufactured under license. Eventually, the laboratory provided testing services for various materials in use in the shipyard, even for non-metal materials, if such tests could be done with the equipment and personnel of the laboratory.

The laboratory gradually became more active, and in 1916, it was given a new building. It is said that the manager of the yard, during a mission to Europe in 1913, was so impressed on seeing the laboratory of the Krupp firm in Germany that he decided that a comparable facility was desirable at Nagasaki [34]. Also, due to increasing naval

Table 6
Number of laboratory employees and reports

	(1)	(2)	(3)	(4)	(5)
1904	3	1	0	2	–
1905	–	–	–	–	–
1906	5	2	1	2	–
1907	–	–	–	–	–
1908	7	3	2	2	–
1909	–	–	–	–	–
1910	–	–	–	–	–
1911	–	–	–	–	–
1912	9	3	2	4	–
1913	–	–	–	–	–
1914	–	–	–	–	–
1915	–	–	–	–	–
1916	20	3	9	8	–
1917	–	5	–	–	–
1918	67	5	9	53	–
1919	150	6	–	–	–
1920	–	14	23	–	–
1921	156	12	16	128	76
1922	107	13	13	81	124
1923	88	8	7	73	87
1924	98	8	8	82	81
1925	–	9	7	–	99
1926	81	10	6	65	118
1927	82	11	5	66	98
1928	89	11	5	73	119
1929	88	11	5	72	76
1930	86	12	7	67	130
1931	85	9	6	70	96
1932	87	9	6	72	95
1933	102	9	6	87	77
1934	106	8	6	92	70
1935	114	8	6	100	94

(1) Total number working in laboratory; (2) Number of engineers; (3) Number of technicians; (4) Others (workers, clerks etc.); (5) Number of reports compiled. *Source:* [8,16,34].

orders, more work was demanded of the laboratory. Since there are no records of R&D expenditures, the extent of research activities of the laboratory, now called *Jikkenba* (literally meaning factory for experiments), has to be estimated by other means. Table 6 shows the number of engineers, technicians and workers (of the years for which records are available) attached to the laboratory between 1904 and 1935. There was a sudden and large increase in staff in the second half of the 1910s. This was due to the deliberate expansion of the laboratory as well as to the war boom. At this time the number of staff of the entire shipyard also increased. In the 1920s, the number initially decreased then quickly stabilized at between 80 and 90.

This indicates that in terms of the number of employees, *Jikkenba* did not suffer continuous contraction in number of staff during the recession which was the fate of most other parts of the yard. Between 1930 and 1933 when the shipyard dismissed a large number of staff and workers, the number working in *Jikkenba* in fact increased. The number of engineers and technicians working in *Jikkenba* changed only slightly between 1923 and 1935. These figures are indicative of the importance the shipyard attached to research during the years of recession. The steadiness of the activities of *Jikkenba* is reflected also in the number of reports written by *Jikkenba* staff between 1921 and 1935. An average of about a hundred reports were produced at a fairly steady pace during these years.

Thus, in quantitative terms, the laboratory at Nagasaki became a very active part of the shipyard during the first 30 years of its existence. *Jikkenba* greatly diversified its functions, especially after 1916. University trained specialists were recruited to follow new lines of research. Chemical analysis which had been concerned mainly with metals and inorganic materials, now started the analysis of organic materials such as coal, fuel and lubricating oils, paints and asbestos. In 1919, a metallurgy laboratory was created and research on electric steel casting was started in 1920 by a newly recruited electrical engineer. This research was in preparation for electric steel casting which the shipyard started a few years later. The metallurgy laboratory also started research on corrosion of metals for which a chemist was employed. A notable achievement in metallurgy research at *Jikkenba* was the development of an alloy, NM (nickel–manganese) bronze. Research on this had been started in the foundry shop, for the purpose of finding suitable brass material for making propellers which was of good strength and easy to cast. Soon Shintaro Sasaki, a metallurgist and the director of *Jikkenba*, joined this effort, and the research culminated in the development of a new alloy in 1920 which was patented. It was further discovered that by changing slightly the ratio of constituent elements, the metal was also suitable for other uses including turbine blades.

Material strength testing which had so far been limited to relatively routine tensile strength testing started other kinds of research. This was done by a mechanical engineer recruited in 1919, who developed methods for strength testing of various metals. A major problem the shipyard faced around 1920 was the frequent failure of turbine blades, caused by fatigue of the materials used for the blades of the turbines built at the yard. The material testing laboratory installed three fatigue testing machines to examine this problem; also, it worked with turbine design and manufacturing sections to develop new material for turbine blades. Whenever accidents occurred or problems arose in the shipyard, and if material strength was in question, they were always brought into this laboratory for investigation.

Another line of research on material strength was pursued by an applied physicist, Katsumi Okuda, recruited in 1925. This was on stress analysis using optical elasticity methods. He was eventually awarded a doctor of engineering degree for his research in this field. Okuda made another significant contribution by involving *Jikkenba* in the research for the development of the MS diesel engine. [7] The experimental set up to study scavenging and fuel injection was built at *Jikkenba*, and members of both the laboratory and diesel design team worked on the set up together.

Electric welding was an area in which *Jikkenba* involved itself deeply. This was due to the efforts of the director, Shintaro Sasaki, who also was responsible for making *Jikkenba* as active and as versatile as it was during the 1920s and 1930s.

[7] This was a new type of marine single acting diesel engine which the shipyard successfully developed in 1932.

Sasaki was concerned not only with doing research on electric welding, but also with promoting the use of the technique in the shipyard and eventually in Japan. In fact, his research interest was in improving the technology so that its value could be recognized and its use promoted. Thus, when IJN prohibited the use of electric welding following a few accidents caused by faulty welding, Sasaki wrote a report [29] analysing the shortcomings of the method employed thus far and how to improve them. IJN did eventually resume its use. *Jikkenba*'s involvement in electric welding was not limited to doing research. It also started the manufacturing of electrodes in 1920, and it organized training courses for electric welders in the shipyard.

Jikkenba was also held responsible for inspecting weights and measures used at the shipyard and was in charge of taking photographs at the time of the launching of ships. There was also a microscope room attached, where microscopic observation of metal sections was done. Thus from 1916 despite the recession in the 1920s and 1930s, *Jikkenba* continued to be a very active part of the shipyard. The shipyard was anxious for new technological developments, since exactly because of the recession, it had to explore new areas of production like diesel engines and turbo generators.

Research activities at the Nagasaki Shipyard were not limited to *Jikkenba*. It was also an integral part of other departments of the shipyard, especially machinery design. Turbine and diesel engines, boilers and auxiliary machinery were all being manufactured under license. The machinery design sections did research to adapt and improve the imported technologies. This was done often in cooperation with *Jikkenba*.

There was another department within the shipyard whose main function was testing and research, the Experimental Tank built in 1908. As steam ships increased in size and speed, accurate estimation of necessary power and the determination of suitable hull shape became an increasingly important question in the second half of 1800s. Model testing technique was first introduced in Britain, and in 1883, Wm. Denny & Bros. shipyard in Dumbarton built a tank. The technology diffused rapidly to other firms and other countries. Nagasaki Shipyard imported the technology from the Denny firm. This decision was taken at the time the Nagasaki Shipyard was building *Tenyo*, the first turbine vessel, and it was expected that vessels would further increase in size and speed.

The model testing technology was imported by sending two engineers to Dumbarton. The equipment and machines were purchased from the firm of M.J. Kelso on the advice of the Denny firm. Denny sent a draftsman, to supervise and instruct model testing methods. Operation was started immediately after the tank was completed in 1908. Nagasaki was the first shipyard to install a tank in Japan. A total of twelve engineers, technicians and workers were stationed at the tank. The number increased to twenty-two by 1922 and remained around that figure. As with *Jikkenba*, the Experimental Tank did not suffer decrease in staff during the 1920s and 1930s.

Significant research work at the Experimental Tank was done by Nobutaro Motora a naval architecture graduate of Tokyo Imperial University. He pursued research on both theoretical and practical aspects of model testing. In 1916, he derived an equation for the analysis of the efficiency of the model propellor [17]. He effected improvements on both the resistance dynamometer and the propellor dynamometer which improved the accuracy of measurement and made the handling of the instruments easier [18]. [8] The most remarkable of his achievements was the invention of the Motora type stabilizer. In 1918 the Nagasaki Shipyard bought the license for the Sperry gyroscope, which was the most popular type of stabilizer at the time. Its performance was good, but the instrument was heavy and complex in structure. Motora, based on a series of experiments on rolling motions of model ships, devised a stabilizer which was better able to eliminate rolling than the Sperry gyroscope and lighter in weight. The invention was patented in Japan in 1920 and subsequently in other countries. The stabilizer was adopted for most large commercial vessels at Nagasaki. For reasons not well documented, the instrument eventually gained more popularity abroad, where improvements were effected on the original invention. The Denny–

[8] Resistance dynamometer is used to measure the resistance of the model ship in the Experimental Tank while propellor dynamometer measures that of the model propellor. These two are the most important tests that are performed in the Experimental Tank.

Brown stabilizer was developed after purchasing patent rights of the Motora type [9,34]. As model tests became standard practice in the second half of the 1920s, the Experimental Tank at Nagasaki became an increasingly active part of the shipyard.

3.2. Mitsubishi Shipbuilding Company Research Institute

Mitsubishi Shipbuilding firm's deep interest in research is reflected in the founding of an institute specializing in research on technology related to shipbuilding, the Mitsubishi Zosen Kabushiki Kaisha Kenkyusho, (Mitsubishi Shipbuilding Company Research Institute, hereafter abbreviated MSCRI) between 1918 and 1933. The idea of founding this institute came from Koyata Iwasaki, the president of the Mitsubishi Goshi Kaisha, the share holding firm of the Mitsubishi concern. This was clearly in line with the national trend at the time to expand scientific and industrial research capability to fill the gap created by the difficulty to import technology during war time. Iwasaki requested Kyoji Suyehiro, professor of naval architecture at Tokyo Imperial University, who was eventually appointed as the director, to draw up a plan for a research institute in 1917. He was sent on a mission to USA to inspect industrial research institutes and to buy machines and instruments for MSCRI. [9] The institute, formally founded in 1918, was expected to achieve innovations that could substitute technology imports. This was to be done mainly through applied research, but the firm expected some basic research to be done, and its newly recruited staff included theoretical physicists and chemists in addition to mechanical, electrical and chemical engineers. Although researchers had quite some freedom to pursue research topics of interest to them, they were also expected to respond to requests from the shipyards.

The main areas of research were machinery and metal engineering. New machines were invented such as a new type of turbine rotor balancing machine, material testing machine, refrigerator, cryptograph machine as well as a machine for analysing wave shapes. In metal research, notable achievements were the development of a new alloy, Iidaka metal, for use as turbine blades, a new method of metal rolling, and arc welding of cast iron. Research on metals using X-ray, which was of importance in advancing electric welding technology, was also pursued. Patents were obtained for many of the inventions. Those that were commercialized were manufactured at Nagasaki or other shipyards and factories [14,35]. Results of research at MSCRI which were not directly related to productive operations of the firm, were published in professional journals in Japan and abroad. MSCRI researchers gave lectures on their research findings at Mitsubishi Technological Research Meetings, and the institute published reports regularly.

There was a close cooperative relationship between MSCRI and the shipyards. Some innovations were joint efforts of the institute and a shipyard; e.g. the refrigerating machine developed in 1924 was invented jointly by an MSCRI researcher and three engineers of the Kobe shipyard. [10] Shipyards manufactured machines and materials when requested by MSCRI; the latter in turn did experiments upon requests from shipyards. There were cooperative research activities with other research institutes; also, MSCRI did research in response to requests from the army and navy. Research on metal rolling was a joint research with *Riken* and the invention of the cryptograph machine was in accordance to the request from the army. As far as the Nagasaki Shipyard is concerned, MSCRI had a rôle to play in the research for improving turbines. The development of Iidaka metal and balancing machines was in response to a request of the Nagasaki Shipyard's turbine design section to search for better material for turbine blades and to devise an improved system for turbine rotor balancing.

Records indicate that in the 1920s, the Mitsubishi Shipbuilding Company attempted to stimulate research efforts of the firm in order to explore new areas of production. In 1924, the firm set aside a part of its annual profit to create a fund for research in the accounts of the head office. [11] This fund was to be used for research

[9] Records indicate that Mitsubishi Shipbuilding Company continued to make substantial investments in machines and facilities for MSCRI until 1923 [11].

[10] Kobe had been manufacturing refrigerators since it obtained license for it from Pulsometer, a firm in Britain, in 1913 [14].

[11] A total of 1,450,000 yen was put into this fund between 1924 and 1927. This account disappears after 1932. The fund seems to have been used up by then [11].

which would contribute to the diversification of the firm's productive activities. The shipyards as well as MSCRI could draw from this fund with the permission of the head office. Unfortunately, records do not indicate on what research the fund was used, but most research activities of MSCRI, and the new product innovations undertaken at shipyards, e.g., MS diesel engine project at Nagasaki, were probably funded by this.

In 1933, MSCRI was donated to *Riken*. The reasons for this are not documented in the records of the firm. The most immediate reason was certainly the recession that hit the firm the hardest in 1931–32, but there must have been other causes for this move, since the firm attempted to expand rather than diminish its research activities despite the unfavourable economic situation in the 1920s. Most probably, the firm decided to consolidate research efforts at the shipyards and concentrate on research of immediate relevance to productive activities. This would explain the sustained activity level of Nagasaki *Jikkenba* despite the crisis faced by the firm in 1931–32. Another possible reason is that the research at MSCRI, although some were commercialized, fell short of achieving major innovations that could replace technology imports, which the institute originally aimed for.

The fact that the Mitsubishi Shipbuilding firm had three specialized research bodies within the firm, *Jikkenba*, the Experimental Tank and MSCRI indicates that in-house research occupied a place of vital importance in the firm. Of these the contrasting fates of *Jikkenba* and MSCRI are striking. The former grew from a small chemical analysis laboratory to a large versatile center for research. Its activities stayed close and responded well to needs of production in the shipyard. Since ships and engines were being built by importing technology, need for research was in problem-solving and trouble shooting in the process of assimilating foreign technology. Indeed, research related to turbine and diesel engines and electric welding, all of which were imported technologies, accounted for a major part of *Jikkenba*'s activities. The indispensability of *Jikkenba* is reflected in the fact that it remained active throughout the recession in the 1920s and 1930s. MSCRI, in contrast, was founded with the aim of achieving major technological breakthroughs; however, actual achievements were more in improving existing technologies than in creating new ones. MSCRI was given up after 15 years of existence. The difference in the course of the evolution of these two in-house research bodies illustrates that the kind of research needed for the firm at this stage of industrial development was in assimilating and improving existing technologies through minor technological changes. Achieving major breakthroughs at this stage was very difficult and not really necessary.

4. Management of research and inventive activities at Nagasaki Shipyard

4.1. Policy toward inventive activities

The research activities often culminated in inventions. In fact, even before the founding of the research laboratory, engineers at Nagasaki started to obtain patents for new instruments, materials and machines they invented in their work. These inventions induced the shipyard to institute rules for inventive activities of its employees in 1907. The rules stipulated that the Mitsubishi firm, not the inventor, was the one entitled to apply for and utilize patent rights obtained for inventions of the employees. The firm was to give compensation and in some cases award for such inventions when they proved to be useful and profitable. This policy was preserved in the revised rules of 1922 which made an addition that any research with an objective to invent something was to receive an advance approval of the firm.

This conservative policy toward inventive activities was subject to dissatisfaction of potential inventor-employees [12] and may have had the effect of depressing inventive activities. In order to stimulate inventions while adhering to the above rules, the firm first introduced a separate award scheme in 1915 for inventions by its employees. This scheme was borrowed from the one in practice at William Denny Bros. Shipyard in Britain. [13]

12 An engineer recounts his dissatisfaction to this policy and says that for some time he only did inventive activities outside working hours [5].

13 At Denny's an award scheme was introduced in 1871. It was to give money awards to a wide range of inventions and improvements in the form of new machines or tools, new methods of work execution, prevention of accidents, avoiding wastes and cutting costs of production [2].

Unfortunately, In the records of the firm, the details of the operation of this scheme do not remain, but there are references to awards actually given for inventions and improvements. For example, a money award was given to foreman in 1915, for inventing tools to make coils which cut both cost and production time [15]. [14] There is little doubt that the intent behind this scheme was to encourage inventive activities as much as possible while preserving the right to patents obtained for inventions of the employees. The scheme was applied to workers as well as to engineers and technicians. Inventions which were patented averaged 30 per year between 1912 and 1935, including several which also secured patent rights abroad [14].

There were inventions and improvements other than those which were subject to patent rights. These were usually those that made tools more efficient, those that improved manufacturing processes or introduced mechanization in production. Workers and technicians improved hand tools they used. Such improvements sometimes were able to replace tools that had thus far been imported. A worker recounts that after his improvement of the hammer he used every day, the importing of the tool was stopped [13]. Construction processes were mechanized during the 1910s by the adoption of pneumatic compressors, gas cutting and electric welding. In the 1920s and 1930s, the need for innovations that would raise productive efficiency and reduce cost came to be acutely felt because of the recession. Numerous improvements and inventions to produce these effects were introduced by engineers as well as workers. The chief engineer in the hull factory arranged assembling operations to be done as much as possible in the factory, for this saved considerably more time than assembling at the construction site [12]. A worker in 1932 devised a method to make most of the pipes on land rather than at the construction site which also economized time [13]. The chief engineer in the machinery factory rearranged the factory layout which facilitated transporting of materials, eased and made safer the access to machines and improved lighting. He also saw to it that the machines and tools were always in good condition and well adjusted so as to insure precision in the manufacturing process [13].

[14] There are a few other articles found in the records of the firm [11,15]. Also, in the memoires, there are quite a few instances of awards received for inventions and improvements cited [12,13].

These inventions and improvements were encouraged not only by the firm but also by the chief engineer of individual factories who sometimes gave awards from their pockets for inventions and improvements of their subordinates. The chief of the power station, carefully examined every invention and improvement and awarded those that were adopted which included fairly trivial items like improvement on lamps and safety signs, which nonetheless contributed to raising the efficiency [13]. The chief of hull factory held marking competitions and awarded the winners which aided the upgrading of the marking skill of the workers [12].

4.2. Technological research committees

In order not only to stimulate research and inventive activities, but also to channel them for the profit of the firm, there was a need for a means to diffuse the fruits of these activities, solve problems and to coordinate them within the firm so that they would continue to be useful. The technological research committees that were set up from 1920 on various topics of interest to the firm served this purpose. The first one remaining in the records is the Electric Welding Research Committee. The committee held a meeting at Kobe in 1920 and at Nagasaki the following year to discuss common research problems and exchange information. The second committee created was on casting technology which met for the first time in 1921. The direct motive for convening this committee was the faulty casting of turbines built for naval cruisers, *Kama* and *Kiso* for which the Nagasaki Shipyard received complaints from IJN. The chairman of the committee, explained the goal of the committee as follows:

> ...to exchange information and discuss common problems in order to improve casting technology... the stage whereby we could learn and borrow everything from abroad has passed ...now it is necessary to do research ourselves... for example we had not succeeded in satisfactory production of man-

ganese–bronze ingots since we bought Stone's license (in 1904), until the stoppage of imports during the first world war forced us to do serious research after which we finally succeeded in its domestic production ... in order to achieve something we need to apply scientific knowledge to eliminate problems encountered... [11, p. 23212].

Such needs and objectives were shared by other committees. The committee on casting met once in a year and a half to discuss problems in casting of various metals, ways to cut cost and increase efficiency of production. According to the rules of the committee adopted in 1935, the committee was to share experiences in casting at different shipyards of the firm and select topics for further survey or research for the purpose of improving casting technology. Members were chosen by the managers of the shipyards and appointed by the president of the firm. Sub-committees were to be set up whenever necessary.

Exchange of information and coordination of research efforts were the main functions of the above two committees. The Internal Combustion Engine Research Committee organized in 1928 was slightly different, in that its main job was to guide the development of a new type of diesel engine. Mitsubishi had purchased the license to manufacture Sulzer marine diesel engines in 1924; however, this engine found itself at a disadvantage in the competition with Burmeister and Wain engines in both cost and performance which were being built by other firms. The committee was set up to look into this problem. At its first meeting, it was decided that research and development for a new type of marine diesel engine was to be launched at the Nagasaki Shipyard. The committee discussed and decided what kind of research was to be done, and the problems encountered at each stage of the development of what was to become the MS diesel engine. The committee met every two to five months until the completion of the MS engine.

These three committees as well as two others were set up before 1935. The records indicate that more were set up afterwards and many of them added several sub-committees; therefore, technological research committees became a well-established means to guide and coordinate research and innovative activities for the further technological development of the firm.

4.3. Mitsubishi Technological Research Meeting and links with universities, other research institutes and professional societies

Mitsubishi Goshi Kaisha, under the initiative of its *Shiryo-ka* (Information department) held a monthly Mitsubishi Technological Research Meeting in 1924–25. [15] This meeting was to promote exchange of information and coordinate research activities of firms within the Mitsubishi concern. It also attempted to encourage communication and cooperation with researchers and research institutes outside Mitsubishi. Topics discussed at these meetings were those of common interest to participants and were chosen from the following areas: fuels, iron and other metals, engines, machines and tools, chemicals, and other topics of interest related to industrial production [10]. Researchers from universities, government research institutes and from research bodies of various firms of the Mitsubishi group were invited to give lectures at these meetings. They usually spoke of the latest developments in their research.

Mitsubishi was anxious to promote exchange of information and cooperation with universities and other research institutes. Engineers and workers were sent on various occasions to naval yards and universities to be trained in specific technologies. Such exchanges on research matters were frequent and often on an informal basis. For example, when *Tenyo-maru*, at the time of its trial run, was found unable to correct its tendency to list, the shipyard asked the professor of naval architecture at Tokyo Imperial University, to examine the problem [12]. Kotaro Honda, professor at the Metal Research Institute of Tohoku Imperial University held a summer lecture session from 1920 to which two or three engineers, technicians or workers from Nagasaki were sent. The memoires of some of these people show that they took advantage of the opportunity to discuss and seek advice on problems they were encountering in their jobs at the shipyard. Many instances of contacts with Professor Honda appear in the memoires of staff. Since the Metal Institute was partially funded by large industrial concerns like

[15] The meetings may have continued after 1925, but there are no records.

Mitsubishi, its link with the industrial sector was probably very close. Another university which had close industrial links was Kyushu Imperial University. Its naval architecture course was set up because of the proximity to Nagasaki and Sasebo, which implied that the university expected to utilize professional resources of the two yards. Nagasaki engineers were invited to take up lectureships one after another for periods of several years. A Nagasaki engineer left the shipyard to take a professorship there. Shintaro Sasaki, the director of *Jikkenba* gave weekly lectures at the university on metallurgy. He influenced some metallurgy students there to come to work at Nagasaki. Some who did, later came to play important rôles in the introduction and development of automatic welding technology as engineers or researchers at Nagasaki. Sasaki was also a member of the *Gakushin* committee on electric welding. Engineers also actively participated in professional societies; the Society of Naval Architects, the Society of Mechanical Engineers and the Welding Society. Many Nagasaki engineers gave lectures at their meetings and/or published in their periodicals. Such professional activities facilitated the exchange of information and enhanced research activities.

5. Concluding remarks

In Japan there already existed a fairly well-defined policy toward research before World War II. It was moulded to be responsive to the needs of industrial development. Research was viewed as a means to enhance industrialization, and government policy gave priority to and concentrated resources on research directly related to industrial development. Industrial research was complemented by the strengthening of government–industry–university–military links. What is noteworthy is that the policy of stimulating industrial research was sustained throughout the recession periods of the 1920s and 1930s. Research was regarded as having the potential to produce innovations that could improve the economy.

In major industrial enterprises, as illustrated in the case of Mitsubishi Nagasaki Shipyard, research was instituted already at the beginning of the century. This was a natural outcome of the need to assimilate and improve foreign technology. *Jikkenba* grew increasingly active as it provided a variety of research services for design and manufacturing departments. No major breakthroughs were achieved, but the minor technological changes that resulted from its research activities facilitated the assimilation and improvement of imported technologies. Therefore, since before the war, industrial research in Japan was clearly and overwhelmingly geared toward "adaptive", research or "improvement engineering".

"Improvement engineering" was encouraged down to the worker's level. It gave rise to technological changes in production engineering that raised efficiency, reduced cost and/or substituted imports. Research activities were enhanced rather than depressed during the recession in the 1920s and 1930s, and the firm instituted measures to make in-house research more effective. Technological research committees coordinated research and facilitated the diffusion of information within the firm. Cooperation with universities and research institutes outside the firm was encouraged through Mitsubishi Technological Research Meeting and other informal links.

There was complementarity between national and firm level efforts in industrial research. The institutionalization of in-house research was itself a result of the fact that national level efforts could not fulfil research requirements of large and complex industries. In-house research in turn benefited from links with government, universities and the military sector which it was a government policy to strengthen. This suggests that what Freeman [4] termed as the "national system of innovation" already existed before the war.

It should be pointed out that the accumulation of technological capabilities through research that started before World War II had significant implications for postwar technological developments at Nagasaki Shipyard. The very rapid absorption of automatic welding technology which played a major rôle in making Japan the leader in ship-

[16] Masao Kanamori, who started his career at *Jikkenba* in 1935, and who eventually held posts of director of *Jikkenba* as well as the chairman of the board of directors of Mitsubishi Heavy Industries until recently, said emphatically that without knowledge and experiences gained through research in various aspects of metallurgy before the war, the mastering of automatic welding after the war would not have been so rapid nor easy. (Dr. Kanamori in interview with the author.)

building by the second half of the 1950s owed very much to prewar experiences in the assimilation of electric welding technology and the accumulation of research on metallurgy. [16] The successful development of UE (uniflow exhaust) diesel engine after the war likewise was due to experiences and knowledge gained through the MS diesel engine project before the war. [17]

The Japanese experience of accumulating technological capabilities examined in this paper strongly suggests that technological learning is a process that can be and should be sustained and elaborated for a long time, and the accumulated capabilities can lead a country's industrial sector to a degree of technological sophistication that is highly competitive in the international economy. This is the implication of the Japanese experience for international transfer of technology whether from developed to developing countries or among these two groups of countries. In any case accumulating technological capabilities is of utmost importance.

For Japan the implications of her experiences are more ambiguous. Although competitiveness was achieved, the long-range efforts have produced a conditioning that equates research with "improvement engineering". This has resulted in the the difficulty Japan is facing now in re-orienting its "national system of innovation" to stress more basic research. The Japanese technological efforts in the next few decades should demonstrate whether long-range technological learning can also lead to strength in basic research.

References

[1] T. Blumenthal, Japan's Technological Strategy, *Journal of Development Economics* 3 (1976) 245–255.

[2] Denny Bros. & Ltd., *Denny-Dumbarton 1844–1932* (Dumbarton, 1932).

[3] M. Fransman and K. King (eds.), *Technological Capability in the Third World* (MacMillan, London, 1984).

[4] C. Freeman, *Technology Policy and Economic Performance – Lessons from Japan* (Pinter, London, 1987).

[5] J. Fukao, *Fukao Junji Gijutsu Kaiso Nanajunen* (Seventy years of recollections on technology) (Tokyo, 1979).

[17] Hideo Fujita, who took part in both projects said that the UE diesel engine project was designed after the MS project and exactly the same procedures were employed. In this case prewar experience in project management was as important as technical knowledge and experience. (Dr. Fujita in interview with the author.)

[6] T. Hiroshige, *Kagaku no Shakaishi* (A social history of science) (Chuokoronsha, Tokyo, 1973).

[7] IDB/ECLA Research Programme on Science and Technology, *Working Papers 1–30* (Buenos Aires, 1977–1978).

[8] Y. Kubo, *Jikkenba Hokoku Mokuji* (List of laboratory reports), 1972.

[9] I. Matsushita, Senkei Shikenjo Jidai no Omoide (Memoires of the days at the experimental tank) in Mitsubishi Zosen Kabushiki Kaisha, *Mitsubishi Zosen no Kenkyu Kikan* (Research departments of Mitsubishi Shipyard) (Nagasaki, 1960).

[10] Mitsubishi Goshi Kaisha Shiryoka, *Shiryoka Nenpo* (Annual report of the Resource Department) (Tokyo, 1924–1925).

[11] Mitsubishi Jukogyo Kaibushiki Kaisha, *Shashi Shiryo* (Materials on the history of the firm) *1916–1948* (1960).

[12] Mitsubishi Jukogyo Kabushiki Kaisha, *Shashi Shiryo Kaikyuroku* (Materials on the history of the firm – recollections) 1&2, (1960).

[13] Mitsubishi Jukogyo Nagasaki Zosensho, *Kaiso no Hyakunen 1–3* (A hundred years of retrospect) (Nagasaki, 1974).

[14] Mitsubishi Jukogyo Kabushiki Kaisha, *Mitsubishi Jukogyo Kabushiki Kaishashi* (History of Mitsubishi Heavy Industries Ltd.) (Tokyo, 1956).

[15] Mitsubishi Shashi Kankokai, *Mitsubishi Shashi 1–40, 1870–1952* (History of Mitsubishi firm) (Tokyo, 1972).

[16] Mitsubishi Zosen Kabushiki Kaisha, *Meibo*, (Staff list).

[17] N. Motora, An Analysis of Model Screw Propeller Experiments, *ZosenKyokai Kaiho* 19 (1916).

[18] N. Motora, Dokeisen renko no baai Kozokusen no ukuru Eikyo ni tsuite (Effect the follower ship receives when being pulled by another ship) *Zosen Kyokai Kaiho* 41 (1927).

[19] D. Mowery, The Emergence and Growth of Industrial Research in American Manufacturing, 1899–1945, unpublished Ph.D. dissertation, Stanford University, 1981.

[20] K. Ohkawa and H. Rosovsky, A Century of Japanese Economic Growth, in: W. Lockwood (ed.), *The State and Economic Enterprise in Japan* (Princeton University Press, Princeton, 1965).

[21] K. Ohkawa and M. Shinohara, *Patterns of Japanese Economic Development – A Quantitative Appraisal* (Yale University Press, New Haven, 1979).

[22] K. Oshima, Research and Development and Economic Growth in Japan, in: B.R. Williams (ed.), *Science and Technology in Economic Growth* (Wiley, New York, 1973).

[23] T. Ozawa, *Japan's Technological Challenge to the West 1950–1974* (MIT Press, Cambridge Mass, 1974).

[24] T. Ozawa, Macroeconomic Factors Affecting Japan's Technology Inflows and Outflows: the Postwar Experience, in: N. Rosenberg and C. Frischtak, *International Technology Transfer: Concepts, Measures and Comparisons* (Praeger, New York, 1985).

[25] M.J. Peck and A. Goto, Technology and Economic Growth: the Case of Japan, *Research Policy* 10 (1981).

[26] M.J. Peck and S. Tamura, Technology, in: H. Patric and H. Rosovsky (eds.), *Asia's New Giant* (Brookings Institution, Washington, 1976).

[27] N. Rosenberg and C. Frischtak, *International Technology Transfer: Concepts, Measures and Comparisons* (Praeger, New York, 1985).

[28] H. Rosovsky, What are the "lessons" of Japanese economic history?, in: A.J. Youngson (ed.), *Economic Development in the Long Run* (St. Martins Press, New York, 1972).
[29] S. Sasaki, Kaizen seraretaru Kohan Denki Yosetsuho (Improved electric welding method of steel plates), *Jikkenby Hokoku* No. 59 (Nagasaki, 1921).
[30] Science and Technology Agency, *Indicators of Science and Technology* (Tokyo, 1980).
[31] Tokkyocho, *Tokkyo Seido Nanajunenshi* (Seventy year history of the patent system) (Tokyo, 1955).
[32] Tokyo Kogyo Shikensho, *Tokyo Kogyo Shikensho Gojunenshi* (Fifty year history of Tokyo Industrial Research Institute) (Tokyo, 1951).
[33] Tsushosangyosho, *Shoko Seisakushi 13 Kogyo Gijutsu* (History of policies on commerce and industry, 13, industrial technology) (Tokyo, 1979).
[34] T. Yano, *Nagasaki Kenkyushoshi Shiryo* (Materials on the history of Nagasaki Research Institute) (Nagasaki, 1966).
[35] Zosen Kyokai (ed.), *Nippon Kinsei Zosenshi – Taisho Jidai* (History of Shipbuilding in Modern Japan – Taisho Era) (Hara Shobo, Tokyo, 1973).

Note:

The author wishes to point out that this article is based on a chapter in the book:

Yukiko Fukasaku (1992), *Technology and Industrial Development in Pre-war Japan – Mitsubishi Nagasaki Shipyard 1884–1934*, London: Routledge.

Excerpt from *National Innovation Systems: A Comparative Analysis*, Richard R. Nelson (ed.)

[23]

3

The Japanese System of Innovation: Past, Present, and Future

HIROYUKI ODAGIRI
AKIRA GOTO

It was commonly believed that the history of the Japanese economy was that of a country desperately trying to catch up with technologically advanced nations. When Japan opened its country around the time of the Meiji Restoration of 1868 following more than two centuries of seclusionism, the leaders had to realize how much Japan was behind Western countries in many aspects of technology. Naturally, the government made efforts to import superior technology, hire engineers from abroad, educate its people, and encourage the entrepreneurs to assimilate foreign technologies and apply them in Japanese factories. In addition, the determination to catch up in terms of military capacity gave the government a strong incentive to support technological advances and make domestic procurement possible. Such a catch-up process has been observed in many late-developing countries, such as Prussia and Russia in the nineteenth century or South Korea and Taiwan in recent decades. Japan's experience was by no means unusual and neither were the policies taken by the government.

However, one cannot attribute Japan's "success" solely to the latecomer advantage or to government policies because neither of these can lead to successful economic development unless the private sector—investors, managers, engineers, workers—is willing and able to respond to the opportunities open to them. In fact, at the time of the Meiji Restoration, entrepreneurs appeared in various industries and most people had been sufficiently educated to read, write, and count. The willingness to start an unfamiliar business and ability to assimilate new technology were present, and these were the main moving forces behind Japan's development. It is also noted that even under the seclusionism during the Tokugawa (Edo) era, technological importation did take place (though not on a large scale) and indigenous technologies and skills were gradually developed. One aim of this chapter is to investigate the conditions necessary for successful catch-up by studying the development of Japanese industries and technologies by means of both a general historical description and industrial cases. Another aim is to study how Japanese businesses have been (and are) behaving and organizing, because such a micro aspect provides an important key to understanding the Japanese innovation system.

The paper is separated into four parts. The first section gives a historical overview of Japan's technological advance from past to present. The second section discusses the technological development of three industries—iron and steel, electrical and communication equipment, and automobiles. The next section discusses the present Japanese management system, since understanding the managerial aspects of Japanese firms is indispensable to understanding why the firms are motivated to innovate and how they maintain efficiency in innovation. The final section concludes the chapter by discussing future prospects.

FROM PAST TO PRESENT

The Tokugawa Era and Before: Up to 1868

Major technology importation before the Tokugawa era (1603–1868) took place twice. The first is from the seventh to ninth century when the emperor's government sent envoys to China and immigration was frequent from China and Korea. The second is the sixteenth century, particularly when guns were first introduced by the Portuguese. This was the period of civil war and the strong demand for guns caused a number of blacksmiths to start producing guns either by copying or with the help of Portuguese gunsmiths. The technological level in guns, ships, and other arms production soon caught up with the West.

The Tokugawa government, fearing that Christians would disobey their rule, secluded the country in 1639 and restricted foreign trade to the Chinese and the Dutch. Thus the only Western contact for Japanese was through a number of Dutchmen allowed to live in Nagasaki in Kyushu, an island in the southwest. The Tokugawa government monopolized trade, though evidence suggests that some of the powerful feudal lords under the Tokugawa rule did trade with the Chinese and other foreigners. With the request of the government the Dutch provided information on foreign affairs and science regularly, some of which were diffused to other lords. The Dutch in Nagasaki were also a source of information on many aspects of science and technology, such as medical science, biology, and geography. Several Dutch books were translated into Japanese, and a number of Japanese studied with Dutch doctors or other scientists living in the Dutch quarters of Nagasaki. Thus the seclusionism by no means implied a complete seclusion from foreign scientific and technological information. Actually, the accumulation of Western knowledge through the Dutch can be assumed to have contributed to a rapid absorption of the Western economic and technological system when the Tokugawa government opened the country to non-Dutch Western countries in 1854. For instance, the first Western-style iron furnace was made by a *samurai* who had studied with a Dutch book.

In terms of indigenous technology too, the Tokugawa era was hardly a static period. Because of the importance of agriculture and mining for the feudal lords' finances, increases in productivity in these areas were encouraged and many improvements were made and diffused. Because of the steep flow of rivers, climatic conditions (rainy season in June and typhoons in summer–autumn), and the importance of water in rice crops, much investment was directed toward improving rivers and making irrigation systems. Consequently, the technological level in civil engineering is believed to have been very high.

Machine engineering was another area for indigenous technology at the time. The innovators were called *karakuri* masters because karakuri, a moving mechanism, is the most essential part of any machine. These mechanisms were applied to many sorts of machines and tools, ranging from dolls and clocks to textile looms and rice-polishing machines.[1] Again, we will show later that there was a continuous development from such indigenous technology to imported Western technology after the Meiji Restoration. H. Tanaka, probably the most important karakuri master toward the end of the Tokugawa era, became a pioneer in the electrical equipment industry, and S. Toyoda (born a year before the Restoration) invented internationally copied looms.

Therefore, although the Japanese failed to invent steam engines, among other things, their technological level was not too much behind the West. This fact should be emphasized because it is in marked difference from the cases of many developing countries today. In addition, the educational level was high—probably higher in elementary education than the United States, Britain, France, and Germany, though lower in higher education because in these countries science education in universities had started by the early nineteenth century.

Basically there were two school systems. The first was the schools owned by feudal local governments, which were usually compulsory to the children of samurais (i.e., the employees of respective local governments). Many of them also admitted the children of wealthier farmers and merchants. The second was private schools, called *terakoya,* since many of them were run by Buddhist temples (*tera*). The length of education was various but, most typically, went from the age of 6 to 12, similar to the present elementary school system. They mainly taught reading, writing, and the use of *soroban* (abacus) to calculate. In addition to these schools for children, there were a number of private or public higher education systems. Some of these taught Japanese or Chinese studies, for instance, Confucianism. However, there were private schools teaching medical science using Dutch books. The best known of the latter case was "Tekijuku," a private school in Osaka taught by K. Ogata, who had learned Dutch medical science in Nagasaki. Among his students were Y. Fukuzawa, who later introduced Western democratic thoughts to Japan and established the first private university, Keio, and M. Ohmura, who later designed the Meiji military system (but was assassinated before the system was completed).

There were not many terakoyas and other schools in the early Tokugawa era but they became more and more popular, and, according to a very rough estimate, there were almost ten thousand such schools in Japan toward the end of the era. Many of them were small with 10 or fewer children, but there were schools with more than 500 in large cities such as Edo (Tokyo) and Kyoto. The ratio of school enrollment (or attendance) varies among estimates and among regions, ranging from about 25 to almost 100% for boys and a lower percentage for girls (Umihara, 1988). These estimates may be misleading because they may include those children enrolled as pupils but who seldom came. Nevertheless, the ratio of, let us say, 50% under the noncompulsory system is surprisingly high and shows the eagerness of parents to invest in child education. This fact suggests that the literacy rate in Japan in the seventeenth and eighteenth century was likely higher than in Europe and America. The introduction of the public education system in the Meiji era would not have been so smooth were it not for this wide educational background.

The Meiji Era: 1868–1911

The end of seclusionism in 1854 and the inauguration of a nonfeudal central government following the Meiji Restoration of 1868 prompted the Japanese government and the public at large to import advanced foreign technology and to catch up with the Western countries economically and militarily. The government thus started an organized effort to "modernize" the country, including the provision of infrastructures for transportation, communication, utilities, education, and finance.

Generally speaking, diffusion of science and technology from one country (or countries) to another can be made by transferring written information (e.g., books, papers, and drawings), people (e.g., hiring foreigners and sending Japanese to study abroad), goods (e.g., importing machines and plants), and capital (i.e., foreign direct investment). All these methods were used in Meiji Japan. Particularly in the case of importing social systems, hire (*yatoi*) of foreign teachers and advisors was common. For instance, when the Meiji government started to establish a national system of education including compulsory elementary education, an American, D. Murray, was hired as an advisor and gave American influences to the system, though these influences were diluted when nationalistic education was later emphasized.

It took about three decades of trial and error before the government could establish a countrywide elementary education system. In 1874 2 years after the government started the effort, there were about twenty thousand schools, less than half of what the government had planned. Some of these schools were converted from terakoyas and taught by former terakoya teachers. The enrollment ratio was 46% for boys and 17% for girls. By 1904 when the compulsory 6-year education system was finally established, the ratio had increased to 99% for boys and 96% for girls. Thus illiteracy among the youth was nearly absent by the beginning of this century. The secondary education system also became commonly available and by 1920 more than half of the children out of elementary schools proceeded to 2-year or 5-year secondary schools (Kaigo, 1971).

For the higher education system, particularly in the field of technology and engineering education, a British influence was introduced. Kobusho (the Ministry of Industries), itself started with the advice of a British railway engineer, hired a British, H. Dyer, to make plans for an engineering college. In 1873, Kogakuryo (the College of Engineering) was established with Dyer's plan and eight more British professors were hired besides Dyer himself who became the head. With the eagerness and high quality of both the instructors and the students, the college was quite successful. It hired more foreign (mostly British) professors and then started to replace them with Japanese graduates from the College. In 1886, the College merged with another college, established by the Tokugawa government to teach science and technology with the help of the Dutch, French, and German, and became the Engineering Department of Imperial University (later renamed the University of Tokyo). The College and the University produced graduates who later founded many of the major Japanese manufacturing companies as we will later show.

It is noteworthy that Dyer later exported his Japanese experience of engineering education to his hometown, Glasgow. He had been recommended to the Japanese government by a professor at the University of Glasgow, and came to Japan at the age of 25 right after receiving a master's degree there. The program he made for Kogakuryo emphasized the interaction between classroom studies and on-site training at the laboratory works he made within Kogakuryo as well as Kobusho's works. This balance between the two aspects of education was lacking in European schools at the time and Dyer's originality should be noted. He left in 1882 to go back to Glasgow and made efforts to introduce a similar engineering education program when a technical college was founded there. Apparently, the Japanese government emphasized engineering education at the time when more developed countries regarded pure science as superior to engineering. This background gave Dyer an opportunity to experiment with his ideas on engineering education and, with his Japanese experience, he persuaded his people of its importance.

During the early Meiji era, in particular the 1870s and early 1880s, the government built and owned plants and factories in industries such as mining, railroad, shipbuilding, machinery, and textile, because it was still difficult for the private sector to finance the required investment and take risks. In addition, personnel with advanced Western technological knowledge were scarce in the private sector. However, the government's investment program in industries was neither quantitatively spectacular compared, say, to that of today's developing countries nor always successful. In fact, most government-owned factories suffered losses and, with the tight government budget, were gradually privatized and sold to emerging private entrepreneurs (mostly merchants and ex-samurais).

The government retained plants in military-related industries, such as shipbuilding, aircraft, munitions, and steel, and in public utilities including telecommunication. The military-related production occupied a significant portion of Japan's economy at the time, because the Meiji government was keen to build up its military capacity to deal with the threat of Russia and other countries that were colonizing China. In 1907, the largest operation in machinery industry (including shipbuilding, vehicles, general machinery, tools, and parts) in terms of the number of blue-collar workers was the navy shipyard in Kure with about 21,500 employees, followed by another navy shipyard and two arsenals (Sawai, 1990). The largest private plant, Mitsubishi's Nagasaki Shipyard, ranked fifth with less than 10,000 employees, whose important customer was again the navy. The military-owned plants were also a center of technological development. They hired a large percentage of scarce engineers and imported advanced machinery from abroad. Their technology was subsequently transferred to the private sector as the engineers and skilled workers moved from the military plants to the private sector, especially during the disarmament period following the Russo-Japanese War of 1904–1905.

The military not only produced goods within its own shipyards and arsenals, but also procured them from the private sector. Since the military, for obvious defense reasons, preferred to procure goods domestically, procurement gave domestic producers in shipbuilding, steel, machines, electrical equipment, and so on, who were under competitive pressure from larger and technically advanced foreign firms, a chance to increase their production and accumulate knowledge through experience. It should be noted that Unequal Treaties the Tokugawa government was forced to sign in 1858 with the United States, the United Kingdom, the Netherlands, France, and Russia deprived the government of the right to set tariffs to imports. In 1902, for instance, the proportion of tariffs in government revenue was 5% in Japan versus 21% in the United Kingdom and 45% in the United States. Thus, until 1911 when the treaties were revised, the government had little means to protect domestic producers from foreign competitors except preferential procurement.

The economy started to grow after about two decades of the Restoration (see Table 3.1). GNP more than doubled in the 30-year period from 1885 to 1914, the year World War I started. In terms of industrial composition, food processing and textile were the largest industries before the turn of the century. Then metal, machinery, chemical, and other heavy industries started to grow fast. During the first four decades of the twentieth century, these heavy industries grew at an annual rate of more than 10% whereas the manufacturing industries as a whole grew at about 6%. Their growth was particularly rapid in the 1920s and the 1930s, and their share in the manufacturing sector exceeded 50% in the 1930s.

Technological progress was an important source of this growth. According to Minami (1981), nearly 70% of the growth of the private mining and manufacturing sector was accounted for by the "residual" factors that include technological progress. This technological progress came from both indigenous (traditional and domestic) technology and the technology imported from advanced countries. Indigenous technology was important not only on its own, particularly in traditional industries, but also in providing the ability to select among the technologies available in developed countries, and in adapting and assimilating them to fit domestic conditions. This fact was most notable in the textile industry, the second largest manufacturing industry at the time (next to food processing) and the largest exporting industry before World War II. The case study of iron and steel industry will provide another example.

This fact notwithstanding, the role of indigenous technology was limited in modern industries, such as metal and machinery, where imported technology played a far greater role. As discussed earlier, technology transfer from abroad was made through many channels. Many foreign engineers and specialists were hired, though they were gradually replaced by the Japanese educated domestically at the above-mentioned College of Engineering and other schools and those who had studied abroad and came back. Other channels of technology transfer were importation of advanced machinery (and reverse engineering), licensing agreements, and foreign direct investment into Japan. The latter two increased since the turn of the century because the government liberalized foreign direct investment and also joined the Paris Convention, though the patent system itself was introduced as early as 1885. The significant role these various means of technology transfer played in Japan's industrial and technological development will be discussed in more detail in the case studies.

The number of patents granted during July 1885 to February 1902 was 4817 (Patent Office, 1955). In comparison, the number was 27,136 in 1902 in the United States, 13,714 in the United Kingdom, 12,026 in France, and 10,610 in Germany. Hence the number of patents in Japan was hardly comparable with the Western countries. Among these 4817 patents, 2175 (45%) were related to machinery, 728 (15%) to chemicals, 52 (1%) to electric equipment, and 1862 (39%) to miscellaneous. Therefore, insofar as we can infer from the number of patents, R&D in the machinery industry seems to have been relatively active around the turn of the century.

Table 3.1. Gross National Expenditures and Production: 1875–1940

	Levels (at 1934–1936 prices)					Annual Growth Rates (%)				Composition (%)				
	1875	1885	1900	1915	1940	1875–1885	1885–1900	1900–1915	1915–1940	1875	1885	1900	1915	1940
Population (thousand)	35036	38176	44056	53110	71933	0.9	1	1.3	1.2					
GNE (million yen)		3852	6238	8522	22848		3.3	2.1	4		100	100	100	100
Personal consumption expenditure		3284	5270	6806	13389		3.2	1.7	2.7		85.3	84.5	79.9	58.6
Government consumption expenditure		283	538	769	3377		4.4	2.4	6.1		7.3	8.6	9	14.8
Gross domestic fixed capital formation		346	703	1176	7070		4.8	3.5	7.4		9	11.3	13.8	30.9
Surplus on current account		−61	−279	−224	−988		10.7	−1.5	6.1		−1.6	−4.5	−2.6	−4.3
Exports and factor income from abroad		68	275	1020	3973		9.8	9.1	5.6		1.8	4.4	12	17.4
Imports and factor income paid abroad		129	554	1244	4961		10.2	5.5	5.7		3.3	8.9	14.6	21.7
GNE per Capita (yen)		101	142	160	318		2.3	0.8	2.8					
Production (million yen)														
Mining	7.1	21.3	90.3	290.6	763.7	11.6	10.1	8.1	3.9					
Coal	4	9.2	53.2	145.5	399.8	8.7	12.4	6.9	4.1					
Manufacturing	742.6	877.9	2101	4029.4	20210.2	1.7	6	4.4	6.7	100	100	100	100	100
Food	457.5	454	1018.5	1356.3	2634.7	−0.1	5.5	1.9	2.7	61.6	51.7	48.5	33.7	13
Textile	64.1	129.7	508.7	1133.6	3454.3	7.3	9.5	5.5	4.6	8.6	14.8	24.2	28.1	17.1
Chemicals	75.6	99.1	186.4	427.6	3342.3	2.7	4.3	5.7	8.6	10.2	11.3	8.9	10.6	16.5
Machines	4.9	12.2	68.9	362.2	5580.5	9.6	12.2	11.7	11.6	0.7	1.4	3.3	9	27.6
Iron and steel	2.3	3.2	5.8	89.1	2494.8	3.4	4	20	14.3	0.3	0.4	0.3	2.2	12.3
Nonferrous metals	6.1	13.8	22.7	145.4	556.2	8.5	3.4	13.2	5.5	0.8	1.6	1.1	3.6	2.8
Others	132.1	165.9	290	515.2	2147.4	2.3	3.8	3.9	5.9	17.8	18.9	13.8	12.8	10.6

Source: Estimates of Long-Term Economic Statistics of Japan since 1868. K. Ohkawa, M. Shinohara, and M. Umemura (eds.). Tokyo: Toyo Keizai Shinposha, 1988.

Between the Two World Wars: 1914–1945

It is reasonable to say that the Japanese economy took off around the time of World War I. As will be shown later, many companies in steel, machinery, chemical, and other heavy industries had been established by the mid-1910s and started growing. Also, the country's scientific and engineering base started to be formed. The education system, which had been expanded by this time to include several universities and other higher education institutions, started to supply many trained engineers. More and more skilled workers capable of handling advanced equipment became available. Some of them spun off with their acquired skills. Academic associations were formed and professional journals started. Access to foreign technological information became easier with the extended availability of foreign books and journals, and through trading companies.

To enhance the scientific and technological base further, various proposals were made by industrialists, policy-makers, the military, and scientists; for instance, the creation of basic research institutions, the increase in national industrial laboratories, and enhanced science and technology education. The industries felt a strong need to upgrade its scientific and technological capability to accommodate the growth of technology-based industries. World War I accelerated this trend, by convincing the military of the importance of high technology and the private sector's capability to contribute to national defense. In addition, the import of equipment and intermediate goods became difficult during the war, giving the private sector an incentive to produce previously imported advanced equipment and materials domestically, such as machine tools, various chemicals, aluminum, and steel. This fact further enhanced the need for advanced technology.

More universities and vocational schools were established by the public as well as the private sector. Several national research institutions were founded. Hiroshige (1973) counts 38 national research laboratories founded during 1914–1930 including those annexed to national universities and to the military. Some of them were founded by reorganizing existing laboratories or testing institutions. For instance, the Electric Research Institution was founded in 1918 by upgrading the laboratory within the Ministry of Communications, which dates back to 1873 when Kobusho started an experimental electrical factory within it.[2] In 1890 it had a staff of about 30 with 7 engineers, the rest being assistants and shop workers. It studied both technologies related to electric power and those related to communication; however, the major part of its duty was to test the equipment procured by the Ministry. Another example is the Industrial Research Institution established in 1900 with a staff of 20. By 1920 the staff had expanded to 220, including 48 researchers and engineers. It had five departments—inspection, chemistry [japan (a type of lacquer), matches, oil, and fat, etc.], ceramics, dye, and (since 1909) electrochemistry. Other research institutions present or established during this period include those studying measures, silk, geography, mining, fuel, and airplanes (within the navy) besides several institutions related to agriculture.

One of the largest and most productive research institutions established during this period was the Institute of Physical and Chemical Research (*Rikagaku Kenkyusho,* known as *Riken*). The proposal for Riken was first made by a chemist, J. Takamine, with the aim of fostering scientific progress and thereby contributing to industries. Thus the aim was not purely academic but also practical. After a long discussion involving businesses, the government, and the academics, it was established in 1917 with roughly half of the funds from the government and half from the private sector, and with a staff of 22, of whom 5 were full-time (including both researchers and supportive staff). In 4 years the staff increased to 63, with 30 full-time members. By 1945 it produced about 800 patents in Japan and 200 overseas. Some of them were developed to products, which Riken's subsidiary companies sold. These include chemical products, such as vitamins and photographic paper, and machinery, such as piston rings and measuring equipment. In terms of the contributions to the national economy, these products were hardly as major as, say, automobiles and electrical products.

To promote more basic research, the Science Council (*Gakujutsu Shinkokai*, known as *Gakushin*) was established in 1933 with funds from both the public and private sector. Its purpose was twofold: first, to increase research funds at universities and other research institutions, and second, to promote efficient research management by, for instance, encouraging interorganizational research collaboration. The research funds were granted both on an individual basis and on a project basis. In the beginning, most grants were given on an individual basis but, by 1942, more than 80% were given on a project basis, with each project pursued by members from a number of institutions. Among the grants given on a project basis in 1942, about two-thirds were given to the projects in the field of engineering, and the rest to medical science, natural science, social science, and humanities. The top three projects in terms of research grants were jet fuel, wireless communication, and atomic nucleus. The emphasis on military-related technology was apparent.[3]

Companies also started their own R&D laboratories. In 1923, there were 162 private R&D laboratories affiliated with companies, cooperatives, and other private foundations. Of these 71 were in chemistry (including pharmaceuticals, dye, paint, rubber, cement, ceramics, and paper), 27 in metals and machinery, and 24 in food. Among the large companies having their own laboratories were Shibaura Seisakusho and Tokyo Denki (later to be merged to become Toshiba), Mitsubishi Shipbuilding (later Mitsubishi Heavy Industries), Nippon Kokan (now NKK), Oji Paper, and Takeda, Tanabe, and Sankyo (all in pharmaceuticals). However, they include testing or development sections within factories and how many of them deserve to be called laboratories in the present sense is unknown. Most of them were small.[4]

After the breakout of the Chinese-Japanese War in 1937, the production of heavy industries increased rapidly as the expansion of armaments speeded up. Compared to the 1934–1936 average, the production of steel more than doubled and machinery almost quadrupled by 1944. The share in total manufacturing production of heavy industries (machinery, metal, and chemical) increased from 33.9% in 1931 to 55.8% in 1937, and then to 70.2% in 1942.[5]

The R&D activity also increased. Although the reliability of the data may be questioned, one survey in 1930 shows the number of research organizations (including departments and institutions within universities, government laboratories, Riken, and laboratories affiliated with private companies) to be 349 in 1930, which spent 30 million yen or 0.22% of the GNP.[6] In 1942, the number of private research organizations was 711, employing a staff of 33,400 and spending 590 million yen or approximately 1% of the GNP. In addition, there were 443 public research organizations (including university departments, etc.) employing a staff of 16,160 and spending 296 million yen.[7]

These increased production and R&D activities enabled some of the Japanese manufacturing industries to start building world-class production facilities and developing advanced products; for instance, large-scale furnace and open hearth for steel production, aircrafts, ships, alloys, and communication equipment. However, even these industries depended on American and European technology in many aspects, and imported technology actively until the late 1930s. The stoppage of technological flow from abroad during World War II had a serious impact, and the increase in R&D efforts at the time reflects Japan's desperate effort to fill the void created by this stoppage. Consequently, despite the increased R&D effort, the technological gap from the West widened in such key munitions industries as aircraft and shipbuilding, which had almost caught up with the world technological frontier before the war. This gap partly explains Japan's defeat in the war.[8]

The Post-World War II Era: 1945 to the early 1970s

The impact of World War II on the Japanese economy was, needless to say, devastating. The production index of the manufacturing industry in 1946 fell to 26% of the prewar peak level in 1934–1936 and the supply of food to 51%. Many of the plants and equipment had been destroyed by bombing during the war and others were destroyed or taken away by the Allied Forces. Still, more than two-thirds of the production capacity had been left intact in most of the heavy industries. The R&D resources that had been increased before and during the war, and the many workers who had worked in the munitions plants and gained production experience returned to civilian production. Of course, not all of these resources were useful. Some of the facilities had been hastily built or were hastily converted with little use after the war, and some of the R&D knowledge or production skills were obsolete. Yet, as Japan's military spending after the war was practically zero, most of these resources could be transferred for the purpose of civilian production.

Owing to various drastic economic measures and the boom during the Korean War, the manufacturing production regained the prewar peak level within 5 years. From the mid-1950s to 1973 (the year of the first oil crisis), the economy grew at an annual rate of approximately 10%. Various growth accounting studies show that this fast growth, which lasted almost two decades, was the result of a high rate of capital accumulation combined with technological progress.

As implied at the end of the previous subsection, after the end of the war Japan had to realize that it had again been left behind. The process of catching up started again. As in the Meiji era, it followed a dual track: encouraging the importation of advanced technology and promoting a domestic technology base. To make an effective technology transfer, the government allocated its scarce foreign currency selectively to those firms capable of adapting and improving imported technology.[9] This gave the government, notably the Ministry of International Trade and Industry (MITI), a strong power over the industries. In fact, most of the firms with adequate technological background were given the allocation of foreign currency. However, a few years delay often occurred and, as Peck and Tamura (1976, p. 553) said, "in a fast-growing economy it can be of importance whether government policy delays or speeds up the use of a particular technology by only three or four years."

Nevertheless, despite the effect this control undoubtedly had on individual firms, it seems unlikely to have affected the overall pattern and pace of technology importation in a major way. A more profound influence, in our view, was provided by the restriction of import and direct investment because the only way for foreign firms to exploit their technological superiority was to sell their technology, even though they might have preferred to export the product or start their own production in Japan. That is, Japan's restriction of trade and investment seems to have made it possible for Japan to import technology only. Therefore, the trade and capital liberalization in the 1960s and 1970s inevitably caused MITI to lose its control over technology and industries in general.

Another important consequence was that the Japanese manufacturing industries remained mostly in the hands of Japanese shareholders, in marked contrast with European countries where many U. S. firms started their own businesses or purchased existing European firms as their subsidiaries. There are exceptions such as IBM, NCR, and a number of petroleum refiners, which established their subsidiaries in Japan using the "grandfather rights" to resume their pre-World War II operations or using the "yen-based stock purchase system" that allowed foreign direct investment into Japan with strict conditions during 1958–1963. However, major firms in the automobile, electric equipment, and steel industries remained Japanese owned and the stakes foreign firms had in Japanese electric and communication equipment companies in fact decreased during the postwar period. Even after the capital liberalization in the 1970s, foreign direct investment into Japan remained low. Thus the management of most Japanese firms has been, and in large part still is, not exposed to the pressure of the international capital market. The consequence of this on corporate motivation will be discussed later.

It is difficult to evaluate these policies. On the one hand, it is fair to assume that they have substantially benefitted domestic firms. On the other hand, they may have lessened market competition by restricting entry of foreign producers either through import or investment. In addition, the restriction of foreign direct investment prevented the inflow of capital when the Japanese economy needed it. Even though the intense competition among domestic firms, and the high saving ratio after the 1950s mitigated these negative effects, the overall effect has to be assessed in a broader context.[10]

We also note that this government policy was feasible only because there were many firms eager to import technology in expectation of a high return. Since these technology-importing firms could start manufacturing those products that had been imported previously, import substitution took place reducing imports and eventually increasing exports, thereby enabling Japan to economize on precious foreign currency. The amount of foreign currency thus saved by far exceeded the payments to technology importation. Through the 1950s and 1960s the limited availability of foreign currency worked as a ceiling to economic growth, that is, economic booms could not be sustained because of acute shortage of foreign currency in boom years when imports increased. Technology importation had the long-run effect of pushing this ceiling upward and making fast economic growth sustainable.

In evaluating the role imported technology played in Japan's development, one should also note the following two facts. First, trade in technology became very active after the war in every country. Although Japan's payment to technology importation (64 billion yen) was the largest among major countries even as recently as 1988, its percentage to GNP, 0.17%, was about the same as in France (0.18%), Germany (0.17%), and the United Kingdom (0.16%), though larger than the United States (0.04%) (STA, 1990). That is, under the GATT-IMF regime, trade in both goods and technology increased worldwide and Japan is hardly peculiar in this regard, though, no doubt, it was one of the major beneficiaries of this trend.

Second, technology transfer would not have succeeded without the prewar technology base and a rapid postwar increase in R&D expenditures. As discussed earlier, heavy industries, such as machines and steel, had a large share in the 1930s and early 1940s, owing partly to the military build-up. When the war ended, a large number of researchers, engineers, and skilled workers were released from such production and could be utilized in civilian production. As for the contribution of domestic R&D effort to technology transfer, empirical findings indicate that those who imported technology also invested in their own R&D.[11] Domestic R&D was essential to enable firms to evaluate, adapt, and improve imported technology.

Emphasis on Own Innovation: Since the 1960s

As Japan became a serious competitor in the international market, the terms of technology importation became less favorable. In addition, as Japanese industries advanced technologically and became world-class, the backlog of technologies available for import decreased. Consequently, during the 1960s when the Japanese economy grew at a high rate and started to compete internationally, the need for increased efforts to develop its own technology became more urgent and government policies to promote domestic R&D began to be emphasized. However, the size of the incentives provided through tax breaks, subsidies, and low-interest loans were modest (see Table 3.2). In total, they amount to a little more than a hundred billion yen in 1983, or 2.6% of industrial R&D expenditures and 2.3% of industrial R&D expenditures plus payment to technology importation. Thus, financial incentives by the government to the private sector has been modest and doubtlessly smaller than in other countries where as much as a one-third of industrial R&D is financed by the government.

Table 3.2 also reveals a clear downward trend in the importance of subsidies: the proportion to industrial R&D was almost 8% in 1960, still lower than in other countries, but gradually declined to 2.6% in 1983. In parallel with the generally decreasing role of industrial policy in Japan's economic growth, government support to industrial R&D has been decreasing rather dramatically in the past two to three decades.

Private firms, under increased competition from other Japanese firms and from American and European firms in foreign and domestic markets (owing to liberated trade and direct investment), felt an urgent need to enhance their technological capability. Their R&D expenditures more than tripled in the latter half of the 1960s. Relative to these increased private efforts, the government policy measures were modest indeed and, in our opinion, most of the R&D projects the government supported would have taken place anyway.

One policy measure that has attracted considerable interest in the West is joint or cooperative research efforts based on the Act on the Mining and Manufacturing Industry Technology Research Association enforced in 1961. During 1961–1987, 87 such research associations (RAs) were formed. In 1983, there were 44 RAs of which 38 received subsidies of 33 billion yen, or about half of their total R&D expenditures

Table 3.2. Government Support to Industrial R&D in Japan[a,b] (in Billion Yen)

Year	(*A*) Total	(*a*) Subsidies and Research Contracts	(*b*) Preferential Tax Treatment	(*c*) Low Interest Rate Loan
1960	9.8	0.7	9.1	—
1965	16.4	3.1	13.3	—
1970	31.0	11.0	19.1	0.9
1975	64.7	29.8	33.0	1.9
1980	101.0	60.8	38.0	2.2
1983	117.7	58.7	57.0	2.0
Year	**(*B*) R&D Expenditure by Industry**	**(*A*)/(*B*) (%)**	**(*C*) Payment for Technology Importation**	**(*A*)/(*B* + *C*) (%)**
1960	124.4	7.88	34.2	6.18
1965	252.4	6.50	59.6	5.26
1970	823.3	3.77	155.1	3.17
1975	1684.8	3.84	211.3	3.41
1980	3142.3	3.21	326.2	2.91
1983	4560.1	2.58	493.8	2.33

[a]Adapted from Goto and Wakasugi (1988), Table 1.
[b]*a* is the amount of subsidies and research contracts. *b* is the amount of tax forgone through the preferential tax treatments to promote R&D. *c* is the interest payments savings due to the low interest rate loan to promote R&D. $A = (a) + (b) + (c)$.
Sources: (*a*) Somu-cho, "Kagaku gijyutsu kenkyu chosa hokoku" (Report on the Survey of Research and Development), each year.
(*b*) Documents submitted to Tax System Council.
(*c*) Estimated from Japan Development Bank documents.

(Wakasugi and Goto, 1985). Since the total government R&D subsidies were about 59 billion yen in the same year (see Table 3.2), we infer that more than one-half of them were supplied through RAs. This reflects an important aspect of RAs: for MITI, RAs have been a convenient way to distribute its subsidies to promote the technology MITI (and participating firms) believed important, most notably semiconductors and computers, and have been used to avoid favoring particular firms and to minimize the cost of supervising the use of subsidies. From this viewpoint, it is not surprising that only two of the 87 associations had joint research facilities; in all other cases, each member firm simply took its share of research funds and carried out the research in its own laboratory. Therefore, how coordinated the research really was among participating firms within each RA is doubtful except for a few cases.[12]

The effectiveness of these RAs in generating new technology is also doubtful. According to Wakasugi and Goto (1985) and Fujishiro (1988), RA's productivity as measured by the number of patents divided by its R&D expenditures was considerably lower than that of industries, though neither of these authors implies that it is the best measure of R&D efficiency.[13] Perhaps RAs were the product of the time of transition: the firms were then feeling a need to promote more long-range and large-scale, if not basic, research but were still unable, technologically and financially, to conduct it individually. Their role has been declining as more and more collaborative research activities are now carried out by research institutions funded jointly by companies or under intercompany technology agreements.

The first oil crisis of 1973 ended the high-growth era. Around the same time, Japan adopted a flexible exchange system, resulting in a major appreciation of yen. The annual growth rate fell from roughly 10% of the high-growth era to less than 5% in the 1970s and 1980s. There were two consequences of these drastic changes on technology. First, developing energy-saving production processes became one of the major targets in R&D for many businesses. Such R&D, together with the accumulation of incremental innovation at production sites, made it possible for Japanese manufacturing industries to increase energy efficiency. For instance, the steel industry developed the so-called oilless steel-making process and reduced energy consumption drastically. This shift in emphasis toward energy R&D was also apparent in government R&D: the proportion of energy R&D among total government R&D funding increased from 7.5% in 1975 to 16.3% in 1985 (National Science Foundation, 1988, p. 55). The Sunshine Project aimed at developing alternative energy-generating technology and the Moonlight Project to develop energy-conserving technology started in 1974 and 1978, respectively.

Second, the entire industrial structure shifted toward an energy-saving, technology-intensive, and high value-added one. Energy-intensive industries scaled down and high-technology industries expanded rapidly. The striking example of the former was the aluminum smelting industry. Japan was the second largest aluminum-producing country in the world in the early 1970s. After a decade and two oil crises, it had been virtually wiped out while the import had soared (Goto, 1988). By contrast, high-technology industries—semiconductors, computers, fine chemicals, and such—grew rapidly.

Through the 1970s to 1980s, therefore, Japanese firms had to cope with the drastically changing environment. Yet, after the two oil crises and two sharp appreciations of yen (65% against the U. S. dollar in 1977–1978, and 92% in 1986–1988), Japanese firms found their share to have considerably increased in the world export market. This increase was most prominent in high-technology products where the Japanese share rose from 7.2% in 1965 to 19.8% in 1986. We may attribute this success in Japan's adaptation to changing environment to two factors. The first is the high rate of investment in R&D and in plants and equipment. The R&D expenditures grew 4.4 times from 1973 to 1987 and, as a proportion to GNP, from 2.0 to 2.8%. The ratio of investment in plants and equipment to GNP fell gradually after the first oil crisis from about 19% in 1970, but remained at 14 to 15%, and then grew in the 1980s to reach 19% again in 1990. The amount of investment in plants and equipment in Japan even surpassed that of the United States, whose economy is 1.8 times larger than Japan's. These active investments in R&D and in plants and equipment reinforced each other, increasing productivity and improving product quality. In addition, as the second factor, the management system of Japanese firms contributed to their adapting to the rapidly changing environment, as we shall discuss later.

With increasing strength of the Japanese firms in the world market, the focus of the science and technology policy is gradually shifting from the promotion of industrial technological capability to emphasize such new issues as the promotion of basic research and the globalization of innovation. These issues will be discussed in the final section.

TECHNOLOGICAL DEVELOPMENT IN THREE INDUSTRIES

In this section, we will describe the development of three industries, iron and steel, electrical and communication equipment, and automobiles, since the Meiji era; this will complement the general historical description previously given.[14]

Iron and Steel

In 1901, the government constructed Yawata Steel Works (and then ran it) wholly dependent on German technology. This case, it has been argued, epitomizes the two salient characteristics of Japan's industrial and technological development—government support and technology importation. Is such an argument justified? Are these two characteristics really evident? A close examination of the history of the iron and steel industry actually gives a different picture.

The first attempt to produce iron and steel using Western technology was made almost a half century before Yawata. This was during the 1850s, still in the Tokugawa era but after the government opened the country to the United States, the United Kingdom, France, and Russia under military threat from these countries. Before this period, iron and steel had been produced from iron sand and charcoal using an indigenous small-scale production method. However, with the threat of Western countries, the Tokugawa government and some of the powerful feudal lords considered it urgent to construct strong cannons. Because this created a demand for quality steel, they constructed reverberating furnaces copying the technology described in a Dutch book (an example of technology import through the Dutch during the Tokugawa era) but entirely with Japanese hands. These furnaces produced wrought iron on a small scale out of pig iron made with indigenous technology. One of the engineers who made these furnaces, T. Ohshima, proposed building a blast furnace to produce pig iron. Ohshima was a samurai of Nambu Clan, northeast of Honshu. Nambu had an iron mine and thus a history of iron making. Adding to this background, Ohshima went to Nagasaki to study Dutch and learned the steel- and cannon-making technology with the book mentioned earlier. He thus knew both indigenous and Western technology, and, with the financial support of private investors, constructed a small blast furnace in Kamaishi, a city near an iron mine in Nambu. Although the production met with difficulties at first (in firebricks, ventilation, etc.), in about a year the production started to go smoothly and several more furnaces were built.

In 1874, the Meiji government decided to build a modern furnace in Kamaishi and hired Ohshima and a German engineer to plan it. The German engineer proposed building a large-scale modern mill (with a railway system to carry the product and iron ore) around two imported furnaces. Ohshima opposed this plan and instead proposed building five relatively small furnaces and using more modest transportation method. The government, which never questioned the superiority of Western technology, rejected Ohshima's plan and imported furnaces, railway cars, and other equipment from Britain and hired British engineers to construct a mill based on the German engineer's plan. It started the operation in 1890, immediately met with difficulties, and was closed down after 196 days of operation. The cause was obvious. At first, the mill used charcoal but, after a short while, the supply of such a large quantity of charcoal was found out to be impossible. It thus decided to switch to coke as fuel but the coke it could obtain did not have a sufficiently high quality. Needless to say, technology cannot be free from social and economic conditions. Apparently, neither the European engineers nor the Meiji government was aware of this simple principle in the belief of superior Western technology. By contrast, Ohshima, who knew the way iron had been produced by that time, was aware of the conditions surrounding iron production in Japan.

Kamaishi restarted its iron production several years after the government closed the mill. C. Tanaka, a private entrepreneur who bought the mill from the government, restarted the production not with the big furnace but with a new small furnace (built after Ohshima's technology). Again several improvements had to be made before the operation became commercially viable, but he could expand the production gradually and in 1894, with the help of K. Noro, a professor at the Imperial University (now University of Tokyo), repaired and improved the British-built big furnaces and succeeded in restarting them.

The lesson of Kamaishi appears to have been forgotten by the time the government built an integrated steel mill in Yawata, Kyushu. The government had been experimenting with steel production within the arsenals using the pig iron produced in Kamaishi or with traditional technology, and imported iron. However, from the military viewpoint it was considered necessary to produce steel using a modern integrated technology and on a large scale. Thus the government decided in 1896 to build a mill in Yawata and let M. Ohshima (T. Ohshima's son) make the plans. Ohshima had studied in Germany and decided to build the mill with German technology, commissioning a German company to make a detailed plan, buying German equipment, and hiring German engineers for the top engineering posts. Completed in 1901, it was a spectacularly modern and large plant. The excitement, however, did not last long. The operational problems mounted and the mill was forced to stop operations in about a year. The major problem was the different quality of coke between Germany and Japan. Yawata was chosen as the site basically because there was a big coal mine nearby, and the mill used the coke produced with this coal. The coke production mill was designed by Germans as well, who never understood the difference between German and Japanese coal and were also under a budget constraint by that time. Consequently, the coke mill was not only inadequate for Japanese coal but also inefficient owing to the use of cheaper obsolete equipment. The coke thus produced was inevitably poor in quality, and made steel production extremely difficult.

Again, like in Kamaishi, Noro was asked to improve the situation. Noro had earlier proposed his own plan for the Yawata mill, which, not surprisingly, suggested a more modest production scale and a gradual expansion. But the government rejected it and he had resigned from government posts for an unrelated reason. Noro made a thorough investigation of Yawata, improved the coke production process, and made several changes to the furnaces to fit them to the attributes of Japanese coke. These changes took a few years but in 1904 the mill restarted its operation successfully.

Yawata Mill was owned and run by the government, mostly because of its role as the supplier of steel to weapon production. Though it was obviously a dominant steel producer, private enterprises were not absent. In addition to Tanaka's Kamaishi mill, Sumitomo, Kawasaki, and Mitsubishi (in occupied Korea) entered into the market and so did Nippon Kokan (now NKK) with the initiative of K. Imaizumi, Noro's student, who had resigned from Yawata when his proposal to privatize Yawata was not accepted. Hence, private profit-seeking entry did take place. Because of the large initial investment, many of the entrants were zaibatsu companies or financially supported by zaibatsus (for instance, Asano, Okura, and other relatively small zaibatsus invested in Nippon Kokan). Some of them built integrated mills, but some imported pig iron and scrap iron from India, America, and other countries. In 1934, a partly government-owned company, Nippon Steel, was established to achieve a stable supply of steel (which the army considered indispensable). Yawata was absorbed into this company and the government urged private steel companies to join. Kamaishi and a few other companies complied with this request but Nippon Kokan, Sumitomo Metal, Kawasaki Steel, and other companies kept their independence.

Since the aim of this subsection is to show that Japan had a certain technological background when it decided to import Western technology and that technology importation alone cannot guarantee successful industrial development, we shall discuss the postwar development only briefly. In terms of innovation, two developments had a significant impact. The first is the importation of the basic oxygen furnace method from Austria and its improvement and adaptation to Japanese steel mills during the 1950s. Japan was ahead of other countries not only in the speed of adoption but also in the extent it improved the technology (see Lynn, 1982). For instance, Yawata invented the oxygen converter gas recovery system as an effective device to save energy use and reduce pollution, which has been used worldwide. The second is the introduction of the continuous casting method and strip mills during the 1950s and 1960s.

Two factors explain why these technologies were most effectively utilized in Japan. One is the strong propensity of Japanese firms to invest in production processes. Since both of these new technologies have to be embodied in capital equipment and require a large-scale integrated production flow, their advantage can be exploited only with new investment. Japanese iron and steel manufacturers in fact invested in a number of new and integrated large plants, and productivity has increased tremendously. For instance, between the two plants of Nippon Steel, a new one at Kimitsu had a labor productivity (steel production per worker) 2.5 times higher than an old one at Yawata in 1973, though Yawata's productivity itself had increased several times in the preceding 30 years (Iida, 1979). The other factor is the constant effort to improve efficiency within the plants. Some of them were conspicuous, for instance, the introduction of computers to control operation; however, some were made more at the shop-floor level based on learning by doing and workers' proposals. As we shall discuss later, Japanese firms have been relatively successful in attaining companywide involvement in productivity increase.

As a result, Japan's productivity has outpaced the other countries and the companies started to export technology, for instance, imparting know-hows in plant construction and operation to many countries including both developing countries such as Brazil and developed countries such as Italy and the United States. Since 1974 its royalty receipts have exceeded its payments, the first among Japanese industries.

Let us conclude this subsection. In terms of industrial policy, the role played by the government and the military in the development of the steel industry was obviously large (as in many other countries). Steel production is inseparable from the military needs, as exemplified by the fact that the first steel production under Western technology was made to make cannons. However, both the cases of Kamaishi and Yawata imply that the government is less qualified than the private sector in selecting appropriate technology. In either case, the government placed too much confidence in the Western technology as it was and disregarded local conditions, such as infrastructure and the accumulated indigenous technology. By contrast, T. Ohshima's and Tanaka's attempts at Kamaishi suggest that private enterprises were better suited for this purpose.

Electrical and Communication Equipment

The electrical and communication equipment industry gives another example of a combination of Japan's indigenous technology or its own R&D efforts with imported Western technology. Such a combination was feasible and effective because, though Japan technologically lagged behind America and Europe (as in other modern industries) in the late nineteenth century when electrical and communication equipment started to be introduced, the lag was not large. Take an example of the electric light. T. W. Swan invented an electric incandescent bulb in 1878 in the United Kingdom and T. A. Edison, in 1879 in the United States. The first incandescent light was exhibited in Japan in 1884, only 5 years after Edison's invention. Needless to say, all the equipment used in this exhibition was imported. However, in a similar exhibition in the following year, a domestically produced generator was already used. This generator was designed by I. Fujioka, a young professor at the College of Engineering (later the Engineering Department of the University of Tokyo, as explained earlier) who had studied with a British professor at the College. Fujioka then undertook to make a bulb. He left the university to join Tokyo Electric Light Company (now Tokyo Electric Power Company), was sent to America and Europe to learn the technology there, bought bulb-manufacturing equipment in the United Kingdom, and came back in 1887. He started the development effort and then, to continue this effort independently from Tokyo Electric Light Company, established a new company, Hakunetsusha, in 1890. Even with the equipment brought from the United Kingdom, the development met with continuous problems. For instance, following Swan's invention, his team first used cotton yarn for filament, but the result was unsatisfactory, and it took some time before they learned about Edison's use of Japanese bamboo and did the same. Hakunetsusha started selling the bulbs in 1890. Thus, although the United States was unquestionably earlier in innovation and was dominant in production size, the start of the industry in Japan did not lag by more than a generation.

As discussed earlier, Japan did not have a means to protect the domestic industries because of Unequal Treaties. Hakunetsusha met with fierce competition from imports. Its bulb cost about 60% more than the imports. The company, however, managed to survive with the booms after the Sino-Japanese war of 1894–1895 and the Russo-Japanese war of 1904–1905, and also because of the international price increase caused by an electric bulb cartel of European manufacturers in 1903. In the meantime, it continued its effort to improve the product and made two innovations, a dual coil bulb and a frosted bulb, that were copied worldwide.

Hakunetsusha consists of one-half of the present Toshiba. The other half came from Tanaka Seisakusho, established by H. Tanaka. Born in 1799, Tanaka was one of the most original and productive inventor–engineers (*karakuri* masters) in the Tokugawa era. At the age of 14, he already invented a loom. At 20 he made a "karakuri doll" with a hydraulic mechanism, which became very popular. The list of his inventions and developments is surprisingly comprehensive—a clock, a torch lamp, a furnace, a pump, a fire extinguisher, a tobacco cutter, an ice maker, a bicycle, an oil press, a rice-polishing machine, a ship, a dredger, guns, and many more. He even made the first steam engine in the country based on Western technology. In 1873, at the age of 74, he was invited by Kobusho (the Ministry of Industries) to come to Tokyo from his hometown in Kyushu to make telegraphs at the ministry's small factory. After this experience, he established his own company, Tanaka Seisakusho, in 1875 and made telegraphs, switches, and miscellaneous electrical and other equipment. For Tanaka, therefore, technological development was continuous, with little discontinuity between indigenous and Western technologies. Tanaka's role in Japan's industrial development was important also because among those under Tanaka's direction in the Kobusho factory were several engineers who later became pioneers themselves. These include S. Miyoshi, who later helped Fujioka in establishing Hakunetsusha, and K. Oki, the founder of the present Oki (to be discussed soon).

Although H. Tanaka died in 1881, his son took over and expanded the company with the production of torpedoes and mines at the request of the navy. In a few years it became one of the largest manufacturing companies at the time, but as the navy started to use competitive bids and then completed its own work, the demand decreased substantially and the company started to lose money. The main creditor to the company, Mitsui Bank, took over the virtually bankrupt company and renamed it Shibaura Engineering Works. The company's main products were steam engines, power generators, and other heavy electrical equipment.

Even though Shibaura had developed its own generators (after copying imported ones), it found its technology still behind the American's. Thus in 1910 it formed an affiliation with GE of America. GE acquired about a quarter of the share and allowed Shibaura the use of GE's technology. For instance, engineers were sent to GE to learn its technology and production methods.

Hakunetsusha, renamed Tokyo Electric in 1898, was also affiliated with GE, with GE acquiring a 55% share. Since it started with the production of electric bulbs, the main business was light electrical equipment and appliances, in contrast to Shibaura, which was biased toward heavy equipment. Thus the two were considered complementary and, with GE's support, the two merged in 1939 to become Tokyo Shibaura Electric (later Toshiba). GE owned 33% and Mitsui owned 15%. The relation with GE continued until the beginning of the war and, after the war, resumed in 1953 with GE's 24% share, though this percentage has decreased substantially since then.

Toshiba's main rival, Hitachi, is rather unique among the Japanese producers in that it relied on its own R&D efforts (together with reverse engineering) without forming any affiliation with foreign companies. It was established by N. Odaira, a graduate of the University of Tokyo's Engineering Department, in 1911 as a plant affiliated to Hitachi Mine, and became independent 9 years later. It started with the production of generators, blowers, transformers, and other heavy electrical equipment to be used in mining, but soon began selling to other customers and expanding the business to include industrial machines and lighter electrical equipment.

Hitachi never entered into any affiliation with foreign companies, in contrast to Toshiba, which affiliated with GE, Mitsubishi Electric, which affiliated in 1923 with Westinghouse (with 10% ownership), and Fuji Electric, which started in 1923 as a joint venture between Siemens (30% ownership) and Furukawa Mine. That Hitachi became the largest among these four main manufacturers must suggest either that Japan's technological lag was not serious or that Japanese engineers were capable of developing the products by themselves with the help of reverse engineering. It must also be noted that even companies with foreign affiliations made a significant amount of its own R&D efforts. Thus the two predecessors of Toshiba, Tokyo Electric and Shibaura Engineering, established their laboratories in 1918 and 1921, respectively, while affiliated with GE. Hitachi established its laboratory in 1934, though it started an independent R&D section much earlier in 1918.

Let us turn to communication equipment. The Meiji government started telegraphs as a government business in 1868 with a British engineer using equipment imported from Britain. Kobusho, the ministry, wished to supply them domestically and, as mentioned above, asked H. Tanaka to undertake it. Tanaka succeeded in making them in Kobusho's experimental factory and later in his own Tanaka Seisakusho. In 1877, only a year after A. G. Bell's invention, Kobusho experimented with the first telephone in Japan using American equipment. The engineers at the Kobusho factory started its domestic production by copying and in the next year came up with its first product, however unsatisfactory the quality was. One of the engineers, K. Oki, felt the business had a bright future and left Kobusho to start his own company named Meikosha (later Oki). The telephone business itself was not established until 1890, partly because of scepticism about its merit (for instance, hiring a messenger was thought to be a cheaper and more reliable way to convey information) and partly because of the controversy within the government as to whether the business should be run by the public or private sector. However, soon after the start (as a government business), the public was easily convinced of its convenience, and the demand surged. Both imported and domestic equipment (receivers, switchboards, wire, etc.) were used and, as the sole domestic producer, Meikosha prospered.

Western Electric (WE) of the United States saw a growing market in Japan and decided to enter through a joint venture. At first, it thought of teaming with Meikosha, but the prospering Oki proposed conditions too severe for WE to accept. WE finally decided to team with K. Iwadare, who had been WE's agent in Japan, and established Nippon Electric Co. (NEC) in 1898 with WE owning 54% of the share and Iwadare owning 33%. To start the business, NEC bought one of the largest manufacturing factories at the time, owned by S. Miyoshi but on the brink of bankruptcy as a consequence of its too rapid expansion and the depression at the time. As mentioned earlier, Miyoshi was another Tanaka disciple at the Kobusho factory and also a collaborator with Fujioka in establishing Hakunetsusha. Hence, one can see a historical connection from Tanaka, the master of pre-Meiji technology, not only to Oki but also to NEC, in addition to Toshiba.

Iwadare's career illustrates another (and newer) type of engineer at the time, that is, university educated. He studied electrical engineering at the College of Engineering a year after Fujioka, worked at Kobusho for 4 years, went to the United States and got a job at GE, came back to Japan after about a year to be an engineering manager of Osaka Electric Lamp Company, and, after 8 years, quit this post to start his own business as an agent for GE and WE. Thus, his career, like Fujioka's, is more "Western" than Oki's or Miyoshi's in that he studied technology in a university designed and taught by Westerners, and then was employed as an engineer in large companies at the time.

In the beginning, NEC not only produced telephone and other communication equipment with WE's technology, but also copied WE's management methods (including the accounting system and work management) and bought WE's supplies ranging from machines and materials to notepaper. However, it gradually replaced them with their own or with products procured domestically. With the expansion of the telephone system, the company expanded rapidly and competed against Oki and new entrants including Toshiba, Hitachi, and Fuji (whose communication equipment business was later separated as Fujitsu). NEC started the R&D department in 1926, which was expanded into a separate laboratory in 1939. From just assimilating WE's technology at first, it gradually increased its own R&D efforts and started producing innovations that surpassed WE's.

The technological and industrial development of Japan's electrical and communication equipment industry shows those characteristics that can be commonly found in many other industries. First, Japan was technologically behind America and Europe but the lag was in a tolerable range—a marked difference from present developing countries. Second, as a consequence, imported technologies were not entirely discontinuous from indigenous technologies. Third, therefore, some of the entrepreneurs/engineers could utilize the skills and knowledge accumulated in the pre-Meiji era. Fourth, at the same time, those engineers educated by the colleges established by the Meiji government and taught by foreign professors greatly contributed to assimilating Western technology. Fifth, foreign technologies were actively imported through joint ventures, technology contracts, reverse engineering, and such. Japanese had an ability to absorb or adapt the imported technologies, and an ability to run the joint ventures successfully. It should also be noted that a strong desire to become independent of foreign dominance of technology—one may call it nationalism—was common among the entrepreneurs/engineers as illustrated most clearly by Hitachi's Odaira. Finally, a strong entrepreneurship and a strong will to enter into uncertain and unknown fields were present. This conclusion does not imply that the government played no role. It did play an important role in establishing science and engineering education systems, introducing new technologies such as the telegraph, telephone, and weaponry, guaranteeing demand through military procurement, and so forth. In our view, however, such policies would have never resulted in Japan's development were it not for the willingness and capability in the private sector.

This conclusion is reinforced by the postwar experience. Although the government continued to affect the industry through its nurturing policy during the 1950s and 1960s, procurement by Nippon Telephone and Telegram, and research done within MITI's Agency of Industrial Science and Technology and NTT laboratories, its significance was more limited than previously because of the minimal defense research and military procurement. As for the nurturing policy, the computer industry was the main subject. Yet subsidies given to the industry, its sizable portion being granted through research associations discussed earlier, were very small compared to the total R&D expenditures of the industry. Apparently, it was the competitive threat from domestic rivals and foreign giants such as IBM that fostered firms to R&D activities. Their R&D expenditures increased more than 40-fold during the 20-year period from 1965 to 1985.

The importation of technology and its improvement within Japanese laboratories and factories continued. After World War II, almost all the firms resumed their relation with American or European firms. Even firms such as Hitachi, which kept its independence before the war, entered into contracts to import technology from companies such as RCA, GE, Phillips, and Western Electric. Royalty payments thus greatly exceeded receipts. However, with continuous innovation by Japanese firms, the receipts have increased much faster than the payment: 13.4 times as opposed to

2.5 times from 1973 to 1988. Hence, though payment still dominates receipt, the difference has narrowed considerably: in 1988, payment was 1.65 times larger.[15]

Automobiles

The history of the development of the automobile industry also shows similar characteristics: the dominance of foreign producers in the beginning, a number of (domestic) entry attempts, the acquisition of technology first with reverse engineering, the gradual accumulation of technologies with learning by doing, the procurement by the military and the transportation authority, support by the government through financial incentives and standard setting, and the entrepreneurs' risk-taking behavior in the form of physical investment and R&D. After several decades of struggle, Japanese car manufacturers have now caught up with the European and American rivals technologically, and are now surpassing them in certain areas. It is therefore worthwhile to trace the Japanese experience in this industry in some detail.

The first car introduced into Japan was an European-made steam-driven one imported in 1897. However, it was in 1903 when two cars (one steam-driven and the other electric-driven) were used as buses in a large exposition that many Japanese people saw cars for the first time. One of the visitors to this exposition saw them and started an effort to make one domestically. This effort produced one steam-driven car, which had so much trouble (particularly in terms of the tires) in the test run that he gave up any commercial application. Four years later, the first domestic production of a motor-driven car was made with the main parts (including, it seems, engines) imported at first but gradually replaced by domestic production. This effort also collapsed after selling 10 cars, 8 of which were bought by the police to add to its stock of 41 European cars and 11 American cars. In the meantime, import of foreign cars increased and completely dominated the market: more than 200 cars had been imported by 1910 and 6800 by 1920. The Great Earthquake in Tokyo of 1923 increased this trend further because all the train services had been disrupted and cars were urgently needed to deliver food and other necessities. Thus by 1925 more than 16,000 cars had been imported, and in this year Ford established an assembly plant in Yokohama. This was followed by GM, which established a similar plant in Osaka 2 years later. Thus the market became dominated by these two American producers.

More than 10 attempts had been made during the Meiji era to produce cars domestically but all failed, and the first significant one was made by a company named Kaishinsha established in 1911 by M. Hashimoto, who had studied engineering in the United States for 3 years. He started the business by importing and repairing British cars, and then with the technology learned from these British cars and his experience with American cars, began to develop a passenger car domestically. His first car was a failure because of lack of the casting technology needed to make a sufficiently durable engine, but he finally succeeded in producing a viable product in 1913. He began selling this car with a brand name "Dat," but, in competition with Ford and GM, could not sell in sufficiently large numbers to make the production commercially viable.

The first government initiative was made by the army who, after a survey of military procurement policy in Britain, France, and Germany, concluded that it was urgent to foster domestic production of trucks and buses for military purposes. In 1918 it started a policy of subsidizing domestic production and civilian purchase of trucks and buses. Although the subsidy itself was modest, this policy gave an expectation of increased procurement and three companies started production. One was Kaishinsha, who, with the difficulty in the passenger car market, decided to enter into the production of trucks; the second was Tokyo Gas Electric (now defunct), which inherited the trial truck production in the arsenal; and the third was Tokyo Ishikawajima Zosensho (now IHI), which, with the profits from shipbuilding during the World War I boom, entered into the production of passenger cars and then trucks, first by disassembling a Fiat car to learn the technology and then by buying technology from Wolseley, a British company. Despite the army's support, domestic production was small. During 1919–1931, these three companies together produced 2575 cars and the market (including trucks and buses) was totally dominated by Ford, GM, and other imports.

The next governmental initiative occurred in 1930 when the Ministry of Railways made a detailed open test of the three domestic producers, and then in 1932 when the Ministry of Commerce and Industry (now MITI), in collaboration with the Ministry of Railways, urged the three companies to collaborate in a joint venture to develop a so-called "standard" car (truck and bus) with the brand name "Isuzu" (the ancestor of current Isuzu). The joint venture lasted for only 2 years with the production of 750 cars, but the experience was utilized in these companies' efforts to produce their own trucks and buses (both for military and civilian purposes).

Although the major part of Kaishinsha was absorbed in this joint venture and later by Isuzu, the business of small passenger cars (called "Datsun" by this time) was retained by G. Ayukawa, an energetic entrepreneur in the process of expanding its Nissan Zaibatsu (one of the so-called new zaibatsus), who had acquired Kaishinsha by that time. Ayukawa established Nissan Motors for this purpose. Another origin of Nissan Motors was a company called Jitsuyo Jidosha, which was started with the technological guidance of American engineers using the engines of American Harley-Davidson motorcycles. Hence, Nissan's technology originated both from a Japanese's effort to copy British and American cars and the technology brought by hired Americans.

The number of entry attempts around this time was surprisingly large. To trucks, Mitsubishi and Kawasaki made some efforts. To motorcycles and three-wheel cars, more than 10 companies made efforts, some of whom, like Daihatsu and Mazda, survived until later, and some of whom eventually exited. A number of efforts were also made in the passenger car market, including Mitsui Shipbuilding, but mostly by small-scale nonzaibatsu companies. Most failed but an important exception was Toyota. Two facts deserve emphasis. The first is the frequency of entry attempts, clearly indicating the presence of risk-taking spirits and the will as well as capability to challenge new technologies. In short, Schumpeterian entrepreneurship was abundant. Second, though this entrepreneurship was present among established zaibatsus, their role was limited. In fact, they gave up rather early. By contrast, independent entrepreneurs such as Ayukawa, Toyoda (Toyota), and Matsuda (Mazda), not to mention Hashimoto and others in the early period, were more persistent.

Sakichi Toyoda was a typical inventor and entrepreneur of pre-World War II Japan, much like Thomas Edison, though in a different field. Born in 1867 in a poor carpenter's family, he invented the first (wooden-made hand-driven) loom in 1890 after several years of struggle. He started selling the looms and making cloth using them, but the business did not prosper. The struggle continued until he invented automatic looms in 1897. This time the business was a success with a strong demand both for the looms and the high-quality cloth made with them. He worked further to improve the loom, and to make a wider cloth faster and more efficiently. The company started exporting the products and the quality of his loom was proved when the world's largest loom manufacturer, Pratt (U. K.), bought the technology for one hundred thousand pounds. It is worth noting that Toyoda's development effort over his lifetime was a purely private and independent enterprise. Though Mitsui Bussan (the largest trading company at the time, now Mitsui Co.) at one time gave financial support to Toyoda's enterprise, Sakichi parted the relationship fairly soon after a disagreement.

Sakichi's son, Kiichiro, plowed back the profits from loom sales and the fee from Pratt into the development of cars. He organized a development team by hiring engineers from fields such as steel, tools, and three-wheel cars from other companies. He visited several professors to learn about technology, some of whom were his former classmates when he studied at the University of Tokyo's Engineering Department. And he sent his engineer to the United States to visit the Ford plant and get ideas on how to make a factory, and purchase necessary machines. The team started by disassembling American cars. Finding that the American technology was far advanced, they decided to imitate Chevrolet's engine and use many of Chevrolet's and Ford's parts. As in the previous experience of Kaishinsha, casting was a problem. Although Toyoda had the casting technology to manufacture looms, they needed a more complex and finer technology to produce an engine. After a seemingly endless process of trial and learning in casting and other aspects of technology, the first viable product, a truck, was introduced in 1935 to sell to the army. A passenger car followed in 1936.

An important policy initiative was taken in 1936 to foster domestic car production. The government restricted the production to licensed companies who received financial and other support. In contrast, the biggest producers at the time, Ford and GM, were banned from increasing their production level in the future. The license was given to Toyota (Toyoda's brand name for cars), Nissan, and (later) Isuzu. Toyota and Nissan thus became the dominant producers and increased production. In 1938, the production level of Toyota and Nissan combined was 6568 for nonmilitary sales versus 12,441 of Ford and GM. For military sales, they were 5930 and 5999, respectively, showing the army's willingness to support domestic producers. Both Ford and GM exited the Japanese market in 1939, and the production level of Toyota and Nissan increased further, each exceeding 15,000 around 1941, with Isuzu following with about half the production level. By this time, most of the demand was for trucks for military purposes.

Without question, therefore, military procurement played an important role in supporting domestic car production. However, the quality of domestically produced cars never caught up with that of Ford and GM before the war. Complaints of breakdowns were frequent, including broken shafts, water dripping out of radiators, and early wear of moving parts.[16]

Except for the short period of the Korean war, the postwar production shifted to civilian uses. The big question at the time was whether Japan should protect and nurture domestic producers following the infant industry theory, or import cheaper and better-quality cars from abroad. The first view was taken mostly by MITI and the producers, whereas the latter was taken by the Ministry of Transport, the Bank of Japan, users such as taxi companies, and dealers of imported cars. The government settled on the first position and restricted imports quantitatively and with high tariffs. This protection guaranteed a growing market to domestic producers, giving them a chance to lower costs and improve quality through scale economies, learning by doing, and imported technology. Compared to this effect, the financial incentives provided by the government, such as the provision of low-interest loans through Japan Development Bank and accelerated depreciation allowed to selected equipment, seem less important. Government support specifically aimed at R&D was very limited.

To catch up technologically, Toyota relied on its own efforts but Nissan, Isuzu, and Hino imported technology from, respectively, Austin (U. K.), Rootes (U. K.), and Renault (France). In the latter case, at first they imported major parts including engines, but gradually replaced them with domestically produced ones. All these companies incorporated the technology they learned from the production of licensed models into the development of their own models such as Datsun, and terminated their relationship with the foreign producers by 1965.

These producers, all established before the war, were not the only car manufacturers. Entry attempts were many, counting nearly 30 companies between 1945 and 1960. These included ex-zaibatsu companies, such as Mitsubishi Heavy Industries, and reentrants, such as Mazda, but the majority were new independents including Honda and Suzuki, both of which started as motorcycle producers. However, except for these companies, few of them survived more than 5 years. These private efforts were against MITI's wish to integrate domestic car production to two groups, Toyota and Nissan, to attain scale economies large enough to compete against American giants after trade and capital liberalization. MITI's guidance bore some fruits in the form of a Nissan–Prince merger and an alliance between Toyota and Hino (and later Daihatsu). As a consequence, market concentration to the two groups increased, but the entry of several powerful independents kept the market highly competitive, which together with the increasing importance of global competition contributed to maintaining and even intensifying the innovation efforts of Japanese car makers.

This experience of the Japanese car industry throws lights on some key issues. How could Japan succeed in establishing its own car industry under the presence (in the prewar period) of technologically far advanced Ford and GM? Two factors appear most important. One is the presence of engineers and entrepreneurs, such as Toyoda and Ayukawa, who were willing to take risks and sustain efforts under adversity. The other is the general ability of engineers to absorb foreign technology and the ability of workers to absorb new production processes. Needless to say, the education system discussed earlier contributed to this effect. Also, the emphasis on engineering education in universities helped, as shown by the episode of Toyoda's visiting professors to ask technical questions.

As in any late-developing countries, reverse engineering was the first source of foreign technology. Visiting advanced factories of foreign producers, hiring foreign engineers, and purchasing foreign technology were other important means of acquiring technology.

The government's role was important in two aspects, military and other procurement, and the protection of domestic producers. These policies, by guaranteeing demand, encouraged investment in capital and R&D (mostly development rather than research, particularly in the early period). Nevertheless, one should recall that many entry attempts were made even before government policies or against MITI's guidance. It is misleading indeed to attribute Japan's success solely to government policies.

Concluding Remarks

This section gave three case studies to show how Japanese industries acquired and developed technology in its broad sense. Since the urgent economic target for Meiji Japan was to acquire more advanced Western technology and build competitive industries, our cases focused on "modern" industries and mostly the pre-World War II era. In these industries, intensive efforts were made to transfer technologies from the Western countries. The means were various. In steel, especially the case of Yawata Work, the basic strategy was to import a whole set—plan, technology, equipment, and personnel. In automobiles and electrical/communication equipment, reverse engineering and technology import through licence or joint ventures were common. The balance between these two means varied between the two industries. Apparently, joint ventures were more common in the electrical/communication equipment industry. The balance also varied across companies within each industry; thus, Toshiba, NEC, and Nissan relied more on imported technology than did Hitachi and Toyota. It is difficult to assess the relative merit of the two strategies.

We also revealed a significant role played by indigenous technology. A right choice of technology to import is easier to make with knowledge of indigenous technology, and imported technology has to be modified with the help of indigenous technology that is compatible with local conditions. The cases of Kamaishi and Yawata Steel Works clearly demonstrate this fact. Reverse engineering needs understanding of the basic mechanism behind products and the skills to imitate, which the masters of indigenous technology, such as H. Tanaka, tended to have.

In addition to these technology-specific factors, there were, first, the persistent efforts and inquisitive minds of industrial leaders and, second, educational background and skills of middle-to-top engineers and key workers. The government's hurried effort to establish a compulsory education system and engineering schools was clearly a contributing factor, though the effort would not have achieved its purpose were it not for the parents' enthusiasm for education and the background of widespread private education in the pre-Meiji era. Besides, education may be necessary but not sufficient to produce entrepreneurial will, which, as Schumpeter has so forcefully argued, is the key factor in economic development. A full inquiry into the factors supporting such will is beyond the scope of this study. Some of them may be similar to the Japanese business system today, to be discussed in the next section; however, some are not. For instance, long-term employment was not a norm before World War I as illustrated by NEC's Iwadare's career.[17]

Some observers stress the role zaibatsus played. Zaibatsus, such as Mitsui, Mitsubishi, and Sumitomo, were big, diversified, and family-owned businesses often with government or military connections, and were relatively strong in finance, mining, and shipbuilding as they acquired these businesses when the government privatized them. Among the three industries studied here, their presence was relatively large in steel because the investment required large funds. Even in this industry, however, they remained as followers to giant Nippon Steel. In the other two, their role was further limited. In fact, it was independent entrepreneurs who made risky investment and persistent efforts to catch up with the West technologically. Some of them had certain relations with zaibatsus, including Ayukawa, whose business expanded into one of the so-called new zaibatsus; Tanaka's company, which became partly owned by Mitsui because of bankruptcy; NEC, which formed a tie with Sumitomo after it intermediated between WE and Sumitomo in WE's entry into the wire business[18]; and Toyota's automatic looms, which were sold by Mitsui's trading company from time to time. In neither of these cases, however, did zaibatsu (that is, established zaibatsus) make a first move or become a dominant player. As Ayukawa's case suggests, zaibatsu is, after all, an enterprise that has successfully expanded and diversified; hence, it hardly differs from large diversified enterprises in other capitalist economies—from conglomerates in the United States and European countries to chaebols in Korea (see Chapter 11 on Korea) and other equivalents in developing countries. Thus, it is misleading to emphasize the peculiarities of Japan's experience in reference to zaibatsu.

Zaibatsus were dissolved after the war and the present Mitsui Group, Mitsubishi Group, or any other group is no more than a very loose federation of basically independent companies, unlike pre-World War II zaibatsu where there was a holding company that served as headquarters. Thus it is now frequently observed that a Mitsui company purchases from or sells to, say, a Mitsubishi company, or gets loans from Mitsubishi Bank. On this difference between prewar zaibatsu and present business groups, see Goto (1982) and Odagiri (1992, Chapter 7).

We have discussed the development of the three industries after World War II only briefly, because the discussion of the prewar experience alone was long enough, and the literature on the postwar experience is relatively plentiful. Komiya et al. (1988), in our view, gives the best discussion both on the postwar industrial policy in general and individual studies on the industries discussed in this section.[19]

We have discussed the general shift among Japanese industries from technology import to their own R&D efforts, and described the postwar R&D policies. More generally, the industrial policy (not particularly specific to R&D) may be summarized as follows.

1. Even in the early years the amount of government funding to industries, including subsidies, tax credits, and low-interest loans, was rather modest, though some industries, such as iron and steel, coal mining, shipping, and electric power, benefitted significantly.
2. There is a clear tendency that such government funding has decreased in amount and importance over time; in fact, it appears minimal and by no means larger than in other industrialized countries in recent years.[20]
3. The restriction on imports and foreign direct investment into Japan was probably the most important policy until the early 1970s. Restricting the growing Japanese market, already the second largest in the capitalist economy in the late 1960s, to Japanese firms who were competing intensively among themselves gave a strong incentive to invest in plants, equipment, and R&D. In addition, because postwar Japan's Peace Constitution meant that the military was no longer a significant customer to businesses, industries such as automobiles, which had been helped by military procurement before the war but was still in its infancy relative to American and European producers, might have been wiped out were the market made open to foreign competition. However, such restriction on trade and investment has

been drastically reduced in the 1960s and the 1970s so that the rates of tariff and the number of manufactured goods under import restriction are among the smallest in the world.

4. Procurement by government corporations such as Nippon Telephone and Telegraph (NTT, privatized in 1985) and Japan National Railways (split into six regional companies and privatized in 1987) helped the communication and electronic equipment industry, and the rolling stock industry, respectively. However, they now buy from foreign suppliers as well.
5. MITI's role in the collection and diffusion of information may have been significant, as it could obtain information on overseas markets through its Japan External Trade Organization (JETRO) and technological information through its Agency of Industry Science and Technology, which has a number of laboratories. However, this role has also declined as firms themselves accumulated international experiences and technological knowledge. In addition, Japanese firms have various other "networks" through, for instance, trading companies (*sogo shosha*), trade associations, and buyer–supplier relationships.

Thus, government policies, especially the protection of domestic markets, played a significant role in Japan's postwar industrial development at least until the early 1970s. However, as argued in the beginning of this chapter, government policies can attain their purposes only if the private sector is willing and able to take the opportunities given to them (or even to overcome the obstacles the policies create, as evidenced by Honda's experience). To understand why Japanese firms were willing and able, and to understand why they continue to be so despite the weakening role of government policies, one has to know how Japanese businesses today make decisions and carry out innovation efforts. This is why a detailed examination will be given in the next section on innovation inside Japanese firms. It will be preceded by descriptions of the current R&D efforts and performance in the aggregate.

INNOVATION IN TODAY'S JAPAN

Overview

Table 3.3 shows Japan's current R&D efforts in comparison to other major developed countries. Clearly, the role of government in funding R&D is smaller in Japan than in the other countries. One reason for this smaller share of government funds is the small defense-related expenditures in Japan. However, even when defense-related R&D expenditures are excluded, the proportion of government funds in total R&D expenditures is the smallest.

The share of government funds in industrial R&D expenditures is a mere 1.2% in Japan; that is, Japanese companies are financing almost all of their R&D out of their own funds. This is in contrast to other countries where 11 to 34% of industrial R&D expenditures are supported by government funds. Again, the gap narrows considerably if defense-related industries are excluded because, without aircraft and missile industries, the comparable percentage for the United States is reduced to 7.7 (Eads and Nelson, 1986). Yet, in comparison with Japan's 1.2%, a substantial gap remains between the two countries.

Table 3.3. R&D Expenditures and Personnel in Five Countries[a]

	Japan (Y)	United States ($)	Germany (DM)	France (F)	United Kingdom (£)
Total R&D expenditures (in billions, local currency)	10909	142.0	64.82	141.9	10334*
Total R&D expenditures (in billion yen)	10909	19596	4757	3068	2359*
Total R&D expenditures/GNP (%)	2.69	2.73	2.89	2.33	2.19*
Nondefense R&D expenditures/ GNP (%)	2.66	1.95	2.75	1.79	1.77*
Government funds/total R&D expenditures (%)	17.1	46.4	33.2	49.3	36.7*
Government funds/nondefense R&D expenditures (%)	16.4	25.1	29.9	33.9	21.6*
Industry expenditures/total R&D expenditures (%)	75.5	72.0	73.0	56.8***	66.4*
Industry funds/industry R&D expenditures (%)	98.6	65.7	86.8	73.0***	71.4*
Government funds/industry R&D expenditures (%)	1.2	34.3	11.3	22.4***	16.5*
R&D personnel (in thousands)	462	949*	166**	115*	102*
R&D personnel/10,000 population	37	39*	27**	21*	18*

[a]Year 1989 except *1988, **1987, and ***1983. R&D expenditures and personnel exclude humanities and social sciences in Japan and the United Kingdom but include them in other countries. Germany refers to the Federal Republic (West) before unification.

Source: Science and Technology Agency (1991).

Table 3.3 also shows the number of R&D personnel. The number of research personnel was about 462,000 in 1989, growing three times in the preceding two decades. Of these 64% of them were working in industry. Although the number of research personnel is about half that of the United States, per capita Japan has about the same number as the United States and more than the European countries. One notable feature of the Japanese research personnel is a relatively larger proportion of engineers (42% of the total research personnel in 1989) than scientists (16%).[21] The same tendency exists for university degrees: the number of Ph.Ds granted was 860 in science and 1404 in engineering in Japan in 1988. In contrast, these were 7438 and 3236 in the United States and 2894 and 1020 in Germany, respectively. The same can be said of the undergraduate students in Japan. In 1988, there were 368,000 students in engineering departments in contrast to 62,000 in science departments. This emphasis on engineering education started in the Meiji era as shown earlier.[22]

An active R&D activity results in patents and, more generally, technological progress. The number of patent application in Japan (by Japanese or foreigners) increased from 131,000 in 1970 to 341,000 in 1987. This is in marked contrast with European countries where the number of application decreased in the 1970s to the early 1980s, and with the United States, where the number increased only gradually to 134,000 in 1987. Although this international comparison of patent numbers may be biased owing to the Japanese system under which patents are granted separately even when inventions behind them were closely related technologically, the fact that the share of Japanese applications in the U. S. patents is also increasing rapidly—from

5.1% in 1970 to 20.3% in 1988—suggests that Japan's increased R&D efforts are indeed bearing fruits (STA, 1990).

Another popular measure of technological progress is the rate of change in total factor productivity (TFP), which is affected not only by R&D but also by other miscellaneous factors, such as demand shocks, and the "distance" from the best practice in the world. Table 3.4 shows an impressive rate of TFP increase in Japan, especially before 1973. The extremely high rate of economic growth in this period is explained mainly by this TFP increase, which, in turn, appears to be explained by the combination of successful technology importation and extensive domestic R&D, among other factors.

Innovation Inside Japanese Firms[23]

Motivations

Inherently, the choice of the level of R&D is a dynamic problem because the costs and returns take place at different and over many periods. Thus the firm's willingness to invest in R&D very much depends on how much the firm is inclined to growth (i.e., long-run objectives) as opposed to short-run objectives. It has often been argued that Japanese firms are more growth oriented than American or European firms. Several institutional factors are important.

It is first noted that the capital market constraints to the firm are presumably not as strong in Japan as in the United States or the United Kingdom. Most of the large

Table 3.4. Annual Productivity and Output and Input Growth in Business Sector (in Percentages)

	OECD Average	United States	Japan	Germany	France	United Kingdom
1960s to 1973						
Output	5.2	3.8	9.7	4.6	6.4	3.2
Factor input	2.4	2.3	3.5	1.8	2.1	1.2
TFP[a]	2.8	1.5	6.1	2.8	4.3	2.0
Labor productivity	4.1	2.2	8.6	4.9	5.9	3.3
Capital productivity	−0.4	0.3	−2.4	−1.1	0.6	−0.7
1973–1979						
Output	2.9	2.8	3.8	2.4	3.5	1.1
Factor input	2.2	2.9	2.0	0.6	1.4	0.9
TFP	0.7	−0.1	1.8	1.8	2.1	0.2
Labor productivity	1.6	0.3	3.2	3.4	3.5	1.3
Capital productivity	−1.4	−0.9	−3.0	−1.1	−1.2	−1.9
1979–1986						
Output	2.3	2.2	3.8	1.6	1.5	1.4
Factor input	1.7	2.2	2.1	0.8	0.2	0.3
TFP	0.6	0.0	1.7	0.8	1.3	1.1
Labor productivity	1.4	0.6	2.8	2.0	2.5	1.9
Capital productivity	−1.3	−1.0	−2.0	−1.3	−1.4	−0.8
Memorandum						
1985 capital share	32.2	34.3	22.6	34.8	30.8	32.5

[a]TFP, total factor productivity.

Source: Steven Englander, A., and Mittelstädt, A. (1988). "Total Factor Productivity: Macroeconomic and Structural Aspects of the Slowdown." *OECD Economic Studies*, No. 10, Spring.

shareholders are banks and other firms that are basically friendly to the firm. Often these shareholdings are reciprocal, that is, the firm also owns their shares.[24] As a consequence, it is unlikely that these shareholders interfere with the management of the firm unless it is in trouble. These shareholders are themselves controlled by the managers sympathetic to the management of this firm.[25]

Hence, Japanese managers tend to have a fair amount of discretionary power. They are typically promoted from within among the pool of those having worked with the company for, say, three decades and gradually climbed up the promotion ladder. They consider themselves, first and foremost, as representing the employees rather than the shareholders. Indeed, if they have worked with the employees all the time since their graduation from the schools, how can they stop feeling closer to the employees than to the shareholders with whom they have probably been acquainted only briefly or they have never met?

That managers tend to pursue growth more than the value of the firm has been put forth by many writers, most notably Marris (1964), for several reasons such as pecuniary and nonpecuniary gains managers receive from growth more than from shareholder wealth. Add to this the managers' strong identification with the employees and the employees' long-term attachment to the firms common in Japan. Whether this attachment should be called "lifetime" employment is doubtful, because it is not explicitly written in any labor contract, and de facto dismissal has taken place more frequently than is usually presumed. Yet it has been and is still regarded as a norm by both employers and employees, and the employer–employee relationship has been more stable in Japan than in any other country.

The result of these considerations is an even stronger motivation toward growth, for the employees are most concerned with the long-term survival and performance of the company and the prospects of promotion. Since the positions for promotion are larger the faster the corporate organization expands, the employees are more growth-oriented than the shareholders, and, with the weak capital market constraints and with the managers identifying themselves with the employees, it appears reasonable to assume that Japanese firms are more growth oriented and place a great emphasis on the acquisition of technological knowledge.

The Managers' Background

There is another difference between Japanese managers and American or British managers—their background. In Japan, the largest proportion of directors (who, unlike in the United States, are mostly full-time and insiders) came from production and technology departments, followed by marketing and export, which together accounted for 50 to 70%, far larger than those with financial and accounting origins, which are merely 5 to 19% (Kono, 1984, p. 33). More recently, according to the report in *Nihon Keizai Shimbun* (May 25, 1987), among the 126 presidents who assumed the post during the first 6 months of 1987 in the firms listed in the Tokyo Stock Exchange, 36% had a marketing origin and 28% had a production/R&D origin. Those with a financial origin again accounted for a mere 11%. This lack of importance of financial background is in contrast to the United States or the United Kingdom where financial experience has been found most helpful in attaining managerial positions.[26]

The better knowledge and experience of Japanese managers in production/R&D provide them with a better understanding of the potentials and limitations of R&D

projects, more accurate evaluation of the outcomes from R&D, and more favorable general attitudes toward R&D. Similarly, their better knowledge and experience in sales/marketing provide them with a keen understanding of what kinds of products are in demand in the markets. These familiarities of Japanese managers with technological seeds and market needs are particularly valuable in technologically rapidly changing markets.

R&D–Production–Sales Linkages

The Japanese internal labor market system is characterized not only by long-term attachment but also by a carefully organized training and rotation scheme. It is not unusual for the firm to provide several months of well-programmed training at the time of entry of the workers, not only so that they can acquire technical knowledge and skills but also to familiarize them with diverse activities of the firm, and then mid-career training every 5 years or so. In addition, rotation of workers from, say R&D to production in the case of engineers, or one shop to another in the case of production workers, is more common than in other countries. These practices help workers acquire a company-wide view and acquire flexibility to changing work environment.

Furthermore, long-term employment naturally leads the workers to develop personal linkages across departments. If you are working in R&D for many years, you will have many opportunities to meet and talk with other people in the company, and opportunities to visit other departments and other plants to discuss problems of mutual concern. Thus a close human relation is created between the R&D department and other departments within the firm. Consequently, the R&D staff will be more familiar with the technological needs arising from production and marketing, and the non-R&D departments will be more familiar with what is going on in the laboratories. Consequently, the company R&D departments tend to be more production based and market oriented. The advantage is that the research undertaken will be more commercially relevant and the introduction of a new product into the production and marketing stages will be faster. On the other hand, truly original and basic research may not be emphasized.

Two other features of Japanese firms make this R&D–production–sales link even more effective. One is the infrequent adoption in Japan of multidivisional forms with substantial discretion within each division. The weaker divisional separation not only fosters easy rotation across divisions (which probably is why a multidivisional form is unpopular) but also facilitates a companywide use of technological knowledge. The other is a close link with suppliers and subcontractors. The stable buyer–supplier relationship with a constant flow of information is usually observed in Japanese companies. As a result they tend to share the threat of market competition as well as the need for innovation. Improvements in the product or production process by the supplier or subcontractor will be noted and rewarded by the parent company and, if possible, will be utilized in other firms in the group. In many occasions, cooperative R&D will be carried out between the parent and the supplier(s). For instance, when an automobile assembler develops a new car model, it is essential that suitable components are developed at the same time. Such development will be done by the supplier(s) with close communication with the assembler's developing team and often with the latter's technical advice. See Asanuma (1985) and Clark and Fujimoto (1991) for a detailed study on the automobile industry, and Imai et al. (1985) for a general discussion.

Introduction of New Process/Product

The introduction of a new process or new product into production and marketing may be smoother in Japanese firms for several reasons. Most important is the close communication between R&D and production as previously discussed. In particular, in the course of developing a product or process, the production staff tends to participate from a rather early stage. This may not be efficient in terms of development itself but, since the views of the production department are reflected in the development of the final product/process, the transfer from development to production is quicker, with little bottleneck or disruption occurring in the process.

It is also not unusual that the engineers in the development team are transferred to the production department and actively involve themselves in applying the new product/process into production (Sakakibara and Westney, 1985). Such a transfer is common in the careers of Japanese engineers because the ability for research tends to reach a peak in the late thirties (or even earlier) and, afterward, the management as well as the engineers themselves tend to use their ability and expertise in more or less administrative and managerial positions in the production and R&D departments. The transfer, therefore, not only smooths the introduction of a particular new product/process in which the engineer has been involved, but also increases the technological knowledge in general in the production department. It also fosters an even tighter personal link between R&D and production.

Even if the production manager is familiar with the new product/process, its introduction to production cannot be smoothly done unless the workers are skilled and flexible. Here again the advantage of Japanese firms in terms of labor management has to be emphasized. In Japan, under the expectation of lifetime employment, the management has an ample incentive to provide training to both white-collar or blue-collar workers. Hence, in-company education, on-the-job training, and rotation are provided to the workers. Their skills are therefore not only high but also flexible and wide, which makes adaptation of a new product/process into shop-floors much easier and faster.

In this adaptation process, it is important for management to have a free hand in reorganizing the work organization. The prevalence of single and companywide trade unions in Japan is indispensable for this purpose. For instance, when automobile manufacturers introduced industrial robots, most of the welding jobs were eliminated and yet none of the Japanese firms discharged the welders for this reason. They were all retrained and transferred to other shops, such as metalwork and assembling; hence, no incidence of grievance or industrial action was reported. If, on the contrary, workers in separate jobs are organized into different unions, such transfers would have been impossible and the introduction of new processes would have caused disputes. The relation between innovation and labor relations can be indeed quite important, however distant they may appear. That Japanese firms starting production in the United Kingdom have all sought a single union representation has to be understood in this light.

Synthesis

In this section, we have looked into the inside of the so-called black box and came up with four major factors in Japanese management favorable in making large R&D efforts, making them efficiently, and applying the outcome swiftly and smoothly into manufacturing and marketing. They were a bias to growth maximization; familiarity of management with research, production, and marketing; close R&D–production–sales links; and the smooth transfer of new processes and products into production. Behind all these factors, we stressed the importance of the human aspects of management. The internal labor practices of Japanese management, particularly, a long-term worker–company attachment with internal training and internal promotion, have been conducive to the growth preference of the managers and employees, interdepartmental personal contacts, and easy adaptability of production processes to new processes and products. Innovation is predominantly a labor-intensive process. However powerful a computer may be, it needs a human brain to start with ideas and make a final evaluation of alternative projects. However labor-saving a new process may be, it needs experienced workers to install and operate it. Creating the human resources and organizations that are most suitable for these needs is the key to successful innovation.

These internal aspects, we note, must be accompanied by external factors. An organization, however capable of being efficient and flexible, need not be so unless external threats of competition and rapid changes in industrial structure exist. Competition among rival firms has been very intensive in many industries in Japan, even where the concentration ratio is not particularly low. Entry into growing industries has been fast and frequent. And shifts in industrial structure have been drastic in the past decades, with some of the top industries now fading into obscurity. These facts provided the firms with a strong sense of crisis and a motivation to be innovative. To maintain the smooth internal labor system, growth and diversification have to be sought internally and the firm has to acquire the necessary technology internally. Consequently, under the threat of competition and of rapidly changing technology and industrial structures, the firm's survival is threatened unless it keeps innovating. Competition in industries, itself a product of the Japanese management system (through the effort to grow internally and diversify, for instance), should never be disregarded as a key to a nation's technical progress.

THE FUTURE?

What shall we expect of the Japanese innovation system in the future? We suggested that the weight of government policies will further decline, particularly because the government is losing most of its control tools through deregulation and liberalization. The industries are less dependent on the government, and government–business personal relations are weakening.

There are a few signs suggesting that a change is inevitable in the Japanese management system. The first is the move by many firms toward increased offshore production. How much this move will make the R&D–production linkage difficult is an important question. The second is the change in financial markets, such as deregulation and an increase in mergers and acquisitions. Again, it is important to know how much impact this change will make on the way Japanese firms make decisions, and particularly on their attitudes toward R&D. The third is the changing environment surrounding labor relations. As the age composition of the Japanese working population is rapidly changing toward that of more senior workers, and as technological change accelerates, it will become more and more difficult for the firms to maintain lifetime employment, and some of the advantages of the Japanese management system may be lost. An increase in offshore production and immigrant workers may further add to this tendency.

The fourth is an increasing need for global research. As international technological race intensifies, and as globalization in production increases, the firm has to be alert to technological development in any part of the world, to technological needs from any production site worldwide, and to international market needs. The need for global research is therefore increasingly felt, and many companies have started establishing research institutions overseas or increasing technological collaborations with foreign firms. Whether the Japanese firm is suitable for this globalization is an interesting question. Sakakibara (1988), for instance, suggests that American managers are more experienced in employing people with different values, personalities, and educational backgrounds, and, therefore, have a higher flexibility and the adaptability needed in international transplantation of technology management. However, the heterogeneity of scientists in the United States and homogeneity in Japan (the difference being not absolute but only relative) may suggest that American scientists working in U. S.-based Japanese firms may adjust more easily than Japanese scientists working in American firms in Japan. To this extent, American firms may be handicapped in obtaining technological knowledge originating from Japan, and the language barrier may further aggravate this handicap. In addition, in view of the R&D–production link in Japanese firms, foreign firms may not be able to carry out research efficiently in Japan unless they have production facilities as well (Westney and Sakakibara, 1986).

Finally, more and more emphasis will be placed on basic research by both private firms and policy-makers. As Japan has caught up with the best practices in the world in many fields and become a leader in some of them, technological seeds that can be imported have become scarce. Even if Japanese firms find technologies they wish to import, it is now common that seller firms attach various restrictive conditions in their offer. Consequently, Japanese firms are now keen to create technology of their own, in particular in the field of basic research. In addition, recent development in high technology has demonstrated the importance of basic research for commercial success.

Many Japanese firms have thus set up basic research laboratories independently of existing research facilities that have been biased toward applied research and development. Although it is sometimes doubtful how basic their version of basic research truly is, the shift of emphasis toward the basic end of technological spectrum is apparent. The government is also emphasizing the need for more basic research, in the belief that, as the status of the Japanese economy in the world rises, its contribution to "international public goods" has to be increased as well. Basic research that contributes to the world stock of knowledge is viewed as such an international public good to which Japan is expected to contribute.

In terms of expenditures on basic research, the increase has been attained far more by industries than universities. During the 10-year period from 1977 to 1987, industries increased the expenditures by 15.8% annually and universities by 6.5%. As a consequence, whereas in 1977 universities undertook 64% of Japan's basic research and companies undertook 20%, the percentages in 1987 were 50 and 34, respectively (STA, 1990). Needless to say, universities have been traditionally assumed to be the prime and most appropriate performers of basic research because industries are unlikely to undertake research with externality and distant commercial applicability. One wonders, therefore, whether the increasing dependence on the private effort can lead to a healthy development of basic research. Insufficient funds in universities is also a problem. In 1987, R&D expenditure per researcher was 25 million yen in companies but only 10 million yen in universities (STA, 1990).

All the above discussions suggest that further changes are inevitable and necessary to the Japanese innovation system. Which way to change can gravely influence future technological performance, and innovation to create a leader's innovation system out of a follower's system, however successful it may have been in the past, is very much in need.

NOTES

We thank Chris Freeman, Richard Nelson, Hugh Patrick, and Nathan Rosenberg for helpful comments to earlier versions.

1. That the Japanese made rice-polishing machines while the West made mills reflects the difference in eating habits. Some authors suggest that this difference caused the Japanese to be good at the machines with reciprocating motion but behind the West at those with rotary motion.

2. We will show later how this factory relates to indigenous *karakuri* technology and to present Toshiba.

3. For national research institutions, Riken, and Gakushin discussed above, see Hiroshige (1973) and Kamatani (1988).

4. The data here are from *Kogyo Chosa Shuho*, 2(2), 1924, reprinted in *Nihon Kagaku Gijutsushi Taikei*, Vol. 3, Tokyo: Daiichi Hoki Shuppan, 1967.

5. See Miyazaki and Ito (1989) for further details of the historical development of the Japanese industry.

6. A survey conducted by Resources Bureau of the Government, quoted in Hiroshige (1973, p. 115).

7. A survey conducted by the Technology Agency of the Cabinet, quoted in the Agency of Industrial Science and Technology (1964, p. 125).

8. Take the case of Zero fighters known for their speed and maneuverability in the beginning. Karasawa (1986) shows that during the 5-year war period, their horsepower could be increased only by 20% whereas Americans and Germans more than doubled the horsepower of their fighters, thereby surpassing Zero fighters. Also, guns on the fighters were difficult to aim and suffered from short range and slowness. The weaker technology in communication and radar was also evident. In the army, too, the trucks and other cars had great trouble, as will be discussed later. Hence, the technological gap between Japan and America seems to have been very large by the end of the war, not to mention America's development of atomic bombs.

9. For Japan's technology importation and the role of government policy up to the mid-1970s, see Peck and Tamura (1976) and Peck and Goto (1981).

10. We will return to the discussion of the postwar industrial policy later.

11. Blumenthal (1976) and Odagiri (1983) found a significantly positive correlation between the amount of royalty payments for imported technology by industries or firms and their R&D expenditures.

12. Some authors, such as Saxonhouse (1985), emphasized the information-disseminating role of cooperative research. It is difficult to estimate how important this role was in RAs. It has been reported that in the much publicized case of the Very-Large-Scale Integrated Circuits

(VLSI) Research Association, it took considerable talent and effort (and drinks!) by the director to have the scientists from different companies mingle (Sakakibara, 1981). Since this RA was quite exceptional in having a common laboratory and a dedicated director from a third party (namely, a government research institution and not companies), we imagine that communication among scientists from different companies must have been infrequent in most other RAs.

13. Two arguments may be made. First, research subjects of the RAs are in principle closer to the basic end, need longer lead time, and have a smaller chance of success than the research subjects of individual firms. Second, when most of the actual and potential rivals are within an RA, there may be little need for protective patents, that is, to apply for patents only to preempt rivals.

14. The discussion in this section is based on miscellaneous Japanese sources including company histories published by respective companies. References will not be given individually.

15. Bureau of Statistics, Management and Coordination Agency, *Report on the Survey of Research and Development.*

16. This fact supports the discussion in footnote 7.

17. See Hirschmeier and Yui (1975).

18. When WE decided to enter the wire business in Japan, NEC recommended Sumitomo, one of the so-called four big zaibatsus before World War II, as the partner because Sumitomo had a big copper business. A new company, Sumitomo Electric Wire (now Sumitomo Electric Industries), was established, jointly owned by WE, Sumitomo, and NEC. In return, Sumitomo acquired a share of NEC.

19. For other views, see Johnson (1982) and Okimoto (1989).

20. This finding agrees with the declining government support on R&D shown in Table 3.2.

21. The number of researchers (or scientists and engineers) is difficult to compare internationally because the definition is different and the necessary data to adjust the Japanese figures to full-time equivalents are lacking. See National Science Foundation (1988).

22. The data on education are from Ministry of Education, *Gakko Kihon Chosa Houkokusho.*

23. For a more detailed discussion of the topic in this section, see Odagiri (1992).

24. As suggested at the end of the previous section, it is incorrect to attribute this ownership pattern to post-zaibatsu business groups. The discussion in the text applies to most firms whether they belong to such groups or not.

25. This reciprocal shareholding is one reason why hostile takeovers are infrequent in Japan. The labor practices in Japan are another reason. See Odagiri (1992, Chapter 5).

26. See Fidler (1981) and Kohn/Ferry (1981) for the United Kingdom and Browne and Motamedi (1977) for the United States.

REFERENCES

Agency of Industrial Science and Technology, the Ministry of International Trade and Industry. (1964). *Gijutsu Kakushin to Nihon no Kogyo: Kogyo Gijutsu-in 15 Nen no Ayumi.* [Innovation and Industries in Japan: The 15 Year History of the Agency of Industrial Science and Technology]. Tokyo: Nikkan Kogyo Shimbunsha.

Asanuma, B. (1985). "The Contractual Framework for Parts Supply in the Japanese Automotive Industry." *Journal of Japanese Studies* Summer, 54–78.

Blumenthal, T. (1976). "Japan's Technological Strategy." *Journal of Development Economics* 3: 245–55.

Browne, W. G., and Motamedi, K. K. (1977). "Transition at the Top." *California Management Review* 20(2): 67–73.

Clark K. B., and Fujimoto, T. (1991). *Product Development Performance: Strategy, Organization, and Management in the World Auto Industry.* Boston: Harvard Business School Press.

Eads, G. C., and Nelson, R. R. (1986). "Japanese

High Technology Policy: What Lessons for the United States?" In Hugh Patrick (ed.), *Japan's High Technology Industries*, 243–69. Seattle: University of Washington Press.

Fidler, J. (1981). *The British Business Elite: Its Attitudes to Class, Status and Power*. London: Routledge and Kegan Paul.

Fujishiro, N. (1988). "Computer Sangyo ni Okeru Kyodo-Kenkyu no Yakuwari." [The Role of Joint R&D in the Computer Industry]. Unpublished master's thesis, University of Tsukuba.

Goto, A. (1982). "Business Groups in a Market Economy." *European Economic Review* 19: 53–70.

Goto, A. (1988) "Japan: A Sunset Industry." In Merton Peck (ed.), *The World Aluminium Industry in a Changing Energy Era*. Washington, D. C.: Resources for the Future.

Goto, A., and Wakasugi, R. (1988). "Technology Policy." In R. Komiya, M. Okuno, and K. Suzumura (eds.), *Industrial Policy of Japan*, 183–204. New York: Academic Press.

Hiroshige, T. (1973). *Kagaku no Shakaishi* [The Social History of Science]. Tokyo: Chuo Koron Sha.

Hirschmeier, J., and Yui, T. (1975). *The Development of Japanese Business, 1600–1973*. London: George Allen & Unwin.

Iida, K. (1979). *Nippon Tekkou Gijutsu Shi* [The History of Iron and Steel Technology in Japan]. Tokyo: Toyo Keizai Shinposha.

Imai, K., Nonaka, I., and Takeuchi, H. (1985). "Managing the New Product Development Process: How Japanese Companies Learn and Unlearn." In K. B. Clark, R. H. Hayes, and C. Lorenz (eds.), *The Uneasy Alliance*, 337–75. Boston: Harvard Business School Press.

Johnson, C. (1982). *MITI and the Japanese Miracle*. Stanford: Stanford University Press.

Kaigo, M. (ed.) (1971). *Nihon Kindai Kyouiku Jiten* [Encyclopedia on Modern Education in Japan]. Tokyo: Heibon Sha.

Kamatani, C. (1988) *Gijutsu Taikoku 100 Nen no Kei: Nihon no Kindaika to Kokuritsu Kenkyuu Kikan* [The 100 Year Strategy to an Innovative Country: Japan's Modernization and the National Research Institutions]. Tokyo: Heibonsha.

Karasawa, E. (1986). "Koku Gijutsu no Hizumi." [Deficiencies in Aircraft Technology]. In K. Hasegawa (ed.), *Nihon Kindai to Senso*, Vol. 6, 193–224. [Modern Japan and Wars]. Kyoto: PHP.

Kohn/Ferry International in Conjunction with the London Business School (1981). *British Corporate Leaders—A Profile*. London: Kohn/Ferry International.

Komiya, R., Okuno, M., and Suzumura, K. (eds.) (1988). *Industrial Policy of Japan*. New York: Academic Press.

Kono, T. (1984). *Strategy and Structure of Japanese Enterprises*. London: Macmillan.

Lynn, L. H. (1982). *How Japan Innovates*. Boulder: Westview Press.

Marris, R. L. (1964). *An Economic Theory of 'Managerial' Capitalism*. London: Macmillan.

Minami, R. (1981). *Nippon no Keizai Hatten* [Economic Development in Japan]. Tokyo: Toyo Keizai Shinposha.

Miyazaki, M., and Ito, O. (1989). "Senji Sengo no Sangyo to Kigyo" [Industries and Firms during and after the War]. In T. Nakamura (ed.), *Nihon Keizaishi 7: Keikakuka to Minshuka*, 165–235. [The History of Japanese Economy, Vol. 7: Planning and Democratization], Tokyo: Iwanami Shoten.

National Science Foundation (1988). *The Science and Technology Resources of Japan: A Comparison with the United States*. Washington, D. C.: U. S. Government Printing Office.

Odagiri, H. (1983). "R&D Expenditures, Royalty Payments, and Sales Growth in Japanese Manufacturing Corporations." *Journal of Industrial Economics* 32(1): 61–71.

Odagiri, H. (1992). *Growth through Competition, Competition through Growth: Strategic Management and the Economy in Japan*. Oxford: Oxford University Press.

Okimoto, D. I. (1989). *Between MITI and the Market: Japanese Industrial Policy for High Technology*. Stanford: Stanford University Press.

Patent Office (1955). *Tokkyo Seido 70 Nen Shi* [The 70 Year History of the Patent System]. Tokyo: Hatsumei Kyokai.

Peck, M. J., and Goto, A. (1981). "Technology and Economic Growth: The Case of Japan." *Research Policy* 10(3): 222–43.

Peck, M. J., and Tamura, S. (1976). "Technology." In H. Patrick and H. Rosovsky (eds.), *Asia's New Giant*, 525–85. Washington, D. C.: The Brookings Institution.

Sakakibara, K. (1981). "Soshiki to Innovation: Jirei Kenkyu, Cho LSI Gijutsu Kenkyu Kumiai" [Organization and Innovation: A Case Study of VLSI Technology Association]. *The Hitotsubashi Review* 86(2): 160–75.

Sakakibara, K. (1988). "Managing Global Innovation: The Challenge of International Transplantation of Technology Management." Working Paper No. 8803, Hitotsubashi University.

Sakakibara, K., and Westney, D. E. (1985). "Comparative Study of the Training, Careers, and Organization of Engineers in the Computer Industry in the United States and Japan." *Hitotsubashi Journal of Commerce & Management* 20(1): 1–20.

Sawai, M. (1990). "Kikai Kogyo." [Machinery Industry]. In S. Nishikawa and T. Abe (eds.), *Nihon Keizaishi 4: Sangyo-ka no Jidai*, 213–53. Tokyo: Iwanami Shoten.

Saxonhouse, G. (1985). "Japanese Cooperative R&D Ventures: A Market Evaluation." Discussion Paper No. 156, Department of Economics, University of Michigan.

Science and Technology Agency (STA) (1990, 1991). *Kagaku Gijutsu Hakusho.* [White Paper on Science and Technology]. Tokyo: Printing Bureau of the Ministry of Finance.

Umihara, T. (1988). *Kinsei no Gakko to Kyouiku* [Schools and Education in the Tokugawa Japan]. Kyoto: Shibunkaku.

Wakasugi, R., and Goto, A. (1985). "Kyodo Kenkyu Kaihatsu to Gijutsu Kakushin." [Joint R&D and Technological Innovation]. In Y. Okamoto and T. Wakasugi (eds.), *Gijutsu Kakushin to Kigyo Kodo,* 193–217. Tokyo: Tokyo University Press.

Westney, D. E., and Sakakibara, K. (1986). "The Role of Japan-Based R&D in Global Technology Strategy." In M. Horwitch (ed.), *Technology in the Modern Corporation: A Strategic Perspective,* 217–32. Oxford: Pergamon Press.

Name Index